VOLUME ONE HUNDRED AND ELEVEN

CURRENT TOPICS IN DEVELOPMENTAL BIOLOGY

Neural Crest and Placodes

CURRENT TOPICS IN DEVELOPMENTAL BIOLOGY

"A meeting-ground for critical review and discussion of developmental processes"
A.A. Moscona and Alberto Monroy (Volume 1, 1966)

VOLUME ONE HUNDRED AND ELEVEN

CURRENT TOPICS IN DEVELOPMENTAL BIOLOGY

Neural Crest and Placodes

Edited by

PAUL A. TRAINOR

Stowers Institute for Medical Research, Kansas City, Missouri and Department of Anatomy and Cell Biology, University of Kansas Medical Center, Kansas City, Kansas, USA

ELSEVIER

AMSTERDAM • BOSTON • HEIDELBERG • LONDON
NEW YORK • OXFORD • PARIS • SAN DIEGO
SAN FRANCISCO • SINGAPORE • SYDNEY • TOKYO
Academic Press is an imprint of Elsevier

AP

Academic Press is an imprint of Elsevier
225 Wyman Street, Waltham, MA 02451, USA
525 B Street, Suite 1800, San Diego, CA 92101-4495, USA
125 London Wall, London, EC2Y 5AS, UK
The Boulevard, Langford Lane, Kidlington, Oxford OX5 1GB, UK

First edition 2015

ISBN: 978-0-12-407759-1
ISSN: 0070-2153

CONTENTS

CONTRIBUTORS

Youngwook Ahn
Stowers Institute for Medical Research, Kansas City, Missouri, USA

Aria C. Attia
Division of Human Genetics, Department of Pediatrics, Cincinnati Children's Hospital Medical Center, Cincinnati, Ohio, USA

Tiziano Barberi
Pluripotent Stem Cell Differentiation Laboratory, Southwest National Primate Research Center, Texas Biomedical Research Institute, San Antonio, Texas, USA, and Department of Anatomy Neuroscience, The University of Melbourne, Parkville, Victoria, Australia

Linda A. Barlow
Department of Cell and Developmental Biology; Graduate Program in Cell Biology, Stem Cells and Development, and Rocky Mountain Taste and Smell Center, University of Colorado School of Medicine, Anschutz Medical Campus, Aurora, Colorado, USA

Onur Birol
Program in Developmental Biology, Baylor College of Medicine, Houston, Texas, USA

Bianca E. Borchin
Pluripotent Stem Cell Differentiation Laboratory, Southwest National Primate Research Center, Texas Biomedical Research Institute, San Antonio, Texas, USA, and Australian Regenerative Medicine Institute, Monash University, Clayton, Victoria, Australia

Samantha A. Brugmann
Division of Plastic Surgery, Department of Surgery, and Division of Developmental Biology, Department of Pediatrics, Cincinnati Children's Hospital Medical Center, Cincinnati, Ohio, USA

Ching-Fang Chang
Division of Plastic Surgery, Department of Surgery, and Division of Developmental Biology, Department of Pediatrics, Cincinnati Children's Hospital Medical Center, Cincinnati, Ohio, USA

Bharesh Chauhan
Division of Pediatric Ophthalmology and Strabismus, Department of Ophthalmology, University of Pittsburgh School of Medicine, Pittsburgh, Pennsylvania, USA

Alwyn Dady*
Laboratoire de Biologie du Développement, Université Pierre et Marie Curie-Paris 6, and CNRS, Laboratoire de Biologie du Développement, Paris, France

*Present address: Children's Hospital of Pittsburgh, Rangos Research Building, Pittsburgh, Pennsylvania, USA

Jean-Loup Duband
Laboratoire de Biologie du Développement, Université Pierre et Marie Curie-Paris 6, and CNRS, Laboratoire de Biologie du Développement, Paris, France

Renée K. Edlund
Program in Developmental Biology, Baylor College of Medicine, Houston, Texas, USA

Katherine A. Fantauzzo
Department of Developmental and Regenerative Biology, Icahn School of Medicine at Mount Sinai, New York, USA

Alessandro Fantin
UCL Institute of Ophthalmology, University College London, London, United Kingdom

Vincent Fleury
Laboratoire Matière et Systèmes Complexes, CNRS et Université Denis-Diderot-Paris 7, Paris, France

Andrew K. Groves
Program in Developmental Biology; Department of Molecular and Human Genetics, and Department of Neuroscience, Baylor College of Medicine, Houston, Texas, USA

Ophir D. Klein
Departments of Orofacial Sciences and Pediatrics; Program in Craniofacial and Mesenchymal Biology, and Institute for Human Genetics, University of California San Francisco, San Francisco, California, USA

Takahiro Kunisada
Department of Tissue and Organ Development, Regeneration and Advanced Medical Science, Gifu University Graduate School of Medicine, Gifu, and Japan Science and Technology Agency (JST), Core Research for Evolutional Science and Technology (CREST), Tokyo, Japan

Anthony-Samuel LaMantia
George Washington University Institute for Neuroscience, and Department of Pharmacology and Physiology, The George Washington University, School of Medicine and Health Sciences, Washington, DC, USA

Richard Lang
The Visual Systems Group, Abrahamson Pediatric Eye Institute, Division of Pediatric Ophthalmology, Department of Ophthalmology, University of Cincinnati and Children's Hospital Research Foundation, Cincinnati, Ohio, USA

Ming Lou
Department of Chemistry and Physics, Lamar University, Beaumont, Texas, USA

Sally A. Moody
Department of Anatomy and Regenerative Biology, The George Washington University, School of Medicine and Health Sciences, and George Washington University Institute for Neuroscience, Washington, DC, USA

Tsutomu Motohashi
Department of Tissue and Organ Development, Regeneration and Advanced Medical Science, Gifu University Graduate School of Medicine, Gifu, and Japan Science and Technology Agency (JST), Core Research for Evolutional Science and Technology (CREST), Tokyo, Japan

William A. Muñoz
Stowers Institute for Medical Research, Kansas City, Missouri, USA

Jason M. Newbern
School of Life Sciences, Arizona State University, Tempe, Arizona, USA

Noriko Osumi
Department of Developmental Neuroscience, Centers for Neuroscience, Tohoku University Graduate School of Medicine, Sendai, Japan

Timothy Plageman
College of Optometry, The Ohio State University, Columbus, Ohio, USA

Alice Plein
UCL Institute of Ophthalmology, University College London, London, United Kingdom

Christiana Ruhrberg
UCL Institute of Ophthalmology, University College London, London, United Kingdom

Gerhard Schlosser
School of Natural Sciences & Regenerative Medicine Institute (REMEDI), National University of Ireland, Galway, Ireland

Elizabeth N. Schock
Division of Plastic Surgery, Department of Surgery, and Division of Developmental Biology, Department of Pediatrics, Cincinnati Children's Hospital Medical Center, Cincinnati, Ohio, USA

Philippe Soriano
Department of Developmental and Regenerative Biology, Icahn School of Medicine at Mount Sinai, New York, USA

Rolf W. Stottmann
Division of Developmental Biology, and Division of Human Genetics, Department of Pediatrics, Cincinnati Children's Hospital Medical Center, Cincinnati, Ohio, USA

Jun Suzuki
Department of Developmental Neuroscience, Centers for Neuroscience, and Department of Otorhinolaryngology-Head and Neck Surgery, Tohoku University Graduate School of Medicine, Sendai, Japan

Paul A. Trainor
Stowers Institute for Medical Research, Missouri, and Department of Anatomy and Cell Biology, University of Kansas Medical Center, Kansas City, Kansas, USA

PREFACE

Neural crest cells and placodes give rise to an extraordinary array of cell types and tissues. Neural crest cells form bone; cartilage; odontoblasts of teeth; connective tissue; cranial and trunk sensory neurons; peripheral autonomic neurons; and glia, smooth muscle, pigment, and endocrine cells. Ectodermal placodes contribute to the major sensory organs including the olfactory epithelium, lens of the eye, inner ear, and teeth and generate most of the cranial sensory neurons, together with hair and mammary glands. Neural crest cells and placodes are essential for embryonic development and adult homeostasis and are increasingly clinically significant. Collectively, they generate many of the defining characteristics of the craniates and have played major roles in vertebrate evolution.

Neural crest cells and placodes were discovered independently in the nineteenth century and in different species. Neural crest cells were first described by His (1868) in chick embryos, while placodes were described a little latter by van Wijhe (1883) in sharks. The study of neural crest cells and placodes exhibits a rich history, serving as important paradigms for vertebrate evolution, cell and tissue induction, epithelial to mesenchymal transformation, migration, and differentiation, while also providing a profound understanding of the underlying pathogenesis of congenital disorders. The persistence of neural crest cells and placodes into adulthood serves as important models of stem cell biology and tissue homeostasis and provides insights into cancer and metastasis.

Recent studies in tunicates and amphioxus point to neural crest cells and placodes having independent evolutionary origins. However, neural crest cells and placodes develop similarly in many respects and are mutually interdependent. This is particularly true with respect to evolution and development of the vertebrate head and more specifically the peripheral nervous system. For example, cranial neural crest cell-derived glia support placode-derived neurons during the formation and function of the cranial sensory ganglia. Furthermore, cranial neural crest cells establish corridors for the proper migration of epibranchial placode-derived neurons. These properties are a reflection of their extensive coevolution.

This issue of *Current Topics and Developmental Biology* highlights the current state of our knowledge concerning the evolution and development of neural crest cells and placodes throughout the entire body. Where and when did these specialized cells occur and how are they governed by signaling pathways and increasingly complex gene regulatory networks? What contributions do these cells make to specific tissues and organs and how are they integrated? The answers to these questions together with the derivation and application of stem cell-derived neural crest and placode cells in regenerative medicine have major implications for understanding and potentially treating congenital disorders.

Paul A. Trainor

If I have seen further it is by standing on the shoulders of Giants.

Isaac Newton

SECTION A

Neural Crest Cells

CHAPTER ONE

Neural Crest Cell Evolution: How and When Did a Neural Crest Cell Become a Neural Crest Cell

William A. Muñoz*, Paul A. Trainor*,†,1

*Stowers Institute for Medical Research, Kansas City, Missouri, USA
†Department of Anatomy and Cell Biology, University of Kansas Medical Center, Kansas City, Kansas, USA
[1]Corresponding author: e-mail address: pat@stowers.org

Contents

Abstract

As vertebrates evolved from protochordates, they shifted to a more predatory lifestyle, and radiated and adapted to most niches of the planet. This process was largely facilitated by the generation of novel vertebrate head structures, which were derived from neural crest cells (NCC). The neural crest is a unique vertebrate cell population that is frequently termed the "fourth germ layer" because it forms in conjunction with the other germ layers and contributes to a diverse array of cell types and tissues including the craniofacial skeleton, the peripheral nervous system, and pigment cells among many other tissues and cell types. NCC are defined by their origin at the neural plate border, via an epithelial-to-mesenchymal transition (EMT), together with multipotency and polarized patterns of migration. These defining characteristics, which evolved independently in the germ layers of invertebrates, were subsequently co-opted through their gene regulatory networks to form NCC in vertebrates. Moreover, recent data suggest that the ability to undergo an EMT was one of the latter features co-opted by NCC. In this review, we discuss the potential origins of NCC and how they evolved to contribute to nearly all tissues and organs throughout the body, based on paleontological evidence together with an evaluation of the evolution of molecules involved in NCC development and their migratory cell paths.

Current Topics in Developmental Biology, Volume 111
ISSN 0070-2153
http://dx.doi.org/10.1016/bs.ctdb.2014.11.001

1. INTRODUCTION

Neural crest cells (NCC) are considered to be a vertebrate innovation that significantly contributed to the ability of chordates to diversify and radiate to most niches on the planet. Originally identified by Wilhelm His in 1868 (Hall, 2000), NCC have been shown to contribute to almost all tissues throughout the body. NCC give rise to neurons, glia, Schwann cells, cartilage, bone, smooth muscles, adipocytes, and melanocytes, among many others (Table 1) (Bronner & LeDouarin, 2012; Dupin, Creuzet, & Le Douarin, 2006; Le Douarin & Dupin, 2012). Interestingly, many of these cell types originally arose from the other germ layers, particularly the mesoderm, in vertebrates and nonvertebrate chordates (Bronner & LeDouarin, 2012; Dupin et al., 2006; Etchevers, Vincent, Le Douarin, & Couly, 2001). The function of the NCC and their diversity of cell and tissue derivatives lent to the idea that NCC constituted a "fourth germ layer" (Hall, 2000).

One of the most significant accomplishments of the NCC was in contributing to evolution of a "new head" with a hinged jaw, special sense

Table 1 Contributions of NCC to tissues throughout the animal

Peripheral nervous system	**Fat tissue**
Cranial sensory ganglia	Adipocytes
Sympathetic ganglia	**Skin and inner ear**
Parasympathetic ganglia	Melanocytes
Sensory dorsal root ganglia	Dermal cells
Schwann cells	**Blood vessels and heart**
Ensheathing olfactory cells	Smooth muscle cells
Satellite cells of PNS ganglia	Pericytes
Central nervous system	Heart conotruncus
Meninges	**Striated muscles**
Enteric nervous system	Connective cells
Ganglia	Tendons
Glial cells	Extraocular muscles
Enteric neurons	**Craniofacial skeleton**
Endocrine system	Odontoblasts
Carotid body cells	Osteocytes
C cells (thyroid gland)	Chondrocytes
Adrenal-medullary cells	

The tissues with NCC contributions and the terminally differentiated NCC-derived cell types that populate the respective tissues are summarized here.

organs, and neural circuitry. These novel, predominately NCC-derived tissues facilitated vertebrates becoming predatory, shifting away from the filtration feeding lifestyle of their *Amphioxus*-like ancestors (Gans & Northcutt, 1983; Northcutt & Gans, 1983). Additionally, NCC have become integral in the organization of the vertebrate brain, possibly facilitating its enhanced growth in vertebrates (Creuzet, Martinez, & Le Douarin, 2006; Le Douarin, Couly, & Creuzet, 2012). Deficiencies in NCC development are known to result in various birth defects including craniofacial and heart anomalies, disorders affecting the bowel and other organs, and loss of pigmentation in the skin and hair. In contrast over-proliferation of NCC can result in several aggressive tumor types (Butler Tjaden & Trainor, 2013; Noack Watt & Trainor, 2014). Therefore, the innovation of NCC is one of the most significant factors contributing to vertebrate evolution and diversity. Understanding the mechanisms controlling the specification, migration, and terminal specification of NCC will provide insights into the evolutionary history of vertebrates and may lead to the development of therapies for treating disorders of NCC development, which are known collectively as neurocristopathies.

2. DEFINING NEURAL CREST CELLS

NCC have been the focus of extensive research since their initial discovery, particularly with respect to the mechanisms underlying their formation, the signals that determine how and where they migrate, and to what cell types and tissues they contribute. NCC are induced to form at the neural plate border, which is the junction between the neural ectoderm and surface ectoderm (Simoes-Costa & Bronner, 2013). During neurulation, the neural ectoderm elevates to form neural folds, which then join to form the neural tube. During this process dorsal neuroepithelial cells lose their intercellular connections, acquire apicobasal polarity, and undergo and epithelial-to-mesenchymal transition (EMT). These processes facilitate the delamination and migration of NCC in streams or in chains (Fig. 1A and B), which then proceed to their terminal sites of differentiation (Fig. 1C) (Baker & Bronner-Fraser, 1997; Groves & LaBonne, 2014; Mayor & Theveneau, 2013).

During their emigration from the neural plate or neural tube, NCC maintain a stem cell-like, multipotent state with the capacity for self-renewal (Bronner-Fraser & Fraser, 1988, 1989; Coelho-Aguiar, Le Douarin, & Dupin, 2013; Crane & Trainor, 2006; Dupin & Sommer, 2012; Le

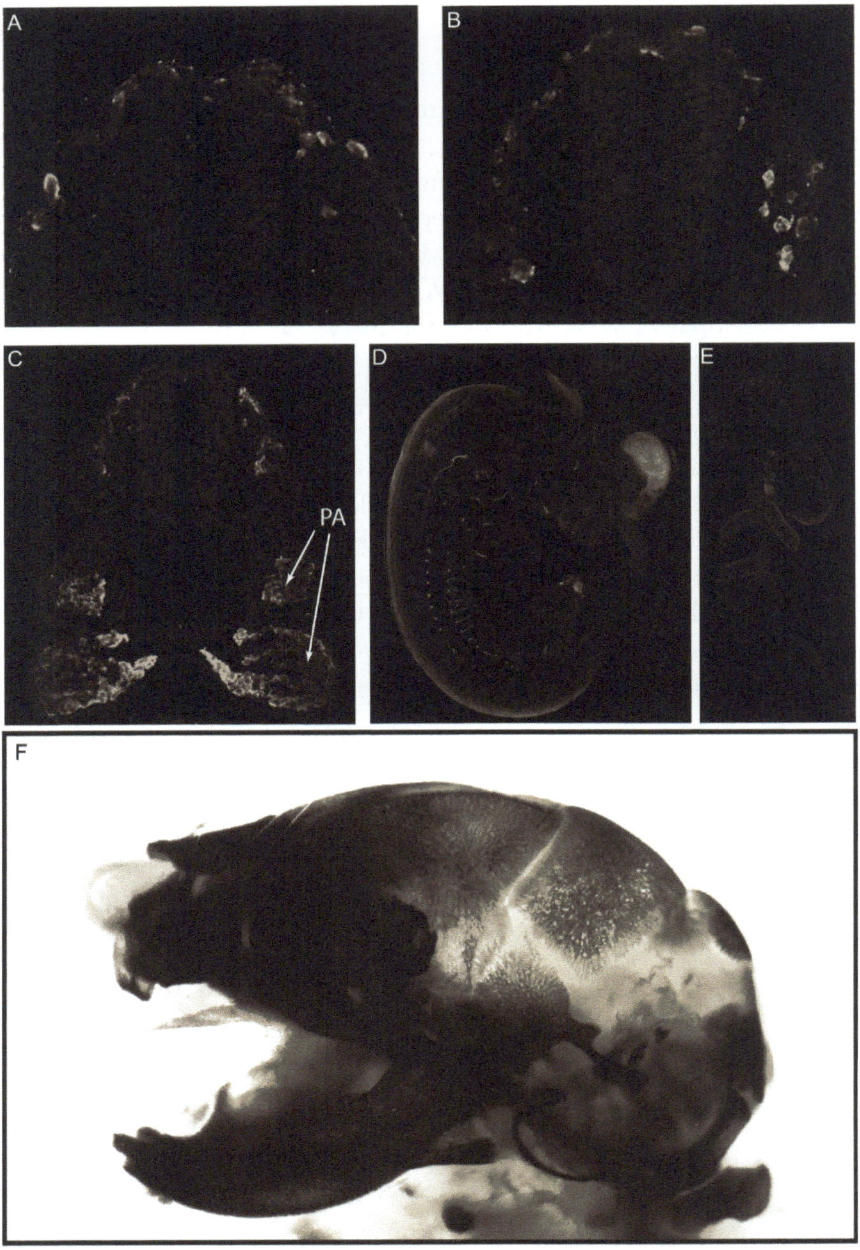

Figure 1 NCC migration and differentiation in mice. (A–C) Wnt1-Cre YFP mouse section at E10.5. Green labels NCC, blue labels DAPI stained nuclei. (A) NCC in the neural tube and start of emigration from the dorsal neural tube. (B) NCC migrating in streams from the neural tube. (C) NCC populating sites of terminal differentiation including pharyngeal arches (PA). (D) Terminally differentiated NCC-derived neurons in the peripheral nervous system in an E11.5 mouse stained for Tuj1 (green) and DAPI (blue). (E) Terminally differentiated NCC-derived neurons in the enteric nervous system of an E13.5 mouse stained for Tuj1 (red) and DAPI (blue). (F) NCC-derived bone, stained with alizarin red, and cartilage, stained with alcian blue, of the craniofacial skeleton of an E18.5 mouse. (See the color plate.)

Douarin, Calloni, & Dupin, 2008; Le Douarin, Creuzet, Couly, & Dupin, 2004; McKinney et al., 2013; Prasad, Sauka-Spengler, & LaBonne, 2012; Trentin, Glavieux-Pardanaud, Le Douarin, & Dupin, 2004). Interestingly, this stemness is partially retained in adult NCC in stem cell niches which can be isolated, purified, cultured, and used in neurodegenerative clinical applications (El-Nachef & Grikscheit, 2014; Greiner et al., 2014; Konig et al., 2014; Sanchez-Lara & Zhao, 2014; Trolle, Konig, Abrahamsson, Vasylovska, & Kozlova, 2014).

As NCC migrate throughout the developing embryo to their final destinations, they respond to various intrinsic and extrinsic signals promoting their proliferation, survival, and terminal differentiation into numerous cell types depending on their axial position (Barlow, Dixon, Dixon, & Trainor, 2012; Bhatt, Diaz, & Trainor, 2013; Walker & Trainor, 2006). NCC can be categorized as cranial, cardiac, vagal, trunk, and sacral based on their axial position of origin together with the cells and tissues they contribute to during terminal differentiation. Cranial NCC give rise to most of the bone and cartilage of the facial skeleton (Fig. 1F) and neurons and glia of the cranial ganglia (Fig. 1D), as well as smooth muscle and pigment cells. Cardiac NCC contribute to the valves, septa, and outflow tract of the heart. The vagal NCC form the enteric nervous system, which innervates the gastrointestinal tract (Fig. 1E). Trunk NCC differentiate to form melanocytes, secretory cells, and neurons and glia of the peripheral nervous system through their formation of dorsal root and sympathetic ganglia. Sacral NCC, which are a component of trunk NCC, also contribute to the enteric nervous system, however to a significantly less extent than the vagal NCC (Trainor, 2014; Simoes-Costa & Bronner, 2013). Despite this compartmentalization, each axial population of NCC maintains a latent capacity to differentiate into lineages derived from other NCC populations when isolated at the start of their migration and cultured *in vitro*, and when heterotypically transplanted *in vivo* (Calloni, Glavieux-Pardanaud, Le Douarin, & Dupin, 2007; Coelho-Aguiar et al., 2013; Le Douarin & Teillet, 1974; Le Lievre, Schweizer, Ziller, & Le Douarin, 1980; McGonnell & Graham, 2002).

Hörstadius posited the multipotency and regionalization of neural crest populations along the body axis in amphibians (Hörstadius, 1950, 1973). This was subsequently refined by Nicole Le Douarin through quail–chick chimeras, which demonstrated that specific regions of NCC would differentiate and contribute to adult tissues, and furthermore that NCC contributed to many more tissues than previously thought (Le Douarin & Kalcheim, 1999). As cell-labeling technologies improved, it was definitively demonstrated that an individual NCC is multipotent and capable of differentiating into very diverse cell types (Bronner-Fraser & Fraser, 1988; Chai et al., 2000; Jiang, Rowitch, Soriano, McMahon, & Sucov, 2000; Matsuoka et al., 2005).

Thus the defining features of NCC are their origin at the neural plate border, multipotency, formation via EMT, and acquisition of polarity and migratory ability, together with their regulation by a conserved gene regulatory network (GRN). These features must be considered collectively and individually in any analysis of the evolutionary origins of NCC.

3. CHORDATE EVOLUTION AND VERTEBRATE ORIGINS

For over a century, the origin of vertebrates and thus NCC has been debated based on shared structural homology between primitive vertebrates and protochordates (urochordates and cephalochordates) (Fig. 2), which had separated by the early Cambrian period (Dupret, Sanchez, Goujet, Tafforeau, & Ahlberg, 2014; Gai, Donoghue, Zhu, Janvier, & Stampanoni, 2011; Hall & Gillis, 2013; Mallatt & Chen, 2003). All animals in these subphyla have a similar body plan at a particular stage in their life, which includes a dorsal hollow neural tube and an axial structure, the

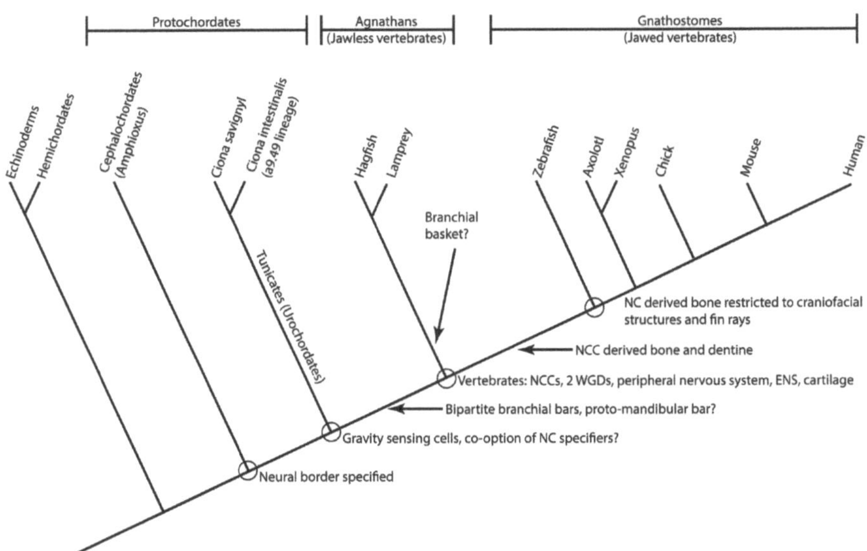

Figure 2 Cladogram showing the position of commonly studied model organisms of NCC development and evolution. Key evolutionary events contributing to NCC formation and function are indicated.

notochord, which collectively defines them as chordates (Hall & Gillis, 2013; Holland & Chen, 2001). Vertebrates are further characterized by the addition of a head and skeletal tissues both of which are absent in protochordates.

Comparative embryology suggests that the cephalochordates are the sister clade to vertebrates, and consistent with this idea, *Amphioxus* has been used extensively as a model to explore the origins of vertebrates (Bertrand & Escriva, 2011; Delsuc, Brinkmann, Chourrout, & Philippe, 2006; Holland, 2013; Holland et al., 2008). Consequently, the epidermal nerve plexus of protochordate-like ancestors resembling amphioxus was hypothesized to have evolved to produce the anterior parts of the head (Gans & Northcutt, 1983). Mechanistically, this was thought to be achieved through additions to the anterior regions of protochordates rather than a transformation resulting in evolution of a "new head." This addition of a "new head" is evident in Amphioxus by a notochord that extends to the most anterior region of the body. In contrast, in vertebrates the forebrain and sense organs lie rostral to the notochord. Additionally, a significant majority of the tissues that constitute the "new head" are derived from NCC (Gans & Northcutt, 1983; Le Douarin & Dupin, 2012). This strongly suggests that NCC were necessary for "new head" formation and therefore vertebrate evolution.

Interestingly, decades of extensive work in *Amphioxus* has not revealed the presence of bona fide primitive NCC-like cells by either morphology or molecular markers (Hall & Gillis, 2013; Yu, 2010). This missing evolutionary link may be partially explained by the recent genomic discovery that urochordates are in fact more closely related to vertebrates than cephalochordates (Fig. 2) (Delsuc et al., 2006). This change in subphyla organization provides new avenues for exploring NCC innovation and vertebrate evolution, as urochordates comprise a very diverse extant group with little known about their embryology. Recently, it was provocatively proposed that rudimentary NCC-like cells may exist in at least one genus of urochordates (Abitua, Wagner, Navarrete, & Levine, 2012).

4. NEURAL CREST CELL ORIGIN

The origin and diversification of NCC and ectodermal placodes were proposed by Gans and Northcutt to be associated with the shift to active predation in vertebrate evolution (Gans & Northcutt, 1983). With the evolution of NCC, vertebrates became quadroblastic with ectoderm, mesoderm, endoderm and NCC contributing to an enormous increase in cell diversity.

The significance of NCC in vertebrate evolution has made the identification of NCC-like structures during development and their derivatives in adults of nonvertebrate chordates of fundamental scientific importance.

Although, significant work has focused on *Amphioxus* in the pursuit of NCC precursors, no such cells have been identified to date. This lack of NCC-like candidates in cephalochordates could be a result of a true lack of this cell type in the subphyla or due to a lack of species diversity with so few extant species within the subphyla. It is highly unlikely that NCC-like cells will be found given the lack of typical NCC derivatives in amphioxus normally found in vertebrates, such as peripheral pigment cells and dentine-, bone-, or cartilage-forming cells. Interestingly, dorsal root nerves in these animals are ensheathed by Schwann cell-like glial cells (Bone, 1960; Peters, 1963). However, the origin for these glial cells has been shown to be ectodermal. Furthermore, numerous invertebrate animals also possess peripheral glial cells despite a complete absence of NCC-like cells (Coles & Abbott, 1996; Hall & Gillis, 2013). This suggests that the glial cell fate of ectodermal cells evolved prior to the divergence of NCC from ectoderm. Furthermore, the GRN for glial cell development may have not arisen separately in NCC, but rather been conserved in ectoderm-derived NCC and lost in the remainder of the ectodermal cells.

Recently, potential rudimentary NCC precursors in the form of pigment cells were identified in the mangrove tunicate, *Ciona intestinalis*, which belongs to the larger urochordate group. These cells originate near, though not necessarily from, the neural plate in blastomere pair *a7.6*. These cells migrate as individual cells either through the mesoderm or between the mesoderm and epidermis, eventually populating the siphon and body wall (Jeffery, Strickler, & Yamamoto, 2004). These cells and their derivatives express HNK-1, a carbohydrate epitope found in migrating NCC in many species together with *Zic3*, a transcription factor expressed in the vertebrate neural plate border but not in migrating NCC. Furthermore, HNK-1 positive cells in ascidians also express enzymes necessary for melanogenesis. This led to speculation that the evolutionary origins of a NCC precursor resided in cells fated to become pigment cells (Jeffery, 2006, 2007). However, there are significant discrepancies within these studies including but not limited to the fact that DiI lineage tracing labeled large populations of cells near the neural plate not just the *a7.6* blastomere pair. Furthermore, the molecular markers used label multiple cell types including neural plate, not just NCC (Jeffery et al., 2004). Blastomere *a7.6* derivatives include a number of mesendodermal cell types (Hirano & Nishida, 1997), and in other animals pigment cells may arise from mesodermal origins (Gibson & Burke, 1985). These studies highlight the difficulties associated with identifying potential NCC-like cells outside the vertebrate subphyla, given NCC mimic some behaviors of the mesoderm-derived cells in vertebrates. For example, NCC migrate significant distances with large protrusions and differentiate into various cell types in numerous tissues, including cell types that prior to the innovation of NCC were derived from the mesoderm such as chondroblasts and osteoblasts. Interestingly, this also suggests that the NCC may have co-opted developmental programs from other cell types during evolution.

More recently, it was found that blastomere *a9.49* found near the neural plate border of *C. intestinalis* expresses various neural plate border markers and perhaps some NCC specification signals (Abitua et al., 2012; Imai, Levine, Satoh, & Satou, 2006; Jeffery et al., 2008; Squarzoni, Parveen, Zanetti, Ristoratore, & Spagnuolo, 2011). Interestingly, cells derived from blastomere *a9.49* give rise to the gravity sensing otolith and the melanocytes in light-sensing tissues (Nishida & Satoh, 1989). Furthermore, ectopic expression of *Twist* in the *a9.49* blastomere endows the lineage with migratory ability. This suggests that a GRN of the type that might be expected in rudimentary NCC was indeed already present in this lineage, and furthermore, that it evolved before the divergence of urochordates and vertebrates. More importantly, the identification and characterization of this NCC-like cell lineage that gives rise to pigment cells in a urochordate suggest that the vertebrate NCC specification GRN was partially achieved by co-opting existing differentiation networks together with an EMT GRN to endow rudimentary NCC-like cells with migratory ability.

The identification of blastomere *a9.49* of *C. intestinalis* as a very strong potential candidate of primitive NCC-like cells is already a major step in the direction of identifying when and where NCC originated. It remains to be seen whether, given the extensive diversity of extant urochordates, that more NCC-like cells will be identified in this subphylum. If there were, it would suggest that NCC began to evolve in the Precambrian period. In addition, primitive NCC-like cells in urochordates may reveal what cell types NCC evolved from, together with their early functions.

5. NEURAL CREST CELL EVOLUTION IN VERTEBRATES

While protochordates have been important in identifying the potential origins of NCC and their influence on evolution and development, it is equally important to characterize how NCC have shaped the vertebrate subphyla. Numerous vertebrate model systems have been used including but not limited to mice, chick, amphibians, zebrafish, and the jawless basal vertebrates, lamprey and hagfish. Despite highly conserved GRNs, vertebrate NCC are able to create many diverse structures from the shell of a turtle to the heart outflow tract of a mammal (Cebra-Thomas et al., 2007; Gilbert, Bender, Betters, Yin, & Cebra-Thomas, 2007; Trainor, 2014).

Analyses of NCC development were first carried out in amphibians and fish, and later in avians and other amniotes (Hörstadius, 1950; Le Douarin, 1969), and much of the early focus centered on the contributions of NCC to the craniofacial skeleton. NCC contribute to the cranium, facial skeleton, connective and adipose tissues, dermis of the face, ventral part of the neck, and several sense organs (Couly, Coltey, & Le Douarin, 1993; Creuzet et al., 2006; Le Douarin & Kalcheim, 1999; Le Lievre et al., 1980). However, it has become apparent that during vertebrate evolution, NCC have evolved to participate in numerous other functions providing specific advantages suited to an animal's environment. The most influential evolutionary effects of the NCC on vertebrates were through contributions that improved metabolism, circulation, and respiration with significant changes from one evolutionary branch to the next (Green & Bronner, 2014; Landacre, 1921; Lievre, 1974; Mongera et al., 2013).

NCC contribute to skeletal structures through their differentiation into chondrocytes that form primary and secondary cartilage, osteoblasts forming dermal and endochondral bone, and odontoblasts producing teeth and scale dentine. Of these cell types only dentine secreting cells are exclusively a NCC derivative. All the other cell types can also arise from the mesoderm (Hall, 2009; Hall & Gillis, 2013). This makes dentine a definitive marker of NCC in fossils and in tissues today. Congruent with studies that NCC-derived bone formation is important for vertebrate evolution, it has been suggested that the first mineralized skeletal tissues of the vertebrate subphylum were of NCC origin (Fig. 2). Early vertebrates, such as pteraspidomorphs agnathans, possessed ossified armor consisting of dentine produced by odontoblasts (Janvier, 1996; Le Douarin & Dupin, 2012; Smith, 1991). Interestingly, these animals are devoid of well-characterized vertebrae suggesting the presence of a cartilaginous endoskeleton most likely derived from the somites (Ota, Fujimoto, Oisi, & Kuratani, 2011). As vertebrate evolution proceeded, gnathostomes began to develop an endoskeleton made of somite-derived cellular bone, while still retaining their dentine-based exoskeleton. Later, however, as the endoskeleton became fully developed, the putatively NCC-derived exoskeleton was lost and only the NCC-derived skull, facial bones, and cartilages remained (Janvier, 2011). Interestingly, while trunk NCC generally no longer contribute to skeletal formation, they still maintain a latent ability *in vitro* to produce mesenchymal derivatives including skeletal tissues from both birds and mammals (Calloni et al., 2007; Coelho-Aguiar et al., 2013; Ido & Ito, 2006; McGonnell & Graham, 2002; Nakamura & Ayer-le Lievre, 1982).

In addition to these skeletal studies, extensive work has focused on the most basal extant vertebrate, the jawless lamprey. The lamprey is used to understand the role of NCC in the progressive evolution of vertebrates after their separation from protochordates. Interestingly, the NCC GRN, which is discussed later in this review, is mostly conserved in lamprey when compared to gnathostomes suggesting that it has an ancient origin (Medeiros, 2013).

Histological studies of lamprey revealed the presence of NCC in embryos and their contributions to numerous tissues in adults (Hardisty, 1979; Medeiros, 2013). Furthermore, transplantation experiments showed that cranial NCC grafted to the flank of lamprey or newts generated neurons, pigment cells, and cartilage nodules, while ablation experiments demonstrated that removal of NCC resulted in reductions of typical NCC derivatives (Langille & Hall, 1986, 1988; Newth, 1951, 1956). Vital dye labeling experiments subsequently confirmed the primary observations and conclusions drawn from NCC transplantations and ablations (Horigome et al., 1999; McCauley & Bronner-Fraser, 2003). However, despite the significant conservation between lamprey and gnathostomes, there are some differences in the NCC GRN, which will be discussed later. Nonetheless, these studies in lamprey show that the NCC have an ancient origin predating the divergence of lamprey and gnathostomes despite a complete lack of jaw structures in lamprey.

6. CRANIAL NEURAL CREST CELL GENE REGULATORY NETWORK

NCC development comprises several distinct steps and events including specification and induction at the neural plate border, delamination through EMT, migration, and finally terminal differentiation. To help understand these complex processes, molecular manipulation has been used to identify, characterize, and describe interactions between a considerable number of genes involved in these processes of NCC development in species spanning the vertebrate subphyla from jawless vertebrates to mice (Ota, Kuraku, & Kuratani, 2007; Sauka-Spengler, Meulemans, Jones, & Bronner-Fraser, 2007). These data have subsequently been assembled into a putative conserved GRN to help explain the complex events of NCC development (Betancur, Bronner-Fraser, & Sauka-Spengler, 2010a, 2010b; Meulemans & Bronner-Fraser, 2004; Sauka-Spengler & Bronner-Fraser, 2008). It is important to note that the proposed cranial GRN cannot account for all NCC processes along the body axis as each region of NCC exhibits a distinct repertoire of cell fates with some variations between species as well. Examples of these variations are exhibited by *Snail* and *Twist*, which are expressed in postmigratory cells in lamprey while they are expressed in premigratory and migratory NCC of gnathostomes (Rahimi, Allmond, Wagner, McCauley, & Langeland, 2009; Sauka-Spengler et al., 2007). Additional variations in the NCC GRN nodes exist between less basal species. Mouse for example does not require Snail1 or Snail2 for NCC formation (Murray & Gridley, 2006) nor Pax3 and Pax7, because individual or combined loss-of-function does not affect NCC development (our unpublished results). Furthermore, in *Xenopus laevis,* Snail and Twist regulate mesoderm formation in addition to NCC formation (Shi, Severson, Yang, Wedlich, & Klymkowsky, 2011).

The NCC GRN is subdivided into a series of temporal steps, with each step governed by a distinct regulatory module. This network is initiated at the neural plate border by several ligands secreted from the neuroectoderm, nonneural ectoderm, and underlying mesoderm, including Wnts, BMPs, FGFs, and retinoic acid. These extracellular signals act during gastrulation, activating a transcriptional program that specifies the neural plate border, making it competent to produce NCC (Groves & LaBonne, 2014; Munoz, Sater, & McCrea, 2014).

BMPs form a gradient in the ectoderm, with low levels specifying the neural plate, high levels the surface ectoderm, and intermediate levels the neural plate border. The intermediate levels of BMP are thought to work together with ectoderm-derived Wnt signaling and underlying mesoderm-derived Fgf signaling, which leads to the activation of transcription factors such as *Pax3/7, Zic1*, and *Msx1* in the neural plate border and the competency to form NCC (Basch, Bronner-Fraser, & Garcia-Castro, 2006; Betancur et al., 2010a; Huang & Saint-Jeannet, 2004; Munoz et al., 2014).

The neural plate border transcription factors are expressed not only in the neural plate border but also in the immediate surrounding areas as well where ectodermal placodes will form. The neural plate border becomes further refined as the extracellular signals continue to enhance the neural plate border specifiers eventually activating "NCC specifiers." Included in these specifiers are conserved transcription factors *Snail/Slug, Id, FoxD3, cMyc*, and *Sox9/10*. These genes will begin to repress epidermal and neural tube markers, such as Sox2 and E-cadherin, while upregulating factors necessary for migration, such as cadherin-7 and various matrix metalloproteinases. This facilitates EMT and the delamination of NCC from the neuroepithelium. These genes also play reiterated roles in the terminal differentiation of several NCC lineages (Bronner & LeDouarin, 2012; Hall & Gillis, 2013; Milet & Monsoro-Burq, 2012).

This presumptive NCC GRN is well conserved across multiple species and continues to evolve. Novel genomic high-throughput methodologies continue to reveal additional transcription factors, as well as epigenetic, posttranscriptional, and posttranslational regulators, broadening the network (Attanasio et al., 2013; Brunskill et al., 2014; Simoes-Costa & Bronner, 2013; Simoes-Costa, Tan-Cabugao, Antoshechkin, Sauka-Spengler, & Bronner, 2014).

7. EVOLUTION OF NEURAL CREST CELL GENE REGULATORY NETWORKS

NCC formation, migration, and differentiation are controlled by a complex GRN consisting of both intrinsic and extrinsic input. How these transcriptional regulators and signals were first incorporated into a GRN during evolution is beginning to elucidate the origin of NCC. In early vertebrate evolution, there were two whole-genome duplication events (WGD), not found in closely related invertebrates (Fig. 2) (Green & Bronner, 2013; Holland, 1999; Ohno, 1999; Simoes-Costa & Bronner, 2013; Taylor & Raes, 2004). These two WGD appear to coincide with the evolutionary origin of NCC that are found in all vertebrates. This suggests that WGD may have contributed to or facilitated the diversification of NCC from the potentially rudimentary neural crest-like cells found in nonvertebrate chordates (Abitua et al., 2012). WGD could have accomplished this through several mechanisms including alterations in gene expression patterns, creation of novel genes through domain shuffling, the appearance of novel sequence motifs in transcription factors, or the co-option of genes and their functions leading to the generation of new cell types (Fig. 3).

Co-option occurs when a feature or molecule that performs a function is repurposed for another use. Despite *Amphioxus* lacking NCC, they possess

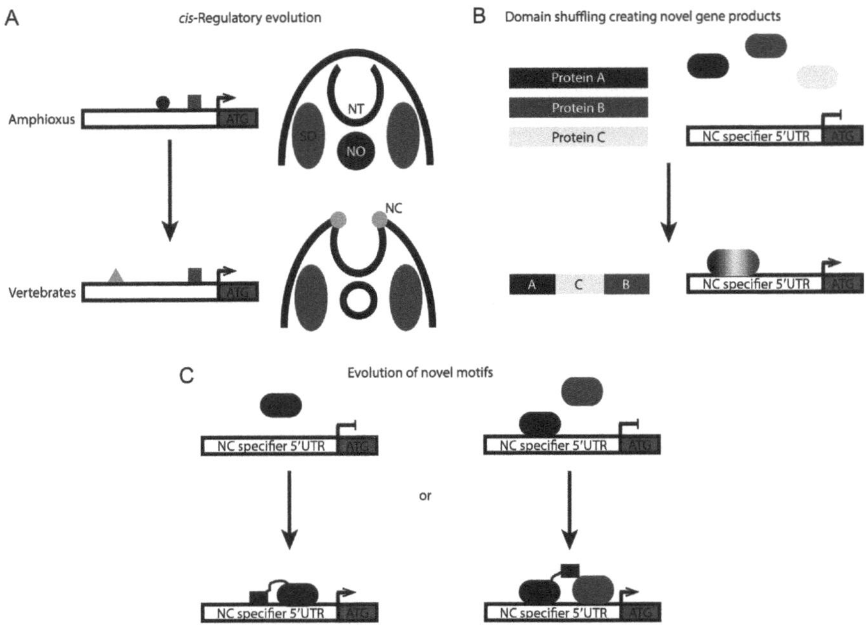

Figure 3 The potential mechanisms contributing to the molecular evolution of NCC GRN components. (A) Evolution of *cis*-regulatory elements of NCC specifiers. Shapes indicate tissue-specific, *cis*-regulatory elements on the DNA with colors corresponding to expression in the tissues of the embryo section on right. As vertebrates evolved from *Amphioxus*, loss-of-function mutations in regulatory elements (loss of circle) resulted in loss of expression in corresponding tissue. Concurrently, a novel *cis*-regulatory element (triangle) evolved to promote expression of the NCC specifier in the evolving NCC. NT, neural tube; NO, notochord; SO, somites. (B) Evolution of novel gene product via domain shuffling of chromosomes. Several gene products unspliced do not bind to regulatory elements of NCC specifier genes. Following domain shuffling, a novel gene, consisting of regions of each original gene, is created that can bind and activate transcription of the NCC specifier promoting NCC development. (C) Evolution of novel protein motifs in existing molecules. Proteins evolve either a novel motif allowing DNA binding or interactions with additional transcription factors promoting transcription of a NCC specifier. (See the color plate.)

homologs of most genes known to be necessary for neural crest specification, migration, and terminal differentiation (i.e., *Snail1/2*, *AP-2*, *FoxD3*, *Twist*, *Id*, etc.). Interestingly however, none of these molecules are expressed at the neural plate border in *Amphioxus* except *Snail* (Holland, 2013; Holland et al., 2008; Meulemans & Bronner-Fraser, 2005, 2007; Meulemans, McCauley, & Bronner-Fraser, 2003). This indicates that NCC evolved in part by co-opting genes into a primitive gene network that was already present in the neural plate border of the ancestral chordate.

It has been assumed that for a gene to be expressed in a new domain, it must have acquired new *cis*-regulatory control enhancers. One study supporting this hypothesis involves the vertebrate transcriptional repressor FoxD3. *FoxD* genes are members of the Fox family of transcription factors with only one *FoxD* gene identified in amphioxus. In amphioxus, *FoxD* is expressed in the forebrain, somites, and notochord, while in vertebrates, five duplicates have independent regions of expression. *FoxD3* is the only duplicate that is expressed in NCC, but *FoxD3* is also expressed in the somites and spinal cord. This demonstrates that FoxD3 acquired a new domain of activity in NCC, where it maintains NCC potential at the neural plate border and prevents melanocyte differentiation (Ignatius, Moose, El-Hodiri, & Henion, 2008; Teng, Muncell, Frist, Wang, & Labosky, 2008; Thomas & Erickson, 2008). To investigate if a regulatory region for *FoxD3* had evolved since a common ancestor, a previously identified regulatory region of *Amphioxus FoxD* (Yu, Holland, & Holland, 2004) was introduced into chick where it directed expression to the somites, notochord, and neural tube, but not to neural crest (Yu, Meulemans, McKeown, & Bronner-Fraser, 2008). These studies demonstrate that evolution of a regulatory region was necessary for *FoxD3* to be expressed in the neural plate border region and potentially become part of the NCC GRN (Fig. 3A). However, this work does not address whether the spatiotemporal expression was sufficient for *FoxD* homologs to contribute to NCC specification.

Given the significant evolutionary distance between vertebrates and *Amphioxus*, it is expected that there would be differences in the protein structure as well as in the regulatory regions. This made it a distinct possibility that FoxD3 had also developed a novel protein function allowing it to regulate neural crest specification beyond *Amphioxus* FoxD activity. To test this idea numerous chimeric versions of *Amphioxus* FoxD with zebrafish FoxD3, which alone induces ectopic NCC in the neural tube, were tested for their ability to induce ectopic NCC. The chimeras and full-length proteins, including the FoxD3 paralogs, were assayed for their ability to induce ectopic NCC when electroporated into chick neural tubes. The FoxD3 paralogs and *Amphioxus* FoxD could not induce ectopic NCC; however, a chimeric construct replacing the first 39 amino acids of *Amphioxus* FoxD with the first 39 amino acids of Xenopus FoxD3 was able to induce ectopic NCC. This study identified a necessary amino-terminal region of FoxD3 that has evolved specifically allowing FoxD3 to induce NCC when expressed in the neural tube (Fig. 3C). These two studies demonstrate that neural crest specifiers may not require just one evolutionary advantage over their cephalochordate homologs.

As NCC specifiers were modified and co-opted into the NCC GRN, various activities distinct from the surrounding ectoderm and neurectoderm were acquired, including the ability to undergo EMT, migrate away from the neural tube and differentiate into novel cell types. Among these, it has been suggested that molecules governing EMT were the last to be co-opted into the GRN. In *Amphioxus*, many of the NCC specifiers are expressed in mesenchymal structures, with only the conserved NCC specifier *Snail* also being expressed at the neural plate border (Imai, Hino, Yagi, Satoh, & Satou, 2004). Interestingly in *Ciona*, the NCC-like cell (blastomere *a9.49*) derivatives that form pigment cells express many of the NCC specification genes, including *Id*, *FoxD*, *Snail*, and *Ets* but do not migrate as mesenchymal cells. However, *a9.49* derivatives can be transformed into mesenchymal-like cells through ectopic expression of *Twist*, which is required for the EMT of mesodermal cells in *Ciona* (Abitua et al., 2012). This demonstrates that much of the molecular machinery to transform these NCC-like cells into mesenchymal cell types is present in *Ciona*, but it still lacks key regulators to promote EMT.

8. CONCLUSIONS AND PERSPECTIVES

In this chapter, we have briefly summarized our knowledge regarding the properties of NCC, a vertebrate innovation that facilitated the extensive evolutionary diversity and expansion of vertebrates. Recent work has begun to identify and characterize the coincident origins of NCC and vertebrates. The data obtained in various species and phyla provide an overview of NCC and how they are defined by their differing regulatory states, position at particular developmental time points, and ability to form a wide variety of cell types. The regulatory state referenced here as the cranial NCC GRN still requires significant future work to further define additional interactions that have not been identified yet and continue to characterize the interactions that have already been established. However, there are still significant gaps in our knowledge of the NCC GRN with little known about the earliest NCC specifiers that commit NCC to their fates. Furthermore, as NCC migrate and undergo differentiation, our understanding of what signals they receive and how they are processed remains incomplete. Within the already defined NCC GRN there remain undefined interactions with recent work describing crosstalk between several known components and potential post-translational regulation that has yet to be considered in these models. It will be important to understand these mechanisms, so comparisons with other networks can be established to illuminate how various cellular pathways were adapted and integrated into the GRN.

Of key importance to furthering our understanding of the NCC regulatory states will be continued efforts to understand the GRN of NCC at each axial level during their formation and migration. This will require comparable studies to those accomplished with the cranial NCC already, across multiple species and at various times of development possibly using the cranial NCC GRN as a guidepost. Overlap with cranial NCC is expected, but so too is significant variance and specificity within each axial region, illustrating a much broader involvement of the genome. It will be important in these studies to monitor and identify changes outside the defined GRN with respect to epigenetic, posttranscriptional, and posttranslational modifications. High-throughput sequencing combined with various molecular perturbations has already begun to provide insight into these questions highlighting large changes in transcriptional profiles as NCC progress through each regulatory state. Epigenetic profiling is a maturing field and can now be easily performed in parallel to RNA-seq experiments to understand the effects of various perturbations on chromatin structure and function. Finally, proteomics can provide another method to monitor whole proteome changes in NCC at distinct axial levels and stages of development. These high-throughput methods will continue to evolve and become major resources for expanding the GRN of NCC.

Aside from the GRN, there is still much to be understood concerning how NCC evolved from primitive NCC-like cells. With a shift in research from cephalochordates to urochordates, it is likely that NCC-like cells will be identified in a wide variety of these diverse extant subphyla. The identification of additional NCC-like cells and variation between the cells, their derivatives, and active pathways will promote our understanding of the steps that were necessary for NCC formation prior to the divergence of vertebrates from protochordates. These steps will hopefully reveal how NCC specifiers came to be expressed in the neural plate border and how they came to regulate gene products necessary for NCC specification and differentiation. Furthermore, they may also help to reveal the factors necessary for the induction of mammalian NCC. Although we understand many of the signals required for NCC formation in avian and aquatic species, the counterparts underpinning mammalian NCC formation remain to be functionally determined. What is clear is that the origins of NCC are closely linked to evolution of the vertebrate lineage, and disruptions of NCC development result in neurocristopathies, demonstrating their significance in vertebrates.

Despite the phenomenal work that has been accomplished to date to understand the origins, mechanisms, and evolution of NCC, the field is still relatively young with numerous questions and gaps in our knowledge. Technological advances will provide exciting opportunities to address these outstanding questions and significantly impact the field of developmental biology.

ACKNOWLEDGMENTS

The authors thank Annita Achilleos, Naomi Tjaden, and George Bugarinovic for images used in Fig. 1A–C, E, and F, respectively. Research in the Trainor laboratory is supported by the Stowers Institute for Medical Research and the National Institute of Dental and Craniofacial Research (DE 016082).

REFERENCES

Abitua, P. B., Wagner, E., Navarrete, I. A., & Levine, M. (2012). Identification of a rudimentary neural crest in a non-vertebrate chordate. *Nature*, *492*(7427), 104–107. http://dx.doi.org/10.1038/nature11589.

Attanasio, C., Nord, A. S., Zhu, Y., Blow, M. J., Li, Z., Liberton, D. K., et al. (2013). Fine tuning of craniofacial morphology by distant-acting enhancers. *Science*, *342*(6157), 1241006. http://dx.doi.org/10.1126/science.1241006.

Baker, C. V., & Bronner-Fraser, M. (1997). The origins of the neural crest. Part I: Embryonic induction. *Mechanisms of Development*, *69*(1–2), 3–11.

Barlow, A. J., Dixon, J., Dixon, M. J., & Trainor, P. A. (2012). Balancing neural crest cell intrinsic processes with those of the microenvironment in Tcof1 haploinsufficient mice enables complete enteric nervous system formation. *Human Molecular Genetics*, *21*(8), 1782–1793. http://dx.doi.org/10.1093/hmg/ddr611.

Basch, M. L., Bronner-Fraser, M., & Garcia-Castro, M. I. (2006). Specification of the neural crest occurs during gastrulation and requires Pax7. *Nature*, *441*(7090), 218–222. http://dx.doi.org/10.1038/nature04684.

Bertrand, S., & Escriva, H. (2011). Evolutionary crossroads in developmental biology: Amphioxus. *Development*, *138*(22), 4819–4830. http://dx.doi.org/10.1242/dev.066720.

Betancur, P., Bronner-Fraser, M., & Sauka-Spengler, T. (2010a). Assembling neural crest regulatory circuits into a gene regulatory network. *Annual Review of Cell and Developmental Biology*, *26*, 581–603. http://dx.doi.org/10.1146/annurev.cellbio.042308.113245.

Betancur, P., Bronner-Fraser, M., & Sauka-Spengler, T. (2010b). Genomic code for Sox10 activation reveals a key regulatory enhancer for cranial neural crest. *Proceedings of the National Academy of Sciences of the United States of America*, *107*(8), 3570–3575. http://dx.doi.org/10.1073/pnas.0906596107.

Bhatt, S., Diaz, R., & Trainor, P. A. (2013). Signals and switches in mammalian neural crest cell differentiation. *Cold Spring Harbor Perspectives in Biology*, *5*(2), 243–262. http://dx.doi.org/10.1101/cshperspect.a008326.

Bone, Q. (1960). The central nervous system in amphioxus. *The Journal of Comparative Neurology*, *115*(1), 27–64. http://dx.doi.org/10.1002/cne.901150105.

Bronner, M. E., & LeDouarin, N. M. (2012). Development and evolution of the neural crest: An overview. *Developmental Biology*, *366*(1), 2–9. http://dx.doi.org/10.1016/j.ydbio.2011.12.042.

Bronner-Fraser, M., & Fraser, S. E. (1988). Cell lineage analysis reveals multipotency of some avian neural crest cells. *Nature, 335*(6186), 161–164. http://dx.doi.org/10.1038/335161a0.

Bronner-Fraser, M., & Fraser, S. (1989). Developmental potential of avian trunk neural crest cells in situ. *Neuron, 3*(6), 755–766.

Brunskill, E. W., Potter, A. S., Distasio, A., Dexheimer, P., Plassard, A., Aronow, B. J., et al. (2014). A gene expression atlas of early craniofacial development. *Developmental Biology, 391*(2), 133–146. http://dx.doi.org/10.1016/j.ydbio.2014.04.016.

Butler Tjaden, N. E., & Trainor, P. A. (2013). The developmental etiology and pathogenesis of Hirschsprung disease. *Translational Research, 162*(1), 1–15. http://dx.doi.org/10.1016/j.trsl.2013.03.001.

Calloni, G. W., Glavieux-Pardanaud, C., Le Douarin, N. M., & Dupin, E. (2007). Sonic Hedgehog promotes the development of multipotent neural crest progenitors endowed with both mesenchymal and neural potentials. *Proceedings of the National Academy of Sciences of the United States of America, 104*(50), 19879–19884. http://dx.doi.org/10.1073/pnas.0708806104.

Cebra-Thomas, J. A., Betters, E., Yin, M., Plafkin, C., McDow, K., & Gilbert, S. F. (2007). Evidence that a late-emerging population of trunk neural crest cells forms the plastron bones in the turtle *Trachemys scripta. Evolution & Development, 9*(3), 267–277. http://dx.doi.org/10.1111/j.1525-142X.2007.00159.x.

Chai, Y., Jiang, X., Ito, Y., Bringas, P., Jr., Han, J., Rowitch, D. H., et al. (2000) Fate of the mammalian cranial neural crest during tooth and mandibular morphogenesis. *Development, 127*(8), 1671–1679.

Coelho-Aguiar, J. M., Le Douarin, N. M., & Dupin, E. (2013). Environmental factors unveil dormant developmental capacities in multipotent progenitors of the trunk neural crest. *Developmental Biology, 384*(1), 13–25. http://dx.doi.org/10.1016/j.ydbio.2013.09.030.

Coles, J. A., & Abbott, N. J. (1996). Signalling from neurones to glial cells in invertebrates. *Trends in Neurosciences, 19*(8), 358–362.

Couly, G. F., Coltey, P. M., & Le Douarin, N. M. (1993). The triple origin of skull in higher vertebrates: A study in quail–chick chimeras. *Development, 117*(2), 409–429.

Crane, J. F., & Trainor, P. A. (2006). Neural crest stem and progenitor cells. *Annual Review of Cell and Developmental Biology, 22*, 267–286. http://dx.doi.org/10.1146/annurev.cellbio.22.010305.103814.

Creuzet, S. E., Martinez, S., & Le Douarin, N. M. (2006). The cephalic neural crest exerts a critical effect on forebrain and midbrain development. *Proceedings of the National Academy of Sciences of the United States of America, 103*(38), 14033–14038. http://dx.doi.org/10.1073/pnas.0605899103.

Delsuc, F., Brinkmann, H., Chourrout, D., & Philippe, H. (2006). Tunicates and not cephalochordates are the closest living relatives of vertebrates. *Nature, 439*(7079), 965–968. http://dx.doi.org/10.1038/nature04336.

Dupin, E., Creuzet, S., & Le Douarin, N. M. (2006). The contribution of the neural crest to the vertebrate body. *Advances in Experimental Medicine and Biology, 589*, 96–119. http://dx.doi.org/10.1007/978-0-387-46954-6_6.

Dupin, E., & Sommer, L. (2012). Neural crest progenitors and stem cells: From early development to adulthood. *Developmental Biology, 366*(1), 83–95. http://dx.doi.org/10.1016/j.ydbio.2012.02.035.

Dupret, V., Sanchez, S., Goujet, D., Tafforeau, P., & Ahlberg, P. E. (2014). A primitive placoderm sheds light on the origin of the jawed vertebrate face. *Nature, 507*(7493), 500–503. http://dx.doi.org/10.1038/nature12980.

El-Nachef, W., & Grikscheit, T. (2014). Enteric nervous system cell replacement therapy for Hirschsprung disease: Beyond tissue-engineered intestine. *European Journal of Pediatric Surgery, 24*(3), 214–218. http://dx.doi.org/10.1055/s-0034-1376311.

Etchevers, H. C., Vincent, C., Le Douarin, N. M., & Couly, G. F. (2001). The cephalic neural crest provides pericytes and smooth muscle cells to all blood vessels of the face and forebrain. *Development, 128*(7), 1059–1068.

Gai, Z., Donoghue, P. C., Zhu, M., Janvier, P., & Stampanoni, M. (2011). Fossil jawless fish from China foreshadows early jawed vertebrate anatomy. *Nature, 476*(7360), 324–327. http://dx.doi.org/10.1038/nature10276.

Gans, C., & Northcutt, R. G. (1983). Neural crest and the origin of vertebrates: A new head. *Science, 220*(4594), 268–273. http://dx.doi.org/10.1126/science.220.4594.268.

Gibson, A. W., & Burke, R. D. (1985). The origin of pigment cells in embryos of the sea urchin *Strongylocentrotus purpuratus. Developmental Biology, 107*(2), 414–419.

Gilbert, S. F., Bender, G., Betters, E., Yin, M., & Cebra-Thomas, J. A. (2007). The contribution of neural crest cells to the nuchal bone and plastron of the turtle shell. *Integrative and Comparative Biology, 47*(3), 401–408. http://dx.doi.org/10.1093/icb/icm020.

Green, S. A., & Bronner, M. E. (2013). Gene duplications and the early evolution of neural crest development. *Seminars in Cell & Developmental Biology, 24*(2), 95–100. http://dx.doi.org/10.1016/j.semcdb.2012.12.006.

Green, S. A., & Bronner, M. E. (2014). The lamprey: A jawless vertebrate model system for examining origin of the neural crest and other vertebrate traits. *Differentiation, 87*(1–2), 44–51. http://dx.doi.org/10.1016/j.diff.2014.02.001.

Greiner, J. F., Grunwald, L. M., Muller, J., Sudhoff, H., Widera, D., Kaltschmidt, C., et al. (2014). Culture bag systems for clinical applications of adult human neural crest-derived stem cells. *Stem Cell Research & Therapy, 5*(2), 34. http://dx.doi.org/10.1186/scrt422.

Groves, A. K., & LaBonne, C. (2014). Setting appropriate boundaries: Fate, patterning and competence at the neural plate border. *Developmental Biology, 389*(1), 2–12. http://dx.doi.org/10.1016/j.ydbio.2013.11.027.

Hall, B. K. (2000). The neural crest as a fourth germ layer and vertebrates as quadroblastic not triploblastic. *Evolution & Development, 2*(1), 3–5.

Hall, B. K. (2009). *The neural crest and neural crest cells in vertebrate development and evolution.* New York: Springer.

Hall, B. K., & Gillis, J. A. (2013). Incremental evolution of the neural crest, neural crest cells and neural crest-derived skeletal tissues. *Journal of Anatomy, 222*(1), 19–31. http://dx.doi.org/10.1111/j.1469-7580.2012.01495.x.

Hardisty, M. W. (1979). *Biology of the cyclostomes.* London/New York: Chapman and Hall/Distributed by Halsted Press.

Hirano, T., & Nishida, H. (1997). Developmental fates of larval tissues after metamorphosis in ascidian *Halocynthia roretzi.* I. Origin of mesodermal tissues of the juvenile. *Developmental Biology, 192*(2), 199–210.

Holland, P. W. (1999). Gene duplication: Past, present and future. *Seminars in Cell & Developmental Biology, 10*(5), 541–547. http://dx.doi.org/10.1006/scdb.1999.0335.

Holland, L. Z. (2013). Evolution of new characters after whole genome duplications: Insights from amphioxus. *Seminars in Cell & Developmental Biology, 24*(2), 101–109. http://dx.doi.org/10.1016/j.semcdb.2012.12.007.

Holland, L. Z., Albalat, R., Azumi, K., Benito-Gutierrez, E., Blow, M. J., Bronner-Fraser, M., et al. (2008). The amphioxus genome illuminates vertebrate origins and cephalochordate biology. *Genome Research, 18*(7), 1100–1111. http://dx.doi.org/10.1101/gr.073676.107.

Holland, N. D., & Chen, J. (2001). Origin and early evolution of the vertebrates: New insights from advances in molecular biology, anatomy, and palaeontology. *Bioessays, 23*(2), 142–151. http://dx.doi.org/10.1002/1521-1878(200102)23:2<142::AID-BIES1021>3.0.CO;2-5.

Horigome, N., Myojin, M., Ueki, T., Hirano, S., Aizawa, S., & Kuratani, S. (1999). Development of cephalic neural crest cells in embryos of *Lampetra japonica,* with special

reference to the evolution of the jaw. *Developmental Biology, 207*(2), 287–303. http://dx.doi.org/10.1006/dbio.1998.9175.
Hörstadius, S. O. (1950). *The neural crest: Its properties and derivatives in the light of experimental research.* London, New York: Oxford University Press.
Hörstadius, S. (1973). *Experimental embryology of echinoderms.* Oxford, England Clarendon Press.
Huang, X., & Saint-Jeannet, J. P. (2004). Induction of the neural crest and the opportunities of life on the edge. *Developmental Biology, 275*(1), 1–11. http://dx.doi.org/10.1016/j.ydbio.2004.07.033.
Ido, A., & Ito, K. (2006). Expression of chondrogenic potential of mouse trunk neural crest cells by FGF2 treatment. *Developmental Dynamics, 235*(2), 361–367. http://dx.doi.org/10.1002/dvdy.20635.
Ignatius, M. S., Moose, H. E., El-Hodiri, H. M., & Henion, P. D. (2008). Colgate/hdac1 repression of foxd3 expression is required to permit mitfa-dependent melanogenesis. *Developmental Biology, 313*(2), 568–583. http://dx.doi.org/10.1016/j.ydbio.2007.10.045.
Imai, K. S., Hino, K., Yagi, K., Satoh, N., & Satou, Y. (2004). Gene expression profiles of transcription factors and signaling molecules in the ascidian embryo: Towards a comprehensive understanding of gene networks. *Development, 131*(16), 4047–4058. http://dx.doi.org/10.1242/dev.01270.
Imai, K. S., Levine, M., Satoh, N., & Satou, Y. (2006). Regulatory blueprint for a chordate embryo. *Science, 312*(5777), 1183–1187. http://dx.doi.org/10.1126/science.1123404.
Janvier, P. (1996). *Early vertebrates.* Oxford/New York: Clarendon Press/Oxford University Press.
Janvier, P. (2011). Comparative anatomy: All vertebrates do have vertebrae. *Current Biology, 21*(17), R661–R663. http://dx.doi.org/10.1016/j.cub.2011.07.014.
Jeffery, W. R. (2006). Ascidian neural crest-like cells: Phylogenetic distribution, relationship to larval complexity, and pigment cell fate. *Journal of Experimental Zoology Part B: Molecular and Developmental Evolution, 306*(5), 470–480. http://dx.doi.org/10.1002/jez.b.21109.
Jeffery, W. R. (2007). Chordate ancestry of the neural crest: New insights from ascidians. *Seminars in Cell & Developmental Biology, 18*(4), 481–491. http://dx.doi.org/10.1016/j.semcdb.2007.04.005.
Jeffery, W. R., Chiba, T., Krajka, F. R., Deyts, C., Satoh, N., & Joly, J. S. (2008). Trunk lateral cells are neural crest-like cells in the ascidian *Ciona intestinalis*: Insights into the ancestry and evolution of the neural crest. *Developmental Biology, 324*(1), 152–160. http://dx.doi.org/10.1016/j.ydbio.2008.08.022.
Jeffery, W. R., Strickler, A. G., & Yamamoto, Y. (2004). Migratory neural crest-like cells form body pigmentation in a urochordate embryo. *Nature, 431*(7009), 696–699. http://dx.doi.org/10.1038/nature02975.
Jiang, X., Rowitch, D. H., Soriano, P., McMahon, A. P., & Sucov, H. M. (2000). Fate of the mammalian cardiac neural crest. *Development, 127*(8), 1607–1616.
Konig, N., Trolle, C., Kapuralin, K., Adameyko, I., Mitrecic, D., Aldskogius, H., et al. (2014). Murine neural crest stem cells and embryonic stem cell-derived neuron precursors survive and differentiate after transplantation in a model of dorsal root avulsion. *Journal of Tissue Engineering and Regenerative Medicine.* http://dx.doi.org/10.1002/term.1893.
Landacre, F. L. (1921). The fate of the neural crest in the head of the urodeles. *The Journal of Comparative Neurology, 33*(1), 1–43. http://dx.doi.org/10.1002/cne.900330102.
Langille, R. M., & Hall, B. K. (1986). Evidence of cranial neural crest contribution to the skeleton of the sea lamprey, *Petromyzon marinus. Progress in Clinical and Biological Research, 217B*, 263–266.
Langille, R. M., & Hall, B. K. (1988). The organ culture and grafting of lamprey cartilage and teeth. *In Vitro Cellular & Developmental Biology, 24*(1), 1–8.

Le Douarin, N. (1969). Details of the interphase nucleus in Japanese quail (*Coturnix coturnix japonica*). *Bulletin Biologique de la France et de la Belgique, 103*(3), 435–452.
Le Douarin, N. M., Calloni, G. W., & Dupin, E. (2008). The stem cells of the neural crest. *Cell Cycle,* 7(8), 1013–1019.
Le Douarin, N. M., Couly, G., & Creuzet, S. E. (2012). The neural crest is a powerful regulator of pre-otic brain development. *Developmental Biology, 366*(1), 74–82. http://dx.doi.org/10.1016/j.ydbio.2012.01.007.
Le Douarin, N. M., Creuzet, S., Couly, G., & Dupin, E. (2004). Neural crest cell plasticity and its limits. *Development, 131*(19), 4637–4650. http://dx.doi.org/10.1242/dev.01350.
Le Douarin, N. M., & Dupin, E. (2012). The neural crest in vertebrate evolution. *Current Opinion in Genetics & Development, 22*(4), 381–389. http://dx.doi.org/10.1016/j.gde.2012.06.001.
Le Douarin, N., & Kalcheim, C. (1999). *The neural crest* (2nd ed.). Cambridge, UK/New York, USA: Cambridge University Press.
Le Douarin, N. M., & Teillet, M. A. (1974). Experimental analysis of the migration and differentiation of neuroblasts of the autonomic nervous system and of neurectodermal mesenchymal derivatives, using a biological cell marking technique. *Developmental Biology, 41*(1), 162–184.
Le Lievre, C. S., Schweizer, G. G., Ziller, C. M., & Le Douarin, N. M. (1980). Restrictions of developmental capabilities in neural crest cell derivatives as tested by in vivo transplantation experiments. *Developmental Biology,* 77(2), 362–378.
Lievre, C. L. (1974). Role of mesectodermal cells arising from the cephalic neural crest in the formation of the branchial arches and visceral skeleton. *Journal of Embryology and Experimental Morphology, 31*(2), 453–477.
Mallatt, J., & Chen, J. Y. (2003). Fossil sister group of craniates: Predicted and found. *Journal of Morphology, 258*(1), 1–31. http://dx.doi.org/10.1002/jmor.10081.
Matsuoka, T., Ahlberg, P. E., Kessaris, N., Iannarelli, P., Dennehy, U., Richardson, W. D., et al. (2005). Neural crest origins of the neck and shoulder. *Nature, 436*(7049), 347–355. http://dx.doi.org/10.1038/nature03837.
Mayor, R., & Theveneau, E. (2013). The neural crest. *Development, 140*(11), 2247–2251. http://dx.doi.org/10.1242/dev.091751.
McCauley, D. W., & Bronner-Fraser, M. (2003). Neural crest contributions to the lamprey head. *Development, 130*(11), 2317–2327.
McGonnell, I. M., & Graham, A. (2002). Trunk neural crest has skeletogenic potential. *Current Biology, 12*(9), 767–771.
McKinney, M. C., Fukatsu, K., Morrison, J., McLennan, R., Bronner, M. E., & Kulesa, P. M. (2013). Evidence for dynamic rearrangements but lack of fate or position restrictions in premigratory avian trunk neural crest. *Development, 140*(4), 820–830. http://dx.doi.org/10.1242/dev.083725.
Medeiros, D. M. (2013). The evolution of the neural crest: New perspectives from lamprey and invertebrate neural crest-like cells. *Wiley Interdisciplinary Reviews. Developmental Biology, 2*(1), 1–15. http://dx.doi.org/10.1002/wdev.85.
Meulemans, D., & Bronner-Fraser, M. (2004). Gene-regulatory interactions in neural crest evolution and development. *Developmental Cell,* 7(3), 291–299. http://dx.doi.org/10.1016/j.devcel.2004.08.007.
Meulemans, D., & Bronner-Fraser, M. (2005). Central role of gene cooption in neural crest evolution. *Journal of Experimental Zoology Part B: Molecular and Developmental Evolution, 304*(4), 298–303. http://dx.doi.org/10.1002/jez.b.21047.
Meulemans, D., & Bronner-Fraser, M. (2007). Insights from amphioxus into the evolution of vertebrate cartilage. *PLoS One, 2*(8), e787. http://dx.doi.org/10.1371/journal.pone.0000787.
Meulemans, D., McCauley, D., & Bronner-Fraser, M. (2003). Id expression in amphioxus and lamprey highlights the role of gene cooption during neural crest evolution. *Developmental Biology, 264*(2), 430–442.

Milet, C., & Monsoro-Burq, A. H. (2012). Neural crest induction at the neural plate border in vertebrates. *Developmental Biology, 366*(1), 22–33. http://dx.doi.org/10.1016/j.ydbio.2012.01.013, S0012-1606(12)00029-2 [pii].

Mongera, A., Singh, A. P., Levesque, M. P., Chen, Y. Y., Konstantinidis, P., & Nusslein-Volhard, C. (2013). Genetic lineage labeling in zebrafish uncovers novel neural crest contributions to the head, including gill pillar cells. *Development, 140*(4), 916–925. http://dx.doi.org/10.1242/dev.091066.

Munoz, W., Sater, A., & McCrea, P. (2014). Development of neural tissues in *Xenopus laevis*. In M. Kloc & J. Kubiak (Eds.), *Xenopus development* (pp. 239–263). Hoboken: John Wiley & Sons, Inc.

Murray, S. A., & Gridley, T. (2006). Snail family genes are required for left-right asymmetry determination, but not neural crest formation, in mice. *Proceedings of the National Academy of Sciences of the United States of America, 103*(27), 10300–10304. http://dx.doi.org/10.1073/pnas.0602234103.

Nakamura, H., & Ayer-le Lievre, C. S. (1982). Mesectodermal capabilities of the trunk neural crest of birds. *Journal of Embryology and Experimental Morphology, 70*, 1–18.

Newth, D. R. (1951). Experiments on the neural crest of the lamprey embryo. *The Journal of Experimental Biology, 28*, 247–260.

Newth, D. R. (1956). On the neural crest of the lamprey embryo. *The Journal of Experimental Biology, 4*, 358–375.

Nishida, H., & Satoh, N. (1989). Determination and regulation in the pigment cell lineage of the ascidian embryo. *Developmental Biology, 132*(2), 355–367.

Noack Watt, K. E., & Trainor, P. A. (2014). Neurocritopathies: The etiology and pathogenesis of disorders arising from defects in neural crest cell development. In P. A. Trainor (Ed.), *Neural crest cells evolutions, development and disease* (1st ed., pp. 361–394) San Diego: Academic Press.

Northcutt, R. G., & Gans, C. (1983). The genesis of neural crest and epidermal placodes: A reinterpretation of vertebrate origins. *The Quarterly Review of Biology, 58*(1), 1–28.

Ohno, S. (1999). Gene duplication and the uniqueness of vertebrate genomes circa 1970–1999. *Seminars in Cell & Developmental Biology, 10*(5), 517–522. http://dx.doi.org/10.1006/scdb.1999.0332.

Ota, K. G., Fujimoto, S., Oisi, Y., & Kuratani, S. (2011). Identification of vertebra-like elements and their possible differentiation from sclerotomes in the hagfish. *Nature Communications, 2*, 373. http://dx.doi.org/10.1038/ncomms1355.

Ota, K. G., Kuraku, S., & Kuratani, S. (2007). Hagfish embryology with reference to the evolution of the neural crest. *Nature, 446*(7136), 672–675. http://dx.doi.org/10.1038/nature05633.

Peters, A. (1963). The structure of the dorsal root nerves of amphioxus an electron microscope study. *The Journal of Comparative Neurology, 121*(2), 287–304. http://dx.doi.org/10.1002/cne.901210210.

Prasad, M. S., Sauka-Spengler, T., & LaBonne, C. (2012). Induction of the neural crest state: Control of stem cell attributes by gene regulatory, post-transcriptional and epigenetic interactions. *Developmental Biology, 366*(1), 10–21. http://dx.doi.org/10.1016/j.ydbio.2012.03.014.

Rahimi, R. A., Allmond, J. J., Wagner, H., McCauley, D. W., & Langeland, J. A. (2009). Lamprey snail highlights conserved and novel patterning roles in vertebrate embryos. *Development Genes and Evolution, 219*(1), 31–36. http://dx.doi.org/10.1007/s00427-008-0258-4.

Sanchez-Lara, P. A., & Zhao, H. (2014). Neural crest stem cell: Tissue regeneration and repair. In P. A. Trainor (Ed.), *Neural crest cells evolution, development and disease* (pp. 413–439). San Diego: Academic Press.

Sauka-Spengler, T., & Bronner-Fraser, M. (2008). A gene regulatory network orchestrates neural crest formation. *Nature Reviews Molecular Cell Biology*, *9*(7), 557–568. http://dx.doi.org/10.1038/nrm2428.

Sauka-Spengler, T., Meulemans, D., Jones, M., & Bronner-Fraser, M. (2007). Ancient evolutionary origin of the neural crest gene regulatory network. *Developmental Cell*, *13*(3), 405–420. http://dx.doi.org/10.1016/j.devcel.2007.08.005.

Shi, J., Severson, C., Yang, J., Wedlich, D., & Klymkowsky, M. W. (2011). Snail2 controls mesodermal BMP/Wnt induction of neural crest. *Development*, *138*(15), 3135–3145. http://dx.doi.org/10.1242/dev.064394.

Simoes-Costa, M., & Bronner, M. E. (2013). Insights into neural crest development and evolution from genomic analysis. *Genome Research*, *23*(7), 1069–1080. http://dx.doi.org/10.1101/gr.157586.113.

Simoes-Costa, M., Tan-Cabugao, J., Antoshechkin, I., Sauka-Spengler, T., & Bronner, M. E. (2014). Transcriptome analysis reveals novel players in the cranial neural crest gene regulatory network. *Genome Research*, *24*(2), 281–290. http://dx.doi.org/10.1101/gr.161182.113.

Smith, M. M. (1991). Putative skeletal neural crest cells in early late Ordovician vertebrates from Colorado. *Science*, *251*(4991), 301–303. http://dx.doi.org/10.1126/science.251.4991.301.

Squarzoni, P., Parveen, F., Zanetti, L., Ristoratore, F., & Spagnuolo, A. (2011). FGF/MAPK/Ets signaling renders pigment cell precursors competent to respond to Wnt signal by directly controlling Ci-Tcf transcription. *Development*, *138*(7), 1421–1432. http://dx.doi.org/10.1242/dev.057323.

Taylor, J. S., & Raes, J. (2004). Duplication and divergence: The evolution of new genes and old ideas. *Annual Review of Genetics*, *38*, 615–643. http://dx.doi.org/10.1146/annurev.genet.38.072902.092831.

Teng, L., Mundell, N. A., Frist, A. Y., Wang, Q., & Labosky, P. A. (2008). Requirement for Foxd3 in the maintenance of neural crest progenitors. *Development*, *135*(9), 1615–1624. http://dx.doi.org/10.1242/dev.012179.

Thomas, A. J., & Erickson, C. A. (2008). The making of a melanocyte: The specification of melanoblasts from the neural crest. *Pigment Cell & Melanoma Research*, *21*(6), 598–610. http://dx.doi.org/10.1111/j.1755-148X.2008.00506.x.

Trainor, P. A. (Ed.). (2014). *Neural crest cells evolution, development and disease* (1st ed.). San Diego: Academic Press.

Trentin, A., Glavieux-Pardanaud, C., Le Douarin, N. M., & Dupin, E. (2004). Self-renewal capacity is a widespread property of various types of neural crest precursor cells. *Proceedings of the National Academy of Sciences of the United States of America*, *101*(13), 4495–4500. http://dx.doi.org/10.1073/pnas.0400629101.

Trolle, C., Konig, N., Abrahamsson, N., Vasylovska, S., & Kozlova, E. N. (2014). Boundary cap neural crest stem cells homotopically implanted to the injured dorsal root transitional zone give rise to different types of neurons and glia in adult rodents. *BMC Neuroscience*, *15*(1), 60. http://dx.doi.org/10.1186/1471-2202-15-60.

Walker, M. B., & Trainor, P. A. (2006). Craniofacial malformations: Intrinsic vs extrinsic neural crest cell defects in Treacher Collins and 22q11 deletion syndromes. *Clinical Genetics*, *69*(6), 471–479. http://dx.doi.org/10.1111/j.0009-9163.2006.00615.x.

Yu, J. K. (2010). The evolutionary origin of the vertebrate neural crest and its developmental gene regulatory network—Insights from amphioxus. *Zoology (Jena, Germany)*, *113*(1), 1–9. http://dx.doi.org/10.1016/j.zool.2009.06.001.

Yu, J. K., Holland, N. D., & Holland, L. Z. (2004). Tissue-specific expression of FoxD reporter constructs in amphioxus embryos. *Developmental Biology*, *274*(2), 452–461. http://dx.doi.org/10.1016/j.ydbio.2004.07.010.

Yu, J. K., Meulemans, D., McKeown, S. J., & Bronner-Fraser, M. (2008). Insights from the amphioxus genome on the origin of vertebrate neural crest. *Genome Research*, *18*(7), 1127–1132. http://dx.doi.org/10.1101/gr.076208.108.

CHAPTER TWO

Resolving Time and Space Constraints During Neural Crest Formation and Delamination

Jean-Loup Duband[*,†,1], **Alwyn Dady**[*,†,2], **Vincent Fleury**[‡]

[*]Laboratoire de Biologie du Développement, Université Pierre et Marie Curie-Paris 6, Paris, France
[†]CNRS, Laboratoire de Biologie du Développement, Paris, France
[‡]Laboratoire Matière et Systèmes Complexes, CNRS et Université Denis-Diderot-Paris 7, Paris, France
[1]Corresponding author: e-mail address: jean-loup.duband@upmc.fr
[2]Present address: Children's Hospital of Pittsburgh, Rangos Research Building, Pittsburgh, Pennsylvania, USA.

Contents

Abstract

A striking feature of neural crest development in vertebrates is that all the specification, delamination, migration, and differentiation steps occur consecutively in distinct areas of the embryo and at different timings of development. The significance and consequences of this partition into clearly separated events are not fully understood yet, but it ought to be related to the necessity of controlling precisely and independently each step, given the wide array of cell types and tissues derived from the neural crest and the long duration of their development spanning almost the entire embryonic life. In this chapter, using the examples of early neural crest induction and delamination, we discuss how time and space constraints influence their development and describe the

Current Topics in Developmental Biology, Volume 111
ISSN 0070-2153
http://dx.doi.org/10.1016/bs.ctdb.2014.11.002

molecular and cellular responses that are employed by cells to adapt. In the first example, we analyze how cell sorting and cell movements cooperate to allow nascent neural crest cells, which are initially mingled with other neurectodermal progenitors after induction, to segregate from the neural tube and ectoderm populations and settle at the apex of the neural tube prior to migration. In the second example, we examine how cadherins drive the entire process of neural crest segregation from the rest of the neurectoderm by their dual role in mediating first cell sorting and cohesion during specification and later in promoting their delamination. In the third example, we describe how the expression and activity of the transcription factors known to drive epithelium-to-mesenchyme transition (EMT) are regulated timely and spatially by the cellular machinery so that they can alternatively and successively regulate neural crest specification and delamination. In the last example, we briefly tackle the problem of how factors triggering EMT may elicit different cell responses in neural tube and neural crest progenitors.

The term neural crest cells (NCC) stems from the fact that, in many species and particularly in some urodele amphibians, its cellular components protrude over the dorsal midline of the embryo after delaminating from the closing neural tube (Le Douarin & Kalcheim, 1999; Trainor, 2013); and as a matter of fact, it is precisely at that time that this cell population individualizes as a distinct entity, fully segregated from the rest of the neurectoderm from which it derives. This morphological feature has long served to define NCC as a population of migrating cells that arise from the dorsal region of the neural tube and populate various sites in the embryo, where they undergo terminal differentiation. However, a variety of cell-tracing studies in chick and *Xenopus* clearly established that NCC are not born in the neural epithelium at the time when they initiate migration but long before, by a mechanism that is separate from that of induction of the neural plate, the primordium of the central nervous system (Aybar & Mayor, 2002; Basch, Bronner-Fraser, & Garcia-Castro, 2006; Betancur, Bronner-Fraser, & Sauka-Spengler, 2010; Ezin, Fraser, & Bronner-Fraser, 2009; Milet & Monsoro-Burq, 2012; Prasad, Sauka-Spengler, & LaBonne, 2012; Stuhlmiller & Garcia-Castro, 2012a). In birds, fishes, and amphibians and possibly also in mammals, NCC progenitors are specified during gastrulation, before its subdivision into neural and non-neural epithelium. As neurulation proceeds, these progenitors are progressively positioned, first in the elevating neural folds, then in the dorsal aspect of the neural tube. Later on, by a mechanism related to epithelium-to-mesenchyme transition (EMT), they delaminate from the neural tube, and owing to a complex code involving multiple guidance cues, they disperse through restricted pathways. Finally, they colonize many different territories throughout the body, where they eventually give rise to a large collection of cell types. Thus, NCC development divides into four well-defined steps: the premigration phase, consisting essentially of the early induction of progenitors and their maturation as a population of cells endowed with the competence to undergo migration and to give rise to multiple lineages; the delamination phase, which allows their physical separation from the neural tube; the migration phase; and finally the ultimate phase of differentiation in the sites of colonization. Intriguingly, these steps all occur in distinct areas of the embryo and at different times during development, the course of NCC ontogeny spanning almost the entire duration of embryonic life: specification in the blastoderm during gastrulation and early neurulation, delamination from the dorsal neural tube along the embryonic midline at the end of neurulation, migration and differentiation throughout the course of organogenesis and late embryogenesis in as diverse locations as the vertebral column, the skin, the aorta, the periphery of the brain, the face, the heart, or the gut. Although not unique, it is infrequent that during embryogenesis, cells execute sequentially their main developmental steps in separate locations. Therefore, time and space are both important parameters that may underlie the entire process of NCC development and may impinge on the regulatory processes involved; and there are now many situations in which mismatches between them may cause severe developmental anomalies. In this chapter, using selected examples taken from specification and delamination, mainly in the chick and frog but also in the zebrafish and mouse, we address the question of how space is integrated during NCC development and how this is regulated over time.

1. INTEGRATING SPACE DURING NCC INDUCTION AND SPECIFICATION: ROLES OF CELL MOVEMENTS AND CADHERIN-MEDIATED CELL SORTING

The generation of NCC progenitors in the blastoderm results from the combined action of various signals mediated by diffusible morphogens of the Wnt, fibroblast growth factor (FGF), and bone morphogenetic protein (BMP) families as well as by retinoic acid, emanating from the direct environment and acting sequentially or in synergy on a restricted cell population of the neurectoderm. In response to these signals, cells express a gradually more complex and specific repertoire of transcription factors assembled into a gene regulatory network, which assigns them their NCC identity and allows them to segregate from the ectoderm and neural plate (Betancur et al., 2010; Groves & LaBonne, 2014; Huang & Saint-Jeannet, 2004; Milet & Monsoro-Burq, 2012; Prasad et al., 2012; Sauka-Spengler & Bronner-Fraser, 2008; Stuhlmiller & Garcia-Castro, 2012a).

The first manifestation of NCC induction during gastrulation is the collective expression of the transcription factors *Pax-7* in chick, *Pax-3* in *Xenopus*, together with *Msx-1/2*, and *Zic-1*. Expression commences in a territory at the border between the prospective neural plate and ectoderm, hence its name neural plate border, and which encompasses progenitors of the future NCC, as well as those of the dorsal neural tube and ectodermal placodes. Although individually these factors are neither restricted to the neural plate border nor sufficient to drive NCC specification, their combined effect is sufficient to restrict progressively the potential of cells to adopt an ectodermal or a neural plate fate, while providing them with the competence to acquire NCC, dorsal neural tube, and placode identities. For this reason, these genes have been regrouped under the term of border specifiers. The second step, named specification, is aimed at consolidating the NCC identity, in particular to acquire unique adhesive, proliferative, migratory, multipotency, and survival traits that enable NCC progenitors to undergo EMT and disperse throughout their migratory routes. This is achieved by the cumulative expression of the NCC specifier genes, including members of the *Snail* and *SoxE* families of transcription factors, as well as a variety of other factors such as *Foxd-3*, *Ets-1*, *AP-2*, and *Id* genes.

Strikingly, the expression patterns of border specifiers that have been resolved in different species reveal that at onset of induction, the neural plate border is not a well-defined and delineated territory (Basch et al., 2006; Ezin et al., 2009; Khudyakov & Bronner-Fraser, 2009; Milet & Monsoro-Burq, 2012; Stuhlmiller & Garcia-Castro, 2012a). By combining the techniques of lipophilic dye tracing and time-lapse imaging, Ezin et al. (2009) performed fate-mapping studies in chick and were able to trace precisely the position of NCC progenitors over time, from early induction to their final position in the dorsal neural tube. They found that at gastrulation, progenitors receiving the induction signals are scattered in the blastoderm in a broad domain corresponding to the BMP-4-producing region and are mingled with other neurectodermal progenitors. During neurulation, the neural plate border undergoes spectacular morphological changes from a short, wide, and flat strip of cells with a horseshoe shape in the anterior blastoderm into two long, narrow, and protruding neural folds that appose and fuse together along the embryonic midline. During this process, ectodermal, placodal, NCC and neural plate progenitors, initially dispersed and mixed in the blastoderm, become spatially segregated and patterned into highly ordered, compact entities along the rostrocaudal and dorsoventral axes. How is this process achieved? Ezin et al. (2009) revealed that the dynamic displacement of the neural plate border and sorting of its different cell components occur as a result of tightly coordinated movements of convergence, extension, reorientations, as well as cell mixing, coincident with the movements of folding, rolling, and bending of the neural plate into a hollow neural tube. These observations therefore uncover the previously unsuspected importance of cell movements and rearrangements in the progressive definition of the NCC territory (and by the way that of the other cell populations derived from the neurectoderm) during neurulation and its confinement to the dorsal neural tube prior to delamination.

Movements of convergence extension cannot, however, explain by themselves the assembly of NCC progenitors and their segregation from other cells with which they are intermingled. Other cellular events, such as cell cohesion and cell sorting, need to be invoked to account for the whole process. Interestingly, recent studies in the zebrafish on the mechanism of specification of ventral neural tube progenitors by the inductive signaling activity of Sonic hedgehog (Shh; Xiong et al., 2013) provide clues about the sequence of events that might be responsible for NCC assembly after induction. Indeed, in zebrafish, instead of being static in a stable epithelium as often assumed in classical models, both the Shh-producing cells in the axial mesoderm and the Shh-responding cells in the neural tube are motile and capable of exchanging neighbors. As a result, morphogen signaling across the tissue is spatially noisy, and neural progenitors being exposed to variable doses of Shh over time and space exhibit a large range of responses and distribute first in a salt-and-pepper specification pattern. It is only later that the rearrangement of the different neural progenitors of the neural tube into sharply bordered domains is achieved by a cell-sorting mechanism operating in a gradually more-static population. These observations indicate that specification and positioning are separate in time and that cell sorting acts to refine spatial patterning by inductive signals. The fact that the blastoderm is subjected to intense cell movements during induction of the neural plate border and that the extents of the BMP-4 and FGF signals occur in broad regions of the blastoderm (Ezin et al., 2009; Stuhlmiller & Garcia-Castro, 2012b) suggests that the same paradigm may apply during NCC formation.

The question now is by which mechanisms NCC would be sorted from the other populations of the neurectoderm? Cell sorting results primarily from the selective recognition and adhesion of cell populations expressing different repertoires or doses of molecules of the cadherin family (Halbleib & Nelson, 2006). In the case of NCC in the chick, until recently the prevailing assumption was that, due to their integration into the dorsal neural tube, they express primarily N-cadherin, whereas in the overlying ectoderm, cells instead express E-cadherin. In addition, it was generally believed that such a situation is established precociously during early neural induction and persists throughout neurulation until delamination (Duband, 2010; Kalcheim, 2000; Powell, Blasky, Britt, & Artinger, 2013; Strobl-Mazzulla & Bronner, 2012a; Taneyhill & Schiffmacher, 2013; Theveneau & Mayor, 2012). However, recent observations show that there is no mutually exclusive partition of E- and N-cadherins in the prospective ectoderm and neural epithelium until they are entirely segregated at completion of neurulation (Dady, Blavet, & Duband, 2012). Indeed, during gastrulation, E-cadherin is uniformly expressed throughout the superficial layer of the blastoderm, while N-cadherin is restricted to the primitive streak, the mesoderm and the notochord. It is only very progressively that E-cadherin is replaced by N-cadherin in the neural plate, with a relatively long time period where both molecules are coexpressed in cells. Thus, rather than defining precociously the embryonic territories fated to become neural, N-cadherin upregulation occurs secondarily as a result of neural induction and early neurulation. Moreover, a detailed survey of the spatiotemporal expression patterns of E- and N-cadherins throughout neurulation revealed a far more intricate pattern than previously thought (Dady et al., 2012). At the most anterior levels, from the forebrain to the hindbrain, the E- to N-cadherin switch is complete at the late stages of neurulation even after neural tube closure, while in the trunk, it occurs earlier during fold elevation. This indicates that this switch is not correlated with the movements of neurulation and fully supports previous genetic studies in the mouse and zebrafish showing that in the absence of N-cadherin, onset of formation of the neural tube occurs normally and that the first signs of malformations are detected relatively late (Hong & Brewster, 2006; Radice et al., 1997). Which morphogenetic events are then driven by the E- to N-cadherin switch? Studies in the zebrafish showed that the lack of N-cadherin prevents cells from establishing stable protrusive activity and causes alterations in their radial intercalation in the neural epithelium (Hong & Brewster, 2006). This suggests that N-cadherin plays a role in cell sorting and in tissue stabilization and implies that changes in the expression patterns of E- and N-cadherins correlate primarily with the segregation of neural and ectodermal progenitors.

Still, neither E-cadherin nor N-cadherin can account for the subdivision of the blastoderm into three distinct territories (and even more when also considering the placodes and the roof plate), and the implication of a third partner has to be invoked. Interestingly, the expression pattern of cadherin-6B in the chick is suggestive of a role for this cadherin in NCC segregation from the rest of the neurectoderm and their assembly at the interface between the ectoderm and neural tube, as originally proposed by Nakagawa and Takeichi (1995, 1998). Far from being induced late in the dorsal neural tube after NCC specification, cadherin-6B appears early in the neural plate border slightly after induction of the border genes *Pax-7* and *Msx-1* but before NCC specifiers. In addition, consistent with a role in cell sorting, cadherin-6B is initially distributed in a punctate pattern that gradually becomes more intense and compact as NCC progenitors accumulate at the apex of the neural tube (Dady & Duband, 2015). Thus, organization of the blastoderm into sharply delineated, contiguous cellular compartments is likely to result from the dynamic interplay between E-cadherin, N-cadherin, and cadherin-6B over time during neurulation. In this regard, it is worthmentioning that cadherin-6B is upregulated by BMP-4 signaling (Sela-Donenfeld & Kalcheim, 1999), while N-cadherin is in contrast under the control of FGF via the Sox-2 transcription factor, an early definitive marker for neural tissues (Linker & Stern, 2004; Matsumata, Uchikawa, Kamachi, & Kondoh, 2005; Uchikawa et al., 2011).

As suggested by the model developed by Xiong et al. (2013) on Shh specification of ventral neural tube cells in the zebrafish, cadherin-mediated cell sorting must be coordinated with dynamic cell movements so that cell populations can first travel and reorganize over long distances to exchange neighbors and then become more static and cohesive to stabilize their mutual associations. How are these movements driven and regulated during NCC induction and specification? In chick, a possibility is that they are synchronized with the movements of intercalation occurring nearby the primitive streak (Voiculescu, Bertocchini, Wolpert, Keller, & Stern, 2007; Ybot-Gonzalez et al., 2007). However, the limited range of these movements along the primitive streak cannot account for the large amplitude of the so-called polonaise movements throughout the blastoderm. Biophysical studies based on the high-resolution microscopy and fine particle tracking in chick have documented cell movements on a large scale over the entire blastoderm, from gastrulation to the end of neurulation, and defined the physical parameters, such as viscoelasticity and mechanical forces, that drive the shaping of the embryo (Fleury, 2012). Prior to gastrulation, the blastula exhibits roughly three rings of cells of different sizes, cells being smaller internally and larger at the periphery. At that time, two types of movements can be observed, a radial one involving cells at the periphery responsible for the expansion of the blastoderm over the yolk and a rotatory one in the center. This creates large vortices forming a quadrupolar flow. As the embryo initiates gastrulation, propagation of the mesoderm exerts a traction on the epiblast, stretching and folding the blastoderm anteroposteriorly. Importantly, folds form at boundaries between cell types and match to the expression patterns of cadherins in the neurectoderm, segregating the tissue into domains of different cell sizes and cell adhesion properties and separated by deep furrows (Fleury, 2012, unpublished). While E-cadherin is progressively replaced by N-cadherin in small cells in the medial part of the blastoderm, cadherin-6B becomes specifically expressed in the folds. Thus, folding in the blastoderm produced by differential cell movements may drive the segregation of the diverse cell populations derived from the neurectoderm.

Collectively, the different studies described earlier suggest a model in which, rather than being *de facto* a well-defined area of progenitors all sharing the same differentiation potential that is passively moved toward the embryonic midline by the movements of neurulation, the neural plate border constitutes an unsteady population that is subjected to intense cellular reorganizations and is progressively partitioned by the interplay between physical events operating in the entire blastoderm and biochemical factors acting locally (Fig. 1). In addition, this model ascribes to cadherins a critical role in the specification and segregation of the different cell lineages arising from the anterior blastoderm. It would be interesting to analyze to which extent and how this model developed for a flat embryo like in chick must be modified to adapt to spheric embryos like in frog and fish.

2. COORDINATING NCC DELAMINATION TIMELY AND SPATIALLY: REGULATION OF CADHERIN ACTIVITY

By analogy to the now-classical models of EMT developed for cultured cell lines and cancer cells (Peinado, Olmeda, & Cano, 2007; Thiery, Acloque, Huang, & Nieto, 2009; Thiery & Sleeman, 2006), a model has been proposed for chick NCC in which EMT would occur through repression of cadherins (namely N-cadherin and cadherin-6B) by a mechanism involving the so-called core EMT regulatory factors, Snail-1, Snail-2, Zeb-2 (also known as SIP-1), and Twist-1, and elicited by

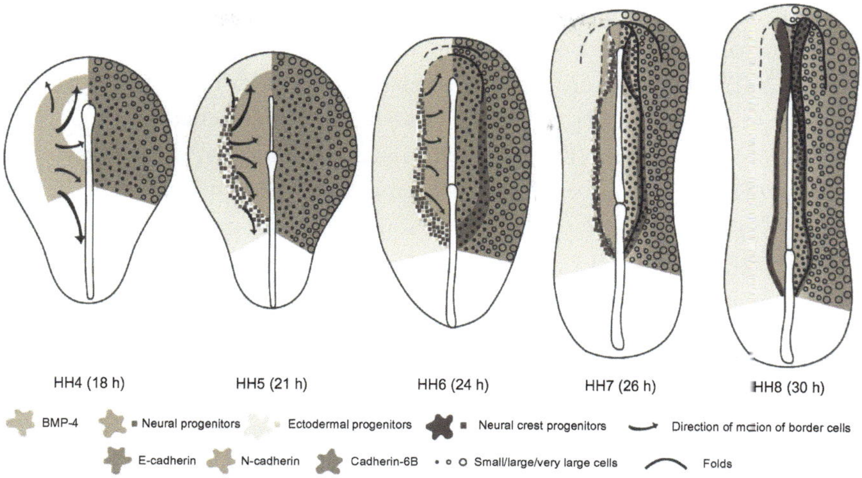

Figure 1 Integrating space during NCC induction and specification. Hypothetical model depicting how, in the chick embryo, surface movements and cadherin-mediated selective cell adhesion can coordinately contribute to the sorting and partition of the ectodermal, NCC, and neural tube lineages from gastrulation to late neurulation. **Schematic representations of the chick blastoderm viewed dorsally at stages 4–8 of Hamburger and Hamilton (1951) showing: on the left, the expression pattern of the inducing signal BMP-4 at stage 4 and the positions of the different populations of progenitors over time at stages 5–8 and, on the right, the distribution patterns of cadherins as well as the positions of cells according to their sizes. Surface movements combined with the distribution of cells of different sizes in concentric rings result in the formation of folding in the blastoderm, responsible for its transformation into a three-dimensional structure. In a strictly overlapping picture, E- and N-cadherins become distributed in the large/medium cells and in the small cells, respectively, while cadherin-6B appears at the border between medium and small cells where folds develop; this corresponds to the neural plate border where NCC are ultimately generated. (See the color plate.)

Wnt-1 and Bmp-4 signals (Duband, 2010; Kalcheim, 2000; Powell et al., 2013; Strobl-Mazzulla & Bronner, 2012a; Taneyhill & Schiffmacher, 2013; Theveneau & Mayor, 2012). However, although widely accepted, this model must be reevaluated in light of progress during the last decade on our comprehension of the intimate molecular processes spatiotemporally regulating the sequence of events leading to EMT.

An important point concerns how cadherins are regulated during delamination: what is the exact timing of their expression and what are their functions in this event? Moreover, an intriguing question has emerged gradually during the last few years: are cadherin-6B and N-cadherin actually involved in NCC delamination in chick? Regarding cadherin-6B, the sharp diminution of their messages at onset of EMT is indicative of a role in cell delamination (Nakagawa & Takeichi, 1995; Taneyhill, Coles, & Bronner-Fraser, 2007). In addition, Snail-2 has been shown to bind specifically to E boxes in the regulatory sequence of the *cadherin-6B* gene and to directly control its repression (Taneyhill et al., 2007). Finally, knockdown of *cadherin-6B* at cranial levels leads to premature NCC emigration, whereas its overexpression blocks their migration (Coles, Taneyhill, & Bronner-Fraser, 2007). Thus, all these data concur with the idea of transcriptional repression of *cadherin-6B* being a major triggering event at onset of EMT. This view is further supported by the observation that αN-catenin, a cytoskeleton-associated partner of cadherins, present in adherens junctions in premigratory NCC is also transcriptionally repressed by Snail-2 upon delamination and that manipulating its expression leads to the same outcomes as for cadherin-6B (Jhingory, Wu, & Taneyhill, 2010). Interestingly, other studies suggest that, in addition to transcriptional regulation, the timing of NCC delamination is also defined by posttranscriptional regulation of cadherin-6B activity (Fairchild & Gammill, 2013). Indeed, tetraspanin-18, a member of the tetraspanin family specific of epithelial cells, is abundantly expressed in premigratory but not migratory cranial NCC. Tetraspanin-18 functions as a stabilizer of cadherin-6B proteins at the cell surface, and its downregulation under the control of Foxd-3 is required for NCC to initiate migration. These data illustrate the complexity of the mechanisms controlling the timing of NCC delamination: while transcription of the *cadherin-6B* gene is under the control of Snail-2, the activity of cadherin-6B proteins at the cell surface is regulated by tetraspanin-18 itself under the transcriptional control of Foxd-3.

Several observations, however, argue in favor of alternative roles for cadherin-6B during delamination. First, its messages persist in the dorsal neural tube after completion of NCC emigration, indicating that its function may not be restricted to the temporal control of intercellular adhesion during EMT (Nakagawa & Takeichi, 1995). In mouse and zebrafish, *cadherin-6*, a close relative to *cadherin-6B*, is not downregulated after NCC delamination, and at trunk levels in the chick, cadherin-6B proteins remain present transiently on the surface of early migrating cells (Clay & Halloran, 2014; Inoue, Chisaka, Matsunami, & Takeichi, 1997; Park & Gumbiner, 2010). Additionally, functional studies suggest that, rather than maintaining epithelial stability among NCC, *cadherin-6B* in chick and *cadherin-6* in fish are required for cells to lose their epithelial junctional polarity and acquire their motile properties (Clay & Halloran, 2014; Park & Gumbiner, 2010). This effect is mediated in chick by noncanonical BMP signaling involving the BMP type II receptor, LIM kinase, and its major target cofilin. In zebrafish, the effect is mediated by controlling Rho GTPase distribution in the cytoplasm. Both signaling pathways ultimately regulate locally the organization of the actin cytoskeleton and promote polarized actomyosin contraction necessary for disruption of apical cell junctions (Clay & Halloran, 2014; Park & Gumbiner, 2010, 2012).

The discrepancy between the data in favor of a role of cadherin-6B in cell–cell adhesion and those supporting intracellular signaling activity regulating actin dynamics has been interpreted in terms of differences between cranial and trunk NCC (see, e.g., Taneyhill & Schiffmacher, 2013), and it is now clearly demonstrated that both populations differ radically in their modes and kinetics of delamination (Duband, 2010). However, there are no objectively good reasons for regional differences in cadherin-6B function and regulation, as many of its partners (BMP receptors, catenins, Rho GTPases, cytoskeletal elements) and regulators (Snail-2, Foxd-3) are found in both cranial and trunk levels. In addition, a question remains in the signaling model as to how cell cohesion is maintained among NCC until delamination if the role of cadherin-6B is solely restricted to the disruption of cell–cell contacts. This is consistent neither with its precocious expression during NCC induction nor with its maintenance in the roof plate after delamination. A possible explanation may reside in the fact that these experiments differ in an important detail, i.e., the timing when cadherin-6B expression was affected, and this might be sufficient to account for the differences observed. Indeed, Coles et al. (2007) modified cadherin-6B expression at cranial levels less than 6 h before delamination, while Park and Gumbiner (2010) affected it at trunk levels more than 12 h before. Because the adhesive and signaling functions described for cadherin-6B are not mutually exclusive, another scenario can be proposed that would reconcile both models (Fig. 2). Both in the head and trunk, cadherin-6B would promote cell sorting during NCC specification and maintain their cohesion until delamination while also activating signaling pathways that would act as a negative feedback loop and set the grounds for future delamination by driving polarized changes in the actin cytoskeleton dynamics. Whether both activities occur concurrently or sequentially prior to delamination is a critical point that remains, however, to be clarified. At cranial levels, additional events such as the stabilizing activity of tetraspanin-18 may complement this process to coordinate EMT into a massive emigration. Thus, according to this scenario, premature inhibition of cadherin-6B signaling

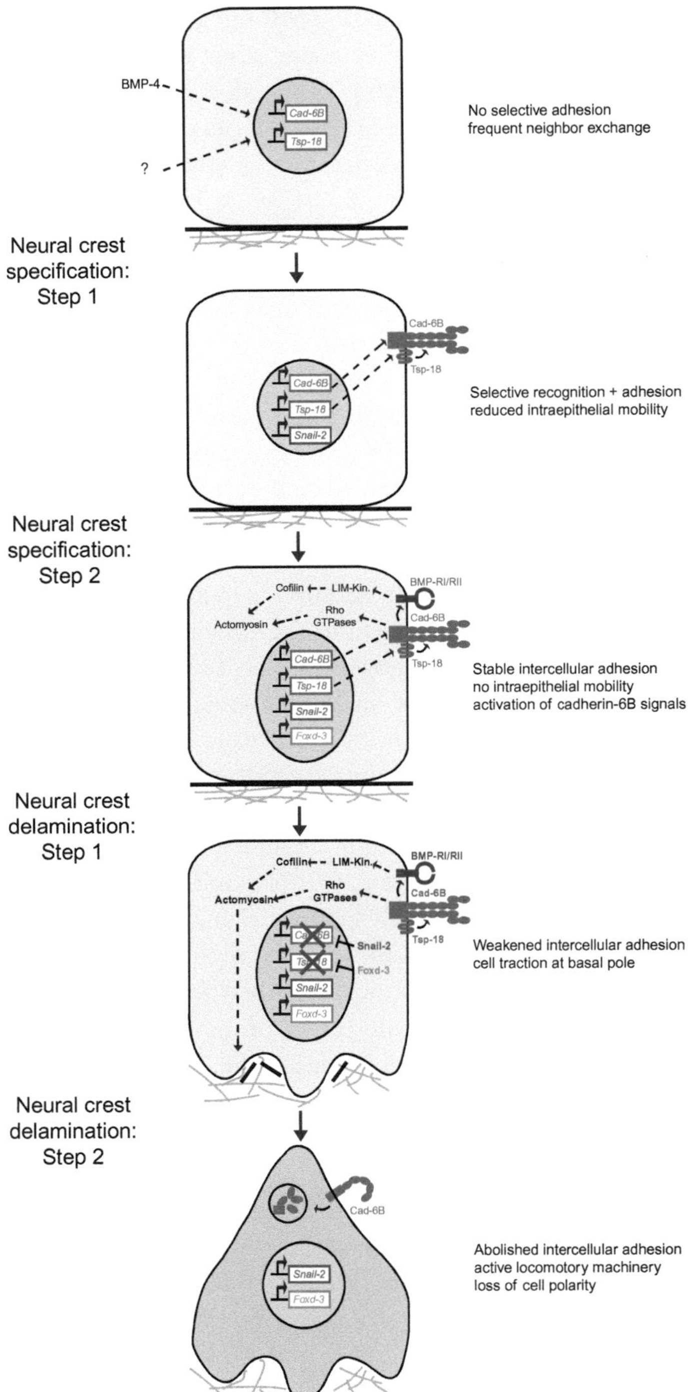

Figure 2 See legend on opposite page.

would block the cascade of events leading to NCC delamination, while late inhibition of its adhesive function at a time when polarized contraction of actomyosin has already been activated would in contrast force their emigration. Conversely, early overexpression of cadherin-6B in NCC would enhance emission of the signals activating cytoskeleton dynamics and therefore promote delamination, whereas its late overexpression can oppose the diminution of endogenous cadherins and block delamination. Regardless of the precise sequence of events controlling its expression and function, cadherin-6B must therefore be considered as a unique element of the developmental program responsible for the segregation of the NCC population from the rest of the neurectoderm by initially mediating their sorting during specification, and then later by inducing EMT. A question remains however as to why roof plate cells do not undergo EMT despite persistent expression of cadherin-6B.

As far as N-cadherin is concerned, it is generally considered the major cadherin to be repressed at the time of NCC delamination. In support to this assumption, it has been demonstrated that at trunk levels in chick, delamination involves cleavage of N-cadherin by the metalloprotease ADAM-10 under the control of BMP-4. In addition, the released cytoplasmic fragment of N-cadherin translocates to the nucleus and stimulates cyclin-D1 transcription and cell delamination (Shoval, Ludwig, & Kalcheim, 2007). As for cadherin-6B, this model is consistent with both the adhesive and signaling activities of cadherins. However, it has not been validated yet for anterior levels, and given the growing evidence for heterogeneity in the modes of NCC delamination along the embryonic axis, it may be true only for the caudal trunk of the embryo. Moreover, it should be stressed that previous analyses of N-cadherin distribution on which this model relies either did not employ markers for premigratory NCC or were not performed systematically at all axial levels and at all stages throughout NCC development (Akitaya & Bronner-Fraser, 1992; Duband, Volberg, Sabanay, Thiery, & Geiger, 1988; Hatta, Takagi, Fujisawa, & Takeichi, 1987; Park & Gumbiner, 2010; Shoval et al., 2007). Thus, several recent

Figure 2 Dual role of cadherin-6B in NCC specification and delamination. Hypothetical model depicting how cadherin-6B expression is regulated during cranial NCC development in chick and how it may control both cell cohesion and initiation of delamination owing to its dual adhesive and signaling activities. Although this model is illustrated for the cranial region, the same principles apply for trunk NCC except for tetraspanin-18 expression and role. (See the color plate.)

studies argue that N-cadherin may not be implicated at all or at least very secondarily in NCC delamination in the rostral half of the embryo. A detailed description of the kinetics of NCC delamination at the midbrain level in chick and mouse (Lee et al., 2013) revealed that a large proportion of the delaminating cells do not express N-cadherin but rather E-cadherin, which is gradually repressed as they separate from the neurectoderm. Interestingly, these cells are located adjacent to the nonneural ectoderm in the neural folds and correspond to the mesectodermal contingent of cranial NCC (i.e., cells at the origin of bones, cartilage, and other nonneural derivatives). Beside these cells, a smaller contingent situated adjacent to the neural part of the neurectoderm expresses both N-cadherin and E-cadherin prior to migration and provides the neuronal derivatives of cranial NCC. These observations suggest that rather than controlling the timing of initiation of cell delamination, N-cadherin expression would constitute an early neuronal signature in the premigratory NCC population. In a systematic analysis of the repertoire of cadherin expression throughout NCC formation until their complete segregation from the neural tube, we found that at least in the head and upper trunk, N-cadherin expression is induced progressively in the neural part of the neurectoderm during neurulation but not in NCC progenitors, including those which adopt a neuronal fate (Dady et al., 2012; Dady & Duband, 2015). Thus, N-cadherin repression may not constitute a prominent event during NCC delamination in the anterior half of the body. As discussed earlier, this role is most likely devoted to cadherin-6B whose repression occurs sharply. The function of E-cadherin in this process is apparently less critical as its repression proceeds at a slow pace and that, depending on the axial level considered, it may be expressed on the surface of migrating cells (though at moderate levels) or down-regulated prior to delamination (Dady et al., 2012; Dady & Duband, 2015; Lee et al., 2013). This view is consistent with the expression patterns and functions during delamination of *Snail-2* and *Zeb-2*, known as bona fide transcriptional repressors of *E-cadherin* and *cadherin-6B* (Peinado et al., 2007; Taneyhill et al., 2007) but not of *N-cadherin* (Dady et al., 2012; Dady & Duband, 2015). Conversely, *Sox-2*, an activator of *N-cadherin* transcription (Uchikawa, Ishida, Takemoto, Kamachi, & Kondoh, 2003), is excluded from the neural plate border where it was found to repress NCC identity (Wakamatsu, Endo, Osumi, & Weston, 2004).

In conclusion, in a new model that differs quite significantly from the previous ones, we propose that, in chick, cadherins play dual roles during

NCC specification and delamination and constitute key elements in a coherent program aimed at progressively segregating these cells from the ectoderm and neural tube, first locally within the neurectodermal epithelium by promoting cell sorting and then out of it by inducing EMT and migration (Fig. 3). An important consequence of this program is that it spatiotemporally restricts EMT and migration only to NCC without interfering with the stability and development of the neighboring neural and ectodermal epithelia. This might explain why EMT can be induced in those tissues only if a NCC identity is forced (Barembaum, Moreno, LaBonne, Sechrist, & Bronner-Fraser, 2000; Cheung et al., 2005; Théveneau, Duband, & Altabef, 2007). It should be noted, however, that this model does not apply to *Xenopus* for several reasons. In this species, cadherin-6 activity has not been observed in premigratory NCC (David & Wedlich, 2000), and at the present time, whether NCC are sorted from the rest of the neurectoderm by a cadherin-dependent mechanism has not been investigated. Furthermore, the precocious expression of N-cadherin in NCC progenitors at onset of neurulation (Nandadasa, Tao, Menon, Heasman, & Wylie, 2009) and its persistent expression during migration (Theveneau & Mayor, 2012) raises the intriguing question as to whether onset of migration of cranial NCC in *Xenopus* involves a complete EMT. Indeed, these cells do not migrate as a population of individual cells as trunk NCC in chick, but collectively as a dense pseudoepithelial sheet, and it has been shown very recently that this intermediate phenotype results from a solid-like-to-fluid-like transition initiated by signaling of the lysophosphatidic acid receptor 2 together with a reduction of surface expression of N-cadherin (Kuriyama et al., 2014).

3. COORDINATING NCC SPECIFICATION AND DELAMINATION: SPATIOTEMPORAL CONTROL OF THE CORE EMT REGULATORY FACTORS

Although the moment when NCC exit the neural tube generally lasts less than an hour (Ahlstrom & Erickson, 2009; Clay & Halloran, 2010), Snail-1/2, Zeb-2, and, in a lesser extent, Twist-1 all appear in their progenitors early during specification long before onset of delamination In addition, as described earlier, their major cadherin targets undergo different temporal patterns of repression. This situation contrasts strikingly with that found in the primitive streak during chick gastrulation, where ingressing mesodermal cells express Snail-2 in a dynamic fashion and immediately

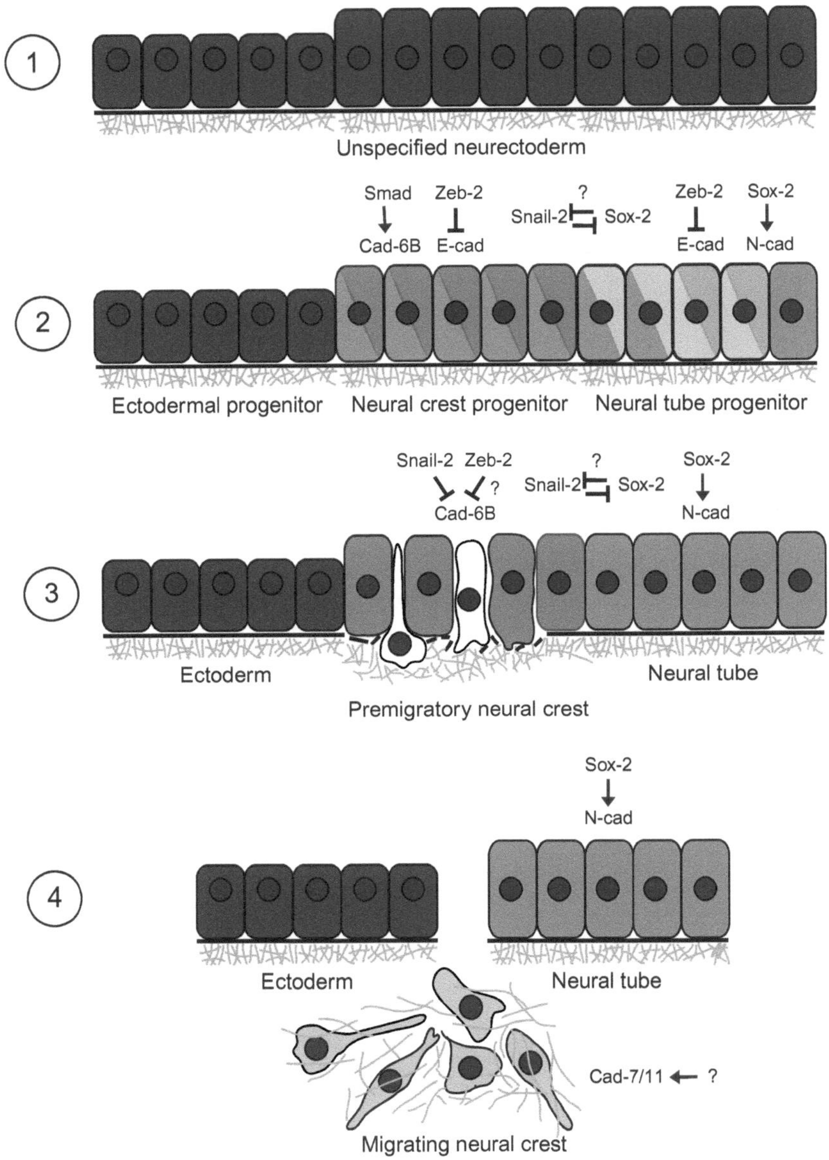

Figure 3 Coordinated regulation of cadherin expression during NCC specification and delamination. Schematic model depicting the sequence of cadherin expression during NCC development in chick: induction (1), specification (2), delamination (3), and migration (4). The model presented corresponds to the situation found in the hindbrain and anterior trunk. The same scenario applies however to the entire rostral half of the embryo, except that the time at which the E- to N-cadherin switch in the neural tube and in NCC is complete differs according to the level considered, being earlier in the trunk than in the midbrain. In this model, we propose that by the time NCC

execute a complete EMT program with breakdown of the basement membrane, loss of cell polarity, and a rapid E- to N-cadherin switch (Acloque et al., 2011; Dady et al., 2012; Nakaya, Sukowati, Wu, & Sheng, 2008). This suggests that, in NCC, the transcription-repressing activity of the core EMT regulatory factors is finely tuned over time and that prior to EMT, these factors may exert additional functions independently of their role in delamination by regulating different sets of "early" and "late" genes. In this respect, *Snail-1/2* have been proposed to play a role in both specification and delamination (LaBonne & Bronner-Fraser, 2000; Mancilla & Mayor, 1996; Sakai, Suzuki, Osumi, & Wakamatsu, 2006), albeit no specific *Snail-1/2* target specification genes have been identified yet. Therefore, knowing how *Snail*, *Zeb*, and *Twist* expressions and activities are regulated may provide insight into their missions throughout NCC development.

EMT is a strictly controlled process driven by multiple, closely interconnected regulatory networks involving a plethora of factors which modulate Snail, Zeb, and Twist expression, cellular localization, stability, and activity (Fig. 4; De Craene & Berx, 2013, for review). The possible implication for a number of them during NCC development has been investigated during the last decade, most often at cranial levels in *Xenopus* or chick embryos, and a temporal sequence of their activity progressively emerges that might account for the time course of cell specification and delamination.

---

become induced, the unspecified neurectoderm encompassing all ectodermal, NCC, and neural tube progenitors is an epithelium uniformly expressing E-cadherin (red). Following induction, while E-cadherin expression is maintained at high levels in the ectoderm, it decreases gradually in the rest of the neurectoderm along a mediolateral gradient. Progenitors of the neural tube express progressively higher levels of N-cadherin (green) and lower levels of E-cadherin under the concerted actions of Zeb-2-repressing E-cadherin and Sox-2-activating N-cadherin. NCC also repress *E-cadherin* under the control of Zeb-2 and start expressing *cadherin-6B* (blue) under the control of Smads rather than N-cadherin, due to the putative mutual repression between Snail-2 and Sox-2. This results in N-cadherin confinement to the neural tube. By the time of NCC delamination, cadherin pattern exhibits a clear-cut partition, with E-cadherin only in the ectoderm, cadherin-6B in premigratory NCC, and N-cadherin in the neural tube. NCC undergo EMT after losing cadherin-6B from their surface under the control of Snail-2 and most likely Zeb-2 and transiently express low cadherin levels (white). As NCC initiate migration after delamination, they express a new cadherin content (cadherin-7 or cadherin-11, beige), while the ectoderm and neural tube become entirely segregated. (See the color plate.)

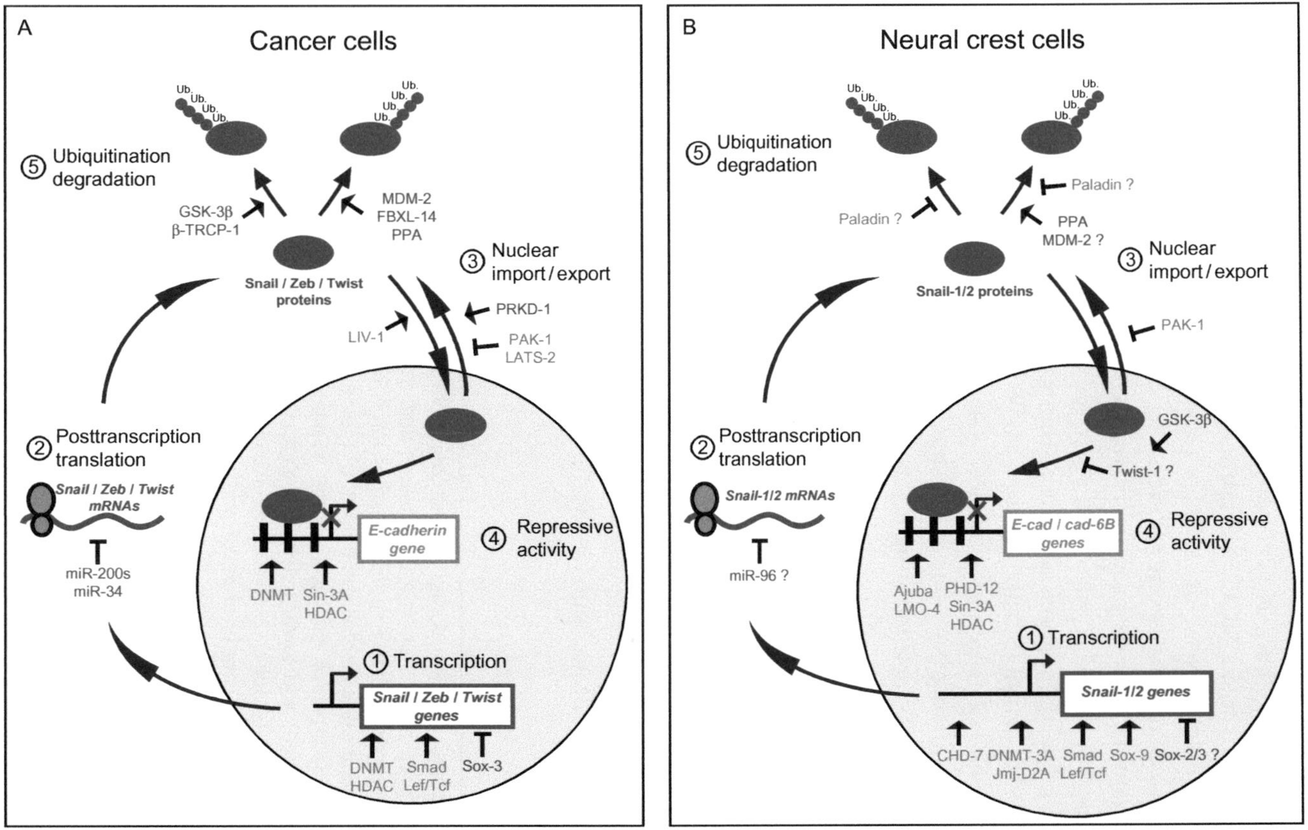

Figure 4 Control mechanism of the expression and activity of the core EMT regulators. Schematic representation of the molecular cascades controlling (1) transcription and (2) translation of Snail-1/2 messages, (3) the nucleocytoplasmic trafficking of Snail-1/2 proteins, (4) their repressing activity on the E-cadherin gene, and (5) their degradation by the cytoplasmic machinery. The main players identified in cancer cells are depicted in panel (A), while those identified in NCC are shown in panel (B). Players whose action results in increased Snail activity and EMT are indicated in green and those involved in reduced Snail activity are in red. We also propose in this schema that in NCC Twist-1 may reduce Snail-2 activity by titrating Snail-2. (See the color plate.)

3.1. Transcriptional and translational controls of the expression of the core EMT regulatory factors

Transcriptional control of expression of core EMT regulatory factors is the primary mode of regulation of the EMT program, and this has been specifically investigated in NCC owing to our understanding of the gene regulatory networks underlying their development (Betancur et al., 2010; Prasad et al., 2012). As in cancer cells (Lim & Thiery, 2012; Thiery & Sleeman, 2006), EMT is triggered in NCC by the BMP-4 and Wnt-signaling pathways. The classical transcription factors recruited by these pathways (e.g., Smads for BMPs and Lef/Tcf for Wnts) have been shown to directly upregulate expression of *Snail-1/2* in these cells (Sakai et al., 2005). However, aside from these factors, additional transcription factors have been identified as potent inducers of core EMT regulators during NCC specification and delamination. Sox-9, for example, is essential for *Snail-2* expression and its effect is mediated through direct binding to the *Snail-2* promoter in synergy with Snail-2 itself, which is able to activate its own expression (Sakai et al., 2006). Other factors, such as the border specifiers, Pax-3/7, and Msx-1, can also upregulate *Snail-2*, but whether they act directly or via other factors has not been fully investigated. Recently, functional experiments in chick have placed Elk-3, a member of the Ets family of transcription factors present in NCC early during specification and migration, downstream of border specifiers and upstream of *Snail-2*, in the genetic cascade accompanying their development (Rogers, Saxena, & Bronner, 2013). How Elk-3, which is known to function as a transcriptional repressor, controls these genes is not known. Likewise, c-Myb, a transcription factor also expressed in premigratory and early migrating NCC, has been found to control expression of *Snail-2* and *Twist-1* (Betancur, Simoes-Costa, Sauka-Spengler, & Bronner, 2014; Karafiat et al., 2005). Whether this cohort of factors is responsible for modulating expression of *Snail-1/2* (as well as *Twist-1* and *Zeb-2*) between specification and delamination has not been formally established, and to our knowledge, no transcription factors whose expression correlates with the initiation of delamination have been identified so far. The only described exception is Ets-1, the prototypic member of the Ets family, which is induced just prior to EMT at cranial levels in chick. Functional studies, however, indicate that Ets-1 acts merely as a coordinator of delamination for cranial NCC and not as a direct modulator of *Snail-2* expression (Théveneau et al., 2007).

Relatively little is known about the transcriptional repressors of *Snail*, *Twist*, and *Zeb*. At variance with what was found in NCC, *Snail-1* has also been reported to bind to an E box in its own promoter and repress its transcription (Peiro et al., 2006). However, a recent study has identified Sox-3 as a potent negative regulator of *Snail-2* expression during chick gastrulation. Sox-3 expression in the neural plate represses *Snail-2* gene, thus protecting neuroepithelial cells from ingressing, while in the primitive streak, Snail-2 conversely represses *Sox-3*, activating EMT and ingression of precursors of the mesoderm (Acloque et al., 2011). Whether this antagonistic relationship also applies to *Sox-2/3* and *Snail-2* in NCC is not known. Although highly plausible, because *Sox-2*, *Sox-3*, and *Snail-2* expressions are mutually exclusive in NCC progenitors and in the neural tube (Uchikawa et al., 2011; Wakamatsu et al., 2004), it remains to be demonstrated. Indeed, Sakai et al. (2006) reported that Sox-2 does not affect *Snail-2* promoter activity in fibroblast cell lines.

Small noncoding RNAs also emerged as potent modulators of the expression of the core EMT regulatory factors. MicroRNAs (miRNAs) typically suppress gene expression by interacting with the 3′-untranslated region of target messengers to repress their translation or degrade them. A large number of miRNAs have been involved in the control of EMT and, among them, the miR-34 and the miR-200 families are strongly associated with the epithelial state and are downregulated upon EMT (De Craene & Berx, 2013, for a review). While miR-34 represses *Snail-1* expression, miR-200s act as silencers for *Zeb-1/2*. Interestingly, both miRNAs contain E boxes in their promoters and are directly regulated by Snail and Zeb, thereby revealing intricate reciprocal feedback loops between miRNAs and their targets (Burk et al., 2008; Siemens et al., 2011). These miRNAs are, on the other hand, under the positive control of p53 (Chang et al., 2011), an important player in NCC development. Contrary to miR-200s and miR-34, miR-9, which is upregulated in breast cancer cells and activated by Myc, has been found to directly target *E-cadherin* messages, leading to increased cell motility and invasiveness in epithelial cell lines (Ma et al., 2010). To date, the presence of these various miRNAs during specification and delamination of NCC has not been reported, yet, miRNAs undoubtedly play an important role in these cells (Mayanil, 2013; Strobl-Mazzulla, Marini, & Buzzi, 2012). Indeed, in mouse, specific deletion in NCC of DICER, the RNase III enzyme required for miRNAs maturation, leads to craniofacial and cardiac anomalies (Zehir, Hua, Maska, Morikawa, & Cserjesi, 2010). Likewise, in *Xenopus*, loss of DICER or of FMR1 (fragile-X mental retardation syndrome-1) and FXR1 (fragile-X related-1), two RNA-binding proteins interacting with the miRNA-induced silencing complex (RISC) also causes craniofacial defects due to strong reduction of NCC migration (Gessert, Bugner, Tecza, Pinker, & Kuhl, 2010). Consistent with its role in repressing *Zeb* expression and preventing EMT, depletion of miR-200b affects development of cranial NCC-derived structures but, unexpectedly, this effect was not found to result from a block in induction and migration. Depletion of another miRNA, miR-96, causes alterations of both induction and migration (Gessert et al., 2010). Clearly, more systematic analyses of the repertoire, function and interplay of miRNAs regulating expression of the core EMT factors during NCC delamination are needed to define precisely how they contribute to its spatiotemporal control.

3.2. Epigenetic control of the expression of the core EMT regulatory factors

Control of *Snail*, *Zeb*, and *Twist* expressions may also be a part of more general cellular programs regulated at the epigenetic level, allowing genome-wide coordinated modulation of gene activity. Though still preliminary, recent studies have highlighted the major transcriptional reprogramming events that accompany NCC specification and delamination and provided insights into the synergistic control of NCC specifiers by the epigenetic machinery (Mayanil, 2013; Strobl-Mazzulla et al., 2012).

DNA methylation by DNA-methyltransferases (DNMTs) is one of the epigenetic modifications resulting in transcriptional repression of genes. Only DNMT-3A/3B, the DNMT responsible for *de novo* methylations, have been studied during NCC development. DNMT-3B is expressed in mouse in premigratory and migrating cells but is not restricted to them, and genetic studies demonstrated that it is largely dispensable in cranial and cardiac NCC for migration and differentiation. DNMT-3B is however required in their host tissues, such as the branchial arches and heart, during the latter steps of their development (Jacques-Fricke, Roffers-Agarwal, & Gammill, 2012). DNMT-3A, in contrast, shows in chick a more-restricted expression pattern than DNMT-3B in mouse as it is initially present in the neural plate border before becoming confined to premigratory and migrating NCC. Knockdown of DNMT-3A results in a severe reduction in the expression of *Snail-2* and of a variety of other NCC specifiers, accompanied by a remarkable increase in *Sox-2/3* activity. Importantly, these two genes appear to be direct targets of DNMT-3A in NCC. These data therefore suggest that DNA methylation acts as a major molecular switch to turn off neural tube transcription factors in neural plate border cells and promote NCC fate (Hu, Strobl-Mazzulla, Sauka-Spengler, & Bronner, 2012). The mechanism by which DNMT-3A is recruited to the promoters of its target genes in prospective NCC is currently unknown, but it has been found to interact with a multiplicity of transcription factors, some of which are highly relevant to NCC formation, i.e., p53, Ets-1, and AP-2 (Hervouet, Vallette, & Cartron, 2009). It would be of interest therefore to establish a complete list of targets of DNMT-3A in NCC and to determine by which factors it is recruited to them.

Beside DNA methylation, histone methylation and acetylation are the major chemical modifications, which influence chromatin structure and regulate gene expression (Kouzarides, 2007). Methylation of histone H3 at specific amino acid residues recruits a variety of chromatin modifiers and transcriptional activators or repressors, resulting in differential effects on gene expression. Schematically, trimethylated lysine 4 of histone H3 (H3K4me3) and H3K36me3 is associated with active transcription, while H3K27me3 and H3K9me3 are repressive marks. Histone methylation is regulated by the interplay between methyltransferases and demethylases, notably the Jumonji proteins (Jmj). To date, only Jmj-D2A has been implicated in the control of *Snail-2* expression during chick NCC development (Strobl-Mazzulla, Sauka-Spengler, & Bronner-Fraser, 2010). It is expressed initially throughout the neural plate and at the border and gradually resolves to premigratory NCC. Of interest, its expression is the highest during early specification and then declines until the onset of migration at which stage it is almost no longer expressed. Furthermore, loss of *Jmj-D2A* expression causes a dramatic downregulation of NCC specifier genes, notably *Snail-2*, *Foxd-3*, and *Sox-10*, but has no effect on neural plate border genes. Additionally, the location and abundance of the H3K9me3 and H3K36me3 epigenetic marks, as revealed by chromatin immunoprecipitation, illustrate the dynamic occupancy of sites in proximity to the transcriptional start site of *Snail-2* and *Sox-10* genes and clearly reflect their transcriptional state. At initiation of specification, *Sox-10* exhibits H3K9me3 repression marks, while *Snail-2* shows a more mitigated pattern with some H3K9me3 repression but H3K36m3 activation marks as well, reflecting an ongoing shift from transcriptional repression to activation. Later, by the time of migration, occupancy of the repressive H3K9me3 marks near the *Sox-10* and *Snail-2* genes is clearly reduced, consistent with high expression of these, while H3K36me3 marks are unchanged. Correlating with these changes, direct interactions between Jmj-D2A and the regulatory regions of *Sox-10* and

Snail-2 genes can be detected at early specification stages but not later during migration. Finally, demethylation of H3K9me3 on the *Sox-10* promoter is inhibited upon knockdown of Jmj-D2A in early NCC. These results therefore indicate that the relief of transcriptional repression by the Jmj-D2A-mediated demethylation of H3K9me3 marks plays an important role in the induction of *Snail-2* and other NCC specifiers.

Other chromatin remodellers, such as CHD-7, a chromodomain helicase DNA-binding protein, have also been shown to control expression of core EMT regulators during *Xenopus* NCC development (Bajpai et al., 2010). *CHD-7* is expressed in the neural ectoderm and in premigratory and migrating NCC. Depletion of *CHD-7* specifically targets NCC specifiers, such as *Snail-2*, *Sox-9*, and *Twist-1*, but not border specifiers, and causes major craniofacial defects. In addition, CHD-7, in association with the PBAF (polybromo- and BRG1-associated factor) complex, binds to a NCC-specific distal enhancer of the *Sox-9* gene and a conserved genomic element located upstream of the *Twist-1* gene. Consistent with these observations, CHD-7- and PBAF-bound regions were preferentially enriched for H3K4me1, a mark previously associated with enhancers. Thus, in a similar manner to Jmj-D2A in chick, CHD-7 promotes expression of NCC specifiers in *Xenopus* by interacting with epigenetic marks on target genes. Whether both activities coexist in either species is not known yet.

3.3. Control of the stability and intracellular location of the core EMT regulatory factors

Snail proteins are by nature highly unstable, including in NCC (Vernon & LaBonne, 2006). Snail turnover in the cytoplasm is tightly regulated by regulatory mechanisms involving posttranslational phosphorylation and ubiquitination (De Craene & Berx, 2013). These regulatory mechanisms can be classified into glycogen synthase kinase-3β (GSK-3β)-dependent and -independent processes (Fig. 4). In the GSK-3β-dependent process, cytoplasmic Snail-1 proteins are phosphorylated at serine–threonine residues first by casein kinase-1 then by GSK-3β and they are subsequently processed for ubiquitination by the E3 ubiquitin ligase β-TRCP-1 for degradation (Vinas-Castells et al., 2010). Thus, signaling events elicited by Wnt and other growth factors causing inactivation of GSK-3β may repress Snail-1 degradation and favor transition toward a mesenchymal phenotype. Given that Wnt-1 signals are activated in NCC progenitors prior to delamination, it is tempting to suggest that their activity may recruit β-TRCP-1 resulting in increased stability of Snail proteins. Interestingly, at least in chick and mouse, Wnt-1 expression is induced late during NCC specification, shortly prior to onset of migration (Burstyn-Cohen, Stanleigh, Sela-Donenfeld, & Kalcheim, 2004), suggesting that Snail stabilization upon inhibition of GSK-3β activity may constitute a late triggering event of NCC delamination. This hypothesis awaits to be tested, however.

Beside β-TRCP-1, GSK-3β-independent ubiquitin ligases, such as MDM-2 and FBXL-14, and the F-box protein partner of paired (PPA) have also been reported to target Snail proteins for degradation (Lander, Nordin, & LaBonne, 2011; Vinas-Castells et al., 2010), and, at least two of them, MDM-2 and PPA, have been implicated in NCC development (Daujat, Neel, & Piette, 2001; Lander et al., 2011; Vernon & LaBonne, 2006). Indeed, it has been found in *Xenopus* that, contrary to Snail-1, Snail-2 does not contain in its sequence the β-TRCP-1 destruction motif. Inhibiting GSK-3β causes only a marginal increase in Snail-1/2 stability, suggesting that in this species the Wnt-1/GSK-3β/β-TRCP-1 axis plays only a minor contribution to Snail regulation (Lander et al., 2011; Vernon & LaBonne, 2006). In contrast, PPA targets not only Snail-1, but also Snail-2, Twist-1, and Zeb-2, making this pathway a major candidate for regulating the timing of NCC delamination (Lander et al., 2011; Vernon & LaBonne, 2006). PPA is induced specifically in premigratory NCC just prior to EMT. In addition, morpholino-depletion of PPA stabilizes Snail-2 protein, whereas its misexpression promotes its turnover. Importantly, Sox-9 and Foxd-3 have been found to repress expression of PPA and oppose its Snail-protein degradation activity when overexpressed in the neural tube. Thus, prior to EMT, Snail-protein levels (and probably those of Zeb-2 and Twist-1) would be constantly maintained through repression of PPA by the combined activities of Sox-9 and Foxd-3. Then, upon EMT, PPA levels would increase, resulting in higher Snail-1/2, Zeb-2, and Twist-1 degradations. Consistent with this idea, PPA inhibition has been found to cause migration defects, indicating that high levels of Snail-2 might be deleterious for efficient NCC migration (Vernon & LaBonne, 2006). However, as neither Sox-9 nor Foxd-3 is downregulated at onset of migration, how *PPA* repression is relieved remains to be found.

Insights into the role of MDM-2 in NCC come essentially from the studies on the regulation of the tumor suppressor p53. p53 is one of the main targets of the MDM-2 ubiquitination activity, an important step in the oncogenic process. In mouse and chick, p53 messengers are expressed in premigratory NCC (although proteins are barely detectable) and decline after delamination, while MDM-2 shows a reverse expression pattern

(Daujat et al., 2001; Jones et al., 2008; Rinon et al., 2011), suggesting that in NCC, MDM-2 also drives p53 to the proteasome. In chick, stabilization of the endogenous p53 protein by an inhibitor of MDM-2 activity reduces *Snail-2* expression and inhibits NCC delamination (Rinon et al., 2011). Likewise, in a mouse model of the Treacher Collins syndrome, a congenital disorder of craniofacial development due to the deficient formation of migrating NCC, levels of p53 proteins have been found to be stabilized. Furthermore, in those animals, inhibition of p53 restores normal NCC migration and suppresses craniofacial abnormalities (Jones et al., 2008). In an apparent contradiction with these data, further experiments in chick show that in the absence of p53, NCC fail to delaminate and colonize their target territories although they are produced in large excess in the neural tube. It should be noted though that a significant proportion of *p53*-null mouse embryos exhibit neural tube defects and a variety of craniofacial abnormalities (Armstrong, Kaufman, Harrison, & Clarke, 1995; Rinon et al., 2011), indicating that p53 levels must be finely tuned in NCC progenitors for correct delamination and migration to occur. Whether MDM-2 is required for this process has not been investigated, however. Interestingly, p53 has been found to control cell invasion in cancer cells by inducing the MDM-2-mediated degradation of Snail-2 (Wang et al., 2009), and Pax-3 has been shown to reduce p53 protein stability in NCC by interacting physically with both p53 and MDM2 (Wang, Morgan, & Loeken, 2011). These data therefore reveal the previously unsuspected importance of a signaling cascade involving Pax-3/MDM-2/p53 and Snail-2 in the triggering of NCC delamination.

As nuclear Snail proteins degrade more slowly than cytoplasmic Snail, there are mechanisms that modulate their nuclear trafficking (Fig. 4). Thus, two kinases, PAK-1 and LATS-2, favor nuclear retention of Snail-1, thereby enhancing its stability and activity (Yang et al., 2005; Zhang et al., 2012). Furthermore, the zinc transporter LIV-1, a STAT-3 target, has been found to drive nuclear import of Snail-1 in the zebrafish gastrula and promote EMT (Yamashita et al., 2004). In contrast, the protein kinase D1 (PRKD-1) by enhancing its nuclear export restricts EMT (Du, Zhang, Hassan, Biswas, & Balaji, 2010). The roles of LATS-2, LIV-1, and PRKD-1 in NCC delamination have not been investigated yet, but a recent report indicates that in *Xenopus*, PAK-1 expression coincides temporally with their migration (Bisson, Wedlich, & Moss, 2012). Consistent with this restricted pattern, expression of a dominant-negative form of PAK-1 blocks migration but does not affect specification. Intriguingly, PAK-1 is able to phosphorylate Snail-1 and Twist-1 but not Snail-2, indicating that during migration requirement for Snail-1 and Snail-2 activities is different.

Recently, a novel level of regulation of Snail-1/2 activity has been uncovered in *Xenopus*, implicating Twist-1. Although it is commonly presented as an important player in NCC specification and delamination (Betancur et al., 2010; Milet & Monsoro-Burq, 2012; Prasad et al., 2012; Sauka-Spengler & Bronner-Fraser, 2008), Twist-1 function in *Xenopus* has long remained elusive, and it is only very recently that it has been reevaluated. Unlike Snail-1/2 and Zeb-2, Twist-1 appears relatively late during the process of NCC specification and it is restricted to cranial regions (Soo et al., 2002). However, elegant studies by Lander et al. (2013) showed that *Twist-1* misexpression or depletion impacts the levels of *Snail-1/2* message as well as other NCC specifiers. Interestingly, this occurs not merely through regulation of gene expression but primarily through direct binding with Snail-1/2 proteins. The C-terminal WR domain of Twist-1 interacts with the N-terminus of Snail via a cluster of GSK-3β-phosphorylated amino acids. This interaction does not interfere with the ability of Twist-1 to bind DNA (Fig. 5). Moreover, Snail-1/2 binding stabilizes Twist-1 by competing with PPA association, suggesting that Snail-1/2 impacts positively on Twist-1 activity. Surprisingly, GSK-3β-phosphorylation of Twist-1 titers Snail-1/2 and diminishes their recruitment to E boxes in the promoters of their target genes. Thus, although Snail-2 is not directly targeted for degradation by GSK-3β-phosphorylation (see earlier), these findings ascribe a critical role for GSK-3β in the control of its activity via Twist-1. In the future, it will be important to determine where and when Twist-1 is phosphorylated by GSK-3β to decipher the mechanisms triggering NCC EMT. Of note, Twist-1 is not expressed in premigratory cranial NCC in chick and mouse and it appears gradually during migration as cells reach the branchial arches and undergo chondrocytic differentiation, thereby excluding any significant involvement in NCC delamination in these species (J.-L. Duband, unpublished; Soo et al., 2002).

Finally, recent preliminary data on paladin, an antiphosphatase, may provide interesting information regarding how the activity of the core EMT regulators can be modulated over time in premigratory NCC (Roffers-Agarwal, Hutt, & Gammill, 2012). Antiphosphatases have been shown to bind phosphorylated residues on target proteins and protect them from dephosphorylation. In chick, paladin is expressed during NCC development both at cranial and trunk levels. Decreasing paladin expression precociously

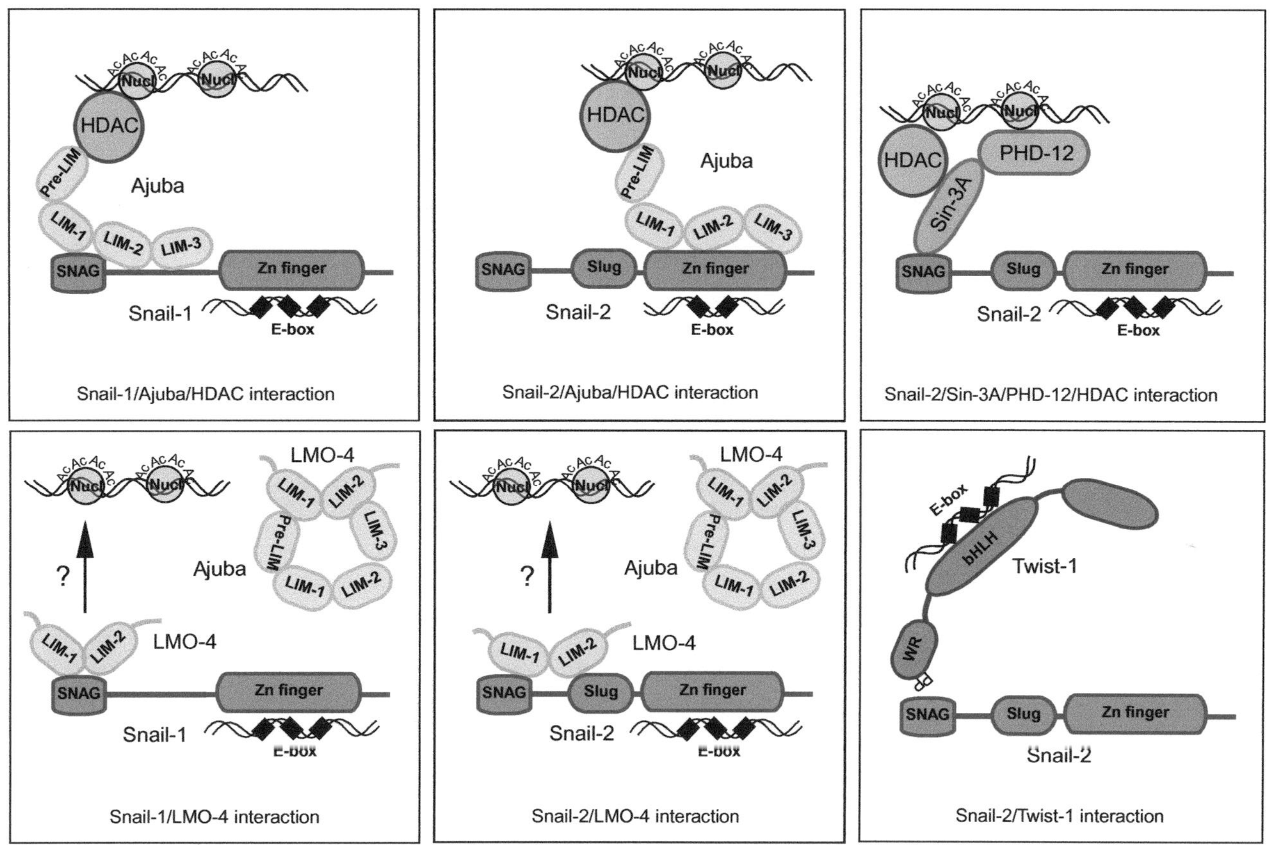

Figure 5 See legend on next page.

using morpholinos results in a substantial but incomplete reduction in *Snail-2* expression at premigratory stages. Interestingly, only *Snail-2* is strongly repressed, as other targets tested were affected either moderately (e.g., *Sox-10*) or not at all (*Foxd-3*, *cadherin-6B*, and *Rho-B*). This indicates that paladin may regulate only a limited set of targets during NCC specification. In light of these data, it is tempting to speculate that control of the level of phosphorylation of Snail-2 by paladin is important for some of its activities and may vary over time during specification and delamination. Further experiments, however, are needed to verify this hypothesis and to better define paladin's role and targets. For example, determining which amino acids on Snail are targeted by paladin and whether Snail stability and trafficking are affected upon manipulation of its expression should help further our understanding of how maintenance of phosphorylation involving paladin contributes to control of NCC delamination.

3.4. Control of the activity of the core EMT regulatory factors

One of the reasons for the ability of the Snail, Zeb, and Twist transcription factors to induce massive phenotypic changes during EMT resides in part in their close interaction with epigenetic modifiers, allowing genomewide changes in gene expression. Indeed, their transcriptional regulatory activity requires the participation of various interacting proteins, which results in strong transcriptional control of EMT, and this has been particularly well studied in the case of the *E-cadherin* gene. For example, DNA methylation of the *E-cadherin* promoter is responsible for its silencing and has been

Figure 5 Snail-1/2 partners in NCC-controlling cadherin gene expression. Snail-1/2 proteins bind to E boxes of the E-cadherin and cadherin-6B promoters via their Zn-finger domain, and their repressing activity involves recruitment of HDAC either through direct binding to the LIM protein Ajuba (as found in *Xenopus*) or via the Sin-3A and PhD-12 adaptors (in chick). Whether Ajuba and PHD-12 exist in both species is not known yet and if so it would be of interest to determine whether they cooperate or compete for Snail activity. Interestingly, Snail-1 and Snail-2 differ from each other by the domain involved in Ajuba binding (the Snail/Gfi (SNAG) domain for Snail-1 and the Zn-finger domain for Snail-2), but whether this may influence the repressing activity of Snail is not known. Snail-1/2 also bind the LIM only protein LMO-4 through their SNAG domain (for Snail-1) or their N-terminal half (for Snail-2). However, as LMO-4 cannot recruit HDAC and that it binds Ajuba, thereby preventing Ajuba–HDAC interaction, it remains to be defined how it functions to mediate Snail-repressing activity. Finally, Snail-2 has been found to bind the GSK-3β-phosphorylated WR domain of Twist-1 and this inhibits its interaction with its target genes. Thus, Twist-1 phosphorylation may function as a negative regulator of Snail-2 activity. (See the color plate.)

correlated with EMT in cancer cells (Lombaerts et al., 2006) and associated with increased *Snail-1* expression (Cheng et al., 2001). More intriguingly, though, it has also been reported that DNMT-1 can repress *E-cadherin* expression in the absence of noticeable changes in DNA methylation patterns in its promoter and that this effect is mediated by direct interaction with Snail-1 (Espada et al., 2011). During NCC development, although DNA methylation by DNMT-3A has been shown to act as a major molecular switch to turn off neural tube transcription factors in neural plate border cells and promote NCC fate (Hu et al., 2012), it is not known yet whether Snail-1/2 and the other core EMT regulators can associate directly or via adaptors with DNMTs to target specific genes.

Beside DNMTs, Snail-1 has been found to associate with a variety of histone modifiers to regulate gene expression (Fig. 5). In particular, it induces histone deacetylation of the *E-cadherin* gene through the recruitment of Sin-3A in association with histone deacetylases (HDAC-1/2; Peinado, Ballestar, Esteller, & Cano, 2004). Interestingly, a recent study uncovered the mechanism by which Snail-2 recruits HDAC to repress *cadherin-6B* expression in NCC in chick (Strobl-Mazzulla & Bronner, 2012b). PHD-12 is an adaptor protein, whose expression is markedly increased in cranial NCC just before EMT. Loss of *PHD-12* function phenocopies *Snail-2* knockdown, inhibiting transcriptional repression of *cadherin-6B* and preventing NCC emigration. PHD-12 and Snail-2 do not bind to each other, but both directly interact with Sin-3A, which in turn complexes with HDAC. PHD-12 is recruited to the *cadherin-6B* promoter during EMT. Consistent with this, lysines on histone H3 at the *cadherin-6B* promoter are hyperacetylated before NCC emigration, correlating with active transcription, but deacetylated during EMT, reflecting a shift to a repressive state. Finally, knockdown of either *PHD-12* or *Snail-2* prevents deacetylation of the *cadherin-6B* promoter. These results therefore suggest a model whereby repression of the *cadherin-6B* gene occurs through binding of PHD-12 and Snail-2 to their transcription start site and E boxes. This allows Sin-3A and HDAC to be recruited and to deacetylate histone H3 at the promoters, resulting in the repression of transcription. Of interest, the authors mention that this model also applies to *E-cadherin* repression. In another study, Murko et al. (2013) showed that the treatment of early chick embryos using a pharmacological blocker of HDAC causes neural tube defects at trunk levels that were associated with an increase in *cadherin-6B* expression and a premature loss of epithelial features among neural epithelial cells.

Among the different adaptors with which Snail interacts in NCC to recruit HDAC and drive transcriptional repression of genes are the LIM proteins, characterized by the presence of two LIM domains in tandem (Fig. 5). The first LIM proteins to be identified were the Ajuba family which function as corepressors of *Snail-1/2* (Langer et al., 2008). Ajuba interacts with the SNAG domain of Snail-1 and this interaction targets the complex to the nucleus and potentiates Snail-1 binding to E boxes of the *E-cadherin* promoter. Interestingly, a recent study by Ochoa, Salvador, and LaBonne (2012) revealed that contrary to Snail-1, Snail-2 binds to Ajuba via its Zn-finger motifs. Expression of Ajuba in *Xenopus* embryos enhances NCC migration in a Snail-1/2-dependent manner, and conversely, its depletion phenocopies *Snail-1/2* knockdown (Langer et al., 2008). This study therefore assigns to Ajuba LIM proteins a critical role in NCC development as Snail-1/2 corepressors. However, their presence in NCC has not been assessed. Interestingly, Ajuba LIM proteins are also components of adherens' junctions and contribute to their assembly or stability (Srichai et al., 2004). Their functional interaction with Snail proteins in the nucleus suggests that they are also important regulators of the dynamics of epithelial cells, linking surface events with nuclear responses. It would then be informative to determine when and where Ajuba LIM proteins are expressed in NCC progenitors to better evaluate their implication in the spatiotemporal control of Snail-1/2 activity.

More recently, studies in *Xenopus* and chick identified the LIM domain only protein 4 (LMO-4) as another LIM protein involved in NCC development (Ferronha et al., 2013; Ochoa et al., 2012). In both species, *LMO-4* is distributed in NCC and neural plate progenitors during early specification and becomes restricted to NCC at the time of their delamination and early migration. In addition, functional studies revealed that LMO-4 is a Snail-1/2-interacting protein essential for NCC development. Morpholino-mediated knockdown of *LMO-4* leads in *Xenopus* to a deficit in the production of NCC progenitors and in chick to a severe reduction in their delamination. Additionally, while misexpression of *LMO-4* in the trunk neural tube in chick is insufficient to induce expression of NCC specifiers but speeds up delamination, in *Xenopus*, excess *LMO-4* leads to ectopic expression of *Snail-1/2*, and to a reduction in the expression of *Foxd-3*, *Sox-8/9/10*, and *Twist-1*. These differences between chick and frog are likely not to result from species specificities but rather to differences in the timing when *LMO-4* was overexpressed (late during NCC development in chick and early during specification in frog), thereby revealing different requirements for LMO-4 throughout NCC development. Detailed analyses of LMO-4-binding specificities further showed that it binds directly to Snail-1/2, but not to other NCC specifiers, and this interaction occurs via the SNAG domain of Snail, like Ajuba LIM proteins (Fig. 5). Intriguingly, Snail-1/2 binding to LMO-4 does not lead to HDAC recruitment, thus raising concerns about its mode of action. The fact that LMO-4 competes with HDAC for Ajuba binding suggests that it may play a key modulatory role of Snail-1/2 interactions with its partners, thereby selectively targeting distinct sets of genes. It will be essential then to determine whether LMO-4 recruits specific proteins to Snail-regulatory complexes on target promoters and to define how Ajuba and LMO-4 are coordinated over time during NCC development.

Twist-1 has also been found to recruit epigenetic modifiers as partners for regulating *E-cadherin* gene expression. Notably, it has been found to associate with Bmi-1, a member of the polycomb-repressive complex, to suppress *E--cadherin* transcription through binding to the E boxes situated in its promoter. Interestingly, Bmi-1 is itself under the direct transcriptional control of Twist-1 (Yang et al., 2010). Knowing the complex relationship between Twist-1 and Snail-1/2 in NCC (see earlier), determining whether Bmi-1 is expressed during their delamination and may contribute to the control of cadherin expression, will be valuable to decipher these interactions and their importance in greater detail.

Although a coherent, complete picture of the regulatory networks controlling expression and activity of the core EMT regulators Snail-1/2, Zeb-2, and Twist-1 is still lacking (in part because studies were performed in different animal models characterized by distinct gene regulatory networks and different cellular features), two major conclusions can be drawn for these studies (Fig. 6). Firstly, a multiplicity of transcription factors and chromatin modifiers and remodellers binds to different domains in the regulatory sequences of *Snail-1/2*, *Twist-1*, and most likely *Zeb-2* (although this has not been documented yet) and is recruited to relieve repressing epigenetic marks and induce their robust expression. In addition, their restricted expression patterns and functions during specification clearly link their activity to the acquisition of NCC identity rather than to delamination. Secondly, the intracellular localization of Snail-1/2, its stability, and its activity are dictated by a plethora of kinases and nuclear corepressors which are induced or activated concurrently just prior to delamination. This suggests that collectively they function as a molecular switch to induce NCC delamination and that this event requires high doses of Snail proteins to regulate a

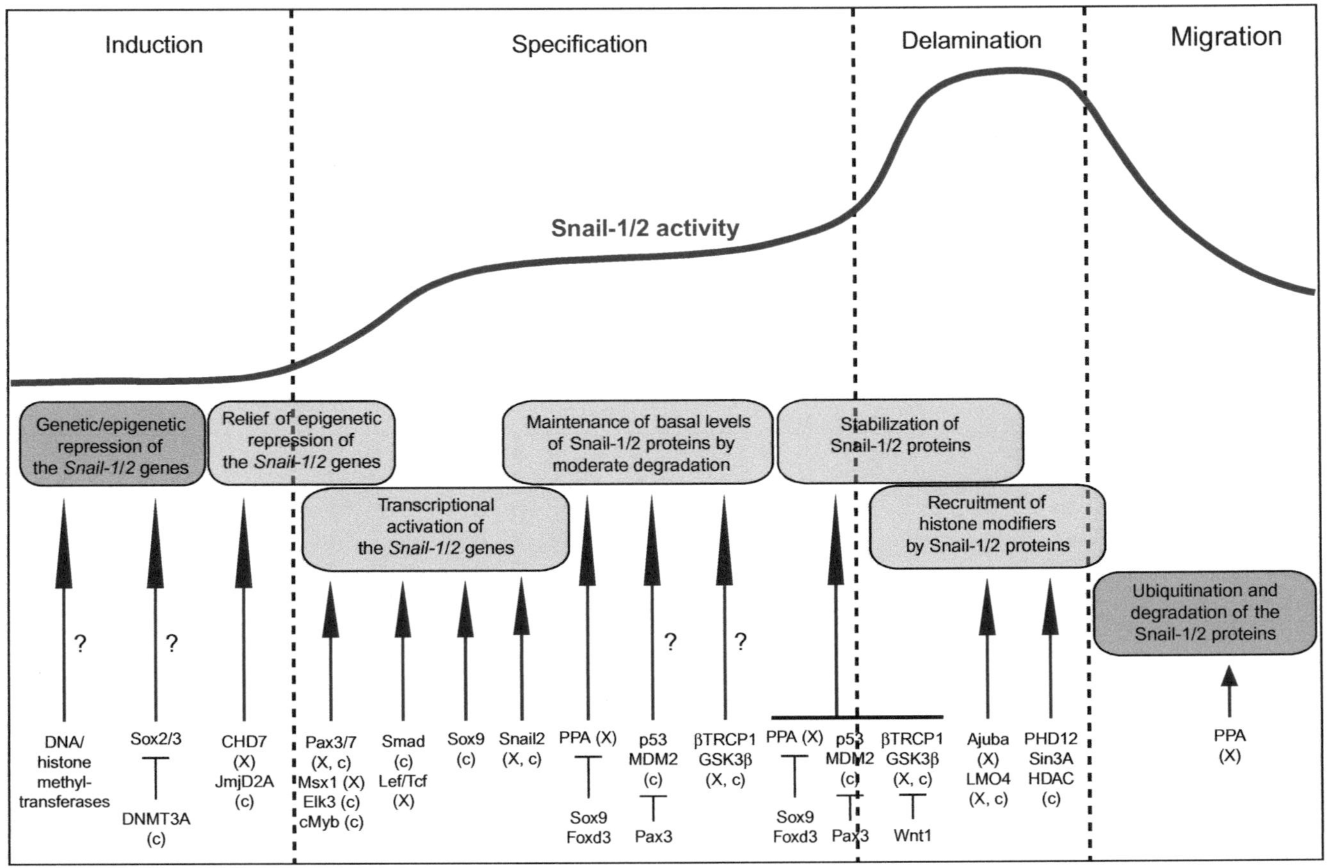

Figure 6 See legend on next page.

large array of effectors within a short period. This model represents, however, only a part of the whole picture. Indeed, it is essentially based on the control of the repressing activities of Snail proteins on its cadherin targets and it has now been demonstrated that they are also capable of activating numerous genes, and in particular those related to the locomotory machinery.

4. SPATIAL CONTROL OF THE ACTIVITY OF THE ZEB-2 TRANSCRIPTION FACTOR

Contrary to the other core EMT regulatory factors, Zeb-2 is not restricted to premigratory NCC but is also expressed by the entire neural plate. *Zeb-2* expression is induced precociously during primary neurulation and correlates with the progressive switch from E- to N-cadherin in neuroepithelial cells (Dady et al., 2012). Knockdown of *Zeb-2* in the neural tube at cranial levels in chick or mouse results in persistent expression of *E-cadherin* in neuroepithelial cells and also in strong inhibition of NCC delamination and migration (Rogers et al., 2013; Van de Putte et al., 2003). Thus, Zeb-2 is clearly an important player in the EMT process by promoting *E-cadherin* silencing. Yet, unlike NCC, neuroepithelial cells retain their epithelial morphology and fail to undergo EMT despite persistent *Zeb-2* expression over long periods. Why the same molecule mediates radically different behaviors in neighboring cells that share a similar origin and numerous cellular features remains unclear. As discussed earlier, an explanation may reside in the fact that neural tube cells do not express NCC specifier genes and therefore lack the appropriate signaling modules that enable them with the ability to undergo EMT and disperse. However,

Figure 6 Coordinated modulation of the expression and activity of Snail-1/2 during NCC specification and delamination. Hypothetical model of the molecular processes controlling Snail-1/2 activity in NCC from induction to early migration. This model combines data obtained either in *Xenopus* (X) or in chick (c), and aside from a few exceptions such as LMO-4, it is not known whether the same players are involved in both species. Question marks indicate that the direct implication of pathway has not formally been established but is plausible. Events causing repression of Snail-1/2 activity are depicted in red (gray in the print version) and those increasing its activity in green (light gray in the print version). From this model, it appears that induction of Snail-1/2 expression during specification occurs essentially at the transcriptional level, while its recruitment during delamination is made possible owing to posttranslational events such as increased stability by the inhibition of the GSK-3β phosphorylation, nuclear import and retention, and interaction with corepressors. At the end of delamination, Snail-2 activity is again reduced to favor migration by mechanisms involving ubiquitination and degradation.

additional reasons may account for the precision by which EMT is restricted to NCC. In particular, it has been found that although *Zeb-2* is evenly expressed at moderate levels in the neural tube and NCC throughout neurulation and during specification, its expression is sharply increased in NCC at the time they undergo EMT. This indicates that execution of the EMT program may require higher doses of Zeb-2 than the sole control of E- to N-cadherin switch (Dady et al., 2012; Rogers et al., 2013; Van de Putte, Francis, Nelles, van Grunsven, & Huylebroeck, 2007). This assumption is supported by a recent study on the genetic network regulating EMT in cancer cells which proposed a model accounting for the three distinct states in which cells can be alternatively found, i.e., epithelial, mesenchymal, and a so-called hybrid state, in which they show mixed mesenchymal and epithelial characteristics (Lu, Jolly, Levine, Onuchic, & Ben-Jacob, 2013). This structural organization of cells has been proposed to be regulated by a circuit composed of two interconnected modules, the miR-34/Snail-2 and the miR-200/Zeb-2 modules, mentioned above. Interestingly, this model suggests that the miR-34/Snail-2 module exists in two mutually exclusive forms: high miR-34/low Snail-2 and low miR-34/high Snail-2. This module is a central regulatory circuit of EMT, acting as an on/off switch responsible for either the epithelial or the mesenchymal states. The miR-200/Zeb-2 module in contrast can be found in three forms: high miR-200/low Zeb-2, low miR-200/high Zeb-2, and medium miR-200/medium Zeb-2, corresponding to the epithelial, mesenchymal, and hybrid states, respectively, and may function as a more-refined modulator of the cohesive state of epithelial cells. Although data on miRNA expression in the neural tube are still lacking, the modest level of *Zeb-2* in neuroepithelial cells associated with the absence of *Snail-2* may explain why, in spite of E- to N-cadherin switch, these cells undergo only partial transformation into a pseudostratified epithelium characterized by elongated cell shapes and loss of tight junctions but are unable to achieve complete EMT. Thus, combination of cadherin switches in the early blastoderm under the tight spatial and temporal controls of the core EMT regulatory factors Snail-2 and Zeb-2 may be responsible for its partition into three distinct portions exhibiting different epithelial properties and capacities to execute the EMT program.

5. FUTURE PROSPECTS

The few examples described earlier illustrate how time and space are coordinated in an exquisite manner to permit a population of progenitor cells to sequentially execute each of the individual steps that eventually lead them to populate a great variety of territories in the embryo and generate an extremely large array of cell types. This is achieved through an intricate combination of genome-wide epigenetic control of gene expression, recruitment of specific transcription factors that act in concert, and activation of chemical modifications that locally regulate the availability and activity of these factors. How these different actors are temporally coordinated and whether there is an hierarchy among them constitutes an important subject of investigation for the future. Clearly, the sole activation of signaling pathways such as BMPs, Wnt, and FGF is far from sufficient to account for the precision by which these events occur, as illustrated in the case of neural plate border induction. Physical processes such as the constraints applied by forces exerted by moving cells or the viscoelastic properties of tissues are likely to play instructive roles at the scale of the whole embryo. We have only just started to address the question of how these physical processes are translated into biological cascades and vice versa how biological mechanisms can modulate and refine ordered and coherent physical events.

ACKNOWLEDGMENTS

The authors wish to thank Paul Trainor for his kind invitation to write this review and for his patience and to express their friendship for all their colleagues working in the field for their endless imagination and stimulating contributions.

REFERENCES

Acloque, H., Ocana, O. H., Matheu, A., Rizzoti, K., Wise, C., Lovell-Badge, R., et al. (2011). Reciprocal repression between Sox3 and snail transcription factors defines embryonic territories at gastrulation. *Developmental Cell, 21*, 546–558.

Ahlstrom, J. D., & Erickson, C. A. (2009). The neural crest epithelial–mesenchymal transition in 4D: A 'tail' of multiple non-obligatory cellular mechanisms. *Development, 136*, 1801–1812.

Akitaya, T., & Bronner-Fraser, M. (1992). Expression of cell adhesion molecules during initiation of and cessation of neural crest cell migration. *Developmental Dynamics, 194*, 12–20.

Armstrong, J. F., Kaufman, M. H., Harrison, D. J., & Clarke, A. R. (1995). High-frequency developmental abnormalities in p53-deficient mice. *Current Biology, 5*, 931–936.

Aybar, M. J., & Mayor, R. (2002). Early induction of neural crest cells: Lessons learned from frog, fish and chick. *Current Opinion in Genetics and Development, 12*, 452–458.

Bajpai, R., Chen, D. A., Rada-Iglesias, A., Zhang, J., Xiong, Y., Helms, J., et al. (2010). CHD7 cooperates with PBAF to control multipotent neural crest formation. *Nature, 463*, 958–962.

Barembaum, M., Moreno, T., LaBonne, C., Sechrist, J., & Bronner-Fraser, M. (2000). Noelin-1 is a secreted glycoprotein involved in generation of the neural crest. *Nature Cell Biology, 2*, 219–2225.

Basch, M. L., Bronner-Fraser, M., & Garcia-Castro, M. I. (2006). Specification of the neural crest occurs during gastrulation and requires Pax7. *Nature, 441*, 218–222.

Betancur, P., Bronner-Fraser, M., & Sauka-Spengler, T. (2010). Assembling neural crest regulatory circuits into a gene regulatory network. *Annual Review of Cell and Developmental Biology*, *26*, 581–603.

Betancur, P., Simoes-Costa, M., Sauka-Spengler, T., & Bronner, M. E. (2014). Expression and function of transcription factor cMyb during cranial neural crest development. *Mechanisms of Development*, *132*, 38–43.

Bisson, N., Wedlich, D., & Moss, T. (2012). The p21-activated kinase Pak1 regulates induction and migration of the neural crest in Xenopus. *Cell Cycle*, *11*, 1316–1324.

Burk, U., Schubert, J., Wellner, U., Schmalhofer, O., Vincan, E., Spaderna, S., et al. (2008). A reciprocal repression between ZEB1 and members of the miR-200 family promotes EMT and invasion in cancer cells. *EMBO Reports*, *9*, 582–589.

Burstyn-Cohen, T., Stanleigh, J., Sela-Donenfeld, D., & Kalcheim, C. (2004). Canonical Wnt activity regulates trunk neural crest delamination linking BMP/noggin signaling with G1/S transition. *Development*, *131*, 5327–5339.

Chang, C. J., Chao, C. H., Xia, W., Yang, J. Y., Xiong, Y., Li, C. W., et al. (2011). p53 regulates epithelial–mesenchymal transition and stem cell properties through modulating miRNAs. *Nature Cell Biology*, *13*, 317–323.

Cheng, C. W., Wu, P. E., Yu, J. C., Huang, C. S., Yue, C. T., Wu, C. W., et al. (2001). Mechanisms of inactivation of E-cadherin in breast carcinoma: Modification of the two-hit hypothesis of tumor suppressor gene. *Oncogene*, *20*, 3814–3823.

Cheung, M., Chaboissier, M.-C., Mynett, A., Hirst, E., Schedl, A., & Briscoe, J. (2005). The transcriptional control of trunk neural crest induction, survival, and delamination. *Developmental Cell*, *8*, 179–192.

Clay, M. R., & Halloran, M. C. (2010). Control of neural crest cell behavior and migration: Insights from live imaging. *Cell Adhesion & Migration*, *4*, 586–594.

Clay, M. R., & Halloran, M. C. (2014). Cadherin 6 promotes neural crest cell detachment via F-actin regulation and influences active Rho distribution during epithelial-to-mesenchymal transition. *Development*, *141*, 2506–2515.

Coles, E. G., Taneyhill, L. A., & Bronner-Fraser, M. (2007). A critical role for Cadherin6B in regulating avian neural crest emigration. *Developmental Biology*, *312*, 533–544.

Dady, A., Blavet, C., & Duband, J. L. (2012). Timing and kinetics of E- to N-cadherin switch during neurulation in the avian embryo. *Developmental Dynamics*, *241*, 1333–1349.

Dady, A., & Duband, J.-L. (2015). Segregation of the neurectoderm into ectoderm, neural crest and neural tube compartments is defined precociously by differential cadherin expressions. In preparation.

Daujat, S., Neel, H., & Piette, J. (2001). Preferential expression of Mdm2 oncogene during the development of neural crest and its derivatives in mouse early embryogenesis. *Mechanisms of Development*, *103*, 163–165.

David, R., & Wedlich, D. (2000). Xenopus cadherin-6 is expressed in the central and peripheral nervous system and in neurogenic placodes. *Mechanisms of Development*, *97*, 187–190.

De Craene, B., & Berx, G. (2013). Regulatory networks defining EMT during cancer initiation and progression. *Nature Reviews. Cancer*, *13*, 97–110.

Du, C., Zhang, C., Hassan, S., Biswas, M. H., & Balaji, K. C. (2010). Protein kinase D1 suppresses epithelial-to-mesenchymal transition through phosphorylation of snail. *Cancer Research*, *70*, 7810–7819.

Duband, J. L. (2010). Diversity in the molecular and cellular strategies of epithelium-to-mesenchyme transitions: Insights from the neural crest. *Cell Adhesion & Migration*, *4*, 458–482.

Duband, J.-L., Volberg, T., Sabanay, I., Thiery, J. P., & Geiger, B. (1988). Spatial and temporal distribution of the adherens-junction-associated adhesion molecule A-CAM during avian embryogenesis. *Development*, *103*, 325–344.

Espada, J., Peinado, H., Lopez-Serra, L., Setien, F., Lopez-Serra, P., Portela, A., et al. (2011). Regulation of SNAIL1 and E-cadherin function by DNMT1 in a DNA methylation-independent context. *Nucleic Acids Research, 39,* 9194–9205.

Ezin, A. M., Fraser, S. E., & Bronner-Fraser, M. (2009). Fate map and morphogenesis of presumptive neural crest and dorsal neural tube. *Developmental Biology, 330,* 221–236.

Fairchild, C. L., & Gammill, L. S. (2013). Tetraspanin18 is a FoxD3-responsive antagonist of cranial neural crest epithelial-to-mesenchymal transition that maintains cadherin-6B protein. *Journal of Cell Science, 126,* 1464–1476.

Ferronha, T., Rabadan, M. A., Gil-Guinon, E., Le Dreau, G., de Torres, C., & Marti, E. (2013). LMO4 is an essential cofactor in the Snail2-mediated epithelial-to-mesenchymal transition of neuroblastoma and neural crest cells. *Journal of Neuroscience, 33,* 2773–2783.

Fleury, V. (2012). Clarifying tetrapod embryogenesis by a dorso-ventral analysis of the tissue flows during early stages of chicken development. *Bio Systems, 109,* 460–474.

Gessert, S., Bugner, V., Tecza, A., Pinker, M., & Kuhl, M. (2010). FMR1/FXR1 and the miRNA pathway are required for eye and neural crest development. *Developmental Biology, 341,* 222–235.

Groves, A. K., & LaBonne, C. (2014). Setting appropriate boundaries: Fate, patterning and competence at the neural plate border. *Developmental Biology, 389,* 2–12.

Halbleib, J. M., & Nelson, W. J. (2006). Cadherins in development: Cell adhesion, sorting, and tissue morphogenesis. *Genes & Development, 20,* 3199–3214.

Hamburger, V., & Hamilton, H. L. (1951). A series of normal stages in the development of the chick embryo. *Journal of Morphology, 88,* 49–92.

Hatta, K., Takagi, S., Fujisawa, H., & Takeichi, M. (1987). Spatial and temporal expression pattern of N-cadherin cell adhesion molecules correlated with morphogenetic processes of chicken embryos. *Developmental Biology, 120,* 215–227.

Hervouet, E., Vallette, F. M., & Cartron, P. F. (2009). Dnmt3/transcription factor interactions as crucial players in targeted DNA methylation. *Epigenetics, 4,* 487–499.

Hong, E., & Brewster, R. (2006). N-cadherin is required for the polarized cell behaviors that drive neurulation in the zebrafish. *Development, 133,* 3895–3905.

Hu, N., Strobl-Mazzulla, P., Sauka-Spengler, T., & Bronner, M. E. (2012). DNA methyltransferase3A as a molecular switch mediating the neural tube-to-neural crest fate transition. *Genes and Development, 26,* 2380–2385.

Huang, X., & Saint-Jeannet, J. P. (2004). Induction of the neural crest and the opportunities of life on the edge. *Developmental Biology, 275,* 1–11.

Inoue, T., Chisaka, O., Matsunami, H., & Takeichi, M. (1997). Cadherin-6 expression transiently delineates specific rhombomeres, other neural tube subdivisions, and neural crest subpopulations in mouse embryos. *Developmental Biology, 183,* 183–194.

Jacques-Fricke, B. T., Roffers-Agarwal, J., & Gammill, L. S. (2012). DNA methyltransferase 3b is dispensable for mouse neural crest development. *PLoS One, 7,* e47794.

Jhingory, S., Wu, C. Y., & Taneyhill, L. A. (2010). Novel insight into the function and regulation of alphaN-catenin by Snail2 during chick neural crest cell migration. *Developmental Biology, 344,* 896–910.

Jones, N. C., Lynn, M. L., Gaudenz, K., Sakai, D., Aoto, K., Rey, J. P., et al. (2008). Prevention of the neurocristopathy Treacher Collins syndrome through inhibition of p53 function. *Nature Medicine, 14,* 125–133.

Kalcheim, C. (2000). Mechanism of early neural crest development: From cell specification to migration. *International Review of Cytology, 200,* 143–196.

Karafiat, V., Dvorakova, M., Krejci, E., Kralova, J., Pajer, P., Snajdr, P., et al. (2005). Transcription factor c-Myb is involved in the regulation of the epithelial–mesenchymal transition in the avian neural crest. *Cellular and Molecular Life Sciences, 62,* 2516–2525.

Khudyakov, J., & Bronner-Fraser, M. (2009). Comprehensive spatiotemporal analysis of early chick neural crest network genes. *Developmental Dynamics, 238*, 716–723.

Kouzarides, T. (2007). Chromatin modifications and their function. *Cell, 128*, 693–705.

Kuriyama, S., Theveneau, E., Benedetto, A., Parsons, M., Tanaka, M., Charras, G., et al. (2014). In vivo collective cell migration requires an LPAR2-dependent increase in tissue fluidity. *Journal of Cell Biology, 206*, 113–127.

LaBonne, C., & Bronner-Fraser, M. (2000). Snail-related transcriptional repressors are required in Xenopus for both the induction of the neural crest and its subsequent migration. *Developmental Biology, 221*, 195–205.

Lander, R., Nordin, K., & LaBonne, C. (2011). The F-box protein Ppa is a common regulator of core EMT factors Twist, Snail, Slug, and Sip1. *Journal of Cell Biology, 194*, 17–25.

Lander, R., Nasr, T., Ochoa, S. D., Nordin, K., Prasad, M. S., & Labonne, C. (2013). Interactions between Twist and other core epithelial-mesenchymal transition factors are controlled by GSK3-mediated phosphorylation. *Nature Communication, 4*, 1542.

Langer, E. M., Feng, Y., Zhaoyuan, H., Rauscher, F. J., III, Kroll, K. L., & Longmore, G. D. (2008). Ajuba LIM proteins are snail/slug corepressors required for neural crest development in Xenopus. *Developmental Cell, 14*, 424–436.

Le Douarin, N. M., & Kalcheim, C. (1999). *The neural crest* (2nd ed.). New York: Cambridge University Press.

Lee, R. T., Nagai, H., Nakaya, Y., Sheng, G., Trainor, P. A., Weston, J. A., et al. (2013). Cell delamination in the mesencephalic neural fold and its implication for the origin of ectomesenchyme. *Development, 140*, 4890–4902.

Lim, J., & Thiery, J. P. (2012). Epithelial–mesenchymal transitions: Insights from development. *Development, 139*, 3471–3486.

Linker, C., & Stern, C. D. (2004). Neural induction requires BMP inhibition only as a late step, and involves signals other than FGF and Wnt antagonists. *Development, 131*, 5671–5681.

Lombaerts, M., van Wezel, T., Philippo, K., Dierssen, J. W., Zimmerman, R. M., Oosting, J., et al. (2006). E-cadherin transcriptional downregulation by promoter methylation but not mutation is related to epithelial-to-mesenchymal transition in breast cancer cell lines. *British Journal of Cancer, 94*, 661–671.

Lu, M., Jolly, M. K., Levine, H., Onuchic, J. N., & Ben-Jacob, E. (2013). MicroRNA-based regulation of epithelial-hybrid-mesenchymal fate determination. *Proceedings of the National Academy of Sciences of the United States of America, 110*, 18144–18149.

Ma, L., Young, J., Prabhala, H., Pan, E., Mestdagh, P., Muth, D., et al. (2010). miR-9, a MYC/MYCN-activated microRNA, regulates E-cadherin and cancer metastasis. *Nature Cell Biology, 12*, 247–256.

Mancilla, A., & Mayor, R. (1996). Neural crest formation in Xenopus laevis: Mechanisms of Xslug induction. *Developmental Biology, 177*, 580–589.

Matsumata, M., Uchikawa, M., Kamachi, Y., & Kondoh, H. (2005). Multiple N-cadherin enhancers identified by systematic functional screening indicate its Group B1 SOX-dependent regulation in neural and placodal development. *Developmental Biology, 286*, 601–617.

Mayanil, C. S. (2013). Transcriptional and epigenetic regulation of neural crest induction during neurulation. *Developmental Neuroscience, 35*, 361–372.

Milet, C., & Monsoro-Burq, A. H. (2012). Neural crest induction at the neural plate border in vertebrates. *Developmental Biology, 366*, 22–33.

Murko, C., Lagger, S., Steiner, M., Seiser, C., Schoefer, C., & Pusch, O. (2013). Histone deacetylase inhibitor Trichostatin A induces neural tube defects and promotes neural crest specification in the chicken neural tube. *Differentiation, 85*, 55–66.

Nakagawa, S., & Takeichi, M. (1995). Neural crest cell–cell adhesion controlled by sequential and subpopulation-specific expression of novel cadherins. *Development, 121*, 1321–1332.

Nakagawa, S., & Takeichi, M. (1998). Neural crest emigration from the neural tube depends on regulated cadherin expression. *Development, 125*, 2963–2971.

Nakaya, Y., Sukowati, E. W., Wu, Y., & Sheng, G. (2008). RhoA and microtubule dynamics control cell–basement membrane interaction in EMT during gastrulation. *Nature Cell Biology, 10*, 765–775.

Nandadasa, S., Tao, Q., Menon, N. R., Heasman, J., & Wylie, C. (2009). N- and E-cadherins in Xenopus are specifically required in the neural and non-neural ectoderm, respectively, for F-actin assembly and morphogenetic movements. *Development, 136*, 1327–1338.

Ochoa, S. D., Salvador, S., & LaBonne, C. (2012). The LIM adaptor protein LMO4 is an essential regulator of neural crest development. *Developmental Biology, 361*, 313–325.

Park, K. S., & Gumbiner, B. M. (2010). Cadherin 6B induces BMP signaling and de-epithelialization during the epithelial mesenchymal transition of the neural crest. *Development, 137*, 2691–2701.

Park, K. S., & Gumbiner, B. M. (2012). Cadherin-6B stimulates an epithelial mesenchymal transition and the delamination of cells from the neural ectoderm via LIMK/cofilin mediated non-canonical BMP receptor signaling. *Developmental Biology, 366*, 232–243.

Peinado, H., Ballestar, E., Esteller, M., & Cano, A. (2004). Snail mediates E-cadherin repression by the recruitment of the Sin3A/histone deacetylase 1 (HDAC1)/HDAC2 complex. *Molecular and Cellular Biology, 24*, 306–319.

Peinado, H., Olmeda, D., & Cano, A. (2007). Snail, Zeb and bHLH factors in tumour progression: An alliance against the epithelial phenotype? *Nature Reviews. Cancer, 7*, 415–428.

Peiro, S., Escriva, M., Puig, I., Barbera, M. J., Dave, N., Herranz, N., et al. (2006). Snail1 transcriptional repressor binds to its own promoter and controls its expression. *Nucleic Acids Research, 34*, 2077–2084.

Powell, D. R., Blasky, A. J., Britt, S. G., & Artinger, K. B. (2013). Riding the crest of the wave: Parallels between the neural crest and cancer in epithelial-to-mesenchymal transition and migration. *Wiley Interdisciplinary Reviews: Systems Biology and Medicine, 5*, 511–522.

Prasad, M. S., Sauka-Spengler, T., & LaBonne, C. (2012). Induction of the neural crest state: Control of stem cell attributes by gene regulatory, post-transcriptional and epigenetic interactions. *Developmental Biology, 366*, 10–21.

Radice, G. L., Rayburn, H., Matsunami, H., Knudsen, K. A., Takeichi, M., & Hynes, R. O. (1997). Developmental defects in mouse embryos lacking N-cadherin. *Developmental Biology, 181*, 64–78.

Rinon, A., Molchadsky, A., Nathan, E., Yovel, G., Rotter, V., Sarig, R., et al. (2011). p53 coordinates cranial neural crest cell growth and epithelial–mesenchymal transition/delamination processes. *Development, 138*, 1827–1838.

Roffers-Agarwal, J., Hutt, K. J., & Gammill, L. S. (2012). Paladin is an antiphosphatase that regulates neural crest cell formation and migration. *Developmental Biology, 371*, 180–190.

Rogers, C. D., Saxena, A., & Bronner, M. E. (2013). Sip1 mediates an E-cadherin-to-N-cadherin switch during cranial neural crest EMT. *Journal of Cell Biology, 203*, 835–847.

Sakai, D., Suzuki, T., Osumi, N., & Wakamatsu, Y. (2006). Cooperative action of Sox9, Snail2 and PKA signaling in early neural crest development. *Development, 133*, 1323–1333.

Sakai, D., Tanaka, Y., Endo, Y., Osumi, N., Okamoto, H., & Wakamatsu, Y. (2005). Regulation of Slug transcription in embryonic ectoderm by beta-catenin-Lef/Tcf and BMP-Smad signaling. *Development, Growth & Differentiation, 47*, 471–482.

Sauka-Spengler, T., & Bronner-Fraser, M. (2008). A gene regulatory network orchestrates neural crest formation. *Nature Reviews. Molecular Cell Biology, 9*, 557–568.

Sela-Donenfeld, D., & Kalcheim, C. (1999). Regulation of the onset of neural crest migration by coordinated activity of BMP4 and Noggin in the dorsal neural tube. *Development, 126*, 4749–4762.

Shoval, I., Ludwig, A., & Kalcheim, C. (2007). Antagonistic roles of full-length N-cadherin and its soluble BMP cleavage product in neural crest delamination. *Development, 134*, 491–501.

Siemens, H., Jackstadt, R., Hunten, S., Kaller, M., Menssen, A., Gotz, U., et al. (2011). miR-34 and SNAIL form a double-negative feedback loop to regulate epithelial–mesenchymal transitions. *Cell Cycle, 10*, 4256–4271.

Soo, K., O'Rourke, M. P., Khoo, P. L., Steiner, K. A., Wong, N., Behringer, R. R., et al. (2002). Twist function is required for the morphogenesis of the cephalic neural tube and the differentiation of the cranial neural crest cells in the mouse embryo. *Developmental Biology, 247*, 251–270.

Srichai, M. B., Konieczkowski, M., Padiyar, A., Konieczkowski, D. J., Mukherjee, A., Hayden, P. S., et al. (2004). A WT1 co-regulator controls podocyte phenotype by shuttling between adhesion structures and nucleus. *Journal of Biological Chemistry, 279*, 14398–14408.

Strobl-Mazzulla, P. H., & Bronner, M. E. (2012a). Epithelial to mesenchymal transition: New and old insights from the classical neural crest model. *Seminars in Cancer Biology, 22*, 411–416.

Strobl-Mazzulla, P. H., & Bronner, M. E. (2012b). A PHD12-Snail2 repressive complex epigenetically mediates neural crest epithelial-to-mesenchymal transition. *Journal of Cell Biology, 198*, 999–1010.

Strobl-Mazzulla, P. H., Marini, M., & Buzzi, A. (2012). Epigenetic landscape and miRNA involvement during neural crest development. *Developmental Dynamics, 241*, 1849–1856.

Strobl-Mazzulla, P. H., Sauka-Spengler, T., & Bronner-Fraser, M. (2010). Histone demethylase JmjD2A regulates neural crest specification. *Developmental Cell, 19*, 460–468.

Stuhlmiller, T. J., & Garcia-Castro, M. I. (2012a). Current perspectives of the signaling pathways directing neural crest induction. *Cellular and Molecular Life Sciences, 69*, 3715–3737.

Stuhlmiller, T. J., & Garcia-Castro, M. I. (2012b). FGF/MAPK signaling is required in the gastrula epiblast for avian neural crest induction. *Development, 139*, 289–300.

Taneyhill, L. A., Coles, E. G., & Bronner-Fraser, M. (2007). Snail2 directly represses cadherin6B during epithelial-to-mesenchymal transitions of the neural crest. *Development, 134*, 1481–1490.

Taneyhill, L. A., & Schiffmacher, A. T. (2013). Cadherin dynamics during neural crest cell ontogeny. *Progress in Molecular Biology and Translational Science, 116*, 291–315.

Théveneau, E., Duband, J.-L., & Altabef, M. (2007). Ets-1 confers cranial features on neural crest delamination. *PLoS One, 2*, e1142.

Theveneau, E., & Mayor, R. (2012). Neural crest delamination and migration: From epithelium-to-mesenchyme transition to collective cell migration. *Developmental Biology, 366*, 34–54.

Thiery, J. P., Acloque, H., Huang, R. Y., & Nieto, M. A. (2009). Epithelial–mesenchymal transitions in development and disease. *Cell, 139*, 871–890.

Thiery, J. P., & Sleeman, J. P. (2006). Complex networks orchestrate epithelial–mesenchymal tansitions. *Nature Reviews. Molecular Cell Biology, 7*, 131–142.

Trainor, P. (2013). *Neural crest cells: Evolution, development and disease*. Academic Press: Elsevier.

Uchikawa, M., Ishida, Y., Takemoto, T., Kamachi, Y., & Kondoh, H. (2003). Functional analysis of chicken Sox2 enhancers highlights an array of diverse regulatory elements that are conserved in mammals. *Developmental Cell, 4*, 509–519.

Uchikawa, M., Yoshida, M., Iwafuchi-Doi, M., Matsuda, K., Ishida, Y., Takemoto, T., et al. (2011). B1 and B2 Sox gene expression during neural plate development in chicken and mouse embryos: Universal versus species-dependent features. *Development, Growth & Differentiation, 53*, 761–771.

Van de Putte, T., Francis, A., Nelles, L., van Grunsven, L. A., & Huylebroeck, D. (2007). Neural crest-specific removal of Zfhx1b in mouse leads to a wide range of neurocristopathies reminiscent of Mowat–Wilson syndrome. *Human Molecular Genetics, 16*, 1423–1436.

Van de Putte, T., Maruhashi, M., Francis, A., Nelles, L., Kondoh, H., Huylebroeck, D., et al. (2003). Mice lacking Zfhx1b, the gene that codes for Smad-interacting protein-1, reveal a role for multiple neural crest cell defects in the ethiology of Hirschsprung disease-mental retardation syndrome. *The American Journal of Human Genetics, 72*, 465–470.

Vernon, A. E., & LaBonne, C. (2006). Slug stability is dynamically regulated during neural crest development by the F-box protein Ppa. *Development, 133*, 3359–3370.

Vinas-Castells, R., Beltran, M., Valls, G., Gomez, I., Garcia, J. M., Montserrat-Sentis, B., et al. (2010). The hypoxia-controlled FBXL14 ubiquitin ligase targets SNAIL1 for proteasome degradation. *Journal of Biological Chemistry, 285*, 3794–3805.

Voiculescu, O., Bertocchini, F., Wolpert, L., Keller, R. E., & Stern, C. D. (2007) The amniote primitive streak is defined by epithelial cell intercalation before gastrulation. *Nature, 449*, 1049–1052.

Wakamatsu, Y., Endo, Y., Osumi, N., & Weston, J. A. (2004). Multiple roles of Sox2, an HMG-box transcription factor in avian neural crest development. *Developmental Dynamics, 229*, 74–86.

Wang, X. D., Morgan, S. C., & Loeken, M. R. (2011). Pax3 stimulates p53 ubiquitination and degradation independent of transcription. *PLoS One, 6*, e29379.

Wang, S. P., Wang, W. L., Chang, Y. L., Wu, C. T., Chao, Y. C., Kao, S. H., et al. (2009). p53 controls cancer cell invasion by inducing the MDM2-mediated degradation of Slug. *Nature Cell Biology, 11*, 694–704.

Xiong, F., Tentner, A. R., Huang, P., Gelas, A., Mosaliganti, K. R., Souhait, L., et al. (2013). Specified neural progenitors sort to form sharp domains after noisy Shh signaling. *Cell, 153*, 550–561.

Yamashita, S., Miyagi, C., Fukada, T., Kagara, N., Che, Y. S., & Hirano, T. (2004). Zinc transporter LIVI controls epithelial–mesenchymal transition in zebrafish gastrula organizer. *Nature, 429*, 298–302.

Yang, M. H., Hsu, D. S., Wang, H. W., Wang, H. J., Lan, H. Y., Yang, W. H., et al. (2010). Bmi1 is essential in Twist1-induced epithelial–mesenchymal transition. *Nature Cell Biology, 12*, 982–992.

Yang, Z., Rayala, S., Nguyen, D., Vadlamudi, R. K., Chen, S., & Kumar, R. (2005). Pak1 phosphorylation of snail, a master regulator of epithelial-to-mesenchyme transition, modulates snail's subcellular localization and functions. *Cancer Research, 65*, 3179–3184.

Ybot-Gonzalez, P., Savery, D., Gerrelli, D., Signore, M., Mitchell, C. E., Faux, C. H., et al. (2007). Convergent extension, planar-cell-polarity signalling and initiation of mouse neural tube closure. *Development, 134*, 789–799.

Zehir, A., Hua, L. L., Maska, E. L., Morikawa, Y., & Cserjesi, P. (2010). Dicer is required for survival of differentiating neural crest cells. *Developmental Biology, 340*, 459–467.

Zhang, K., Rodriguez-Aznar, E., Yabuta, N., Owen, R. J., Mingot, J. M., Nojima, H., et al. (2012). Lats2 kinase potentiates Snail1 activity by promoting nuclear retention upon phosphorylation. *EMBO Journal, 31*, 29–43.

CHAPTER THREE

Extended Multipotency of Neural Crest Cells and Neural Crest-Derived Cells

Tsutomu Motohashi[*,†,1], **Takahiro Kunisada**[*,†]
[*]Department of Tissue and Organ Development, Regeneration and Advanced Medical Science, Gifu University Graduate School of Medicine, Gifu, Japan
[†]Japan Science and Technology Agency (JST), Core Research for Evolutional Science and Technology (CREST), Tokyo, Japan
[1]Corresponding author: e-mail address: tmotohas@gifu-u.ac.jp

Contents

Abstract

Neural crest cells (NCC) are migratory multipotent cells that give rise to diverse derivatives. They generate various cell types during embryonic development, including neurons and glial cells of the peripheral sensory and autonomic ganglia, Schwann cells, melanocytes, endocrine cells, smooth muscle, and skeletal and connective tissue cells of the craniofacial complex. The multipotency of NCC is thought to be transient at the early stage of NCC generation; once NCC emerge from the neural tube, they change into lineage-restricted precursors. Although many studies have described the clear segregation of NCC lineages right after their delamination from the neural tube, recent reports suggest that multipotent neural crest stem cells (NCSC) are present not only in migrating NCC in the embryo, but also in their target tissues in the fetus and adult.

Current Topics in Developmental Biology, Volume 111
ISSN 0070-2153
http://dx.doi.org/10.1016/bs.ctdb.2014.11.003

Furthermore, fully differentiated NCC-derived cells such as glial cells and melanocytes have been shown to dedifferentiate or transdifferentiate into other NCC derivatives. The multipotency of migratory and postmigratory NCC-derived cells was found to be similar to that of NCSC. Collectively, these findings support the multipotency or plasticity of NCC and NCC-derived cells.

1. INTRODUCTION

Neural crest cells (NCC) are migratory multipotent cells that give rise to diverse derivatives. NCC respond to various environmental factors throughout their development, including bone morphogenetic proteins, epidermal growth factor (EGF), Wnt proteins, stem cell factor, and endothelin 3 (EDN3) (Lee et al., 2004; Shah & Anderson, 1997), and differentiate into many cell types, including neurons and glial cells of the peripheral sensory and autonomic ganglia, Schwann cells, melanocytes, endocrine cells, smooth muscle, and skeletal and connective tissue cells of the craniofacial complex (Le Douarin & Kalcheim, 1999). Because of these multipotent characteristics, NCC play critical roles in mammalian embryogenesis.

NCC emerge from the dorsal edge of the neural folds, delaminate from the surrounding tissues of the fusing neural tube, and then migrate throughout the embryo. In the trunk, NCC exit dorsally from a closed dorsal neural tube (Milet & Monsoro-Burq, 2012) and migrate along two major routes: the ventromedial pathway between the neural tube and somites, along which NCC give rise to the autonomic ganglia (sympathetic ganglia and parasympathetic ganglia), dorsal root ganglia (DRG), glial cells, and chromaffin cells; and a dorsolateral pathway between the overlying ectoderm and somites, along which NCC give rise to melanocytes (Le Douarin & Kalcheim, 1999; Le Douarin & Teillet, 1974).

NCC are thought to lose their multipotency before or soon after their delamination from the neural tube and become lineage-restricted depending on the direction of their migration. Ventrally migrating NCC are restricted to neural and glial cell fates, whereas dorsolaterally migrating NCC are restricted to a melanocyte fate (Le Douarin & Kalcheim, 1999). Consistent with this idea, NCC isolated from the ventral pathway do not differentiate into melanocytes in cell culture (Reedy, Faraco, & Erickson, 1998). Furthermore, a naturally occurring chick mutant whose pigment cells migrated ventrally instead of dorsolaterally did not develop ectopic neurogenesis (Faraco, Vaz, Pastor, & Erickson, 2001).

A number of studies have attempted to determine when and where NCC lose their multipotency and become restricted in their fate. A clonal analysis of NCC revealed that almost half of the initial NCC population exists as a lineage-restricted population, generating clones of only a single cell type (Henion & Weston, 1997), while the other half comprises bipotent precursors (Lahav et al., 1998). A subpopulation of early NCC that express the receptor tyrosine kinase Kit exclusively express melanocyte-lineage markers in rodents. Moreover, avian Kit-expressing NCC invariably gave rise to clones containing only melanocytes (Henion & Weston, 1997; Luo, Gao, Wehrle-Haller, & Henion, 2003; Wilson, Richards, Ford-Perriss, Panthier, & Murphy, 2004). This suggests that the expression of Kit is an early determinant of lineage fate in dorsolaterally migrating NCC. Consistent with this idea, none of the ventrally migrating NCC were found to express Kit (Wilson et al., 2004) or differentiate into melanocytes even under conditions that favored the development of melanocytes (Reedy et al., 1998). These studies suggested that migratory NCC might comprise a population of fate-restricted precursors that give rise to single or dual cell types.

However, multipotent neural crest stem cells (NSCS) are maintained throughout adult tissue homeostasis. NCSC have been identified in the adult sciatic nerve, gut, skin, and carotid bodies to which they contribute, and there NCC differentiate into functional tissue components such as neurons, glial, or melanocytes (Fernandes et al., 2004; Kruger et al., 2002; Morrison, White, Zock, & Anderson, 1999; Pardal, Ortega-Saenz, Duran, & Lopez-Barneo, 2007; Stemple & Anderson, 1992). Furthermore, cells derived from NCC exhibit multipotential cell fate. For example, pigmented cells isolated from the skin of quail embryos, which morphologically reflect fully differentiated dendritic melanocytes, characteristic other derivatives of NCC such as glia and myofibroblastic cells, when exposed to EDN3 (Dupin, Glavieux, Vaigot, & Le Douarin, 2000; Real, Glavieux-Pardanaud, Le Douarin, & Dupin, 2006; Trentin, Glavieux-Pardanaud, Le Douarin, & Dupin, 2004). Moreover, we also demonstrated that purified melanoblasts derived from NCC maintained their multipotency, under certain culture conditions (Motohashi, Yamanaka, Chiba, Aoki, & Kunisada, 2009). Schwann cells isolated from quail embryonic nerves can generate myofibroblastic cells (Real, Glavieux-Pardanaud, Vaigot, Le-Douarin, & Dupin, 2005), and NCC-derived cells in the thymus have the potential to differentiate into melanocytes (Yamazaki et al., 2005). Furthermore, we showed that Sox10-positive, but Kit-negative NCC, which demarcate ventromedially migrating NCC, were able to differentiate into melanocytes as well as neurons and glial cells (Motohashi et al., 2011). In addition, NCC-derived cells in embryonic skin, DRG, and the inner ear maintained their multipotency during gestation (Motohashi, Kitagawa, Watanabe, Wakaoka, & Kunisada, 2014). These findings indicate that NCC-derived cells are perhaps more flexible in their cell fate or lineage fate capabilities than previously considered.

2. POSTMIGRATORY NCSC: MULTIPOTENT CELLS ARE MAINTAINED IN THE TISSUES

NCSC research was pioneered based on the study of Stemple and Anderson who established the identification and isolation of NCSC from the trunk neural tube of rat embryos (Stemple & Anderson, 1992). NCSC are molecularly recognized by the expression of the low-affinity nerve growth factor receptor (P75) and can be enriched from cultured neural tubes by a sorting with a monoclonal antibody against P75. Stemple and Anderson then established culture conditions for the maintenance of proliferating NCSC without a loss in their differentiation potential. Furthermore, they showed that single P75-positive (P75+) NCC were capable of self-renewal and multipotency in differentiating into neurons, glia, and smooth muscle cells, thereby confirming P75+ NCC as stem cells. These findings profoundly influenced subsequent NCSC research establishing the quest to identify tissues in which NCC colonized and remained as a resident pool of multipotential NCSC during embryogenesis and even into adulthood (Table 1). Consequently, NCSC have been detected in adult rat peripheral nerves, sciatic nerves, and the DRG (Hagedorn et al., 1999; Morrison et al., 1999). NCSC in sciatic nerves were identified as cells expressing P75, but not peripheral myelin protein, Protein-0 (P0), and their multipotency was confirmed through serial passaging, indicating that they also exhibit potent self-renewal (Morrison et al., 1999). NCSC were subsequently identified in the fetal and adult gut as a cell population expressing P75 and integlin $\alpha 4$, and these cells also exhibited multipotency and self-renewal activities (Bixby et al., 2002; Kruger et al., 2002). Contrary to sciatic nerve NCSC, NCSC in DRG were characterized as two peripheral myelin protein marker-positive cells. P0 and PMP22 double-positive cells gave rise to glia, neurons, and smooth muscle-like cells in response to instructive extracellular cues (Hagedorn et al., 1999). These findings clearly established the

Table 1 Postmigratory NCSC

References	Tissue	Marker	Embryo (E), adult (A)	Self-renewal	Potency
Morrison et al. (1999)	Sciatic nerve	P75, P0-negative	Rat E	+	Neurons, glial cells, myofibroblasts
Hagedorn, Suter, and Sommer (1999)	DRG	P0, PMP22	Rat E	?	Neurons, glial cells, smooth muscle cells
Bixby, Kruger, Mosher, Joseph, and Morrison (2002), Kruger et al. (2002)	Gut	P75, integlin α4	Rat E, A	+	Neurons, glial cells, myofibroblasts
Pardal et al. (2007)	Carotid body	GFAP, P75-negative	Mouse A	+	Neurons, smooth muscle cells
Nagoshi et al. (2008)	BM, DRG, whisker pad	Sox10, P75, nestin, musashi	Mouse juvenile, A	+	Neurons, glial cells, myofibroblasts
Tomita et al. (2005)	Heart	Nestin, musashi, CD29, CD44	Mouse neonate, A	?	Neurons, glial cells, smooth muscle cells, cardiomyocytes
Widera et al. (2009)	Palatum	Nestin, CD133, Sox2, P75	Rat A	+	Neurons, glial cells
Widera et al. (2007)	Periodontal ligament	Nestin, Sox2	Human A	?	Neurons
Techawattanawisal et al. (2007)	Periodontal ligament	Nestin, GFAP, vimentin, Sox2	Rat A	?	Neurons, glial cells, myotubes
Sasaki et al. (2008)	Dental pulp	CD81, nestin	Rat A	+	Neurons
Yoshida et al. (2006)	Cornea	Nestin, Notch1, musashi-1, ABCG2	Mouse A	+	Neurons, glial cells, adipocytes, chondrocytes

existence of multipotent NCSC in adult tissues, contradicting the prior assumption that NCC lose their multipotency after their migration to their target tissues.

NCSC-like cells have also been detected in the adult carotid body (CB), an oxygen-sensing organ of the sympathoadrenal lineage that grows under chronic hypoxemia (Pardal et al., 2007). The CB is derived from the cells of the sympathoadrenal lineage that stems from NCC. The stem-like cells in the CB were detected in the P75-negative population by flow cytometry in contrast to the abovedescribed NCSC. However, the lineage tracking experiments showed that stem cells existed in the glial lineage population of the CB. The CB stem cells exhibit unique characteristics in association with their hypoxic environment. Under these conditions, GFAP+ cells are converted into nestin+ intermediate progenitors and proliferate, which gives rise to TH+ glomus cells and smooth muscle actin+ cells. However, this proliferation and differentiation ceases with the return to normoxia. In the CB, stem-like cells continue to be induced by reductions in arterial O_2 concentrations throughout life.

Recent advances in gene technologies have simplified the isolation and identification of NCSC from multiple distinct tissues. NCSC were detected in unexpected tissues in P0-promoter-Cre/Floxed-EGFP and Wnt1-promoter-Cre/Floxed-EGFP mice, which lineage label NCC and their derivatives (Nagoshi et al., 2008; Takashima et al., 2007). Although P0 is originally a Schwann cell-specific myelin protein, it was previously shown to be transiently activated in migrating NCC in early chick embryos (Lemke, Lamar, & Patterson, 1988). Wnt1 is known to be expressed specifically in the dorsal neuroepithlium which encompasses the premigratory population of NCC (Jiang, Rowitch, Soriano, McMahon, & Sucov, 2000). These mice have facilitated the identification of NCC-derived cells in embryonic and adult tissues and uncovered some surprises such as in the case of bone marrow (BM), which was previously not considered to be colonized by NCC (Nagoshi et al., 2008; Takashima et al., 2007). EGFP-positive NCC-derived cells isolated from the BM form sphere-like structures characteristic of stem cells, and express NCSC-related genes, such as *p75* and *sox10*. Furthermore, these cells can differentiate into neurons, glial cells, and myofibroblasts similar to NCSC isolated from the DRG and the whisker pad (Nagoshi et al., 2008). Purified NCSC were also found to differentiate into mesenchymal cell lineages including osteocytes, chondrocytes, and adipocytes. These findings suggest that so-called mesenchymal stem cells in BM may, at least partly, be NCSC.

Multipotent NCC have also been identified in the neonatal and adult heart (Tomita et al., 2005). NCC are known to contribute to the construction of the aorticopulmonary septum of the outflow tract in the heart (Hall, 2009). Multipotent cardiac multipotent NCC have also been detected in the side population (SP) of the heart following FACS. SP cells were isolated by their ability to actively efflux the fluorescent DNA-binding dye Hoechst 33342 in flow cytometric analyses (Goodell et al., 1997; Matsuzaki, Kinjo, Mulligan, & Okano, 2004). SP cells are generally considered to be tissue-specific progenitor cells and SP cells isolated from the heart generated spheres and differentiated into cells with NCC characteristics, including peripheral nervous system-type neurons, glial cells, and smooth muscle cells in addition to cardiomyocytes (Tomita et al., 2005). They also migrated to typical NCC-targeted tissues, including DRG, spinal nerves, and aortic smooth muscle after transplantation back into chick embryos.

Corneal keratocytes, which form the stroma of the cornea, can heal the cornea throughout life. Their developmental origin was found to reside in cranial NCC (Johnston, Noden, Hazelton, Coulombre, & Coulombre, 1979; Trainor & Tam, 1995); furthermore, corneal keratocytes maintain plasticity and multipotency during late stages of avian embryonic development and even into adults (Lwigale, Cressy, & Bronner-Fraser, 2005). Multipotent keratocyte precursor cells isolated from adult mouse corneas form sphere structures and differentiate into keratocytes, fibroblasts, myofibroblasts, adipocytes, and chondrocytes (Yoshida et al., 2006).

NCSC-like cells have also been identified in the adult rat palatum and periodontal ligament through a sphere-forming culture system that was originally utilized to isolate neuronal stem cells (Techawattanawisal et al., 2007; Widera et al., 2007, 2009). NCSC-like cells formed sphere structures even after serial passages and expressed the NCC-related genes, *p75*, *sox2*, and *nestin*. This sphere-forming activity depended on the presence of fibroblast growth factor (FGF) and EGF. Moreover, the sphere-forming cells differentiated into multinucleated myotubes, Tuj1-positive neurons, and GFAP-positive glial cells in adherent culture, similar to NCSC isolated from the other tissues. These results demonstrate that the sphere-forming cells are multipotent NCSC.

Dental pulp tissue originates from NCC-derived mesenchyme and has also been shown to contain sphere-forming NCSC-like cells that depend on basic FGF (bFGF) and EGF similar to those from the palatum and periodontal ligament (Sasaki et al., 2008). Sphere-forming NCSC-like cells are thought to be the descendants of cranial NCC retained in the oral appendages; however, recent studies suggested that some of the NCSC-like cells were derived from myelinated Schwann cells within the palatal ridges (Widera et al., 2011). Thus, further experiments are needed to fully understand the nature of these oral NCSC.

3. CULTURED NCSC-LIKE CELLS ORIGINATING FROM THE HAIR FOLLICLE AND DERMIS

The induction of multipotent NCSC from the skin was a valuable finding because the skin is easily accessible and the separation and preparation of the reported skin-derived multipotent NCSC-like cells only requires a simple culture system.

The first of these kinds of multipotent NCSC-like cells was established from the mouse dermis (Toma et al., 2001). Cells termed skin-derived precursors (SKPs) were established in a suspension culture supplemented with EGF and bFGF that has frequently been used for neural stem cell cultures. Small floating spheres formed after 7 days and these spheres could be cultured in long-term suspension culture without losing the capacity to generate multiple diverse cell types. Plating these spheres onto a polylysine-coated surface triggered their differentiation into neurons and glia in the absence of exogenous growth factors. They could also form smooth muscle cells and adipocytes following the addition of 3% and 10% serum, respectively, similar to multipotent NCSC. SKPs express the intermediate filament protein Nestin and also the extracellular matrix glycoprotein fibronectin; however, NCSC markers such as P75 and the polysialylated form of neural cell adhesion molecule (PSA-NCAM) have not been detected on the surface of SKPs. Similar cells could be isolated from neonatal human foreskin tissue. These SKPs can be maintained in culture for extended periods without losing their multipotency to differentiate into neurons, glia, and smooth muscle cells (Toma, McKenzie, Bagli, & Miller, 2005). SKPs appear to be distinct from NCSC isolated from the neural tube or sciatic nerve plexus because these previously reported NCSCs did not generate spheres under the same conditions and they express P75 which contrasts with the properties of SKPs (Toma et al., 2001). However, NCC lineage tracing in Wnt1-promoter-Cre/R26R mice indicated that almost all SKP isolated from the dermal papilla of whisker hair follicles in facial skin were found to express Wnt1-Cre-activated β-galactosidase, as well as NCC-specific transcription factors such as *slug*, *snail*, and *twist*. Furthermore, the transplantation of SKPs into chick embryos revealed that the SKPs could migrate along endogenous NCC migration routes and express neuron or glial markers in their target tissues. These findings indicated that SKPs from facial skin were in fact derived from NCC (Fernandes et al., 2004). In contrast to SKPs from facial skin, the origin of dorsal (trunk) skin-derived SKPs remains obscure. A previous study demonstrated that trunk SKP originated from dermal mesenchymal stem cells based on the fact that Sox2^{+} EGFP-positive cells isolated from trunk skins of mice generated SKPs (Biernaskie et al., 2009). Sox2-driven EGFP was expressed in the trunk dermal papilla cells, dermal sheath, and dermis, which are all considered to be derived from mesenchymal dermal stem cells (Driskell, Giangreco, Jensen, Mulder, & Watt, 2009).

Other studies have reported the existence of multipotent cells in the skins of mice, pigs, and humans. Cells called epidermal NCSC (EPI-NCSC) were isolated from hair follicles of adult mice (Sieber-Blum, Grim, Hu, & Szeder, 2004). Explants of the bulge region of adult mouse whisker follicles yielded migratory EPI-NCSC that differentiated into neurons, smooth muscle cells, Schwann cells, and melanocytes, as well as demonstrating self-renewal capability. Wnt1 promoter-Cre/R26R lineage tracing revealed that EPI-NCSC were derived from NCC (Sieber-Blum et al., 2004). Furthermore, transplanted EPI-NCSC survived and expressed the GABAergic neuron marker GAD67, oligodendrocyte marker RIP, or myelin basic protein (MBP) in lesioned spinal cords of mice (Sieber-Blum et al., 2006). The bulge cells of hair follicles expressing Nestin and the stem cell marker CD34 differentiated into neuron, glial cells, keratinocytes, smooth muscle cells, and melanocytes *in vitro* (Amoh, Li, Katsuoka, Penman, & Hoffman, 2005). Nestin-positive stem cells isolated from hair follicles were found to differentiate into functional Schwann cells after being transplanted into severed tibial nerves (Amoh, Li, Campillo, et al., 2005). The Nestin-positive stem cells isolated from hair follicles were shown to be distinct from EPI-NCSC because EPI-NCSC did not express CD34 (Hu, Zhang, & Sieber-Blum, 2006). The same type of multipotent stem cells were isolated and characterized from porcine skin (Dyce, Zhu, Craig, & Li, 2004). Dissociated porcine skin cells formed spheres in the presence of EGF and bFGF, and these spheres generated cells with the characteristics of neurons, astrocytes, and adipocyte-like cells *in vitro*.

Another distinct type of multipotent skin-derived stem cell was established from human tissues (Belicchi et al., 2004, Joannides et al., 2004, Shih et al., 2005). These cells form sphere structures in the presence of EGF and bFGF and exhibit similar mesenchymal and neural differentiation potentials. In particular, AC133-positive cell population, isolated from human skin, generated immature neurons, and well-differentiated astrocytes after transplantation to the brain (Belicchi et al., 2004). These kinds of stem cells isolated from various skin tissues with different methods are likely to be originated from the same cell lineage because of their similarity for the growth factor dependency and differentiation potential.

Interestingly, sphere-forming cells from adult mice and human skin differ from SKP (Wong et al., 2006). Culturing skin cells under slightly different conditions to SKP (except the addition EGF and FGF) revealed that P75/Sox10 expressing spheres were formed after 7 days. Both P75 and Sox10 are immature NCC markers and are not expressed in SKP. The P75/Sox10-expressing spheres were found to be differentiated into NCC lineages just like SKP. Furthermore, that these sphere-forming cells were distributed in all whisker follicle components such as the dermal papilla, capsule, upper dermal sheath, and lower part of the dermal sheath, as evidenced through lineage tracing with Wnt1 promoter-Cre/R26R reporter mice and human tissue plasminogen activator (Ht-PA)-Cre/R26R mice, in which Cre recombinase was expressed in NCC lineages. In addition, trunk skin from desert hedgehog promoter-Cre/R26R-EYFP double-transgenic mice and dopachrome tautomerase (Dct) promoter-Cre/R26R-EYFP mice revealed that peripheral glial lineage cells and melanocyte-lineage cells, respectively, contained sphere-forming cells (Wong et al., 2006). These finding suggested that NCSC in skin may not be confined to a particular location but can be obtained from multiple cell types and tissues derived from NCC.

4. RETENTION OF MULTIPOTENCY IN DIFFERENTIATED NCC-DERIVED CELLS

NCC are thought to lose their multipotency soon after leaving the neural tube and restrict their differentiated cell fates in target tissues to ensure proper tissue development and homeostasis (Le Douarin & Kalcheim, 1999). In addition, NCC that migrate aberrantly are often subjected to apoptosis. For example, when neuronal-lineage NCC that endogenously migrate via the ventromedial pathway are rerouted to the dorsolateral pathway, they undergo apoptosis. Only melanocyte-fated NCC appear to be permitted to migrate along the dorsolateral pathway (Wakamatsu, Mochii, Vogel, & Weston, 1998). However, recent findings that are discussed below have revealed the possibility of multipotency and plasticity in cell fate determination in NCC even after tissue contribution.

4.1. Plasticity of NCC-derived glial cells

NCC-derived glial cells are known for their cell lineage plasticity. Quail embryonic peripheral nerves tissues, which do not normally contain melanocytes, generate pigmented melanocytes after being cultured in the presence of the phorbol ester 12-O-tetradecanoyl phorbol-13-acetate (TPA) and bFGF (Ciment, Glimelius, Nelson, & Weston, 1986; Stocker, Sherman, Rees, & Ciment, 1991). Furthermore, immunostaining clearly showed that the NCC-derived cells were P0-expressing Schwann cell precursors (SCP) (Sherman, Stocker, Morrison, & Ciment, 1993). Since melanocytes and Schwann cells are known to be derived from a common bipotent intermediate cell type in the NCC lineage (Nichols & Weston, 1977), Schwann cells or their precursors may not completely lose their potential to differentiate into melanocytes even after migration to their target tissue. Schwann cell myelin protein (SMP)-positive glial cells isolated from the sciatic nerve of a quail embryo were also reported to differentiate into pigmented melanocytes in cell culture containing highly concentrated EDN3 (Dupin, Real, Glavieux-Pardanaud, Vaigot, & Le Douarin, 2003). Moreover, quail Schwann cells expressing SMP were reported to differentiate into α smooth muscle actin (αSMA)-expressing myofibroblasts as well as melanocytes in cultures containing transforming growth factor β1 (TGF β1) (Real et al., 2005). Furthermore, Schwann cells implanted into chick embryos were found to contribute to perivascular smooth muscle cells. A clonal analysis of the Schwann cells in these cultures revealed that the differentiation of myofibroblasts from Schwann cells may involve reversal to a common bipotent progenitor cell for glial and myofibroblastic cell types and reversal to pluripotent progenitor cells for glial, myofibroblastic cell, and melanocyte types (Real et al., 2005). These findings in avian embryos suggest that lineage-restricted NCC-derived cells exhibit plasticity with respect to differentiation. Their fate changed according to the presence of secreted factors or the local tissue environment and these environmental cues may lead to dedifferentiation of NCC-derived cells to more immature bipotent or multipotent progenitor cell states.

Schwann cells appear to possess considerable plasticity in mammals. Severing the adult mouse sciatic nerve plexus was found to generate pigmented cells around the nerve fascicles (Rizvi et al., 2002). Furthermore, the transplantation of dye-labeled Schwann cells into wounded nerves demonstrated that pigment cells arose from the grafted Schwann cells. Further analyses revealed that this pigmentation was enhanced in the wounded nerves of GM-CSF-, IL5-, and IL3-common β chain receptor (βc) null mice or neurofibromin (a downstream molecule of βc) heterozygous mutant mice, suggesting that environment factors including cytokines suppress this pigmentation (Rizvi et al., 2002). Mammalian Schwann cells in damaged nerves dedifferentiated into immature Schwann cells similar to avian Schwann cells (Jessen & Mirsky, 2008). A previous study reported that Notch signaling play a crucial role during the regeneration process, and this was accompanied by the demyelination of mature Schwann cells, which is crucial for dedifferentiation (Woodhoo et al., 2009). In addition, c-Jun was shown to be upregulated after nerve injury and promoted the dedifferentiation of Schwann cells (Arthur-Farraj et al., 2012; Parkinson et al., 2008).

NCSC in palatal tissues were reported to originate from dedifferentiated myelinated Schwann cells (Widera et al., 2011). Immunocytochemistry, reverse transcription–polymerase chain reaction, and transmission electron microscopy revealed that myelinated Schwann cells within palatal tissues could dedifferentiate into multipotent Nestin+, P75+, and S100+ NCSC in the presence of bFGF and EGF.

In summary, glial cells thus far thought to be fully restricted for their differentiation are actually sustained the plasticity even in the nerve tissue; however, their plasticity is likely to be suppressed by the environmental factors. However, when the suppressive environment was disturbed, for example, by the wound healing process, the plasticity of glial cells could be exerted and differentiate into other NCC derivatives *in situ*.

4.2. Plasticity of glial cells plays a significant role in embryogenesis

Glial cell plasticity is expected to be responsible for the regeneration of injured tissues; however, this has not yet been examined in detail. What kind of biological roles would be expected for the plasticity of glial cells? In quails, Schwann cells can differentiated into αSMA-expressing myofibroblasts even under growth factor-free conditions, suggesting that a physiological merit of this transdifferentiation is that glial cells repair damaged blood vessels by their direct differentiation into myofibroblasts in wound regions (Real et al., 2005). However, SCP are developmentally programmed to differentiate into functional melanocytes (Adameyko et al., 2009). In avian embryos, melanocyte-lineage-specific Mitf-expressing cells emerge from Sox10+ SCP associated with neural cells that elongated from DRG to cutaneous tissues. Furthermore, using PLP-CreERT2 mice which genetically trace the progeny of PLP+SCP revealed that SCP were unequivocally ancestral to melanocytes during mouse development. Moreover, detachment of SCP from the nerve is critical for melanocyte differentiation and SCP acquire a melanocyte fate in the absence of signals provided by the nerve. Thus, a higher number of Mitf-expressing cells is observed at the end of dorsal spinal nerves in receptor tyrosine kinase erbB3 mutant mouse, in which SCP cannot respond to neuregulin-1 provided by nerve. Therefore, glial cells retain the ability to dedifferentiate and reenter the cell cycle and give rise to adult melanocytes (Kaucka & Adameyko, 2014). However, from a physiological standpoint, it remains unknown why glial cells dedifferentiate and differentiate into melanocytes.

4.3. Plasticity of lineage-restricted melanocytes or their precursors

All the melanocytes in the body are thought to be derivatives of the NCC that migrated along the dorsolateral pathway during embryogenesis. Melanocytes synthesize melanin to form melanosomes for serving the functions of hair coloration and protection against UV radiation (Hearing, 2005). It has been reported that terminally differentiated pigmented melanocytes have the ability to differentiate into other NCC lineages in response to exogenous factors. For example, quail epidermal melanocytes were purified by flow cytometry for their dense pigmentation from E7.5 dorsal skin and then cultured in the presence of a high concentration of EDN3, which is known to promote the differentiation and proliferation of melanocytes. Under this condition, colorless cell colonies were occasionally detected among pigmented melanocyte colonies (Dupin et al., 2000). The cells forming the nonpigmented colony were SMP- and P0-expressing glial cells, indicating that pigmented melanocytes retained the ability to differentiate into glial cells even after colonizing the skin. Since EDN3 is considered to promote the proliferation and differentiation of melanocytes, these results suggested that a high concentration of EDN3 markedly enhanced the proliferation of melanocytes, thereby increasing the probability of dedifferentiation or transdifferentiation of melanocytes to glial cells. Furthermore, SMP-positive glial cells isolated from the sciatic nerve of a quail embryo differentiated into pigmented melanocytes when cultured in the presence of high concentrations of EDN3 (Dupin et al., 2003). Clonal cell culture analyses revealed that some SMP-positive cells express the melanocyte marker MelEM during this differentiation process. Thus, mutual transdifferentiation in the form of fully differentiated SMP-positive glial cells being directly converted to melanocytes was observed. A previous study also reported that quail melanocytes dedifferentiated into SMP-positive glial cells and myofibroblastic cells in the absence of EDN3 (Real et al., 2006). In this clonal analysis, melanin granules disappeared from individual melanocytes and, along with this depigmentation process, markers of undifferentiated NCC such as FoxD3, Pax3, Slug, and HNK-1 were expressed in the depigmented cells. These findings indicated that the melanocytes were transformed into immature NCC.

Thus, fully differentiated and functional melanocytes in the skin maintain the ability to dedifferentiate into immature NCC, and then redifferentiate into other NCC derivatives such as glial cells. Since matured glial cells also generate melanocytes, glial to melanocyte differentiation and vice versa is held in an equilibrium state. It should be noted that dedifferentiated immature NCC are the only bipotential NCC capable of differentiating into either glial cells or melanocytes and have not yet been shown to differentiate into neurons.

5. MAINTAINED MULTIPOTENCY IN POSTMIGRATING NCC AND NCC DERIVATIVES

Above we described the existence of multipotential NCSC or cultured NCSC-like cells in adult tissues as well as the dedifferentiation or transdifferentiation of NCC derivatives to other NCC lineages via a transient NCC precursor state. Here, we discuss an unexpected feature of fate-restricted NCC precursors, the ability to retain a multipotential differentiation capacity.

5.1. Multipotency of NC-derived lineage-restricted melanoblasts

All melanocytes are thought to be generated from cells with a restricted ability to differentiate only into melanocytes and, these cells have been designated melanoblasts. In the mouse, the melanocyte-restricted NCC that will take the dorsolateral pathway do not migrate from the neural tube until 6–12 h later than the other NCC from the same axial level (Henion & Weston, 1997; Reedy et al., 1998). Expression of the receptor tyrosine kinase Kit is upregulated in these melanocyte-restricted NCC prior to their dorsolateral migration (Manova & Bachvarova, 1991; Wilson et al., 2004). Thus, a group of cells residing in the most dorsomedial region of the neural tube was found to express Kit, and these cells exclusively migrated toward the developing dermis to become melanoblasts that expressed definitive markers for the melanocyte lineage (Wilson et al., 2004). In quail embryos, Kit only begins to be expressed in melanoblasts after they have migrated from the neural tube (Luo et al., 2003). In any case, melanoblasts expressing Kit migrate from the neural tube and are considered to be fate-restricted precursors that differentiate only into melanocytes. However, in the course of *in vitro* differentiation of melanocytes from purified melanoblasts, we noticed that Kit+/CD45- cells purified from mouse embryonic skin as melanoblasts maintain their multipotency, similar to NCC, even after colonizing the entire skin including the hair follicles (Motohashi et al., 2009). Kit+/CD45- cells isolated from E12.5 skin were placed into a ST2 coculture system to promote melanocyte differentiation. Unexpectedly, mixtures of neurons, glial cells, and smooth muscle cells together with melanocytes were observed in the colonies that formed *in vitro*. This was true even when the colony was derived from a single Kit+/CD45- cell. Kit+/CD45- cells were found to express *mitf-m, sox10, trp-1,* and *trp-2,* all of which are melanoblast markers (Motohashi et al., 2009). In this study, the direct differentiation of Kit+/CD45− cells into neurons was observed after several round of cell division. Subsequently, other cell types such as glia and melanocytes appeared, which strongly implies that melanoblasts themselves have multipotency in spite of being considered as fate-restricted precursors for melanocytes. It is important to emphasize that with respect to the previously presented idea that high concentrations of EDN3 may promote mouse Kit+/CD45- melanoblasts to dedifferentiate into NCSC-like cells, which can differentiate into various NCC derivatives, might not be consistent with our findings (Dupin et al., 2000; Real et al., 2006; Trentin et al., 2004). In fact, the multipotency of Kit+/CD45- cells was maintained in a culture even in the presence of BQ788, an antagonist against endothelin receptor B. Furthermore, Kit+/CD45- melanoblasts show limited, but significant self-renewal capacity *in vitro*. These findings suggest that NCC-derived lineage-restricted cells such as melanoblasts do not lose their multipotency. Kit+/CD45- melanoblasts exhibited their multipotency only in an *in vitro* coculture system with ST2 stromal cells and not *in vivo*. It also should be noted that melanoblasts only differentiated into melanocytes and not into other cells in the coculture system with skin keratinocytes, which may better recapitulate their *in vivo* environments (T. Motohashi, unpublished data). The skin microenvironment may therefore continuously suppress the melanoblast multipotency to forcefully guide the proper development of melanoblasts.

5.2. Duration of the multipotency of NCC-derived cells after delamination from the neural tube

It remains unknown how long the multipotency of NCC-derived melanoblasts or other cell lineages can be maintained. Conventional thinking suggests that NCC lose their multipotency before or soon after emerging from the neural tube, and their fate is gradually restricted according to their migration pathways. For example, ventromedially migrating NCC become restricted to neural and glial cell fates, while dorsolaterally migrating cells become restricted to melanocytes.

To evaluate the differentiation potential of these very early-stage NCC, we took advantage of the mouse embryonic stem (ES) cell culture system to replicate the development of NCC (Motohashi, Aoki, Chiba, Yoshimura, & Kunisada, 2007). For this purpose, we used mouse ES cells designed to express green fluorescent protein under the control of the NCC marker *Sox10* (Sox10-IRES-GFP ES cell) in order to efficiently isolate ES cell-derived NCC-like cells (Motohashi et al., 2011). Purified Sox10 and Kit double-positive (Sox10+/Kit+) NCC-like cells equivalent to dorsolaterally migrating NCC were able to differentiate into neurons, glial cells, and melanocytes. Unexpectedly, Sox10+/Kit- NCC-like cells, which were equivalent to the ventromedially migrating NCC, were also able to differentiate into melanocytes as well as neurons and glial cells. Thus, multipotential NCC were obtained from ES cells.

To further extend our findings, cloned mice were established from Sox10-IRES-GFP ES cells. Sox10+/Kit+ NCC and Sox10+/Kit- NCC were clonally isolated from E9.5 Sox10-IRES-GFP mice and exhibited the potential to differentiate into neurons, glial cells, and melanocytes. These findings indicated that NCC maintain their multipotency even after emerging from the neural tube and also that NCC maintained their multipotency, regardless of their migratory pathway (Fig. 1; Motohashi et al., 2011).

Another issue that needed to be addressed was when NCC lose their multipotency. The multipotency of NCC-derived cells after their arrival at each target tissue, including embryonic skin, DRG, the inner ear, and intestines, was evaluated using Sox10-IRES-GFP embryos (Motohashi et al., 2014). The fraction of multipotential cells in harvested NCC varied according to the target tissues and their development stage. We separated NCC into Sox10+/Kit+ and Sox10+/Kit- cells to evaluate their differentiation potency. In embryonic skin and the inner ear, Sox10+/Kit+ cells maintained

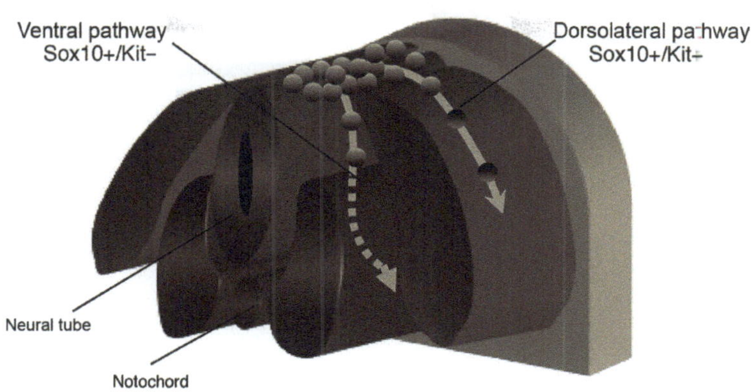

	Sox10+/Kit+	Sox10+/Kit−
E9.5	N, G	Multipotency
E10.5	Multipotency	Multipotency
E12.5	Multipotency	Multipotency

Figure 1 NCC maintained their multipotency after emerging from the neural tube. Schematic representation of the migrating pathway of NCC. Sox10+/Kit+ and Sox10+/Kit- NCC were isolated from tissues around the neural tube of each developmental stage embryo. The table shows the differentiation potencies of Sox10−/Kit+ and Sox10+/Kit- NCC. N, neuron; G, glial cells.

their multipotency throughout embryogenesis, whereas Sox10+/Kit- cells lost the multipotency. In the developing DRG, Sox10+/Kit+ cells maintained their multipotency up until E12.5–13.5 after which they become bipotent with an ability to differentiate only into neurons and glial cells. Sox10+/Kit- cells in DRG exhibit neuron and glial cell bipotency throughout embryogenesis. Sox10+/Kit- cells, but not Sox10+/Kit+ cells, were present in the developing intestines; however, Sox10+/Kit- cells did not display multipotency in this system. Collectively, Sox10+/Kit+ NCC-derived cells maintain their multipotency even after their migration and colonization of their target tissues (Fig. 2). It should be noted that the multipotency of these NCC-derived cells was only detected in an *in vitro* culture system using a monolayer of ST2 stromal cells, conditions which would never be encountered *in vivo*. Other fate-restricted NCC-derived cell populations that migrate to various tissues may also maintain this latent multipotency. The potential of NCC-derived cells to differentiate into multilineage cells may be well controlled or suppressed in a given tissue environment.

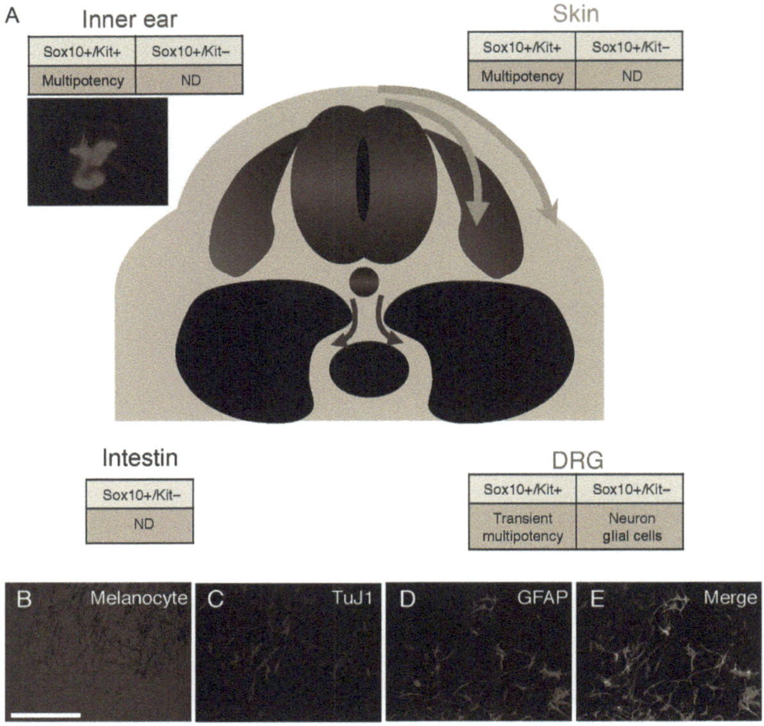

Figure 2 NCC-derived cells maintained their multipotency after entry into the target tissues. (A) NCC-derived cells maintain their multipotency after entering into the target tissues. This unexpected differentiation potency of these cells appears to have been restricted by the tissue environment. The scheme shows a transverse section of the embryonic trunk. The arrow shows the NCC migration pathway. The picture shows the inner ear of E12.5 Sox10-IRES-GFP mouse embryo. The table shows the differentiation potencies of Sox10+/Kit+ and Sox10+/Kit- NCC-derived cells in each tissue. ND, not detected. (B–E) Typical multipotent colonies generated from E12.5 skin Sox10+/Kit+ cells. All pictures were taken from a single typical colony. The same visual field is shown in "B–E." Scale bar = 200 μm. (See the color plate.)

6. REMARKS ON THE ORIGIN OF MELANOMAS

Melanoma is a tumor of transformed melanocytes originated from the NCC. In melanomas, melanocyte-specific markers are frequently down-regulated and stem cell-like characteristics are often observed (Bittner et al., 2000; Hendrix, Seftor, Hess, & Seftor, 2003; Seftor et al., 2002). Established melanoma cell lines can differentiate *in vitro* into a wide range of cell lineages, including neural and glial cells (Brocker, Magiera, & Herlyn, 1991;

Fang et al., 2001; Reed, Finnerty, & Albino, 1999) much like NCSC. Human metastatic melanoma transplanted *in ovo* adjacent to host chick premigratory NCC, distribute in a pattern similar to host NCC migratory pathways, and go onto populate the branchial arches, dorsal root, and sympathetic ganglia, and a small number of the transplanted melanoma cells expressed the neuronal marker TuJ1 (Kulesa et al., 2006). Moreover, in adoptive cell transfer therapies for patients with metastatic melanoma, melanoma cells were found to dedifferentiate into mesenchymal-like cells exhibiting a mesenchymal-like gene expression profile with downregulated melanocyte-related genes and strongly upregulated NCC marker genes such as P75 (Landsberg et al., 2012). The presence of cancer stem cells in melanomas have been reported recently by several groups. One group claimed the successful culture of melanoma stem cells by using a nonadherent sphere-forming culture (Fang et al., 2005). The established melanoma-derived spheres displayed a self-renewal capability and differentiated into melanocytic, adipocytic, osteocytic, and chondrocytic cell lineages, and appeared to recapitulate the plasticity of NCSC. Another group also established a melanoma tumor stem cell line expressing P75 (Boiko et al., 2010). These findings suggest that so-called melanoma stem cells may originate from NCSC residing in the body. The multipotency of NCC derivatives such as melanoblasts also suggests that these cells, as a direct origin of melanomas or even fully differentiated melanocytes, may be dedifferentiated into an immature stem cell state and then transform into melanoma stem cells.

Some melanoma cell lines exhibit considerably high Notch signaling activity. When Notch was constitutively activated in normal melanocytes by overexpressing the Notch inner cellular domain (NICD), these melanocytes displayed biological features resembling melanoma, including increased proliferative capacity and dysregulated cell adhesion and migration (Pinnix et al., 2009). Furthermore, NICD reprogrammed melanocytes directly to NCSC-like cells with migratory characteristics and the ability to differentiate into mesenchymal and neuronal lineages (Zabierowski et al., 2011). Epidermal cell-specific Notch activation in Keratin-14-promoter-CreER/NICD-ER transgenic mice caused the accumulation of cells resembling mature melanocytes, smooth muscle cells, and peripheral nerve cells in their dermis (Ambler & Watt, 2010).

The expression of *Sox10* was recently detected in a large set of human primary and metastatic melanoma (Agnarsdottir et al., 2010; Bakos et al., 2010; Civenni et al., 2011; Flammiger et al., 2009; Nonaka, Chiriboga, & Rubin, 2008), and also in giant congenital naevi which is associated with a high risk of melanomas (Shakhova et al., 2012). Notch signaling may be required for the continuous expression of *Sox10* (Okamura & Saga, 2008), a principal constituent of NCC regulatory gene networks (Kim, Lo, Dormand, & Anderson, 2003; Paratore, Goerich, Suter, Wegner, & Sommer, 2001). Thus, aberrant external signals such as Notch may stimulate otherwise dormant NCSC or their precursors or even terminally differentiated NCC derivatives to develop NCC-like characteristics and eventually become melanoma stem cells.

7. CONCLUDING REMARKS

NCSC are not just a transient cell population generated during early embryogenesis. They persist in the tissues they colonize throughout late embryogenesis and even into adulthood. The origin of NCSC remains obscure. One possibility is that such a specialized cell population is present and maintained *in vivo* in the form of NCSC from the moment of NCC induction. Alternatively, migratory NCC colonize and reside in stem cell niches in each tissue or organ. The latter possibility appears to be more likely because *in vitro*, delaminated NCC from dissected neural tubes were non-selectively cultured and expanded as NCSC (Stemple & Anderson, 1992), and fate-restricted precursors derived from NCC also retained their multipotency, as discussed in this chapter. Looking at the melanocyte cell lineage as a model, migrating melanoblasts in the embryonic skin are a uniform cell population and those that reached the hair bulge, a niche within the hair follicle, are by chance preserved as melanocyte stem cells that produce pigmented melanocytes throughout life. However, the presence of a specialized migrating melanoblast population destined for the stem cell niche is still possible. Although technically hard to perform and confirm, single cell transplantation may resolve this question. Other NCSC-like cells harvested and cultured from adult skin such as SKPs and EPI-NCSC have the same ambiguity for their origin.

In any case, these NCSC or NCSC-like cells can be isolated from adult tissues and expanded enough to be transplanted back to living adults and their organs and tissues. Therefore, these cells represent an attractive candidate for cell therapy. Consistent with this idea, EPI-NCSC grafted into the lesioned spinal cord of mice, survived and differentiated into functional neurons expressing the GABAergic neuron marker GAD67, oligodendrocyte marker RIP, or MBP (Sieber-Blum et al., 2006). Furthermore, human and rodent SKPs were transplanted into either the peripheral nerve or CNS of *shiverer* mutant mice, which are genetically deficient for the synthesis of MBP, and the defective phenotype was found to be recovered by Schwann cells derived from transplanted SKPs (McKenzie, Biernaskie, Toma, Midha, & Miller, 2006). On the other hand, multipotent cells are a double-edged sword. They are versatile for the regenerative maintenance of tissues, but are accompanied by the risk of malignant transformation.

It has yet to be determined why NCC-derived cells so widely maintain their multipotency, unlike other cell lineages in the adult body. The answer may be uncovered by studying on our ancestors. In tunicate urochordates, the sister clade of vertebrates (Jeffery, 2006; Jeffery, Strickler, & Yamamoto, 2004), the melanocyte-lineage cell a9.49 expresses *Mitf* together with NCC specification genes, such as *Id*, *Snail*, *Ets*, and *FoxD*. This cell differentiates into the gravity sensing otolith and melanocyte of the light-detecting ocellus, suggesting that the a9.46 lineage may represent rudimentary or putative NCC (Abitua, Wagner, Navarrete, & Levine, 2012). This tunicate study clearly showed that melanocytes, which are the derivatives of NCC in vertebrates, have already played indispensable roles similar to NCC in our ancestral animals. NCC-derived cells may not have additionally acquired multipotency through evolution, but may just have inherited their multipotency even after having differentiated into specialized NCC derivatives.

ACKNOWLEDGMENTS

We thank the other members in our laboratory for their discussions and critical reading of the manuscript. This study was supported by a Grant-in-aid for CREST-JST and by a grant from the program Grants-in-Aid for Scientific Research (C) from the Japan Society for Promotion for Science.

REFERENCES

Abitua, P. B., Wagner, E., Navarrete, I. A., & Levine, M. (2012). Identification of a rudimentary neural crest in a non-vertebrate chordate. *Nature*, *492*, 104–107.

Adameyko, I., Lallemend, F., Aquino, J. B., Pereira, J. A., Topilko, P., Muller, T., et al. (2009). Schwann cell precursors from nerve innervation are a cellular origin of melanocytes in skin. *Cell*, *139*, 366–379.

Agnarsdottir, M., Sooman, L., Bolander, A., Stromberg, S., Rexhepaj, E., Bergqvist, M., et al. (2010). SOX10 expression in superficial spreading and nodular malignant melanomas. *Melanoma Research*, *20*, 468–478.

Ambler, C. A., & Watt, F. M. (2010). Adult epidermal Notch activity induces dermal accumulation of T cells and neural crest derivatives through upregulation of jagged 1. *Development*, *137*, 3569–3579.

Amoh, Y., Li, L., Campillo, R., Kawahara, K., Katsuoka, K., Penman, S., et al. (2005). Implanted hair follicle stem cells form Schwann cells that support repair of severed peripheral nerves. *Proceedings of the National Academy of Sciences of the United States of America*, *102*, 17734–17738.

Amoh, Y., Li, L., Katsuoka, K., Penman, S., & Hoffman, R. M. (2005). Multipotent nestin-positive, keratin-negative hair-follicle bulge stem cells can form neurons. *Proceedings of the National Academy of Sciences of the United States of America, 102*, 5530–5534.

Arthur-Farraj, P. J., Latouche, M., Wilton, D. K., Quintes, S., Chabrol, E., Banerjee, A., et al. (2012). c-Jun reprograms Schwann cells of injured nerves to generate a repair cell essential for regeneration. *Neuron, 75*, 633–647.

Bakos, R. M., Maier, T., Besch, R., Mestel, D. S., Ruzicka, T., Sturm, R. A., et al. (2010). Nestin and SOX9 and SOX10 transcription factors are coexpressed in melanoma. *Experimental Dermatology, 19*, e89–e94.

Belicchi, M., Pisati, F., Lopa, R., Porretti, L., Fortunato, F., Sironi, M., et al. (2004). Human skin-derived stem cells migrate throughout forebrain and differentiate into astrocytes after injection into adult mouse brain. *Journal of Neuroscience Research, 77*, 475–486.

Biernaskie, J., Paris, M., Morozova, O., Fagan, B. M., Marra, M., Pevny, L., et al. (2009). SKPs derive from hair follicle precursors and exhibit properties of adult dermal stem cells. *Cell Stem Cell, 5*, 610–623.

Bittner, M., Meltzer, P., Chen, Y., Jiang, Y., Seftor, E., Hendrix, M., et al. (2000). Molecular classification of cutaneous malignant melanoma by gene expression profiling. *Nature, 406*, 536–540.

Bixby, S., Kruger, G. M., Mosher, J. T., Joseph, N. M., & Morrison, S. J. (2002). Cell-intrinsic differences between stem cells from different regions of the peripheral nervous system regulate the generation of neural diversity. *Neuron, 35*, 643–656.

Boiko, A. D., Razorenova, O. V., van de Rijn, M., Swetter, S. M., Johnson, D. L., Ly, D. P., et al. (2010). Human melanoma-initiating cells express neural crest nerve growth factor receptor CD271. *Nature, 466*, 133–137.

Brocker, E. B., Magiera, H., & Herlyn, M. (1991). Nerve growth and expression of receptors for nerve growth factor in tumors of melanocyte origin. *The Journal of Investigative Dermatology, 96*, 662–665.

Ciment, G., Glimelius, B., Nelson, D. M., & Weston, J. A. (1986). Reversal of a developmental restriction in neural crest-derived cells of avian embryos by a phorbol ester drug. *Developmental Biology, 118*, 392–398.

Civenni, G., Walter, A., Kobert, N., Mihic-Probst, D., Zipser, M., Belloni, B., et al. (2011). Human CD271-positive melanoma stem cells associated with metastasis establish tumor heterogeneity and long-term growth. *Cancer Research, 71*, 3098–3109.

Driskell, R. R., Giangreco, A., Jensen, K. B., Mulder, K. W., & Watt, F. M. (2009). Sox2-positive dermal papilla cells specify hair follicle type in mammalian epidermis. *Development, 136*, 2815–2823.

Dupin, E., Glavieux, C., Vaigot, P., & Le Douarin, N. M. (2000). Endothelin 3 induces the reversion of melanocytes to glia through a neural crest-derived glial-melanocytic progenitor. *Proceedings of the National Academy of Sciences of the United States of America, 97*, 7882–7887.

Dupin, E., Real, C., Glavieux-Pardanaud, C., Vaigot, P., & Le Douarin, N. M. (2003). Reversal of developmental restrictions in neural crest lineages: Transition from Schwann cells to glial-melanocytic precursors in vitro. *Proceedings of the National Academy of Sciences of the United States of America, 100*, 5229–5233.

Dyce, P. W., Zhu, H., Craig, J., & Li, J. (2004). Stem cells with multilineage potential derived from porcine skin. *Biochemical and Biophysical Research Communications, 316*, 651–658.

Fang, D., Hallman, J., Sangha, N., Kute, T. E., Hammarback, J. A., White, W. L., et al. (2001). Expression of microtubule-associated protein 2 in benign and malignant melanocytes: Implications for differentiation and progression of cutaneous melanoma. *The American Journal of Pathology, 158*, 2107–2115.

Fang, D., Nguyen, T. K., Leishear, K., Finko, R., Kulp, A. N., Hotz, S., et al. (2005). A tumorigenic subpopulation with stem cell properties in melanomas. *Cancer Research, 65*, 9328–9337.

Faraco, C. D., Vaz, S. A., Pastor, M. V., & Erickson, C. A. (2001). Hyperpigmentation in the Silkie fowl correlates with abnormal migration of fate-restricted melanoblasts and loss of environmental barrier molecules. *Developmental Dynamics, 220*, 212–225.

Fernandes, K. J., McKenzie, I. A., Mill, P., Smith, K. M., Akhavan, M., Barnabe-Heider, F., et al. (2004). A dermal niche for multipotent adult skin-derived precursor cells. *Nature Cell Biology, 6*, 1082–1093.

Flammiger, A., Besch, R., Cook, A. L., Maier, T., Sturm, R. A., & Berking, C. (2009). SOX9 and SOX10 but not BRN2 are required for nestin expression in human melanoma cells. *The Journal of Investigative Dermatology, 129*, 945–953.

Goodell, M. A., Rosenzweig, M., Kim, H., Marks, D. F., DeMaria, M., Paradis, G., et al. (1997). Dye efflux studies suggest that hematopoietic stem cells expressing low or undetectable levels of CD34 antigen exist in multiple species. *Nature Medicine, 3*, 1337–1345.

Hagedorn, L., Suter, U., & Sommer, L. (1999). P0 and PMP22 mark a multipotent neural crest-derived cell type that displays community effects in response to TGF-beta family factors. *Development, 126*, 3781–3794.

Hall, B. K. (2009). *The neural crest and neural crest cells in vertebrate development and evolution.* New York: Springer.

Hearing, V. J. (2005). Biogenesis of pigment granules: A sensitive way to regulate melanocyte function. *Journal of Dermatological Science, 37*, 3–14.

Hendrix, M. J., Seftor, E. A., Hess, A. R., & Seftor, R. E. (2003). Vasculogenic mimicry and tumour-cell plasticity: Lessons from melanoma. *Nature Reviews. Cancer, 3*, 411–421.

Henion, P. D., & Weston, J. A. (1997). Timing and pattern of cell fate restrictions in the neural crest lineage. *Development, 124*, 4351–4359.

Hu, Y. F., Zhang, Z. J., & Sieber-Blum, M. (2006). An epidermal neural crest stem cell (EPI-NCSC) molecular signature. *Stem Cells, 24*, 2692–2702.

Jeffery, W. R. (2006). Ascidian neural crest-like cells: Phylogenetic distribution, relationship to larval complexity, and pigment cell fate. *Journal of Experimental Zoology. Part B, Molecular and Developmental Evolution, 306*, 470–480.

Jeffery, W. R., Strickler, A. G., & Yamamoto, Y. (2004). Migratory neural crest-like cells form body pigmentation in a urochordate embryo. *Nature, 431*, 696–699.

Jessen, K. R., & Mirsky, R. (2008). Negative regulation of myelination: Relevance for development, injury, and demyelinating disease. *Glia, 56*, 1552–1565.

Jiang, X., Rowitch, D. H., Soriano, P., McMahon, A. P., & Sucov, H. M. (2000) Fate of the mammalian cardiac neural crest. *Development, 127*, 1607–1616.

Joannides, A., Gaughwin, P., Schwiening, C., Majed, H., Sterling, J., Compston, A., et al. (2004). Efficient generation of neural precursors from adult human skin: Astrocytes promote neurogenesis from skin-derived stem cells. *Lancet, 364*, 172–178.

Johnston, M. C., Noden, D. M., Hazelton, R. D., Coulombre, J. L., & Coulombre, A. J. (1979). Origins of avian ocular and periocular tissues. *Experimental Eye Research, 29*, 27–43.

Kaucka, M., & Adameyko, I. (2014). Non-canonical functions of the peripheral nerve. *Experimental Cell Research, 321*, 17–24.

Kim, J., Lo, L., Dormand, E., & Anderson, D. J. (2003). SOX10 maintains multipotency and inhibits neuronal differentiation of neural crest stem cells. *Neuron, 38*, 17–31.

Kruger, G. M., Mosher, J. T., Bixby, S., Joseph, N., Iwashita, T., & Morrison, S. J. (2002). Neural crest stem cells persist in the adult gut but undergo changes in self-renewal, neuronal subtype potential, and factor responsiveness. *Neuron, 35*, 657–669.

Kulesa, P. M., Kasemeier-Kulesa, J. C., Teddy, J. M., Margaryan, N. V., Seftor, E. A., Seftor, R. E., et al. (2006). Reprogramming metastatic melanoma cells to assume a neural crest cell-like phenotype in an embryonic microenvironment. *Proceedings of the National Academy of Sciences of the United States of America, 103*, 3752–3757.

Lahav, R., Dupin, E., Lecoin, L., Glavieux, C., Champeval, D., Ziller, C., et al. (1998). Endothelin 3 selectively promotes survival and proliferation of neural crest-derived glial and melanocytic precursors in vitro. *Proceedings of the National Academy of Sciences of the United States of America, 95*, 14214–14219.

Landsberg, J., Kohlmeyer, J., Renn, M., Bald, T., Rogava, M., Cron, M., et al. (2012). Melanomas resist T-cell therapy through inflammation-induced reversible dedifferentiation. *Nature, 490*, 412–416.

Le Douarin, N. M., & Kalcheim, C. (1999). *The neural crest*. Cambridge: Cambridge University Press.

Le Douarin, N. M., & Teillet, M. A. (1974). Experimental analysis of the migration and differentiation of neuroblasts of the autonomic nervous system and of neuroectodermal mesenchymal derivatives, using a biological cell marking technique. *Developmental Biology, 41*, 162–184.

Lee, H. Y., Kleber, M., Hari, L., Brault, V., Suter, U., Taketo, M. M., et al. (2004). Instructive role of Wnt/beta-catenin in sensory fate specification in neural crest stem cells. *Science, 303*, 1020–1023.

Lemke, G., Lamar, E., & Patterson, J. (1988). Isolation and analysis of the gene encoding peripheral myelin protein zero. *Neuron, 1*, 73–83.

Luo, R., Gao, J., Wehrle-Haller, B., & Henion, P. D. (2003). Molecular identification of distinct neurogenic and melanogenic neural crest sublineages. *Development, 130*, 321–330.

Lwigale, P. Y., Cressy, P. A., & Bronner-Fraser, M. (2005). Corneal keratocytes retain neural crest progenitor cell properties. *Developmental Biology, 288*, 284–293.

Manova, K., & Bachvarova, R. F. (1991). Expression of c-kit encoded at the W locus of mice in developing embryonic germ cells and presumptive melanoblasts. *Developmental Biology, 146*, 312–324.

Matsuzaki, Y., Kinjo, K., Mulligan, R. C., & Okano, H. (2004). Unexpectedly efficient homing capacity of purified murine hematopoietic stem cells. *Immunity, 20*, 87–93.

McKenzie, I. A., Biernaskie, J., Toma, J. G., Midha, R., & Miller, F. D. (2006). Skin-derived precursors generate myelinating Schwann cells for the injured and dysmyelinated nervous system. *The Journal of Neuroscience, 26*, 6651–6660.

Milet, C., & Monsoro-Burq, A. H. (2012). Neural crest induction at the neural plate border in vertebrates. *Developmental Biology, 366*, 22–33.

Morrison, S. J., White, P. M., Zock, C., & Anderson, D. J. (1999). Prospective identification, isolation by flow cytometry, and in vivo self-renewal of multipotent mammalian neural crest stem cells. *Cell, 96*, 737–749.

Motohashi, T., Aoki, H., Chiba, K., Yoshimura, N., & Kunisada, T. (2007). Multipotent cell fate of neural crest-like cells derived from embryonic stem cells. *Stem Cells, 25*, 402–410.

Motohashi, T., Kitagawa, D., Watanabe, N., Wakaoka, T., & Kunisada, T. (2014). Neural crest-derived cells sustain their multipotency even after entry into their target tissues. *Developmental Dynamics, 243*, 368–380.

Motohashi, T., Yamanaka, K., Chiba, K., Aoki, H., & Kunisada, T. (2009). Unexpected multipotency of melanoblasts isolated from murine skin. *Stem Cells, 27*, 888–897.

Motohashi, T., Yamanaka, K., Chiba, K., Miyajima, K., Aoki, H., Hirobe, T., et al. (2011). Neural crest cells retain their capability for multipotential differentiation even after lineage-restricted stages. *Developmental Dynamics, 240*, 1681–1693.

Nagoshi, N., Shibata, S., Kubota, Y., Nakamura, M., Nagai, Y., Satoh, E., et al. (2008). Ontogeny and multipotency of neural crest-derived stem cells in mouse bone marrow, dorsal root ganglia, and whisker pad. *Cell Stem Cell, 2*, 392–403.

Nichols, D. H., & Weston, J. A. (1977). Melanogenesis in cultures of peripheral nervous tissue. I. The origin and prospective fate of cells giving rise to melanocytes. *Developmental Biology*, *60*, 217–225.
Nonaka, D., Chiriboga, L., & Rubin, B. P. (2008). Sox10: A pan-schwannian and melanocytic marker. *The American Journal of Surgical Pathology*, *32*, 1291–1298.
Okamura, Y., & Saga, Y. (2008). Notch signaling is required for the maintenance of enteric neural crest progenitors. *Development*, *135*, 3555–3565.
Paratore, C., Goerich, D. E., Suter, U., Wegner, M., & Sommer, L. (2001). Survival and glial fate acquisition of neural crest cells are regulated by an interplay between the transcription factor Sox10 and extrinsic combinatorial signaling. *Development*, *128*, 3949–3961.
Pardal, R., Ortega-Saenz, P., Duran, R., & Lopez-Barneo, J. (2007). Glia-like stem cells sustain physiologic neurogenesis in the adult mammalian carotid body. *Cell*, *131*, 364–377.
Parkinson, D. B., Bhaskaran, A., Arthur-Farraj, P., Noon, L. A., Woodhoo, A., Lloyd, A. C., et al. (2008). c-Jun is a negative regulator of myelination. *The Journal of Cell Biology*, *181*, 625–637.
Pinnix, C. C., Lee, J. T., Liu, Z. J., McDaid, R., Balint, K., Beverly, L. J., et al. (2009). Active Notch1 confers a transformed phenotype to primary human melanocytes. *Cancer Research*, *69*, 5312–5320.
Real, C., Glavieux-Pardanaud, C., Le Douarin, N. M., & Dupin, E. (2006). Clonally cultured differentiated pigment cells can dedifferentiate and generate multipotent progenitors with self-renewing potential. *Developmental Biology*, *300*, 656–669.
Real, C., Glavieux-Pardanaud, C., Vaigot, P., Le-Douarin, N., & Dupin, E. (2005). The instability of the neural crest phenotypes: Schwann cells can differentiate into myofibroblasts. *The International Journal of Developmental Biology*, *49*, 151–159.
Reed, J. A., Finnerty, B., & Albino, A. P. (1999). Divergent cellular differentiation pathways during the invasive stage of cutaneous malignant melanoma progression. *The American Journal of Pathology*, *155*, 549–555.
Reedy, M. V., Faraco, C. D., & Erickson, C. A. (1998). The delayed entry of thoracic neural crest cells into the dorsolateral path is a consequence of the late emigration of melanogenic neural crest cells from the neural tube. *Developmental Biology*, *200*, 234–246.
Rizvi, T. A., Huang, Y., Sidani, A., Atit, R., Largaespada, D. A., Boissy, R. E., et al. (2002). A novel cytokine pathway suppresses glial cell melanogenesis after injury to adult nerve. *The Journal of Neuroscience*, *22*, 9831–9840.
Sasaki, R., Aoki, S., Yamato, M., Uchiyama, H., Wada, K., Okano, T., et al. (2008). Neurosphere generation from dental pulp of adult rat incisor. *The European Journal of Neuroscience*, *27*, 538–548.
Seftor, E. A., Meltzer, P. S., Schatteman, G. C., Gruman, L. M., Hess, A. R., Kirschmann, D. A., et al. (2002). Expression of multiple molecular phenotypes by aggressive melanoma tumor cells: Role in vasculogenic mimicry. *Critical Reviews in Oncology/Hematology*, *44*, 17–27.
Shah, N. M., & Anderson, D. J. (1997). Integration of multiple instructive cues by neural crest stem cells reveals cell-intrinsic biases in relative growth factor responsiveness. *Proceedings of the National Academy of Sciences of the United States of America*, *94*, 11369–11374.
Shakhova, O., Zingg, D., Schaefer, S. M., Hari, L., Civenni, G., Blunschi, J., et al. (2012). Sox10 promotes the formation and maintenance of giant congenital naevi and melanoma. *Nature Cell Biology*, *14*, 882–890.
Sherman, L., Stocker, K. M., Morrison, R., & Ciment, G. (1993). Basic fibroblast growth factor (bFGF) acts intracellularly to cause the transdifferentiation of avian neural crest-derived Schwann cell precursors into melanocytes. *Development*, *118*, 1313–1326.
Shih, D. T., Lee, D. C., Chen, S. C., Tsai, R. Y., Huang, C. T., Tsai, C. C., et al. (2005). Isolation and characterization of neurogenic mesenchymal stem cells in human scalp tissue. *Stem Cells*, *23*, 1012–1020.

Sieber-Blum, M., Grim, M., Hu, Y. F., & Szeder, V. (2004). Pluripotent neural crest stem cells in the adult hair follicle. *Developmental Dynamics*, *231*, 258–269.

Sieber-Blum, M., Schnell, L., Grim, M., Hu, Y. F., Schneider, R., & Schwab, M. E. (2006). Characterization of epidermal neural crest stem cell (EPI-NCSC) grafts in the lesioned spinal cord. *Molecular and Cellular Neurosciences*, *32*, 67–81.

Stemple, D. L., & Anderson, D. J. (1992). Isolation of a stem cell for neurons and glia from the mammalian neural crest. *Cell*, *71*, 973–985.

Stocker, K. M., Sherman, L., Rees, S., & Ciment, G. (1991). Basic FGF and TGF-beta 1 influence commitment to melanogenesis in neural crest-derived cells of avian embryos. *Development*, *111*, 635–645.

Takashima, Y., Era, T., Nakao, K., Kondo, S., Kasuga, M., Smith, A. G., et al. (2007). Neuroepithelial cells supply an initial transient wave of MSC differentiation. *Cell*, *129*, 1377–1388.

Techawattanawisal, W., Nakahama, K., Komaki, M., Abe, M., Takagi, Y., & Morita, I. (2007). Isolation of multipotent stem cells from adult rat periodontal ligament by neurosphere-forming culture system. *Biochemical and Biophysical Research Communications*, *357*, 917–923.

Toma, J. G., Akhavan, M., Fernandes, K. J., Barnabe-Heider, F., Sadikot, A., Kaplan, D. R., et al. (2001). Isolation of multipotent adult stem cells from the dermis of mammalian skin. *Nature Cell Biology*, *3*, 778–784.

Toma, J. G., McKenzie, I. A., Bagli, D., & Miller, F. D. (2005). Isolation and characterization of multipotent skin-derived precursors from human skin. *Stem Cells*, *23*, 727–737.

Tomita, Y., Matsumura, K., Wakamatsu, Y., Matsuzaki, Y., Shibuya, I., Kawaguchi, H., et al. (2005). Cardiac neural crest cells contribute to the dormant multipotent stem cell in the mammalian heart. *The Journal of Cell Biology*, *170*, 1135–1146.

Trainor, P. A., & Tam, P. P. (1995). Cranial paraxial mesoderm and neural crest cells of the mouse embryo: Co-distribution in the craniofacial mesenchyme but distinct segregation in branchial arches. *Development*, *121*, 2569–2582.

Trentin, A., Glavieux-Pardanaud, C., Le Douarin, N. M., & Dupin, E. (2004). Self-renewal capacity is a widespread property of various types of neural crest precursor cells. *Proceedings of the National Academy of Sciences of the United States of America*, *101*, 4495–4500.

Wakamatsu, Y., Mochii, M., Vogel, K. S., & Weston, J. A. (1998). Avian neural crest-derived neurogenic precursors undergo apoptosis on the lateral migration pathway. *Development*, *125*, 4205–4213.

Widera, D., Grimm, W. D., Moebius, J. M., Mikenberg, I., Piechaczek, C., Gassmann, G., et al. (2007). Highly efficient neural differentiation of human somatic stem cells, isolated by minimally invasive periodontal surgery. *Stem Cells and Development*, *16*, 447–460.

Widera, D., Heimann, P., Zander, C., Imielski, Y., Heidbreder, M., Heilemann, M., et al. (2011). Schwann cells can be reprogrammed to multipotency by culture. *Stem Cells and Development*, *20*, 2053–2064.

Widera, D., Zander, C., Heidbreder, M., Kasperek, Y., Noll, T., Seitz, O., et al. (2009). Adult palatum as a novel source of neural crest-related stem cells. *Stem Cells*, *27*, 1899–1910.

Wilson, Y. M., Richards, K. L., Ford-Perriss, M. L., Panthier, J. J., & Murphy, M. (2004). Neural crest cell lineage segregation in the mouse neural tube. *Development*, *131*, 6153–6162.

Wong, C. E., Paratore, C., Dours-Zimmermann, M. T., Rochat, A., Pietri, T., Suter, U., et al. (2006). Neural crest-derived cells with stem cell features can be traced back to multiple lineages in the adult skin. *The Journal of Cell Biology*, *175*, 1005–1015.

Woodhoo, A., Alonso, M. B., Droggiti, A., Turmaine, M., D'Antonio, M., Parkinson, D. B., et al. (2009). Notch controls embryonic Schwann cell differentiation, postnatal myelination and adult plasticity. *Nature Neuroscience*, *12*, 839–847.

Yamazaki, H., Sakata, E., Yamane, T., Yanagisawa, A., Abe, K., Yamamura, K., et al. (2005). Presence and distribution of neural crest-derived cells in the murine developing thymus and their potential for differentiation. *International Immunology*, *17*, 549–558.
Yoshida, S., Shimmura, S., Nagoshi, N., Fukuda, K., Matsuzaki, Y., Okano, H., et al. (2006). Isolation of multipotent neural crest-derived stem cells from the adult mouse cornea. *Stem Cells*, *24*, 2714–2722.
Zabierowski, S. E., Baubet, V., Himes, B., Li, L., Fukunaga-Kalabis, M., Patel, S., et al. (2011). Direct reprogramming of melanocytes to neural crest stem-like cells by one defined factor. *Stem Cells*, *29*, 1752–1762.

CHAPTER FOUR

The Ciliary Baton: Orchestrating Neural Crest Cell Development

Ching-Fang Chang[*,†], Elizabeth N. Schock[*,†], Aria C. Attia[‡], Rolf W. Stottmann[†,‡], Samantha A. Brugmann[*,†,1]

[*]Division of Plastic Surgery, Department of Surgery, Cincinnati Children's Hospital Medical Center, Cincinnati, Ohio, USA
[†]Division of Developmental Biology, Department of Pediatrics, Cincinnati Children's Hospital Medical Center, Cincinnati, Ohio, USA
[‡]Division of Human Genetics, Department of Pediatrics, Cincinnati Children's Hospital Medical Center, Cincinnati, Ohio, USA
[1]Corresponding author: e-mail address: samantha.brugmann@cchmc.org

Contents

Current Topics in Developmental Biology, Volume 111
ISSN 0070-2153
http://dx.doi.org/10.1016/bs.ctdb.2014.11.004

Abstract

Primary cilia are cell surface, microtubule-based organelles that dynamically extend from cells to receive and process molecular and mechanical signaling cues. In the last decade, this organelle has gained increasing popularity due to its ability to act as a cellular antenna, receive molecular stimuli, and respond to the cell's environment. A growing field of data suggests that various tissues utilize and interpret the loss of cilia in different ways. Thus, careful examination of the role of cilia on individual cell types and tissues is necessary. Neural crest cells (NCCs) are an excellent example of cells that survey their environment for developmental cues. In this review, we discuss how NCCs utilize primary cilia during their ontogenic development, paying special attention to the role primary cilia play in processing developmental signals required for NCC specification, migration, proliferation, and differentiation. We also discuss how the loss of functional cilia on cranial and trunk NCCs affects the development of various organ systems to which they contribute. A deeper understanding of ciliary function could contribute greatly to understanding the molecular mechanisms guiding NCC development and differentiation. Furthermore, superimposing the ciliary contribution on our current understanding of NCC development identifies new avenues for therapeutic intervention in neurocristopathies.

1. INTRODUCTION

Over 100 years ago, biologists observed a thin, singular eyelash-type extension on various cell types (Zimmermann, 1898). These scientists, without any of the molecular knowledge or tools we have today, hypothesized that this "cilium" had a sensory function within the cell (Zimmermann, 1898). Sadly, after this initial discovery the field of ciliary biology lay essentially dormant for the next 80 years. Now, in the past two decades, the field of ciliary biology has undergone a renaissance due to improved molecular and cellular tools and a better understanding of what these nearly ubiquitous organelles are doing during various cellular processes. This review will discuss what a cilium is, the process of ciliogenesis and ciliary function. We will highlight what is currently known regarding how neural crest cells (NCCs) utilize primary cilia during development and the disorders affecting the NCC derivatives that can arise when ciliary structure or function is disrupted.

2. THE PRIMARY CILIUM: DEFINING THE ORGANELLE

2.1. Structure equals function

A cilium is a microtubule-based cellular projection (Fig. 1). There are several types of cilia, which are typically classified by their microtubule arrangement, capacity for motility, and number per cell. Generally, and for the purposes of this review, a primary cilium is defined by three criteria, (1) a 9+0 microtubule arrangement, (2) lack of motility, and (3) cellular singularity. Establishing how these three criteria define the organelle is important to understand primary cilia function during various cellular processes, including NCC development. Furthermore, understanding how variations in these key characteristics affect ciliary function is also important for gaining a better understanding of how cilia, primary and otherwise, function within a cell.

The 9+0 conformation of a primary cilium, as the name implies, refers to an arrangement of axonemal microtubules in an outer ring of 9 microtubule doublets with an absence of any inner microtubule pair (Fig. 1). Other types of cilia can have a 9+2 arrangement consisting of an outer ring of 9 microtubule doublets and a central core of 2 microtubules. Even 9+4 cilia have been reported in some species (Feistel & Blum, 2006).

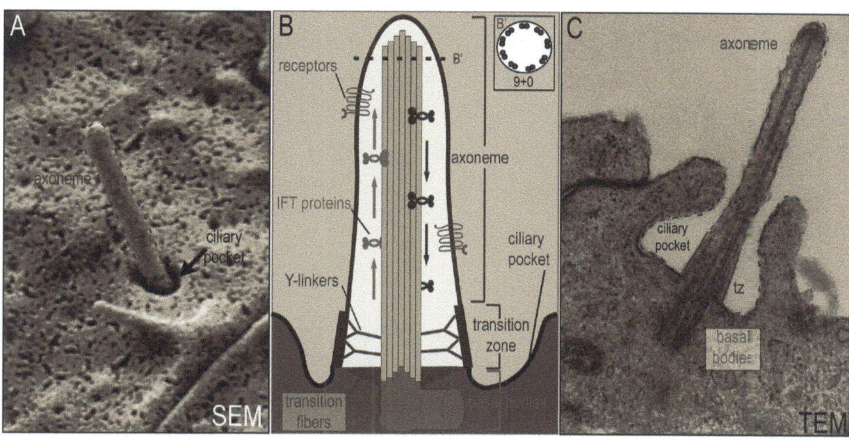

Figure 1 Structure of the primary cilium. (A) Scanning electron microscopy (SEM) of the primary cilium. The axoneme is outlined in purple and the ciliary pocket is indicated by an arrow. (B) Schematic diagram of the primary cilium. The axoneme (purple) extends out from the basal bodies (light green) into the extracellular space (light gray). Receptors (red) for various signaling pathways localized to the membrane of the axoneme. (B′) Schematic cross-section of the axoneme reveals the microtubules in a 9+0 conformation. Intraflagellar transport proteins travel along microtubules, moving cargo up (light blue) and back down (dark blue) the axoneme. The more proximal portion of the primary cilium contains the transition fibers (blue), transition zone (dark green), and Y-linkers (pink). Collectively, these structures are important for ciliogenesis and act as a selective barrier between the intracellular space (dark gray) and the ciliary compartment (white). (C) Transmission electron microscopy (TEM) of the primary cilium. The axoneme (purple dotted lines), transition zone (tz; dark green lines), basal bodies (light green dotted lines), and ciliary pocket (black dotted lines) have all been highlighted in the image. (See the color plate.)

The second feature common to primary cilia is their nonmotile status. Microtubule arrangement broadly correlates with cilium motility: 9 + 0 cilia are canonically not motile and 9 + 2 cilia are motile. While this is a general rule to go by, it is more accurate to use the presence or absence of dynein arms to determine motility of the cilium (Afzelius, 1976; Blum, Hayes, Whisnant, & Rosen, 1977). Dyneins are large motor protein complexes that generate force, causing the movement of eukaryotic cilia and flagella. The dynein protein complexes directly connect to the outer ring of microtubules (Neesen et al., 2001). Generally, 9 + 0 cilia and nonmotile 9 + 2 cilia lack dynein arms (Bloodgood, 2009).

Finally, primary cilia are reported to be solitary cellular extensions (i.e., each cell only extends one primary cilium). Similar to motility, cilia number has also been linked to microtubule arrangement. Monociliated cells are generally 9 + 0 and nonmotile, whereas multiciliated cells are generally 9 +2 and motile. There are, however, several exceptions to this generality, thus it should not be considered a strict rule. Nodal cilia are perhaps the best example of an exception to this general classification. Current data suggest that the node may possess a mixture of both motile and nonmotile monocilia (McGrath, Somlo, Makova, Tian, & Brueckner, 2003) as well as a mixture of 9 + 2 and 9 + 0 monocilia (Caspary, Larkins, & Anderson, 2007).

2.2. Widespread and dynamic

The last 10 years have witnessed increased interest in the primary cilia (Beales & Jackson, 2012). The major causes of this resurgence are twofold; cilia are omnipresent and highly dynamic. First, as with most organelles, they are present in almost all cells rendering their function relevant to just about every tissue and organ system. However, there are a handful of cells and organisms that lack primary cilia. Yeast, fungi, and higher plants do not extend cilia. Of the "model organisms" classically used in biomedical research, the invertebrate *C. elegans* extend cilia only on sensory neurons, and none of the cilia in the nematode are motile. Similarly, *Drosophila* also extend cilia on sensory cells, but they are not utilized in the same way as mammalian cilia (Avidor-Reiss et al., 2004; Han, Kwok, & Kernan, 2003; Ray et al., 1999). Primary cilia are much better understood in vertebrates. Still, there are some mammalian cells that lack primary cilia including hepatocytes, acinar cells, and lymphocytes and granulocytes in the hematopoietic lineage (Wheatley, 1969; Wheatley, Wang, & Strugnell, 1996).

Second, unlike many organelles, primary cilia are highly dynamic in their extension, size, and specialization. Primary cilia are extended during times of quiescence (G_0) and begin to retract during cell cycle re-entry. This cyclical extension and retraction suggests a mutually exclusive relationship between the cell cycle and ciliogenesis. Indeed, various studies have shown that cell cycle progression requires the continued suppression of primary cilia formation (Goto, Inoko, & Inagaki, 2013). Furthermore, links have been established between ciliary extension and cancer, with some cancers being devoid of primary cilia and some depending on ciliary signaling (Basten & Giles, 2013; Molla-Herman et al., 2010; Wong et al., 2009). Although a bona fide understanding of the exact nature of the relationship between cilia and the cell cycle remains elusive, the current data open the door to the possibility of therapeutic intervention via ciliary targeting, in diseases of aberrant cellular proliferation (Basten & Giles, 2013).

In addition to their dynamic extension and retraction, primary cilia can also vary in length. Under most circumstances, primary cilia are between 1 and 10 μm long, but they can extend to over 20 μm. Various factors can influence the length of cilia including injury/hypoxia (Verghese, Weidenfeld, Bertram, Ricardo, & Deane, 2008; Verghese, Zhuang, Saiti, Ricardo, & Deane, 2011), treatment with monovalent cation chlorides (LiCl, NaCl, and KCl) (Miyoshi, Kasahara, Miyazaki, & Asanuma, 2009; Ou et al., 2009), cytoskeletal dynamics (Miyoshi, Kasahara, Miyazaki, & Asanuma, 2011), and autophagy (Tang et al., 2013; Tang, Zhu, & Zhong, 2014). It is hypothesized that with all other variables being constant, increased length of the cilium is directly related to the sensitivity of the organelle. Understanding the factors that regulate cilium length could determine how sensitive a cell is to the molecular and mechanical stimuli in the surrounding environment. Thus, a feedback system dictating primary cilium length could be an important yet underappreciated factor regulating development (Miyoshi et al., 2011).

2.3. Ciliogenesis: Building the cilium

The cilium itself is composed of various functional domains required for proper ciliary function: basal bodies (centrosomes), transition fibers, transition zone, intraflagellar transport (IFT) machinery, axoneme, and a specialized ciliary membrane (Fig. 1). In cycling cells, two centrioles (mother and daughter) together with pericentriolar material form the centrosome, which serves as microtubule organization center. Once cells enter G_1/G_0 phase, the mother centriole differentiates into a basal body before ciliogenesis initiates (Kobayashi & Dynlacht, 2011). Similar to the mother centriole, a basal body is comprised of 9 microtubule triplets arranging in a barrel shape, a subdistal appendage (basal foot), and a distal appendage (transition fiber, Fig. 1).

Depending on cell type, ciliogenesis can occur via two distinct mechanisms: (1) extracellular ciliogenesis, where the basal body docks and fuses to the plasma membrane directly or (2) intracellular ciliogenesis, where the basal body first fuses to a ciliary vesicle in the cytoplasm before docking to the plasma membrane (Molla-Herman et al., 2010; Reiter, Blacque, & Leroux, 2012; Sorokin, 1962; Fig. 2). In both pathways, the initial docking is facilitated by the interaction between the membrane and transition fibers. The primary ciliary vesicle adds additional membrane by recruiting secondary vesicles. The enlarged vesicle soon deforms into an invaginated sac,

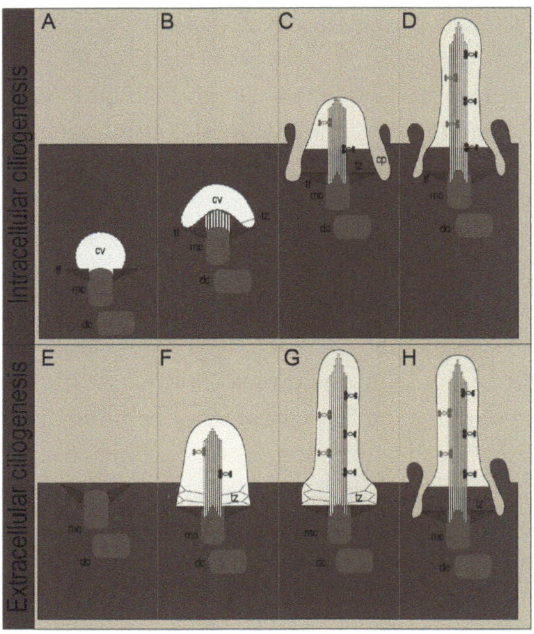

Figure 2 Models of intracellular and extracellular ciliogenesis. *Steps of intracellular ciliogenesis.* (A) A ciliary vesicle (cv) binds to the distal end of the mother centriole (mc) via associations with transition fibers (tf). (B) Microtubule and transition zone (tz) outgrowth protrudes, causing the cv to invaginate. (C) Docking of the centriolar/cv complex to the plasma membrane. (D) Axonemal outgrowth. *Steps of extracellular ciliogenesis.* (E) The mother centriole (mc) docks directly to the plasma membrane via associations with transition fibers (tf). (F–G) Axonemal outgrowth occurs. (H) Ciliary pocket forms. dc, daughter centriole; pc, primary cilium; cp, ciliary pocket. *Figure modified from Reiter et al. (2012) and Molla-Herman et al. (2010).*

which forms a double membrane sheath (later becoming the ciliary pocket) surrounding the apical end of basal body (Fig. 2). At this point, the transition zone begins to emerge as microtubule doublets start to assemble within the vesicle sac.

While the transition zone is growing, the basal body–ciliary vesicle migrates to the membrane through a process dependent on both the actin cytoskeleton and membrane-associated components in the transition zone (Dawe, Farr, & Gull, 2007). As the basal body–ciliary vesicle complex migrates to, and fuses with, the plasma membrane, the transition zone matures and is now able to function as a ciliary gate. The transition zone is characterized by distinctive Y-shaped fibers, which connect microtubule doublets to the ciliary membrane. Structurally, the transition zone stabilizes the basal body–ciliary vesicle after the initial docking. Functionally, all these components collectively serve as the ciliary gate, a domain required for establishing a distinct ciliary compartment separate from the rest of the cell body.

After formation of the ciliary compartment, the axoneme starts to extend via microtubule lengthening. There is a distinct set of posttranslational modifications on axonemal tubulins as the microtubules are acetylated, detyrosinated, polyglutamylated, and polyglycylated (Konno, Setou, & Ikegami, 2012). These modifications are important for axonemal stability. The microtubule extension occurs through a process called IFT. In IFT, the anterograde transport is carried out by IFT-B complexes and kinesin II motors. IFT builds the microtubule scaffold and carries essential ciliary proteins to the distal tip of the axoneme. Finally, numerous receptors and ion channels localize to the specialized membrane surrounding the fully extended axoneme. With the axoneme extended and functional, cilia are now capable of surveying the cell's molecular environment. Various cell types utilize these cilia in different ways. In Section 3, we focus on how NCCs utilize primary cilia during their development and differentiation.

3. THE ROLE OF PRIMARY CILIA DURING NCC ONTOGENY

During development, NCCs progress through multiple phases: specification, migration, proliferation, and differentiation. While there are entire fields of study on each phase, little is known regarding primary cilia function during these phases. In this portion of the review, we summarize what is known about primary cilia with respect to each phase.

3.1. Primary cilia and NCC specification

NCC specification occurs at the boundary between the neural plate and the nonneural ectoderm. At this junction Wnt, BMP, and fibroblast growth factor (FGF) signaling initiates a transcriptional program (e.g., *Pax3*, *Pax7*, *Msx*, and *Zic1*) that will define the neural plate boarder (Garcia-Castro, Marcelle, & Bronner-Fraser, 2002; LaBonne & Bronner-Fraser, 1999; Marchant, Linker, Ruiz, Guerrero, & Mayor, 1998; Mayor, Guerrero, & Martinez, 1997; Monsoro-Burq, Fletcher, & Harland, 2003). Within this region, both NCCs and placodal cells are specified; however, NCCs express a unique combination of transcription factors including *Snail2*, *Sox10*, *Sox9*, and *FoxD3* (Bronner & LeDouarin, 2012). While no in-depth study has been conducted specifically examining if primary cilia are required for NCC specification, postmigration data can be used to infer the answer to this question. Various conditional knockouts of ciliary proteins, using NCC drivers, still exhibit the formation of NCCs (Brugmann et al., 2010; Brugmann, Stottmann, unpublished). This suggests that NCCs are specified, to some appreciable extent, in the absence of normal primary cilia function. This conclusion is further supported by data from zebrafish *bbs* morphant embryos, which exhibit normal domains of *foxd3* or *sox10* expression suggesting that NCCs are specified properly (Tobin et al., 2008).

While data from both of these model systems support the hypothesis that primary cilia are not absolutely required for proper NCC specification, data from *Fuz*$^{-/-}$ murine mutants suggests the cilia regulate NCC number. In these mutants, lineage tracing of NCCs indicates an expansion in the domain and number of NCCs in the midbrain, hindbrain, and maxillary process (Gray et al., 2009; Tabler et al., 2013). Taken together, these studies suggest a number of interesting possibilities regarding NCCs and primary cilia. First, perhaps only certain ciliary proteins play a role in NCC specification. Whereas, both Kif3a and the Bardet–Biedl syndrome (BBS) proteins are required for ciliogenesis and localize to the cilia and basal body (Nachury et al., 2007; Yamazaki, Nakata, Okada, & Hirokawa, 1995), respectively, Fuzzy is a cilia-associated protein that functions in the cytoplasm as an effector of the planar cell polarity (PCP) pathway (Brooks & Wallingford, 2012; Park, Haigo, & Wallingford, 2006). Thus, *Fuz*$^{-/-}$ mutants may have defects in NCC specification due to a cilia-independent mechanism. Second, since *Fuz*$^{-/-}$ is a null mutant (rather than the conditional knockout targeting ciliary function specifically in NCCs), it is possible that the loss of cilia in surrounding tissues negatively affects the molecular crosstalk required for NCC specification. Continued examination of NCC specification in various ciliary mutants is necessary to clarify this issue.

3.2. Primary cilia and NCC migration

Following specification, NCCs will undergo an epithelial to mesenchymal transition, delaminate from the neural tube and migrate in a highly regulated process. Broadly, NCC migration requires two main types of interactions: (1) between NCCs and the surrounding tissues and (2) between individual migrating NCCs. These interactions are dependent upon NCCs receiving and interpreting either long-range or short-range cellular signals (Teddy & Kulesa, 2004). For migration to efficiently occur, NCCs must be capable of responding to attractants/repellants and creating discrete, directional migration streams (Kuriyama & Mayor, 2008; Teddy & Kulesa, 2004) Chemotaxis, the movement of a cell toward a directive substance, is also a guiding force of NCC migration (Theveneau & Mayor, 2012). Several chemotactic molecules essential for cranial NCC migration have been identified including SDF-1, PDGF-AA, and VEGFA (Theveneau & Mayor, 2012). The receptors for some of these molecules are found on primary cilia (Schneider et al., 2005).

Several ciliary mutants exhibit aberrant NCC cell migration. For example, *bbs* zebrafish morphants have a significant reduction in the migratory streams of both cranial and trunk NCCs. Through transplantation studies, it was determined that *bbs8* is required cell-autonomously for proper NCC migration (Tobin et al., 2008). *Fto* zebrafish morphants also have NCC migration defects as evidenced by diffuse and reduced *Sox10* expression in the head (Osborn et al., 2014). More caudal NCC migration is also severely disrupted in these mutants. NCCs posterior to the seventh somite fail to migrate away from the dorsal neural tube and melanocytes are mislocalized (Osborn et al., 2014). In support of these data from animal models, fibroblasts from ciliopathy patients also fail to migrate normally and exhibit disruptions in actin cytoskeletal architecture (Hernandez-Hernandez et al., 2013; Madhivanan, Mukherjee, & Aguilar, 2012; Tobin et al., 2008).

While, it is clear that aberrant ciliary function can negatively impact NCC migration, we currently lack an understanding of *how* primary cilia regulate migration. One potential mechanism is platelet-derived growth factor (PDGF)-dependent chemotaxis as PDGF is a known NCC chemoattractant, PDGFRα localizes to the axoneme, and PDGF signaling is transduced through the primary cilium (Schneider et al., 2005).

Additionally, studies in both fibroblasts and mouse embryonic fibroblasts (MEFs) demonstrated primary cilia are required for directed migration toward a PDGF-AA source (Schneider et al., 2005, 2010). Interestingly, loss of PDGFRα on cranial NCCs yields aberrant cranial NCC migration (He & Soriano, 2013). Another possible mechanism to affect NCC migration is via regulation of the PCP pathway. The PCP proteins Vangl2 and Dishevelled have been shown to localize to the primary cilium and have genetic interactions with both IFT and basal body proteins (Jones et al., 2008; Ross et al., 2005; Veland et al., 2013). Additionally, ciliopathies such as Lowe, Joubert, and Bardet–Biedl syndromes have defects in RhoA or Rac1 localization or activity, and RhoA has also been show to localize to the basal body (Hernandez-Hernandez et al., 2013; Madhivanan et al., 2012; Valente et al., 2010). All of these parallel findings support the hypothesis that primary cilia play an important role during NCC migration, and we suggest this is an area ripe for continued investigation.

3.3. Primary cilia and NCC proliferation

After NCCs have finished migrating, they will continue to proliferate and greatly expand in their target tissues. Primary cilia extension and the cell cycle are tightly linked together. For example, IFT88 is associated with the centrosome throughout the cell cycle (Robert et al., 2007) and is required for spindle orientation in mitosis (Delaval, Bright, Lawson, & Doxsey, 2011). Furthermore, overexpression of IFT88 prevents the G_1–S transition and induces apoptotic cell death, whereas knockdown of IFT88 promotes cell cycle progression (Robert et al., 2007). Data from mouse ciliopathic mutants indicate that primary cilia on NCCs are essential for regulating proliferation, especially in the developing face. *Kif3a*$^{f/f}$;*Wnt1-Cre* mutants have increased NCC proliferation. This increase is most notable at the facial midline of embryos and correlates with an expansion of *Gli1* expression. Phenotypically, the medial increase in proliferation manifests as an expanded midline in these mutants (Brugmann et al., 2010; Fig. 3).

Although the link to proliferation is clear in *Kif3a*$^{f/f}$;*Wnt1-Cre* mutants, the effect that loss of other ciliary proteins has on proliferation is highly variable. *Fuz*$^{-/-}$ mutants have hyperplastic maxillary prominences that result in a high, arched palate, yet NCC proliferation is actually decreased (Tabler et al., 2013). Furthermore, there are other ciliary mutants that do not show any indication of alterations in NCC proliferation or midline development, such as *Ttc21b*$^{f/null}$;*Wnt1-Cre* (Fig. 3, Stottmann unpublished). These data

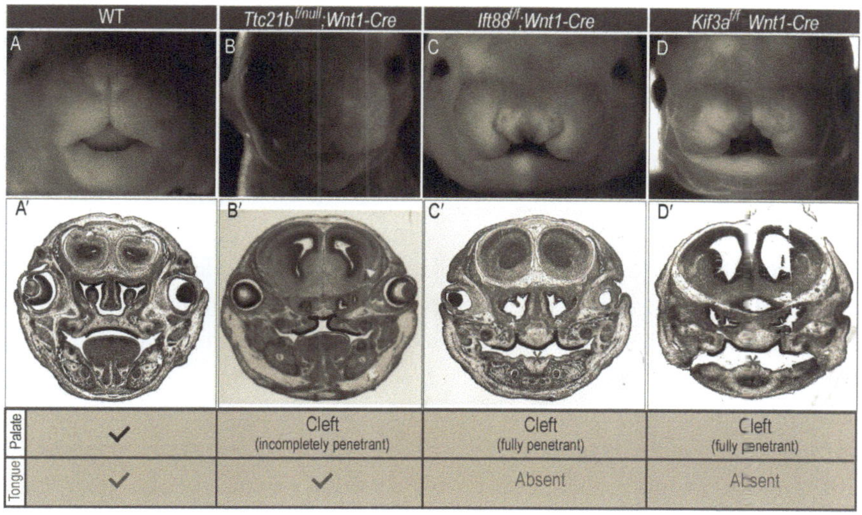

Figure 3 Craniofacial phenotypes of ciliary mutants. Whole-mount images of e14.5 (A) Wild-type, (B) *Ttc21b*$^{f/null}$;*Wnt1-Cre*, (C) *Ift88*$^{f/f}$;*Wnt1-Cre*, and (D) *Kif3a*$^{f/f}$;*Wnt1-Cre* mice. (A′–D′) Frontal sections of (A–D). (A′) Wild-type mice have a fused palatal shelves (purple outline) and a tongue (blue outline). In contrast, all ciliary mutants have (B′–D′) some degree of cleft palate (purple outline) and *Ift88*$^{f/f}$;*Wnt1-Cre* and *Kif3a*$^{f/f}$;*Wnt1-Cre* mutants have aglossia (C′, D′). Chart indicates craniofacial phenotype for each mouse. (See the color plate.)

reiterate that cilia have a role in proliferation, but also suggest that the exact role the cilia play during proliferation is unclear. Currently, the most accurate conclusion to draw from these data is that individual ciliary proteins play specific roles during proliferation.

3.4. Primary cilia and NCC differentiation

NCCs give rise to a wide variety of cell types including bone and cartilage cells in the facial skeleton. Both patients with ciliopathies and ciliopathic animal models often have dysmorphic craniofacial skeletons. In zebrafish, *bbs*, *ift*, and *ofd1* morphants have dysmorphic craniofacial cartilage (Ferrante et al., 2009; Lunt, Haynes, & Perkins, 2009; Tobin et al., 2008). Murine ciliopathic mutants also frequently have skeletal abnormalities. *Msk*^{krc} mutants exhibit misshapen skulls and a decrease in frontal and parietal bone ossification (Weatherbee, Niswander, & Anderson, 2009). Humans with Meckel syndrome, often caused by mutations in the *Msk* gene, have skeletal malformation of the cranial base (Kjaer, Hansen, Keeling, Nolting, & Kjaer, 1999). Thus, taken together, these data demonstrate that cranioskeletal defects are a common feature of human ciliopathies and further suggest that primary cilia may play an important role in NCC differentiation into skeletal elements.

In addition to the importance of primary cilia for initial skeletal differentiation, they are also critical for postnatal changes in the skeleton. Mechanosensation is another cellular process that helps to shape the architecture of skeletal elements. Primary cilia are able to respond to mechanical signals (Hoey, Downs, & Jacobs, 2012). Specific subsets of ciliary proteins have been shown to be involved in these mechanical-based signaling responses. For example, PKD2 localizes to the primary cilium and is essential for fluid-flow sensation (Nauli et al., 2003). *Pkd2*$^{f/f}$;*Wnt1-Cre* postnatal mutants exhibit dome-shaped skull defects, anterior–posterior snout compression, and abnormal fusion in the face. These abnormalities are not detected in embryos, suggesting that these phenotypes are the result of an inability of NCC derivatives to properly respond to postnatal mechanical forces. These data suggest that primary cilia continue to play a role in development and remodeling of NCC-derived structures even after cell differentiation.

4. CRANIOFACIAL PHENOTYPES IN ANIMAL MODELS AND HUMAN PATIENTS SUPPORT A ROLE FOR PRIMARY CILIA IN NCC DEVELOPMENT

4.1. Insights from animal models

In addition to the ciliary mutants in which early stages of NCC development have been analyzed, there are a significant number of ciliary mouse mutants with craniofacial phenotypes suggesting aberrant NCC behavior (Table 1). For example, several mutants have cleft lip and/or palate, including the *Ift172*^{avc1} allele with cleft secondary palate (Friedland-Little et al., 2011), *Rpgrip1l* with cleft lip and hypoplastic mandible (Delous et al., 2007), *Mks1* with cleft lip/palate and pointy snout (Weatherbee et al., 2009), and *Kif7* with cleft lip/palate (direct submission by Cecilia Lo to MGI). Other mutants are reported to have phenotypes in the arches, without clefting defects, including *Dync2h1* (formerly *Dnchc2*), which exhibits micrognathia (C Lo direct submission to MGI), *Ift57* (*Hippi*), which displays hypertelorism and small arches (Houde et al., 2006), *Ift52*^{hypo} with fused maxillary prominences (Liu et al., 2005), *Tmem67* with mandibular hypoplasia (Abdelhamed et al., 2013), and *Ift122* mutants with enlarged branchial arches (Cortellino et al., 2009). Furthermore, an in-depth morphometric

Table 1 Animal models for craniofacial ciliopathies

Model system	Mutant/morphant name	Gene	Craniofacial phenotype	References
Mouse	*Aln*	*Ttc21b* (*Ift39*)	Delayed eye and forebrain development; neural tube defects	Tran et al. (2008)
	$Bbs4^{-/-}$	*Bbs4*	Mid-facial shortening due to premaxillary and maxillary hypoplasia; shortened snouts	Tobin et al. (2008)
	$Bbs6^{-/-}$	*Bbs6*	Mid-facial shortening due to premaxillary and maxillary hypoplasia; shortened snouts	Tobin et al. (2008)
	$Dync2h1^{b2b414Clo}$	*Dync2h1*	Micrognathia	Cecilia Lo to MGI
	$Evc^{-/-}$	*Evc*	Chondroplasia, hypodontia	Ruiz-Perez et al. (2007)
	$Fuz^{-/-}$	*Fuz*	Exencephaly; anophthalmia; encephalocoele; mandibular defects; micrognathia; high arched palate; enlarged maxillary processes	Tabler et al. (2013) and Gray et al. (2009)
	$Hippi^{-/-}$	*Ift57*	Exencephaly; abnormal cranial flexure; hypertelorism; abnormal maxillary processes; small arches	Houde et al. (2006)
	$Ift52^{hypo}$	*Ift52*	Single median nostril; fused maxillary prominences; exencephaly	Liu, Wang, and Niswander (2005)
	$Ift122^{-/-}$	*Ift122*	Exencephaly; anencephaly; altered eye and branchial arch development; defects of the ventral portion of the head	Cortellino et al. (2009)
	$Ift172^{avc1}$	*Ift172*	Hydrocephalus; abnormal facies; shortened muzzle; cleft lip and palate	Friedland-Little et al. (2011)

Continued

Table 1 Animal models for craniofacial ciliopathies—cont'd

Model system	Mutant/morphant name	Gene	Craniofacial phenotype	References
	Kif3a$^{f/f}$; *Wnt1-Cre*	*Kif3a* in cranial NCCs	Anterior cranium occultum; bifid nasal septum; aglossia; severe hypertelorism; dental abnormalities	Brugmann et al. (2010)
	Kif7$^{-/-}$	*Kif7*	Cleft lip and palate	Cecilia Lo to MGI
	Mks$^{-/-}$	*Mks1*	Occipital meningoencephalocele; hypomineralization and/or splitting of the frontal, parietal, and superaoccipital bones; hydrocephaly; cleft palate; pointy snout	Weatherbee et al. (2009)
	Ofd1$^{-/-}$	*Ofd1*	Severe cleft palate; shortened skull and facial regions	Ferrante et al. (2006)
	Orpk	*Ift88*	Dental abnormalities; micrognathia; increased gonial angle; poorly differentiated branchial arches; minor clefting; disorganized/patent frontal sutures	Zhang et al. (2003), Murcia et al. (2000), and Ohazama et al. (2009)
	Rpgrip1l$^{-/-}$	*Rpgrip1l/Ftm*	Exencephaly, microphthalmia; rounded skull; cleft upper lip; micrognathia	Delous et al. (2007)
	Tmem67$^{-/-}$	*Tmem67*	Hypoplastic mandible, semilobar HPE, occipital meningocele, midbrain and/or forebrain exencephaly	Abdelhamed et al. (2013)
Chicken	*talpid*^{2}	*c2cd3*	Short, wide upper beak; cleft lip/palate; hypo- or aglossia	Chang et al. (2014)
	talpid^{3}	*talpid3*	Ocular hypotelorism; frontonasal hypoplasia; fused maxillary prominences; micrognathia	Yin et al. (2009)

Zebrafish	*Bbs morphants*	*bbs4, bbs6, bbs8*	Mandibular dysplasia; severe craniofacial reduction; anteroneurocranium hypoplasia	Tobin et al. (2008)
	Ift mutants	*ift57, ift88, ift172*	Mild dysmorphology of craniofacial cartilages	Lunt et al. (2009)
	Ift80 morphant	*ift80*	Abnormal anterior neurocranial development; midline fusion	Beales et al. (2007)
	Ift122 morphant	*ift122*	Reduced ocular development; distended cranium with hydrocephalus; otolith defects	Walczak-Sztulpa et al. (2010)
	Ofd1 morphant	*ofd1*	Abnormal blunting of the jaw; disorganization of Meckel's cartilage cells	Ferrante et al. (2009)

Ciliopathic models are grouped according to model system.

analysis was performed on BBS mouse models and both *Bbs4* and *Bbs6* homozygous mutants have larger ratios of mid-face width to height and shortened snouts (Tobin et al., 2008). These craniofacial defects strongly suggest NCC patterning and/or development is frequently affected in ciliary mutants.

Other mutations seem to reveal broader requirements for some ciliary genes. The *Ift88*^{Tg737Rpw} mice have been described with poorly differentiated branchial arches, minor clefting, supernumerary teeth, and disorganized and patent frontal sutures (Murcia et al., 2000; Zhang et al., 2003). Loss of *Ofd1* leads to both severe cleft palate and a shortened skull (Ferrante et al., 2006). *Evc* mutants have abnormal incisors (Ruiz-Perez et al., 2007). These findings would suggest that ciliary proteins execute a number of roles in craniofacial development, but that loss of particular ciliary genes leads to specific phenotypes. This latter conclusion resonates with our model that cilia in distinct regions of an embryo have possibly unique roles in specific developmental signaling events and not all cilia have the same complement of proteins. However, we must also consider that mouse experiments on different genetic backgrounds may incorporate different modifier genes, which further complicate comparisons between ciliary gene loss-of-function phenotypes. This is consistent with the known multi-allelic effects in ciliopathies (Davis et al., 2011).

Additional insight into the role of cilia in NCC development can be gleaned from a recent study of the *Fuzzy* mouse (Tabler et al., 2013). This is a unique mutant that exhibits a high arched palate, a phenotype identified in a high proportion of BBS patients (Beales, Elcioglu, Woolf, Parker, & Flinter, 1999). A detailed developmental analysis presented in this study uncovered an enlarged maxillary process and aberrant patterns of NCC migration in E9.5 embryos. Interestingly, a NCC specific ablation of *Fuzzy* with *Wnt1-Cre* resulted in a true cleft phenotype without the associated maxillary hypoplasia or hypoplastic tongue (Tabler et al., 2013). Together, these findings are consistent with our model that cilia are actually performing discrete tasks in different regions of the embryo and future studies will need to incorporate this idea to fully explain the role of cilia in craniofacial NCCs.

While mouse models allow for insights into mammalian ciliary function, both fish and chicken models have proven very valuable, especially when evaluating early NCC development. Several fish mutants and morphants exist allowing for rapid evaluation of NCC development, migration, and differentiation (Beales et al., 2007; Ferrante et al., 2009; Lunt et al., 2009; Tobin et al., 2008; Walczak-Sztulpa et al., 2010; Table 1). In addition, two separate, naturally occurring ciliopathic models have been identified in the chicken. The causal genetic elements for both the *talpid*^{2} and *talpid*^{3} mutants have recently been identified (Chang et al., 2014; Yin et al., 2009). The accessibility of the chicken embryo, and its availability for culturing and transplanting tissues, provides promise for careful analysis of how primary cilia participate in each individual step of NCC development.

4.2. Human craniofacial ciliopathies

Given the data generated in animal models, it is not surprising there is a growing class of human craniofacial ciliopathies with features indicative of aberrant NCC development. Some years ago, Baker and Beales reviewed 102 human conditions of known or possible ciliopathies. Of the 102 conditions, the craniofacial complex was primarily affected in up to 30% of cases (Zaghloul & Brugmann, 2011). Although primary cilia extend from almost all cells in the body, the clinical features observed in patients with ciliopathies highlight the importance of primary cilia in development of certain types of tissue. The phenotypes associated with human craniofacial ciliopathies were similar to those observed in animal models: midline, cephalic, oral, and dental anomalies (Zaghloul & Brugmann, 2011; Table 2). Three such examples of human craniofacial ciliopathies include Meckel–Gruber syndrome (MKS), BBS, and Oro-facial-digital syndrome (OFD). MKS presents with cleft lip in 20% of cases and cleft lip and palate in 45% of cases. Clefting is accompanied by tongue malformations including ankyloglossia, bifid tongue, and anterior marginal hamartomas (Fried, Liban, Lurie, Friedman, & Reisner, 1971; Moerman, Verbeken, Fryns, Goddeeris, & Lauweryns, 1982; Rehder & Labbe, 1981). BBS is associated with oligodontia, hypodontia, and numerous other dental abnormalities (Lofterod, Riise, Skuseth, & Storhaug, 1990). Additional studies have reported characteristic midfacial defects hypothesized to be due to improper hedgehog induced NCC migration (Tobin et al., 2008). A host of other defects related to NCC development/migration including hypopigmentation, heart defects, middle ear defects, choanal atresia, and Hirschsprung disease have also been reported in BBS patients (Gorlin, Cohen, & Levin, 1990). OFD is characterized by a wide nose with a broad, flat nasal bridge, hypertelorism, cleft lip and palate, glossal deformities, hyperplastic frenula, the growth of noncancerous tumors and hypodontia. OFD has long been linked to mutations in *OFD1* (Ferrante et al., 2001; Gurrieri, Franco, Toriello, & Neri, 2007). Recently however, a new

Table 2 Human craniofacial ciliopathies and associated phenotypes

Syndrome	Gene	Craniofacial phenotype
Bardet–Biedl syndrome	*BBS1-12, MKS1, MKS3*	Mid-face shortening and flattening; nasal bridge hypoplasia; reduced length/bulbosity of the nasal tip; mild retrognathia
Cranioectodermal dysplasia/ Sensenbrenner syndrome	*IFT122, IFT144, WDR35*	Sagittal craniosynostosis; epicanthal folds; hypodontia or microdontia; everted lip; multiple oral frenula; high arched palate; skeletal and ectodermal anomalies
Ellis–van Creveld syndrome	*EVC1, EVC2*	Hypertrophy labiogingival frenulum; upper lip abnormalities; presence of teeth at birth; microdontic teeth
Joubert syndrome	*MKS3, NPHP6, RPGRIP1L, AH1, ARL13B, CSPP1*	Prominent forehead; high rounded eyebrows; epicanthal folds; ptosis; upturned nose with evident nostrils; hypertelorism; open mouth and tongue protrusion with rhythmic tongue motions
Lowe syndrome	*OCRL1*	Frontal bossing; retrognathia; high arched palate; deep set eyes; full cheeks; maxillary prognathism; malocclusion
Meckel–Gruber syndrome	*MKS1, MKS3, TMEM231, TCTN2, CC2D2A*	Encephalocele; cleft lip and palate
Oro-facial-digital syndrome	*OFD1, C2CD3*	Facial asymmetry; hypertelorism; micrognathia; broadened nasal ridge; hypoplasia of the malar bones and nasal alar cartilages; frontal bossing; pseudocleft; cleft palate; hamartomas of the tongue; bifid tongue; hyperplastic oral frenula; up-slanting palpebral fissures

subtype of OFD has been identified and linked to mutations in *C2CD3* (Thauvin-Robinet et al., 2014); the same gene recently found responsible for the *talpid*^{2} avian mutant (Chang et al., 2014). The rapid identification of new ciliary genes and ciliopathic models suggests that many yet to be classified syndromes may in fact be ciliopathies. Furthermore, the high frequency of dysmorphic faces suggests NCCs are indeed highly sensitive to the loss of primary cilia.

5. BEYOND THE FACE: TRUNK NCCs ARE ALSO AFFECTED BY THE LOSS OF PRIMARY CILIA

Although NCCs make substantial and vital contributions to the face, they also contribute to many other lineages and primary cilia have been implicated in proper development of many of these populations as well. The best-studied NCC lineage outside the face is arguably the cardiac NCC. Cardiac NCCs normally make a significant contribution to the outflow tract of the heart, which separates the pulmonary and systemic circulations.

An elegant example of a role for primary cilia in cardiac development is the zebrafish *cobblestone* (*cbs*) mutant. *Cbs* is a hypomorphic allele of *Ift88* and homozygous mutants present with a number of the most common congenital heart defects seen in newborn humans including: ventricular and atrial septal defects, and persistent truncus arteriosus (failure of outflow tract to divide). Further study of this mutation showed that the cardiac NCCs migrated normally into the heart, but the NCC-derived outflow tract exhibited a significant reduction in expression of *Bmp4* (Willaredt, Gorgas, Gardner, & Tucker, 2012). In our own studies, we have noted frequent edema in late-embryonic stage *Ttc21b*^{aln} mutants and abnormal patterns of NCC migration into the heart (Fig. 4). Thus, primary cilia are important for the molecular signaling that guides outflow tract development.

Another NCC derivative affected by loss of cilia is the enteric nervous system, the main division of the autonomic nervous system that innervates the gut. A failure of NCC migration to the gut or differentiation into enteric ganglia results in Hirschsprung disease (aganglionic megacolon; McKeown, Stamp, Hao, & Young, 2013), a disease in which variable portions of the gut lose the ability to relax. Hirschsprung disease is frequently present in both BBS patients (Amiel et al., 2001) and ciliopathic animal models. Zebrafish *bbs8* morphants have decreased vagal NCC and incomplete enteric nervous

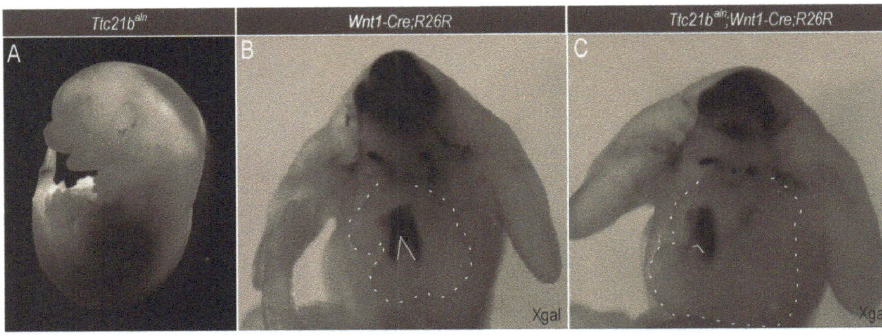

Figure 4 Loss of primary cilia affects cardiac NCCs. *Ttc21b*^{aln} mutant has cardiac NCC defects. (A) Whole-mount image of an e14.5 *Ttc21b*^{aln} mutant with systemic edema suggestive of cardiopulmonary dysfunction. (B) Xgal staining in an e11.5 *Wnt1-Cre;R26R* embryo to trace the NCC lineage. (C) Xgal staining in an e11.5 *Ttc21b*^{aln}*Wnt1-Cre; R26R* embryo. Dotted white lines outline the heart. Solid white lines indicate the degree of cardiac NCC migration into the septating outflow tract.

system formation leading to disordered gut motility (Tobin et al., 2008). Furthermore, reduction of *ADP ribosylation factor-like 6 interacting protein 1* (*Arl6ip1*) in zebrafish produces disorganized cilia, dampened *sonic hedgehog* (*Shh*) signaling and subsequent reduction in the number of enteric ganglia (Tu, Yang, Huang, & Tsai, 2012).

Various other NCC derivatives extend cilia and are affected by a loss of primary cilia. Melanocytes extend cilia, both *in vitro* (Le Coz, Benmerah, & Larue, 2014) and *in vivo* (Wandel, Steigleder, & Bodeux, 1984) and zebrafish *bbs8* morphants exhibit a pigmentation phenotype (Tobin et al., 2008). Primary cilia have been identified on corneal endothelial cells in a number of mammals (Collin & Barry Collin, 2004; Doughty, 1998) and Polaris mutants have disorganized patterning of the corneal endothelium (Blitzer et al., 2011). Although no studies have reported defects in the dorsal root ganglia or sympathetic ganglia, we have noted defects in both of these lineages in the *Ttc21b*^{aln} mutants (Stottmann, unpublished).

6. NCCs UTILIZE PRIMARY CILIA FOR TISSUE–TISSUE INTERACTIONS

6.1. Structures that require reciprocal signaling are disrupted in ciliary mutants

Throughout the course of development NCCs interact with cells from all three germ layers to form epithelial specializations or discrete organs. Many structures that require reciprocal signaling with NCCs for their formation are disrupted in ciliary mutants. Prime examples of structures or tissues that require molecular input from NCCs are teeth, whiskers, and the tongue.

Odontogenesis, or tooth formation, requires reciprocal interactions between the oral ectoderm and the underlying NCC-derived mesenchyme (Thesleff, 2006). Many human ciliopathies including Ellis–van Creveld syndrome, OFD syndrome, and Cranioectodermal dysplasia exhibit abnormal dentition (Gurrieri et al., 2007; Mostafa, Temtamy, el-Gammal, & Mazen, 2005; Walczak-Sztulpa et al., 2010; Table 2). Studies in mice provide the genetic advantage of tissue-specific conditional deletions of primary cilia. *Ift88* hypomorphs develop supernumerary teeth accompanied by an expansion of *Gli1* expression; however, when a keratin (K)-5-Cre driver is used to delete *Ift88* from epithelial cells, no dental abnormalities are observed (Ohazama et al., 2009). This suggests that cilia are required specifically on NCCs for proper tooth formation. This hypothesis is supported by data from *Kif3a*$^{f/f}$;*Wnt1-Cre* mutants where Shh and Wnt activity in the dental lamina is aberrant (Brugmann et al., 2010), and tooth development is abnormal (Fig. 5). Together, these data suggest that primary cilia are essential mediators of NCC-epithelial signaling during odontogenesis.

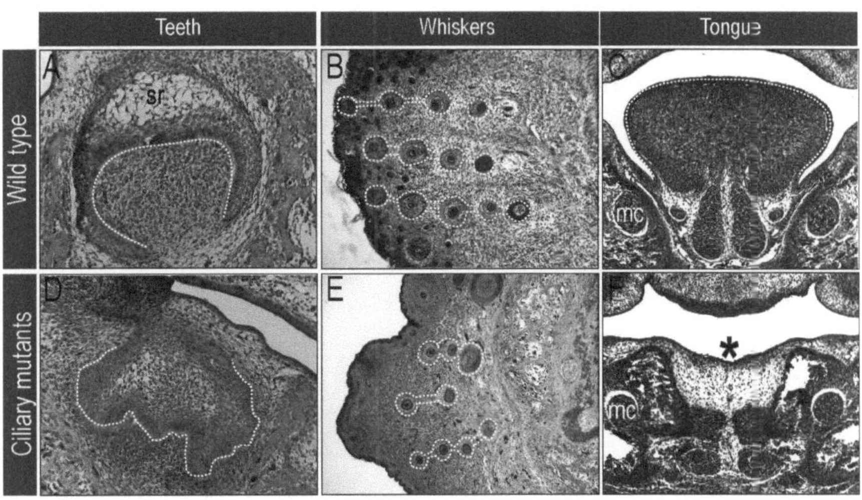

Figure 5 Defects in craniofacial specialization. Images of wild-type (A–C) and ciliary mutant (D–F) craniofacial structures. Ciliary mutants are unable to properly form teeth (compare dotted white lines in A to D), whiskers are abnormally spaced and have aberrant morphology (compare dotted white lines in B to E) and some ciliary mutants can have aglossia (*) (compare dotted white lines in C to asterisk in F). sr, stellate reticulum; mc, Meckel's cartilage. (See the color plate.)

A curious example of a ciliopathy causing dental abnormalities is the avian ciliopathic mutant, *talpid*^{2}, which develops crocodilian-like teeth (Brugmann et al., 2010; Chang et al., 2014; Harris, Hasso, Ferguson, & Fallon, 2006). Modern birds lack teeth. However, some studies suggest that the avian oral ectoderm still possesses instructive signals for forming teeth. Thus, it was hypothesized that avian dentition does not form because the competent NCC-derived mesenchyme is not in a position to receive instructive signaling from the ectoderm (Kollar & Fisher, 1980; Kollar & Mina, 1991). In *talpid*^{2}, the oral/aboral boundary is shifted such that competent mesenchyme is in a position to receive instructive signal for the ectoderm, promoting specification of dentition (Harris et al., 2006). It is unclear the exact role primary cilia play in this context, as this study was performed prior to the identification of *talpid*^{2} as ciliopathic mutant, but these data further suggest that reciprocal signaling between oral ectoderm and NCC-derived mesenchyme require primary cilia.

Similar to ondontogenesis, whisker formation requires reciprocal interactions between NCC-derived mesenchyme and the overlying ectoderm (Chiang et al., 1999). When NCCs lack primary cilia, as in the *Kif3a*$^{f/f}$; *Wnt1-Cre* mutants, whiskers still form; however, they are reduced in number and disorganized (Brugmann et al., 2010). Instead of forming in a distinct spatial pattern, they are irregularly spaced (Fig. 5). While an in-depth analysis of whisker formation has not been carried out in cilia mutants, one might expect defects in primary cilia-dependent reciprocal signaling between overlying ectoderm and NCC-derived mesenchyme to at least partially explain this anomaly.

The developing tongue is comprised of three different cell types: epithelial cells encase the tongue, NCC-derived mesenchyme contribute to the vasculature and connective tissue of the tongue, and the mesodermal-derived myoblasts give rise to the muscles of the tongue (Parada, Han, & Chai, 2012). Mesoderm-derived myoblast progenitors associate with NCC-derived mesenchyme once they both populate the tongue anlage (Huang, Zhi, Izpisua-Belmonte, Christ, & Patel, 1999). In order for muscle differentiation to occur, there must be reciprocal interaction between NCCs and the mesoderm. Signaling molecules such as Fgf4, Fgf6, and BMP4 are important for myogenic cell proliferation and differentiation in the tongue and when BMP signaling is reduced intrinsically in NCCs, as is the case for *Alk5*$^{f/f}$;*Wnt1-Cre* mutants, tongue muscles are disrupted (Han, Zhao, Li, Pelikan, & Chai, 2014). Similarly, Shh signaling is also essential for tongue development and both *Gli3*$^{-/-}$ and *Smo*$^{n/c}$;*Wnt1-Cre* mice display abnormal tongue development (Huang, Goudy, Ketova, Litingtung, & Chiang, 2008; Jeong, Mao, Tenzen, Kottmann, & McMahon, 2004).

Several human ciliopathies and animal models encompass tongue malformations. Human ciliopathies that often present with glossal defects include OFD syndrome and Joubert syndrome (Table 2). Patients with OFD syndrome will frequently have a bifid tongue, sometimes accompanied by hamartomas (Gurrieri et al., 2007). With Joubert syndrome, a small percentage of patients will develop soft tissue tumors of the tongue (Pellegrino, Lensch, Muenke, & Chance, 1997). Others exhibit a groove in the midline of the tongue or frequently protrude their tongue rhythmically (Dahlstrom, Cookman, & Jain, 2000; Parisi, 2009). Numerous animal ciliopathic models have glossal defects. The avian *talpid*^{2} mutant exhibits hypoglossia/aglossia (Chang et al., 2014) and both the *Kif3a*$^{f/f}$;*Wnt1-Cre* and *Ift88*$^{f/f}$;*Wnt1-Cre* mouse mutants display aglossia (Brugmann et al., 2010; Fig. 3). It is tempting to hypothesize that primary cilia on NCCs are required to interpret the Shh signal from the epithelium or other molecular signals from the mesoderm important for specification/patterning of the tongue. It should be noted, however, that other ciliary mutants such as *Ttc21b*$^{f/f}$;*Wnt1-Cre* have normal tongues (Fig. 3). Further investigation is needed to understand why loss of certain ciliary proteins results in aglossia, while others do not.

In addition to craniofacial defects, many ciliopathic patients also have brain defects, including anencephaly, exencephaly, and hydrocephaly (Carter et al., 2012). Similar brain phenotypes are observed in several ciliary mutants including *Dynch2*^{GT}, *Hippi*$^{-/-}$, *hennin*, *Ift122*$^{-/-}$, *Msk*, and *Fuz* murine mutants (Caspary et al., 2007; Cortellino et al., 2009; Gray et al., 2009; Houde et al., 2006; May et al., 2005; Weatherbee et al., 2009). Interestingly, brain abnormalities can also occur when primary cilia are lost on NCCs. *Kif3a*$^{f/f}$;*Wnt1-Cre* mutants present with agenesis of the corpus callosum (Brugmann et al., 2010). In this case, the expression pattern of *Wnt1* in dorsal neural stem cells does not preclude that these phenotypes are due to loss of cilia within the neural plate. However, the possibility remains that loss of cilia affecting NCC derivatives, such as smooth muscle cells or pericytes, could contribute to the aberrant brain development observed in these mutants.

7. THE ROLE FOR PRIMARY CILIA IN MOLECULAR SIGNAL TRANSDUCTION

The data outlined above show NCCs require primary cilia for several developmental processes and intracellular interactions. As previously discussed, primary cilia are essential for the reception and processing of multiple signaling pathways. Below, we briefly summarize what is known regarding the role of primary cilia and transduction of various signaling pathways important in NCC development.

7.1. Sonic hedgehog

NCC survival, proliferation, and patterning are dependent on Shh signaling (Ahlgren & Bronner-Fraser, 1999; Ahlgren, Thakur, & Bronner-Fraser, 2002; Brito, Teillet, & Le Douarin, 2008; Marcucio & Helms, 2002; Wada et al., 2005) and defects in these processes are reflected in the tissues of the face in ciliopathy patients (Brugmann et al., 2010; Tobin et al., 2008). Identification of the primary cilium as a signaling hub for the Hedgehog pathway came from seminal experiments reporting that anterograde and retrograde IFT proteins in the cilium are required for Shh signal propagation (Huangfu & Anderson, 2005; Huangfu et al., 2003). Subsequent analyses showed that Ptch1 and Smo receptors localize to the primary cilium and that in the presence of ligand, Ptch1 is sequestered from the cilium facilitating an accumulation of Smo and activation of signaling (Corbit et al., 2005; Rohatgi, Milenkovic, & Scott, 2007). For a number of years, it was believed that there was a strictly linear relationship between Shh signaling and primary cilia (e.g., the loss of functional cilia would result in a loss of Shh function). In some mutants, however, the exact opposite is the case; loss of cilia causes a gain of Shh signaling (Brugmann et al., 2010; Cortellino et al., 2009; Tran et al., 2008). How might Shh signaling be activated as a consequence of losing ciliary proteins? To begin to hypothesize how this seemingly paradoxical event could occur, we must first understand the relationship between the cilia and components of the Shh pathway.

The most well-known links between primary cilia and the Shh pathway are (1) that the receptors of the pathway are localized to the cilium and (2) that IFT proteins are involved in trafficking and processing of Gli2 and Gli3 from full-length isoforms, into either activator or repressor forms (Haycraft et al., 2005). In the absence of Shh, Ptch1 is present in cilia and Gli3 (and Gli2) may translocate at low levels in and out of cilia, being phosphorylated by PKA. Presumably, IFT proteins are then involved in shuttling Gli3FL to the pericentrosomal proteasome where it is processed into Gli3R and subsequently translocated to the nucleus to inhibit *Gli1* and *Gli2* (Wen et al., 2010). In the presence of Shh, Smo displaces Ptch1 in cilia, thereby inhibiting PKA phosphorylation and Gli3FL (and Gli2FL) accumulates at the distal tips. Thus, one mechanism by which a ciliary defect could result in a gain of Shh signaling is via aberrant proteolytic processing of Gli3 (the predominant repressor of the pathway). Deregulation of Gli3R production would result in an increased Gli3FL:Gli3R ratio and subsequent increase in Shh activity. This scenario has been observed in numerous IFT mouse models (Haycraft et al., 2005; Huangfu & Anderson, 2005; Liu et al., 2005; May et al., 2005; Tran et al., 2008; Willaredt et al., 2008).

Whereas Gli processing is the most commonly examined aspect in ciliary mutants, the identification of novel ciliary proteins implicates additional mechanisms that could explain how the loss of cilia could cause a gain of Shh activity. For example, recent work identified G-protein-coupled receptor 161 (Gpr161) as a ciliary protein that negatively regulates Shh activity via promoting PKA-mediated generation of GliR (Mukhopadhyay et al., 2013). With respect to Gpr161 function, loss of cilia could alleviate negative regulation of the pathway and thus also cause a gain of Shh activity. Furthermore, some studies have suggested "noncanonical" and ciliary -independent mechanisms for Shh signaling that do not function through Gli activity (Robbins, Fei, & Riobo, 2012). In light of this rapidly emerging data, caution must be taken when making generalizations about the role of primary cilia and a signaling pathway because many of these mechanisms appear to be tissue-specific (Zaghloul & Brugmann, 2011).

7.2. Wnt

Wnt signaling is required throughout NCC development. Various Wnt ligands are expressed in developing neural tissue at stages that strongly suggest a role for Wnt proteins in NCC induction in several species (Rogers, Jayasena, Nie, & Bronner, 2012). Furthermore, Wnts also have reiterative roles later in NCC development during migration (Mayor & Theveneau, 2014), proliferation (Brugmann et al., 2007), and differentiation (Brault et al., 2001). Initial evidence for the involvement of the canonical *Wnt* pathway in ciliary function came from *in vitro* experiments demonstrating that *Inversin* localized to the cilia and physically interacted with *Wnt* pathway component *Dishevelled* (*Dvl*) (Otto et al., 2003; Simons et al., 2005; Watanabe et al., 2003). One study reported that a knockdown of Kif3a led to increased Wnt signaling, suggesting that the primary cilium restricts canonical Wnt signaling in mouse embryos, primary fibroblasts, and embryonic stem cells (Corbit et al., 2008). Other studies focused on different ciliary proteins have supported this hypothesis. MEFs isolated from retrograde protein *Ttc21b* mutant mice showed an approximately fivefold increase in reporter activity in response to *Wnt3a*-conditioned media (Tran et al., 2014). Similarly, loss of primary cilia in the mammary ducts of *Ift88* mutant mice resulted in increased Wnt signaling (McDermott, Liu, Tlsty, & Pazour, 2010).

Despite this plethora of data suggesting a role for cilia in Wnt signal transduction, there is significant evidence supporting the contrary. Zebrafish *ift88* mutants fail to make cilia and display characteristic Shh-related phenotypes, but expression of Wnt targets is unchanged (Huang & Schier, 2009). Another study found that *Axin2* expression levels and readouts from Wnt reporters were indistinguishable between wild-type and a number of ciliary mutants (Ocbina, Tuson, & Anderson, 2009). Furthermore, wild-type and ciliary mutant MEFs displayed no difference in response to exposure to both Wnt3a and Wnt5a ligands (Ocbina et al., 2009). In NCCs specifically, the loss of Kif3a did not appear to alter Wnt activity in early facial mesenchyme (Brugmann et al., 2010). Despite these opposing findings, several studies still support a role for ciliary proteins in restraining Wnt signaling. In light of these seemingly contradictory data, one possibility to consider is that the perturbations of the Wnt pathway observed in ciliary mutants may not be direct. Gli proteins can bind to and directly affect transcription of Wnt effector genes. Thus, ciliary-dependent disruptions of posttranslational Gli processing could secondarily affect Wnt activity. Clearly the complexity of the issue requires further investigation regarding Wnt signaling and cilia involvement in NCCs and other cell types.

7.3. Fibroblast growth factor

The FGF signaling pathway has been associated with growth and development of the facial prominences, the facial skeleton, and craniofacial musculature (Creuzet, Schuler, Couly, & Le Douarin, 2004; Macatee et al., 2003; Szabo-Rogers, Geetha-Loganathan, Nimmagadda, Fu, & Richman, 2008; Trumpp, Depew, Rubenstein, Bishop, & Martin, 1999). Similar to Wnts, many FGFs are expressed in spatiotemporal patterns that suggest a role in NCC induction and patterning (Rogers et al., 2012). More recently, a novel role for FGF signaling in ciliary assembly has been hypothesized (Neugebauer, Amack, Peterson, Bisgrove, & Yost, 2009). *Fgfr1* morphant zebrafish exhibited reduced expression of the ciliary genes *foxj1* and *rfx2* as well as diminished *ift88* expression, indicating that FGF signaling may exert its effects on cilia by regulating IFT (Neugebauer et al., 2009).

Furthermore, cilia length was affected in *fgf24/fgf8* double mutants, suggesting that cilia length may be regulated by functional redundancy across FGF ligands. Additional support for this hypothesis comes from studies that report decreased length and number of cilia as a result of knocking down FGF target genes *ier2* and *fibp1* (Hong & Dawid, 2009). These experiments raise the possibility that FGF-mediated craniofacial malformations could be cilia-related, and in fact, recent studies have begun to substantiate this mechanism (Tabler et al., 2013).

7.4. Platelet-derived growth factor

Aberrant *PDGF* signaling has been linked with cleft lip/palate in humans (Smith & Tallquist, 2010). *Pdgfa/Pdgfc* double mutants develop cleft face and cranial bone defects (Ding et al., 2004). Mice deficient in *Pdgfra* in the neural crest lineage display palatal fusion defects as well as nasal septation and various other anomalies of bone and cartilage (Morrison-Graham, Schatteman, Bork, Bowen-Pope, & Weston, 1992; Soriano, 1997; Tallquist & Soriano, 2003). Primary cilia are required for PDGFα signaling (Schneider et al., 2005, 2009). PDGFα receptor localizes to primary cilium in growth-arrested NIH3T3 cells. Upon signaling activation, the downstream protein Mek1/2 is phosphorylated and localizes to primary cilium. This behavior is lost in cilia mutant $Ift88^{orpk}$ fibroblasts. Moreover, the primary cilium is required for PDGF-AA induced directional cell migration during wound healing (Schneider et al., 2010).

7.5. Notch

Experiments across several species have implicated a role for Notch signaling in NCC induction (Cornell & Eisen, 2002, 2005; Endo, Osumi, & Wakamatsu, 2002, 2003; Ezratty et al., 2011; Glavic, Silva, Aybar, Bastidas, & Mayor, 2004; Leitch, Lodh, Prieto-Echague, Badano, & Zaghloul, 2014; Mead & Yutzey, 2012; Noisa et al., 2014). Notch signaling is required for the induction of NCC-specifier genes in premigratory neural crest-like cells (pNCC) derived from human pluripotent stem cells (hPSC) (Noisa et al., 2014). Activated *Notch1* intracellular domain (*NICD1*) directly binds to and induces expression of NCC-specifier genes. Interestingly, Notch activity is also required for the maintenance of the pNCC fate and the suppression of Notch signaling led to the generation of NCC-derived neurons (Noisa et al., 2014). Furthermore, in avian models, gain and loss of Notch both result in loss of NCC markers (Endo et al., 2002, 2003).

The link between Notch signaling and primary cilia is just beginning to emerge. The Notch3 receptor is enriched in the ciliary membrane while Presenilin-2 (the enzyme responsible for Notch cleavage and activation) is localized to the ciliary basal body. *bbs1* and *bbs4* morphant zebrafish have increased expression of Notch targets (Leitch et al., 2014), and loss of *Bbs1* and *Bbs4* caused Notch receptor to accumulate in late endosomes and subsequent lysosomal degradation of Notch receptor was impaired. These experiments strongly suggest that the Notch pathway utilizes the cilium for signal transduction. Uncovering the relationship between Notch signaling, cilia, and NCC development holds great promise for identifying novel molecular mechanism of ciliopathies and possible avenues for therapeutic intervention.

8. CONCLUSIONS

The resurgence of interest in the primary cilium has been impressive and offers new perspectives on how cells sense and respond to their molecular environments. Continued study of primary cilia is vital to both basic science and biomedicine. For basic research, more insight into cilia function will be necessary for truly understanding the molecular mechanisms of signal pathway transduction. For the biomedical community, ciliary research will identify both the basis for ciliopathies as well as potential avenues for therapeutic intervention. Relative to the field of NCC biology, understanding the role primary cilia play during ontogenic development of NCCs could provide novel insights into the etiology of specific neurocristopathies.

Although the field has made significant strides in the last decade, there is still much to learn. For example, why do ciliopathies have differing phenotypes? Why do some ciliopathies have severe craniofacial phenotypes while others do not? Are primary cilia unique depending upon which type of cell they extend from or which signaling pathway they are geared to transduce? In order to address these questions, cilia must be carefully examined both at different time-points during development and on different cell types (Irigoin & Badano, 2011).

The data summarized here emphasize the shift in the current paradigm that primary cilia are homogeneous organelles and that ciliary function can be uniformly defined across multiple tissues and organ systems. Although a plethora of phenotypes have been reported in various ciliopathic animal models (Zaghloul & Brugmann, 2011; Table 1), the perception still exists that disruption of cilia equates simply to a loss of Shh function. Our work, as well as that of many others, suggests that not only will different tissues interpret the loss of cilia in a unique manner, but also the same tissue in different molecular environments will have a unique response to ciliary disruption. It will be of great interest to multiple fields to see what is in store for primary cilia research in the next decade.

REFERENCES

Abdelhamed, Z. A., Wheway, G., Szymanska, K., Natarajan, S., Toomes, C., Inglehearn, C., et al. (2013). Variable expressivity of ciliopathy neurological phenotypes that encompass Meckel-Gruber syndrome and Joubert syndrome is caused by complex de-regulated ciliogenesis, Shh and Wnt signalling defects. *Human Molecular Genetics*, *22*(7), 1358–1372.

Afzelius, B. A. (1976). A human syndrome caused by immotile cilia. *Science*, *193*(4250), 317–319.

Ahlgren, S. C., & Bronner-Fraser, M. (1999). Inhibition of sonic hedgehog signaling in vivo results in craniofacial neural crest cell death. *Current Biology*, *9*(22), 1304–1314.

Ahlgren, S. C., Thakur, V., & Bronner-Fraser, M. (2002). Sonic hedgehog rescues cranial neural crest from cell death induced by ethanol exposure. *Proceedings of the National Academy of Sciences of the United States of America*, *99*(16), 10476–10481.

Amiel, J., Espinosa-Parrilla, Y., Steffann, J., Gosset, P., Pelet, A., Prieur, M., et al. (2001). Large-scale deletions and SMADIP1 truncating mutations in syndromic Hirschsprung disease with involvement of midline structures. *American Journal of Human Genetics*, *69*(6), 1370–1377.

Avidor-Reiss, T., Maer, A. M., Koundakjian, E., Polyanovsky, A., Keil, T., Subramaniam, S., et al. (2004). Decoding cilia function: Defining specialized genes required for compartmentalized cilia biogenesis. *Cell*, *117*(4), 527–539.

Basten, S. G., & Giles, R. H. (2013). Functional aspects of primary cilia in signaling, cell cycle and tumorigenesis. *Cilia*, *2*(1), 6.

Beales, P. L., Bland, E., Tobin, J. L., Bacchelli, C., Tuysuz, B., Hill, J., et al. (2007). IFT80, which encodes a conserved intraflagellar transport protein, is mutated in Jeune asphyxiating thoracic dystrophy. *Nature Genetics*, *39*(6), 727–729.

Beales, P. L., Elcioglu, N., Woolf, A. S., Parker, D., & Flinter, F. A. (1999). New criteria for improved diagnosis of Bardet-Biedl syndrome: Results of a population survey. *Journal of Medical Genetics*, *36*(6), 437–446.

Beales, P., & Jackson, P. K. (2012). Cilia—The prodigal organelle. *Cilia*, *1*(1), 1.

Blitzer, A. L., Panagis, L., Gusella, G. L., Danias, J., Mlodzik, M., & Iomini, C. (2011). Primary cilia dynamics instruct tissue patterning and repair of corneal endothelium. *Proceedings of the National Academy of Sciences of the United States of America*, *108*(7), 2819–2824.

Bloodgood, R. A. (2009). From central to rudimentary to primary: The history of an underappreciated organelle whose time has come. The primary cilium. *Methods in Cell Biology*, *94*, 3–52.

Blum, J. J., Hayes, A., Whisnant, C. C., & Rosen, G. (1977). Effect of spin-labeled maleimide on 14S and 30S dyneins in solution and on demembranated ciliary axonemes. *Biochemistry*, *16*(9), 1937–1943.

Brault, V., Moore, R., Kutsch, S., Ishibashi, M., Rowitch, D. H., McMahon, A. P., et al. (2001). Inactivation of the beta-catenin gene by Wnt1-Cre-mediated deletion results in dramatic brain malformation and failure of craniofacial development. *Development*, *128*(8), 1253–1264.

Brito, J. M., Teillet, M. A., & Le Douarin, N. M. (2008). Induction of mirror-image supernumerary jaws in chicken mandibular mesenchyme by Sonic Hedgehog-producing cells. *Development*, *135*(13), 2311–2319.

Bronner, M. E., & LeDouarin, N. M. (2012). Development and evolution of the neural crest: An overview. *Developmental Biology*, *366*(1), 2–9.

Brooks, E. R., & Wallingford, J. B. (2012). Control of vertebrate intraflagellar transport by the planar cell polarity effector Fuz. *The Journal of Cell Biology*, *198*(1), 37–45.

Brugmann, S. A., Allen, N. C., James, A. W., Mekonnen, Z., Madan, E., & Helms, J. A. (2010). A primary cilia-dependent etiology for midline facial disorders. *Human Molecular Genetics*, *19*(8), 1577–1592.

Brugmann, S. A., Moore, R., Kutsch, S., Ishibashi, M., Rowitch, D. H., McMahon, A. P., et al. (2007). Wnt signaling mediates regional specification in the vertebrate face. *Development*, *134*(18), 3283–3295.

Carter, C. S., Vogel, T. W., Zhang, Q., Seo, S., Swiderski, R. E., Moninger, T. O., et al. (2012). Abnormal development of NG2 + PDGFR-alpha + neural progenitor cells leads to neonatal hydrocephalus in a ciliopathy mouse model. *Nature Medicine*, *18*(12), 1797–1804.

Caspary, T., Larkins, C. E., & Anderson, K. V. (2007). The graded response to Sonic Hedgehog depends on cilia architecture. *Developmental Cell*, *12*(5), 767–778.

Chang, C. F., Schock, E. N., O'Hare, E. A., Dodgson, J., Cheng, H. H., Muir, W. M., et al. (2014). The cellular and molecular etiology of the craniofacial defects in the avian ciliopathic mutant talpid2. *Development*, *141*(15), 3003–3012.

Chiang, C., Swan, R. Z., Grachtchouk, M., Bolinger, M., Litingtung, Y., Robertson, E. K., et al. (1999). Essential role for Sonic hedgehog during hair follicle morphogenesis. *Developmental Biology*, *205*(1), 1–9.

Collin, S. P., & Barry Collin, H. (2004). Primary cilia in vertebrate corneal endothelial cells. *Cell Biology International*, *28*(2), 125–130.

Corbit, K. C., Aanstad, P., Singla, V., Norman, A. R., Stainier, D. Y., & Reiter, J. F. (2005). Vertebrate smoothened functions at the primary cilium. *Nature*, *437*(7061), 1018–1021.

Corbit, K. C., Shyer, A. E., Dowdle, W. E., Gaulden, J., Singla, V., Chen, M. H., et al. (2008). Kif3a constrains beta-catenin-dependent Wnt signalling through dual ciliary and non-ciliary mechanisms. *Nature Cell Biology*, *10*(1), 70–76.

Cornell, R. A., & Eisen, J. S. (2002). Delta/Notch signaling promotes formation of zebrafish neural crest by repressing Neurogenin 1 function. *Development*, *129*(11), 2639–2648.

Cornell, R. A., & Eisen, J. S. (2005). Notch in the pathway: The roles of notch signaling in neural crest development. *Seminars in Cell & Developmental Biology*, *16*(6), 663–672.

Cortellino, S., Wang, C., Wang, B., Bassi, M. R., Caretti, E., Champeval, D., et al. (2009). Defective ciliogenesis, embryonic lethality and severe impairment of the Sonic Hedgehog pathway caused by inactivation of the mouse complex A intraflagellar transport gene Ift122/Wdr10, partially overlapping with the DNA repair gene Med1/Mbd4. *Developmental Biology*, *325*(1), 225–237.

Creuzet, S., Schuler, B., Couly, G., & Le Douarin, N. M. (2004). Reciprocal relationships between Fgf8 and neural crest cells in facial and forebrain development. *Proceedings of the National Academy of Sciences of the United States of America*, *101*(14), 4843–4847.

Dahlstrom, J. E., Cookman, J., & Jain, S. (2000). Joubert syndrome: An affected female with bilateral colobomata. *Pathology*, *32*(4), 283–285.

Davis, E. E., Zhang, Q., Liu, Q., Diplas, B. H., Davey, L. M., Hartley, J., et al. (2011). TTC21B contributes both causal and modifying alleles across the ciliopathy spectrum. *Nature Genetics*, *43*(3), 189–196.

Dawe, H. R., Farr, H., & Gull, K. (2007). Centriole/basal body morphogenesis and migration during ciliogenesis in animal cells. *Journal of Cell Science*, *120*(Pt. 1), 7–15.

Delaval, B., Bright, A., Lawson, N. D., & Doxsey, S. (2011). The cilia protein IFT88 is required for spindle orientation in mitosis. *Nature Cell Biology*, *13*(4), 461–468.

Delous, M., Baala, L., Salomon, R., Laclef, C., Vierkotten, J., Tory, K., et al. (2007). The ciliary gene RPGRIP1L is mutated in cerebello-oculo-renal syndrome (Joubert syndrome type B) and Meckel syndrome. *Nature Genetics*, *39*(7), 875–881.

Ding, H., Wu, X., Boström, H., Kim, I., Wong, N., Tsoi, B., et al. (2004). A specific requirement for PDGF-C in palate formation and PDGFR-alpha signaling. *Nature Genetics*, *36*(10), 1111–1116.

Doughty, M. J. (1998). Changes in cell surface primary cilia and microvilli concurrent with measurements of fluid flow across the rabbit corneal endothelium ex vivo. *Tissue & Cell*, *30*(6), 634–643.

Endo, Y., Osumi, N., & Wakamatsu, Y. (2002). Bimodal functions of Notch-mediated signaling are involved in neural crest formation during avian ectoderm development. *Development*, *129*(4), 863–873.

Endo, Y., Osumi, N., & Wakamatsu, Y. (2003). Deltex/Dtx mediates NOTCH signaling in regulation of Bmp4 expression in cranial neural crest formation during avian development. *Development, Growth & Differentiation*, *45*(3), 241–248.

Ezratty, E. J., Stokes, N., Chai, S., Shah, A. S., Williams, S. E., & Fuchs, E. (2011). A role for the primary cilium in Notch signaling and epidermal differentiation during skin development. *Cell*, *145*(7), 1129–1141.

Feistel, K., & Blum, M. (2006). Three types of cilia including a novel 9+4 axoneme on the notochordal plate of the rabbit embryo. *Developmental Dynamics*, *235*(12), 3348–3358.

Ferrante, M. I., Giorgio, G., Feather, S. A., Bulfone, A., Wright, V., Ghiani, M., et al. (2001). Identification of the gene for oral-facial-digital type I syndrome. *American Journal of Human Genetics*, *68*(3), 569–576.

Ferrante, M. I., Romio, L., Castro, S., Collins, J. E., Goulding, D. A., Stemple, D. L., et al. (2009). Convergent extension movements and ciliary function are mediated by ofd1, a zebrafish orthologue of the human oral-facial-digital type 1 syndrome gene. *Human Molecular Genetics*, *18*(2), 289–303.

Ferrante, M. I., Zullo, A., Barra, A., Bimonte, S., Messaddeq, N., Studer, M., et al. (2006). Oral-facial-digital type I protein is required for primary cilia formation and left-right axis specification. *Nature Genetics*, *38*(1), 112–117.

Fried, K., Liban, E., Lurie, M., Friedman, S., & Reisner, S. H. (1971). Polycystic kidneys associated with malformations of the brain, polydactyly, and other birth defects in newborn sibs. A lethal syndrome showing the autosomal-recessive pattern of inheritance. *Journal of Medical Genetics*, *8*(3), 285–290.

Friedland-Little, J. M., Hoffmann, A. D., Ocbina, P. J., Peterson, M. A., Bosman, J. D., Chen, Y., et al. (2011). A novel murine allele of Intraflagellar Transport Protein 172 causes a syndrome including VACTERL-like features with hydrocephalus. *Human Molecular Genetics*, *20*(19), 3725–3737.

Garcia-Castro, M. I., Marcelle, C., & Bronner-Fraser, M. (2002). Ectodermal Wnt function as a neural crest inducer. *Science*, *13*, 13.

Glavic, A., Silva, F., Aybar, M. J., Bastidas, F., & Mayor, R. (2004). Interplay between Notch signaling and the homeoprotein Xiro1 is required for neural crest induction in Xenopus embryos. *Development*, *131*(2), 347–359.

Gorlin, R. J., Cohen, M. M., & Levin, L. S. (Eds.), (1990). Syndromes of the head and neck. *Oxfordmonographs on medical genetics*, *1* (3rd ed.). New York: Oxford University Press.

Goto, H., Inoko, A., & Inagaki, M. (2013). Cell cycle progression by the represion of primary cilia formation in proliferating cells. *Cellular and Molecular Life Sciences*, *70*(20), 3893–3905.

Gray, R. S., Abitua, P. B., Wlodarczyk, B. J., Szabo-Rogers, H. L., Blanchard, O., Lee, I., et al. (2009). The planar cell polarity effector Fuz is essential for targeted membrane trafficking, ciliogenesis and mouse embryonic development. *Nature Cell Biology*, *11*(10), 1225–1232.

Gurrieri, F., Franco, B., Toriello, H., & Neri, G. (2007). Oral-facial-digital syndromes: Review and diagnostic guidelines. *American Journal of Medical Genetics Part A*, *143A*(24), 3314–3323.

Han, Y. G., Kwok, B. H., & Kernan, M. J. (2003). Intraflagellar transport is required in Drosophila to differentiate sensory cilia but not sperm. *Current Biology*, *13*(19), 1679–1686.

Han, A., Zhao, H., Li, J., Pelikan, R., & Chai, Y. (2014). ALK5-mediated TGF-beta signaling in neural crest cells controls craniofacial muscle development via tissue-tissue interactions. *Molecular and Cellular Biology*, *34*, 3120–3131.

Harris, M. P., Hasso, S. M., Ferguson, M. W., & Fallon, J. F. (2006). The development of archosaurian first-generation teeth in a chicken mutant. *Current Biology*, *16*(4), 371–377.

Haycraft, C. J., Banizs, B., Aydin-Son, Y., Zhang, Q., Michaud, E. J., & Yoder, B. K. (2005). Gli2 and Gli3 localize to cilia and require the intraflagellar transport protein polaris for processing and function. *PLoS Genetics*, *1*(4), e53.

He, F., & Soriano, P. (2013). A critical role for PDGFRalpha signaling in medial nasal process development. *PLoS Genetics*, *9*(9), e1003851.

Hernandez-Hernandez, V., Pravincumar, P., Diaz-Font, A., May-Simera, H., Jenkins, D., Knight, M., et al. (2013). Bardet-Biedl syndrome proteins control the cilia length through regulation of actin polymerization. *Human Molecular Genetics*, *22*(19), 3858–3868.

Hoey, D. A., Downs, M. E., & Jacobs, C. R. (2012). The mechanics of the primary cilium: An intricate structure with complex function. *Journal of Biomechanics*, *45*(1), 17–26.

Hong, S. K., & Dawid, I. B. (2009). FGF-dependent left-right asymmetry patterning in zebrafish is mediated by Ier2 and Fibp1. *Proceedings of the National Academy of Sciences of the United States of America*, *106*(7), 2230–2235.

Houde, C., Dickinson, R. J., Houtzager, V. M., Cullum, R., Montpetit, R., Metzler, M., et al. (2006). Hippi is essential for node cilia assembly and Sonic hedgehog signaling. *Developmental Biology*, *300*(2), 523–533.

Huang, X., Goudy, S. L., Ketova, T., Litingtung, Y., & Chiang, C. (2008). Gli3-deficient mice exhibit cleft palate associated with abnormal tongue development. *Developmental Dynamics*, *237*(10), 3079–3087.

Huang, P., & Schier, A. F. (2009). Dampened hedgehog signaling but normal Wnt signaling in zebrafish without cilia. *Development*, *136*(18), 3089–3098.

Huang, R., Zhi, Q., Izpisua-Belmonte, J. C., Christ, B., & Patel, K. (1999). Origin and development of the avian tongue muscles. *Anatomy and Embryology*, *200*(2), 137–152.

Huangfu, D., & Anderson, K. V. (2005). Cilia and Hedgehog responsiveness in the mouse. *Proceedings of the National Academy of Sciences of the United States of America*, *102*(32), 11325–11330.

Huangfu, D., Liu, A., Rakeman, A. S., Murcia, N. S., Niswander, L., & Anderson, K. V. (2003). Hedgehog signalling in the mouse requires intraflagellar transport proteins. *Nature*, *426*(6962), 83–87.

Irigoin, F., & Badano, J. L. (2011). Keeping the balance between proliferation and differentiation: The primary cilium. *Current Genomics*, *12*(4), 285–297.

Jeong, J., Mao, J., Tenzen, T., Kottmann, A. H., & McMahon, A. P. (2004). Hedgehog signaling in the neural crest cells regulates the patterning and growth of facial primordia. *Genes & Development*, *18*(8), 937–951.

Jones, C., Roper, V. C., Foucher, I., Qian, D., Banizs, B., Petit, C., et al. (2008). Ciliary proteins link basal body polarization to planar cell polarity regulation. *Nature Genetics*, *40*(1), 69–77.

Kjaer, K. W., Hansen, B. F., Keeling, J. W., Nolting, D., & Kjaer, I. (1999). Malformations of cranial base structures and pituitary gland in prenatal Meckel syndrome. *APMIS, 107*(10), 937–944.

Kobayashi, T., & Dynlacht, B. D. (2011). Regulating the transition from centriole to basal body. *The Journal of Cell Biology, 193*(3), 435–444.

Kollar, E. J., & Fisher, C. (1980). Tooth induction in chick epithelium: Expresion of quiescent genes for enamel synthesis. *Science, 207*(4434), 993–995.

Kollar, E. J., & Mina, M. (1991). Role of the early epithelium in the patterning of the teeth and Meckel's cartilage. *Journal of Craniofacial Genetics and Developmental Biology, 11*(4), 223–228.

Konno, A., Setou, M., & Ikegami, K. (2012). Ciliary and flagellar structure and function—their regulations by posttranslational modifications of axonemal tubulin. *International Review of Cell and Molecular Biology, 294*, 133–170.

Kuriyama, S., & Mayor, R. (2008). Molecular analysis of neural crest migration. *Philosophical Transactions of the Royal Society of London, Series B: Biological Sciences, 363*(1495), 1349–1362.

LaBonne, C., & Bronner-Fraser, M. (1999). Molecular mechanisms of neural crest formation. *Annual Review of Cell and Developmental Biology, 15*, 81–112.

Le Coz, M., Benmerah, A., & Larue, L. (2014). Quiescent melanocytes form primary cilia. *Experimental Dermatology, 23*(6), 426–427.

Leitch, C. C., Lodh, S., Prieto-Echague, V., Badano, J. L., & Zaghloul, N. A. (2014). Basal body proteins regulate Notch signaling through endosomal trafficking. *Journal of Cell Science, 127*(Pt. 11), 2407–2419.

Liu, A., Wang, B., & Niswander, L. A. (2005). Mouse intraflagellar transport proteins regulate both the activator and repressor functions of Gli transcription factors. *Development, 132*(13), 3103–3111.

Lofterod, B., Riise, R., Skuseth, T., & Storhaug, K. (1990). Laurence-Moon-Bardet-Biedl syndrome. *Nordisk Medicin, 105*(5), 146–148.

Lunt, S. C., Haynes, T., & Perkins, B. D. (2009). Zebrafish ift57, ift88, and ift172 intraflagellar transport mutants disrupt cilia but do not affect hedgehog signaling. *Developmental Dynamics, 238*(7), 1744–1759.

Macatee, T. L., Macatee, T. L., Hammond, B. P., Arenkiel, B. R., Francis, L., Frank, D. U., et al. (2003). Ablation of specific expression domains reveals discrete functions of ectoderm- and endoderm-derived FGF8 during cardiovascular and pharyngeal development. *Development, 130*(25), 6361–6374.

Madhivanan, K., Mukherjee, D., & Aguilar, R. C. (2012). Lowe syndrome: Between primary cilia assembly and Rac1-mediated membrane remodeling. *Communicative & Integrative Biology, 5*(6), 641–644.

Marchant, L., Linker, C., Ruiz, P., Guerrero, N., & Mayor, R. (1998). The inductive properties of mesoderm suggest that the neural crest cells are specified by a BMP gradient. *Developmental Biology, 198*(2), 319–329.

Marcucio, R., & Helms, J. A. (2002). Shh-dependent signals from forebrain direct facial development. Unpublished observations.

May, S. R., Ashique, A. M., Karlen, M., Wang, B., Shen, Y., Zarbalis, K., et al. (2005). Loss of the retrograde motor for IFT disrupts localization of Smo to cilia and prevents the expression of both activator and repressor functions of Gli. *Developmental Biology, 287*(2), 378–389.

Mayor, R., Guerrero, N., & Martinez, C. (1997). Role of FGF and noggin in neural crest induction. *Developmental Biology, 189*(1), 1–12.

Mayor, R., & Theveneau, E. (2014). The role of the non-canonical Wnt-planar cell polarity pathway in neural crest migration. *The Biochemical Journal, 457*(1), 19–26.

McDermott, K. M., Liu, B. Y., Tlsty, T. D., & Pazour, G. J. (2010). Primary cilia regulate branching morphogenesis during mammary gland development. *Current Biology*, *20*(8), 731–737.

McGrath, J., Somlo, S., Makova, S., Tian, X., & Brueckner, M. (2003). Two populations of node monocilia initiate left-right asymmetry in the mouse. *Cell*, *114*(1), 61–73.

McKeown, S. J., Stamp, L., Hao, M. M., & Young, H. M. (2013). Hirschsprung disease: A developmental disorder of the enteric nervous system. *Wiley Interdisciplinary Reviews Developmental Biology*, *2*(1), 113–129.

Mead, T. J., & Yutzey, K. E. (2012). Notch pathway regulation of neural crest cell development in vivo. *Developmental Dynamics*, *241*(2), 376–389.

Miyoshi, K., Kasahara, K., Miyazaki, I., & Asanuma, M. (2009). Lithium treatment elongates primary cilia in the mouse brain and in cultured cells. *Biochemical and Biophysical Research Communications*, *388*(4), 757–762.

Miyoshi, K., Kasahara, K., Miyazaki, I., & Asanuma, M. (2011). Factors that influence primary cilium length. *Acta Medica Okayama*, *65*(5), 279–285.

Moerman, P., Verbeken, E., Fryns, J. P., Goddeeris, P., & Lauweryns, J. M. (1982). The Meckel syndrome. Pathological and cytogenetic observations in eight cases. *Human Genetics*, *62*(3), 240–245.

Molla-Herman, A., Ghossoub, R., Blisnick, T., Meunier, A., Serres, C., Silbermann, F., et al. (2010). The ciliary pocket: An endocytic membrane domain at the base of primary and motile cilia. *Journal of Cell Science*, *123*(Pt. 10), 1785–1795.

Monsoro-Burq, A. H., Fletcher, R. B., & Harland, R. M. (2003). Neural crest induction by paraxial mesoderm in Xenopus embryos requires FGF signals. *Development*, *130*(14), 3111–3124.

Morrison-Graham, K., Schatteman, G. C., Bork, T., Bowen-Pope, D. F., & Weston, J. A. (1992). A PDGF receptor mutation in the mouse (Patch) perturbs the development of a non-neuronal subset of neural crest-derived cells. *Development*, *115*(1), 133–142.

Mostafa, M. I., Temtamy, S. A., el-Gammal, M. A., & Mazen, I. M. (2005). Unusual pattern of inheritance and orodental changes in the Ellis-van Creveld syndrome. *Genetic Counseling*, *16*(1), 75–83.

Mukhopadhyay, S., Wen, X., Ratti, N., Loktev, A., Rangell, L., Scales, S. J., et al. (2013). The ciliary G-protein-coupled receptor Gpr161 negatively regulates the Sonic hedgehog pathway via cAMP signaling. *Cell*, *152*(1–2), 210–223.

Murcia, N. S., Richards, W. G., Yoder, B. K., Mucenski, M. L., Dunlap, J. R., & Woychik, R. P. (2000). The Oak Ridge Polycystic Kidney (orpk) disease gene is required for left-right axis determination. *Development*, *127*(11), 2347–2355.

Nachury, M. V., Loktev, A. V., Zhang, Q., Westlake, C. J., Peränen, J., Merdes, A., et al. (2007). A core complex of BBS proteins cooperates with the GTPase Rab8 to promote ciliary membrane biogenesis. *Cell*, *129*(6), 1201–1213.

Nauli, S. M., Alenghat, F. J., Luo, Y., Williams, E., Vassilev, P., Li, X., et al. (2003). Polycystins 1 and 2 mediate mechanosensation in the primary cilium of kidney cells. *Nature Genetics*, *33*(2), 129–137.

Neesen, J., Kirschner, R., Ochs, M., Schmiedl, A., Habermann, B., Mueller, C., et al. (2001). Disruption of an inner arm dynein heavy chain gene results in asthenozoospermia and reduced ciliary beat frequency. *Human Molecular Genetics*, *10*(11), 1117–1128.

Neugebauer, J. M., Amack, J. D., Peterson, A. G., Bisgrove, B. W., & Yost, H. J. (2009). FGF signalling during embryo development regulates cilia length in diverse epithelia. *Nature*, *458*(7238), 651–654.

Noisa, P., Lund, C., Kanduri, K., Lund, R., Lähdesmäki, H., Lahesmaa, R., et al. (2014). Notch signaling regulates the differentiation of neural crest from human pluripotent stem cells. *Journal of Cell Science*, *127*(Pt. 9), 2083–2094.

Ocbina, P. J., Tuson, M., & Anderson, K. V. (2009). Primary cilia are not required for normal canonical Wnt signaling in the mouse embryo. *PLoS One*, *4*(8), e6839.

Ohazama, A., Haycraft, C. J., Seppala, M., Blackburn, J., Ghafoor, S., Cobourne, M., et al. (2009). Primary cilia regulate Shh activity in the control of molar tooth number. *Development*, *136*(6), 897–903.

Osborn, D. P., Roccasecca, R. M., McMurray, F., Hernandez-Hernandez, V., Mukherjee, S., Barroso, I., et al. (2014). Loss of FTO antagonises Wnt signaling and leads to developmental defects associated with ciliopathies. *PLoS One*, *9*(2), e87662.

Otto, E. A., Schermer, B., Obara, T., O'Toole, J. F., Hiller, K. S., Mueller, A. M., et al. (2003). Mutations in INVS encoding inversin cause nephronophthisis type 2, linking renal cystic disease to the function of primary cilia and left-right axis determination. *Nature Genetics*, *34*(4), 413–420.

Ou, Y., Ruan, Y., Cheng, M., Moser, J. J., Rattner, J. B., & van der Hoorn, F. A. (2009). Adenylate cyclase regulates elongation of mammalian primary cilia. *Experimental Cell Research*, *315*(16), 2802–2817.

Parada, C., Han, D., & Chai, Y. (2012). Molecular and cellular regulatory mechanisms of tongue myogenesis. *Journal of Dental Research*, *91*(6), 528–535.

Parisi, M. A. (2009). Clinical and molecular features of Joubert syndrome and related disorders. *American Journal of Medical Genetics Part C: Seminars in Medical Genetics*, *151C*(4), 326–340.

Park, T. J., Haigo, S. L., & Wallingford, J. B. (2006). Ciliogenesis defects in embryos lacking inturned or fuzzy function are associated with failure of planar cell polarity and Hedgehog signaling. *Nature Genetics*, *38*(3), 303–311.

Pellegrino, J. E., Lensch, M. W., Muenke, M., & Chance, P. F. (1997). Clinical and molecular analysis in Joubert syndrome. *American Journal of Medical Genetics*, *72*(1), 59–62.

Ray, K., Perez, S. E., Yang, Z., Xu, J., Ritchings, B. W., Steller, H., et al. (1999). Kinesin-II is required for axonal transport of choline acetyltransferase in Drosophila. *The Journal of Cell Biology*, *147*(3), 507–518.

Rehder, H., & Labbe, F. (1981). Prenatal morphology in Meckel's syndrome. *Prenatal Diagnosis*, *1*(3), 161–171.

Reiter, J. F., Blacque, O. E., & Leroux, M. R. (2012). The base of the cilium: Roles for transition fibres and the transition zone in ciliary formation, maintenance and compartmentalization. *EMBO Reports*, *13*(7), 608–618.

Robbins, D. J., Fei, D. L., & Riobo, N. A. (2012). The Hedgehog signal transduction network. *Science Signaling*, *5*(246), re6.

Robert, A., Margall-Ducos, G., Guidotti, J. E., Brégerie, O., Celati, C., Bréchot, C., et al. (2007). The intraflagellar transport component IFT88/polaris is a centrosomal protein regulating G1-S transition in non-ciliated cells. *Journal of Cell Science*, *120*(Pt. 4), 628–637.

Rogers, C. D., Jayasena, C. S., Nie, S., & Bronner, M. E. (2012). Neural crest specification: Tissues, signals, and transcription factors. *Wiley Interdisciplinary Review Developmental Biology*, *1*(1), 52–68.

Rohatgi, R., Milenkovic, L., & Scott, M. P. (2007). Patched1 regulates hedgehog signaling at the primary cilium. *Science*, *317*(5836), 372–376.

Ross, A. J., May-Simera, H., Eichers, E. R., Kai, M., Hill, J., Jagger, D. J., et al. (2005). Disruption of Bardet-Biedl syndrome ciliary proteins perturbs planar cell polarity in vertebrates. *Nature Genetics*, *37*(10), 1135–1140.

Ruiz-Perez, V. L., Blair, H. J., Rodriguez-Andres, M. E., Blanco, M. J., Wilson, A., Liu, Y. N., et al. (2007). Evc is a positive mediator of Ihh-regulated bone growth that localises at the base of chondrocyte cilia. *Development*, *134*(16), 2903–2912

Schneider, L., Cammer, M., Lehman, J., Nielsen, S. K., Guerra, C. F., Veland, I. R., et al. (2010). Directional cell migration and chemotaxis in wound healing response to

PDGF-AA are coordinated by the primary cilium in fibroblasts. *Cellular Physiology and Biochemistry*, *25*(2–3), 279–292.

Schneider, L., Clement, C. A., Teilmann, S. C., Pazour, G. J., Hoffmann, E. K., Satir, P., et al. (2005). PDGFRalphaalpha signaling is regulated through the primary cilium in fibroblasts. *Current Biology*, *15*(20), 1861–1866.

Schneider, L., Stock, C. M., Dieterich, P., Jensen, B. H., Pedersen, L. B., Satir, P., et al. (2009). The Na+/H+ exchanger NHE1 is required for directional migration stimulated via PDGFR-alpha in the primary cilium. *The Journal of Cell Biology*, *185*(1), 163–176.

Simons, M., Gloy, J., Ganner, A., Bullerkotte, A., Bashkurov, M., Krönig, C., et al. (2005). Inversin, the gene product mutated in nephronophthisis type II, functions as a molecular switch between Wnt signaling pathways. *Nature Genetics*, *37*(5), 537–543.

Smith, C. L., & Tallquist, M. D. (2010). PDGF function in diverse neural crest cell populations. *Cell Adhesion & Migration*, *4*(4), 561–566.

Soriano, P. (1997). The PDGF alpha receptor is required for neural crest cell development and for normal patterning of the somites. *Development*, *124*(14), 2691–2700.

Sorokin, S. (1962). Centrioles and the formation of rudimentary cilia by fibroblasts and smooth muscle cells. *The Journal of Cell Biology*, *15*, 363–377.

Szabo-Rogers, H. L., Geetha-Loganathan, P., Nimmagadda, S., Fu, K. K., & Richman, J. M. (2008). FGF signals from the nasal pit are necessary for normal facial morphogenesis. *Developmental Biology*, *318*(2), 289–302.

Tabler, J. M., Barrell, W. B., Szabo-Rogers, H. L., Healy, C., Yeung, Y., Perdiguero, E. G., et al. (2013). Fuz mutant mice reveal shared mechanisms between ciliopathies and FGF-related syndromes. *Developmental Cell*, *25*(6), 623–635.

Tallquist, M. D., & Soriano, P. (2003). Cell autonomous requirement for PDGFRalpha in populations of cranial and cardiac neural crest cells. *Development*, *130*(3), 507–518.

Tang, Z., Lin, M. G., Stowe, T. R., Chen, S., Zhu, M., Stearns, T., et al. (2013). Autophagy promotes primary ciliogenesis by removing OFD1 from centriolar satellites. *Nature*, *502*(7470), 254–257.

Tang, Z., Zhu, M., & Zhong, Q. (2014). Self-eating to remove cilia roadblock. *Autophagy*, *10*(2), 379–381.

Teddy, J. M., & Kulesa, P. M. (2004). In vivo evidence for short- and long-range cell communication in cranial neural crest cells. *Development*, *131*(24), 6141–6151.

Thauvin-Robinet, C., Lee, J. S., Lopez, E., Herranz-Pérez, V., Shida, T., Franco, B., et al. (2014). The oral-facial-digital syndrome gene C2CD3 encodes a positive regulator of centriole elongation. *Nature Genetics*, *46*(8), 905–911.

Thesleff, I. (2006). The genetic basis of tooth development and dental defects. *American Journal of Medical Genetics Part A*, *140*(23), 2530–2535.

Theveneau, E., & Mayor, R. (2012). Neural crest migration: Interplay between chemorepellents, chemoattractants, contact inhibition, epithelial-mesenchymal transition, and collective cell migration. *Wiley Interdisciplinary Reviews Developmental Biology*, *1*(3), 435–445.

Tobin, J. L., Di Franco, M., Eichers, E., May-Simera, H., Garcia, M., Yan, J., et al. (2008). Inhibition of neural crest migration underlies craniofacial dysmorphology and Hirschsprung's disease in Bardet-Biedl syndrome. *Proceedings of the National Academy of Sciences of the United States of America*, *105*(18), 6714–6719.

Tran, P. V., Talbott, G. C., Turbe-Doan, A., Jacobs, D. T., Schonfeld, M. P., Silva, L. M., et al. (2014). Downregulating hedgehog signaling reduces renal cystogenic potential of mouse models. *Journal of the American Society of Nephrology*, *25*(10), 2201–2212.

Tran, P. V., Haycraft, C. J., Besschetnova, T. Y., Turbe-Doan, A., Stottmann, R. W., Herron, B. J., et al. (2008). THM1 negatively modulates mouse sonic hedgehog signal transduction and affects retrograde intraflagellar transport in cilia. *Nature Genetics*, *40*(4), 403–410.

Trumpp, A., Depew, M. J., Rubenstein, J. L., Bishop, J. M., & Martin, G. R. (1999). Cre-mediated gene inactivation demonstrates that FGF8 is required for cell survival and patterning of the first branchial arch. *Genes and Development*, *13*(23), 3136–3148.

Tu, C. T., Yang, T. C., Huang, H. Y., & Tsai, H. J. (2012). Zebrafish arl6ip1 is required for neural crest development during embryogenesis. *PLoS One*, 7(3), e32899.

Valente, E. M., Logan, C. V., Mougou-Zerelli, S., Lee, J. H., Silhavy, J. L., Brancati, F., et al. (2010). Mutations in TMEM216 perturb ciliogenesis and cause Joubert, Meckel and related syndromes. *Nature Genetics*, *42*(7), 619–625.

Veland, I. R., Montjean, R., Eley, L., Pedersen, L. B., Schwab, A., Goodship, J., et al. (2013). Inversin/Nephrocystin-2 is required for fibroblast polarity and directional cell migration. *PLoS One*, *8*(4), e60193.

Verghese, E., Weidenfeld, R., Bertram, J. F., Ricardo, S. D., & Deane, J. A. (2008). Renal cilia display length alterations following tubular injury and are present early in epithelial repair. *Nephrology, Dialysis, Transplantation*, *23*(3), 834–841.

Verghese, E., Zhuang, J., Saiti, D., Ricardo, S. D., & Deane, J. A. (2011). In vitro investigation of renal epithelial injury suggests that primary cilium length is regulated by hypoxia-inducible mechanisms. *Cell Biology International*, *35*(9), 909–913.

Wada, N., Javidan, Y., Nelson, S., Carney, T. J., Kelsh, R. N., & Schilling, T. F. (2005). Hedgehog signaling is required for cranial neural crest morphogenesis and chondrogenesis at the midline in the zebrafish skull. *Development*, *132*(17), 3977–3988.

Walczak-Sztulpa, J., Eggenschwiler, J., Osborn, D., Brown, D. A., Emma, F., Klingenberg, C., et al. (2010). Cranioectodermal Dysplasia, Sensenbrenner syndrome, is a ciliopathy caused by mutations in the IFT122 gene. *American Journal of Human Genetics*, *86*(6), 949–956.

Wandel, A., Steigleder, G. K., & Bodeux, E. (1984). Primary cilia in cells of the epidermis and dermis. *Zeitschrift für Hautkrankheiten*, *59*(6), 389–392.

Watanabe, D., Saijoh, Y., Nonaka, S., Sasaki, G., Ikawa, Y., Yokoyama, T., et al. (2003). The left-right determinant Inversin is a component of node monocilia and other 9+0 cilia. *Development*, *130*(9), 1725–1734.

Weatherbee, S. D., Niswander, L. A., & Anderson, K. V. (2009). A mouse model for Meckel syndrome reveals Mks1 is required for ciliogenesis and Hedgehog signaling. *Human Molecular Genetics*, *18*(23), 4565–4575.

Wen, X., Lai, C. K., Evangelista, M., Hongo, J. A., de Sauvage, F. J., & Scales, S. J. (2010). Kinetics of hedgehog-dependent full-length Gli3 accumulation in primary cilia and subsequent degradation. *Molecular and Cellular Biology*, *30*(8), 1910–1922.

Wheatley, D. N. (1969). Cilia in cell-cultured fibroblasts. I. On their occurrence and relative frequencies in primary cultures and established cell lines. *Journal of Anatomy*, *105*(Pt. 2), 351–362.

Wheatley, D. N., Wang, A. M., & Strugnell, G. E. (1996). Expression of primary cilia in mammalian cells. *Cell Biology International*, *20*(1), 73–81.

Willaredt, M. A., Gorgas, K., Gardner, H. A., & Tucker, K. L. (2012). Multiple essential roles for primary cilia in heart development. *Cilia*, *1*(1), 23.

Willaredt, M. A., Hasenpusch-Theil, K., Gardner, H. A., Kitanovic, I., Hirschfeld-Warneken, V. C., Gojak, C. P., et al. (2008). A crucial role for primary cilia in cortical morphogenesis. *The Journal of Neuroscience*, *28*(48), 12887–12900.

Wong, S. Y., Seol, A. D., So, P. L., Ermilov, A. N., Bichakjian, C. K., Epstein E. H., Jr., et al. (2009). Primary cilia can both mediate and suppress Hedgehog pathway-dependent tumorigenesis. *Nature Medicine*, *15*(9), 1055–1061.

Yamazaki, H., Nakata, T., Okada, Y., & Hirokawa, N. (1995). KIF3A/B: A heterodimeric kinesin superfamily protein that works as a microtubule plus end-directed motor for membrane organelle transport. *The Journal of Cell Biology*, *130*(6), 1387–1399.

Yin, Y., Bangs, F., Paton, I. R., Prescott, A., James, J., Davey, M. G., et al. (2009). The Talpid3 gene (KIAA0586) encodes a centrosomal protein that is essential for primary cilia formation. *Development*, *136*(4), 655–664.

Zaghloul, N. A., & Brugmann, S. A. (2011). The emerging face of primary cilia. *Genesis*, *49*(4), 231–246.

Zhang, Q., Murcia, N. S., Chittenden, L. R., Richards, W. G., Michaud, E. J., Woychik, R. P., et al. (2003). Loss of the Tg737 protein results in skeletal patterning defects. *Developmental Dynamics*, *227*(1), 78–90.

Zimmermann, K. W. (1898). Beitrage zur Kenntnis einiger Drusen und Epithelien. In *52. Arch. mikr. Anat. u Entwick* (p. 552).

CHAPTER FIVE

Receptor Tyrosine Kinase Signaling: Regulating Neural Crest Development One Phosphate at a Time

Katherine A. Fantauzzo[1], Philippe Soriano
Department of Developmental and Regenerative Biology, Icahn School of Medicine at Mount Sinai, New York, USA
[1]Corresponding author: e-mail address: katherine.fantauzzo@mssm.edu

Contents

Abstract

Receptor tyrosine kinases (RTKs) bind to a subset of growth factors on the surface of cells and elicit responses with broad roles in developmental and postnatal cellular processes. Receptors in this subclass consist of an extracellular ligand-binding domain, a single

Current Topics in Developmental Biology, Volume 111
ISSN 0070-2153
http://dx.doi.org/10.1016/bs.ctdb.2014.11.005

transmembrane domain, and an intracellular domain harboring a catalytic tyrosine kinase and regulatory sequences that are phosphorylated either by the receptor itself or by various interacting proteins. Once activated, RTKs bind signaling molecules and recruit effector proteins to mediate downstream cellular responses through various intracellular signaling pathways. In this chapter, we highlight the role of a subset of RTK families in regulating the activity of neural crest cells (NCCs) and the development of their derivatives in mammalian systems. NCCs are migratory, multipotent cells that can be subdivided into four axial populations, cranial, cardiac, vagal, and trunk. These cells migrate throughout the vertebrate embryo along defined pathways and give rise to unique cell types and structures. Interestingly, individual RTK families often have specific functions in a subpopulation of NCCs that contribute to the diversity of these cells and their derivatives in the mammalian embryo. We additionally discuss current methods used to investigate RTK signaling, including genetic, biochemical, large-scale proteomic, and biosensor approaches, which can be applied to study intracellular signaling pathways active downstream of this receptor subclass during NCC development.

1. INTRODUCTION

Growth factors encompass an expansive range of proteins, cytokines, and hormones that bind to receptors on the surface of cells and stimulate various cellular activities. A subset of these factors signal through membrane-spanning receptor tyrosine kinases (RTKs) and elicit wide-ranging responses with broad roles in developmental and postnatal cellular processes, such as proliferation, growth, survival, apoptosis, adhesion, migration, differentiation, metabolism, cell-cycle progression, transcription, RNA processing, protein synthesis, vesicle trafficking, and autophagy. All RTKs share a basic structural framework consisting of an extracellular ligand-binding domain, a single transmembrane domain, and an intracellular domain harboring a catalytic tyrosine kinase (Fig. 1) and regulatory sequences that are phosphorylated either by the receptor itself or by various interacting proteins. RTKs are activated by ligand binding that, in the case of most receptors, induces receptor dimerization and promotes tyrosine kinase activity. In some instances, such as for the fibroblast growth factor (FGF) receptor, interactions with additional extracellular proteins including heparan sulfate proteoglycans are required for maximal receptor dimerization and activation (reviewed in Lemmon & Schlessinger, 2010; Schlessinger, 2000).

Tyrosine autophosphorylation sites in the noncatalytic portions of the receptor intracellular domains act as docking sites for signaling molecules

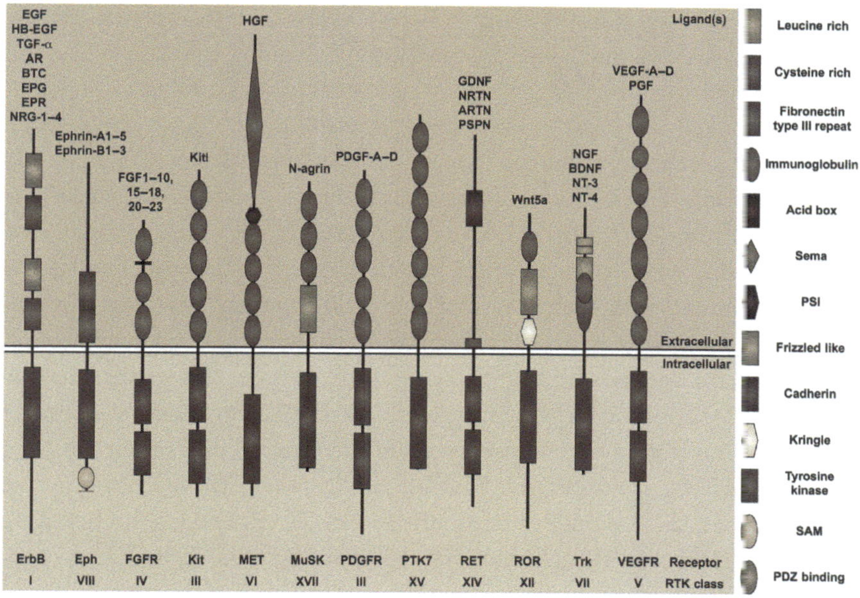

Figure 1 Schematic representation of the 12 RTK families involved in mammalian NCC development. The family name and RTK class are listed below each receptor, while the interacting ligands are listed above. Structural domains in the extracellular and intracellular portions of the receptors are depicted according to the key at right. (See the color plate.)

containing phosphotyrosine recognition motifs such as Src homology 2 (SH2) or phosphotyrosine-binding (PTB) domains. SH2 domain-containing interacting proteins include a subset with intrinsic enzymatic activity, including Src family tyrosine kinases, phosphatidylinositol 3-kinase (PI3K), the tyrosine phosphatase SHP-2, phospholipase Cγ (PLCγ), and GTPase-activating protein (GAP); the signal transducer and activator of transcription (Stat) family of transcription factors; and adaptor proteins such as Shc, growth factor receptor-bound proteins (Grb) 2/7/10/14, Crk/Crkl, and Nck. PTB domain-containing proteins that bind RTKs include FGF receptor substrate (Frs) 2/3 and insulin receptor substrate (IRS) 1/2. Once bound, these molecules employ various additional domains (pleckstrin homology, FYVE, SH3, WW, PDZ, etc.) to establish multiprotein signaling complexes. Upon recruitment, effector proteins are activated by several mechanisms, such as membrane translocation, conformational changes, and/or tyrosine phosphorylation, and subsequently mediate downstream cellular responses through a variety of intracellular signaling pathways (reviewed in Lemmon & Schlessinger, 2010; Schlessinger, 2000).

Despite the fact that the various RTKs interact with similar subsets of signaling molecules and utilize an overlapping network of intracellular signaling pathways to stimulate a range of cellular activities, biological specificity is introduced at several levels to generate a unique response downstream of receptor activation. First, the expression of individual growth factors and their corresponding RTKs is neither ubiquitous nor continuous within the embryo, thereby localizing their activity to specific tissues and/or timepoints during development. Similarly, the inherent binding properties of individual RTKs combined with enrichment of interacting proteins and their effectors at particular sites together dictate the unique combination of signaling molecules that are engaged by the various RTKs. Second, the requirement in some cases for additional interactions with extracellular proteins to optimize receptor dimerization and activation adds an extra layer of constraint for a subset of RTKs, for example, FGFR. Additionally, the ability of some RTK families, such as the erythroblastic leukemia viral oncogene homolog (ErbB) receptor and the platelet-derived growth factor (PDGF) receptor families, to induce signaling downstream of both homodimeric and heterodimeric receptor complexes may impart distinct effects on cellular behavior. Finally, differences in the strength and duration of signaling pathway activation induced by various RTKs have been shown to alter downstream biological responses (reviewed in Lemmon & Schlessinger, 2010; Schlessinger, 2000).

Here, we will discuss the role of a subset of RTK families in mediating the activity of neural crest cells (NCCs) and the development of their derivatives in mammalian systems, with a particular emphasis on their role in the mouse embryo (Table 1). NCCs are migratory, multipotent cells that play a critical role in vertebrate development. During mammalian embryogenesis, NCCs arise at the border of the neural ectoderm, undergo an epithelial to mesenchymal transition, and subsequently delaminate from the cranial neural folds or dorsal neural tube. They can be subdivided into four axial populations, cranial, cardiac, vagal, and trunk, which migrate throughout the embryo along defined pathways and contribute to diverse derivatives (Fig. 2). Cranial, or cephalic, NCCs originate from the forebrain to the hindbrain, which is segmented into seven transient neuroepithelial rhombomeres, and populate the frontonasal prominence and pharyngeal arches 1–4. These cells give rise to the bone and cartilage of the frontonasal skeleton and cartilages of the jaw, middle ear, hyoid, and thyroid. Cranial NCCs additionally generate smooth muscle, tendons, connective tissue, melanocytes, and cranial sensory ganglia of the peripheral nervous system

Table 1 RTKs contributing to murine NCC development

Family	Receptor	Mouse model NCC phenotype(s)	Function
ErbB receptors	EGFR	Craniofacial abnormalities (cleft palate, misshapen snout, micrognathia, and abnormal Meckel's cartilage development)	
		Defective cardiac semilunar valvulogenesis	
	ErbB2	Hypoplastic cranial sensory and sympathetic ganglia, loss of sensory and motor neurons, and absent Schwann cells in the peripheral nerves	Migration
		Absence of cardiac ventricular trabeculae	
	ErbB3	Hypoplastic cranial, dorsal root, sympathetic, and enteric ganglia; absent Schwann cells in sensory and motor neurons	
		Absence of mesenchyme in cardiac cushions; hypoplastic cardiac valves	
	ErbB4	Misprojections of cranial sensory ganglia	Migration
		Absence of cardiac ventricular trabeculae	
Eph receptors	EphA4	Coronal synostosis	Migration
FGF receptors	FGFR1	Craniofacial abnormalities (cleft palate, cleft lip, micrognathia, abnormal tooth bud development, inner ear malformations, and hypoplastic hyoid)	Survival and migration
	FGFR2	Craniofacial abnormalities (domed head, dysmorphic cranial base, craniosynostosis, cleft palate, misshapen snout, abnormal tooth bud development, and inner ear malformations)	Survival and migration
Kit receptor	Kit	Defective pigmentation	Survival and migration
		Absent small- and medium-diameter sensory neurons in dorsal root ganglia	Differentiation

Continued

Table 1 RTKs contributing to murine NCC development—cont'd

Family	Receptor	Mouse model NCC phenotype(s)	Function
MET receptor	MET	Ectopic melanoblasts	Survival and differentiation
MuSK receptor	MuSK	Impaired segmental trunk NCC migration	Migration
PDGF receptors	PDGFRα	Craniofacial bone hypoplasia (cleft palate; cleft lip; shortening of the premaxilla; and malformations of the basisphenoid, alisphenoid, and pterygoid bones)	Migration and proliferation
		Thymus hypoplasia	
		Cardiac outflow tract defects	Migration
		Defective pigmentation	
	PDGFRβ	Cardiac ventricular septal defects	Migration
PTK7 receptor	PTK7	Misshapen cranial and dorsal root ganglia	Distribution
		Cardiac outflow tract defects; cardiac ventricular septal defects	
RET receptor	RET	Absent enteric ganglia	Survival, migration, proliferation, and differentiation
		Neuronal loss in dorsal root and sympathetic ganglia	
ROR receptors	Ror2	Craniofacial bone hypoplasia (cleft palate, short snout, and micrognathia)	Migration and proliferation
		Cardiac ventricular septal defects	
Trk receptors	TrkA	Neuronal loss in trigeminal, dorsal root, and sympathetic ganglia	
		Defective thymus development	Differentiation
	TrkB	Neuronal loss in trigeminal and dorsal root ganglia	Survival
		Defective enteric glial cell development	
	TrkC	Neuronal loss in trigeminal and dorsal root ganglia	
		Cardiac atrial and ventricular septal defects; valvular defects	

The RTK families and individual receptors are listed, along with the NCC phenotype(s) of their corresponding mouse models and their functional role(s), if known, in regulating NCC activity.

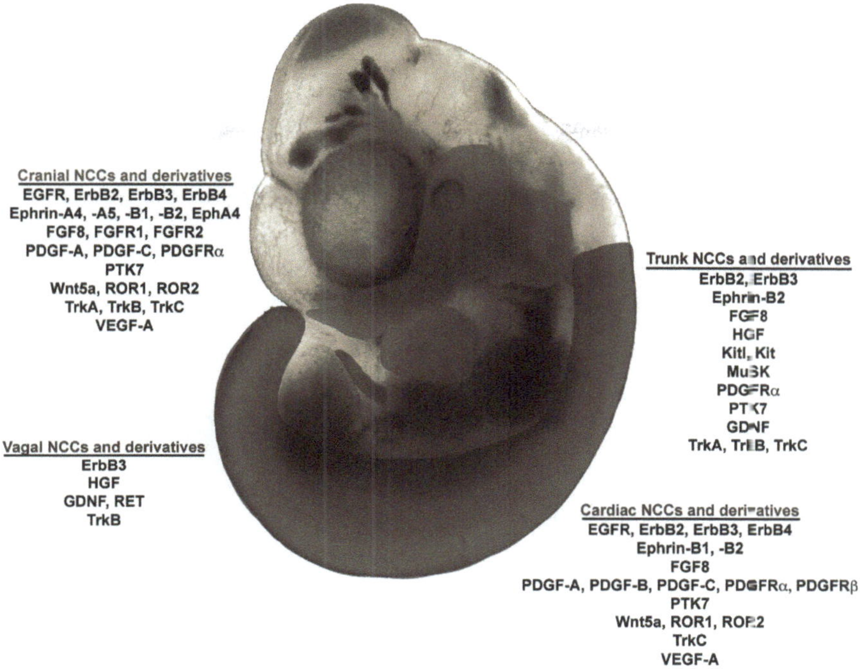

Figure 2 Schematic diagram of an E11.5 mouse embryo in which the target tissues of the four axial NCC populations are broadly highlighted. Ligands and corresponding RTKs contributing to the development of each population and its derivatives are listed. (See the color plate.)

as well as contribute to the formation of the eye, teeth, thyroid gland, parathyroid gland, and thymus. Cardiac NCCs are a subpopulation of cranial NCCs that arise as far rostrally as the otic vesicle and contribute to the aorticopulmonary septum and the caudal pharyngeal arch arteries. Vagal and sacral NCCs generate the enteric ganglia of the gut peripheral nervous system. Finally, trunk NCCs, which originate caudally to the cranial NCC domain, give rise to melanocytes, the dorsal root and sympathetic ganglia of the peripheral nervous system, Schwann cells, and the adrenal medulla (reviewed in Mayor & Theveneau, 2013; Trainor, 2005).

In humans, diseases stemming from defects in NCC activity are collectively referred to as neurocristopathies (Bolande, 1974). These diseases fall under two broad categories: congenital malformations and neoplasms. Dysgenetic neurocristopathies encompass craniofacial malformations; pigmentary disorders; diseases of the peripheral nervous system, such as Hirschsprung's disease; and syndromes affecting multiple sites throughout the body, such as DiGeorge, Kallmann, and craniofrontonasal (reviewed in Bolande, 1997; Etchevers, Amiel, & Lyonnet, 2006).

While not all RTK families have been shown to play a role in mammalian NCC development, those that do often have specific functions in a subpopulation of NCCs that contribute to the diversity of these cells and their derivatives in the mammalian embryo (Table 1 and Fig. 2). The combined application of genetic, proteomic, and *in vivo* biosensor approaches to investigate RTK signaling promises to shed further light on the intracellular signaling pathways active downstream of this receptor subclass during NCC development.

2. RTK SIGNALING IN MAMMALIAN NCC DEVELOPMENT

2.1. ErbB receptors

In mammals, the ErbB family is composed of 11 ligands, epidermal growth factor (EGF), heparin-binding EGF-like growth factor (HB-EGF), transforming growth factor-α (TGF-α), amphiregulin, betacellulin, epigen, epiregulin, and neuregulins 1–4, which variously bind and activate 3 receptors, ErbB1 (also known as Her1, EGFR), ErbB3 (Her3), and ErbB4 (Her4). A fourth receptor, ErbB2 (Her2, Neu), does not directly bind ligands (Stein & Staros, 2000). The ErbB receptors are composed of an extracellular region harboring four subdomains organized as a tandem repeat of homologous domains, leucine-rich 1 (LR1), cysteine-rich 1 (CR1), LR2 and CR2, and a cytoplasmic tyrosine kinase domain (Bajaj, Waterfield, Schlessinger, Taylor, & Blundell, 1987; Ullrich et al., 1984; Fig. 1). While the neuregulins mainly activate ErbB3 and ErbB4, the remaining ligands in the family primarily activate EGFR (Leahy, 2004). ErbB2, which lacks a known ligand, and ErbB3, which lacks an active kinase domain (Guy, Platko, Cantley, Cerione, & Carraway, 1994), are incapable of signaling on their own and heterodimerize with other receptors in the family to potentiate a signal (Citri, Skaria, & Yarden, 2003; Klapper et al., 1999).

EGFR is expressed in various epithelial tissues throughout the developing embryo (Sibilia & Wagner, 1995). Homozygous null mice show strain-dependent phenotypes ranging from peri-implantation lethality stemming from inner cell mass defects to midgestation lethality owing to placental defects and perinatal lethality approximately 3 weeks after birth (Sibilia & Wagner, 1995; Threadgill et al., 1995). In the latter case, mice display abnormalities in the development of several organs, including the brain, eye, lung, kidney, liver, gastrointestinal tract, skin, and hair follicles (Miettinen et al., 1995; Sibilia & Wagner, 1995; Threadgill et al., 1995). Homozygous null neonates additionally exhibit defects in NCC-derived structures in the face and heart. These include craniofacial abnormalities such as cleft palate, misshapen snouts, micrognathia, and abnormal Meckel's cartilage development, which are caused, at least in part, by decreased matrix metalloproteinase secretion (Miettinen et al., 1999), as well as defects in semilunar valvulogenesis mediated through signaling of the tyrosine phosphatase SHP-2 (Chen et al., 2000).

Targeted disruption of *Erbb2*, *Erbb3*, or *Erbb4* receptors in mice results in embryonic lethality during midgestation and a subset of overlapping NCC phenotypes (Erickson et al., 1997; Gassmann et al., 1995; Lee et al., 1995; Riethmacher et al., 1997). ErbB2 is expressed in the mouse nervous system and cardiac myocytes during development, and *Erbb2* homozygous null embryos display defects in cranial sensory ganglia, sympathetic ganglia, motor nerve, and heart development, due in part to defects in NCC migration (Britsch et al., 1998; Lee et al., 1995). Genetic rescue of the cardiac defects of *Erbb2* mutant mice revealed an absence of NCC-derived Schwann cells in the peripheral nerves and loss of motor and sensory neurons (Morris et al., 1999; Woldeyesus et al., 1999). Embryos homozygous for a kinase-dead *Erbb2* allele recapitulate the null phenotypes, indicating that the catalytic activity of ErbB2 is required for proper development (Chan, Hardy, Laing, Hardy, & Muller, 2002). Moreover, analysis of an allelic series of autophosphorylation mutant knock-in mice at the *Erbb2* locus revealed a role for Shc adaptor signaling in mediating the activity of the receptor in cutaneous sensory neurons (Chan, Hardy, Dankort, Laing, & Muller, 2004). Consistent with expression of *Erbb3* in murine NCCs, brain, Schwann cells, various ganglia, and heart, among other sites (Britsch et al., 1998; Meyer et al., 1997), *Erbb3* homozygous null embryos exhibit brain defects, a lack of Schwann cell precursors accompanying sensory and motor neurons, defects in cranial ganglia, dorsal root ganglia, sympathetic ganglia and enteric ganglia, and heart abnormalities (Erickson et al., 1997; Riethmacher et al., 1997). *Erbb4* is expressed in the brain and myocardium, and *Erbb4* homozygous null embryos accordingly display defects in innervation of the hindbrain and heart development (Gassmann et al., 1995). Moreover, abnormal migration of hindbrain-derived cranial NCCs in *Erbb4* mutant embryos results in misprojections, and in some cases fusions, of the cranial sensory ganglia (Golding, Trainor, Krumlauf, & Gassmann, 2000).

2.2. Eph receptors

The erythropoietin-producing hepatocellular carcinoma (Eph) receptors are the largest subfamily of RTKs in vertebrates and are subdivided into two classes, A-type and B-type, based on homology and ligand-binding affinities (Gale et al., 1996). In mammals, the family consists of 8 Eph receptor-interacting (ephrin) proteins, ephrin-A1–5 and ephrin-B1–3, and 14 Eph receptors, EphA1–8, EphA10, EphB1–4, and EphB6, which are capable of bidirectional signaling. Signaling downstream of Ephs upon ephrin binding constitutes forward signaling, while activation of signal transduction pathways downstream of ephrins upon interaction with Eph receptors comprises reverse signaling (Brückner, Pasquale, & Klein, 1997; Davy & Soriano, 2005; Holland et al., 1996). Ephrin-A proteins are tethered to the membrane via a glycosylphosphatidyl inositol anchor, while ephrin-B proteins possess a transmembrane domain and a cytoplasmic domain harboring a PDZ-domain-binding motif (Davis et al., 1994; Lin, Gish, Songyang, & Pawson, 1999). Eph receptors are composed of an extracellular region with a globular domain, a cysteine-rich region and two fibronectin type III repeats, and an intracellular region consisting of a tyrosine kinase domain, a SAM domain, and a carboxyl-terminal PDZ-domain-binding motif (Hirai, Maru, Hagiwara, Nishida, & Takaku, 1987; Fig. 1). Ephrin proteins and Eph receptors typically interact within their subclass (Gale et al., 1996), with few exceptions: ephrin-A5 can also bind EphB2 and EphA4 is also capable of binding to all B-type ephrins.

Despite the considerable number of proteins in the Eph receptor family and their extensive roles during mammalian development, relatively few, ephrins-A4, -A5, -B1, -B2, and EphA4, have been demonstrated to regulate NCC activity in the embryo. The majority of the functional studies in mice have addressed the roles of ephrins-B1 and -B2 in controlling the migration of various NCC populations. Ephrin-B1 and its cognate receptors are expressed in cranial NCCs. *Efnb1* null embryos as well as those with conditional inactivation of the gene in NCCs using the *Wnt1-Cre* driver exhibit abnormalities in cranial and cardiac NCC migration and partially penetrant defects in the palate and tympanic ring (Davy, Aubin, & Soriano, 2004). Mice harboring point mutations in *Efnb1* that disrupt reverse signaling pathways do not display craniofacial abnormalities, indicating that the NCC defects observed in *Efnb1* null mutants are due to impaired forward signaling (Bush & Soriano, 2009). Similarly, Ephrin-B2 forward signaling from the branchial arch surface ectoderm through its cognate receptors in cranial

NCCs is required for proper cranial NCC migration and subsequent branchial arch morphogenesis. Whereas *Efnb2* null embryos variably exhibit hypoplastic first and second branchial arches and reduced aortic arches and fifth (trigeminal) cranial ganglia due to impaired NCC migration (Adams et al., 2001; Davy & Soriano, 2007), these defects are rescued in embryos hemizygous for a null allele in combination with an allele encoding an ephrin-B2 protein lacking tyrosine resides required for reverse signaling (Davy & Soriano, 2007). Ephrin-B2 is also expressed in the posterior half of somites (Davy & Soriano, 2007; Wang & Anderson, 1997), while its cognate receptors are expressed in migrating trunk NCCs (Wang & Anderson, 1997). As such, *Efnb2* null embryos additionally exhibit trunk NCC migration defects with an invasion of NCCs into the posterior half of somites (Davy & Soriano, 2007). Furthermore, conditional ablation of *Efnb2* in NCCs using the *Wnt1-Cre* driver impairs migration of the thymus into the thoracic cavity due to defects in the motility of NCC-derived thymic mesenchymal cells (Foster et al., 2010). *Efnb1/Efnb2* compound heterozygous embryos display further abnormalities in tissues composed of NCC-derived mesenchyme, including the salivary gland and eyelid, indicating that the two proteins act redundantly at these sites (Davy & Soriano, 2007), though their exact mechanism of action on these tissues is unknown.

Additionally, both ephrin-A4 and EphA4 have been shown to play a role in establishing calvarial boundaries. In a *Twist1* haploinsufficiency model of coronal synostosis, typified by premature fusion of the suture between the NCC-derived frontal bone and the mesoderm-derived parietal bone, reduced ephrin-A4 distribution in the suture accompanies a boundary defect wherein the NCCs invade the undifferentiated mesoderm (Merrill et al., 2006). This same coronal synostosis phenotype is observed in *Epha4* null mice and is associated with reduced phospho-Erk1/2 expression in the ectocranial layer of the suture (Ting et al., 2009). Consistent with the mouse model data, heterozygous germline mutations of *EFNA4* in humans are associated with several cases of nonsyndromic coronal synostosis (Merrill et al., 2006).

Functional roles for ephrin-A5 and ephrin-B1 in regulating survival of cranial NCCs and proliferation of cranial NCC-derived mesenchyme, respectively, have also been demonstrated. *Wnt1-Cre*-mediated expression of ephrin-A5-Fc along the dorsal midline of the mouse midbrain results in craniofacial hypoplasia. Accompanying NCC explant experiments revealed that ephrin-A5-Fc hinders cranial NCC survival in the dorsal midline (Noh, Park, & Park, 2014), providing a potential mechanism for the phenotype. In the case of X-linked ephrin-B1, heterozygous loss of the gene in mice results in sorting out of ephrin-B1-expressing and -nonexpressing cells due to X inactivation, disruption of cell proliferation in the anterior palatal shelf mesenchyme, and perturbations in downstream Erk/MAPK signaling. These defects result in a cleft palate phenotype that mirrors a subset of craniofacial abnormalities in human X-linked craniofrontonasal syndrome (Bush & Soriano, 2010) caused by heterozygous loss-of-function mutations in *EFNB1* (Twigg et al., 2004).

Finally, additional analyses of *Efnb1*$^{+/-}$ mice revealed that the observed calvarial defects in this model stem from inhibition of gap junction communication and impaired differentiation of NCCs into osteogenic precursors (Davy, Bush, & Soriano, 2006).

2.3. FGF receptors

The mammalian FGF family consists of 22 FGF proteins, 18 of which variously signal through 4 receptors, FGFR1–4. Most FGF ligands additionally bind heparin or heparan sulfate proteoglycans, an interaction that serves to increase the affinity between the ligands and FGFRs and contribute to receptor dimerization and activation (Rapraeger, Krufka, & Olwin, 1991; Spivak-Kroizman et al., 1994; Yayon, Klagsbrun, Esko, Leder, & Ornitz, 1991). The FGFRs are composed of an extracellular portion, containing three immunoglobulin-like domains (D1–D3) and an acid box, and a cytoplasmic portion with a split tyrosine kinase domain (Lee, Johnson, Cousens, Fried, & Williams, 1989; Fig. 1). The FGFRs are subject to extensive alternative splicing which produce, among other forms, FGFR1–3 isoforms containing an alternatively spliced C-terminal half of D3 depending on the inclusion of exon 8 ("b" isoforms) or exon 9 ("c" isoforms; Johnson, Lu, Chen, Werner, & Williams, 1991; Miki et al., 1992; Yayon et al., 1992). Importantly, this alternative splicing produces receptor isoforms with different tissue-specific expression as well as unique ligand-binding properties (Miki et al., 1992; Yayon et al., 1992).

The FGFRIIIb isoforms are generally expressed by the epithelia during development, while the FGFRIIIc isoforms commonly localize to the mesenchyme (Avivi, Yayon, & Givol, 1993; Orr-Urtreger et al., 1993), with their respective ligands expressed in the adjacent compartment (Ornitz et al., 1996). To date, only a handful of FGF family members—the ligand FGF8 and the receptors FGFR1 and FGFR2—have been shown to regulate mammalian NCC activity *in vivo*. *Fgf8* is expressed in the craniofacial, central nervous system, and limb bud epithelia as well as the pharyngeal arch ectoderm and endoderm during development and binds to FGFR isoforms expressed in the mesenchyme (MacArthur et al., 1995). Targeted disruption of *Fgf8* in mice results in embryonic lethality before E10.5 and a loss of all mesoderm-derived structures (Meyers, Lewandoski, & Martin, 1998). Mice homozygous for an *Fgf8* hypomorphic allele die perinatally and exhibit impaired development of the midbrain, cerebellum and olfactory bulbs (Meyers et al., 1998). Compound heterozygous mice harboring one null and one hypomorphic *Fgf8* allele have variable defects, a subset of which phenocopy human 22q11 deletion syndromes, including DiGeorge syndrome. These defects include abnormalities in craniofacial, pharyngeal gland, brain, cardiac, and posterior axis development (Abu-Issa, Smyth, Smoak, Yamamura, & Meyers, 2002; Frank et al., 2002; Meyers et al., 1998). Analyses of mutant mouse models have revealed that FGF8 regulates several aspects of NCC activity during development of various craniofacial, pharyngeal, cardiac, and neural structures. The ligand promotes survival of the NCC-derived ectomesenchymal cells of the first branchial arch (Trumpp, Depew, Rubenstein, Bishop, & Martin, 1999) and pharyngeal and cardiac NCCs (Abu-Issa et al., 2002; Frank et al., 2002). FGF8 expression from the ectoderm establishes and maintains rostral–caudal polarity in the ectomesenchymal cells of the first branchial arch (Tucker, Yamada, Grigoriou, Pachnis, & Sharpe, 1999) and is required for outgrowth of this structure (Trumpp et al., 1999). Furthermore, upregulation of craniofacial *Fgf8* expression has been shown to result in an expansion of cranial NCCs leading to an enlarged first branchial arch, maxillary hyperplasia, and a high arched palate in two ciliopathic mutant mouse models (Tabler et al., 2013). In pharyngeal arches 3–6, FGF8 expression in both the ectoderm and endoderm is required for proper thymus, parathyroid, and cardiac development (Macatee et al., 2003; Park et al., 2006). Moreover, a requirement for FGF8 has been demonstrated in promoting the survival of NCCs that will give rise to postganglionic neurons (Chen, Moon, & Gaufo, 2012).

Targeted disruption of *Fgfr1* results in embryonic lethality between E9.5 and E12.5, defects in cell migration out of the posterior primitive streak and defective patterning of axial structures (Deng et al., 1994; Yamaguchi, Harpal, Henkemeyer, & Rossant, 1994). While mice homozygous for an inactivating mutation in the FGFR1IIIb isoform are viable and fertile, those deficient in the FGFR1IIIc isoform phenocopy the *Fgfr1* null mutant (Partanen, Schwartz, & Rossant, 1998). Mice homozygous for *Fgfr1* hypomorphic alleles die perinatally and exhibit defects in craniofacial and limb development as well as abnormalities in the formation of the anterior–posterior axis (Partanen et al., 1998). Further analyses of the *Fgfr1* hypomorphic allele in combination with conditional deletion of the receptor in NCCs revealed that FGFR1 is required for the entry of NCCs into the second branchial arch and for proper lip and secondary palate formation (Trokovic, Trokovic, Mai, & Partanen, 2003; Wang et al., 2013). While the former requirement for the receptor is non-cell-autonomous in NCCs (Trokovic et al., 2003), the latter is cell-autonomous in the palatal mesenchyme and non-cell-autonomous in the palatal and medial edge epithelia (Wang et al., 2013).

Targeted disruption of *Fgfr2* results in embryonic lethality at E10–E11 and defects in placenta and limb bud formation (Xu et al., 1998). Mice lacking FGFR2IIIb die perinatally and exhibit defects in the skull, palate, teeth, inner ear, salivary gland, anterior pituitary gland, limb, lung, and skin (De Moerlooze et al., 2000), while those deficient in the FGFR2IIIc isoform are viable and display delayed ossification, craniosynostosis, and dwarfism of the long bones and axial skeleton (Eswarakumar et al., 2002). Mice heterozygous for an *Fgfr2* gain-of-function allele expressed exclusively in NCCs exhibit prematurely fused cranial sutures and dysmorphologies affecting the snout, cranial base, and cranial vault (Heuzé et al., 2014).

Additional studies have revealed roles for various adaptor proteins, specifically Crkl and Frs2, in mediating NCC activity downstream of FGFR activation. FGF8 binding induces the phosphorylation of both FGFR1 and FGFR2 and their subsequent binding to Crkl, which in turn promotes NCC survival and migration (Moon et al., 2006). Similar to mutant mouse models of *Fgf8*, targeted disruption of the gene encoding the Crkl adaptor protein also results in a combination of craniofacial, glandular, and cardiovascular phenotypes reminiscent of human 22q11 deletion syndromes (Guris, Fantes, Tara, Druker, & Imamoto, 2001), suggesting that Crkl signaling downstream of FGF8-mediated FGFR activation regulates NCC activity in several locations throughout the embryo. In regard to Frs2, homozygous mutant mice in which the Frs2/3 binding site on FGFR1 is deleted die during late embryogenesis with defects in spinal neural tube closure and development of the second pharyngeal arch and tail bud (Hoch & Soriano, 2006). The pharyngeal arch defect was shown to stem from impaired NCC migration into the first and second arches and ectopic cell death along the migratory path, and further resulted in hypoplastic second pharyngeal arch derivatives in the middle ear and hyoid (Hoch & Soriano, 2006).

Point mutations in *FGFR1*, *FGFR2*, and *FGFR3* underlie a series of human syndromes characterized by craniosynostosis and/or midface hypoplasia, among other associated defects (reviewed in Kelleher, O'Sullivan, Smyth, McDermott, & Viterbo, 2013). A recent study using mesoderm- or NCC-specific *Cre* drivers in combination with a *Cre*-inducible activating mutation in *Fgfr2* revealed that mesodermal expression of the mutation is necessary and sufficient to generate a craniosynostosis phenotype (Holmes & Basilico, 2012). Finally, heterozygous mutations in *FGFR1* and *FGF8* cause different forms of Kallmann syndrome typified by hypogonadotropic hypogonadism and a variable frequency of cleft palate, sensorineural hearing loss, and anosmia (Dodé et al., 2003; Falardeau et al., 2008).

2.4. Kit receptor

The Kit family is composed of the Kit ligand (Kitl, also known as MGF, SCF, Steel factor) and its receptor, Kit (Huang et al., 1990; Williams et al., 1990; Zsebo et al., 1990). *Kitl* is alternatively spliced to generate membrane-bound and soluble ligands, both of which are able to bind and activate Kit (Anderson et al., 1991). The Kit receptor consists of an extracellular portion harboring five immunoglobulin-like domains and an intracellular portion containing a split tyrosine kinase domain (Qui et al., 1988; Yarden et al., 1987; Fig. 1).

Kitl is encoded at the *Sl* (*Steel*) locus (Copeland et al., 1990; Huang et al., 1990; Zsebo et al., 1990) and is expressed in the mesenchyme lining the migratory pathways and destinations of melanoblast precursors, hematopoietic cells, and germ cells, with additional expression in the brain and spinal cord (Matsui, Zsebo, & Hogan, 1990). *Kit* is encoded at the *W* (*Dominant white spotting*) locus (Chabot, Stephenson, Chapman, Besmer, & Bernstein, 1988; Geissler, Ryan, & Housman, 1988) and is reciprocally expressed in melanoblasts, hematopoietic cells, and germ cells, as well as in ectoderm-derived structures of the central nervous system and craniofacial region, and the endoderm-derived intestinal tract (Orr-Urtreger et al., 1990). Mutations at both loci result in defects in pigmentation, hematopoiesis, and reproduction (Bennett, 1956; Dunn, 1937; Geissler, McFarland, & Russell, 1981; Gruneberg, 1942; McCoshen & McCallion, 1975; Sarvella & Russell, 1956). As the names of their loci suggest, these mutants were prized for their coat color and collected in Asia (Tokuda, 1935) and eventually the European "Mouse Fancy," ultimately contributing to the establishment of inbred mouse strains.

Signaling through the Kit receptor has been shown to regulate the survival and migration of NCC-derived melanoblast precursors in mice. Membrane-bound Kitl and Kit kinase activity are required for melanoblast survival, as melanoblasts of *Kitl*$^{Sl\text{-}d}$ mutants lacking both the transmembrane and cytoplasmic domains of the ligand and *Kitl*$^{Sl\text{-}17H}$ mutants with reduced cell surface expression of the membrane-bound ligand, as well as those of *Kit*$^{W\text{-}v}$ and *Kit*$^{W\text{-}41J}$ mutants carrying point mutations in the RTK domain of the receptor, are observed along migratory pathways but are subsequently absent from their target sites in the inner ear and dermis (Cable, Jackson, & Steel, 1995; MacKenzie, Jordan, Budd, & Jackson, 1997; Steel, Davidson, & Jackson, 1992; Wehrle-Haller & Weston, 1995, 1999). Conversely, melanoblast precursors are absent from the migratory pathways of embryos homozygous for the *Kitl*^{Sl} null allele, indicating that soluble ligand is necessary for melanoblast precursor dispersal (Wehrle-Haller & Weston, 1995). By rescuing melanoblast precursor survival in *Kitl*^{Sl} null embryos, Kit signaling was shown to regulate the entrance and migration of these cells along a dorsolateral pathway (Wehrle-Haller, Meller, & Weston, 2001). The role of Kit in melanogenesis is dependent, at least in part, on signaling through Src family kinases, SHP-1 and/or SHP-2, as mice homozygous for two tyrosine to phenylalanine mutations at residues known to mediate association of the receptor with these proteins have a white coat devoid of melanocytes. Further analysis revealed that these mutations result in reduced receptor autophosphorylation and attenuated downstream Erk/MAPK signaling (Kimura et al., 2004). Examination of an additional mouse model harboring a targeted mutation in one of these residues demonstrated that homozygous mutant animals exhibit variably penetrant loss of ventral pigment that is exacerbated in hemizygous animals carrying one copy of a *Kit* null allele, confirming a role for Src signaling downstream of the receptor in pigmentation (Agosti et al., 2004). These authors also revealed a potential role for PI3K-mediated Kit signaling in regulating melanogenesis, as mice hemizygous for an allele abrogating binding of the receptor to PI3K in combination with a *Kit* null allele have a mild pigmentation phenotype (Agosti et al., 2004). Finally, signaling through the Kit receptor has also been shown to contribute to the development of a subset of primary sensory neurons, as both small- and medium-diameter sensory neurons are absent in the dorsal root ganglia of *Kit*^{W} mutant mice (Zhang & Sieber-Blum, 2009).

2.5. MET receptor

The hepatocyte growth factor (HGF) family consists of the HGF/scatter factor (SF) ligand and its receptor, MET (Bottaro et al., 1991). The MET receptor consists of an extracellular portion composed of a Sema domain, a cysteine-rich PSI domain, and four immunoglobulin-like repeat domains, as well as an intracellular portion containing a tyrosine kinase domain and a carboxy-terminal binding domain (Graveel, Tolbert, & Vande Woude, 2013; Fig. 1).

HGF/SF is generally expressed in the mesenchyme of various organs in the developing mouse embryo and the MET receptor is reciprocally expressed in the adjacent epithelia, yet expression of the two transcripts overlaps in NCCs, among other sites (Andermarcher, Surani, & Gherardi, 1996; Sonnenberg, Meyer, Weidner, & Birchmeier, 1993). Both *Hgf* and *Met* homozygous null embryos die during midgestation with liver and placental defects, and *Met* null embryos additionally exhibit skeletal muscle abnormalities (Bladt, Riethmacher, Isenmann, Aguzzi, & Birchmeier, 1995; Schmidt et al., 1995; Uehara et al., 1995). While neither knockout model has a NCC phenotype, analyses of transgenic mice ubiquitously overexpressing HGF/SF uncovered a role for this signaling pathway in NCC derivatives. Overexpression of HGF/SF induces the presence of ectopic melanoblasts in the embryonic neural tube and dorsal root ganglia, as well as ectopic melanocyte formation in the adult central nervous system and skin (Kos et al., 1999; Takayama, La Rochelle, Anver, Bockman, & Merlino, 1996). Additionally, HGF/SF was shown to promote melanoblast survival and melanoctye differentiation in NCC explant cultures (Kos et al., 1999). Finally, HGF/SF transgenic mice have a high incidence of gastrointestinal obstruction, which may stem from abnormal development of the enteric ganglia, thus pointing to a potential additional role for this pathway in regulating NCC derivatives (Takayama et al., 1996).

2.6. MuSK receptor

The mammalian muscle-specific kinase (MuSK) family consists of one *bona fide* ligand, the heparan sulfate proteoglycan N-agrin, which activates the MuSK receptor (Glass et al., 1996). The receptor is composed of an extracellular portion harboring three immunoglobulin-like domains and a Frizzled-like cysteine-rich domain, and an intracellular portion containing a tyrosine kinase domain (Masiakowski & Yancopoulos, 1998; Valenzuela et al., 1995; Xu & Nusse, 1998; Fig. 1). While Wnt11r, the zebrafish orthologue of the mammalian secreted glycoprotein Wnt11, has been shown to bind the MuSK receptor via its cysteine-rich domain (Jing, Lefebvre, Gordon, & Granato, 2009), N-agrin does not bind MuSK, but instead interacts with MuSK-bound LRP4 to enhance the LRP4–MuSK association and activate MuSK (Kim et al., 2008; Zhang et al., 2008).

MuSK is expressed in developing muscle, at the neuromuscular junction, in the brain, and in sperm (Garcia-Osta et al., 2006; Kumar, Ferns, & Meizel, 2006; Valenzuela et al., 1995), and mutant mouse models of both MuSK and N-agrin die perinatally and exhibit defective neuromuscular synaptogenesis (DeChiara et al., 1996; Gautam et al., 1996). While studies of MuSK function during murine development have primarily focused on its role in neuromuscular junction formation, a recent study revealed an additional requirement for the receptor in maintaining segmental NCC migration. In *Musk* homozygous null mouse embryos, trunk NCCs fail to be restricted to the anterior somite and instead spread throughout the entire somite (Banerjee et al., 2011). In zebrafish, the identical role for MuSK is mediated through the Wnt11r ligand and Disheveled signaling downstream of the receptor (Banerjee et al., 2011).

2.7. PDGF receptors

The mammalian PDGF family is composed of four ligands, PDGF-A–D, which variously signal through two receptors, PDGFRα and PDGFRβ. The PDGF receptors consist of five extracellular immunoglobulin-like loops and a split intracellular tyrosine kinase domain (Williams, 1989; Fig. 1). While numerous ligand and receptor interactions have been demonstrated *in vitro*, relatively few functional interactions have been demonstrated *in vivo*. The homodimers PDGF-AA and PDGF-CC have been shown to exclusively activate PDGFRα signaling during mammalian development (Boström et al., 1996; Ding et al., 2004; Soriano, 1997), while PDGF-BB solely activates PDGFRβ signaling (Levéen et al., 1994; Soriano, 1994). The role of PDGF-DD has yet to be investigated using mouse models.

To date, PDGFRα signaling has been shown to contribute to both cranial and cardiac NCC development, while a more restricted role in cardiac NCC development has been demonstrated for PDGFRβ. PDGFRα is expressed in the embryonic mesenchyme, specifically in the non-neuronal derivatives of NCCs, while its ligands PDGF-A and PDGF-C are reciprocally expressed in the surface ectoderm and epithelium (Ding et al., 2000; Morrison-Graham, Schatteman, Bork, Bowen-Pope, & Weston, 1992; Orr-Urtreger & Lonai, 1992). Targeted disruption of *Pdgfra* in mice results in embryonic lethality during midgestation, with homozygous null embryos exhibiting a cleft face, subepidermal blebbing, edema, hemorrhaging, cardiac outflow tract defects, abnormalities in neural tube development, abnormally patterned somites, and skeletal defects (Soriano, 1997). These defects are phenocopied in mice lacking both PDGF-A and PDGF-C ligands (Ding et al., 2004). Conditional ablation of *Pdgfra* in NCCs using the *Wnt1-Cre* driver results in a subset of the null phenotypes, specifically facial clefting, midline hemorrhaging, aortic arch defects, and thymus hypoplasia (Tallquist & Soriano, 2003). Additional analyses exploring the cellular mechanism of the facial clefting phenotype demonstrated that $Pdgfra^{fl/fl}$; *Wnt1-Cre* embryos exhibit a delay in the migration of NCCs into the frontonasal prominence and decreased proliferation in this structure (He & Soriano, 2013). PDGFRβ is also expressed in the embryonic mesenchyme, with high expression levels in the heart, among other sites (Soriano, 1994). *Pdgfrb*-deficient mice die perinatally and exhibit hemorrhaging, thrombocytopenia, anemia, and kidney defects (Soriano, 1994). Furthermore, both *Pdgfrb* and *Pdgfb* null mouse embryos exhibit cardiac defects associated with impaired cardiac NCC development (Richarte, Mead, & Tallquist, 2007; Van den Akker et al., 2008). Conditional ablation of both *Pdgfra* and *Pdgfrb* in NCCs using the *Wnt1-Cre* driver results in defects in multiple cardiac NCC derivatives that are more severe than those observed in either single conditional homozygous mutant alone, stemming from impaired cardiac NCC migration into the outflow tract (Richarte et al., 2007).

Analysis of an allelic series of autophosphorylation mutant knock-in mice at the *Pdgfra* locus identified PI3K signaling as the main intracellular pathway downstream of PDGFRα signaling during embryogenesis in the mouse (Klinghoffer, Hamilton, Hoch, & Soriano, 2002). Embryos homozygous for an allele ($Pdgfra^{PI3K}$) harboring two tyrosine to phenylalanine mutations at residues that mediate the ability of PDGFRα to bind PI3K (Yu et al., 1991) die perinatally and exhibit a cleft palate, among other defects (Klinghoffer et al., 2002). Moreover, the full range of $Pdgfra^{-/-}$ phenotypes, including complete facial clefting, is observed in $Pdgfra^{PI3K/PI3K}$;$Pdgfrb^{PI3K/PI3K}$ double homozygous mutant embryos in which PI3K signaling cannot be engaged through PDGFRα/β heterodimers (Klinghoffer et al., 2002). We recently employed a mass spectrometry-based phosphoproteomic approach to identify intracellular effectors downstream of PI3K/Akt-mediated PDGFRα signaling in mouse embryonic NCC-derived palatal mesenchyme. Our analysis identified both established and novel Akt phosphorylation target proteins, a subset of which we demonstrated regulate cell survival and proliferation downstream of PDGFRα activation in part through modulation of p53 activity (Fantauzzo & Soriano, 2014).

While PDGFRβ does not have an established role in cranial NCC development, limited evidence, including the phenotypes of *Pdgfra*$^{fl/fl}$;*Pdgfrb*$^{fl/fl}$; *Wnt1-Cre* and *Pdgfra*$^{PI3K/PI3K}$;*Pdgfrb*$^{PI3K/PI3K}$ double homozygous mutant embryos described above (Klinghoffer et al., 2002; Richarte et al., 2007), indicate that the two receptors may be able to form functional heterodimers *in vivo*. Previous *in vitro* studies suggest that PDGFRα/β heterodimers have distinct properties from homodimeric receptor complexes. For example, PDGFRα is phosphorylated at Y754 exclusively upon heterodimer formation, resulting in preferential binding of SHP-2 (Rupp et al., 1994). Conversely, phosphorylation of PDGFRβ at Y771 is decreased in heterodimer receptor complexes, leading to reduced association with Ras-GAP and prolonged activation of Ras and MAP kinase (Ekman, Thuresson, Heldin, & Rönnstrand, 1999). Furthermore, heterodimers have been shown to generate an enhanced mitogenic response over that of either homodimer receptor complex (Rupp et al., 1994). The prevalence of these heterodimers, as well as their effect on cellular behavior during development, has yet to be determined.

Finally, PDGFRα may also play a cell-autonomous role in melanocyte development. Chimeric mice derived from embryonic stem cells heterozygous for either the *Pdgfra* null or *Pdgfra*^{PI3K} allele exhibit poor coat color chimerism, indicating a particular dosage requirement for PI3K-mediated PDGFRα signaling in contributing to coat color pigmentation (Klinghoffer et al., 2002; Soriano, 1997).

2.8. PTK7 receptor

Mammalian protein tyrosine kinase 7 (PTK7) is an orphan receptor composed of an extracellular portion containing seven immunoglobulin-like loops and an intracellular portion with an inactive tyrosine kinase domain (Park, Lee, & Lee, 1996; Fig. 1). *Ptk7* transcripts are detected in the primitive streak, craniofacial region, heart, gut, lung, somites, limbs, and tail in the mouse embryo (Jung, Shin, Song, & Lee, 2004; Paudyal et al., 2010), while the protein has been shown to localize specifically to the neuroepithelium and mesenchyme during early development (Paudyal et al., 2010). Mice homozygous for a gene trap insertion in the second intron of the gene die perinatally and display defects in neural tube closure and stereociliary bundle orientation in the ear (Lu et al., 2004). *Chuzhoi* mice, which are homozygous for an ENU-induced splice site mutation in the *Ptk7* gene, also die perinatally and similarly exhibit severe neural tube defects and altered planar cell polarity in the ear, as well as cardiac outflow tract and ventricular septal defects, omphalocele, abnormal lung development, and skeletal defects affecting the ribs and limbs (Paudyal et al., 2010). *Chuzhoi* mutant embryos additionally have abnormal NCC distribution, displaying misshapen cranial and dorsal root ganglia (Paudyal et al., 2010).

2.9. RET receptor

Signaling through the rearranged during transfection (RET) receptor involves the formation of a multicomponent receptor complex consisting of a glial cell line-derived neurotrophic factor (GDNF) family ligand (GDNF, neurturin (NRTN), artemin (ARTN), or persephin (PSPN)), a ligand-binding glycosyl-phosphatidylinositol (GPI)-anchored coreceptor (GFRα1–4), and the RET receptor tyrosine kinase. The RET receptor consists of an extracellular domain composed of a cadherin-related motif and a cysteine-rich region, and an intracellular domain harboring a split tyrosine kinase domain (Iwamoto et al., 1993; Fig. 1). GDNF, NRTN, ARTN, and PSPN primarily bind and signal through GFRα1, GFRα2, GFRα3, and GFRα4, respectively (Baloh et al., 1998; Buj-Bello et al., 1997; Enokido et al., 1998; Jing et al., 1996; Klein et al., 1997; Treanor et al., 1996).

In the developing mouse embryo, *Gdnf* is expressed in the anterior neuroectoderm during early neurogenesis and later localizes to the mesenchyme at several sites throughout the embryo, including the gastrointestinal tract, kidney, testes, facial prominences, eye, tongue, tooth primordia, vibrissae (whisker) follicles, ear, paravertebral mesenchyme, and limbs (Hellmich, Kos, Cho, Mahon, & Zimmer, 1996). *Ret* is expressed in NCCs; several lineages of the peripheral and central nervous systems, including the cranial, autonomic, dorsal root, and enteric ganglia; the nephric duct; the epithelia of the ureteric bud; and the renal collecting ducts (Pachnis, Mankoo, & Costantini, 1993). Targeted disruption of both *Gdnf* and *Ret* in mice results in perinatal lethality, with homozygous null embryos lacking enteric neurons and displaying renal agenesis due to defective induction of the ureteric bud, among other defects (Moore et al., 1996; Pichel et al., 1996; Sánchez et al., 1996; Schuchardt, D'Agati, Larsson-Blomberg, Costantini, & Pachnis, 1994). Conditional ablation of *Ret* in NCCs using the *Wnt1-Cre* driver similarly results in intestinal aganglionosis (Luo et al., 2007). *Gdnf* null neonates additionally exhibit neuronal loss in dorsal root and sympathetic ganglia (Moore et al., 1996).

Studies using rodent models have demonstrated that perturbations of the GDNF-GFRα1-RET signaling pathway disrupt several aspects of enteric NCC development, including survival, migration, proliferation, and/or neuronal differentiation, which contribute to the pathology of the intestinal aganglionosis phenotype in mammals (Heuckeroth, Lampe, Johnson, & Milbrandt, 1998; Taravira et al., 1999; Uesaka & Enomoto, 2010; Uesaka, Nagashimada, Yonemura, & Enomoto, 2008). Furthermore, analyses of RET phosphorylation mutant knock-in mice have revealed roles for the MAPK, PI3K/Akt, and JNK signaling cascades downstream of Shc adaptor and cAMP-dependent PKA interactions in mediating the role of the receptor during enteric nervous system development (Asai et al., 2006; Jain et al., 2010; Jijiwa et al., 2004; Wong et al., 2005).

In humans, heterozygous germline mutations in *GDNF* and/or *RET* underlie a significant subset of cases of Hirschsprung's disease (Edery et al., 1994; Eketjäll & Ibáñez, 2002; Romeo et al., 1994), characterized by a congenital absence of enteric ganglia in a portion of the gastrointestinal tract.

2.10. ROR receptors

The mammalian RTK-like orphan receptor (ROR) family consists of two receptors, Ror1 and Ror2. While a ligand has not been identified for Ror1, the secreted glycoprotein Wnt5a has been shown to bind and signal through Ror2 to activate the noncanonical Wnt pathway, thereby inducing cell migration (He et al., 2008; Nishita et al., 2006; Nomachi et al., 2008; Oishi et al., 2003; Yamamoto, Yoo, Nishita, Kikuchi, & Minami, 2007) and regulating cell proliferation (He et al., 2008). The extracellular portions of the receptors are composed of an immunoglobulin-like domain, Frizzled-like cysteine-rich domain, and a kringle domain, while the intracellular portions contain a tyrosine kinase domain and a carboxy-terminal proline-rich domain (Masiakowski & Carroll, 1992; Oishi et al., 1999; Fig. 1).

In the developing mouse embryo, *Ror1* is expressed in the anterior part of the embryo at E7.5, while *Ror2* is detected throughout the primitive streak. One day later, *Ror1* transcripts localize to the cephalic mesenchyme, with particularly high expression levels in NCCs, while *Ror2* is more broadly expressed in neural and non-neural tissues, including cephalic NCCs. At later stages, the two receptors have largely overlapping expression patterns at several sites throughout the embryo, with particularly high transcript levels in NCCs and their derivatives in the face and heart (Al-Shawi, Ashton, Underwood, & Simons, 2001; Matsuda et al., 2001; Oishi et al., 1999). Similar to *Ror2*, *Wnt5a* is expressed in the primitive streak and later in the facial primordia and heart, among other sites (Schleiffarth et al., 2007; Yamaguchi, Bradley, McMahon, & Jones, 1999).

Both *Ror1* and *Ror2* homozygous null embryos die perinatally with respiratory defects (DeChiara et al., 2000; Nomi et al., 2001; Takeuchi et al., 2000). *Ror2*-deficient mice additionally exhibit widespread skeletal defects, including craniofacial bone hypoplasia, and ventricular septal defects of the heart (DeChiara et al., 2000; Takeuchi et al., 2000). Double homozygous mutant mice also die perinatally and exhibit enhanced skeletal phenotypes over those observed in *Ror2* null mice, as well as transposition of the great arteries, indicating that the two receptors interact genetically during skeletal and cardiac development (Nomi et al., 2001). *Wnt5a* homozygous null embryos also display craniofacial bone truncations and cardiac outflow tract abnormalities, among other defects (Schleiffarth et al., 2007; Yamaguchi et al., 1999). Furthermore, $Ror2^{+/-};Wnt5a^{+/-}$ double heterozygous embryos exhibit a cleft palate, confirming an interaction between this ligand and receptor pair in the palatal mesenchyme (He et al., 2008).

In humans, a subset of mutations in *ROR2* result in autosomal recessive Robinow syndrome, characterized by skeletal dysplasia affecting the craniofacial bones, limbs, and vertebra, as well as genital hypoplasia (Afzal et al., 2000; van Bokhoven et al., 2000), while the phenotypically less severe autosomal dominant form of the syndrome stems from mutations in *WNT5A* (Person et al., 2010).

2.11. Trk receptors

The mammalian tropomyosin-related kinase (Trk) receptor family is composed of four ligands, nerve growth factor (NGF), brain-derived neurotrophic factor (BDNF), neurotrophin-3 (NT-3), and NT-4, which variously bind and activate three receptors, TrkA, TrkB, and TrkC. The vertebrate Trk receptors consist of an extracellular portion harboring a cysteine-rich cluster, three leucine-rich repeats, a second cysteine-rich cluster, and two immunoglobulin-like domains, as well as an intracellular portion containing a tyrosine kinase domain (Benito-Gutiérrez, Garcia-Fernàndez, &

Comella, 2006; Fig. 1). NGF binds to TrkA (Kaplan, Hempstead, Martin-Zanca, Chao, & Parada, 1991; Klein, Jing, Nanduri, O'Rourke, & Barbacid, 1991); BDNF and NT-4 interact with TrkB (Klein, Nanduri, et al., 1991; Squinto et al., 1991); and NT-3 binds with high affinity to TrkC and with lower affinity to other receptors of the family (Lamballe, Klein, & Barbacid, 1991).

Ntrk1, encoding TrkA, is expressed exclusively in the NCC-derived dorsal root and cranial sensory ganglia of the peripheral nervous system during murine development (Martin-Zanca, Barbacid, & Parada, 1990). *Ntrk2* and *Ntrk3*, encoding TrkB and TrkC, respectively, are expressed at the same sites as TrkA, with broader expression in the brain, the spinal cord, the non-sensory cranial ganglia and placode-derived sensory ganglia of the peripheral nervous system, and the dorsal aorta, among other non-neural sites (Klein, Martin-Zanca, Barbacid, & Parada, 1990; Tessarollo et al., 1993). *Ntrk2* and *Ntrk3* are additionally expressed in the NCC-derived enteric nervous system, with *Ntrk2* expressed in the enteric glial cells (Levanti et al., 2009) and *Ntrk3* expressed in the enteric ganglia (Tessarollo et al., 1993). The expression of specific Trk receptors in various neuronal populations appears to specify their subtype, as expression of *Ntrk3* from the *Ntrk1* locus alters the fate of a subset of dorsal root ganglia neurons (Moqrich et al., 2004).

Targeted disruption of *Ntrk1* in mice results in premature death, typically within the first month after birth, with homozygous null mice exhibiting neuronal loss in dorsal root, trigeminal, and sympathetic ganglia (Smeyne et al., 1994), as well as defects in thymus development (García-Suárez et al., 2000). Both the *Ntrk2* and *Ntrk3* loci undergo alternative splicing to generate either a full-length receptor or a truncated receptor lacking the tyrosine kinase domain. Mice homozygous for a germline mutation in the tyrosine kinase domain of *Ntrk2* die within the first 2 weeks after birth and display neuronal loss in both the central and peripheral nervous systems, including in the trigeminal and dorsal root ganglia, among other defects (Klein et al., 1993). Mutant mice lacking all isoforms of *Ntrk2* exhibit similar timing of premature death, yet less severe neuronal loss in sensory ganglia, indicating that the truncated receptor may negatively regulate neuron survival (Luikart, Nef, Shipman, & Parada, 2003). *Ntrk2*-deficient mice additionally display defects in enteric glial cell development (Levanti et al., 2009). Mice homozygous for a targeted mutation in the tyrosine kinase domain of *Ntrk3* usually die within the first month of birth and lack a subset of dorsal root ganglia neurons (Klein et al., 1994). Mutant mice lacking all isoforms of *Ntrk3* die within 1 week of birth and display more severe neuronal phenotypes affecting the trigeminal, vestibular, cochlear, petrosal–nodose, and dorsal root ganglia. These mice also exhibit cardiac defects such as atrial and ventricular septal defects and valvular defects, among other abnormalities, suggesting a kinase-independent role for TrkC during mouse development (Tessarollo et al., 1997). Analysis of mice harboring a mutation of the Shc binding site in TrkB or TrkC receptors revealed a role for this pathway in regulating target innervation of sensory neurons downstream of the TrkB, but not the TrkC, receptor (Postigo et al., 2002).

In humans, mutations in *NTRK1* underlie the autosomal recessive disorder congenital insensitivity to pain with anhidrosis (Indo et al., 1996), which is characterized by defective NCC differentiation into a subset of sensory neurons as well as neuronal loss in the sympathetic ganglia. Furthermore, polymorphisms in *NTRK3* have been found in several patients with Hirschsprung's disease, which is typified by defective NCC activity during development of the enteric nervous system, though a causal role for these variants in the disease phenotype has not yet been demonstrated (Fernández et al., 2009; Ruiz-Ferrer, Fernandez, Antiñolo, Lopez-Alonso, & Borrego, 2008).

2.12. VEGF receptors

The mammalian vascular endothelial growth factor (VEGF) family consists of five ligands that are subject to alternative splicing and/or processing, VEGF-A–D and placental growth factor (PGF), which variously signal through three receptors, VEGFR1 (also known as Flt-1), VEGFR2 (KDR/Flk-1), and VEGFR-3 (Flt-4), as well as two neuropilin (Nrp) coreceptors, Nrp1 and Nrp2. The VEGF receptors consist of an extracellular portion with seven immunoglobulin-like domains and an intracellular portion with a split tyrosine kinase domain (Shibuya et al., 1990; Fig. 1). The NRP coreceptors, which also bind semaphorins (He & Tessier-Lavigne, 1997; Kolodkin et al., 1997), are quite distinct from the VEGF receptors and contain an extracellular portion with three interaction domains designated a1/a2, b1/b2, and c and a negligible cytoplasmic domain that lacks catalytic function (Kawakami, Kitsukawa, Takagi, & Fujisawa, 1996). Binding of VEGF ligand to a VEGF receptor induces receptor homo- or heterodimerization. VEGF-A binds VEGFR1 and VEGFR2 homodimers, VEGFR1/2 and VEGFR2/3 heterodimers, as well as Nrp1 homodimers; VEGF-B and PGF bind VEGFR1 homodimers, VEGFR1/2 heterodimers, and Nrp1 homodimers; and VEGF-C and VEGF-D bind VEGFR2 and

VEGFR3 homodimers, VEGFR2/3 heterodimers, and Nrp2 homodimers (reviewed in Koch & Claesson-Welsh, 2012).

While all members of the family function in vascular development, only the interaction of VEGF-A with Nrp1 has been implicated in NCC biology. *Vegfa* is widely expressed by parenchymal cells throughout the embryo, including the cardiac outflow tract, pharyngeal arch endoderm, thymus, facial prominences, and palate, among other sites, while Nrp1 is expressed in neighboring, often endothelial, cells at each of these sites (Stalmans et al., 2003). Consistent with its expression, mouse embryos devoid of the major, Nrp1-binding isoform of *Vegfa* exhibit cardiac outflow tract, pharyngeal arch artery, thymic, parathyroid, and craniofacial defects, reminiscent of human DiGeorge syndrome (Stalmans et al., 2003). Moreover, endothelial-specific disruption of *Nrp1* similarly results in a combination of phenotypes typical of DiGeorge syndrome, such as defects in the cardiac outflow tract (Gu et al., 2003; Zhou, Pashmforoush, & Sucov, 2012), pharyngeal organ hypoplasia, and cleft palate (Zhou et al., 2012). The defects observed in the above mouse models are not due to defective NCC migration, but have instead been attributed to vascular dysgenesis and endothelial cell dysfunction (Stalmans et al., 2003; Zhou et al., 2012). In contrast, a study in chick revealed that VEGFR2 and Nrp1 are expressed in cranial NCCs, while VEGF-A is expressed in the surface ectoderm adjacent to the rhombomere 4 NCC migratory route, and furthermore, that the VEGF-A–Nrp1 interaction was required for proper cranial NCC invasion from the rhombomere 4 migratory stream into branchial arch 2 (McLennan, Teddy, Kasemeier-Kulesa, Romine, & Kulesa, 2010).

3. CURRENT METHODS TO INVESTIGATE RTK SIGNALING

3.1. Receptor allelic series

As highlighted above, beyond the analysis of null mouse models, the use of conditional, floxed alleles in conjunction with NCC-specific *Cre* driver alleles has allowed researchers to examine the roles of various receptors and the signaling proteins with which they interact exclusively in NCCs. This approach has been utilized with a *Wnt1-Cre* driver (Danielian, Muccino, Rowitch, Michael, & McMahon, 1998) in combination with *Efnb1*, *Efnb2*, *Fgfr1*, *Pdgfra*, and *Ret* conditional alleles to demonstrate cell-autonomous functions of these receptors in NCCs (Davy et al., 2004; Foster et al., 2010; He & Soriano, 2013; Luo et al., 2007;

Tallquist & Soriano, 2003; Wang et al., 2013). While these studies have provided critical data on the roles of each of these RTKs in NCCs, it should be noted that the original *Wnt1-Cre* driver (Danielian et al., 1998) ectopically activates Wnt signaling, resulting in defects in midbrain development in heterozygous animals which are even more severe in *Wnt1-Cre*$^{Tg/Tg}$ mice (Lewis, Vasudevan, O'Neill, Soriano, & Bush, 2013). However, the development of a new tool, the *Wnt1-Cre2* transgenic mouse line (Lewis et al., 2013), circumvents these issues and will likely be of considerable use to the field going forward. Further NCC-specific *Cre* drivers include the *P0-Cre* (Yamauchi et al., 1999), *P3Pro-Cre* (Li, Chen, & Epstein, 2000), *Ht-PA-Cre* (Pietri, Eder, Blanche, Thiery, & Dufour, 2003), and *S4F:Cre* (Stine et al., 2009) alleles. Moreover, by employing additional, tissue-specific *Cre* drivers active in NCC target sites, the cell-autonomous role of a particular protein can be assessed in the various layers of tissues populated by NCCs. Using the pharyngeal arch as an example, the *Foxg1-Cre* transgene (Hébert & McConnell, 2000) can be used to inactivate gene expression throughout the arch, while *Crect* (Reid, Yang, Melvin, Taketo, & Williams, 2011), *Foxa2*^{mcm} (Park et al., 2008), and *Myf5-Cre* (Tallquist, Weismann, Hellström, & Soriano, 2000) drivers can be used to specifically target the pharyngeal arch ectoderm, pharyngeal pouch endoderm, and paraxial mesoderm, respectively (Tavares et al., 2012). Further tissue-specific *Cre* drivers of potential interest include *Ap2-Cre* alleles, which drive expression in the pharyngeal arch ectoderm (Macatee et al., 2003) or frontonasal process (Nelson & Williams, 2004); the *Mesp1-Cre* allele, targeting the cranial mesoderm and myocardium of the heart tube (Saga et al., 1999); and the *Tyr-Cre* allele, which drives expression in the melanocytes and peripheral nerves (Delmas, Martinozzi, Bourgeois, Holzenberger, & Larue, 2003; Tonks et al., 2003). Lastly, it is possible to perform tissue-specific, *in vivo* lineage tracing by combining *Cre* drivers with *lacZ* (Soriano, 1999) or fluorescent (Muzumdar, Tasic, Miyamichi, Li, & Luo, 2007; Prigge et al., 2013) *Cre* reporter alleles, such that all cells of a particular lineage are permanently marked for detection.

One approach that has yielded a wealth of functional information for a subset of RTK families to which it has been applied is the use of homologous recombination to generate series of knock-in alleles that disrupt either particular domains of a receptor or individual tyrosine autophosphorylation sites necessary for interaction with specific proteins. In contrast to the analysis of null alleles for these signaling molecules and their effectors, which can be obscured by the fact that multiple RTKs may utilize the same molecule as well as redundancy in intracellular signaling networks, the knock-in approach addresses the role of these molecules downstream of an individual receptor. As discussed earlier, such studies have revealed roles for Shc adaptor signaling in mediating the activity of ErbB2 in cutaneous sensory neurons (Chan et al., 2004); Frs2 downstream of FGFR1 in regulating NCC migration and survival in the second pharyngeal arch (Hoch & Soriano, 2006); Src family kinases and PI3K downstream of Kit in regulating melanogenesis in the coat (Agosti et al., 2004; Kimura et al., 2004); PI3K-mediated PDGFRα signaling in contributing to NCC-derived skeletal development (Klinghoffer et al., 2002); MAPK, PI3K, and JNK signaling cascades downstream of Shc adaptor and PKA interactions in mediating the role of RET during enteric nervous system development (Asai et al., 2006; Jain et al., 2010; Jijiwa et al., 2004; Wong et al., 2005); and Shc in regulating target innervation of sensory neurons downstream of TrkB activation (Postigo et al., 2002).

Finally, genetic knock-in approaches in which the domain of one RTK is replaced with that of another have begun to address questions of receptor functional specificity during mouse development. For example, swapping of the PDGFRα and PDGFRβ intracellular signaling domains revealed that signaling downstream of the two receptors is largely conserved, despite differences in expression and ligand binding affinities, such that null phenotypes are mostly rescued in both knock-in lines. Interestingly, however, this analysis demonstrated that sustained MAPK signaling specifically downstream of PDGFRβ is required for proper vascular development (Klinghoffer, Mueting-Nelsen, Faerman, Shani, & Soriano, 2001). A second study using a knock-in approach to fuse the extracellular domain of PDGFRα to the intracellular domain of either the *Drosophila* RTK Torso or mouse FGFR1 revealed that neither replacement can completely rescue proper development, due to alterations in MAPK and/or PI3K signaling, further indicating that strict regulation of downstream signaling pathways is required in some instances to mediate the specific biological function of individual RTKs (Hamilton, Klinghoffer, Corrin, & Soriano, 2003).

3.2. Phospho-specific reagents

The development of phospho-specific reagents, particularly antibodies recognizing individual phosphorylated residues within signaling molecules or directed against phosphorylated consensus recognition motifs, has greatly enhanced the biochemical analysis of intracellular events downstream of RTK activation. Within the embryo, whole mount or section immunohistochemistry can be performed using phospho-specific antibodies to assess the spatiotemporal expression of activated downstream effector proteins. This approach has revealed localization of several such molecules to discrete domains during murine development, indicating a role for these signaling molecules at particular sites and/or timepoints during embryogenesis (Corson, Yamanaka, Lai, & Rossant, 2003; Fantauzzo & Soriano, 2014).

For *in vitro* studies, cells can be serum-starved, stimulated with ligand and a particular receptor immunoprecipitated from the cell lysates. Western blotting can then be performed using an antibody recognizing phosphorylated tyrosine residues such as 4G10 or pY20 (Druker, Mamon, & Roberts, 1989; Glenney, Zokas, & Kamps, 1988) to examine receptor activation, or antibodies specific to signaling proteins known to interact with the receptor, such as those against Src, p85, SHP-2, PLCγ, Ras-GAP, Stat3, Shc, Grb2, Crk, Nck, Frs2, and IRS-1. Similarly, whole cell lysates can be subjected to Western blotting with antibodies directed against various activated downstream effector proteins, such as phospho-SAPK/JNK, phospho-Akt, phospho-p44/42 (Erk1/2), and phospho-PLCγ, to assess the status of these pathways downstream of receptor stimulation.

Furthermore, phosphosubstrate-specific antibodies have allowed for detection of phosphorylation events downstream of particular intracellular signaling molecules in response to RTK activation. One widely used example is the anti-Akt-phosphosubstrate antibody (Manning, Tee, Logsdon, Blenis, & Cantley, 2002) generated against the phosphorylated AGC kinase family consensus recognition motif RXRXXS/T recognized by Akt, RSK, and p70 S6 kinases (Alessi, Caudwell, Andjelkovic, Hemmings, & Cohen, 1996; Obata et al., 2000). This antibody has been used to identify individual Akt substrates downstream of RTK signaling using standard immunoprecipitation and Western blotting techniques (Manning et al., 2002), as well as to perform larger, mass spectrometry-based screens in response to growth factor stimulation in both cancer (Moritz et al., 2010) and primary (Fantauzzo & Soriano, 2014) cell lines.

Notably, these approaches can be combined with the use of pharmacological inhibitors that allow researchers to target RTK signaling pathway components at several levels, with the caveat that several of these inhibitors target more than one protein. Using the PDGFRα pathway as an example, Gleevec (imatinib mesylate) can be used to inhibit the receptor itself (Buchdunger et al., 1996), LY294002 can be used to inhibit the receptor-binding protein PI3K (Vlahos, Matter, Hui, & Brown, 1994), and rapamycin can be used to inhibit the intracellular signaling molecule mTOR (Brown et al., 1994).

Finally, a recent study has combined several techniques to investigate signaling networks commonly engaged downstream of individual RTKs (Wagner et al., 2013). Six isogenic transformed human embryonic kidney cell lines expressing EGFR, FGFR1, IGF-1R, MET, PDGFRβ, or TRKB were used in combination with lentiviral shRNA expression vectors to alter the levels of intracellular signaling proteins. Upon stimulation with relevant growth factors, phosphorylation of downstream proteins was evaluated at several timepoints by probing lysate microarrays with phospho-specific antibodies. Intriguingly, analysis of the resulting data highlighted three distinct RTK classes with conserved downstream signaling networks: (1) EGFR, FGFR1, and MET; (2) IGF-1R and TRKB; and (3) PDGFRβ (Wagner et al., 2013). Such multifaceted approaches will no doubt identify further commonalities and differences in signaling downstream of the various RTK families.

3.3. Proteomics

Mass spectrometry-based proteomic approaches have been employed to identify protein phosphorylation targets downstream of growth factor stimulation and new technologies have allowed for the quantification of these post-translational modifications. One shotgun proteomics strategy that has been used in two studies examining phosphorylation targets downstream of RTK signaling in NCC-derived primary mouse embryonic palatal mesenchyme cells is the immunoprecipitation of target proteins from whole cell lysates using either an anti-phosphotyrosine or anti-Akt-phosphosubstrate antibody, analysis of the tryptic peptides by nanoliquid chromatography coupled to tandem mass spectrometry (nano-LC–MS/MS), and assessment of phosphorylation changes in response to ephrin-B1-Fc or PDGF-AA treatment, respectively, by spectral counting (Bush & Soriano, 2010; Fantauzzo & Soriano, 2014). In these scenarios, summing the number of tandem mass spectra obtained for a given protein, a process known as spectral counting, approximates the abundance of the protein in the sample within over two orders of magnitude (Liu, Sadygov, & Yates, 2004).

Alternative isotope labeling approaches have been more commonly used with transformed or cancer cell lines in the RTK field and allow for quantitative proteomics analyses. One such technique, iTRAQ (isobaric tag for relative and absolute quantitation; Ross et al., 2004), has successfully been used to investigate, for example, the dynamics of tyrosine phosphorylation in response to EGF treatment in a transformed human mammary epithelial cell line (Zhang et al., 2005). For this study, tryptic peptides from four growth factor stimulation timepoints were separately labeled with one of four covalent tags of the same mass, mixed, immunoprecipitated with an anti-phosphotyrosine antibody and analyzed by LC–MS/MS (Zhang et al., 2005). In the case of iTRAQ, individual peptides are quantitated by comparing the relative ratios of reporter ions generated by fragmentation of the covalent tags in tandem mass spectrometry (Ross et al., 2004). Two additional studies used a related approach, SILAC (stable isotope labeling with amino acids in cell culture; Ong et al., 2002), to identify phosphorylation targets downstream of EGFR, MET, and/or PDGFRα signaling in various human cancer cell lines (Moritz et al., 2010; Olsen et al., 2006). Here, cells were grown in the presence of isotope-substituted forms of arginine and lysine, stimulated with growth factor or treated with various inhibitors, and mixed. Tryptic peptides were then enriched for phosphopeptides and analyzed by LC–MS/MS (Moritz et al., 2010; Olsen et al., 2006). With SILAC, peptides are subsequently quantitated by assessing the relative intensities of isotopic forms detected by mass spectrometry (Ong et al., 2002). Importantly, each of the mass spectrometry-based proteomics techniques discussed here has unique benefits and drawbacks (reviewed in Ahmad & Lamond, 2014; Brewis & Brennan, 2010) that should be considered when designing a relevant experimental strategy.

3.4. Biosensors

Lastly, various biosensors have been utilized both *in vitro* and *in vivo* to examine the spatiotemporal dynamics of RTK signaling. Bioluminescence resonance energy transfer (BRET) involves the transfer of energy from a luminescent donor (such as *Renilla* luciferase) to a fluorescent acceptor (such as GFP or EYFP). Upon co-expression of fusion molecules in live cells, protein–protein interactions or conformational changes can be assessed by measuring the ratio of emissions from the donor and acceptor (reviewed in Siddiqui, Cong, Daimon, Martin, & Maudsley, 2013). One such study utilizing this technology examined the interactions between RTKs of the ErbB, Kit, PDGF, Trk, and VEGF receptor families with the signaling molecules Grb2, p85, Stat5a, Shc46, and PLCγ1 in transformed human embryonic kidney cells, revealing specific receptor–signaling molecule interactions in response to growth factor treatment (Tan et al., 2007). Additional studies have employed BRET to examine receptor conformational changes upon ligand treatment. For example, BRET assays conducted in Chinese hamster ovary cells demonstrated that the association between TrkB and Shc is constitutive and that the complex undergoes a conformational rearrangement in response to BDNF stimulation (De Vries et al., 2010).

More recently, biosensor mouse models have been developed that allow for the assessment of intracellular signaling molecule activity downstream of RTK signaling *in vivo*. To date, a single study has employed this technology in the examination of neural crest-derived cell activity, using transgenic mouse lines expressing Förster (or fluorescence) resonance energy transfer (FRET) biosensors in conjunction with live imaging by two-photon excitation microscopy (Goto et al., 2013). The authors used transgenic lines harboring PKA, Erk, Rac1, Cdc42, and JNK FRET biosensors (Goto et al., 2013; Kamioka et al., 2012; Komatsu et al., 2011) to demonstrate that PKA activity in migrating enteric neural crest-derived cells is positively correlated with the distribution of GDNF and inversely correlated with Rac1 and Cdc42 activity (Goto et al., 2013). Similar application of *in vivo* biosensors will likely provide a profusion of information on the activity of signaling molecules downstream of RTK induction during NCC development, migration, and differentiation.

4. CONCLUDING REMARKS

Over the past two decades, numerous advances have been made in the growth factor signaling field using biochemical, expression, and genetic knockout approaches that have highlighted the mechanism and function of RTK signaling during murine embryogenesis. A role for several of these receptor families has thus been demonstrated in regulating NCC activity and the development of their derivatives in mammalian embryogenesis. The application of additional techniques, including receptor allelic series, large-scale, quantitative proteomics, and biosensor imaging, promises to reveal novel aspects of RTK signaling during development. Furthermore, the *in vivo* analysis of transcriptional readout in response to individual RTK stimulation will likely provide a wealth of knowledge on the mechanisms by which extracellular growth factors mediate diverse cellular activities.

ACKNOWLEDGMENTS

We thank our laboratory colleagues for their helpful discussions and comments on this chapter. We apologize to authors whose work we were unable to cite due to space limitations. Work in the Soriano laboratory is supported by National Institutes of Health/National

Institute of Dental and Craniofacial Research (NIH/NIDCR) Grants R01DE022363 and R01DE022778 and NYSTEM Grant IIRP N11G-131 to P. S. K. A. F. is additionally supported by NIH/NIDCR Ruth L. Kirschstein NRSA Individual Postdoctoral Fellowship F32DE022719.

REFERENCES

Abu-Issa, R., Smyth, G., Smoak, I., Yamamura, K., & Meyers, E. N. (2002). Fgf8 is required for pharyngeal arch and cardiovascular development in the mouse. *Development, 129*, 4613–4625.

Adams, R. H., Diella, F., Hennig, S., Helmbacher, F., Deutsch, U., & Klein, R. (2001). The cytoplasmic domain of the ligand ephrinB2 is required for vascular morphogenesis but not cranial neural crest migration. *Cell, 104*, 57–69.

Afzal, A. R., Rajab, A., Fenske, C. D., Oldridge, M., Elanko, N., Ternes-Pereira, E., et al. (2000). Recessive Robinow syndrome, allelic to dominant brachydactyly type B, is caused by mutation of ROR2. *Nature Genetics, 25*, 419–422.

Agosti, V., Corbacioglu, S., Ehlers, I., Waskow, C., Sommer, G., Berrozpe, G., et al. (2004). Critical role of Kit-mediated Src kinase but not PI 3-kinase signaling in pro T and pro B cell development. *Journal of Experimental Medicine, 199*, 867–878.

Ahmad, Y., & Lamond, A. I. (2014). A perspective on proteomics in cell biology. *Trends in Cell Biology, 24*, 257–264.

Alessi, D. R., Caudwell, F. B., Andjelkovic, M., Hemmings, B. A., & Cohen, P. (1996). Molecular basis for the substrate specificity of protein kinase B; comparison with MAPKAP kinase-1 and p70 S6 kinase. *FEBS Letters, 399*, 333–338.

Al-Shawi, R., Ashton, S. V., Underwood, C., & Simons, J. P. (2001). Expression of the Ror1 and Ror2 receptor tyrosine kinase genes during mouse development. *Development Genes and Evolution, 211*, 161–171.

Andermarcher, E., Surani, M. A., & Gherardi, E. (1996). Co-expression of the HGF/SF and c-met genes during early mouse embryogenesis precedes reciprocal expression in adjacent tissues during organogenesis. *Developmental Genetics, 18*, 254–266.

Anderson, D. M., Williams, D. E., Tushinski, R., Gimpel, S., Eisenman, J., Cannizzaro, L. A., et al. (1991). Alternate splicing of mRNAs encoding human mast cell growth factor and localization of the gene to chromosome 12q22-q24. *Cell Growth & Differentiation, 2*, 373–378.

Asai, N., Fukuda, T., Wu, Z., Enomoto, A., Pachnis, V., Takahashi, M., et al. (2006). Targeted mutation of serine 697 in the Ret tyrosine kinase causes migration defect of enteric neural crest cells. *Development, 133*, 4507–4516.

Avivi, A., Yayon, A., & Givol, D. (1993). A novel form of FGF receptor-3 using an alternative exon in the immunoglobulin domain III. *FEBS Letters, 330*, 249–252.

Bajaj, M., Waterfield, M. D., Schlessinger, J., Taylor, W. R., & Blundell, T. (1987). On the tertiary structure of the extracellular domains of the epidermal growth factor and insulin receptors. *Biochimica et Biophysica Acta, 916*, 220–226.

Baloh, R. H., Tansey, M. G., Lampe, P. A., Fahrner, T. J., Enomoto, H., Simburger, K. S., et al. (1998). Artemin, a novel member of the GDNF ligand family, supports peripheral and central neurons and signaling through the GFRalpha3-RET receptor complex. *Neuron, 21*, 1291–1302.

Banerjee, S., Gordon, L., Donn, T. M., Berti, C., Moens, C. B., Burden, S. J., et al. (2011). A novel role for MuSK and non-canonical Wnt signaling during segmental neural crest cell migration. *Development, 138*, 3287–3296.

Benito-Gutiérrez, E., Garcia-Fernàndez, J., & Comella, J. X. (2006). Origin and evolution of the Trk family of neurotrophic receptors. *Molecular and Cellular Neuroscience, 31*, 179–192.

Bennett, D. (1956). Developmental analysis of a mutation with pleiotropic effects in the mouse. *Journal of Morphology, 98*, 199–233.
Bladt, F., Riethmacher, D., Isenmann, S., Aguzzi, A., & Birchmeier, C. (1995). Essential role for the c-met receptor in the migration of myogenic precursor cells into the limb bud. *Nature, 376*, 768–771.
Bolande, R. P. (1974). The neurocristopathies: A unifying concept of disease arising in neural crest maldevelopment. *Human Pathology, 5*, 409–429.
Bolande, R. P. (1997). Neurocristopathy: Its growth and development in 20 years. *Pediatric Pathology & Laboratory Medicine, 17*, 1–25.
Boström, H., Willetts, K., Pekny, M., Levéen, P., Lindahl, P., Hedstrand, H., et al. (1996). PDGF-A signaling is a critical event in lung alveolar myofibroblast development and alveogenesis. *Cell, 85*, 863–873.
Bottaro, D. P., Rubin, J. S., Faletto, D. L., Chan, A. M., Kmiecik, T. E., Vande Woude, G. F., et al. (1991). Identification of the hepatocyte growth factor receptor as the c-met proto-oncogene product. *Science, 251*, 802–804.
Brewis, I. A., & Brennan, P. (2010). Proteomics technologies for the global identification and quantification of proteins. *Advances in Protein Chemistry and Structural Biology, 80*, 1–44.
Britsch, S., Li, L., Kirchhoff, S., Theuring, F., Brinkmann, V., Birchmeier, C., et al. (1998). The ErbB2 and ErbB3 receptors and their ligand, neuregulin-1, are essential for development of the sympathetic nervous system. *Genes & Development, 12*, 1825–1836.
Brown, E. J., Albers, M. W., Shin, T. B., Ichikawa, K., Keith, C. T., Lane, W. S., et al. (1994). A mammalian protein targeted by G1-arresting rapamycin–receptor complex. *Nature, 369*, 756–758.
Brückner, K., Pasquale, E. B., & Klein, R. (1997). Tyrosine phosphorylation of transmembrane ligands for Eph receptors. *Science, 275*, 1640–1643.
Buchdunger, E., Zimmermann, J., Mett, H., Meyer, T., Müller, M., Druker, B. J., et al. (1996). Inhibition of the Abl protein-tyrosine kinase in vitro and in vivo by a 2-phenylaminopyrimidine derivative. *Cancer Research, 56*, 100–104.
Buj-Bello, A., Adu, J., Piñón, L. G., Horton, A., Thompson, J., Rosenthal, A., et al. (1997). Neurturin responsiveness requires a GPI-linked receptor and the Ret receptor tyrosine kinase. *Nature, 387*, 721–724.
Bush, J. O., & Soriano, P. (2009). Ephrin-B1 regulates axon guidance by reverse signaling through a PDZ-dependent mechanism. *Genes & Development, 23*, 1586–1599.
Bush, J. O., & Soriano, P. (2010). Ephrin-B1 forward signaling regulates craniofacial morphogenesis by controlling cell proliferation across Eph-ephrin boundaries. *Genes & Development, 24*, 2068–2080.
Cable, J., Jackson, I. J., & Steel, K. P. (1995). Mutations at the W locus affect survival of neural crest-derived melanocytes in the mouse. *Mechanisms of Development, 50*, 139–150.
Chabot, B., Stephenson, D. A., Chapman, V. M., Besmer, P., & Bernstein, A. (1988). The proto-oncogene c-kit encoding a transmembrane tyrosine kinase receptor maps to the mouse W locus. *Nature, 335*, 88–89.
Chan, R., Hardy, W. R., Dankort, D., Laing, M. A., & Muller, W. J. (2004). Modulation of Erbb2 signaling during development: A threshold level of Erbb2 signaling is required for development. *Development, 131*, 5551–5560.
Chan, R., Hardy, W. R., Laing, M. A., Hardy, S. E., & Muller, W. J. (2002). The catalytic activity of the ErbB-2 receptor tyrosine kinase is essential for embryonic development. *Molecular and Cellular Biology, 22*, 1073–1078.
Chen, B., Bronson, R. T., Klaman, L. D., Hampton, T. G., Wang, J. F., Green, P. J., et al. (2000). Mice mutant for Egfr and Shp2 have defective cardiac semilunar valvulogenesis. *Nature Genetics, 24*, 296–299.

Chen, Y., Moon, A. M., & Gaufo, G. O. (2012). Influence of mesodermal Fgf8 on the differentiation of neural crest-derived postganglionic neurons. *Developmental Biology*, *361*, 125–136.

Citri, A., Skaria, K. B., & Yarden, Y. (2003). The deaf and the dumb: The biology of ErbB-2 and ErbB-3. *Experimental Cell Research*, *284*, 54–65.

Copeland, N. G., Gilbert, D. J., Cho, B. C., Donovan, P. J., Jenkins, N. A., Cosman, D., et al. (1990). Mast cell growth factor maps near the steel locus on mouse chromosome 10 and is deleted in a number of steel alleles. *Cell*, *63*, 175–183.

Corson, L. B., Yamanaka, Y., Lai, K. M., & Rossant, J. (2003). Spatial and temporal patterns of ERK signaling during mouse embryogenesis. *Development*, *130*, 4527–4537.

Danielian, P. S., Muccino, D., Rowitch, D. H., Michael, S. K., & McMahon, A. P. (1998). Modification of gene activity in mouse embryos in utero by a tamoxifen-inducible form of Cre recombinase. *Current Biology*, *8*, 1323–1326.

Davis, S., Gale, N. W., Aldrich, T. H., Maisonpierre, P. C., Lhotak, V., Pawson, T., et al. (1994). Ligands for EPH-related receptor tyrosine kinases that require membrane attachment or clustering for activity. *Science*, *266*, 816–819.

Davy, A., Aubin, J., & Soriano, P. (2004). Ephrin-B1 forward and reverse signaling are required during mouse development. *Genes & Development*, *18*, 572–583.

Davy, A., Bush, J. O., & Soriano, P. (2006). Inhibition of gap junction communication at ectopic Eph/ephrin boundaries underlies craniofrontonasal syndrome. *PLoS Biology*, *4*, e315.

Davy, A., & Soriano, P. (2005). Ephrin signaling in vivo: Look both ways. *Developmental Dynamics*, *232*, 1–10.

Davy, A., & Soriano, P. (2007). Ephrin-B2 forward signaling regulates somite patterning and neural crest cell development. *Developmental Biology*, *304*, 182–193.

DeChiara, T. M., Bowen, D. C., Valenzuela, D. M., Simmons, M. V., Poueymirou, W. T., Thomas, S., et al. (1996). The receptor tyrosine kinase MuSK is required for neuromuscular junction formation in vivo. *Cell*, *85*, 501–512.

DeChiara, T. M., Kimble, R. B., Poueymirou, W. T., Rojas, J., Masiakowski, P., Valenzuela, D. M., et al. (2000). Ror2, encoding a receptor-like tyrosine kinase, is required for cartilage and growth plate development. *Nature Genetics*, *24*, 271–274.

Delmas, V., Martinozzi, S., Bourgeois, Y., Holzenberger, M., & Larue, L. (2003). Cre-mediated recombination in the skin melanocyte lineage. *Genesis*, *36*, 73–80.

De Moerlooze, L., Spencer-Dene, B., Revest, J.-M., Hajihosseini, M., Rosewell, I., & Dickson, C. (2000). An important role for the IIIb isoform of fibroblast growth factor receptor 2 (FGFR2) in mesenchymal–epithelial signaling during mouse organogenesis. *Development*, *127*, 483–492.

Deng, C.-X., Wynshaw-Boris, A., Shen, M. M., Daugherty, C., Ornitz, D. M., & Leder, P. (1994). Murine FGFR-1 is required for early postimplantation growth and axial organization. *Genes & Development*, *8*, 3045–3057.

De Vries, L., Finana, F., Cachoux, F., Vacher, B., Sokoloff, P., & Cussac, D. (2010). Cellular BRET assay suggests a conformational rearrangement of preformed TrkB/Shc complexes following BDNF-dependent activation. *Cellular Signalling*, *22*, 158–165.

Ding, H., Wu, X., Boström, H., Kim, I., Wong, N., Tsoi, B., et al. (2004). A specific requirement for PDGF-C in palate formation and PDGFR-alpha signaling. *Nature Genetics*, *36*, 1111–1116.

Ding, H., Wu, X., Kim, I., Tam, P. P., Koh, G. Y., & Nagy, A. (2000). The mouse Pdgfc gene: Dynamic expression in embryonic tissues during organogenesis. *Mechanisms of Development*, *96*, 209–213.

Dodé, C., Levilliers, J., Dupont, J. M., De Paepe, A., Le Dû, N., Soussi-Yanicostas, N., et al. (2003). Loss-of-function mutations in FGFR1 cause autosomal dominant Kallmann syndrome. *Nature Genetics*, *33*, 463–465.

Druker, B. J., Mamon, B. S., & Roberts, T. M. (1989). Oncogenes, growth factors, and signal transduction. *The New England Journal of Medicine, 321*, 1383–1391.

Dunn, L. C. (1937). Studies on spotting patterns: II. Genetic analysis of variegated spotting in the house mouse. *Genetics, 22*, 43–64.

Edery, P., Lyonnet, S., Mulligan, L. M., Pelet, A., Dow, E., Abel, L., et al. (1994). Mutations of the RET proto-oncogene in Hirschsprung's disease. *Nature, 367*, 378–380.

Eketjäll, S., & Ibáñez, C. F. (2002). Functional characterization of mutations in the GDNF gene of patients with Hirschsprung disease. *Human Molecular Genetics, 11*, 325–329.

Ekman, S., Thuresson, E. R., Heldin, C. H., & Rönnstrand, L. (1999). Increased mitogenicity of an alphabeta heterodimeric PDGF receptor complex correlates with lack of RasGAP binding. *Oncogene, 18*, 2481–2488.

Enokido, Y., de Sauvage, F., Hongo, J. A., Ninkina, N., Rosenthal, A., & Buchman, V. L. (1998). GFR alpha-4 and the tyrosine kinase Ret form a functional receptor complex for persephin. *Current Biology, 8*, 1019–1022.

Erickson, S. L., O'Shea, K. S., Ghaboosi, N., Loverro, L., Frantz, G., Bauer, M., et al. (1997). ErbB3 is required for normal cerebellar and cardiac development: A comparison with ErbB2- and heregulin-deficient mice. *Development, 124*, 4999–5011.

Eswarakumar, V. P., Monsonego-Ornan, E., Pines, M., Antonopoulou, I., Morriss-Kay, G. M., & Lonai, P. (2002). The IIIc alternative of Fgfr2 is a positive regulator of bone formation. *Development, 129*, 3783–3793.

Etchevers, H. C., Amiel, J., & Lyonnet, S. (2006). Molecular basis of human neurocristopathies. *Advances in Experimental Medicine and Biology, 589*, 213–234.

Falardeau, J., Chung, W. C., Beenken, A., Raivio, T., Plummer, L., Sidis, Y., et al. (2008). Decreased FGF8 signaling causes deficiency of gonadotropin-releasing hormone in humans and mice. *Journal of Clinical Investigation, 118*, 2822–2831.

Fantauzzo, K. A., & Soriano, P. (2014). PI3K-mediated PDGFRα signaling regulates survival and proliferation in skeletal development through p53-dependent intracellular pathways. *Genes & Development, 28*, 1005–1017.

Fernández, R. M., Sánchez-Mejías, A., Mena, M. D., Ruiz-Ferrer, M., López-Alonso, M., Antiñolo, G., et al. (2009). A novel point variant in NTRK3, R645C, suggests a role for this gene in the pathogenesis of Hirschsprung disease. *Annals of Human Genetics, 73*, 19–25.

Foster, K. E., Gordon, J., Cardenas, K., Veiga-Fernandes, H., Makinen, T., Grigorieva, E., et al. (2010). EphB–ephrin-B2 interactions are required for thymus migration during organogenesis. *Proceedings of the National Academy of Sciences of the United States of America, 107*, 13414–13419.

Frank, D. U., Fotheringham, L. K., Brewer, J. A., Muglia, L. J., Tristani-Firouzi, M., Capecchi, M. R., et al. (2002). An Fgf8 mouse mutant phenocopies human 22q11 deletion syndrome. *Development, 129*, 4591–4603.

Gale, N. W., Holland, S. J., Valenzuela, D. M., Flenniken, A., Pan, L., Ryan, T. E., et al. (1996). Eph receptors and ligands comprise two major specificity subclasses and are reciprocally compartmentalized during embryogenesis. *Neuron, 17*, 9–19.

Garcia-Osta, A., Tsokas, P., Pollonini, G., Landau, E. M., Blitzer, R., & Alberini, C. M. (2006). MuSK expressed in the brain mediates cholinergic responses, synaptic plasticity, and memory formation. *Journal of Neuroscience, 26*, 7919–7932.

García-Suárez, O., Germanà, A., Hannestad, J., Ciriaco, E., Laurà, R., Naves, J., et al. (2000). TrkA is necessary for the normal development of the murine thymus. *Journal of Neuroimmunology, 108*, 11–21.

Gassmann, M., Casagranda, F., Orioli, D., Simon, H., Lai, C., Klein, R., et al. (1995). Aberrant neural and cardiac development in mice lacking the ErbB4 neuregulin receptor. *Nature, 378*, 390–394.

Gautam, M., Noakes, P. G., Moscoso, L., Rupp, F., Scheller, R. H., Merlie J. P., et al. (1996). Defective neuromuscular synaptogenesis in agrin-deficient mutant mice. *Cell, 85*, 525–535.

Geissler, E. N., McFarland, E. C., & Russell, E. S. (1981). Analysis of pleiotropism at the dominant white-spotting (W) locus of the house mouse: A description of ten new W alleles. *Genetics, 97*, 337–361.

Geissler, E. N., Ryan, M. A., & Housman, D. E. (1988). The dominant-white spotting (W) locus of the mouse encodes the c-kit proto-oncogene. *Cell, 55*, 185–192.

Glass, D. J., Bowen, D. C., Stitt, T. N., Radziejewski, C., Bruno, J., Ryan, T. E., et al. (1996). Agrin acts via a MuSK receptor complex. *Cell, 85*, 513–523.

Glenney, J. R., Jr., Zokas, L., & Kamps, M. P. (1988). Monoclonal antibodies to phosphotyrosine. *Journal of Immunological Methods, 109*, 277–285.

Golding, J. P., Trainor, P., Krumlauf, R., & Gassmann, M. (2000). Defects in pathfinding by cranial neural crest *cells* in mice lacking the neuregulin receptor ErbB4. *Nature Cell Biology, 2*, 103–109.

Goto, A., Sumiyama, K., Kamioka, Y., Nakasyo, E., Ito, K., Iwasaki, M., et al. (2013). GDNF and endothelin 3 regulate migration of enteric neural crest-derived cells via protein kinase A and Rac1. *Journal of Neuroscience, 33*, 4901–4912.

Graveel, C. R., Tolbert, D., & Vande Woude, G. F. (2013). MET: A critical player in tumorigenesis and therapeutic target. *Cold Spring Harbor Perspectives in Biology, 5*, a009209.

Gruneberg, H. (1942). Inherited macrocytic anaemias in the house mouse. II. Dominance relationships. *Journal of Genetics, 43*, 285–293.

Gu, C., Rodriguez, E. R., Reimert, D. V., Shu, T., Fritzsch, B., Richards, L. J., et al. (2003). Neuropilin-1 conveys semaphorin and VEGF signaling during neural and cardiovascular development. *Developmental Cell, 5*, 45–57.

Guris, D. L., Fantes, J., Tara, D., Druker, B. J., & Imamoto, A. (2001). Mice lacking the homologue of the human 22q11.2 gene CRKL phenocopy neurocristopathies of DiGeorge syndrome. *Nature Genetics, 27*, 293–298.

Guy, P. M., Platko, J. V., Cantley, L. C., Cerione, R. A., & Carraway, K. L., 3rd. (1994). Insect cell-expressed p180erbB3 possesses an impaired tyrosine kinase activity. *Proceedings of the National Academy of Sciences of the United States of America, 91*, 8132–8136.

Hamilton, T. G., Klinghoffer, R. A., Corrin, P. D., & Soriano, P. (2003). Evolutionary divergence of platelet-derived growth factor alpha receptor signaling mechanisms. *Molecular and Cellular Biology, 23*, 4013–4025.

He, F., & Soriano, P. (2013). A critical role for PDGFRα signaling in medial nasal process development. *PLoS Genetics, 9*, e1003851.

He, Z., & Tessier-Lavigne, M. (1997). Neuropilin is a receptor for the axonal chemorepellent Semaphorin II. *Cell, 90*, 739–751.

He, F., Xiong, W., Yu, X., Espinoza-Lewis, R., Liu, C., Gu, S., et al. (2008). Wnt5a regulates directional cell migration and cell proliferation via Ror2-mediated noncanonical pathway in mammalian palate development. *Development, 135*, 3871–3879

Hébert, J. M., & McConnell, S. K. (2000). Targeting of cre to the Foxg1 (BF-1) locus mediates loxP recombination in the telencephalon and other developing head structures. *Developmental Biology, 222*, 296–306.

Hellmich, H. L., Kos, L., Cho, E. S., Mahon, K. A., & Zimmer, A. (1996). Embryonic expression of glial cell-line derived neurotrophic factor (GDNF) suggests multiple developmental roles in neural differentiation and epithelial-mesenchymal interactions. *Mechanisms of Development, 54*, 95–105.

Heuckeroth, R. O., Lampe, P. A., Johnson, E. M., & Milbrandt, J. (1998). Neurturin and GDNF promote proliferation and survival of enteric neuron and glial progenitors in vitro. *Developmental Biology, 200*, 116–129.

Heuzé, Y., Singh, N., Basilico, C., Jabs, E. W., Holmes, G., & Richtsmeier, J. T. (2014). Morphological comparison of the craniofacial phenotypes of mouse models expressing the Apert FGFR2 S252W mutation in neural crest- or mesoderm-derived tissues. *Bone*, *63*, 101–109.

Hirai, H., Maru, Y., Hagiwara, K., Nishida, J., & Takaku, F. (1987). A novel putative tyrosine kinase receptor encoded by the eph gene. *Science*, *238*, 1717–1720.

Hoch, R. V., & Soriano, P. (2006). Context-specific requirements for Fgfr1 signaling through Frs2 and Frs3 during mouse development. *Development*, *133*, 663–673.

Holland, S. J., Gale, N. W., Mbamalu, G., Yancopoulos, G. D., Henkemeyer, M., & Pawson, T. (1996). Bidirectional signaling through the EPH-family receptor Nuk and its transmembrane ligands. *Nature*, *383*, 722–725.

Holmes, G., & Basilico, C. (2012). Mesodermal expression of Fgfr2S252W is necessary and sufficient to induce craniosynostosis in a mouse model of Apert syndrome. *Developmental Biology*, *368*, 283–293.

Huang, E., Nocka, K., Beier, D. R., Chu, T.-Y., Buck, J., Lahm, H.-W., et al. (1990). The hematopoietic growth factor KL is encoded by the Sl locus and is the ligand for the c-kit receptor, the gene product of the W locus. *Cell*, *63*, 225–233.

Indo, Y., Tsuruta, M., Hayashida, Y., Karim, M. A., Ohta, K., Kawano, T., et al. (1996). Mutations in the TRKA/NGF receptor gene in patients with congenital insensitivity to pain with anhidrosis. *Nature Genetics*, *13*, 485–488.

Iwamoto, T., Taniguchi, M., Asai, N., Ohkusu, K., Nakashima, I., & Takahashi, M. (1993). cDNA cloning of mouse ret proto-oncogene and its sequence similarity to the cadherin superfamily. *Oncogene*, *8*, 1087–1091.

Jain, S., Knoten, A., Hoshi, M., Wang, H., Vohra, B., Heuckeroth, R. O., et al. (2010). Organotypic specificity of key RET adaptor-docking sites in the pathogenesis of neurocristopathies and renal malformations in mice. *Journal of Clinical Investigation*, *120*, 778–790.

Jijiwa, M., Fukuda, T., Kawai, K., Nakamura, A., Kurokawa, K., Murakumo, Y., et al. (2004). A targeting mutation of tyrosine 1062 in Ret causes a marked decrease of enteric neurons and renal hypoplasia. *Molecular and Cellular Biology*, *24*, 8026–8036.

Jing, L., Lefebvre, J. L., Gordon, L. R., & Granato, M. (2009). Wnt signals organize synaptic prepattern and axon guidance through the zebrafish unplugged/MuSK receptor. *Neuron*, *61*, 721–733.

Jing, S., Wen, D., Yu, Y., Hoist, P. L., Luo, Y., Fang, M., et al. (1996). GDNF-induced activation of the ret protein tyrosine kinase is mediated by GDNFR-alpha, a novel receptor for GDNF. *Cell*, *85*, 1113–1124.

Johnson, D. E., Lu, J., Chen, H., Werner, S., & Williams, L. T. (1991). The human fibroblast growth factor receptor genes: A common structural arrangement underlies the mechanisms for generating receptor forms that differ in their third immunoglobulin domain. *Molecular and Cellular Biology*, *11*, 4627–4634.

Jung, J. W., Shin, W. S., Song, J., & Lee, S. T. (2004). Cloning and characterization of the full-length mouse Ptk7 cDNA encoding a defective receptor protein tyrosine kinase. *Gene*, *328*, 75–84.

Kamioka, Y., Sumiyama, K., Mizuno, R., Sakai, Y., Hirata, E., Kiyokawa, E., et al. (2012). Live imaging of protein kinase activities in transgenic mice expressing FRET biosensors. *Cell Structure and Function*, *37*, 65–73.

Kaplan, D. R., Hempstead, B. L., Martin-Zanca, D., Chao, M. V., & Parada, L. F. (1991). The trk proto-oncogene product: A signal transducing receptor for nerve growth factor. *Science*, *252*, 554–558.

Kawakami, A., Kitsukawa, T., Takagi, S., & Fujisawa, H. (1996). Developmentally regulated expression of a cell surface protein, neuropilin, in the mouse nervous system. *Journal of Neurobiology*, *29*, 1–17.

Kelleher, F. C., O'Sullivan, H., Smyth, E., McDermott, R., & Viterbo, A. (2013). Fibroblast growth factor receptors, developmental corruption and malignant disease. *Carcinogenesis, 34*, 2198–2205.

Kim, N., Stiegler, A. L., Cameron, T. O., Hallock, P. T., Gomez, A. M., Huang, J. H., et al. (2008). Lrp4 is a receptor for Agrin and forms a complex with MuSK. *Cell, 135*(2), 334–342.

Kimura, Y., Jones, N., Klüppel, M., Hirashima, M., Tachibana, K., Cohn, J. B., et al. (2004). Targeted mutations of the juxtamembrane tyrosines in the Kit receptor tyrosine kinase selectively affect multiple cell lineages. *Proceedings of the National Academy of Sciences of the United States of America, 101*, 6015–6020.

Klapper, L. N., Glathe, S., Vaisman, N., Hynes, N. E., Andrews, G. C., Sela, M., et al. (1999). The ErbB-2/HER2 oncoprotein of human carcinomas may function solely as a shared coreceptor for multiple stroma-derived growth factors. *Proceedings of the National Academy of Sciences of the United States of America, 96*, 4995–5000.

Klein, R., Jing, S. Q., Nanduri, V., O'Rourke, E., & Barbacid, M. (1991). The trk proto-oncogene encodes a receptor for nerve growth factor. *Cell, 65*, 189–197.

Klein, R., Martin-Zanca, D., Barbacid, M., & Parada, L. F. (1990). Expression of the tyrosine kinase receptor gene trkB is confined to the murine embryonic and adult nervous system. *Development, 109*, 845–850.

Klein, R., Nanduri, V., Jing, S. A., Lamballe, F., Tapley, P., Bryant, S., et al. (1991). The trkB tyrosine protein kinase is a receptor for brain-derived neurotrophic factor and neurotrophin-3. *Cell, 66*, 395–403.

Klein, R. D., Sherman, D., Ho, W. H., Stone, D., Bennett, G. L., Moffat, B., et al. (1997). A GPI-liked protein that interacts with Ret to form a candidate neurturin receptor. *Nature, 387*, 717–721.

Klein, R., Silos-Santiago, I., Smeyne, R. J., Lira, S. A., Brambilla, R., Bryant, S., et al. (1994). Disruption of the neurotrophin-3 receptor gene trkC eliminates Ia muscle afferents and results in abnormal movements. *Nature, 368*, 249–251.

Klein, R., Smeyne, R. J., Wurst, W., Long, L. K., Auerbach, B. A., Joyner, A. L., et al. (1993). Targeted disruption of the trkB neurotrophin receptor gene results in nervous system lesions and neonatal death. *Cell, 75*, 113–122.

Klinghoffer, R. A., Hamilton, T. G., Hoch, R., & Soriano, P. (2002). An allelic series at the PDGFalphaR locus indicates unequal contributions of distinct signaling pathways during development. *Developmental Cell, 2*, 103–113.

Klinghoffer, R. A., Mueting-Nelsen, P. F., Faerman, A., Shani, M., & Soriano, P. (2001). The two PDGF receptors maintain conserved signaling in vivo despite divergent embryological functions. *Molecular Cell, 7*, 343–354.

Koch, S., & Claesson-Welsh, L. (2012). Signal transduction by vascular endothelial growth factor receptors. *Cold Spring Harbor Perspectives in Medicine, 2*, a006502.

Kolodkin, A. L., Levengood, D. V., Rowe, E. G., Tai, Y. T., Giger, R. J., & Ginty, D. D. (1997). Neuropilin is a semaphorin III receptor. *Cell, 90*, 753–762.

Komatsu, N., Aoki, K., Yamada, M., Yukinaga, H., Fujita, Y., Kamioka, Y., et al. (2011). Development of an optimized backbone of FRET biosensors for kinases and GTPases. *Molecular Biology of the Cell, 22*, 4647–4656.

Kos, L., Aronzon, A., Takayama, H., Maina, F., Ponzetto, C., Merlino, G., et al. (1999). Hepatocyte growth factor/scatter factor-MET signaling in neural crest-derived melanocyte development. *Pigment Cell Research, 12*, 13–21.

Kumar, P., Ferns, M. J., & Meizel, S. (2006). Identification of agrinSN isoform and muscle-specific receptor tyrosine kinase (MuSK) in sperm. *Biochemical and Biophysical Research Communications, 342*, 522–528.

Lamballe, F., Klein, R., & Barbacid, M. (1991). trkC, a new member of the trk family of tyrosine protein kinases, is a receptor for neurotrophin-3. *Cell, 66*, 967–979.

Leahy, D. J. (2004). Structure and function of the epidermal growth factor (EGF/ErbB) family of receptors. *Advances in Protein Chemistry*, *68*, 1–27.

Lee, P. L., Johnson, D. E., Cousens, L. S., Fried, V. A., & Williams, L. T. (1989). Purification and complementary DNA cloning of a receptor for basic fibroblast growth factor. *Science*, *245*, 57–60.

Lee, K. F., Simon, H., Chen, H., Bates, B., Hung, M. C., & Hauser, C. (1995). Requirement for neuregulin receptor erbB2 in neural and cardiac development. *Nature*, *378*, 394–398.

Lemmon, M. A., & Schlessinger, J. (2010). Cell signaling by receptor tyrosine kinases. *Cell*, *141*, 1117–1134.

Levanti, M. B., Esteban, I., Ciriaco, E., Pérez-Piñera, P., Cabo, R., García-Suarez, O., et al. (2009). Enteric glial cells express full-length TrkB and depend on TrkB expression for normal development. *Neuroscience Letters*, *454*, 16–21.

Levéen, P., Pekny, M., Gebre-Medhin, S., Swolin, B., Larsson, E., & Betsholtz, C. (1994). Mice deficient for PDGF B show renal, cardiovascular, and hematological abnormalities. *Genes & Development*, *8*, 1875–1887.

Lewis, A. E., Vasudevan, H. N., O'Neill, A. K., Soriano, P., & Bush, J. O. (2013). The widely used Wnt1-Cre transgene causes developmental phenotypes by ectopic activation of Wnt signaling. *Developmental Biology*, *379*, 229–234.

Li, J., Chen, F., & Epstein, J. A. (2000). Neural crest expression of Cre recombinase directed by the proximal Pax3 promoter in transgenic mice. *Genesis*, *26*, 162–164.

Lin, D., Gish, G. D., Songyang, Z., & Pawson, T. (1999). The carboxyl terminus of B class ephrins constitutes a PDZ domain binding motif. *Journal of Biological Chemistry*, *274*, 3726–3733.

Liu, H., Sadygov, R. G., & Yates, J. R., 3rd. (2004). A model for random sampling and estimation of relative protein abundance in shotgun proteomics. *Analytical Chemistry*, *76*, 4193–4201.

Lu, X., Borchers, A. G., Jolicoeur, C., Rayburn, H., Baker, J. C., & Tessier-Lavigne, M. (2004). PTK7/CCK-4 is a novel regulator of planar cell polarity in vertebrates. *Nature*, *430*, 93–98.

Luikart, B. W., Nef, S., Shipman, T., & Parada, L. F. (2003). In vivo role of truncated trkb receptors during sensory ganglion neurogenesis. *Neuroscience*, *117*, 847–858.

Luo, W., Wickramasinghe, S. R., Savitt, J. M., Griffin, J. W., Dawson, T. M., & Ginty, D. D. (2007). A hierarchical NGF signaling cascade controls Ret-dependent and Ret-independent events during development of nonpeptidergic DRG neurons. *Neuron*, *54*, 739–754.

MacArthur, C. A., Lawshé, A., Xu, J., Santos-Ocampo, S., Heikinheimo, M., Chellaiah, A. T., et al. (1995). FGF-8 isoforms activate receptor splice forms that are expressed in mesenchymal regions of mouse development. *Development*, *121*, 3603–3613.

Macatee, T. L., Hammond, B. P., Arenkiel, B. R., Frances, L., Frank, D. U., & Moon, A. M. (2003). Ablation of specific expression domains reveals discrete functions of ectoderm- and endoderm-derived FGF8 during cardiovascular and pharyngeal development. *Development*, *130*, 6361–6374.

MacKenzie, M. A. F., Jordan, S. A., Budd, P. S., & Jackson, I. J. (1997). Activation of the receptor tyrosine kinase Kit is required for the proliferation of melanoblasts in the mouse embryo. *Developmental Biology*, *192*, 99–107.

Manning, B. D., Tee, A. R., Logsdon, M. N., Blenis, J., & Cantley, L. C. (2002). Identification of the tuberous-sclerosis complex-2 tumor suppressor gene product tuberin as a target of the phosphoinositide 3-kinase/Akt pathway. *Molecular Cell*, *10*, 151–162.

Martin-Zanca, D., Barbacid, M., & Parada, L. F. (1990). Expression of the trk proto-oncogene is restricted to the sensory cranial and spinal ganglia of neural crest origin in mouse development. *Genes & Development*, *4*, 683–694.

Masiakowski, P., & Carroll, R. D. (1992). A novel family of cell surface receptors with tyrosine kinase-like domain. *Journal of Biological Chemistry*, *267*, 26181–26190.
Masiakowski, P., & Yancopoulos, G. D. (1998). The Wnt receptor CRD domain is also found in MuSK and related orphan receptor tyrosine kinases. *Current Biology*, *8*, R407.
Matsuda, T., Nomi, M., Ikeya, M., Kani, S., Oishi, I., Terashima, T., et al. (2001). Expression of the receptor tyrosine kinase genes, Ror1 and Ror2, during mouse development. *Mechanisms of Development*, *105*, 153–156.
Matsui, Y., Zsebo, K. M., & Hogan, B. L. (1990). Embryonic expression of a haematopoietic growth factor encoded by the Sl locus and the ligand for c-kit. *Nature*, *347*, 667–669.
Mayor, R., & Theveneau, E. (2013). The neural crest. *Development*, *140*, 2247–2251.
McCoshen, J. A., & McCallion, D. J. (1975). A study of the primordial germ cells during their migratory phase in Steel mutant mice. *Experientia*, *31*, 589–590.
McLennan, R., Teddy, J. M., Kasemeier-Kulesa, J. C., Romine, M. H., & Kulesa, P. M. (2010). Vascular endothelial growth factor (VEGF) regulates cranial neural crest migration in vivo. *Developmental Biology*, *339*, 114–125.
Merrill, A. E., Bochukova, E. G., Brugger, S. M., Ishii, M., Pilz, D. T., Wall, S. A., et al. (2006). Cell mixing at a neural crest-mesoderm boundary and deficient ephrin-Eph signaling in the pathogenesis of craniosynostosis. *Human Molecular Genetics*, *15*, 1319–1328.
Meyer, D., Yamaai, T., Garratt, A., Riethmacher-Sonnenberg, E., Kane, D., Theill, L. E., et al. (1997). Isoform-specific expression and function of neuregulin. *Development*, *124*, 3575–3586.
Meyers, E. N., Lewandoski, M., & Martin, G. R. (1998). An Fgf8 mutant allelic series generated by Cre- and Flp-mediated recombination. *Nature Genetics*, *18*, 136–141.
Miettinen, P. J., Berger, J. E., Meneses, J., Phung, Y., Pedersen, R. A., Werb, Z., et al. (1995). Epithelial immaturity and multiorgan failure in mice lacking epidermal growth factor receptor. *Nature*, *376*, 337–341.
Miettinen, P. J., Chin, J. R., Shum, L., Slavkin, H. C., Shuler, C. F., Derynck, R., et al. (1999). Epidermal growth factor receptor function is necessary for normal craniofacial development and palate closure. *Nature Genetics*, *22*, 69–73.
Miki, T., Bottaro, D. P., Fleming, T. P., Smith, C. L., Burgess, W. H., Chan, A. M., et al. (1992). Determination of ligand-binding specificity by alternative splicing: Two distinct growth factor receptors encoded by a single gene. *Proceedings of the National Academy of Sciences of the United States of America*, *89*, 246–250.
Moon, A. M., Guris, D. L., Seo, J., Li, L., Hammond, J., Talbot, A., et al. (2006). Crkl deficiency disrupts Fgf8 signaling in a mouse model of 22q11 deletion syndromes. *Developmental Cell*, *10*, 71–80.
Moore, M. W., Klein, R. D., Fariñas, I., Sauer, H., Armanini, M., Phillips, H., et al. (1996). Renal and neuronal abnormalities in mice lacking GDNF. *Nature*, *382*, 76–79.
Moqrich, A., Earley, T. J., Watson, J., Andahazy, M., Backus, C., Martin-Zanca, D., et al. (2004). Expressing TrkC from the TrkA locus causes a subset of dorsal root ganglia neurons to switch fate. *Nature Neuroscience*, *7*, 812–818.
Moritz, A., Li, Y., Gui, A., Villén, J., Wang, Y., MacNeill, J., et al. (2010). Akt-RSK-S6 kinase signaling networks activated by oncogenic receptor tyrosine kinases. *Science Signaling*, *3*, ra64.
Morris, J. K., Lin, W., Hauser, C., Marchuk, Y., Getman, D., & Lee, K. F. (1999). Rescue of the cardiac defect in ErbB2 mutant mice reveals essential roles of ErbB2 in peripheral nervous system development. *Neuron*, *23*, 273–283.
Morrison-Graham, K., Schatteman, G. C., Bork, T., Bowen-Pope, D. F., & Weston, J. A. (1992). A PDGF receptor mutation in the mouse (Patch) perturbs the development of a non-neuronal subset of neural crest-derived cells. *Development*, *115*, 133–142.
Muzumdar, M. D., Tasic, B., Miyamichi, K., Li, L., & Luo, L. (2007). A global double-fluorescent Cre reporter mouse. *Genesis*, *45*, 593–605.

Nelson, D. K., & Williams, T. (2004). Frontonasal process-specific disruption of AP-2alpha results in postnatal midfacial hypoplasia, vascular anomalies, and nasal cavity defects. *Developmental Biology, 267*, 72–92.
Nishita, M., Yoo, S. K., Nomachi, A., Kani, S., Sougawa, N., Ohta, Y., et al. (2006). Filopodia formation mediated by receptor tyrosine kinase Ror2 is required for Wnt5a-induced cell migration. *Journal of Cell Biology, 175*, 555–562.
Noh, H., Park, E., & Park, S. (2014). In vivo expression of ephrinA5-Fc in mice results in cephalic neural crest agenesis and craniofacial abnormalities. *Molecules and Cells, 37*, 59–65.
Nomachi, A., Nishita, M., Inaba, D., Enomoto, M., Hamasaki, M., & Minami, Y. (2008). Receptor tyrosine kinase Ror2 mediates Wnt5a-induced polarized cell migration by activating c-Jun N-terminal kinase via actin-binding protein filamin A. *Journal of Biological Chemistry, 283*, 27973–27981.
Nomi, M., Oishi, I., Kani, S., Suzuki, H., Matsuda, T., Yoda, A., et al. (2001). Loss of mRor1 enhances the heart and skeletal abnormalities in mRor2-deficient mice: Redundant and pleiotropic functions of mRor1 and mRor2 receptor tyrosine kinases. *Molecular and Cellular Biology, 21*, 8329–8335.
Obata, T., Yaffe, M. B., Leparc, G. G., Piro, E. T., Maegawa, H., Kashiwagi, A., et al. (2000). Peptide and protein library screening defines optimal substrate motifs for AKT/PKB. *Journal of Biological Chemistry, 17*, 36108–36115.
Oishi, I., Suzuki, H., Onishi, N., Takada, R., Kani, S., Ohkaware, B., et al. (2003). The receptor tyrosine kinase Ror2 is involved in non-canonical Wnt5a/JNK signaling pathway. *Genes to Cells, 8*, 645–654.
Oishi, I., Takeuchi, S., Hashimoto, R., Nagabukuro, A., Ueda, T., Liu, Z. J., et al. (1999). Spatio-temporally regulated expression of receptor tyrosine kinases, mRor1, mRor2, during mouse development: Implications in development and function of the nervous system. *Genes to Cells, 4*, 41–56.
Olsen, J. V., Blagoev, B., Gnad, F., Macek, B., Kumar, C., Mortensen, P., et al. (2006). Global, in vivo, and site-specific phosphorylation dynamics in signaling networks. *Cell, 127*, 635–648.
Ong, S. E., Blagoev, B., Kratchmarovak, I., Kristensen, D. B., Steen, H., Pandey, A., et al. (2002). Stable isotope labeling by amino acids in cell culture, SILAC, as a simple and accurate approach to expression proteomics. *Molecular & Cellular Proteomics, 1*, 376–386.
Ornitz, D. M., Xu, J., Colvin, J. S., McEwen, D. G., MacArthur, C. A., Coulier, F., et al. (1996). Receptor specificity of the fibroblast growth factor family. *Journal of Biological Chemistry, 271*, 15292–15297.
Orr-Urtreger, A., Avivi, A., Zimmer, Y., Givol, D., Yarden, Y., & Lonai, P. (1990). Developmental expression of c-kit, a proto-oncogene encoded by the W locus. *Development, 109*, 911–923.
Orr-Urtreger, A., Bedford, M. T., Burakova, T., Arman, E., Zimmer, Y., Yayon, A., et al. (1993). Developmental localization of the splicing alternatives of fibroblast growth factor receptor-2 (FGFR2). *Developmental Biology, 158*, 475–486.
Orr-Urtreger, A., & Lonai, P. (1992). Platelet-derived growth factor-A and its receptor are expressed in separate, but adjacent cells layers of the mouse embryo. *Development, 115*, 1045–1058.
Pachnis, V., Mankoo, B., & Costantini, F. (1993). Expression of the c-ret proto-oncogene during mouse embryogenesis. *Development, 119*, 1005–1017.
Park, S. K., Lee, H. S., & Lee, S. T. (1996). Characterization of the human full-length PTK7 cDNA encoding a receptor protein tyrosine kinase-like molecule closely related to chick KLG. *Journal of Biochemistry, 119*, 235–239.
Park, E. J., Ogden, L. A., Talbot, A., Evans, S., Cai, C.-L., Black, B. L., et al. (2006). Required, tissue-specific roles for Fgf8 in outflow tract formation and remodeling. *Development, 133*, 2419–2433.

Park, E. J., Sun, X., Nichol, P., Saijoh, Y., Martin, J. F., & Moon, A. M. (2008). System for tamoxifen-inducible expression of cre-recombinase from the Foxa2 locus in mice. *Developmental Dynamics, 237*, 447–453.

Partanen, J., Schwartz, L., & Rossant, J. (1998). Opposite phenotypes of hypomorphic and Y766 phosphorylation site mutations reveal a function for Fgfr1 in anteroposterior patterning of mouse embryos. *Genes & Development, 12*, 2332–2344.

Paudyal, A., Damrau, C., Patterson, V. L., Ermakov, A., Formstone, C., Lalanne, Z., et al. (2010). The novel mouse mutant, chuzhoi, has disruption of Ptk7 protein and exhibits defects in neural tube, heart and lung development and abnormal planar cell polarity in the ear. *BMC Developmental Biology, 10*, 87.

Person, A. D., Beiraghi, S., Sieben, C. M., Hermanson, S., Neumann, A. N., Robu, M. E., et al. (2010). WNT5A mutations in patients with autosomal dominant Robinow syndrome. *Developmental Dynamics, 239*, 327–337.

Pichel, J. G., Shen, L., Sheng, H. Z., Granholm, A.-C., Drago, J., Grinberg, A., et al. (1996). Defects in enteric innervation and kidney development in mice lacking GDNF. *Nature, 382*, 73–76.

Pietri, T., Eder, O., Blanche, M., Thiery, J. P., & Dufour, S. (2003). The human tissue plasminogen activator-Cre mouse: A new tool for targeting specifically neural crest cells and their derivatives in vivo. *Developmental Biology, 259*, 176–187.

Postigo, A., Calella, A. M., Fritzsch, B., Knipper, M., Katz, D., Ellers, A., et al. (2002). Distinct requirements for TrkB and TrkC signaling in target innervation by sensory neurons. *Genes & Development, 16*, 633–645.

Prigge, J. R., Wiley, J. A., Talago, E. A., Young, E. M., Johns, L. L., Kundert J. A., et al. (2013). Nuclear double-fluorescent reporter for in vivo and ex vivo analyses of biological transitions in mouse nuclei. *Mammalian Genome, 24*, 389–399.

Qui, F. H., Ray, P., Brown, K., Barker, P. E., Jhanwar, S., Ruddle, F. H., et al. (1988). Primary structure of c-kit: Relationship with the CSF-1/PDGF receptor kinase family—Oncogenic activation of v-kit involves deletion of extracellular domain and C terminus. *EMBO Journal, 7*, 1003–1011.

Rapraeger, A. C., Krufka, A., & Olwin, B. B. (1991). Requirement of heparan sulfate for bFGF-mediated fibroblast growth and myoblast differentiation. *Science, 252*, 1705–1708.

Reid, B. S., Yang, H., Melvin, V. S., Taketo, M. M., & Williams, T. (2011). Ectodermal Wnt/β-catenin signaling shapes the mouse face. *Developmental Biology, 349*, 261–269.

Richarte, A. M., Mead, H. B., & Tallquist, M. D. (2007). Cooperation between the PDGF receptors in cardiac neural crest cell migration. *Developmental Biology, 306*, 785–796.

Riethmacher, D., Sonnenberg-Riethmacher, E., Brinkmann, V., Yamaai, T., Lewin, G. R., & Birchmeier, C. (1997). Severe neuropathies in mice with targeted mutations in the ErbB3 receptor. *Nature, 389*, 725–730.

Romeo, G., Ronchetto, P., Luo, Y., Barone, V., Seri, M., Ceccherini, I., et al. (1994). Point mutations affecting the tyrosine kinase domain of the RET proto-oncogene in Hirschsprung's disease. *Nature, 367*, 377–378.

Ross, P. L., Huang, Y. N., Marchese, J. N., Williamson, B., Parker, K., Hattan, S., et al. (2004). Multiplexed protein quantitation in Saccharomyces cerevisiae using amine-reactive isobaric tagging reagents. *Molecular & Cellular Proteomics, 3*, 1154–1169.

Ruiz-Ferrer, M., Fernandez, R. M., Antiñolo, G., Lopez-Alonso, M., & Borrego, S. (2008). NTF-3, a gene involved in the enteric nervous system development, as a candidate gene for Hirschsprung disease. *Journal of Pediatric Surgery, 43*, 1308–1311.

Rupp, E., Siegbahn, A., Rönnstrand, L., Wernstedt, C., Claesson-Welsh, L., & Heldin, C. H. (1994). A unique autophosphorylation site in the platelet-derived growth factor alpha receptor from a heterodimeric receptor complex. *European Journal of Biochemistry, 225*, 29–41.

Saga, Y., Miyagawa-Tomita, S., Takagi, A., Kitajima, S., Miyazaki, J., & Inoue, T. (1999). MesP1 is expressed in the heart precursor cells and required for the formation of a single heart tube. *Development, 126*, 3437–3447.

Sánchez, M. P., Silos-Santiago, I., Frisén, J., He, B., Lira, S. A., & Barbacid, M. (1996). Renal agenesis and the absence of enteric neurons in mice lacking GDNF. *Nature, 382*, 70–73.

Sarvella, P. A., & Russell, L. B. (1956). Steel, a new dominant gene in the house mouse. *Journal of Heredity, 47*, 123–128.

Schleiffarth, J. R., Person, A. D., Martinsen, B. J., Sukovich, D. J., Neumann, A., Baker, C. V., et al. (2007). Wnt5a is required for cardiac outflow tract septation in mice. *Pediatric Research, 61*, 386–391.

Schlessinger, J. (2000). Cell signaling by receptor tyrosine kinases. *Cell, 103*, 211–225.

Schmidt, C., Bladt, F., Goedecke, S., Brinkmann, V., Zschiesche, W., Sharpe, M., et al. (1995). Scatter factor/hepatocyte growth factor is essential for liver development. *Nature, 373*, 699–702.

Schuchardt, A., D'Agati, V., Larsson-Blomberg, L., Costantini, F., & Pachnis, V. (1994). Defects in the kidney and enteric nervous system of mice lacking the tyrosine kinase receptor Ret. *Nature, 367*, 380–383.

Shibuya, M., Yamaguchi, S., Yamane, A., Ikeda, T., Tojo, A., Matsushime, H., et al. (1990). Nucleotide sequence and expression of a novel human receptor-type tyrosine kinase gene (flt) closely related to the fms family. *Oncogene, 5*, 519–524.

Sibilia, M., & Wagner, E. F. (1995). Strain-dependent epithelial defects in mice lacking the EGF receptor. *Science, 269*, 234–238.

Siddiqui, S., Cong, W.-N., Daimon, C. M., Martin, B., & Maudsley, S. (2013). BRET biosensor analysis of receptor tyrosine kinase functionality. *Frontiers in Endocrinology, 4*, 46.

Smeyne, R. J., Klein, R., Schnapp, A., Long, L. K., Bryant, S., Lewin, A., et al. (1994). Severe sensory and sympathetic neuropathies in mice carrying a disrupted Trk/NGF receptor gene. *Nature, 368*, 246–249.

Sonnenberg, E., Meyer, D., Weidner, K. M., & Birchmeier, C. (1993). Scatter factor/hepatocyte growth factor and its receptor, the c-met tyrosine kinase, can mediate a signal exchange between mesenchyme and epithelia during mouse development. *Journal of Cell Biology, 123*, 223–235.

Soriano, P. (1994). Abnormal kidney development and hematological disorders in PDGF beta-receptor mutant mice. *Genes & Development, 8*, 1888–1896.

Soriano, P. (1997). The PDGF alpha receptor is required for neural crest cell development and for normal patterning of the somites. *Development, 124*, 2691–2700.

Soriano, P. (1999). Generalized lacZ expression with the ROSA26 Cre reporter strain. *Nature Genetics, 21*, 70–71.

Spivak-Kroizman, T., Lemmon, M. A., Dikic, I., Ladbury, J. E., Pinchasi, D., Huang, J., et al. (1994). Heparin-induced oligomerization of FGF molecules is responsible for FGF receptor dimerization, activation, and cell proliferation. *Cell, 79*, 1015–1024.

Squinto, S. P., Stitt, T. N., Aldrich, T. H., Davis, S., Bianco, S. M., Radziejewski, C., et al. (1991). trkB encodes a functional receptor for brain-derived neurotrophic factor and neurotrophin-3 but not nerve growth factor. *Cell, 65*, 885–893.

Stalmans, I., Lambrechts, D., De Smet, F., Jansen, S., Wang, J., Maity, S., et al. (2003). VEGF: A modifier of the del22q11 (DiGeorge) syndrome? *Nature Medicine, 9*, 173–182.

Steel, K. P., Davidson, D. R., & Jackson, I. J. (1992). TRP-2/DT, a new early melanoblast marker, shows that steel growth factor (c-kit ligand) is a survival factor. *Development, 115*, 1111–1119.

Stein, R. A., & Staros, J. V. (2000). Evolutionary analysis of the ErbB receptor and ligand families. *Journal of Molecular Evolution, 50*, 397–412.

Stine, Z. E., Huynh, J. L., Loftus, S. K., Gorkin, D. U., Salmasi, A. H., Novak, T., et al. (2009). Oligodendroglial and pan-neural crest expression of Cre recombinase directed by Sox10 enhancer. *Genesis, 47*, 765–770.

Tabler, J. M., Barrell, W. B., Szabo-Rogers, H. L., Healy, C., Yeung, Y., Perdiguero, E. G., et al. (2013). Fuz mutant mice reveal shared mechanisms between ciliopathies and FGF-related syndromes. *Developmental Cell, 25*, 623–635.

Takayama, H., La Rochelle, W. J., Anver, M., Bockman, D. E., & Merlino, G. (1996). Scatter factor/hepatocyte growth factor as a regulator of skeletal muscle and neural crest development. *Proceedings of the National Academy of Sciences of the United States of America, 93*, 5866–5871.

Takeuchi, S., Takeda, K., Oishi, I., Nomi, M., Ikeya, M., Itoh, K., et al. (2000). Mouse Ror2 receptor tyrosine kinase is required for the heart development and limb formation. *Genes to Cells, 5*, 71–78.

Tallquist, M. D., & Soriano, P. (2003). Cell autonomous requirement for PDGFRalpha in populations of cranial and cardiac neural crest cells. *Development, 130*, 507–518.

Tallquist, M. D., Weismann, K. E., Hellström, M., & Soriano, P. (2000). Early myotome specification regulates PDGFA expression and axial skeleton development. *Development, 127*, 5059–5070.

Tan, P. K., Wang, J., littler, P.-L. H., Wong, K. K., Sweetnam, T. A., Keefe, W., et al. (2007). Monitoring interactions between receptor tyrosine kinases and their downstream effector proteins in living cells using bioluminescence resonance energy transfer. *Molecular Pharmacology, 72*, 1440–1446.

Taravira, S., Marcos-Gutierrez, C. V., Durbec, P., Jani, H., Grigoriou, M., Sukumaran, M., et al. (1999). Signaling by the RET receptor tyrosine kinase and its role in the development of the mammalian enteric nervous system. *Development, 126*, 2785–2797.

Tavares, A. L., Garcia, E. L., Kuhn, K., Woods, C. M., Williams, T., & Clouthier, D. E. (2012). Ectodermal-derived Endothelin1 is required for patterning the distal and intermediate domains of the mouse mandibular arch. *Developmental Biology, 371*, 47–56.

Tessarollo, L., Tsoulfas, P., Donovan, M. J., Palko, M. E., Blair-Flynn, J., Hempstead, B. L., et al. (1997). Targeted deletion of all isoforms of the trkC gene suggests the use of alternate receptors by its ligand neurotrophin-3 in neuronal development and implicates trkC in normal cardiogenesis. *Proceedings of the National Academy of Sciences of the United States of America, 94*, 14776–14781.

Tessarollo, L., Tsoulfas, P., Martin-Zanca, D., Gilbert, D. J., Jenkins, N. A., Copeland, N. G., et al. (1993). trkC, a receptor for neurotropin-3, is widely expressed in the developing nervous system and in non-neuronal tissues. *Development, 118*, 463–475.

Threadgill, D. W., Dlugosz, A. A., Hansen, L. A., Tennenbaum, T., Lichti, U., Yee, D., et al. (1995). Targeted disruption of mouse EGF receptor: Effect of genetic background on mutant phenotype. *Science, 269*, 230–234.

Ting, M.-C., Wu, N. L., Roybal, P. G., Sun, J., Lui, L., Yen, Y., et al. (2009). EphA4 as an effector of Twist1 in the guidance of osteogenic precursor cells during calvarial bone growth and in craniosynostosis. *Development, 136*, 855–864.

Tokuda, M. (1935). An eighteenth century Japanese guide-book on mouse-breeding. *Journal of Heredity, 26*, 481–484.

Tonks, I. D., Nurcombe, V., Paterson, C., Zournazi, A., Prather, C., Mould, A. W., et al. (2003). Tyrosinase-Cre mice for tissue-specific gene ablation in neural crest and neuroepithelial-derived tissues. *Genesis, 37*, 131–138.

Trainor, P. A. (2005). Specification of neural crest cell formation and migration in mouse embryos. *Seminars in Cell & Developmental Biology, 16*, 683–693.

Treanor, J. J., Goodman, L., de Sauvage, F., Stone, D. M., Poulsen, K. T., Beck, C. D., et al. (1996). Characterization of a multicomponent receptor for GDNF. *Nature, 382*, 80–83.

Trokovic, N., Trokovic, R., Mai, P., & Partanen, J. (2003). Fgfr1 regulates patterning of the pharyngeal region. *Genes & Development, 17*, 141–153.

Trumpp, A., Depew, M. J., Rubenstein, J. L. R., Bishop, J. M., & Martin, G. R. (1999). Cre-mediated gene inactivation demonstrates that FGF8 is required for cell survival and patterning of the first branchial arch. *Genes & Development, 13*, 3136–3148.

Tucker, A. S., Yamada, G., Grigoriou, M., Pachnis, V., & Sharpe, P. T. (1999). Fgf-8 determines rostral-caudal polarity in the first branchial arch. *Development, 126*, 51–61.

Twigg, S. R. F., Kan, R., Babbs, C., Bochukova, E. G., Robertson, S. P., Wall, S. A., et al. (2004). Mutations of ephrin-B1 (EFNB1), a marker of tissue boundary formation, cause craniofrontonasal syndrome. *Proceedings of the National Academy of Sciences of the United States of America, 101*, 8652–8657.

Uehara, Y., Minowa, O., Mori, C., Shiota, K., Kuno, J., Noda, T., et al. (1995). Placental defect and embryonic lethality in mice lacking hepatocyte growth factor/scatter factor. *Nature, 373*, 702–705.

Uesaka, T., & Enomoto, H. (2010). Neural precursor death is central to the pathogenesis of intestinal aganglionosis in Ret hypomorphic mice. *Journal of Neuroscience, 30*, 5211–5218.

Uesaka, T., Nagashimada, M., Yonemura, S., & Enomoto, H. (2008). Diminished Ret expression compromises neuronal survival in the colon and causes intestinal aganglionosis in mice. *Journal of Clinical Investigation, 118*, 1890–1898.

Ullrich, A., Coussens, L., Hayflick, J. S., Dull, T. J., Gray, A., Tam, A., et al. (1984). Human epidermal growth factor receptor cDNA sequence and aberrant expression of the amplified gene in A431 epidermoid carcinoma cells. *Nature, 309*, 418–425.

Valenzuela, D. M., Stitt, T. N., DiStefano, P. S., Rojas, E., Mattsson, K., Compton, D. L., et al. (1995). Receptor tyrosine kinase specific for the skeletal muscle lineage: Expression in embryonic muscle, at the neuromuscular junction, and after injury. *Neuron, 15*, 573–584.

van Bokhoven, H., Celli, J., Kayserili, H., van Beusekom, E., Balci, S., Brussel, W., et al. (2000). Mutation of the gene encoding the ROR2 tyrosine kinase causes autosomal recessive Robinow syndrome. *Nature Genetics, 25*, 423–426.

Van den Akker, N. M., Winkel, L. C., Nisancioglu, M. H., Maas, S., Wisse, L. J., Armulik, A., et al. (2008). PDGF-B signaling is important for murine cardiac development: Its role in developing atrioventricular valves, coronaries, and cardiac innervation. *Developmental Dynamics, 237*, 494–503.

Vlahos, C. J., Matter, W. F., Hui, K. Y., & Brown, R. F. (1994). A specific inhibitor of phosphatidylinositol 3-kinase, 2-(4-morpholinyl)-8-phenyl-4H-1-benzopyran-4-one (LY294002). *Journal of Biological Chemistry, 269*, 5241–5248.

Wagner, J. P., Wolf-Yadlin, A., Sevecka, M., Grenier, J. K., Root, D. E., Lauffenburger, D. A., et al. (2013). Receptor tyrosine kinases fall into distinct classes based on their inferred signaling networks. *Science Signaling, 6*, ra58.

Wang, H. U., & Anderson, D. J. (1997). Eph family transmembrane ligands can mediate repulsive guidance of trunk neural crest migration and motor axon outgrowth. *Neuron, 18*, 383–396.

Wang, C., Chang, J. Y. F., Yang, C., Huang, Y., Liu, J., You, P., et al. (2013). Type 1 fibroblast growth factor receptor in cranial neural crest cell-derived mesenchyme is required for palatogenesis. *Journal of Biological Chemistry, 288*, 22174–22183.

Wehrle-Haller, B., Meller, M., & Weston, J. A. (2001). Analysis of melanocyte precursors in Nf1 mutants reveals that MGF/KIT signaling promotes directed cell migration independent of its function in cell survival. *Developmental Biology, 232*, 471–483.

Wehrle-Haller, B., & Weston, J. A. (1995). Soluble and cell-bound forms of steel factor activity play distinct roles in melanocyte precursor dispersal and survival on the lateral neural crest migration pathway. *Development, 121*, 731–742.
Wehrle-Haller, B., & Weston, J. A. (1999). Altered cell-surface targeting of stem cell factor causes loss of melanocytes precursors in Steel17H mutant mice. *Developmental Biology, 210*, 71–86.
Williams, L. T. (1989). Signal transduction by the platelet-derived growth factor receptor. *Science, 243*, 1564–1570.
Williams, D. E., Eisenman, J., Baird, A., Rauch, C., Van Ness, K., March, C. J., et al. (1990). Identification of a ligand for the c-kit proto-oncogene. *Cell, 63*, 167–174.
Woldeyesus, M. T., Britsch, S., Riethmacher, D., Xu, L., Sonnenberg-Riethmacher, E., Abou-Rebyeh, F., et al. (1999). Peripheral nervous system defects in erbB2 mutants following genetic rescue of heart development. *Genes & Development, 13*, 2538–2548.
Wong, A., Bogni, S., Kotka, P., de Graaff, E., D'Agati, V., Costantini, F., et al. (2005). Phosphotyrosine 1062 is critical for the in vivo activity of the Ret9 receptor tyrosine kinase isoform. *Molecular and Cellular Biology, 25*, 9661–9673.
Xu, Y. K., & Nusse, R. (1998). The Frizzled CRD domain is conserved in diverse proteins including several receptor tyrosine kinases. *Current Biology, 8*, R405–R406.
Xu, X., Weinstein, M., Li, C., Naski, M., Cohen, R. I., Ornitz, D. M., et al. (1998). Fibroblast growth factor receptor 2 (FGFR2)-mediated reciprocal regulation loop between FGF8 and FGF10 is essential for limb induction. *Development, 125*, 753–765.
Yamaguchi, T. P., Bradley, A., McMahon, A. P., & Jones, S. (1999). A Wnt5a pathway underlies outgrowth of multiple structures in the vertebrate embryo. *Development, 126*, 1211–1223.
Yamaguchi, T. P., Harpal, K., Henkemeyer, M., & Rossant, J. (1994). fgfr-1 is required for embryonic growth and mesodermal patterning during mouse gastrulation. *Genes & Development, 8*, 3032–3044.
Yamamoto, H., Yoo, S. K., Nishita, M., Kikuchi, A., & Minami, Y. (2007). Wnt5a modulates glycogen synthase kinase 3 to induce phosphorylation of receptor tyrosine kinase Ror2. *Genes to Cells, 12*, 1215–1223.
Yamauchi, Y., Abe, K., Mantani, A., Hitoshi, Y., Suzuki, M., Osuzu, F., et al. (1999). A novel transgenic technique that allows specific marking of the neural crest cell lineage in mice. *Developmental Biology, 212*, 191–203.
Yarden, Y., Kuang, W.-J., Yang-Feng, T., Coussens, L., Munemitsu, S., Dull, T. J., et al. (1987). Human proto-oncogene c-kit: A new cell surface receptor tyrosine kinase for an unidentified ligand. *EMBO Journal, 6*, 3341–3351.
Yayon, A., Klagsbrun, M., Esko, J. D., Leder, P., & Ornitz, D. M. (1991). Cell surface, heparin-like molecules are required for binding of basic fibroblast growth factor to its high affinity receptor. *Cell, 64*, 841–848.
Yayon, A., Zimmer, Y., Shen, G. H., Avivi, A., Yarden, Y., & Givol, D. (1992). A confined variable region confers ligand specificity on fibroblast growth factor receptors: Implications for the origin of the immunoglobulin fold. *EMBO Journal, 11*, 1885–1890.
Yu, J. C., Heidaran, M. A., Pierce, J. H., Gutkind, J. S., Lombardi, D., Ruggiero, M., et al. (1991). Tyrosine mutations within the alpha platelet-derived growth factor receptor kinase insert domain abrogate receptor-associated phosphatidylinositol-3 kinase activity without affecting mitogenic or chemotactic signal transduction. *Molecular and Cellular Biology, 11*, 3780–3785.
Zhang, B., Luo, S., Wang, Q., Suzuki, T., Xiong, W. C., & Mei, L. (2008). LRP4 serves as a coreceptor of agrin. *Neuron, 60*, 285–297.
Zhang, Z.-J., & Sieber-Blum, M. (2009). Essential role of stem cell factor signaling in primary sensory neuron development. *Developmental Neuroscience, 31*, 202–211.

Zhang, Y., Wolf-Yadlin, A., Ross, P. L., Pappin, D. J., Rush, J., Lauffenburger, D. A., et al. (2005). Time-resolved mass spectrometry of tyrosine phosphorylation sites in the epidermal growth factor receptor signaling network reveals dynamic modules. *Molecular & Cellular Proteomics*, *4*, 1240–1250.

Zhou, J., Pashmforoush, M., & Sucov, H. M. (2012). Endothelial neuropilin disruption in mice causes DiGeorge syndrome-like malformations via mechanisms distinct to those caused by loss of Tbx1. *PLoS One*, *7*, e32429.

Zsebo, K. M., Williams, D. A., Geissler, E. N., Broudy, V. C., Martin, F. H., Atkins, H. L., et al. (1990). Stem cell factor is encoded at the Sl locus of the mouse and is the ligand for the c-kit tyrosine kinase receptor. *Cell*, *63*, 213–224.

CHAPTER SIX

Neural Crest Cells in Cardiovascular Development

Alice Plein, Alessandro Fantin, Christiana Ruhrberg[1]
UCL Institute of Ophthalmology, University College London, London, United Kingdom
[1]Corresponding author: e-mail address: c.ruhrberg@ucl.ac.uk

Contents

Abstract

Cardiac neural crest cells (NCCs) are a transient, migratory cell population exclusive to vertebrate embryos. Ablation, transplantation, and lineage-tracing experiments in chick and mouse have demonstrated their essential role in the remodeling of the initially bilateral and symmetric pharyngeal artery pairs into an aortic arch and for the septation of the cardiac outflow tract into the base of the pulmonary artery and aorta. Accordingly, defective cardiac NCC function is a common cause of congenital birth defects. Here, we review our current understanding of cardiac NCC-mediated vascular remodeling and signaling pathways important for this process. We additionally discuss their contribution to the cardiac valves as well as the still contentious role of cardiac NCCs in the development of the myocardium and conductive system of the heart.

1. INTRODUCTION

Cardiac neural crest cells (NCCs), like other NCC populations, are a vertebrate-specific cell type that is derived from the dorsal part of the embryonic neural tube through epithelial-to-mesenchymal transition. Even

Current Topics in Developmental Biology, Volume 111
ISSN 0070-2153
http://dx.doi.org/10.1016/bs.ctdb.2014.11.006

though NCCs had already been described in 1868 by Wilhelm His as the "Zwischenstrang" (intermediate cord) and were subsequently renamed "neural crest" by Arthur Marshall in 1879, NCCs with specific functions in heart development were not discovered until the 1980s by Kirby and colleagues through lineage tracing, ablation, and transplantation experiments in avian embryos (Creazzo, Godt, Leatherbury, Conway, & Kirby, 1998; Kirby, 1987; Kirby, Gale, & Stewart, 1983). Thus, Kirby and colleagues described how NCCs delaminating from the neural tube between the otic placode and the third somite colonize the embryonic pharyngeal arches and outflow tract (OFT) of the heart (Fig. 1A and B). They further demonstrated that ablation of this NCC subset causes developmental defects that resemble congenital heart defects in patients, such as a common arterial trunk (CAT; also known as persistent truncus arteriosus). This groundbreaking research initiated an intense era of research to identify the molecular and cellular mechanisms that govern NCC-induced cardiovascular development and the role of NCCs in congenital heart disease (reviewed by Keyte & Hutson, 2012). Here, we review the role of cardiac NCCs in remodeling the heart-associated vasculature and their contribution to nonvascular heart tissues. We also describe several key genetic pathways involved in cardiac NCC function during development and disease.

2. CARDIAC NCCs ENABLE PHARYNGEAL ARCH ARTERY REMODELING

After cardiac NCCs have delaminated from the neural tube, they migrate ventrally into the circumpharyngeal ridge. Here, they pause while the pharyngeal arch arteries (PAAs) form by vasculogenesis to generate a bilateral series of artery pairs, which in the mouse occurs between embryonic day (E) 8.5 and 9.5 and in the chick between Hamburger and Hamilton (HH) stages 12 and 13 (Hamburger & Hamilton, 1951; Hiruma & Hirakow, 1995). As the first and second artery pairs remodel into the mandibular and hyoid arteries, respectively, the cardiac NCCs invade the pharyngeal arches in three streams to sequentially associate with the third, the fourth, and the sixth PAA pairs, reflecting the order of artery formation along the rostrocaudal axis (Hiruma, Nakajima, & Nakamura, 2002) (Fig. 1A). Notably, current PAA nomenclature refers to the six artery pairs that exist in fish, even though the presence of a fifth PAA pair in mammals is contentious (Bamforth et al., 2013).

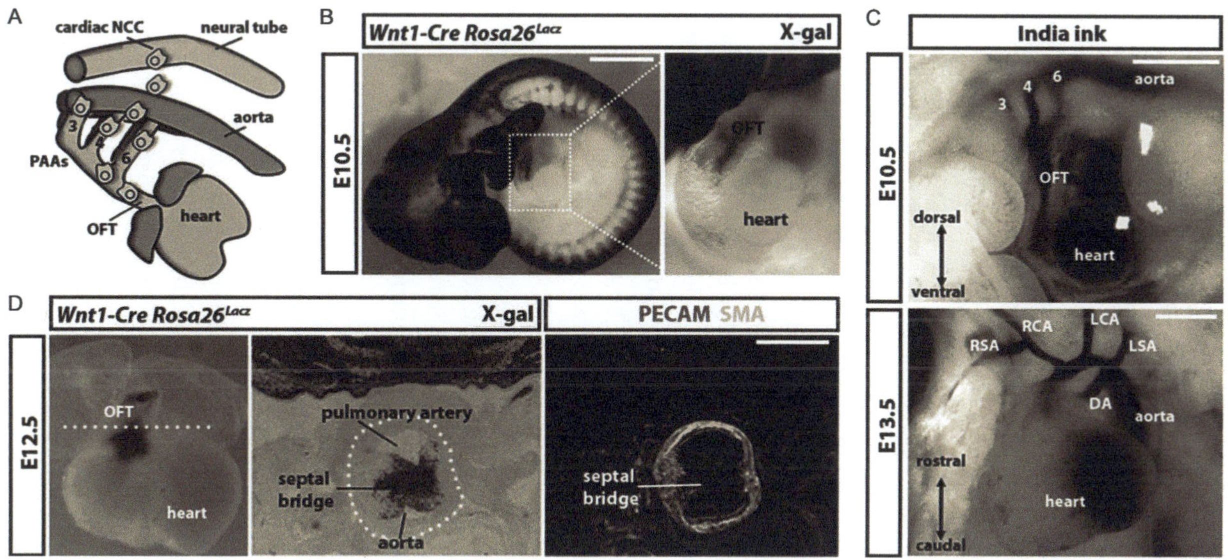

Figure 1 Cardiac NCC contribution to murine OFT and PAA development. (A) Schematic representation of cardiac NCC migration. Cardiac NCCs delaminate from the neural tube at E8.5 and associate with the PAAs to differentiate into the SMCs of the arterial tunica media or continue to the OFT where they contribute to septal bridge formation. (B) Visualization of NCC and their derivatives in an E10.5 *Wnt1-Cre;Rosa26^{Lacz}* mouse embryo. The X-gal stain illustrates NCC contribution to trunk, head, and pharyngeal arch tissues. The higher magnification image shows two prongs of cardiac NCCs in the OFT. (C) PAA remodeling into the aortic arch. Ink injections into embryonic hearts show the third, fourth, and sixth PAAs at E10.5 and their remodeled derivatives at E13.5. (D) Cardiac NCC contribution to the OFT septal bridge. X-gal staining of an E12.5 *Wnt1-Cre;Rosa26^{Lacz}* heart shows accumulation of cells from the cardiac NCC lineage in the OFT; the dotted line indicates the level at which a section was taken to illustrate the position of the cardiac NCCs within the OFT in the adjacent panel. Double staining with the endothelial marker PECAM and the SMC marker SMA illustrates OFT anatomy at E12.5. Scale bars: B—1 mm; C and D—500 μm. (See the color plate.)

Having colonized the PAAs, which initially consist only of a sheath of endothelial cells, the cardiac NCCs differentiate into smooth muscle cells (SMCs) (Bergwerff, Verberne, DeRuiter, Poelmann, & Gittenberger-de Groot, 1998). This process is thought to be important for subsequent PAA remodeling, which occurs between E10.5 and E13.5 in the mouse and between HH14 and HH28 in the chick (Hamburger & Hamilton, 1951; Hiruma & Hirakow, 1995) (Fig. 1C).

The third PAAs extend to form the right common carotid and basal part of the left internal carotid in both birds and mammals. In mammals, the right fourth artery regresses and becomes part of the right subclavian artery, whereas the left fourth artery persists and forms the segment of the aortic arch that connects the aortic sac to the descending aorta. Moreover, in mammals, the proximal right sixth PAA contributes to the base of the pulmonary artery, whereas the distal segment regresses (Schneider & Moore, 2006). The left sixth PAA gives rise to the ductus arteriosus, an embryonic structure that connects the pulmonary artery with the descending aorta; this shunt allows blood from the right ventricle to bypass the lungs because the fetal blood is oxygenated through the placenta. At birth, the ductus arteriosus collapses due to a rise in p_{O_2}, and deoxygenated blood is now able to enter the pulmonary circulation (Leonhardt et al., 2003). The nonfunctional vestige of the ductus arteriosus persists through life and is called ligamentum arteriosus.

Ultimately, the asymmetric remodeling of the PAAs in mammals gives rise to a vascular tree in which the aortic sac connects via the left fourth PAA to the descending aorta to form the left-looping aortic arch. In birds, remodeling of the fourth and sixth PAAs occurs in the reverse configuration: the right fourth artery remodels into the definitive aortic arch, the left sixth PAA forms the proximal segment of the pulmonary artery, and the right sixth PAA forms the ductus arteriosus, which collapses at hatching. In contrast to mammals, the right fourth PAA persists and the aortic arch therefore curves to the right.

In both birds and mammals, formation of an asymmetric aortic arch and pulmonary artery is a prerequisite for blood from the ventricles to enter two distinct circulations. While the aortic arch directs the oxygenated blood into the systemic circulation, the pulmonary artery directs the blood into the pulmonary circulation. Accordingly, defects in the remodeling process are often lethal or severely impair cardiovascular performance. When cardiac NCCs are ablated in chick or mouse embryos, the PAAs form normally, but regress or persist inappropriately (Kirby et al., 1983; Porras & Brown, 2008; Waldo, Kumiski, & Kirby, 1996). For example, a common defect in mouse mutants with abnormal cardiac NCC behavior is the regression of the left fourth PAA. This defect results in an interrupted aortic arch (IAA) and is referred to as a type b interruption. Despite a multitude of mouse models with defective cardiac NCC behavior, only a limited number of the signaling pathways involved have been discovered, and the exact cellular interactions by which cardiac NCCs orchestrate PAA remodeling and survival remain incompletely understood (see below).

3. CARDIAC NCCs ARE ESSENTIAL FOR CARDIAC OFT SEPTATION

Having colonized the PAAs, a subset of cardiac NCCs continue to migrate via the aortic sac into the cardiac OFT at the anterior pole of the heart (Gitler, Brown, Kochilas, Li, & Epstein, 2002) (Fig. 1A, B, and D). The OFT provides a conduit for blood leaving the ventricles into the PAAs and later their derivatives. Concomitantly with PAA rearrangement during embryogenesis, the OFT undergoes a complex remodeling process that begins around E9.5 in the mouse and HH14 in the chick and gives rise to the base of the aorta and pulmonary artery (reviewed in Webb, Qayyum, Anderson, Lamers, & Richardson, 2003) (Fig. 1D). The OFT of mammals and birds thus remodels into a completely septated vessel to allow blood from the ventricles to enter the systemic versus pulmonary circulation. In contrast, the OFT of amphibians and reptiles septates only in its distal part, which results in a partial mixing of the blood before it enters the systemic or pulmonary circulation.

Initially, the OFT is a solitary tube of endocardial cells that is covered by a myocardial layer. For remodeling to occur, the interstitial space between endothelium and smooth muscle is filled with an acellular "cardiac jelly" consisting of extracellular matrix (Eisenberg & Markwald, 1995; Markwald, Krook, Kitten, & Runyan, 1981). This jelly gives rise to the endocardial cushions on either side of the central OFT canal. Starting at E10.5 in the mouse, the cardiac NCCs migrate in bilateral advancing columns into these cushions (Fig. 1B), populating first the distal and then the proximal region of the OFT to fill the length of the OFT by E12.5 (Fig. 1D). While the proximal cushions are mostly invaded by cardiac NCCs, the distal cushions are invaded both by cardiac NCCs and by cells derived from the pharyngeal mesoderm (Ward, Stadt, Hutson, & Kirby, 2005). In addition, the proximal region is invaded by the progeny of endothelial cells that have undergone an endothelial-to-mesenchymal transition (endoMT) in the OFT between E10.5 and E13.5 in mouse (reviewed by von Gise & Pu, 2012). This mobilization of endothelial cells coincides with the migration of the cardiac NCCs into the OFT; however, it remains to be established whether cardiac NCCs play a role in inducing endoMT (reviewed by Sugishita, Watanabe, & Fisher, 2004). Importantly, cell invasion causes the cushions to swell, which is thought to facilitate the apposition of endothelium in the central OFT canal. Coincident with cushion swelling, the cardiac NCCs in the opposing endocardial cushions come together to form the aorticopulmonary septum, a tissue bridge that segregates the single OFT vessel into two endothelial tubes (Waldo, Miyagawa-Tomita, Kumiski, & Kirby, 1998) (Fig. 1D).

Ablation of cardiac NCCs in mouse or chick prevents OFT septation and causes abnormal patterning of the aortic arch and ventricular septal defects (VSDs) (Kirby et al., 1983; Porras & Brown, 2008). In contrast, ablation of cardiac NCCs does not impair OFT development in amphibian species, where the OFT only septates distally and a complete segregation of the blood from the right and left ventricle is not required (Lee & Saint-Jeannet, 2011). Thus, it appears that NCCs were recruited during vertebrate evolution to enable the formation of a septal bridge in the proximal OFT for complete separation of the pulmonary and systemic circulation. This anatomical advance is an essential prerequisite for postnatal life both in birds and in mammals, presumably because both species are endothermic and therefore have a high metabolic demand. Thus, segregating the deoxygenated blood from the arterial blood improves the oxygen supply to all tissues. Cardiac NCCs have therefore likely played a key role in the evolution of higher vertebrates.

4. SIGNALING PATHWAYS IN CARDIAC NCC-MEDIATED VASCULAR REMODELING

A number of genes involved in cardiac NCC-induced PAA remodeling and OFT septation have been uncovered through the analysis of mouse mutants with congenital heart defects (reviewed in Gruber & Epstein, 2004). These genes regulate cardiac NCC induction or their migration, survival, and differentiation. Below, we will discuss several key genetic pathways in cardiac NCC function that are activated by members of the transforming growth factor β (TGFβ), fibroblast growth factor (FGF), vascular endothelial growth factor (VEGF), and class 3 semaphorin (SEMA3) families of secreted signaling molecules.

The TGFβ superfamily comprises a large group of secreted polypeptides that includes the bona fide TGFβs and the bone morphogenetic proteins (BMPs), with both types of proteins shown to play a role in cardiac NCC-regulated vascular remodeling (reviewed in Keyte & Hutson, 2012). For instance, the NCC-specific deletion of the murine BMP receptor ALK2 showed that it is required by cardiac NCCs for their migration into the PAAs and the OFT to enable correct remodeling and prevent CAT and IAA (Kaartinen et al., 2004). In contrast, the NCC-specific deletion of the alternative BMP receptor BMPR1A does not affect cardiac NCC migration in the mouse, but impairs endocardial cushion formation and consequently results in an unseptated OFT (Stottmann, Choi, Mishina, Meyers, & Klingensmith, 2004). The authors also proposed that cardiac NCCs within the epicardium promote ventricular development by stimulating myocardial cell proliferation.

TGFβ2 appears to be particularly important for cardiac NCC-mediated vascular remodeling. The addition of exogenous TGFβ2 to whole mouse embryo cultures causes OFT and aortic arch remodeling defects (Kubalak, Hutson, Scott, & Shannon, 2002). Similarly, TGFβ2 knockout mice have defects in the development of several tissues, including the OFT and the aortic arch. In contrast, similar defects are not observed in TGFβ1 and TGFβ3 knockouts (Sanford et al., 1997). TGFβ signaling may also have a cell autonomous role in cardiac NCCs. The loss of TGFβRII in cardiac NCCs prevents their differentiation into SMCs and therefore perturbs remodeling of both the OFT and PAAs in mice (Wurdak et al., 2005). Furthermore, loss of the TGFβ-receptor ALK5 in the NCC lineage also causes PAA and OFT defects by impairing postmigratory cardiac NCC survival (Wang et al., 2006).

To date, 20 FGFs have been described, most of which are small, heparin-binding secreted proteins (however, FGF11–14 are not secreted). Of these, only FGF8 has to date been implicated in cardiac NCC development (Frank et al., 2002). FGF8 deletion causes embryonic death at midgastrulation (Sun, Meyers, Lewandoski, & Martin, 1999); however, FGF8 hypomorphic mice secrete enough FGF8 to survive gastrulation and have defective cardiac NCC migration into the pharyngeal arches (Frank et al., 2002) FGF8 is secreted by the pharyngeal ectoderm and endoderm as a guidance cue for migrating cardiac NCCs (Sato et al., 2011) and promotes the survival of the cardiac NCCs that colonize the fourth PAAs (Abu-Issa, Smyth, Smoak, Yamamura, & Meyers, 2002; Macatee et al., 2003). In particular, the conditional deletion of FGF8 in pharyngeal ectoderm impaired fourth PAA formation, while conditional deletion in the endoderm resulted in a failure of OFT septation (Park et al., 2006).

The vascular endothelial growth factor VEGF-A is a secreted polypeptide most commonly studied in regard to its critical role in the processes of vasculogenesis, angiogenesis, and vascular permeability (reviewed in Ruhrberg, 2003). In addition, VEGF-A has been reported to act as a guidance cue for cranial NCCs that invade the second PAAs to contribute to craniofacial tissues (McLennan, Teddy, Kasemeier-Kulesa, Romine, & Kulesa, 2010). In contrast, VEGF-A is thought to not play a major role in cardiac NCC migration (Kirby & Hutson, 2010; Stalmans et al., 2003). Nevertheless, genetic studies in the mouse have implicated specific isoforms of VEGF-A in OFT and PAA remodeling.

The VEGF-A polypeptide exists in three major isoforms, which are generated by alternative splicing and differ in their molecular mass (reviewed in Plein, Fantin, & Ruhrberg, 2014). Reflecting the length of their amino acid chain, the shortest isoform is termed $VEGF_{121}$, as it is 121 amino acids long, and the larger isoforms are referred to as $VEGF_{165}$ and $VEGF_{189}$. These isoforms differ in their affinity for heparin *in vitro*, and this property is thought to reflect a differential affinity for heparan sulfate proteoglycans in the extracellular matrix *in vivo*. Thus, $VEGF_{121}$ is the most diffusible and $VEGF_{189}$ the least diffusible isoform in cell culture studies (Park, Keller, & Ferrara, 1993), and they cooperate to form extracellular VEGF-A gradients to regulate blood vessel morphogenesis *in vivo* (Ruhrberg et al., 2002). The isoforms also differ in their affinity for the shared VEGF-A/SEMA3 receptor neuropilin 1 (NRP1), which preferentially binds $VEGF_{165}$ over $VEGF_{121}$ (Soker, Takashima, Miao, Neufeld, & Klagsbrun, 1998).

Even though loss of VEGF-A results in early embryonic lethality in mice, the exclusive expression of $VEGF_{120}$ (corresponding to human $VEGF_{121}$) at the expense of $VEGF_{164}$ and $VEGF_{188}$ (corresponding to human $VEGF_{165}$ and $VEGF_{189}$, respectively) in *Vegfa*$^{120/120}$ mice, or the exclusive expression of $VEGF_{188}$ at the expense of $VEGF_{120}$ and $VEGF_{164}$ in *Vegfa*$^{188/188}$ mice, enables embryo survival to birth and has revealed an essential role for $VEGF_{164}$ in OFT and PAA remodeling (Stalmans et al., 2003). Thus, similar cardiac and aortic arch malformations were observed in *Vegfa*$^{120/120}$ and *Vegfa*$^{188/188}$ mice. Aortic arch malformations included type b interruption of the aortic arch, double aortic arch, right-sided aortic arch, and different aberrations in the formation of the carotid and subclavian arteries, while cardiac defects arising from defective OFT septation included Tetralogy of Fallot, CAT, hypoplasia of the pulmonary trunk, and VSDs (Stalmans et al., 2003).

Given that cardiac NCC migration is thought to be at least grossly normal in VEGF-A isoform-deficient mice (Kirby & Hutson, 2010; Stalmans et al., 2003), the mechanistic reasons for OFT and PAA defects in these mice are poorly understood. The prevailing model suggests that these defects are due to the loss of $VEGF_{164}$ signaling through NRP1 in the OFT endothelium, because the complete absence of NRP1, or the genetic ablation of endothelial NRP1, impairs OFT and PAA remodeling (Gu et al., 2003; Kawasaki et al., 1999). However, we recently demonstrated the viability of mice with a mutated NRP1 receptor that is unable to bind VEGF-A (Fantin et al., 2014). This observation is not compatible with the hypothesis that VEGF-A signals through NRP1 in endothelium to enable OFT septation. It is therefore likely that VEGF-A signals through another receptor, such as VEGFR2, and that endothelial NRP1 serves as a receptor for another ligand. However, neither hypothesis has been experimentally tested.

SEMA3 proteins are secreted glycoproteins best known for their role in axon guidance (Rohm, Ottemeyer, Lohrum, & Puschel, 2000; Ruediger et al., 2013). One particular family member termed SEMA3C is secreted within the OFT and PAAs at the time when cardiac NCCs migrate into these tissues, and loss of SEMA3C causes CAT and IAA (Feiner et al., 2001). It has been proposed that SEMA3C acts as an attractive signal for cardiac NCCs, as knockdown of NRP1 in cardiac NCCs perturbs their migration in chick (Toyofuku et al., 2008). Agreeing with a similar role for SEMA3C in mammals, mice lacking semaphorin signaling through both NRP1 and NRP2 also have aortic arch and OFT defects (Gu et al., 2003). However, experimental proof for a role of NRPs in SEMA3C-induced cardiac NCCs in mammals is still lacking.

5. CARDIAC NCCs CONTRIBUTION TO THE CARDIAC VALVES

OFT endothelial cells that have undergone endoMT are thought to give rise to the bulk of the semilunar valves, which form within the aorta and pulmonary artery, respectively, to prevent the backflow of blood into the ventricles (de Lange et al., 2004). In addition, lineage trace studies with *Wnt1-Cre* showed that cardiac NCCs also colonize the semilunar valves, where they mainly contribute to the two leaflets adjacent to the aorticopulmonary septum, with their progeny persisting into adulthood (Nakamura, Colbert, & Robbins, 2006). In contrast, an earlier report using a similar genetic approach suggested that NCC derivatives only marginally contribute to these structures in postnatal life (Jiang, Rowitch, Soriano, McMahon, & Sucov, 2000). Cells of the NCC lineage have also been found to contribute to the atrioventricular valves, consisting of the bicuspid (mitral) valve and the tricuspid valve, which are located between the upper atria and the lower ventricles. In particular, NCC progeny were found in the septal leaflets, where they persisted into adulthood (Hildreth et al., 2008; Nakamura et al., 2006). A cardiac NCC contribution to cardiac valve development agrees with the association of bicuspid aortic valve disease (Hoffman & Kaplan, 2002) with craniofacial defects, which are often caused by defective cranial NCC development (Kappetein et al., 1991). However, further experiments are required to unequivocally link cardiac valve defects to impaired cardiac NCC development.

6. POSSIBLE ROLES FOR CARDIAC NCCs IN MYOCARDIAL DEVELOPMENT

In zebrafish, cardiac NCCs migrate into the ventricles of the heart and then differentiate into myocardial cells (Sato & Yost, 2003). In contrast, the contribution of cardiac NCCs to myocardial development in amniotes has remained contentious. To date, most lineage-tracing studies in mouse and chick did not observe NCC derivatives within the myocardium or epicardium (e.g., Jiang et al., 2000; Lo et al., 1997). However, other reports provided evidence of NCC progeny on the surface of or within the heart ventricles of embryonic mice, either by fate mapping (Brown et al., 2001; Stottmann et al., 2004; Tomita et al., 2005) or by expression of *Plxna2*, a known marker of cardiac NCCs (Brown et al., 2001). In particular, cardiac NCCs have been suggested to contribute to dormant multipotent stem cells in the mammalian heart that can differentiate into cardiomyocytes (Tomita et al., 2005). Moreover, the cell type-specific knockout of several different genes with a *Cre* transgene active in cardiac NCCs impairs myocardial development. For instance, deletion of the BMP receptor BMPR1A with *Wnt1-Cre* results in a thin ventricular myocardium (Stottmann et al., 2004). Deletion of PAX3, which is specifically expressed by cardiac NCCs, but not myocardium, also leads to a thin ventricular myocardium (Engleka et al., 2005). These observations may indicate that cardiac NCCs contribute to the epicardial layer of the heart and regulate myocyte proliferation. Alternatively, ventricular phenotypes may be explained by ectopic expression of the *Wnt1-Cre* transgene, even though two independent studies failed to observe any expression of the *Wnt1* gene or the *Cre* from the *Wnt1-Cre* transgene outside of the neural tube, suggesting that the *Wnt1-Cre* transgene does not get activated after NCC delamination (Nakamura et al., 2006; Stottmann et al., 2004).

7. POSSIBLE ROLES FOR CARDIAC NCCs IN THE DEVELOPMENT OF THE CARDIAC CONDUCTION SYSTEMS

The progeny of cranial, vagal, and trunk NCCs contribute to autonomic nervous system innervation of the heart and are therefore involved in modulating cardiac function (Hildreth et al., 2008). In addition, several studies have suggested a role for cardiac NCCs in the development of the cardiac conduction system, which consists of pace-making tissues (i.e., nodes) and fast-conducting bundles, such as the bundle of His (reviewed in Gourdie et al., 2003). For instance, deletion of the transcription factor HF1B with the NCC-specific *Pax3* promoter causes cardiac conduction defects and reduces the expression of the neurotrophin receptor TRKC within cardiac neurons of the mouse (St Amand et al., 2006). Moreover, the ablation of cardiac NCCs in the chick affects the conductive properties of the heart by impairing the compaction of the bundle of His (Gurjarpadhye et al., 2007). Lineage tracing in mice with a *Wnt1-Cre* reporter confirmed that cardiac NCCs contribute to the cardiac conduction system, including the bundle of His and the posterior internodal tract (Nakamura et al., 2006). However, a more recent study using the same transgenic mouse strain did not confirm a NCC contribution to these structures (Hildreth et al., 2008). Further work is therefore required to resolve whether cardiac NCCs contribute directly to the cardiac conductive system, or if they instead provide transient inductive signals to other cell types that contribute to or pattern the cardiac neural network.

8. CONGENITAL ABNORMALITIES CAUSED BY DEFECTIVE CARDIAC NCC DEVELOPMENT

Congenital heart defects are among the most common major birth defects in humans, with OFT and aortic arch remodeling defects affecting 6 per 1000 live births (Hoffman & Kaplan, 2002). The most common types of abnormalities associated with defective cardiac NCC-induced vascular remodeling are discussed below. In most human syndromes, the aforementioned signaling pathways have not been found to be the underlying cause. Instead, chromosomal deletions and epigenetic changes are common factors.

DiGeorge syndrome (DGS) is a condition that affects the development of many tissues that are patterned by or derived from NCCs. Thus, patients have variable types of craniofacial defects, aplasia or hypoplasia of the thymus and parathyroid glands, and OFT and aortic arch defects. The genetic causes of this condition have been partly uncovered. For instance, a hemizygous deletion within chromosome band 22q11.2 has been found in 25% of DGS patients (Driscoll et al., 1993). Complicating prognosis, this deletion causes phenotypes of varying severity in different patients, and there are even cases of monozygotic twins with identical mutations, but a different spectrum of abnormalities, suggesting a role for epigenetics in these syndromes (Singh, Murphy, & O'Reilly, 2002). In other cases, different deletions or translocations affecting chromosome 22q11 or other chromosomes have been described, suggesting that DGS is a multigenic disorder (Scambler, 2010).

To elucidate which genes on human chromosome 22q11 are important for cardiac NCC remodeling, it is possible to study mice lacking single genes or chromosomal regions that are syntenic with 22q11. For example, mice haploinsufficient for the gene encoding the transcription factor TBX1 recapitulate many cardiovascular phenotypes of DGS (Merscher et al., 2001). *Tbx1* is expressed in tissues the cardiac NCCs migrate through, and loss of TBX1-mediated transcriptional regulation is thought to modify the cardiac NCC microenvironment and thus affect cardiac NCC behavior (Calmont et al., 2009; Vitelli, Morishima, Taddei, Lindsay, & Baldini, 2002). However, in humans, mutations in *TBX1* are only a rare cause of DGS (Prescott et al., 2005). This suggests that additional genes interact with TBX1, such as VEGF-A (Stalmans et al., 2003), or alternatively, that other genes residing in 22q11 locus are important for cardiac NCC function in humans.

CHARGE syndrome derives its name from a characteristic combination of congenital defects that comprises *c*oloboma, *h*eart anomaly, *a*tresia of choanae, *r*etardation of physical and mental development, *g*enital hypoplasia, and *e*ar anomalies (Siebert, Graham, & MacDonald, 1985). In addition, malformations of the foregut, kidneys, limbs, lung, and liver have been described in infants with CHARGE syndrome. The gene most commonly affected in patients with CHARGE syndrome is CHD7, which encodes a DNA binding protein involved in chromatin remodeling

(Janssen et al., 2012). Knockdown of *Chd7* in Xenopus (Bajpai et al., 2010) or mouse (Bosman et al., 2005) embryos causes developmental defects that resemble CHARGE. Defective OFT and PAA remodeling in CHARGE patients suggests a role for CHD7 in regulating cardiac NCC behavior. A recent study demonstrated that genes associated with NCC migration and axon guidance, such as semaphorins and ephrin receptor genes, are misregulated in mice with a heterozygous *Chd7* mutation (Schulz et al., 2014).

The link between the chromatin-remodeling protein CHD7 with cardiac NCC-associated defects suggests that epigenetic regulation is important for genes controlling NCC function. In agreement with this idea, environmental factors such as micronutrients, which can affect histone methylation, have been described to also affect cardiac NCC behavior. For instance, maternal mutations affecting the folate pathway can give rise to congenital heart defects reminiscent of abnormal cardiac NCC behavior, and mothers of children with OFT and aortic arch defects have increased blood levels of homocysteine, an indicator of folate deficiency (Hobbs et al., 2006; Kapusta et al., 1999). In addition, medication affecting maternal retinoid levels can cause CAT and aortic arch defects (Lammer et al., 1985; Zile, 2001).

Several factors may contribute to the high frequency of congenital cardiac NCC-related disease. Both the overexpression and haploinsufficiency of some of the genes described above, such as *Tgfβ2* and *Nrp1*, impair PAA and OFT remodeling, suggesting a tight dose-dependency for signals regulating cardiac NCCs and therefore increasing the likelihood for misregulation. Additionally, the complexity of the signaling pathways and cell types involved may increase the susceptibility to developmental perturbation in the cardiac NCC-mediated remodeling process. Furthermore, NCCs are a cell type found only in vertebrates, raising the possibility that evolutionary time may not have been sufficient to establish robust regulatory networks and backup pathways for processes that rely on cardiac NCCs, in contrast to inductive processes in other tissues.

9. OUTSTANDING QUESTIONS

Despite extensive research efforts over the past decades, the exact role of cardiac NCCs in vascular remodeling is incompletely understood. For instance, cardiac NCCs colonize the third, fourth, and sixth PAAs and differentiate into SMCs, but it is currently not known how this process promotes asymmetric vessel remodeling. Moreover, in the OFT, cardiac NCCs contribute to the aorticopulmonary septum, but the inductive signal relays by which cardiac NCCs enable endothelial fusion and myocardialization remain to be established. Finally, the contribution of cardiac NCCs to the myocardium and conduction system of the heart remains contentious. Thus, more work is required to fully understand the role of cardiac NCCs in the development of these tissues. For example, it should be possible to place the numerous signaling pathways implicated in cardiac NCC-mediated developmental processes into a functional hierarchy by cross-examination of expression patterns and by defining cellular behaviors in previously published tissue-specific mouse knockout models with cardiac NCC defects.

REFERENCES

Abu-Issa, R., Smyth, G., Smoak, I., Yamamura, K., & Meyers, E. N. (2002). Fgf8 is required for pharyngeal arch and cardiovascular development in the mouse. *Development*, *129*, 4613–4625.

Bajpai, R., Chen, D. A., Rada-Iglesias, A., Zhang, J., Xiong, Y., Helms, J., et al. (2010). CHD7 cooperates with PBAF to control multipotent neural crest formation. *Nature*, *463*, 958–962.

Bamforth, S. D., Chaudhry, B., Bennett, M., Wilson, R., Mohun, T. J., Van Mierop, L. H., et al. (2013). Clarification of the identity of the mammalian fifth pharyngeal arch artery. *Clinical Anatomy*, *26*, 173–182.

Bergwerff, M., Verberne, M. E., DeRuiter, M. C., Poelmann, R. E., & Gittenberger-de Groot, A. C. (1998). Neural crest cell contribution to the developing circulatory system: Implications for vascular morphology? *Circulation Research*, *82*, 221–231.

Bosman, E. A., Penn, A. C., Ambrose, J. C., Kettleborough, R., Stemple, D. L., & Steel, K. P. (2005). Multiple mutations in mouse Chd7 provide models for CHARGE syndrome. *Human Molecular Genetics*, *14*, 3463–3476.

Brown, C. B., Feiner, L., Lu, M. M., Li, J., Ma, X., Webber, A. L., et al. (2001). PlexinA2 and semaphorin signaling during cardiac neural crest development. *Development*, *128*, 3071–3080.

Calmont, A., Ivins, S., Van Bueren, K. L., Papangeli, I., Kyriakopoulou, V., Andrews, W. D., et al. (2009). Tbx1 controls cardiac neural crest cell migration during arch artery development by regulating Gbx2 expression in the pharyngeal ectoderm. *Development*, *136*, 3173–3183.

Creazzo, T. L., Godt, R. E., Leatherbury, L., Conway, S. J., & Kirby, M. L. (1998). Role of cardiac neural crest cells in cardiovascular development. *Annual Review of Physiology*, *60*, 267–286.

de Lange, F. J., Moorman, A. F., Anderson, R. H., Manner, J., Soufan, A. T., de Gier-de Vries, C., et al. (2004). Lineage and morphogenetic analysis of the cardiac valves. *Circulation Research*, *95*, 645–654.

Driscoll, D. A., Salvin, J., Sellinger, B., Budarf, M. L., McDonald-McGinn, D. M., Zackai, E. H., et al. (1993). Prevalence of 22q11 microdeletions in DiGeorge and velocardiofacial syndromes: Implications for genetic counselling and prenatal diagnosis. *Journal of Medical Genetics*, *30*, 813–817.

Eisenberg, L. M., & Markwald, R. R. (1995). Molecular regulation of atrioventricular valvuloseptal morphogenesis. *Circulation Research*, *77*, 1–6.

Engleka, K. A., Gitler, A. D., Zhang, M., Zhou, D. D., High, F. A., & Epstein, J. A. (2005). Insertion of Cre into the Pax3 locus creates a new allele of Splotch and identifies unexpected Pax3 derivatives. *Developmental Biology, 280*, 396–406.

Fantin, A., Herzog, B., Mahmoud, M., Yamaji, M., Plein, A., Denti, L., et al. (2014). Neuropilin 1 (NRP1) hypomorphism combined with defective VEGF-A binding reveals novel roles for NRP1 in developmental and pathological angiogenesis. *Development, 141*, 556–562.

Feiner, L., Webber, A. L., Brown, C. B., Lu, M. M., Jia, L., Feinstein, P., et al. (2001). Targeted disruption of semaphorin 3C leads to persistent truncus arteriosus and aortic arch interruption. *Development, 128*, 3061–3070.

Frank, D. U., Fotheringham, L. K., Brewer, J. A., Muglia, L. J., Tristani-Firouzi, M., Capecchi, M. R., et al. (2002). An Fgf8 mouse mutant phenocopies human 22q11 deletion syndrome. *Development, 129*, 4591–4603.

Gitler, A. D., Brown, C. B., Kochilas, L., Li, J., & Epstein, J. A. (2002). Neural crest migration and mouse models of congenital heart disease. *Cold Spring Harbor Symposia on Quantitative Biology, 67*, 57–62.

Gourdie, R. G., Harris, B. S., Bond, J., Edmondson, A. M., Cheng, G., Sedmera, D., et al. (2003). His-Purkinje lineages and development. *Novartis Foundation Symposium, 250*, 110–122, discussion 122–114, 276–119.

Gruber, P. J., & Epstein, J. A. (2004). Development gone awry: Congenital heart disease. *Circulation Research, 94*, 273–283.

Gu, C., Rodriguez, E. R., Reimert, D. V., Shu, T., Fritzsch, B., Richards, L. J., et al. (2003). Neuropilin-1 conveys semaphorin and VEGF signaling during neural and cardiovascular development. *Developmental Cell, 5*, 45–57.

Gurjarpadhye, A., Hewett, K. W., Justus, C., Wen, X., Stadt, H., Kirby, M. L., et al. (2007). Cardiac neural crest ablation inhibits compaction and electrical function of conduction system bundles. *American Journal of Physiology. Heart and Circulatory Physiology, 292*, H1291–H1300.

Hamburger, V., & Hamilton, H. L. (1951). A series of normal stages in the development of the chick embryo. *Journal of Morphology, 88*, 49–92.

Hildreth, V., Webb, S., Bradshaw, L., Brown, N. A., Anderson, R. H., & Henderson, D. J. (2008). Cells migrating from the neural crest contribute to the innervation of the venous pole of the heart. *Journal of Anatomy, 212*, 1–11.

Hiruma, T., & Hirakow, R. (1995). Formation of the pharyngeal arch arteries in the chick embryo. Observations of corrosion casts by scanning electron microscopy. *Anatomy and Embryology (Berlin), 191*, 415–423.

Hiruma, T., Nakajima, Y., & Nakamura, H. (2002). Development of pharyngeal arch arteries in early mouse embryo. *Journal of Anatomy, 201*, 15–29.

Hobbs, C. A., Malik, S., Zhao, W., James, S. J., Melnyk, S., & Cleves, M. A. (2006). Maternal homocysteine and congenital heart defects. *Journal of the American College of Cardiology, 47*, 683–685.

Hoffman, J. I., & Kaplan, S. (2002). The incidence of congenital heart disease. *Journal of the American College of Cardiology, 39*, 1890–1900.

Janssen, N., Bergman, J. E., Swertz, M. A., Tranebjaerg, L., Lodahl, M., Schoots, J., et al. (2012). Mutation update on the CHD7 gene involved in CHARGE syndrome. *Human Mutation, 33*, 1149–1160.

Jiang, X., Rowitch, D. H., Soriano, P., McMahon, A. P., & Sucov, H. M. (2000). Fate of the mammalian cardiac neural crest. *Development, 127*, 1607–1616.

Kaartinen, V., Dudas, M., Nagy, A., Sridurongrit, S., Lu, M. M., & Epstein, J. A. (2004). Cardiac outflow tract defects in mice lacking ALK2 in neural crest cells. *Development, 131*, 3481–3490.

Kappetein, A. P., Gittenberger-de Groot, A. C., Zwinderman, A. H., Rohmer, J., Poelmann, R. E., & Huysmans, H. A. (1991). The neural crest as a possible pathogenetic factor in coarctation of the aorta and bicuspid aortic valve. *The Journal of Thoracic and Cardiovascular Surgery*, *102*, 830–836.

Kapusta, L., Haagmans, M. L., Steegers, E. A., Cuypers, M. H., Blom, H. J., & Eskes, T. K. (1999). Congenital heart defects and maternal derangement of homocysteine metabolism. *The Journal of Pediatrics*, *135*, 773–774.

Kawasaki, T., Kitsukawa, T., Bekku, Y., Matsuda, Y., Sanbo, M., Yagi, T., et al. (1999). A requirement for neuropilin-1 in embryonic vessel formation. *Development*, *126*, 4895–4902.

Keyte, A., & Hutson, M. R. (2012). The neural crest in cardiac congenital anomalies. *Differentiation; Research in Biological Diversity*, *84*, 25–40.

Kirby, M. L. (1987). Cardiac morphogenesis—Recent research advances. *Pediatric Research*, *21*, 219–224.

Kirby, M. L., Gale, T. F., & Stewart, D. E. (1983). Neural crest cells contribute to normal aorticopulmonary septation. *Science*, *220*, 1059–1061.

Kirby, M. L., & Hutson, M. R. (2010). Factors controlling cardiac neural crest cell migration. *Cell Adhesion & Migration*, *4*, 609–621.

Kubalak, S. W., Hutson, D. R., Scott, K. K., & Shannon, R. A. (2002). Elevated transforming growth factor beta2 enhances apoptosis and contributes to abnormal outflow tract and aortic sac development in retinoic X receptor alpha knockout embryos. *Development*, *129*, 733–746.

Lammer, E. J., Chen, D. T., Hoar, R. M., Agnish, N. D., Benke, P. J., Braun, J. T., et al. (1985). Retinoic acid embryopathy. *The New England Journal of Medicine*, *313*, 837–841.

Lee, Y. H., & Saint-Jeannet, J. P. (2011). Cardiac neural crest is dispensable for outflow tract septation in Xenopus. *Development*, *138*, 2025–2034.

Leonhardt, A., Glaser, A., Wegmann, M., Schranz, D., Seyberth, H., & Nusing, R. (2003). Expression of prostanoid receptors in human ductus arteriosus. *British Journal of Pharmacology*, *138*, 655–659.

Lo, C. W., Cohen, M. F., Huang, G. Y., Lazatin, B. O., Patel, N., Sullivan, R., et al. (1997). Cx43 gap junction gene expression and gap junctional communication in mouse neural crest cells. *Developmental Genetics*, *20*, 119–132.

Macatee, T. L., Hammond, B. P., Arenkiel, B. R., Francis, L., Frank, D. U., & Moon, A. M. (2003). Ablation of specific expression domains reveals discrete functions of ectoderm- and endoderm-derived FGF8 during cardiovascular and pharyngeal development. *Development*, *130*, 6361–6374.

Markwald, R. R., Krook, J. M., Kitten, G. T., & Runyan, R. B. (1981). Endocardial cushion tissue development: Structural analyses on the attachment of extracellular matrix to migrating mesenchymal cell surfaces. *Scanning Electron Microscopy*, (Pt. 2), 261–274.

McLennan, R., Teddy, J. M., Kasemeier-Kulesa, J. C., Romine, M. H., & Kulesa, P. M. (2010). Vascular endothelial growth factor (VEGF) regulates cranial neural crest migration in vivo. *Developmental Biology*, *339*, 114–125.

Merscher, S., Funke, B., Epstein, J. A., Heyer, J., Puech, A., Lu, M. M., et al. (2001). TBX1 is responsible for cardiovascular defects in velo-cardio-facial/DiGeorge syndrome. *Cell*, *104*, 619–629.

Nakamura, T., Colbert, M. C., & Robbins, J. (2006). Neural crest cells retain multipotential characteristics in the developing valves and label the cardiac conduction system. *Circulation Research*, *98*, 1547–1554.

Park, J. E., Keller, G. A., & Ferrara, N. (1993). The vascular endothelial growth factor (VEGF) isoforms: Differential deposition into the subepithelial extracellular matrix and bioactivity of extracellular matrix-bound VEGF. *Molecular Biology of the Cell*, *4*, 1317–1326.

Park, E. J., Ogden, L. A., Talbot, A., Evans, S., Cai, C. L., Black, B. L., et al. (2006). Required, tissue-specific roles for Fgf8 in outflow tract formation and remodeling. *Development*, *133*, 2419–2433.

Plein, A., Fantin, A., & Ruhrberg, C. (2014). Neuropilin regulation of angiogenesis, arteriogenesis, and vascular permeability. *Microcirculation*, *21*, 315–323.

Porras, D., & Brown, C. B. (2008). Temporal–spatial ablation of neural crest in the mouse results in cardiovascular defects. *Developmental Dynamics*, *237*, 153–162.

Prescott, K., Woodfine, K., Stubbs, P., Super, M., Kerr, B., Palmer, R., et al. (2005). A novel 5q11.2 deletion detected by microarray comparative genomic hybridisation in a child referred as a case of suspected 22q11 deletion syndrome. *Human Genetics*, *116*, 83–90.

Rohm, B., Ottemeyer, A., Lohrum, M., & Puschel, A. W. (2000). Plexin/neuropilin complexes mediate repulsion by the axonal guidance signal semaphorin 3A. *Mechanisms of Development*, *93*, 95–104.

Ruediger, T., Zimmer, G., Barchmann, S., Castellani, V., Bagnard, D., & Bolz, J. (2013). Integration of opposing semaphorin guidance cues in cortical axons. *Cerebral Cortex*, *23*, 604–614.

Ruhrberg, C. (2003). Growing and shaping the vascular tree: Multiple roles for VEGF. *BioEssays*, *25*, 1052–1060.

Ruhrberg, C., Gerhardt, H., Golding, M., Watson, R., Ioannidou, S., Fujisawa, H., et al. (2002). Spatially restricted patterning cues provided by heparin-binding VEGF-A control blood vessel branching morphogenesis. *Genes & Development*, *16*, 2684–2698.

Sanford, L. P., Ormsby, I., Gittenberger-de Groot, A. C., Sariola, H., Friedman, R., Boivin, G. P., et al. (1997). TGFbeta2 knockout mice have multiple developmental defects that are non-overlapping with other TGFbeta knockout phenotypes. *Development*, *124*, 2659–2670.

Sato, A., Scholl, A. M., Kuhn, E. N., Stadt, H. A., Decker, J. R., Pegram, K., et al. (2011). FGF8 signaling is chemotactic for cardiac neural crest cells. *Developmental Biology*, *354*, 18–30.

Sato, M., & Yost, H. J. (2003). Cardiac neural crest contributes to cardiomyogenesis in zebrafish. *Developmental Biology*, *257*, 127–139.

Scambler, P. J. (2010). 22q11 deletion syndrome: A role for TBX1 in pharyngeal and cardiovascular development. *Pediatric Cardiology*, *31*, 378–390.

Schneider, D. J., & Moore, J. W. (2006). Patent ductus arteriosus. *Circulation*, *114*, 1873–1882.

Schulz, Y., Wehner, P., Opitz, L., Salinas-Riester, G., Bongers, E. M., van Ravenswaaij-Arts, C. M., et al. (2014). CHD7, the gene mutated in CHARGE syndrome, regulates genes involved in neural crest cell guidance. *Human Genetics*, *133*, 997–1009.

Siebert, J. R., Graham, J. M., Jr., & MacDonald, C. (1985). Pathologic features of the CHARGE association: Support for involvement of the neural crest. *Teratology*, *31*, 331–336.

Singh, S. M., Murphy, B., & O'Reilly, R. (2002). Monozygotic twins with chromosome 22q11 deletion and discordant phenotypes: Updates with an epigenetic hypothesis. *Journal of Medical Genetics*, *39*, e71.

Soker, S., Takashima, S., Miao, H. Q., Neufeld, G., & Klagsbrun, M. (1998). Neuropilin-1 is expressed by endothelial and tumor cells as an isoform-specific receptor for vascular endothelial growth factor. *Cell*, *92*, 735–745.

Stalmans, I., Lambrechts, D., De Smet, F., Jansen, S., Wang, J., Maity, S., et al. (2003). VEGF: A modifier of the del22q11 (DiGeorge) syndrome? *Nature Medicine*, *9*, 173–182.

St Amand, T. R., Lu, J. T., Zamora, M., Gu, Y., Stricker, J., Hoshijima, M., et al. (2006). Distinct roles of HF-1b/Sp4 in ventricular and neural crest cells lineages affect cardiac conduction system development. *Developmental Biology*, *291*, 208–217.

Stottmann, R. W., Choi, M., Mishina, Y., Meyers, E. N., & Klingensmith, J. (2004). BMP receptor IA is required in mammalian neural crest cells for development of the cardiac outflow tract and ventricular myocardium. *Development, 131*, 2205–2218.
Sugishita, Y., Watanabe, M., & Fisher, S. A. (2004). The development of the embryonic outflow tract provides novel insights into cardiac differentiation and remodeling. *Trends in Cardiovascular Medicine, 14*, 235–241.
Sun, X., Meyers, E. N., Lewandoski, M., & Martin, G. R. (1999). Targeted disruption of Fgf8 causes failure of cell migration in the gastrulating mouse embryo. *Genes & Development, 13*, 1834–1846.
Tomita, Y., Matsumura, K., Wakamatsu, Y., Matsuzaki, Y., Shibuya, I., Kawaguchi, H., et al. (2005). Cardiac neural crest cells contribute to the dormant multipotent stem cell in the mammalian heart. *The Journal of Cell Biology, 170*, 1135–1146.
Toyofuku, T., Yoshida, J., Sugimoto, T., Yamamoto, M., Makino, N., Takamatsu, H., et al. (2008). Repulsive and attractive semaphorins cooperate to direct the navigation of cardiac neural crest cells. *Developmental Biology, 321*, 251–262.
Vitelli, F., Morishima, M., Taddei, I., Lindsay, E. A., & Baldini, A. (2002). Tbx1 mutation causes multiple cardiovascular defects and disrupts neural crest and cranial nerve migratory pathways. *Human Molecular Genetics, 11*, 915–922.
von Gise, A., & Pu, W. T. (2012). Endocardial and epicardial epithelial to mesenchymal transitions in heart development and disease. *Circulation Research, 110*, 1628–1645.
Waldo, K. L., Kumiski, D., & Kirby, M. L. (1996). Cardiac neural crest is essential for the persistence rather than the formation of an arch artery. *Developmental Dynamics, 205*, 281–292.
Waldo, K., Miyagawa-Tomita, S., Kumiski, D., & Kirby, M. L. (1998). Cardiac neural crest cells provide new insight into septation of the cardiac outflow tract: Aortic sac to ventricular septal closure. *Developmental Biology, 196*, 129–144.
Wang, J., Nagy, A., Larsson, J., Dudas, M., Sucov, H. M., & Kaartinen, V. (2006). Defective ALK5 signaling in the neural crest leads to increased postmigratory neural crest cell apoptosis and severe outflow tract defects. *BMC Developmental Biology, 6*, 51.
Ward, C., Stadt, H., Hutson, M., & Kirby, M. L. (2005). Ablation of the secondary heart field leads to tetralogy of Fallot and pulmonary atresia. *Developmental Biology, 284*, 72–83.
Webb, S., Qayyum, S. R., Anderson, R. H., Lamers, W. H., & Richardson, M. K. (2003). Septation and separation within the outflow tract of the developing heart. *Journal of Anatomy, 202*, 327–342.
Wurdak, H., Ittner, L. M., Lang, K. S., Leveen, P., Suter, U., Fischer, J. A., et al. (2005). Inactivation of TGFbeta signaling in neural crest stem cells leads to multiple defects reminiscent of DiGeorge syndrome. *Genes & Development, 19*, 530–535.
Zile, M. H. (2001). Function of vitamin A in vertebrate embryonic development. *The Journal of Nutrition, 131*, 705–708.

CHAPTER SEVEN

Molecular Control of the Neural Crest and Peripheral Nervous System Development

Jason M. Newbern[1]
School of Life Sciences, Arizona State University, Tempe, Arizona, USA
[1]Corresponding author: e-mail address: jason.newbern@asu.edu

Contents

Abstract

A transient and unique population of multipotent stem cells, known as neural crest cells (NCCs), generate a bewildering array of cell types during vertebrate development. An attractive model among developmental biologists, the study of NCC biology has provided a wealth of knowledge regarding the cellular and molecular mechanisms important for embryogenesis. Studies in numerous species have defined how distinct phases of NCC specification, proliferation, migration, and survival contribute to the formation of multiple functionally distinct organ systems. NCC contributions to the peripheral nervous system (PNS) are well known. Critical developmental processes have been defined that provide outstanding models for understanding how extracellular stimuli, cell–cell interactions, and transcriptional networks cooperate to direct cellular diversification and PNS morphogenesis. Dissecting the complex extracellular and intracellular mechanisms that mediate the formation of the PNS from NCCs may have important therapeutic implications for neurocristopathies, neuropathies, and certain forms of cancer.

Current Topics in Developmental Biology, Volume 111
ISSN 0070-2153
http://dx.doi.org/10.1016/bs.ctdb.2014.11.007

1. INTRODUCTION

Neural crest cells (NCCs) are a stem-cell population that generate much of the peripheral nervous system (PNS) during development (Le Douarin & Kalcheim, 1999; Le Douarin & Smith, 1988). A tightly regulated balance between extrinsically derived cues and intrinsic regulators is required for the appropriate specification, growth, and function of NCCs during PNS formation. Evidence suggests that the early NCC population is comprised of both fate-restricted and multipotent progenitors (Bronner-Fraser & Fraser, 1988; Coelho-Aguiar, Le Douarin, & Dupin, 2013; Crane & Trainor, 2006; Fraser & Bronner-Fraser, 1991; Greenwood, Turner, & Anderson, 1999; Krispin, Nitzan, & Kalcheim, 2010; Le Douarin & Dupin, 2003; Ziller, Dupin, Brazeau, Paulin, & Le Douarin, 1983). During the course of development *in vivo* most NCCs undergo progressive fate restriction. However, some derivatives retain a level of plasticity and self-renewal potential and neural crest-like stem cells have been extracted from the sciatic nerve and dorsal root ganglia (DRG) of adult organisms (Bixby, Kruger, Mosher, Joseph, & Morrison, 2002; Greenwood et al., 1999; Li, Say, & Zhou, 2007; Morrison, White, Zock, & Anderson, 1999; Nagoshi et al., 2008; Stemple & Anderson, 1992; White et al., 2001). Since cranial PNS structures are derived from both NCCs and placode cells, the focus of this review is primarily on the development of the DRG and the peripheral nerves which are derived solely from trunk NCCs. The study of PNS development continues to shed light on the role of distinct molecular mediators of complex cell and tissue interactions.

2. NEURAL CREST SPECIFICATION

NCC arises from the dorsal lip of the developing neural tube at early stages of embryogenesis. Briefly, extracellular cues derived from the ectoderm, mesoderm, and adjacent neuroepithelium play an active role in the process of NCC specification. The inductive cues and fate potentials of NCCs along the neuraxis are diverse and a number of canonical patterning systems participate in this process, including Wnt/β-catenin, FGFs, BMPs, retinoic acid, and Delta/Notch signaling (Cheung et al., 2005; Mead & Yutzey, 2012; Milet & Monsoro-Burq, 2012; Stuhlmiller & Garcia-Castro, 2012). Many of these same signals act at later stages of NCC and

PNS differentiation as well. The tightly regulated expression of various transcription factors is important during this transition; Pax7, Snail/Slug, FoxD3, and Sox9 are but a few that are especially critical at this early stage (Betancur, Bronner-Fraser, & Sauka-Spengler, 2010; Bhatt, Diaz, & Trainor, 2013).

Once specified, NCCs separate from the neuroepithelium and undergo an epithelial to mesenchymal transition (EMT) before initiating migration toward distant sites (Lim & Thiery, 2012). Live cell imaging has revealed significant heterogeneity in the sequence of detachment, division, polarization, and migration during EMT, indicating that highly complex and plastic interactions between multiple cellular subprograms regulate this process (Ahlstrom & Erickson, 2009). Modulation of cadherins, integrins, and multiple extracellular matrix (ECM) components is vital for modulating NCC delamination (Perris & Perissinotto, 2000). For example, a regulated switch from N-cadherin to cadherin-6 expression and noncanonical Wnt/planar cell polarity signaling play a key role in delamination and early migration (Carmona-Fontaine, Matthews, & Mayor, 2008; Clay & Halloran, 2014; De Calisto, Araya, Marchant, Riaz, & Mayor, 2005; Mayor & Theveneau, 2014; Nakagawa & Takeichi, 1995, 1998; Ulmer et al., 2013). Wnt/β-catenin signaling also acts as a potent instructive cue that promotes PNS specification. Activation of β-catenin drives the formation of DRG sensory neurons at the expense of many other NCC derivatives, while inhibition of Wnt or β-catenin attenuates DRG and sympathetic ganglia (SG) formation (Armstrong, Ryu, Chieco, & Kuruvilla, 2011; Hari et al., 2002; Ikeya, Lee, Johnson, McMahon, & Takada, 1997; Lee et al., 2004). The effect of Wnt/β-catenin signaling on DRG fate is most effective at the premigratory stage; however, Wnts continue to have important functions during later stages of neuronal development (Bodmer, Levine-Wilkinson, Richmond, Hirsh, & Kuruvilla, 2009; Hari et al., 2012).

NCCs are induced along the entire neuraxis and can be divided into specific groups with distinct migratory routes and competencies. The PNS arises primarily from trunk NCC, which is derived from the neural tube caudal to the fourth somite. Unlike cranial NCCs, trunk NCCs are generally restricted from generating ectomesenchymal tissues such as bone and cartilage *in vivo* (Coelho-Aguiar et al., 2013). However, exceptions have been observed in turtle carapace and plastron development (Cebra-Thomas et al., 2013). NCCs from the vagal and sacral regions generate the enteric nervous system (ENS), while the cranial and sacral NCCs make important contributions to the parasympathetic nervous system. Outstanding advances have been made in defining mechanisms of ENS morphogenesis that are reviewed elsewhere (Sasselli, Pachnis, & Burns, 2012).

3. MIGRATORY PATTERNS OF TRUNK NEURAL CREST

NCCs exiting the dorsal neural tube first migrate ventrally in a non-segmental fashion before traveling along a set of well-characterized routes (Fig. 1). Cooperative cell interactions, control of cytoskeletal activity, and an array of positive and negative cues directly influence the complex pattern of NCC migration (Friedl & Gilmour, 2009; Mayor & Carmona-Fontaine, 2010). NCCs often migrate in chains, an interaction that is critical for regulating the directionality of migration (Erickson, 1985; Rorth, 2009). Migratory routes are lined with a number of permissive ECM components, such as laminins, versican, and fibronectin, which help guide the path of NCCs (Dutt, Kleber, Matasci, Sommer, & Zimmermann, 2006; Perris & Perissinotto, 2000; Rorth, 2009). Typically, trunk NCCs migrate ipsilaterally; however, some NCCs are capable of crossing the dorsal midline and migrating into the contralateral DRG (George, Chaverra, Todd, Lansford, & Lefcort, 2007).

The timing and choice of migratory pathway is tightly linked to subsequent fate decisions. An early bifurcation occurs when migratory NCCs choose a dorsolateral path along the ectoderm or a ventromedial course in between the neural tube and developing somites (Gammill & Roffers-Agarwal, 2010; Serbedzija, Bronner-Fraser, & Fraser, 1989; Thiery, Duband, & Delouvee, 1982). Trunk NCCs that enter the ventromedial pathway contribute to the peripheral and autonomic nervous system in addition to other trunk derivatives, such as adrenal chromaffin cells, while the dorsolateral pathway mostly generates the pigment cell lineage including melanocytes (Kelsh, Harris, Colanesi, & Erickson, 2009; Serbedzija, Fraser, & Bronner-Fraser, 1990; Shtukmaster et al., 2013). The choice of pathway is also related to the timing of emigration; early NCCs primarily enter the ventromedial pathway, while later waves of NCCs are biased toward the dorsolateral pathway. Interestingly, late-born NCCs transplanted into younger embryos still enter the dorsolateral pathway, showing that the timing of NCC birth is critical for subsequent migratory path and fate choices (Erickson & Goins, 1995; Reedy, Faraco, & Erickson, 1998). NCCs generally show a bias toward populating target organs in a ventral to dorsal order, though variation between chick and mouse has been observed

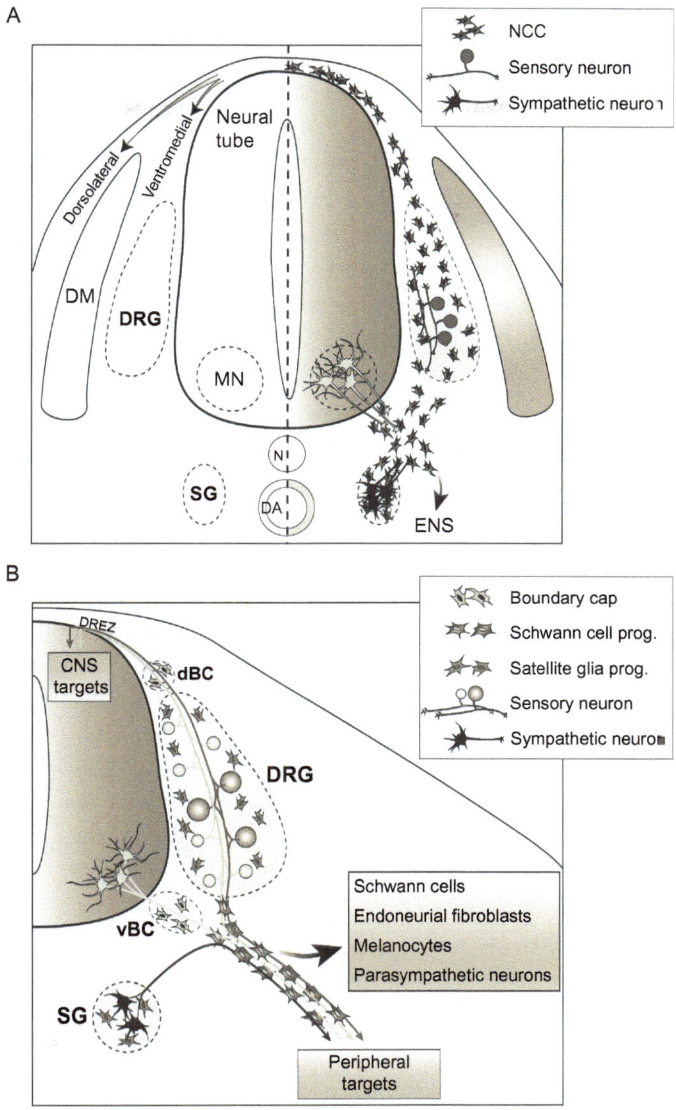

Figure 1 The initial formation of the PNS from NCCs. (A) After undergoing EMT in the roof plate of the neural tube, migratory trunk NCCs are guided by a combination of attractive, repulsive, and instructive cues derived from the developing dermomyotome (DM), sclerotome, dorsal aorta (DA), and notochord (N). NCCs that generate the PNS migrate ventromedially between the neural tube and developing somite, while the dorsomedial NCCs primarily generate melanocytes. Some NCCs migrate to distant sites in the trunk, such as the enteric nervous system (ENS) and adrenal glands. A population of NCCs cease migration at sites of peripheral ganglia formation and enter a phase of neurogenesis that produces sensory neurons in the dorsal root ganglia (DRG) and

(Continued)

(Krispin, Nitzan, Kassem, & Kalcheim, 2010; Serbedzija, Bronner-Fraser, & Fraser, 1994, Serbedzija et al., 1989).

The developing somites provide an additional critical source of patterning cues that initiate segmental migration and direct the metameric organization of the developing DRG, SG, and peripheral nerves (Bronner-Fraser, 1986; Bronner-Fraser & Stern, 1991; Keynes & Stern, 1984; Krull, 2001). Rotation or ablation of the early somites leads to aberrant PNS segmentation and altered NCC migratory patterns (Bronner-Fraser & Stern, 1991; Kalcheim & Teillet, 1989). The early wave of trunk NCCs migrates ventrally along intersomitic blood vessels in between the somites (Bronner-Fraser, 1986; Schwarz, Maden, Davidson, & Ruhrberg, 2009; Thiery et al., 1982). NCCs entering the intersomitic path will generate neurons and glia within the SG and are stimulated by chemoattractant and instructive cues from the dorsal aorta (DA), such as SDF1/CXCR4, BMPs, and neuregulin-1 (Nrg1), as well as blood vessel-derived artemin (Belmadani et al., 2005; Britsch et al., 1998; Honma et al., 2002; Kasemeier-Kulesa, McLennan, Romine, Kulesa, & Lefcort, 2010; Reissmann et al., 1996; Saito, Takase, Murai, & Takahashi, 2012; Schneider, Wicht, Enderich, Wegner, & Rohrer, 1999; Shah, Groves, & Anderson, 1996; Yip, 1986). As the somite differentiates into the sclerotome and dermomyotome (DM), another wave of NCCs migrates segmentally into the space between the developing structures, often along the basement membrane of the DM (Krull, 2001; Tosney, Dehnbostel, & Erickson, 1994). Importantly, the caudal somite produces factors that repel migrating NCCs, while the rostral half

Figure 1—Cont'd sympathetic neurons in the sympathetic ganglia (SG). (B) Axons from NCC-derived SG and DRG neurons and neuroectodermally derived lower motor neurons begin growing into the periphery. The sites where axons enter and exit the spinal cord are populated by the boundary cap, a transient NCC-derived stem-cell niche. The ventral boundary cap (vBC) is localized along outgrowing lower motor neurons axons at CNS/PNS boundary, while the dorsal boundary cap (dBC) is found along fibers near the dorsal root entry zone (DREZ) where sensory afferents enter the spinal cord. TrkC expressing, large-diameter proprioceptive neurons (green) are among the first neurons to be produced in the nascent DRG, followed by small-diameter, TrkA-expressing DRG neurons (red). Subsequent to the onset of neurogenesis, NCCs and boundary cap generate satellite glia progenitors that reside in the ganglia and Schwann cell progenitors (SCPs) that migrate along axons in the developing nerve. SCPs can ultimately differentiate into various cell types that contribute to peripheral nerve function, including myelinating and nonmyelinating Schwann cells and endoneurial fibroblasts. SCPs also generate melanocytes and have even been shown to produce parasympathetic neurons in cranial nerves. (See the color plate.)

provides attractive cues (Bronner-Fraser & Stern, 1991; Goldstein, Teillet, & Kalcheim, 1990; Koblar et al., 2000; Krull et al., 1997; Wang & Anderson, 1997). A subpopulation of NCCs migrate through the developing sclerotome and provide an additional source of SG progenitors. NCCs that arrest migration adjacent to the neural tube generate the DRG and subsequent derivatives.

4. MOLECULAR REGULATORS OF NEURAL CREST MIGRATION

A number of secreted factors act in conjunction with intrinsic regulators to control NCC migration, proliferation, and multipotency during the early migratory stage of development. Extracellular signaling through secreted trophic factors helps promote migratory NCC survival and/or proliferation (Britsch et al., 1998; Kalcheim, 1996; Meyer & Birchmeier, 1995; Murphy, Reid, Ford, Furness, & Bartlett, 1994; Shah, Marchionni, Isaacs, Stroobant, & Anderson, 1994; Sommer, 2006). FGF2, Nrg1, the neurotrophin-3 (NT-3) receptor TrkC, and the thrombospondin/EGF domain-containing factor NELL2 can be detected in a subset of migrating trunk NCCs, all of which promote NCC proliferation and may act as instructive cues (Henion, Garner, Large, & Weston, 1995; Kahane & Kalcheim, 1994; Kalcheim, Carmeli, & Rosenthal, 1992; Murphy et al., 1994; Nelson, Claes, Todd, Chaverra, & Lefcort, 2004; Rifkin, Todd, Anderson, & Lefcort, 2000). The transcription factors Sox2, Sox10, and FoxD3 play well-defined roles in maintaining the stem cell-like features and self-renewal capacity of early migratory NCCs (Kim, Lo, Dormand, & Anderson, 2003; Mundell & Labosky, 2011; Sonnenberg-Riethmacher et al., 2001; Southard-Smith, Kos, & Pavan, 1998; Teng, Mundell, Frist, Wang, & Labosky, 2008). Sox10 increases the expression of the neuregulin receptor ErbB3, providing a specific mechanistic example of a precise intrinsic cue that modulates extrinsic responsiveness (Britsch et al., 2001; Paratore, Goerich, Suter, Wegner, & Sommer, 2001; Prasad et al., 2011). The mechanism of interaction between many other critical extrinsic and intrinsic cues has yet to be fully elucidated.

Importantly, somite-derived factors that direct NCC migration have been defined. The caudal half of the developing somite provides local cues that inhibit NCC migration, while the rostral half appears to produce attractive and mitogenic factors (Goldstein et al., 1990; Koblar et al., 2000; Krull, 2001; Krull et al., 1997; Wang & Anderson, 1997). Repulsion from the caudal somite is mediated by semaphorins and ephrins that act in concert with Neuropilin and Eph-expressing neural crest (Gammill, Gonzalez, Gu, & Bronner-Fraser, 2006; Kawasaki et al., 2002; Krull, 2001; Krull et al., 1997; Maden et al., 2012; Schwarz et al., 2009; Wang & Anderson, 1997). These signaling cues are critical for directing the segmental migration and final pattern of PNS morphogenesis. F-spondin expression in the caudal sclerotome provides an additional repulsive cue for migrating NCCs while thrombospondin in the rostral domain has been shown to act as an attractant (Debby-Brafman, Burstyn-Cohen, Klar, & Kalcheim, 1999; Tucker et al., 1999). Furthermore, a similar role for Delta expression in the caudal somite has been proposed (Bettenhausen, Hrabe de Angelis, Simon, Guenet, & Gossler, 1995). Deletion of *Delta1* results in disruption of the metameric pattern of DRG formation; however, a reduced number of progenitors indicate multiple functions for Delta/Notch signaling that clearly extend beyond strict migratory control (De Bellard, Ching, Gossler, & Bronner-Fraser, 2002; Hrabe de Angelis, McIntyre, & Gossler, 1997; Mead & Yutzey, 2012).

Long-range, local, and contact-dependent molecules have been identified that regulate diverse aspects of NCC migration. These signals are capable of activating numerous intracellular pathways; however, convergent regulation of common downstream components, such as Rho and Rac, serves as a key integration point (Berndt, Clay, Langenberg, & Halloran, 2008; Clay & Halloran, 2014; Liu & Jessell, 1998; Shoval & Kalcheim, 2012; Theveneau & Mayor, 2012). Newly developed high-resolution imaging techniques and genetic tools will continue to provide unique insight into how entire populations of cells are guided into distinct migratory routes and destinations during embryogenesis (Clay & Halloran, 2010).

In vivo clonal analyses suggest that early migratory NCCs contain both multipotent progenitors capable of generating cells within the DRG, SG, and nerve, in addition to progenitors restricted to a specific lineage (Bronner-Fraser & Fraser, 1988, 1989; Frank & Sanes, 1991; Krispin, Nitzan, Kassem, et al., 2010; Serbedzija et al., 1989; Shtukmaster et al., 2013). A disruption in migratory guidance occurs in *Nrp1* and *EdnRB2* mutants that leads precociously misrouting of NCCs into the dorsolateral pathway (Krispin, Nitzan, Kassem, et al., 2010; Schwarz et al., 2009). Interestingly, neuronal markers are detected in the dorsolateral pathway of these mutants. Thus, NCCs can be specified to the neurogenic lineage in the absence of interactions with sclerotome-derived signals (Krispin, Nitzan, Kassem, et al., 2010; Schwarz et al., 2009). Even though DRG-restricted NCCs have been identified, these cells produce both neurons and glia *in vivo* (Greenwood et al., 1999; Zirlinger, Lo, McMahon, McMahon, & Anderson, 2002). Overall, these data provide support for the notion that some migratory NCCs are specified to the DRG or the autonomic lineage prior to choosing between a neuronal or glial fate (Anderson, 2000; Crane & Trainor, 2006; Krispin, Nitzan, Kassem, et al., 2010; Morrison et al., 1999).

5. BOUNDARY CAP

An intermediate population of NCC-derived stem cells, known as the boundary cap, form on the border of the spinal cord and DRG along both the dorsal and ventral roots, known as the dorsal root entry zone (DREZ) and the motor exit point (MEP), respectively (Altman & Bayer, 1984; Golding & Cohen, 1997). NCCs migrating along the ventromedial pathway generate the boundary cap after the initial wave of DRG progenitors is established (Niederlander & Lumsden, 1996). The boundary cap progenitors form a critical boundary between the CNS and the PNS (Bron et al., 2007; Coulpier et al., 2010, 2009; Hjerling-Leffler et al., 2005; Maro et al., 2004; Mauti, Domanitskaya, Andermatt, Sadhu, & Stoeckli, 2007; Vermeren et al., 2003). This cell impermeable barrier relies, in part, on boundary cap-derived, membrane-bound Semaphorin6A (Sema6A; Bron et al., 2007; Mauti et al., 2007). *Sema6A* expression may play two roles in this process, the first being to appropriately aggregate boundary cap cells along the CNS/PNS boundary and the second to inhibit PlexinA or Neuropilin2-expressing CNS-derived cell types from migrating into the PNS (Bron et al., 2007; Kucenas, Wang, Knapik, & Appel, 2009; Mauti et al., 2007). Ablation of the boundary cap by multiple techniques has been shown to result in the ectopic presence of CNS-derived motor neurons and oligodendrocytes in the proximal peripheral nerve (Bron et al., 2007; Mauti et al., 2007; Vermeren et al., 2003). DREZ- and MEP-associated boundary cap cells have distinct molecular profiles and slightly different temporal relationships with outgrowing axons (Coulpier et al., 2009; Fraher, Dockery, O'Donoghue, Riedewald, & O'Leary, 2007). These data suggest that potentially distinct specific functions of the boundary cap at these two sites have yet to be discovered.

During normal development, boundary cap progenitors produce a small subset of neurons in the DRG followed by the production of satellite glia and Schwann cells (Aquino et al., 2006; Hjerling-Leffler et al., 2005; Maro et al., 2004). *Egr2/Krox-20* serves as an important molecular identifier *in vivo* and is required for boundary cap barrier functions, as is *Sox10* expression (Coulpier et al., 2010; Frob et al., 2012; Maro et al., 2004; Vermeren et al., 2003; Wilkinson, Bhatt, Chavrier, Bravo, & Charnay, 1989). Boundary cap progenitors maintain a state of pluripotency somewhere between that of early NCCs and a Schwann cell progenitor (SCP), though *in vitro* studies have shown that these cells can even generate multiple CNS subtypes (Coulpier et al., 2009; Zujovic et al., 2010, 2011). These characteristics have led to a number of studies seeking to utilize boundary cap progenitor transplantation in spinal cord, peripheral nerve, and dorsal root injury paradigms (Aldskogius et al., 2009; Aquino et al., 2006; Trolle, Konig, Abrahamsson, Vasylovska, & Kozlova, 2014; Zujovic et al., 2010, 2011).

6. SENSORY NEUROGENESIS IN THE DRG

Sensory neurons in the PNS relay information into the CNS from a number of specific exteroceptive, proprioceptive, and interoceptive structures, including Merkel's discs, Meissner's and Pacinian corpuscles, Ruffini's end organs, Golgi tendon organs, muscle spindles, and free nerve endings in the skin. Dedicated neurons transmit information of distinct somatosensory modalities; proprioceptive neurons provide spatial information regarding limb position, mechanoreceptive neurons mediate touch, nociceptive neurons respond to painful stimuli or itch, and thermoreceptive neurons relay information regarding temperature (Liu & Ma, 2011; Marmigere & Ernfors, 2007). The importance of trophic factor signaling during the development of PNS neurons has long been recognized, particularly the neurotrophin ligand/receptor components NGF/TrkA, BDNF/TrkB, and NT-3/TrkC (Cowan, 2001; Ernsberger, 2009). The discovery that functionally related neuronal subtypes require specific neurotrophic factors has provided a crucial molecular handle for analyses of PNS development. Dozens of different neuronal subtypes have been characterized based on the expression of specific molecular components and peripheral/central innervation targets (Abraira & Ginty, 2013; Li et al., 2011; Liu & Ma, 2011).

DRG sensory neurons are generated in a number of waves that derive from temporally distinct NCC populations (Carr & Simpson, 1978; Frank & Sanes, 1991; Lawson & Biscoe, 1979; Marmigere & Ernfors, 2007; Rifkin et al., 2000). The initial production of sensory neurons from postmigratory NCCs follows a stereotyped pattern where large-diameter $TrkC/TrkB^{+}$ proprio- and mechanoreceptive neurons are produced first, while small-diameter $TrkA^{+}$ nociceptive neurons are subsequently generated (Carr & Simpson, 1978; Lawson & Biscoe, 1979; Liu & Ma, 2011; Marmigere & Ernfors, 2007). Boundary cap progenitors and contralaterally migrating NCCs also generate a small population of *TrkA*$^+$ nociceptive sensory neurons that populate the DRG (George et al., 2007; Maro et al., 2004). NCCs that first migrate into the nascent DRG generate a core domain of differentiated postmitotic sensory neurons, while subsequent NCCs tend to encapsulate and proliferate in the perimeter region surrounding the core (George, Kasemeier-Kulesa, Nelson, Koyano-Nakagawa, & Lefcort, 2010). Activity-dependent BDNF production from active neurons in the core and protocadherin-1 expression in the perimeter are necessary for proper DRG formation (Bononi, Cole, Tewson, Schumacher, & Bradley, 2008; Wright & Ribera, 2010). Inhibition of either mechanism leads to less NCCs localizing within the DRG and an increase in ventrally migrating NCCs that expand the SG. Lastly, contact-mediated interactions between immature neurons in the core domain and undifferentiated NCCs regulate neuronal specification and subsequent lineage diversification, in part through Delta/Notch signaling (Hagedorn, Suter, & Sommer, 1999; Maynard, Wakamatsu, & Weston, 2000; Wakamatsu, Maynard, & Weston, 2000).

The sequence of transcriptional changes that occurs during sensory neuron specification has been well studied (Lallemend & Ernfors, 2012). The downregulation of factors that maintain NCC multipotency, such as Sox10, Sox2, and FoxD3, is important for NCC differentiation into postmitotic neurons (Montelius et al., 2007; Nitzan et al., 2013; Wakamatsu, Endo, Osumi, & Weston, 2004). The coordinated upregulation of proneural transcription factors, *Neurogenin-1* and *-2*, can be detected in a subset of migrating NCCs shortly after exiting the neural tube (Greenwood et al., 1999; Ma, Fode, Guillemot, & Anderson, 1999; Perez, Rebelo, & Anderson, 1999). *Neurogenins* are potent promoters of DRG specification; however, they do not necessarily drive NCCs toward a specific subtype of sensory neuron or glia (Zirlinger et al., 2002). The subsequent upregulation of neuron-specific transcriptional regulators, *Brn3a* and *Islet1*, is involved in the transition of neurogenic progenitors into sensory neurons (Dykes, Tempest, Lee, & Turner, 2011; Fedtsova & Turner, 1995; McEvilly et al., 1996; Sun et al., 2008). *Brn3a* and *Islet1* also direct the expression of factors important for sensory neuron maturation, such as the Runx family of transcription factors and specific neurotrophin receptors (Chen et al., 2006; Dykes et al., 2011; Kramer et al., 2006; Marmigere et al., 2006). Loss of *Brn3a* leads to an increased number of aberrantly differentiated sensory neurons that express multiple neurotrophin receptors and decreased levels of *Runx1* (Zou, Li, Klein, & Xiang, 2012). Runx1 is critical for the continuing differentiation of nociceptive neurons, while Runx3 primarily regulates proprioceptive maturation (Chen et al., 2006; Inoue et al., 2007; Kramer et al., 2006; Lallemend et al., 2012).

7. NEUROTROPHIC FACTORS IN SENSORY NEURON DEVELOPMENT

Peripheral innervation targets, central neurons, and associated glia produce neurotrophic cues that direct the development of receptive neuronal subtypes at distinct stages (Davies, Thoenen, & Barde, 1986; Kawaja et al., 2011; Lumsden & Davies, 1983; Patapoutian, Backus, Kispert, & Reichardt, 1999; Usui et al., 2012). Neurotrophic factor responsiveness is highly dynamic during development. This mechanism is likely important for generating diverse neuronal characteristics that are necessary for responding to a wide range of sensory stimuli. The transient pan-neuronal expression of *TrkC* is rapidly restricted to a small subset of proprioceptive neurons, while *TrkA* and *TrkB* expression is upregulated in nociceptive and mechanoreceptive neurons, respectively (Farinas, Wilkinson, Backus, Reichardt, & Patapoutian, 1998; Lefcort, Clary, Rusoff, & Reichardt, 1996; Martin-Zanca, Barbacid, & Parada, 1990; Mu, Silos-Santiago, Carroll, & Snider, 1993; Rifkin et al., 2000; Wright & Snider, 1995). Genetic deletion mutants have clearly demonstrated that NT-3/TrkC is critical for the survival of large-diameter proprioceptive neurons, while NGF/TrkA maintains small-diameter nociceptive neuron number (Crowley et al., 1994; Ernfors, Lee, Kucera, & Jaenisch, 1994; Farinas, Jones, Backus, Wang, & Reichardt, 1994; Klein et al., 1994; Ruit, Elliott, Osborne, Yan, & Snider, 1992; Smeyne et al., 1994; Tessarollo, Vogel, Palko, Reid, & Parada, 1994). As embryogenesis continues, a subset of *TrkA*-expressing nociceptive neurons develop responsiveness to GDNF by upregulating the GDNF receptors *Ret/GFRα* (Molliver & Snider, 1997; Molliver et al., 1997).

The p75 low-affinity neurotrophin receptor ($p75^{NTR}$) is also activated by a number of trophic factors (Simi & Ibanez, 2010). $p75^{NTR}$ can bind all of the neurotrophins, but when compared to the Trks, striking differences in structure and intracellular signal transduction have been discovered (Charalampopoulos et al., 2012). Different deletion mutants of $p75^{NTR}$ exhibit complex sensory and sympathetic abnormalities that vary depending on the precise mutation (Davies, Lee, & Jaenisch, 1993; Dhanoa, Krol, Jahed, Crutcher, & Kawaja, 2006; Lee et al., 1992; Majdan, Walsh, Aloyz, & Miller, 2001; Petrie et al., 2013; von Schack et al., 2001). Conditional NCC-specific $p75^{NTR}$ mutants show effects consistent with a disruption in PNS development (Bogenmann et al., 2011). Many of these studies have focused upon the role of $p75^{NTR}$ in neuronal survival and innervation. However, the onset of $p75^{NTR}$ expression occurs in premigratory neural crest and $p75^{NTR}$ has been used to isolate neural crest stem cells (Stemple & Anderson, 1992; Wilson, Richards, Ford-Perriss, Panthier, & Murphy, 2004). It will be interesting to further evaluate whether $p75^{NTR}$ modulates early neural crest migration or patterning events that might also contribute to PNS phenotypes (Hapner, Boeshore, Large, & Lefcort, 1998).

As neurons transition into a postmitotic state, they begin to grow neurites that fasciculate with outgrowing spinal motor axons in the forming ventral root *en route* to the periphery or project centrally into the spinal cord via the dorsal root and innervate CNS targets. Once again somite-derived patterning cues direct the stereotyped position of early sensorimotor projections into the periphery and coordinate alignment with the developing vertebrae (Keynes & Stern, 1984; Koblar et al., 2000; Krull, 2010). Trophic factor regulation of postmitotic sensory and motor neuron survival is well known. Mouse mutants that block the neuronal death associated with trophic factor deletion, by simultaneously deleting the prodeath Bcl-2 family member, *Bax*, are an important tool for defining the additional vital functions of trophic factors in neurons (Deckwerth et al., 1996; Patel, Jackman, Rice, Kucera, & Snider, 2000). For example, mouse mutants that lack both *Bax* and *NGF/TrkA* do not exhibit a loss of sensory neurons; however, nociceptive neurons fail to innervate the peripheral cutaneous field (Patel et al., 2000). With this approach, the multifunctional effects of various trophic factors on target innervation, subtype specification, and synapse formation have been definitively evaluated (Deppmann et al., 2008; Glebova & Ginty, 2004; Guo et al., 2011; Luo et al., 2007; Patel et al., 2003; Sharma et al., 2010).

8. GLIOGENESIS IN THE PNS

After neurogenesis has commenced, a subset of NCCs begin to generate distinct populations of nonneuronal cells. These include satellite glia within the peripheral and enteric ganglia, in addition to SCPs in the developing peripheral nerve. Satellite glia in the DRG can be detected prior to

SCPs in the nerve (Woodhoo, Dean, Droggiti, Mirsky, & Jessen, 2004). Moreover, the satellite glia lineage shows a number of differences from the SCP lineage, such as the expression of *Erm* (Hagedorn et al., 2000). SCPs maintain a close association with developing axons in the nerve and undergo additional lineage diversification into nonmyelinating and myelinating glial subtypes (Jessen & Mirsky, 2005). Myelinating Schwann cells form a myelin sheath around a single axon crucial for nerve transmission, while nonmyelinating Schwann cells ensheath multiple axons in a Remak bundle.

Delta/Notch signaling acts as a critical module for driving gliogenesis in undifferentiated and neurogenic NCCs. Evidence suggests that newly born DRG neurons in the core domain upregulate *Delta1*, which acts on neighboring *Notch*-expressing NCCs to promote the onset of gliogenesis and maintenance of gliogenic precursors (Morrison et al., 2000; Tsarovina, Schellenberger, Schneider, & Rohrer, 2008; Wakamatsu et al., 2004). NCC-specific deletion of *Notch* or the canonical downstream effector, *Rbpj*, results in a profound reduction in gliogenic precursors in the DRG, while *Notch* overactivation drives premature and increased gliogenesis *in vivo* and *in vitro* (Hu et al., 2011; Mead & Yutzey, 2012; Morrison et al., 2000; Taylor, Yeager, & Morrison, 2007). *Sox2* is a critical intrinsic factor important for gliogenesis that is regulated by Notch (Wakamatsu et al., 2004). Sox2 is required for maintaining the gliogenic state of SCPs while also preventing melanocyte specification (Adameyko et al., 2012; Wakamatsu et al., 2004). As in migratory NCCs, Sox10 continues to be vital for maintaining the SCP pool and glial differentiation (Britsch et al., 2001; Kim et al., 2003; Paratore et al., 2001). The functional requirement for Sox10 and Notch persists in developing Schwann cells; both regulate later stages of Schwann cell differentiation and development (Bremer et al., 2011; Britsch et al., 2001; Finzsch et al., 2010; Paratore et al., 2001).

SCPs are distinct from migrating NCCs in that they are dependent on axonal-derived cues for survival (Jessen et al., 1994; Woodhoo et al., 2004). Axonally derived Nrg-1 is a crucial component of the neuron-derived signal that instructs gliogenic neural crest toward a glial fate, promotes SCP survival, and is required for lineage progression and myelination (Dong et al., 1999; Meyer et al., 1997; Michailov et al., 2004; Shah et al., 1994; Taveggia et al., 2005). *ErbB2*, *ErbB3*, and *Nrg1* mutant mice exhibit a near complete absence of SCPs in the developing peripheral nerve (Lin et al., 2000; Meyer & Birchmeier, 1995; Morris et al., 1999; Riethmacher et al., 1997; Woldeyesus et al., 1999). Importantly, these mutants also exhibit profound sensory and motor neuron death and abnormally fasciculated axons in the peripheral nerve. Moreover, SCP-derived trophic factors have been found to be potent stimulators of Nrg1 release from neurons (Esper & Loeb, 2004, 2009; Hapner et al., 2006; Ma, Wang, Song, & Loeb, 2011). These data suggest that Schwann cells and axons form reciprocal trophic feedback loops that support the development of the neuroglial unit and appropriate nerve function.

Nrg1/ErbB and Delta/Notch are critical extracellular modulators of a core transcriptional network necessary for subsequent Schwann cell development and myelination. These transcriptional regulators exhibit complex interactions with some factors promoting (*Sox10, Oct6, Egr2/Krox-20, YY1, NF-κB*) and others inhibiting (*Sox2, Nab, c-Jun, Id2*) lineage progression in developing Schwann cells (Pereira, Lebrun-Julien, & Suter, 2012). Unlike terminally differentiated neurons, mature Schwann cells can dedifferentiate into a progenitor-like state following nerve injury and help promote efficient peripheral nerve regeneration (Glenn & Talbot, 2013; Napoli et al., 2012). Developmental regulators of lineage progression often continue to act as important factors in Schwann cell dedifferentiation and remyelination.

It is important to note that SCPs generate cell types other than Schwann cells. SCPs have been shown to generate melanocytes and endoneurial fibroblasts that line the peripheral nerve sheath (Adameyko et al., 2009; Joseph et al., 2004). Recent exciting work has shown that SCPs can even generate neurons *in vivo* (Dyachuk et al., 2014; Espinosa-Medina et al., 2014). In these studies, elegant whole mount labeling and 3D imaging demonstrate that parasympathetic neurons are derived from SCPs in the developing cranial nerves (Dyachuk et al., 2014; Espinosa-Medina et al., 2014). Thus, the developing cranial nerve appears to serve as both a guide and source of progenitors for the parasympathetic ganglia it will eventually innervate. Further research is clearly necessary to precisely evaluate the mechanisms that balance fate restriction and multipotency in the SCP pool.

9. TROPHIC SIGNALING MECHANISMS DURING PNS DEVELOPMENT

The study of trophic factor functions in PNS neurons and glia has served as a classic system for dissecting the biochemical pathways that mediate cellular development (Cowan, 2001; Dekkers, Nikoletopoulou, & Barde, 2013; Harrington & Ginty, 2013). The intracellular signaling cascades downstream of ErbBs and Trks have been well studied and provide a model for other receptor tyrosine kinases (RTKs; Lemmon & Schlessinger,

2010). Even though there are dozens of RTKs, a number of common core pathways are repeatedly implicated, including extracellular signal-regulated kinase 1/2 (Erk1/2), phosphatidylinositol-3-kinase (PI3K), phospholipase C, and protein kinase C (Lemmon & Schlessinger, 2010). Tight control of the activity of signaling pathways is likely an important mechanism to obtain specific responses in certain neural crest populations. For example, substantial gene dose-dependent defects in cranial and cardiac neural crest derivatives are observed following deletion of *Erk1* and *Erk2*, whereas the initial formation of the DRG from trunk neural crest is relatively intact (Newbern et al., 2008). At later stages of PNS development, neurons appear to utilize distinct intracellular pathways to achieve precise patterns of innervation. Signaling through ERK1/2 is critical for promoting nociceptive cutaneous innervation *in vivo*, possibly via disruption of SRF and ETS family transcription factors downstream of NGF (Arber, Ladle, Lin, Frank, & Jessell, 2000; Fontanet, Irala, Alsina, Paratcha, & Ledda, 2013; Newbern et al., 2011; Patel et al., 2003; Wickramasinghe et al., 2008). In contrast, SAD kinase signaling has little effect on NGF-dependent nociceptive neurons, but strongly regulates NT-3-dependent proprioceptive innervation (Lilley, Pan, & Sanes, 2013).

The study of glial development has provided important insight into the functional requirement and regulatory features of trophic signaling mechanisms. The PI3K/Akt pathway has repeatedly been implicated in the control of Schwann cell myelination in response to Nrg1 and ECM signaling (Heller et al., 2014; Maurel & Salzer, 2000). A number of findings suggest that Nrg1-mediated activation of the ERK1/2 pathway is also crucial for development of the Schwann cell lineage *in vivo*. Neural crest-specific deletion of *Shp2* or *Erk1/2* led to a profound absence of SCPs in the developing mouse peripheral nerve without a substantial alteration in the initial stages of neurogenesis (Grossmann et al., 2009; Newbern et al., 2011). Moreover, hyperactivation of ERK1/2 signaling is sufficient to rescue mature Schwann cell defects in *ErbB3* mutants and even results in hypermyelination (Ishii, Furusho, & Bansal, 2013; Sheean et al., 2014). Interestingly, robust reactivation of the ERK1/2 cascade in adult myelinating Schwann cells following injury induces reversion to a SCP-like state *in vivo* (Napoli et al., 2012). Thus, the level of ERK1/2 kinase activity appears to be tightly linked to the state of glial progenitor differentiation.

In the traditional model, neurotrophic receptors activate intracellular signaling pathways after ligand binding and often support neuronal survival in the PNS (Lemmon & Schlessinger, 2010; Reichardt, 2006). Recent findings have shown that in the *absence* of ligand, some receptors promote death. These receptors have thus been termed "dependence receptors." TrkA and TrkC have been shown to act as dependence receptors in the developing nervous system (Dekkers et al., 2013; Nikoletopoulou et al., 2010; Tauszig-Delamasure et al., 2007). DRG and spinal cord neurons induced to overexpress TrkA and TrkC will undergo death unless the associated ligands, NGF or NT-3, are simultaneously increased (Nikoletopoulou et al., 2010; Tauszig-Delamasure et al., 2007). Furthermore, a comparison of $TrkA^{-/-}$ and $NGF^{-/-}$ mutant mouse embryos revealed that deletion of TrkA protects NGF-dependent E11.5 DRG neurons from death *in vivo* (Nikoletopoulou et al., 2010). Current findings suggest that the death-promoting effect of dependence receptors in the PNS involves complex interactions with p75 and possibly the generation of proapoptotic receptor fragments (Dekkers et al., 2013; Ichim et al., 2013; Nikoletopoulou et al., 2010; Tauszig-Delamasure et al., 2007). Notably, the sensory and sympathetic neuron loss in E13.5 $TrkA^{-/-}$ mutants can be significantly rescued by simultaneous inhibition of $p75^{NTR}$ (Majdan et al., 2001; Nikoletopoulou et al., 2010). It is not yet clear why some receptors act as dependence receptors and others do not. For example, the GDNF receptor, c-Ret, can act as a dependence receptor, while TrkB does not appear to share this property (Bordeaux et al., 2000; Canibano et al., 2007; Nikoletopoulou et al., 2010). Indeed, the rules governing the death-promoting effect of dependence receptors deserve further attention. Future studies will undoubtedly illuminate additional critical functions of RTK signaling and dependence receptors in NCC and PNS development.

10. CONCLUSIONS

The formation of the PNS from trunk NCCs provides a rich developmental process to study how cell–cell interactions, secreted cues, and transcriptional networks contribute to embryogenesis. Additional molecules that regulate key cellular events during NCC development certainly await discovery. Nonetheless, many extracellular cues and transcription factors have been characterized that are necessary for specific stages of trunk NCC development. It will be important to continue to define the intracellular signaling mechanisms that link these two fundamental processes. Relative to the extremely complex repertoire of cellular and subcellular changes in the developing trunk NCCs, the number of known extracellular regulatory cues might seem limiting. Furthermore, many of these cues act at multiple stages of development. The mechanism of cellular response specificity likely depends upon the interaction between distinct canonical cues (Finelli, Murphy, Chen, & Zou, 2013). Dissecting these and many other key issues will yield important insight into the control of NCC development and assist in defining the pathogenesis of various developmental abnormalities.

ACKNOWLEDGMENT

J. N. is supported by R00-NS076661 from NIH.

REFERENCES

Abraira, V. E., & Ginty, D. D. (2013). The sensory neurons of touch. *Neuron, 79*(4), 618–639.

Adameyko, I., Lallemend, F., Aquino, J. B., Pereira, J. A., Topilko, P., Muller, T., et al. (2009). Schwann cell precursors from nerve innervation are a cellular origin of melanocytes in skin. *Cell, 139*(2), 366–379.

Adameyko, I., Lallemend, F., Furlan, A., Zinin, N., Aranda, S., Kitambi, S. S., et al. (2012). Sox2 and Mitf cross-regulatory interactions consolidate progenitor and melanocyte lineages in the cranial neural crest. *Development, 139*(2), 397–410.

Ahlstrom, J. D., & Erickson, C. A. (2009). The neural crest epithelial-mesenchymal transition in 4D: A 'tail' of multiple non-obligatory cellular mechanisms. *Development, 136*(11), 1801–1812.

Aldskogius, H., Berens, C., Kanaykina, N., Liakhovitskaia, A., Medvinsky, A., Sandelin, M., et al. (2009). Regulation of boundary cap neural crest stem cell differentiation after transplantation. *Stem Cells, 27*(7), 1592–1603.

Altman, J., & Bayer, S. A. (1984). The development of the rat spinal cord. *Advances in Anatomy, Embryology, and Cell Biology, 85*, 1–164.

Anderson, D. J. (2000). Genes, lineages and the neural crest: A speculative review. *Philosophical Transactions of the Royal Society of London. Series B: Biological Sciences, 355*(1399), 953–964.

Aquino, J. B., Hjerling-Leffler, J., Koltzenburg, M., Edlund, T., Villar, M. J., & Ernfors, P. (2006). In vitro and in vivo differentiation of boundary cap neural crest stem cells into mature Schwann cells. *Experimental Neurology, 198*(2), 438–449.

Arber, S., Ladle, D. R., Lin, J. H., Frank, E., & Jessell, T. M. (2000). ETS gene Er81 controls the formation of functional connections between group Ia sensory afferents and motor neurons. *Cell, 101*(5), 485–498.

Armstrong, A., Ryu, Y. K., Chieco, D., & Kuruvilla, R. (2011). Frizzled3 is required for neurogenesis and target innervation during sympathetic nervous system development. *Journal of Neuroscience, 31*(7), 2371–2381.

Belmadani, A., Tran, P. B., Ren, D., Assimacopoulos, S., Grove, E. A., & Miller, R. J. (2005). The chemokine stromal cell-derived factor-1 regulates the migration of sensory neuron progenitors. *Journal of Neuroscience, 25*(16), 3995–4003.

Berndt, J. D., Clay, M. R., Langenberg, T., & Halloran, M. C. (2008). Rho-kinase and myosin II affect dynamic neural crest cell behaviors during epithelial to mesenchymal transition in vivo. *Developmental Biology, 324*(2), 236–244.

Betancur, P., Bronner-Fraser, M., & Sauka-Spengler, T. (2010). Assembling neural crest regulatory circuits into a gene regulatory network. *Annual Review of Cell and Developmental Biology, 26*, 581–603.

Bettenhausen, B., Hrabe de Angelis, M., Simon, D., Guenet, J. L., & Gossler, A. (1995). Transient and restricted expression during mouse embryogenesis of Dll1, a murine gene closely related to Drosophila Delta. *Development, 121*(8), 2407–2418.

Bhatt, S., Diaz, R., & Trainor, P. A. (2013). Signals and switches in Mammalian neural crest cell differentiation. *Cold Spring Harbor Perspectives in Biology, 5*(2), http://dx.doi.org/10.1101/cshperspect.a008326.

Bixby, S., Kruger, G. M., Mosher, J. T., Joseph, N. M., & Morrison, S. J. (2002). Cell-intrinsic differences between stem cells from different regions of the peripheral nervous system regulate the generation of neural diversity. *Neuron, 35*(4), 643–656.

Bodmer, D., Levine-Wilkinson, S., Richmond, A., Hirsh, S., & Kuruvilla, R. (2009). Wnt5a mediates nerve growth factor-dependent axonal branching and growth in developing sympathetic neurons. *Journal of Neuroscience, 29*(23), 7569–7581.

Bogenmann, E., Thomas, P. S., Li, Q., Kim, J., Yang, L.-T., Pierchala, B., et al. (2011). Generation of mice with a conditional allele for the p75(NTR) neurotrophin receptor gene. *Genesis, 49*(11), 862–869.

Bononi, J., Cole, A., Tewson, P., Schumacher, A., & Bradley, R. (2008). Chicken protocadherin-1 functions to localize neural crest cells to the dorsal root ganglia during PNS formation. *Mechanisms of Development, 125*(11–12), 1033–1047.

Bordeaux, M. C., Forcet, C., Granger, L., Corset, V., Bidaud, C., Billaud, M., et al. (2000). The RET proto-oncogene induces apoptosis: A novel mechanism for Hirschsprung disease. *The EMBO Journal, 19*(15), 4056–4063.

Bremer, M., Frob, F., Kichko, T., Reeh, P., Tamm, E. R., Suter, U., et al. (2011). Sox10 is required for Schwann-cell homeostasis and myelin maintenance in the adult peripheral nerve. *Glia, 59*(7), 1022–1032.

Britsch, S., Goerich, D. E., Riethmacher, D., Peirano, R. I., Rossner, M., Nave, K. A., et al. (2001). The transcription factor Sox10 is a key regulator of peripheral glial development. *Genes and Development, 15*(1), 66–78.

Britsch, S., Li, L., Kirchhoff, S., Theuring, F., Brinkmann, V., Birchmeier, C., et al. (1998). The ErbB2 and ErbB3 receptors and their ligand, neuregulin-1, are essential for development of the sympathetic nervous system. *Genes and Development, 12*(12), 1825–1836.

Bron, R., Vermeren, M., Kokot, N., Andrews, W., Little, G. E., Mitchell, K. J., et al. (2007). Boundary cap cells constrain spinal motor neuron somal migration at motor exit points by a semaphorin-plexin mechanism. *Neural Development, 2*, 21.

Bronner-Fraser, M. (1986). Analysis of the early stages of trunk neural crest migration in avian embryos using monoclonal antibody HNK-1. *Developmental Biology, 115*(1), 44–55.

Bronner-Fraser, M., & Fraser, S. E. (1988). Cell lineage analysis reveals multipotency of some avian neural crest cells. *Nature, 335*(6186), 161–164.

Bronner-Fraser, M., & Fraser, S. (1989). Developmental potential of avian trunk neural crest cells in situ. *Neuron, 3*(6), 755–766.

Bronner-Fraser, M., & Stern, C. (1991). Effects of mesodermal tissues on avian neural crest cell migration. *Developmental Biology, 143*(2), 213–217.

Canibano, C., Rodriguez, N. L., Saez, C., Tovar, S., Garcia-Lavandeira, M., Borrello, M. G., et al. (2007). The dependence receptor Ret induces apoptosis in somatotrophs through a Pit-1/p53 pathway, preventing tumor growth. *The EMBO Journal, 26*(8), 2015–2028.

Carmona-Fontaine, C., Matthews, H., & Mayor, R. (2008). Directional cell migration in vivo: Wnt at the crest. *Cell Adhesion & Migration, 2*(4), 240–242.

Carr, V. M., & Simpson, S. B. (1978). Proliferative and degenerative events in the early development of chick dorsal root ganglia. II. Responses to altered peripheral fields. *Journal of Comparative Neurology, 182*(4), 741–755.

Cebra-Thomas, J. A., Terrell, A., Branyan, K., Shah, S., Rice, R., Gyi, L., et al. (2013). Late-emigrating trunk neural crest cells in turtle embryos generate an osteogenic ectomesenchyme in the plastron. *Developmental Dynamics, 242*(11), 1223–1235.

Charalampopoulos, I., Vicario, A., Pediaditakis, I., Gravanis, A., Simi, A., & Ibanez, C. F. (2012). Genetic dissection of neurotrophin signaling through the p75 neurotrophin receptor. *Cell Reports*, *2*(6), 1563–1570.

Chen, C.-L., Broom, D. C., Liu, Y., de Nooij, J. C., Li, Z., Cen, C., et al. (2006). Runx1 determines nociceptive sensory neuron phenotype and is required for thermal and neuropathic pain. *Neuron*, *49*(3), 365–377.

Cheung, M., Chaboissier, M.-C., Mynett, A., Hirst, E., Schedl, A., & Briscoe, J. (2005). The transcriptional control of trunk neural crest induction, survival, and delamination. *Developmental Cell*, *8*(2), 179–192.

Clay, M. R., & Halloran, M. C. (2010). Control of neural crest cell behavior and migration: Insights from live imaging. *Cell Adhesion & Migration*, *4*(4), 586–594.

Clay, M. R., & Halloran, M. C. (2014). Cadherin 6 promotes neural crest cell detachment via F-actin regulation and influences active Rho distribution during epithelial-to-mesenchymal transition. *Development*, *141*(12), 2506–2515.

Coelho-Aguiar, J. M., Le Douarin, N. M., & Dupin, E. (2013). Environmental factors unveil dormant developmental capacities in multipotent progenitors of the trunk neural crest. *Developmental Biology*, *384*(1), 13–25.

Coulpier, F., Decker, L., Funalot, B., Vallat, J.-M., Garcia-Bragado, F., Charnay, P., et al. (2010). CNS/PNS boundary transgression by central glia in the absence of Schwann cells or Krox20/Egr2 function. *Journal of Neuroscience*, *30*(17), 5958–5967.

Coulpier, F., Le Crom, S., Maro, G. S., Manent, J., Giovannini, M., Maciorowski, Z., et al. (2009). Novel features of boundary cap cells revealed by the analysis of newly identified molecular markers. *Glia*, *57*(13), 1450–1457.

Cowan, W. M. (2001). Viktor Hamburger and Rita Levi-Montalcini: The path to the discovery of nerve growth factor. *Annual Review of Neuroscience*, *24*, 551–600.

Crane, J. F., & Trainor, P. A. (2006). Neural crest stem and progenitor cells. *Annual Review of Cell and Developmental Biology*, *22*, 267–286.

Crowley, C., Spencer, S. D., Nishimura, M. C., Chen, K. S., Pitts-Meek, S., Armanini, M. P., et al. (1994). Mice lacking nerve growth factor display perinatal loss of sensory and sympathetic neurons yet develop basal forebrain cholinergic neurons. *Cell*, *76*(6), 1001–1011.

Davies, A. M., Lee, K. F., & Jaenisch, R. (1993). p75-deficient trigeminal sensory neurons have an altered response to NGF but not to other neurotrophins. *Neuron*, *11*(4), 565–574.

Davies, A. M., Thoenen, H., & Barde, Y. A. (1986). Different factors from the central nervous system and periphery regulate the survival of sensory neurones. *Nature*, *319*(6053), 497–499.

Debby-Brafman, A., Burstyn-Cohen, T., Klar, A., & Kalcheim, C. (1999). F-Spondin, expressed in somite regions avoided by neural crest cells, mediates inhibition of distinct somite domains to neural crest migration. *Neuron*, *22*(3), 475–488.

De Bellard, M. E., Ching, W., Gossler, A., & Bronner-Fraser, M. (2002). Disruption of segmental neural crest migration and ephrin expression in delta-1 null mice. *Developmental Biology*, *249*(1), 121–130.

De Calisto, J., Araya, C., Marchant, L., Riaz, C. F., & Mayor, R. (2005). Essential role of non-canonical Wnt signalling in neural crest migration. *Development*, *132*(11), 2587–2597.

Deckwerth, T. L., Elliott, J. L., Knudson, C. M., Johnson, E. M., Snider, W. D., & Korsmeyer, S. J. (1996). BAX is required for neuronal death after trophic factor deprivation and during development. *Neuron*, *17*(3), 401–411.

Dekkers, M. P. J., Nikoletopoulou, V., & Barde, Y.-A. (2013). Cell biology in neuroscience: Death of developing neurons: New insights and implications for connectivity. *Journal of Cell Biology*, *203*(3), 385–393.

Deppmann, C. D., Mihalas, S., Sharma, N., Lonze, B. E., Niebur, E., & Ginty, D. D. (2008). A model for neuronal competition during development. *Science*, *320*(5874), 369–373.

Dhanoa, N. K., Krol, K. M., Jahed, A., Crutcher, K. A., & Kawaja, M. D. (2006). Null mutations for exon III and exon IV of the p75 neurotrophin receptor gene enhance sympathetic sprouting in response to elevated levels of nerve growth factor in transgenic mice. *Experimental Neurology*, *198*(2), 416–426.

Dong, Z., Sinanan, A., Parkinson, D., Parmantier, E., Mirsky, R., & Jessen, K. R. (1999). Schwann cell development in embryonic mouse nerves. *Journal of Neuroscience Research*, *56*(4), 334–348.

Dutt, S., Kleber, M., Matasci, M., Sommer, L., & Zimmermann, D. R. (2006). Versican V0 and V1 guide migratory neural crest cells. *Journal of Biological Chemistry*, *281*(17), 12123–12131.

Dyachuk, V., Furlan, A., Shahidi, M. K., Giovenco, M., Kaukua, N., Konstantinidou, C., et al. (2014). Neurodevelopment. Parasympathetic neurons originate from nerve-associated peripheral glial progenitors. *Science*, *345*(6192), 82–87.

Dykes, I. M., Tempest, L., Lee, S.-I., & Turner, E. E. (2011). Brn3a and Islet1 act epistatically to regulate the gene expression program of sensory differentiation. *Journal of Neuroscience*, *31*(27), 9789–9799.

Erickson, C. A. (1985). Control of neural crest cell dispersion in the trunk of the avian embryo. *Developmental Biology*, *111*(1), 138–157.

Erickson, C. A., & Goins, T. L. (1995). Avian neural crest cells can migrate in the dorsolateral path only if they are specified as melanocytes. *Development*, *121*(3), 915–924.

Ernfors, P., Lee, K. F., Kucera, J., & Jaenisch, R. (1994). Lack of neurotrophin-3 leads to deficiencies in the peripheral nervous system and loss of limb proprioceptive afferents. *Cell*, *77*(4), 503–512.

Ernsberger, U. (2009). Role of neurotrophin signalling in the differentiation of neurons from dorsal root ganglia and sympathetic ganglia. *Cell and Tissue Research*, *336*(3), 349–384.

Esper, R. M., & Loeb, J. A. (2004). Rapid axoglial signaling mediated by neuregulin and neurotrophic factors. *Journal of Neuroscience*, *24*(27), 6218–6227.

Esper, R. M., & Loeb, J. A. (2009). Neurotrophins induce neuregulin release through protein kinase Cdelta activation. *Journal of Biological Chemistry*, *284*(39), 26251–26260.

Espinosa-Medina, I., Outin, E., Picard, C. A., Chettouh, Z., Dymecki, S., Consalez, G. G., et al. (2014). Neurodevelopment. Parasympathetic ganglia derive from Schwann cell precursors. *Science*, *345*(6192), 87–90.

Farinas, I., Jones, K. R., Backus, C., Wang, X. Y., & Reichardt, L. F. (1994). Severe sensory and sympathetic deficits in mice lacking neurotrophin-3. *Nature*, *369*(6482), 658–661.

Farinas, I., Wilkinson, G. A., Backus, C., Reichardt, L. F., & Patapoutian, A. (1998). Characterization of neurotrophin and Trk receptor functions in developing sensory ganglia: Direct NT-3 activation of TrkB neurons in vivo. *Neuron*, *21*(2), 325–334.

Fedtsova, N. G., & Turner, E. E. (1995). Brn-3.0 expression identifies early post-mitotic CNS neurons and sensory neural precursors. *Mechanisms of Development*, *53*(3), 291–304.

Finelli, M. J., Murphy, K. J., Chen, L., & Zou, H. (2013). Differential phosphorylation of Smad1 integrates BMP and neurotrophin pathways through Erk/Dusp in axon development. *Cell Reports*, *3*(5), 1592–1606.

Finzsch, M., Schreiner, S., Kichko, T., Reeh, P., Tamm, E. R., Bosl, M. R., et al. (2010). Sox10 is required for Schwann cell identity and progression beyond the immature Schwann cell stage. *Journal of Cell Biology*, *189*(4), 701–712.

Fontanet, P., Irala, D., Alsina, F. C., Paratcha, G., & Ledda, F. (2013). Pea3 transcription factor family members Etv4 and Etv5 mediate retrograde signaling and axonal growth of DRG sensory neurons in response to NGF. *Journal of Neuroscience*, *33*(40), 15940–15951.

Fraher, J. P., Dockery, P., O'Donoghue, O., Riedewald, B., & O'Leary, D. (2007). Initial motor axon outgrowth from the developing central nervous system. *Journal of Anatomy, 211*(5), 600–611.

Frank, E., & Sanes, J. R. (1991). Lineage of neurons and glia in chick dorsal root ganglia: Analysis in vivo with a recombinant retrovirus. *Development, 111*(4), 895–908.

Fraser, S. E., & Bronner-Fraser, M. (1991). Migrating neural crest cells in the trunk of the avian embryo are multipotent. *Development, 112*(4), 913–920.

Friedl, P., & Gilmour, D. (2009). Collective cell migration in morphogenesis, regeneration and cancer. *Nature Reviews. Molecular Cell Biology, 10*(7), 445–457.

Frob, F., Bremer, M., Finzsch, M., Kichko, T., Reeh, P., Tamm, E. R., et al. (2012). Establishment of myelinating Schwann cells and barrier integrity between central and peripheral nervous systems depend on Sox10. *Glia, 60*(5), 806–819.

Gammill, L. S., Gonzalez, C., Gu, C., & Bronner-Fraser, M. (2006). Guidance of trunk neural crest migration requires neuropilin 2/semaphorin 3F signaling. *Development, 133*(1), 99–9106.

Gammill, L. S., & Roffers-Agarwal, J. (2010). Division of labor during trunk neural crest development. *Developmental Biology, 344*(2), 555–565.

George, L., Chaverra, M., Todd, V., Lansford, R., & Lefcort, F. (2007). Nociceptive sensory neurons derive from contralaterally migrating, fate-restricted neural crest cells. *Nature Neuroscience, 10*(10), 1287–1293.

George, L., Kasemeier-Kulesa, J., Nelson, B. R., Koyano-Nakagawa, N., & Lefcort, F. (2010). Patterned assembly and neurogenesis in the chick dorsal root ganglion. *Journal of Comparative Neurology, 518*(4), 405–422.

Glebova, N. O., & Ginty, D. D. (2004). Heterogeneous requirement of NGF for sympathetic target innervation in vivo. *Journal of Neuroscience, 24*(3), 743–751.

Glenn, T. D., & Talbot, W. S. (2013). Signals regulating myelination in peripheral nerves and the Schwann cell response to injury. *Current Opinion in Neurobiology, 23*(6), 1041–1048.

Golding, J. P., & Cohen, J. (1997). Border controls at the mammalian spinal cord: Late-surviving neural crest boundary cap cells at dorsal root entry sites may regulate sensory afferent ingrowth and entry zone morphogenesis. *Molecular and Cellular Neuroscience, 9*(5–6), 381–396.

Goldstein, R. S., Teillet, M. A., & Kalcheim, C. (1990). The microenvironment created by grafting rostral half-somites is mitogenic for neural crest cells. *Proceedings of the National Academy of Sciences of the United States of America, 87*(12), 4476–4480.

Greenwood, A. L., Turner, E. E., & Anderson, D. J. (1999). Identification of dividing, determined sensory neuron precursors in the mammalian neural crest. *Development, 126*(16), 3545–3559.

Grossmann, K. S., Wende, H., Paul, F. E., Cheret, C., Garratt, A. N., Zurborg, S., et al. (2009). The tyrosine phosphatase Shp2 (PTPN11) directs Neuregulin-1/ErbB signaling throughout Schwann cell development. *Proceedings of the National Academy of Sciences of the United States of America, 106*(39), 16704–16709.

Guo, T., Mandai, K., Condie, B. G., Wickramasinghe, S. R., Capecchi, M. R., & Ginty, D. D. (2011). An evolving NGF-Hoxd1 signaling pathway mediates development of divergent neural circuits in vertebrates. *Nature Neuroscience, 14*(1), 31–36.

Hagedorn, L., Paratore, C., Brugnoli, G., Baert, J. L., Mercader, N., Suter, U., et al. (2000). The ETS domain transcription factor Erm distinguishes rat satellite glia from Schwann cells and is regulated in satellite cells by neuregulin signaling. *Developmental Biology, 219*(1), 44–58.

Hagedorn, L., Suter, U., & Sommer, L. (1999). P0 and PMP22 mark a multipotent neural crest-derived cell type that displays community effects in response to TGF-beta family factors. *Development, 126*(17), 3781–3794.

Hapner, S. J., Boeshore, K. L., Large, T. H., & Lefcort, F. (1998). Neural differentiation promoted by truncated trkC receptors in collaboration with p75(NTR). *Developmental Biology, 201*(1), 90–9100.

Hapner, S. J., Nielsen, K. M., Chaverra, M., Esper, R. M., Loeb, J. A., & Lefcort, F. (2006). NT-3 and CNTF exert dose-dependent, pleiotropic effects on cells in the immature dorsal root ganglion: Neuregulin-mediated proliferation of progenitor cells and neuronal differentiation. *Developmental Biology, 297*(1), 182–197.

Hari, L., Brault, V., Kleber, M., Lee, H.-Y., Ille, F., Leimeroth, R., et al. (2002). Lineage-specific requirements of beta-catenin in neural crest development. *Journal of Cell Biology, 159*(5), 867–880.

Hari, L., Miescher, I., Shakhova, O., Suter, U., Chin, L., Taketo, M., et al. (2012). Temporal control of neural crest lineage generation by Wnt/β-catenin signaling. *Development, 139*(12), 2107–2117.

Harrington, A. W., & Ginty, D. D. (2013). Long-distance retrograde neurotrophic factor signalling in neurons. *Nature Reviews. Neuroscience, 14*(3), 177–187.

Heller, B. A., Ghidinelli, M., Voelkl, J., Einheber, S., Smith, R., Grund, E., et al. (2014). Functionally distinct PI 3-kinase pathways regulate myelination in the peripheral nervous system. *Journal of Cell Biology, 204*(7), 1219–1236.

Henion, P. D., Garner, A. S., Large, T. H., & Weston, J. A. (1995). trkC-mediated NT-3 signaling is required for the early development of a subpopulation of neurogenic neural crest cells. *Developmental Biology, 172*(2), 602–613.

Hjerling-Leffler, J., Marmigere, F., Heglind, M., Cederberg, A., Koltzenburg, M., Enerback, S., et al. (2005). The boundary cap: A source of neural crest stem cells that generate multiple sensory neuron subtypes. *Development, 132*(11), 2623–2632.

Honma, Y., Araki, T., Gianino, S., Bruce, A., Heuckeroth, R., Johnson, E., et al. (2002). Artemin is a vascular-derived neurotropic factor for developing sympathetic neurons. *Neuron, 35*(2), 267–282.

Hrabe de Angelis, M., McIntyre, J., 2nd., & Gossler, A. (1997). Maintenance of somite borders in mice requires the Delta homologue DII1. *Nature, 386*(6626), 717–721.

Hu, Z.-L., Shi, M., Huang, Y., Zheng, M.-H., Pei, Z., Chen, J.-Y., et al. (2011). The role of the transcription factor Rbpj in the development of dorsal root ganglia. *Neural Development, 6*, 14.

Ichim, G., Genevois, A.-L., Menard, M., Yu, L.-Y., Coelho-Aguiar, J. M., Llambi, F., et al. (2013). The dependence receptor TrkC triggers mitochondria-dependent apoptosis upon Cobra-1 recruitment. *Molecular Cell, 51*(5), 632–646.

Ikeya, M., Lee, S. M., Johnson, J. E., McMahon, A. P., & Takada, S. (1997). Wnt signalling required for expansion of neural crest and CNS progenitors. *Nature 389*(6654), 966–970.

Inoue, K.-i, Ito, K., Osato, M., Lee, B., Bae, S.-C., & Ito, Y. (2007). The transcription factor Runx3 represses the neurotrophin receptor TrkB during lineage commitment of dorsal root ganglion neurons. *The Journal of Biological Chemistry, 282*(33), 24175–24184.

Ishii, A., Furusho, M., & Bansal, R. (2013). Sustained activation of ERK1/2 MAPK in oligodendrocytes and Schwann cells enhances myelin growth and stimulates oligodendrocyte progenitor expansion. *Journal of Neuroscience, 33*(1), 175–186.

Jessen, K. R., Brennan, A., Morgan, L., Mirsky, R., Kent, A., Hashimoto, Y., et al. (1994). The Schwann cell precursor and its fate: A study of cell death and differentiation during gliogenesis in rat embryonic nerves. *Neuron, 12*(3), 509–527.

Jessen, K. R., & Mirsky, R. (2005). The origin and development of glial cells in peripheral nerves. *Nature Reviews. Neuroscience, 6*(9), 671–682.

Joseph, N. M., Mukouyama, Y.-S., Mosher, J. T., Jaegle, M., Crone, S. A., Dormand, E.-L., et al. (2004). Neural crest stem cells undergo multilineage differentiation in developing

peripheral nerves to generate endoneurial fibroblasts in addition to Schwann cells. *Development, 131*(22), 5599–5612.
Kahane, N., & Kalcheim, C. (1994). Expression of trkC receptor mRNA during development of the avian nervous system. *Journal of Neurobiology, 25*(5), 571–584.
Kalcheim, C. (1996). The role of neurotrophins in development of neural-crest cells that become sensory ganglia. *Philosophical Transactions of the Royal Society of London. Series B: Biological Sciences, 351*(1338), 375–381.
Kalcheim, C., Carmeli, C., & Rosenthal, A. (1992). Neurotrophin 3 is a mitogen for cultured neural crest cells. *Proceedings of the National Academy of Sciences of the United States of America, 89*(5), 1661–1665.
Kalcheim, C., & Teillet, M. A. (1989). Consequences of somite manipulation on the pattern of dorsal root ganglion development. *Development, 106*(1), 85–93.
Kasemeier-Kulesa, J. C., McLennan, R., Romine, M. H., Kulesa, P. M., & Lefcort, F. (2010). CXCR4 controls ventral migration of sympathetic precursor cells. *Journal of Neuroscience, 30*(39), 13078–13088.
Kawaja, M. D., Smithson, L. J., Elliott, J., Trinh, G., Crotty, A.-M., Michalski, B., et al. (2011). Nerve growth factor promoter activity revealed in mice expressing enhanced green fluorescent protein. *Journal of Comparative Neurology, 519*(13), 2522–2545.
Kawasaki, T., Bekku, Y., Suto, F., Kitsukawa, T., Taniguchi, M., Nagatsu, I., et al. (2002). Requirement of neuropilin 1-mediated Sema3A signals in patterning of the sympathetic nervous system. *Development, 129*(3), 671–680.
Kelsh, R. N., Harris, M. L., Colanesi, S., & Erickson, C. A. (2009). Stripes and belly-spots—A review of pigment cell morphogenesis in vertebrates. *Seminars in Cell and Developmental Biology, 20*(1), 90–104.
Keynes, R. J., & Stern, C. D. (1984). Segmentation in the vertebrate nervous system. *Nature, 310*(5980), 786–789.
Kim, J., Lo, L., Dormand, E., & Anderson, D. J. (2003). SOX10 maintains multipotency and inhibits neuronal differentiation of neural crest stem cells. *Neuron, 38*(1), 17–31.
Klein, R., Silos-Santiago, I., Smeyne, R. J., Lira, S. A., Brambilla, R., Bryant, S., et al. (1994). Disruption of the neurotrophin-3 receptor gene trkC eliminates la muscle afferents and results in abnormal movements. *Nature, 368*(6468), 249–251.
Koblar, S. A., Krull, C. E., Pasquale, E. B., McLennan, R., Peale, F. D., Cerretti, D. P., et al. (2000). Spinal motor axons and neural crest cells use different molecular guides for segmental migration through the rostral half-somite. *Journal of Neurobiology, 42*(4), 437–447.
Kramer, I., Sigrist, M., de Nooij, J. C., Taniuchi, I., Jessell, T. M., & Arber, S. (2006). A role for Runx transcription factor signaling in dorsal root ganglion sensory neuron diversification. *Neuron, 49*(3), 379–393.
Krispin, S., Nitzan, E., & Kalcheim, C. (2010). The dorsal neural tube: A dynamic setting for cell fate decisions. *Developmental Neurobiology, 70*(12), 796–812.
Krispin, S., Nitzan, E., Kassem, Y., & Kalcheim, C. (2010). Evidence for a dynamic spatiotemporal fate map and early fate restrictions of premigratory avian neural crest. *Development, 137*(4), 585–595.
Krull, C. E. (2001). Segmental organization of neural crest migration. *Mechanisms of Development, 105*(1–2), 37–45.
Krull, C. E. (2010). Neural crest cells and motor axons in avians: Common and distinct migratory molecules. *Cell Adhesion & Migration, 4*(4), 631–634.
Krull, C. E., Lansford, R., Gale, N. W., Collazo, A., Marcelle, C., Yancopoulos, G. D., et al. (1997). Interactions of Eph-related receptors and ligands confer rostrocaudal pattern to trunk neural crest migration. *Current Biology,* 7(8), 571–580.
Kucenas, S., Wang, W.-D., Knapik, E. W., & Appel, B. (2009). A selective glial barrier at motor axon exit points prevents oligodendrocyte migration from the spinal cord. *Journal of Neuroscience, 29*(48), 15187–15194.

Lallemend, F., & Ernfors, P. (2012). Molecular interactions underlying the specification of sensory neurons. *Trends in Neurosciences*, *35*(6), 373–381.
Lallemend, F., Sterzenbach, U., Hadjab-Lallemend, S., Aquino, J. B., Castelo-Branco, G., Sinha, I., et al. (2012). Positional differences of axon growth rates between sensory neurons encoded by Runx3. *The EMBO Journal*, *31*(18), 3718–3729.
Lawson, S. N., & Biscoe, T. J. (1979). Development of mouse dorsal root ganglia: An autoradiographic and quantitative study. *Journal of Neurocytology*, *8*(3), 265–274.
Le Douarin, N. M., & Dupin, E. (2003). Multipotentiality of the neural crest. *Current Opinion in Genetics and Development*, *13*(5), 529–536.
Le Douarin, N., & Kalcheim, C. (1999). *The neural crest* (2nd ed.). Cambridge, UK; New York, NY: Cambridge University Press.
Le Douarin, N. M., & Smith, J. (1988). Development of the peripheral nervous system from the neural crest. *Annual Review of Cell Biology*, *4*, 375–404.
Lee, H.-Y., Kleber, M., Hari, L., Brault, V., Suter, U., Taketo, M. M., et al. (2004). Instructive role of Wnt/beta-catenin in sensory fate specification in neural crest stem cells. *Science*, *303*(5660), 1020–1023.
Lee, K. F., Li, E., Huber, L. J., Landis, S. C., Sharpe, A. H., Chao, M. V., et al. (1992). Targeted mutation of the gene encoding the low affinity NGF receptor p75 leads to deficits in the peripheral sensory nervous system. *Cell*, *69*(5), 737–749.
Lefcort, F., Clary, D. O., Rusoff, A. C., & Reichardt, L. F. (1996). Inhibition of the NT-3 receptor TrkC, early in chick embryogenesis, results in severe reductions in multiple neuronal subpopulations in the dorsal root ganglia. *Journal of Neuroscience*, *16*(11), 3704–3713.
Lemmon, M. A., & Schlessinger, J. (2010). Cell signaling by receptor tyrosine kinases. *Cell*, *141*(7), 1117–1134.
Li, L., Rutlin, M., Abraira, V. E., Cassidy, C., Kus, L., Gong, S., et al. (2011). The functional organization of cutaneous low-threshold mechanosensory neurons. *Cell*, *147*(7), 1615–1627.
Li, H.-Y., Say, E. H. M., & Zhou, X.-F. (2007). Isolation and characterization of neural crest progenitors from adult dorsal root ganglia. *Stem Cells*, *25*(8), 2053–2065.
Lilley, B. N., Pan, Y. A., & Sanes, J. R. (2013). SAD kinases sculpt axonal arbors of sensory neurons through long- and short-term responses to neurotrophin signals. *Neuron*, *79*(1), 39–53.
Lim, J., & Thiery, J. P. (2012). Epithelial-mesenchymal transitions: Insights from development. *Development*, *139*(19), 3471–3486.
Lin, W., Sanchez, H. B., Deerinck, T., Morris, J. K., Ellisman, M., & Lee, K. F. (2000). Aberrant development of motor axons and neuromuscular synapses in erbB2-deficient mice. *Proceedings of the National Academy of Sciences of the United States of America*, *97*(3), 1299–1304.
Liu, J. P., & Jessell, T. M. (1998). A role for rhoB in the delamination of neural crest cells from the dorsal neural tube. *Development*, *125*(24), 5055–5067.
Liu, Y., & Ma, Q. (2011). Generation of somatic sensory neuron diversity and implications on sensory coding. *Current Opinion in Neurobiology*, *21*(1), 52–60.
Lumsden, A. G., & Davies, A. M. (1983). Earliest sensory nerve fibres are guided to peripheral targets by attractants other than nerve growth factor. *Nature*, *306*(5945), 786–788.
Luo, W., Wickramasinghe, S. R., Savitt, J. M., Griffin, J. W., Dawson, T. M., & Ginty, D. D. (2007). A hierarchical NGF signaling cascade controls Ret-dependent and Ret-independent events during development of nonpeptidergic DRG neurons. *Neuron*, *54*(5), 739–754.
Ma, Q., Fode, C., Guillemot, F., & Anderson, D. J. (1999). Neurogenin1 and neurogenin2 control two distinct waves of neurogenesis in developing dorsal root ganglia. *Genes and Development*, *13*(13), 1717–1728.

Ma, Z., Wang, J., Song, F., & Loeb, J. A. (2011). Critical period of axoglial signaling between neuregulin-1 and brain-derived neurotrophic factor required for early Schwann cell survival and differentiation. *Journal of Neuroscience, 31*(26), 9630–9640.

Maden, C. H., Gomes, J., Schwarz, Q., Davidson, K., Tinker, A., & Ruhrberg, C. (2012). NRP1 and NRP2 cooperate to regulate gangliogenesis, axon guidance and target innervation in the sympathetic nervous system. *Developmental Biology, 369*(2), 277–285.

Majdan, M., Walsh, G. S., Aloyz, R., & Miller, F. D. (2001). TrkA mediates developmental sympathetic neuron survival in vivo by silencing an ongoing p75NTR-mediated death signal. *Journal of Cell Biology, 155*(7), 1275–1285.

Marmigere, F., & Ernfors, P. (2007). Specification and connectivity of neuronal subtypes in the sensory lineage. *Nature Reviews. Neuroscience, 8*(2), 114–127.

Marmigere, F., Montelius, A., Wegner, M., Groner, Y., Reichardt, L. F., & Ernfors, P. (2006). The Runx1/AML1 transcription factor selectively regulates development and survival of TrkA nociceptive sensory neurons. *Nature Neuroscience, 9*(2), 180–187.

Maro, G. S., Vermeren, M., Voiculescu, O., Melton, L., Cohen, J., Charnay, P., et al. (2004). Neural crest boundary cap cells constitute a source of neuronal and glial cells of the PNS. *Nature Neuroscience,* 7(9), 930–938.

Martin-Zanca, D., Barbacid, M., & Parada, L. F. (1990). Expression of the trk proto-oncogene is restricted to the sensory cranial and spinal ganglia of neural crest origin in mouse development. *Genes and Development, 4*(5), 683–694.

Maurel, P., & Salzer, J. L. (2000). Axonal regulation of Schwann cell proliferation and survival and the initial events of myelination requires PI 3-kinase activity. *Journal of Neuroscience, 20*(12), 4635–4645.

Mauti, O., Domanitskaya, E., Andermatt, I., Sadhu, R., & Stoeckli, E. T. (2007). Semaphorin6A acts as a gate keeper between the central and the peripheral nervous system. *Neural Development, 2*, 28.

Maynard, T. M., Wakamatsu, Y., & Weston, J. A. (2000). Cell interactions within nascent neural crest cell populations transiently promote death of neurogenic precursors. *Development, 127*(21), 4561–4572.

Mayor, R., & Carmona-Fontaine, C. (2010). Keeping in touch with contact inhibition of locomotion. *Trends in Cell Biology, 20*(6), 319–328.

Mayor, R., & Theveneau, E. (2014). The role of the non-canonical Wnt-planar cell polarity pathway in neural crest migration. *Biochemistry Journal, 457*(1), 19–26.

McEvilly, R. J., Erkman, L., Luo, L., Sawchenko, P. E., Ryan, A. F., & Rosenfeld, M. G. (1996). Requirement for Brn-3.0 in differentiation and survival of sensory and motor neurons. *Nature, 384*(6609), 574–577.

Mead, T. J., & Yutzey, K. E. (2012). Notch pathway regulation of neural crest cell development in vivo. *Developmental Dynamics, 241*(2), 376–389.

Meyer, D., & Birchmeier, C. (1995). Multiple essential functions of neuregulin in development. *Nature, 378*(6555), 386–390.

Meyer, D., Yamaai, T., Garratt, A., Riethmacher-Sonnenberg, E., Kane, D., Theill, L. E., et al. (1997). Isoform-specific expression and function of neuregulin. *Development, 124*(18), 3575–3586.

Michailov, G. V., Sereda, M. W., Brinkmann, B. G., Fischer, T. M., Haug, B., Birchmeier, C., et al. (2004). Axonal neuregulin-1 regulates myelin sheath thickness. *Science, 304*(5671), 700–703.

Milet, C., & Monsoro-Burq, A. H. (2012). Embryonic stem cell strategies to explore neural crest development in human embryos. *Developmental Biology, 366*(1), 96–99.

Molliver, D. C., & Snider, W. D. (1997). Nerve growth factor receptor TrkA is down-regulated during postnatal development by a subset of dorsal root ganglion neurons. *Journal of Comparative Neurology, 381*(4), 428–438.

Molliver, D. C., Wright, D. E., Leitner, M. L., Parsadanian, A. S., Doster, K., Wen, D., et al. (1997). IB4-binding DRG neurons switch from NGF to GDNF dependence in early postnatal life. *Neuron, 19*(4), 849–861.

Montelius, A., Marmigere, F., Baudet, C., Aquino, J. B., Enerback, S., & Ernfors, P. (2007). Emergence of the sensory nervous system as defined by Foxs1 expression. *Differentiation, 75*(5), 404–417.

Morris, J. K., Lin, W., Hauser, C., Marchuk, Y., Getman, D., & Lee, K. F. (1999). Rescue of the cardiac defect in ErbB2 mutant mice reveals essential roles of ErbB2 in peripheral nervous system development. *Neuron, 23*(2), 273–283.

Morrison, S. J., Perez, S. E., Qiao, Z., Verdi, J. M., Hicks, C., Weinmaster, G., et al. (2000). Transient Notch activation initiates an irreversible switch from neurogenesis to gliogenesis by neural crest stem cells. *Cell, 101*(5), 499–510.

Morrison, S. J., White, P. M., Zock, C., & Anderson, D. J. (1999). Prospective identification, isolation by flow cytometry, and in vivo self-renewal of multipotent mammalian neural crest stem cells. *Cell, 96*(5), 737–749.

Mu, X., Silos-Santiago, I., Carroll, S. L., & Snider, W. D. (1993). Neurotrophin receptor genes are expressed in distinct patterns in developing dorsal root ganglia. *Journal of Neuroscience, 13*(9), 4029–4041.

Mundell, N. A., & Labosky, P. A. (2011). Neural crest stem cell multipotency requires Foxd3 to maintain neural potential and repress mesenchymal fates. *Development, 138*(4), 641–652.

Murphy, M., Reid, K., Ford, M., Furness, J. B., & Bartlett, P. F. (1994). FGF2 regulates proliferation of neural crest cells, with subsequent neuronal differentiation regulated by LIF or related factors. *Development, 120*(12), 3519–3528.

Nagoshi, N., Shibata, S., Kubota, Y., Nakamura, M., Nagai, Y., Satoh, E., et al. (2008). Ontogeny and multipotency of neural crest-derived stem cells in mouse bone marrow, dorsal root ganglia, and whisker pad. *Cell Stem Cell, 2*(4), 392–403.

Nakagawa, S., & Takeichi, M. (1995). Neural crest cell–cell adhesion controlled by sequential and subpopulation-specific expression of novel cadherins. *Development, 121*(5), 1321–1332.

Nakagawa, S., & Takeichi, M. (1998). Neural crest emigration from the neural tube depends on regulated cadherin expression. *Development, 125*(15), 2963–2971.

Napoli, I., Noon, L. A., Ribeiro, S., Kerai, A. P., Parrinello, S., Rosenberg, L. H., et al. (2012). A central role for the ERK-signaling pathway in controlling Schwann cell plasticity and peripheral nerve regeneration in vivo. *Neuron, 73*(4), 729–742.

Nelson, B. R., Claes, K., Todd, V., Chaverra, M., & Lefcort, F. (2004). NELL2 promotes motor and sensory neuron differentiation and stimulates mitogenesis in DRG in vivo. *Developmental Biology, 270*(2), 322–335.

Newbern, J. M., Li, X., Shoemaker, S. E., Zhou, J., Zhong, J., Wu, Y., et al. (2011). Specific functions for ERK/MAPK signaling during PNS development. *Neuron, 69*(1), 91–9105.

Newbern, J., Zhong, J., Wickramasinghe, R. S., Li, X., Wu, Y., Samuels, I., et al. (2008). Mouse and human phenotypes indicate a critical conserved role for ERK2 signaling in neural crest development. *Proceedings of the National Academy of Sciences of the United States of America, 105*(44), 17115–17120.

Niederlander, C., & Lumsden, A. (1996). Late emigrating neural crest cells migrate specifically to the exit points of cranial branchiomotor nerves. *Development, 122*(8), 2367–2374.

Nikoletopoulou, V., Lickert, H., Frade, J. M., Rencurel, C., Giallonardo, P., Zhang, L., et al. (2010). Neurotrophin receptors TrkA and TrkC cause neuronal death whereas TrkB does not. *Nature, 467*(7311), 59–63.

Nitzan, E., Krispin, S., Pfaltzgraff, E. R., Klar, A., Labosky, P. A., & Kalcheim, C. (2013). A dynamic code of dorsal neural tube genes regulates the segregation between neurogenic and melanogenic neural crest cells. *Development*, *140*(11), 2269–2279.

Paratore, C., Goerich, D. E., Suter, U., Wegner, M., & Sommer, L. (2001). Survival and glial fate acquisition of neural crest cells are regulated by an interplay between the transcription factor Sox10 and extrinsic combinatorial signaling. *Development*, *128*(20), 3949–3961.

Patapoutian, A., Backus, C., Kispert, A., & Reichardt, L. F. (1999). Regulation of neurotrophin-3 expression by epithelial–mesenchymal interactions: The role of Wnt factors. *Science*, *283*(5405), 1180–1183.

Patel, T. D., Jackman, A., Rice, F. L., Kucera, J., & Snider, W. D. (2000). Development of sensory neurons in the absence of NGF/TrkA signaling in vivo. *Neuron*, *25*(2), 345–357.

Patel, T. D., Kramer, I., Kucera, J., Niederkofler, V., Jessell, T. M., Arber, S., et al. (2003). Peripheral NT3 signaling is required for ETS protein expression and central patterning of proprioceptive sensory afferents. *Neuron*, *38*(3), 403–416.

Pereira, J. A., Lebrun-Julien, F., & Suter, U. (2012). Molecular mechanisms regulating myelination in the peripheral nervous system. *Trends in Neurosciences*, *35*(2), 123–134.

Perez, S. E., Rebelo, S., & Anderson, D. J. (1999). Early specification of sensory neuron fate revealed by expression and function of neurogenins in the chick embryo. *Development*, *126*(8), 1715–1728.

Perris, R., & Perissinotto, D. (2000). Role of the extracellular matrix during neural crest cell migration. *Mechanisms of Development*, *95*(1–2), 3–21.

Petrie, C. N., Smithson, L. J., Crotty, A.-M., Michalski, B., Fahnestock, M., & Kawaja, M. D. (2013). Overexpression of nerve growth factor by murine smooth muscle cells: Role of the p75 neurotrophin receptor on sympathetic and sensory sprouting. *Journal of Comparative Neurology*, *521*(11), 2621–2643.

Prasad, M. K., Reed, X., Gorkin, D. U., Cronin, J. C., McAdow, A. R., Chain, K., et al. (2011). SOX10 directly modulates ERBB3 transcription via an intronic neural crest enhancer. *BMC Developmental Biology*, *11*, 40.

Reedy, M. V., Faraco, C. D., & Erickson, C. A. (1998). The delayed entry of thoracic neural crest cells into the dorsolateral path is a consequence of the late emigration of melanogenic neural crest cells from the neural tube. *Developmental Biology*, *200*(2), 234–246.

Reichardt, L. F. (2006). Neurotrophin-regulated signalling pathways. *Philosophical Transactions of the Royal Society of London. Series B: Biological Sciences*, *361*(1473), 1545–1564.

Reissmann, E., Ernsberger, U., Francis-West, P. H., Rueger, D., Brickell, P. M., & Rohrer, H. (1996). Involvement of bone morphogenetic protein-4 and bone morphogenetic protein-7 in the differentiation of the adrenergic phenotype in developing sympathetic neurons. *Development*, *122*(7), 2079–2088.

Riethmacher, D., Sonnenberg-Riethmacher, E., Brinkmann, V., Yamaai, T., Lewin, G. R., & Birchmeier, C. (1997). Severe neuropathies in mice with targeted mutations in the ErbB3 receptor. *Nature*, *389*(6652), 725–730.

Rifkin, J. T., Todd, V. J., Anderson, L. W., & Lefcort, F. (2000). Dynamic expression of neurotrophin receptors during sensory neuron genesis and differentiation. *Developmental Biology*, *227*(2), 465–480.

Rorth, P. (2009). Collective cell migration. *Annual Review of Cell and Developmental Biology*, *25*, 407–429.

Ruit, K. G., Elliott, J. L., Osborne, P. A., Yan, Q., & Snider, W. D. (1992). Selective dependence of mammalian dorsal root ganglion neurons on nerve growth factor during embryonic development. *Neuron*, *8*(3), 573–587.

Saito, D., Takase, Y., Murai, H., & Takahashi, Y. (2012). The dorsal aorta initiates a molecular cascade that instructs sympatho-adrenal specification. *Science*, *336*(6088), 1578–1581.

Sasselli, V., Pachnis, V., & Burns, A. J. (2012). The enteric nervous system. *Developmental Biology, 366*(1), 64–73.
Schneider, C., Wicht, H., Enderich, J., Wegner, M., & Rohrer, H. (1999). Bone morphogenetic proteins are required in vivo for the generation of sympathetic neurons. *Neuron, 24*(4), 861–870.
Schwarz, Q., Maden, C. H., Davidson, K., & Ruhrberg, C. (2009). Neuropilin-mediated neural crest cell guidance is essential to organise sensory neurons into segmented dorsal root ganglia. *Development, 136*(11), 1785–1789.
Serbedzija, G. N., Bronner-Fraser, M., & Fraser, S. E. (1989). A vital dye analysis of the timing and pathways of avian trunk neural crest cell migration. *Development, 106*(4), 809–816.
Serbedzija, G. N., Bronner-Fraser, M., & Fraser, S. E. (1994). Developmental potential of trunk neural crest cells in the mouse. *Development, 120*(7), 1709–1718.
Serbedzija, G. N., Fraser, S. E., & Bronner-Fraser, M. (1990). Pathways of trunk neural crest cell migration in the mouse embryo as revealed by vital dye labelling. *Development, 108*(4), 605–612.
Shah, N. M., Groves, A. K., & Anderson, D. J. (1996). Alternative neural crest cell fates are instructively promoted by TGFbeta superfamily members. *Cell, 85*(3), 331–343.
Shah, N. M., Marchionni, M. A., Isaacs, I., Stroobant, P., & Anderson, D. J. (1994). Glial growth factor restricts mammalian neural crest stem cells to a glial fate. *Cell, 77*(3), 349–360.
Sharma, N., Deppmann, C. D., Harrington, A. W., St Hillaire, C., Chen, Z.-Y., Lee, F. S., et al. (2010). Long-distance control of synapse assembly by target-derived NGF. *Neuron, 67*(3), 422–434.
Sheean, M. E., McShane, E., Cheret, C., Walcher, J., Muller, T., Wulf-Goldenberg, A., et al. (2014). Activation of MAPK overrides the termination of myelin growth and replaces Nrg1/ErbB3 signals during Schwann cell development and myelination. *Genes and Development, 28*(3), 290–303.
Shoval, I., & Kalcheim, C. (2012). Antagonistic activities of Rho and Rac GTPases underlie the transition from neural crest delamination to migration. *Developmental Dynamics, 241*(7), 1155–1168.
Shtukmaster, S., Schier, M. C., Huber, K., Krispin, S., Kalcheim, C., & Unsicker, K. (2013). Sympathetic neurons and chromaffin cells share a common progenitor in the neural crest in vivo. *Neural Development, 8*, 12.
Simi, A., & Ibanez, C. F. (2010). Assembly and activation of neurotrophic factor receptor complexes. *Developmental Neurobiology, 70*(5), 323–331.
Smeyne, R. J., Klein, R., Schnapp, A., Long, L. K., Bryant, S., Lewin, A., et al. (1994). Severe sensory and sympathetic neuropathies in mice carrying a disrupted Trk/NGF receptor gene. *Nature, 368*(6468), 246–249.
Sommer, L. (2006). Growth factors regulating neural crest cell fate decisions. *Advances in Experimental Medicine and Biology, 589*, 197–205.
Sonnenberg-Riethmacher, E., Miehe, M., Stolt, C. C., Goerich, D. E., Wegner, M., & Riethmacher, D. (2001). Development and degeneration of dorsal root ganglia in the absence of the HMG-domain transcription factor Sox10. *Mechanisms of Development, 109*(2), 253–265.
Southard-Smith, E. M., Kos, L., & Pavan, W. J. (1998). Sox10 mutation disrupts neural crest development in Dom Hirschsprung mouse model. *Nature Genetics, 18*(1), 60–64.
Stemple, D. L., & Anderson, D. J. (1992). Isolation of a stem cell for neurons and glia from the mammalian neural crest. *Cell, 71*(6), 973–985.
Stuhlmiller, T. J., & Garcia-Castro, M. I. (2012). FGF/MAPK signaling is required in the gastrula epiblast for avian neural crest induction. *Development, 139*(2), 289–300.
Sun, Y., Dykes, I. M., Liang, X., Eng, S. R., Evans, S. M., & Turner, E. E. (2008). A central role for Islet1 in sensory neuron development linking sensory and spinal gene regulatory programs. *Nature Neuroscience, 11*(11), 1283–1293.

Tauszig-Delamasure, S., Yu, L. Y., Cabrera, J. R., Bouzas-Rodriguez, J., Mermet-Bouvier, C., Guix, C., et al. (2007). The TrkC receptor induces apoptosis when the dependence receptor notion meets the neurotrophin paradigm. *Proceedings of the National Academy of Sciences of the United States of America*, *104*(33), 13361–13366.
Taveggia, C., Zanazzi, G., Petrylak, A., Yano, H., Rosenbluth, J., Einheber, S., et al. (2005). Neuregulin-1 type III determines the ensheathment fate of axons. *Neuron*, *47*(5), 681–694.
Taylor, M. K., Yeager, K., & Morrison, S. J. (2007). Physiological Notch signaling promotes gliogenesis in the developing peripheral and central nervous systems. *Development*, *134*(13), 2435–2447.
Teng, L., Mundell, N. A., Frist, A. Y., Wang, Q., & Labosky, P. A. (2008). Requirement for Foxd3 in the maintenance of neural crest progenitors. *Development*, *135*(9), 1615–1624.
Tessarollo, L., Vogel, K. S., Palko, M. E., Reid, S. W., & Parada, L. F. (1994). Targeted mutation in the neurotrophin-3 gene results in loss of muscle sensory neurons. *Proceedings of the National Academy of Sciences of the United States of America*, *91*(25), 11844–11848.
Theveneau, E., & Mayor, R. (2012). Neural crest delamination and migration: From epithelium-to-mesenchyme transition to collective cell migration. *Developmental Biology*, *366*(1), 34–54.
Thiery, J. P., Duband, J. L., & Delouvee, A. (1982). Pathways and mechanisms of avian trunk neural crest cell migration and localization. *Developmental Biology*, *93*(2), 324–343.
Tosney, K. W., Dehnbostel, D. B., & Erickson, C. A. (1994). Neural crest cells prefer the myotome's basal lamina over the sclerotome as a substratum. *Developmental Biology*, *163*(2), 389–406.
Trolle, C., Konig, N., Abrahamsson, N., Vasylovska, S., & Kozlova, E. N. (2014). Boundary cap neural crest stem cells homotopically implanted to the injured dorsal root transitional zone give rise to different types of neurons and glia in adult rodents. *BMC Neuroscience*, *15*, 60.
Tsarovina, K., Schellenberger, J., Schneider, C., & Rohrer, H. (2008). Progenitor cell maintenance and neurogenesis in sympathetic ganglia involves Notch signaling. *Molecular and Cellular Neuroscience*, *37*(1), 20–31.
Tucker, R. P., Hagios, C., Chiquet-Ehrismann, R., Lawler, J., Hall, R. J., & Erickson, C. A. (1999). Thrombospondin-1 and neural crest cell migration. *Developmental Dynamics*, *214*(4), 312–322.
Ulmer, B., Hagenlocher, C., Schmalholz, S., Kurz, S., Schweickert, A., Kohl, A., et al. (2013). Calponin 2 acts as an effector of noncanonical Wnt-mediated cell polarization during neural crest cell migration. *Cell Reports*, *3*(3), 615–621.
Usui, N., Watanabe, K., Ono, K., Tomita, K., Tamamaki, N., Ikenaka, K., et al. (2012). Role of motoneuron-derived neurotrophin 3 in survival and axonal projection of sensory neurons during neural circuit formation. *Development*, *139*(6), 1125–1132.
Vermeren, M., Maro, G. S., Bron, R., McGonnell, I. M., Charnay, P., Topilko, P., et al. (2003). Integrity of developing spinal motor columns is regulated by neural crest derivatives at motor exit points. *Neuron*, *37*(3), 403–415.
von Schack, D., Casademunt, E., Schweigreiter, R., Meyer, M., Bibel, M., & Dechant, G. (2001). Complete ablation of the neurotrophin receptor p75NTR causes defects both in the nervous and the vascular system. *Nature Neuroscience*, *4*(10), 977–978.
Wakamatsu, Y., Endo, Y., Osumi, N., & Weston, J. A. (2004). Multiple roles of Sox2, an HMG-box transcription factor in avian neural crest development. *Developmental Dynamics*, *229*(1), 74–86.
Wakamatsu, Y., Maynard, T. M., & Weston, J. A. (2000). Fate determination of neural crest cells by NOTCH-mediated lateral inhibition and asymmetrical cell division during gangliogenesis. *Development*, *127*(13), 2811–2821.

Wang, H. U., & Anderson, D. J. (1997). Eph family transmembrane ligands can mediate repulsive guidance of trunk neural crest migration and motor axon outgrowth. *Neuron, 18*(3), 383–396.
White, P. M., Morrison, S. J., Orimoto, K., Kubu, C. J., Verdi, J. M., & Anderson, D. J. (2001). Neural crest stem cells undergo cell-intrinsic developmental changes in sensitivity to instructive differentiation signals. *Neuron, 29*(1), 57–71.
Wickramasinghe, S. R., Alvania, R. S., Ramanan, N., Wood, J. N., Mandai, K., & Ginty, D. D. (2008). Serum response factor mediates NGF-dependent target innervation by embryonic DRG sensory neurons. *Neuron, 58*(4), 532–545.
Wilkinson, D. G., Bhatt, S., Chavrier, P., Bravo, R., & Charnay, P. (1989). Segment-specific expression of a zinc-finger gene in the developing nervous system of the mouse. *Nature, 337*(6206), 461–464.
Wilson, Y. M., Richards, K. L., Ford-Perriss, M. L., Panthier, J.-J., & Murphy, M. (2004). Neural crest cell lineage segregation in the mouse neural tube. *Development, 131*(24), 6153–6162.
Woldeyesus, M. T., Britsch, S., Riethmacher, D., Xu, L., Sonnenberg-Riethmacher, E., Abou-Rebyeh, F., et al. (1999). Peripheral nervous system defects in erbB2 mutants following genetic rescue of heart development. *Genes and Development, 13*(19), 2538–2548.
Woodhoo, A., Dean, C. H., Droggiti, A., Mirsky, R., & Jessen, K. R. (2004). The trunk neural crest and its early glial derivatives: A study of survival responses, developmental schedules and autocrine mechanisms. *Molecular and Cellular Neuroscience, 25*(1), 30–41.
Wright, M. A., & Ribera, A. B. (2010). Brain-derived neurotrophic factor mediates non-cell-autonomous regulation of sensory neuron position and identity. *Journal of Neuroscience, 30*(43), 14513–14521.
Wright, D. E., & Snider, W. D. (1995). Neurotrophin receptor mRNA expression defines distinct populations of neurons in rat dorsal root ganglia. *Journal of Comparative Neurology, 351*(3), 329–338.
Yip, J. W. (1986). Migratory patterns of sympathetic ganglioblasts and other neural crest derivatives in chick embryos. *Journal of Neuroscience, 6*(12), 3465–3473.
Ziller, C., Dupin, E., Brazeau, P., Paulin, D., & Le Douarin, N. M. (1983). Early segregation of a neuronal precursor cell line in the neural crest as revealed by culture in a chemically defined medium. *Cell, 32*(2), 627–638.
Zirlinger, M., Lo, L., McMahon, J., McMahon, A. P., & Anderson, D. J. (2002). Transient expression of the bHLH factor neurogenin-2 marks a subpopulation of neural crest cells biased for a sensory but not a neuronal fate. *Proceedings of the National Academy of Sciences of the United States of America, 99*(12), 8084–8089.
Zou, M., Li, S., Klein, W. H., & Xiang, M. (2012). Brn3a/Pou4f1 regulates dorsal root ganglion sensory neuron specification and axonal projection into the spinal cord. *Developmental Biology, 364*(2), 114–127.
Zujovic, V., Thibaud, J., Bachelin, C., Vidal, M., Coulpier, F., Charnay, P., et al. (2010). Boundary cap cells are highly competitive for CNS remyelination: Fast migration and efficient differentiation in PNS and CNS myelin-forming cells. *Stem Cells, 28*(3), 470–479.
Zujovic, V., Thibaud, J., Bachelin, C., Vidal, M., Deboux, C., Coulpier, F., et al. (2011). Boundary cap cells are peripheral nervous system stem cells that can be redirected into central nervous system lineages. *Proceedings of the National Academy of Sciences of the United States of America, 108*(26), 10714–10719.

SECTION B

Placodes

CHAPTER EIGHT

Vertebrate Cranial Placodes as Evolutionary Innovations—The Ancestor's Tale

Gerhard Schlosser[1]
School of Natural Sciences & Regenerative Medicine Institute (REMEDI), National University of Ireland, Galway, Ireland
[1]Corresponding author: e-mail address: gerhard.schlosser@nuigalway.ie

Contents

Current Topics in Developmental Biology, Volume 111
ISSN 0070-2153
http://dx.doi.org/10.1016/bs.ctdb.2014.11.008

Abstract

Evolutionary innovations often arise by tinkering with preexisting components building new regulatory networks by the rewiring of old parts. The cranial placodes of vertebrates, ectodermal thickenings that give rise to many of the cranial sense organs (ear, nose, lateral line) and ganglia, originated as such novel structures, when vertebrate ancestors elaborated their head in support of a more active and exploratory life style. This review addresses the question of how cranial placodes evolved by tinkering with ectodermal patterning mechanisms and sensory and neurosecretory cell types that have their own evolutionary history. With phylogenetic relationships among the major branches of metazoans now relatively well established, a comparative approach is used to infer, which structures evolved in which lineages and allows us to trace the origin of placodes and their components back from ancestor to ancestor. Some of the core networks of ectodermal patterning and sensory and neurosecretory differentiation were already established in the common ancestor of cnidarians and bilaterians and were greatly elaborated in the bilaterian ancestor (with BMP- and Wnt-dependent patterning of dorsoventral and anteroposterior ectoderm and multiple neurosecretory and sensory cell types). Rostral and caudal protoplacodal domains, giving rise to some neurosecretory and sensory cells, were then established in the ectoderm of the chordate and tunicate–vertebrate ancestor, respectively. However, proper cranial placodes as clusters of proliferating progenitors producing high-density arrays of neurosecretory and sensory cells only evolved and diversified in the ancestors of vertebrates.

1. INTRODUCTION

How novelties arise in evolution is one of the most important but least understood questions of Evolutionary Developmental Biology (Hallgrimsson et al., 2012; Moczek, 2008; Peterson & Müller, 2013). In his lucid essay on this problem, François Jacob has pointed out that evolution works like a tinkerer not like an engineer (Jacob, 1977). It cannot design new characters from scratch but has to modify what is already there. It also cannot dismantle the organism, while it rebuilds it: the organism has to keep running, while it is overhauled and redesigned step by step. Evolution is thus like rebuilding ships while at sea: novel characters are fashioned from cobbling together old parts while keeping the vessel afloat. It should therefore not come as a surprise that evolutionary novelties are built from parts that are not new at all. What is new in a novelty are not the parts but the way they interact with each other. "Parts" here may refer to components of various types including genes, cell types, or smaller regulatory networks of interacting proteins and genes. In organisms that occupy different branches of a phylogenetic tree, we call those characters homologous that are derived from the same characters in their last common ancestor (Wagner, 2007). When a novelty arises in one branch of a phylogenetic tree as a new network of interactions between old parts, we therefore expect to find corresponding (homologous) parts in other branches but no corresponding (homologous) network. Thus, even though novelties do not have homologues in other lineages (Müller & Wagner, 1991), their parts do.

We have to be careful to keep these distinctions in mind and recognize that homology at one level of a hierarchy (the parts) does not imply homology at another level (Roth, 1991; Striedter & Northcutt, 1991). Recently, novel structures that are built from components that have clear homologues in other lineages have been said to show "deep homology" (Shubin, Tabin, & Carroll, 2009), but this terminology somewhat obscures the insight that homology is level specific and that nonhomologous structures can be built from homologous parts. The reverse is also true: A character may preserve its integrity even though its components change, a phenomenon called "genetic piracy" or "developmental systems drift" (Roth, 1988; True & Haag, 2001), but this cannot be explored here further.

Researchers studying the evolution of development have been fascinated by the recognition that the same genes, genetic pathways, or cell types are employed in building very diverse body plans. But in this preoccupation with tracing homologies of genes and pathways to ever more distant lineages lies the danger of mistaking homologies of the components for homology of the new structures that are built from them, leaving the origin of innovations out of sight.

Here I will address the question how vertebrate cranial placodes originated as novel structures in the vertebrate lineage by tinkering with ectodermal patterning mechanisms and cell types that have their own evolutionary history. Cranial placodes are thickenings of the embryonic ectoderm that give rise to many cranial sense organs and ganglia. Together with the neural crest, they contribute to many of the evolutionary novelties of the vertebrate head (Northcutt & Gans, 1983). This "new head" of vertebrates probably evolved to support an increasingly active and exploratory life style during the evolution of vertebrates from their filter feeding chordate ancestors (Northcutt & Gans, 1983). Unfortunately, the fossil record of early stem vertebrates is sparse and the phylogenetic position of some of the key fossils like *Haikouella* is contentious (Northcutt, 2005). Therefore, fossils currently provide little information about the sequence in which placodes evolved and also, inherently, allow only limited inference about the cellular and molecular mechanisms at work in these early vertebrates. To reconstruct the evolutionary history of placodes, we will therefore have to rely on comparisons with other living animals. The phylogenetic relationships among the metazoans are now relatively well resolved even though some uncertainties persist. These known phylogenetic relationships will help us to infer, in which lineage new structures evolved, using outgroup comparison as a mode of reasoning (Fig. 1).

Here I briefly describe metazoan phylogeny and then trace the origin of placodes and their components back from ancestor to ancestor, an idea borrowed from Richard Dawkins' book "The ancestor's tale" (Dawkins, 2004). The first section on vertebrates will introduce the placodes and discuss mechanisms of ectodermal patterning underlying placode formation and the cell types arising from placodes. The second section ("The tunicate–vertebrate ancestor") will then discuss ectodermal patterning and cell type differentiation in the sister group of vertebrates, the tunicates, to identify which mechanisms are shared with vertebrates and thus were most likely already present in the vertebrate–tunicate ancestor, and which ones evolved as novelties in the vertebrate lineage. Each subsequent section will similarly use outgroup comparisons to reconstruct the ancestors of more and more inclusive groups (Chordates, Deuterostomes, Bilaterians, Eumetazoans, Metazoans) with conclusions becoming necessarily more and more sketchy and tentative. Due to space constraints, it will not be possible to give a comprehensive overview of all the literature or discuss distant (prechordate) ancestors in detail and the reader is referred to previous reviews (Baker & Bronner-Fraser, 1997; Baker, O'Neill, & McCole, 2008; Fritzsch, Beisel,

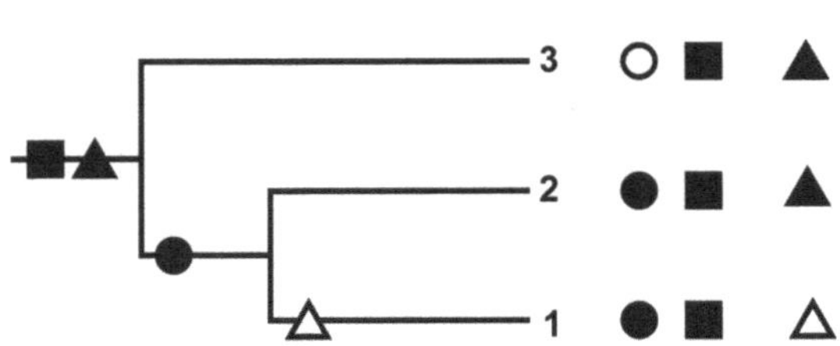

Figure 1 Outgroup comparison. When characters are shared between two sister groups 1 and 2 (circles and squares), comparison with the outgroup 3 allows to determine whether these are shared derived characters which evolved in the last common ancestor of the [1,2] clade (circle) or whether they are shared primitive characters which evolved already in earlier ancestors (square). For characters which differ between two sister groups 1 and 2 (triangles), outgroup comparison may help to clarify, which character is primitive (closed triangle) and was inherited from a common ancestor and which character is derived (open triangle).

Pauley, & Soukup, 2007; Patthey, Schlosser, & Shimeld, 2014; Schlosser, 2005; Schlosser, Patthey, & Shimeld, 2014).

2. A BRIEF PRIMER ON METAZOAN PHYLOGENY

Based on a plethora of morphological similarities, cephalochordates (amphioxus) were long considered to be the closest living relatives of vertebrates with tunicates being more distantly related. However, well-supported molecular phylogenies now show urochordates (tunicates) to be the sister group of vertebrates (Bourlat et al., 2006; Delsuc, Brinkmann, Chourrout, & Philippe, 2006) (Fig. 2). Tunicates are a very rapidly evolving lineage, which have evolved a very specialized life style and have drastically reorganized both their genome as well as their developmental strategies (Holland, 2014; Paps, Holland, & Shimeld, 2012; Putnam et al., 2008). They are therefore overall very poor models for the last common tunicate–vertebrate ancestor. Despite their divergent body plan, however, they share some developmental characteristics with vertebrates not found in amphioxus, which provide insights into placode evolution (see below). Cephalochordates and the tunicate–vertebrate clade (also known somewhat misleadingly as "Olfactores") together comprise the chordates, defined by their notochord, a dorsal hollow nerve cord, segmented muscles, and a post-anal tail. Apart from some peculiar left–right asymmetries in their early embryogenesis, cephalochordates resemble vertebrates in many respects

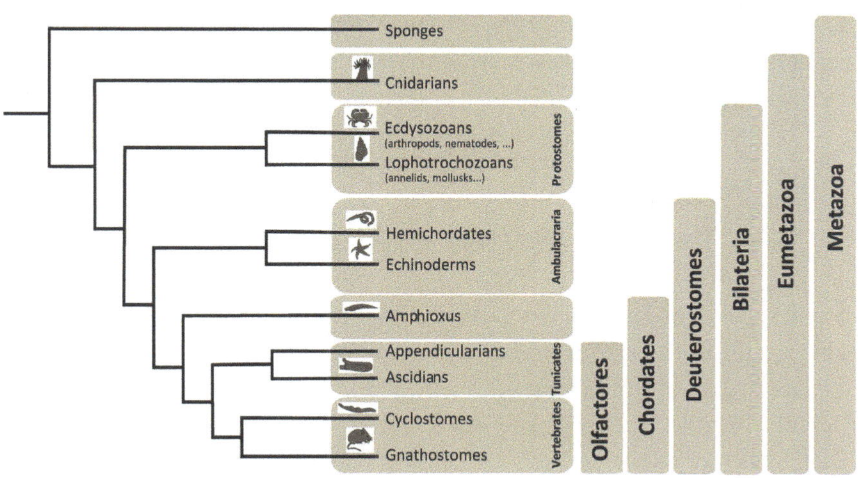

Figure 2 Metazoan phylogeny. Only major lineages are shown. See text for detail. *Modified from Patthey et al. (2014).*

and are thought to retain many primitive characters of the chordate ancestor (Holland, 2014; Holland et al., 2008; Putnam et al., 2008).

The sister group of the chordates are the ambulacrarians, comprising hemichordates and echinoderms. Together with the chordates (and possibly some minor clades such as the Xenoturbellomorpha), these form the deuterostomes (Bourlat et al., 2006; Philippe et al., 2011). While the larval stages of echinoderms and hemichordates are very similar, echinoderms have evolved a highly divergent adult body plan with pentaradial symmetry, which probably retains few traces of how the deuterostome ancestor looked like. Hemichordates are, therefore, thought to provide more insights into the deuterostome ancestor although important controversies remain about whether some of their traits—in particular, their relatively diffuse nervous system—were inherited from the deuterostome ancestor or are due to secondary simplification.

The sister group of the deuterostomes, the protostomes unites two large clades, the lophotrochozoans and the ecdysozoans (Philippe, Lartillot, & Brinkmann, 2005). Lophotrochozoans are characterized by their spiral cleavage pattern and trochophora larvae and include the annelids and molluscs among others. Ecdysozoans are a morphologically heterogeneous group of animals which have a cuticle and grow by molting and include the arthropods and nematodes. Protostomes and deuterostomes together comprise the bilaterians with cnidarians as an outgroup. Cnidarians and bilaterians form the Eumetazoa and the latter together with the sponges are collectively termed metazoans. The position of ctenophores is still controversial. Traditionally thought to be closely related to the cnidarians, recent molecular data suggest instead that they may be the most basal metazoan lineage (Moroz et al., 2014; Ryan et al., 2013).

3. VERTEBRATES

3.1. The cranial placodes of vertebrates and their derivatives

Placodes develop as specialized, usually thickened areas of proliferating progenitor cells in the ectoderm of early vertebrate embryos. Several different types of placodes can be distinguished. This has been reviewed extensively elsewhere (Baker & Bronner-Fraser, 2001; Grocott, Tambalo, & Streit, 2012; Schlosser, 2010) so I will only give a very brief overview here (Fig. 3).

The *adenohypophyseal placode* forms in the anteriormost ectoderm between the mouth and the anterior neural plate. It invaginates to give rise

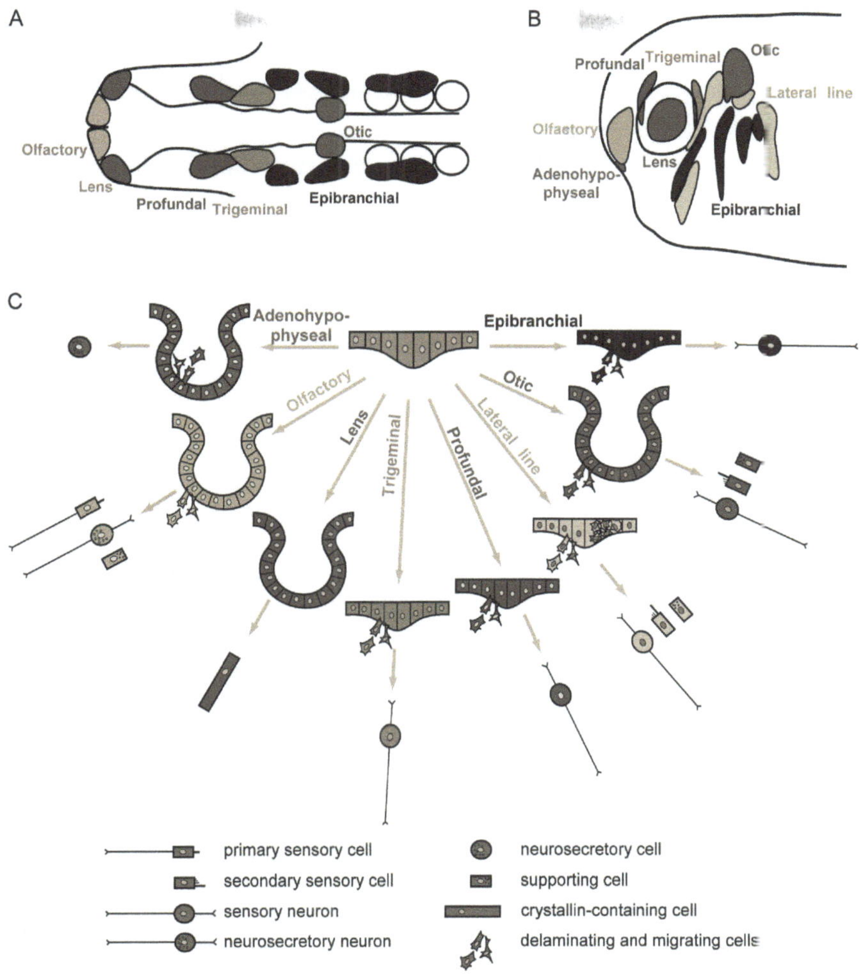

Figure 3 The cranial placodes of vertebrates. (A) Chick embryo. (B) *Xenopus* embryo. (C) Developmental fates and derivative cell types of different cranial placodes. *Panel (A): Modified from Streit (2004). Panel (B): Modified from Schlosser and Northcutt (2000). Panel (C): Modified from Schlosser (2005).* (See the color plate.)

to the anterior pituitary with six types of neurosecretory cells: corticotropes (adrenocorticotropic hormone—ACTH), lactotropes (prolactin—PRL), thyrotropes (thyroid-stimulating hormone—TSH), gonadotropes (luteinizing hormone—LH—and follicle-stimulating hormone—FSH), somatotropes (growth hormone—GH), and melanotropes (melanocyte-stimulating hormone—MSH). The *olfactory placode* gives rise to the olfactory and vomeronasal epithelia with chemoreceptive primary sensory cells (with an axon). Their axons form the olfactory and vomeronasal nerves projecting to the forebrain. In addition, the olfactory placode gives rise to migratory cells producing neuropeptides such as gonadotropin-releasing hormone (GnRH) and neuropeptide Y, which migrate into the forebrain (preoptic area, hypothalamus), where they control gonadotropin release from the anterior pituitary. The glia cells of the olfactory nerve (olfactory ensheathing cells), which were previously thought to be of placodal origin are now known to be derived from the neural crest (Barraud et al., 2010). Whether the neural crest also contributes subpopulations of olfactory receptor cells and GnRH cells is still controversial (Forni, Taylor-Burds, Melvin, Williams, & Wray, 2011; Sabado, Barraud, Baker, & Streit, 2012; Saxena, Peng, & Bronner, 2013; Whitlock, Wolf, & Boyce, 2003). The *lens placode* forms the lens of the eye.

Profundal and *trigeminal placodes* (also known as the ophthalmic and maxillomandibular placode of the trigeminal nerve in amniotes) produce some of the somatosensory neurons of the Vth (trigeminal) cranial nerve—another subpopulation is neural crest derived—which mediate pain, touch, and temperature from the skin of the head and the mouth cavity.

The *otic placode* develops into the inner ear with mechanosensory cells (hair cells) concentrated in several sensory areas dedicated to detection of vestibular (gravity, angular acceleration) and auditory stimuli. Hair cells are secondary sensory cells (without an axon) and are innervated by the somatosensory neurons of the ganglion of the VIIIth (vestibulocochlear) cranial nerve, which also originate from the otic placode. The *lateral line placodes,* which flank the otic placode anteriorly and posteriorly, also give rise to mechanosensory hair cells and the sensory neurons innervating them. These are involved in the detection of water movements. In some groups, modified hair cells function as electroreceptors. A series of *epibranchial placodes* develop at the dorsal (proximal) part of pharyngeal pouches and forms viscerosensory neurons in the distal ganglia of the VIIth (facial), IXth (glossopharyngeal), and Xth (vagal) cranial nerves. These innervate taste buds and mediate chemo- and mechanosensation from the gut and inner organs.

All placodes described so far are shared between different vertebrate groups, although lateral line placodes have been lost in amniotes. Moreover, in lampreys and hagfishes, the adenohypophysis and olfactory epithelium arise from a single unpaired nasohypophyseal placode (Oisi, Ota, Kuraku, Fujimoto, & Kuratani, 2013; Uchida, Murakami, Kuraku, Hirano, & Kuratani, 2003) and only a reduced complement of adenohypophyseal cell types is present (see below). Finally, there are some smaller placodes, which are only found in some vertebrates such as the *hypobranchial placodes* of frogs which produce viscerosensory neurons of unknown function (Schlosser, 2003; Schlosser & Northcutt, 2000) and the *paratympanic placode* of birds, which forms the mechanoreceptors of the paratympanic organ and the sensory neurons innervating it (O'Neill, Mak, Fritzsch, Ladher, & Baker, 2012).

3.2. Origin and patterning of cranial placodes

Fate mapping of the ectoderm in various vertebrates has now firmly established that all placodes arise from a common precursor region, the preplacodal ectoderm (PPE) located around the anterior neural plate and neural crest (Bhattacharyya, Bailey, Bronner-Fraser, & Streit, 2004; Pieper, Eagleson, Wosniok, & Schlosser, 2011; Streit, 2002; Xu, Dude, & Baker, 2008) (Fig. 4). This region is defined by the expression of transcription factors (TFs) of the Six1/2 and Six4/5 families and their coactivators of the Eya family (Grocott et al., 2012; Jemc & Rebay, 2007; Kumar, 2009; Schlosser, 2010; Tadjuidje & Hegde, 2013). Eya proteins also have phosphatase activity and additional functions in the cytoplasm, which are still poorly understood. Six and Eya synergize to promote placodal development at various levels (Ahmed, Wong, et al., 2012; Ahmed, Xu, & Xu, 2012; Kozlowski, Whitfield, Hukriede, Lam, & Weinberg, 2005; Laclef, Souil, Demignon, & Maire, 2003; Schlosser et al., 2008; Xu et al., 1999; Zheng et al., 2003; Zou, Silvius, Fritzsch, & Xu, 2004). At early embryonic stages, they promote the expression of other preplacodal markers. Subsequently, they are required for proper proliferation of neuronal and sensory progenitors and their subsequent differentiation as well as morphogenesis of placodes. Among their direct target genes are cell cycle control genes, *SoxB1* genes (*Sox2*, *Sox3*), which promote neuronal progenitor states and determination genes such as Atoh1, but how the functions of Six and Eya in promoting proliferation of progenitors and differentiation of neurons and sensory cells are coordinated is not understood (Ahmed, Wong, et al., 2012; Li et al., 2003; Schlosser et al., 2008).

The development of the PPE is intimately associated with the origin of other ectodermal territories such as the epidermis, neural plate, and neural crest (reviewed in Grocott et al., 2012; Groves & LaBonne, 2014; Ozair, Kintner, & Brivanlou, 2013; Saint-Jeannet & Moody, 2014; Schlosser, 2010, 2014). During gastrulation, signals including inhibitors of bone morphogenetic proteins (BMPs), inhibitors of wingless/integrated proteins (Wnts), and fibroblast growth factors (FGFs) from the organizer, a signaling

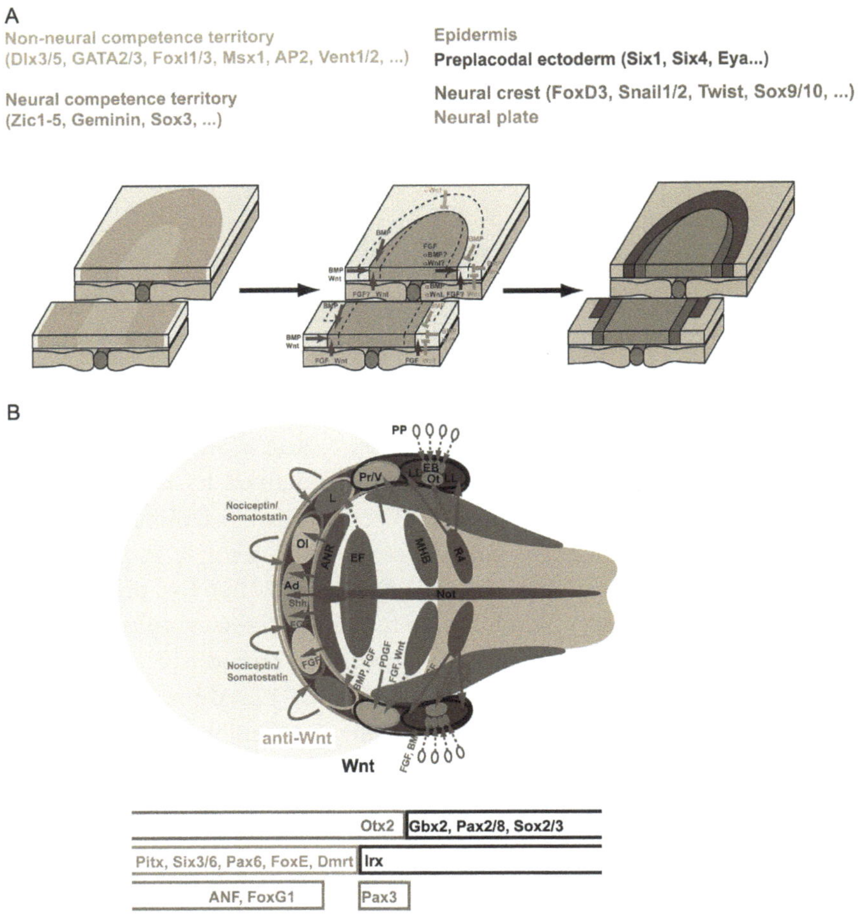

Figure 4 Ectodermal patterning and placode induction. (A) Dorsoventral patterning. Dorsally (green) and ventrally (yellow) restricted transcription factors overlap during gastrulation but subsequently resolve into mutually exclusive neural and nonneural competence territories, respectively. The preplacodal ectoderm is then induced at the border of the nonneural territory by FGF, BMP, and Wnt inhibitors (red), while the neural crest is induced at the border of the neural territory by FGF, BMP, and Wnt (blue). (B) Anteroposterior patterning. The preplacodal ectoderm (red) is subdivided into individual placodes by posteriorly restricted Wnt signals and signalling centers in neural plate and mesoderm. These induce transcription factors, which specify multiplacodal areas (colored outlines) and individual placodes (colored ovals). Some examples of transcription factors are listed below. Ad, adenohypophyseal placode; ANR, anterior neural ridge; EB, epibranchial placodes; EF, eye field; L, lens placode; LL, lateral line placodes; MHB, midbrain–hindbrain boundary; Not, notochord; Ol, olfactory placode; Ot, otic placode; PP, pharyngeal pouches; Pr/V, profundal/trigeminal placode; R4, rhombomere 4. *Panel (A): Modified from Schlosser (2006). Panel (B): Modified from Schlosser (2010).* (See the color plate.)

center in the dorsal mesoderm, establish differences in the expression pattern of TFs between dorsal (neural) and ventral (nonneural) ectoderm (Fig. 4A). Most ventrally (or in many amniotes laterally) restricted TFs like Dlx3/5, Msx1, GATA2/3, AP2, FoxI1/3, and Vent1/2 are activated by BMP, while dorsally (medially) restricted ones like Sox2, Sox3, Geminin, and Zic are repressed by BMP (Feledy et al., 1999; Friedle & Knöchel, 2002; Kwon, Bhat, Sweet, Cornell, & Riley, 2010; Mizuseki, Kishi, Matsui, Nakanishi, & Sasai, 1998; Pera, Stein, & Kessel, 1999; Suzuki, Ueno, & Hemmati-Brivanlou, 1997). The region of overlap between dorsally and ventrally restricted TFs will give rise to the neural plate border region with the neural crest forming medially and the PPE laterally. This region of overlap becomes smaller during gastrulation (partly due to crossrepressive interaction between TFs) until the expression of some of the TFs resolves into mutually complementary nonneural (Dlx3/5, GATA2/3, FoxI) and neural (Zic) domains. Recent experiments in *Xenopus* suggest that the competence to form PPE and neural crest is restricted to the nonneural and neural ectoderm, respectively (Pieper, Ahrens, Rink, Peter, & Schlosser, 2012).

In the nonneural ectoderm, Dlx3/5, GATA2/3, FoxI, and AP2 are required for development of either epidermis or PPE in a signaling-dependent manner and, thus, define a state of nonneural ectodermal competence (Bhat, Kwon, & Riley, 2012; Kwon et al., 2010; Pieper et al., 2012). While epidermis develops as a default state in this region, signals from the anterior neural plate and the dorsolateral endomesoderm, which include BMP inhibitors, Wnt inhibitors, and FGF, induce the PPE in the dorsal part of the nonneural ectoderm at neural plate and fold stages (Ahrens & Schlosser, 2005; Brugmann, Pandur, Kenyon, Pignoni, & Moody, 2004; Kwon et al., 2010; Litsiou, Hanson, & Streit, 2005).

In the neural ectoderm, Zic family TFs similarly seem to act as neural competence factors that are required for neural plate and neural crest induction in the absence and presence of Wnt, respectively, while Sox2/3 recedes further dorsally and defines the neural plate proper (Marchal, Luxardi, Thome, & Kodjabachian, 2009; Pieper et al., 2012; Sato, Sasai, & Sasai, 2005). Zic expression overlaps with expression of AP2, Msx1, and Vent, a group of ventrally restricted TFs which extend further dorsally than Dlx, GATA, and FoxI. In this region of overlap, other TFs (c-Myc, Id, Hairy2, Pax3) will be upregulated in response to Wnt, FGF, and BMP signals and ultimately neural crest specifier genes such as FoxD3, Snail1/2, Twist, and Sox9/10 will be induced (Betancur, Bronner-Fraser, & Sauka-Spengler, 2010; Grocott et al., 2012; Milet & Monsoro-Burq, 2012; Schlosser, 2010, 2014).

The PPE subdivides into individual placodes lined up along the anteroposterior axis in several steps and this process is linked to general mechanisms of anteroposterior patterning (Fig. 4B). Wnt, FGF, and retinoic acid (RA) have all been identified as factors which promote expression of posterior and inhibit expression of anteriorly expressed TFs (reviewed in Grocott et al., 2012; Saint-Jeannet & Moody, 2014; Schlosser, 2010, 2014). For example, direct Wnt targets such as Gbx2 and Irx1 are induced in the posterior and their anterior expression boundary is subsequently sharpened by cross-repressive interactions with anterior TFs Otx2 and Fezf, respectively (Broccoli, Boncinelli, & Wurst, 1999; Martinez-Barbera et al., 2001; Millet et al., 1999; Rodriguez-Seguel, Alarcon, & Gomez-Skarmeta, 2009). While this has first been described for the neural ectoderm, these genes are similarly expressed and regulated in the nonneural ectoderm. Moreover, Otx2 and Gbx2 have been shown to be required for development of anterior (olfactory, lens, trigeminal) and posterior (otic, epibranchial) placodes, respectively (Steventon, Mayor, & Streit, 2012).

Signals from the adjacent mesoderm and neural plate then further subdivide the PPE. First, relatively broad domains of TFs are established, which define multiplacodal areas. These are then further subdivided into individual placodes by more localized signals (Grocott et al., 2012; Ladher, O'Neill, & Begbie, 2010; Saint-Jeannet & Moody, 2014; Schlosser, 2006, 2010; Toro & Varga, 2007). Rostrally, neuropeptides (somatostatin, nociceptin) from the rostral endomesoderm and the PPE itself induce an extended anterior placodal area defined by Six3, Pitx, Pax6, Anf, FoxE, and Dmrt expression (Lleras-Forero et al., 2013). This area later gives rise to the adenohypophyseal, olfactory, and lens placodes in a signaling-dependent manner with sonic hedgehog favoring adenohypophyseal, FGF olfactory, and BMP lens fates (Bailey, Bhattacharyya, Bronner-Fraser, & Streit, 2006; Dutta et al., 2005; Karlstrom, Talbot, & Schier, 1999; Sjödal, Edlund, & Gunhaga, 2007; Treier et al., 2001). Caudally, FGFs from the neural plate and endomesoderm induce a posterior placodal area defined by Pax2, Pax8, Sox2, and Sox3 expression. This area will give rise to the otic, lateral line, and epibranchial placodes, with otic placodes being favored by Wnt and epibranchial placodes by persistent FGF signaling (Freter, Muta, Mak, Rinkwitz, & Ladher, 2008; Ladher, Anakwe, Gurney, Schoenwolf, & Francis-West, 2000; Nechiporuk, Linbo, Poss, & Raible, 2007; Ohyama, Mohamed, Taketo, Dufort, & Groves, 2006). The profundal placode marked by Pax3 expression and the trigeminal placode are induced in between by various signals (Wnt, FGF, PDGF) from the neural plate (Canning, Lee, Luo, Graham, & Jones, 2008; Lassiter et al., 2007; McCabe & Bronner-Fraser, 2008).

3.3. Development of neurosecretory and sensory placodal cell types

After individual placodes have been specified by a unique code of TFs, they segregate into discrete patches of ectoderm, composed of proliferating stem and progenitor cells, which produce high-density arrays of specialized cell types, which differ from placode to placode. In many vertebrates, some stem/progenitor cells persist in placode-derived structures in the adult allowing turnover and regeneration of these structures throughout life, but in mammals, this is only true for the olfactory placode (Maier, Saxena, Alsina, Bronner, & Whitfield, 2014). How the proliferation of progenitors is regulated in the different placodes is still poorly understood but evidence that loss of function in Six and Eya genes compromises proliferation and progenitor formation in most placodes suggests that a common mechanisms may underly this process in different placodes (Chen, Kim, & Xu, 2009; Li et al., 2003; Schlosser et al., 2008; Zheng et al., 2003).

The cell types derived from placodes include neurosecretory cells, various types of sensory cells and neurons, as well as an assortment of specialized epithelial and supporting cells, such as the mucus producing cells of the olfactory epithelium and endolymph producing cells of the inner ear (reviewed in Patthey et al., 2014; Schlosser, 2005).

Neurosecretory cells are mostly found in the adenohypophyseal and olfactory placodes. The adenohypophyseal hormones can be grouped into three classes: (1) peptide hormones (ACTH, MSH) which are produced by proteolytic processing of larger precursor proteins (e.g., pro-opiomelanocortin (POMC), which gives rise to ACTH, MSH and the opioid β-endorphin); (2) dimeric glycoprotein hormones (TSH, LH, FSH) with a common alpha subunit and specific beta subunits; and (3) four-helix cytokine-like proteins (GH, PRL) (Campbell, Satoh, & Degnan, 2004). While neurosecretory cells producing these hormones are concentrated in the adenohypophysis, similar cell types have also been reported in the brain and other tissues (Bicknell, 2008; Murphy & Harvey, 2001; So, Kwok, & Ge, 2005).

In lampreys and hagfishes, only one representative of the adenohypophyseal glycoprotein beta subunits (GTHβ) and only GH but not PRL have been identified (Kawauchi & Sower, 2006; Nozaki, 2008; Sower, Freamat, & Kavanaugh, 2009; Sower et al., 2006; Uchida et al., 2010). This suggests that the vertebrate ancestor had only four adenohypophyseal cell types and that TSH/LH/FSH and PRL arose by duplication of GTHβ and GH genes, respectively, in early gnathostomes.

Differentiation of adenohypophyseal cell types is driven by a complex network of TFs (reviewed in Kelberman, Rizzoti, Lovell-Badge, Robinson, & Dattani, 2009). Cells producing related hormones tend to be regulated by the same TFs, for example, thyrotropes and gonadotropes by GATA2, melanotropes and corticotropes by Tbx19 (T-pit) and NeuroD1, and somatotropes and lactotropes by POU1f1 (Pit1). However, the latter is also required for thyrotropes, and additional TFs (Islet1, Nr5a1) contribute to lineage diversification. TFs of the basic helix loop helix superfamily such as the achaete-scute-related TF Ascl1, which play central roles in sensory/neuronal differentiation (see below), have also been shown to be required for adenohypophyseal cytodifferentiation in zebrafish (Pogoda et al., 2006) and in addition are required for differentiation of various other neurosecretory cells, some of them endoderm derived (Borges et al., 1997; Huber, Combs, Ernsberger, Kalcheim, & Unsicker, 2002).

Additional types of neurosecretory cells are produced by the olfactory placode, most notably the cells migrating into the forebrain and producing neuropeptides including GnRH. There are three forms of GnRH in gnathostomes each encoded by a different gene. Only GnRH1 and GnRH3 are produced in olfactory placode-derived cells (and the latter is only found in teleosts), while GnRH2 cells are neural tube derived (Roch, Busby, & Sherwood, 2011). Fate-mapping studies in zebrafish and putative neural crest-specific reporter lines in mouse suggested that the adenohypophyseal placode and neural crest may also contribute GnRH neurons (Forni et al., 2011; Whitlock et al., 2003), but the tissue selectivity of these experiments has been called into question (Sabado et al., 2012). While lampreys also have three GnRH genes, these are orthologous to GnRH2 (lamprey GnRH II) and GnRH3 (lamprey GnRH I and III) only, and no GnRH1 orthologue has yet been identified (Decatur, Hall, Smith, Li, & Sower, 2013).

Different types of neurons and sensory cells each with their unique specializations are produced by almost all placodes. The chemosensory *olfactory and vomeronasal receptor cells* generated by the olfactory placode have either cilia, microvilli, or both. The prevalence of different types varies for different vertebrate taxa, but the significance of this is still unclear (Eisthen, 1997; Elsaesser & Paysan, 2007). Each olfactory receptor cell expresses one out of many odorant receptors (variable between vertebrate species but typically several hundred), while vomeronasal receptor cells express the pheromone receptors VR1 or VR2. Odorant and pheromone receptor cells belong to two different chordate-specific families of G-protein-coupled receptors (GPCRs) and use different signaling pathways to activate channels of the cyclic nucleotide-gated (CNG) or transient receptor potential (TRP) family, respectively (Kaupp, 2010).

The mechanosensory *hair cells* generated by otic and lateral line placodes have a single nonmotile cilium (kinocilium) eccentrically positioned next to a bundle of microvilli (somewhat misleadingly known as "stereocilia") in a staircase-like arrangement. Different stereocilia are connected by protein filaments (tip links) and connected to myosin VIIa intracellularly (Fritzsch et al., 2007). These tip links are thought to mechanically open ion channels upon deflection of microvilli. Channels of the transmembrane channel-like family have recently been identified as good candidates for the hair cell mechanotransduction channel (Pan et al., 2013) although additional channels, e.g., of the TRP family may also be involved (Eijkelkamp, Quick, & Wood, 2013).

Hair cells as secondary sensory cells are innervated by *somatosensory neurons*, which are derived from the same placodes and may share common progenitors (Satoh & Fekete, 2005). Other somatosensory neurons, which mediate pain, temperature, and touch sensation from the skin, originate from the profundal and trigeminal placodes, but these form either free nerve endings or innervate sensory cells that are not placode-derived such as Merkel cells. The neural crest gives rise to very similar somatosensory neurons, which contribute to the profundal and trigeminal ganglia and form the proximal ganglia of the glossopharyngeal and vagal nerves and the dorsal root ganglia of the spinal nerves. A large diversity of mechanisms for sensory transduction has been identified in these neurons, but many of those involve channels of the TRP family (Lumpkin & Caterina, 2007). The *viscerosensory neurons* derived from the epibranchial placodes also supply sensory cells that are not placode-derived such as taste buds and other visceroreceptors.

The differentiation of sensory cells and neurons from placodes depends on TFs of the basic helix loop helix (bHLH) superfamily. TFs of the Neurogenin (Ngn), Atonal (Atoh), and Achaete-scute (Ascl) families act as neuronal determination (or proneural) genes. Ascl1 (Mash1) plays a central role in differentiation of olfactory receptor cells, while Atoh1 drives hair cell differentiation in the ear and lateral line (Bermingham et al., 1999; Cau, Casarosa, & Guillemot, 2002; Cau, Gradwohl, Fode, & Guillemot, 1997; Chen, Johnson, Zoghbi, & Segil, 2002). Ngn1 and Ngn2, in turn, control the differentiation of sensory neurons derived from the profundal/

trigeminal, otic, lateral line, and epibranchial placodes (Andermann, Ungos, & Raible, 2002; Fode et al., 1998; Ma, Anderson, & Fritzsch, 2000; Ma, Chen, Barrantes, de la Pompa, & Anderson, 1998). These bHLH TFs turn on a cassette of neuronal differentiation genes, promote cell cycle exit, and usually inhibit neuronal differentiation in adjacent cells by activating the Notch signaling pathway (lateral inhibition) (Bertrand, Castro, & Guillemot, 2002). However, neuronal lineages derived from the neural plate or neural crest as well as sensory or neuroendocrine cells derived from either ectoderm or endoderm also are regulated by related bHLH TFs (e.g., Atoh1 for epidermally derived Merkel cells) (Borges et al., 1997; Cau & Wilson, 2003; Huber et al., 2002; Leonard et al., 2002; Li, Ray, Singh, Johnston, & Leiter, 2011). Many other TFs including COE-type bHLH TFs and LIM- (islet1), Paired like- (Phox2a, Phox2b), and POU4-type (POU4f1/Brn3a, POU4f3/Brn3c) homeodomain TFs cooperate with proneural factors in the determination of specific sensory or neuronal cell types (reviewed in Alsina, Giraldez, & Pujades, 2009; Fritzsch et al., 2007; Maier et al., 2014; Schlosser, 2006). Whereas somatosensory neurons are characterized by POU4f1 (Brn3a) and Islet1 expression (Dykes, Tempest, Lee, & Turner, 2011; Eng, Dykes, Lanier, Fedtsova, & Turner, 2007), viscerosensory neurons express Phox2, a TF expressed in all neurons, sensory or motor, innervating the viscera (D'Autreaux, Coppola, Hirsch, Birchmeier, & Brunet, 2011).

Finally, the elongated lens fiber cells are a highly specialized cell type, which loses its nuclei and other cell organelles and accumulates high concentrations of crystallin proteins to become transparent (Cvekl & Duncan, 2007). A diverse group of proteins including heat-shock proteins and metabolic enzymes serve as crystallins (Piatigorsky, 1998). While some of these (α, β, γ) are found throughout the vertebrates, others are taxon specific. Lens fiber cells have several neuronal characteristics (polarized intracellular vesicle transport, dendrite-like protrusions, and shared gene expression) suggesting that they may be derived from neurons (Frederikse, Kasinathan, & Kleiman, 2012).

3.4. The last common vertebrate ancestor

Mechanisms of dorsoventral and anteroposterior patterning are widely conserved among vertebrates. Moreover, with exception of a few neurosecretory cell types of the adenohypophysis, the same types of placode-derived neurosecretory, sensory, and neuronal cells are found throughout vertebrates. Also, with exception of a few small placodes (paratympanic, hypobranchial), all different types of placodes are shared between all vertebrates (unless they are secondarily lost). This indicates that the last common ancestor was already equipped with an essentially full set of placodes and all key events underlying the origin and diversification of placodes happened prior to the radiation of extant vertebrates. Comparisons among living vertebrates will, therefore, not provide us with any insights into the sequence of changes during placode evolution.

4. THE TUNICATE–VERTEBRATE CLADE

To determine which of the patterning mechanisms and cellular derivatives of vertebrate placodes evolved as novelties in stem vertebrates and which ones were inherited from their last common ancestor with tunicates, we have to survey whether corresponding ectodermal patterning mechanisms and cell types exist in tunicates (Figs. 5 and 6). In case of their absence in tunicates, we may also have to look in more distantly related outgroups to account for the possibility of tunicate-specific losses. This task is made more difficult by the highly derived pattern formation mechanisms in tunicates, which have abolished the ancestral chordate mode of development (retained in amphioxus and vertebrates) dependent on morphogens and long-range inductions and have replaced this largely by lineage-dependent segregation of cell fate determinants and local cell interactions (Lemaire, 2009 Lemaire, Smith, & Nishida, 2008). Most of our knowledge of tunicate development comes from ascidians (including *Ciona* and *Halocynthia*) a group characterized by a motile larva and a sessile filter feeding adult, while much less is known about the other tunicate lineages (appendicularians, thaliaceans).

4.1. Ectodermal patterning

In contrast to other chordates, ascidians do not employ a BMP morphogen gradient to set up distinct neural and nonneural ectodermal territories along the dorsoventral axis (Darras & Nishida, 2001a, 2001b; Lemaire et al., 2008). The so-called neural plate of ascidians is composed of six rows of cells (row I-VI from posterior to anterior) at the mid-gastrula stage, but only rows I–IV contribute to the central nervous system (CNS) (Nishida, 1987) (Fig. 5). Rows V and VI give rise to the rostral palps, an adhesive and sensory organ, and to epidermis. The centralmost cells of rows III and IV contribute to the oral siphon; these cells therefore contribute to nonneural (non-CNS) rather than neural ectoderm (Fig. 5A).

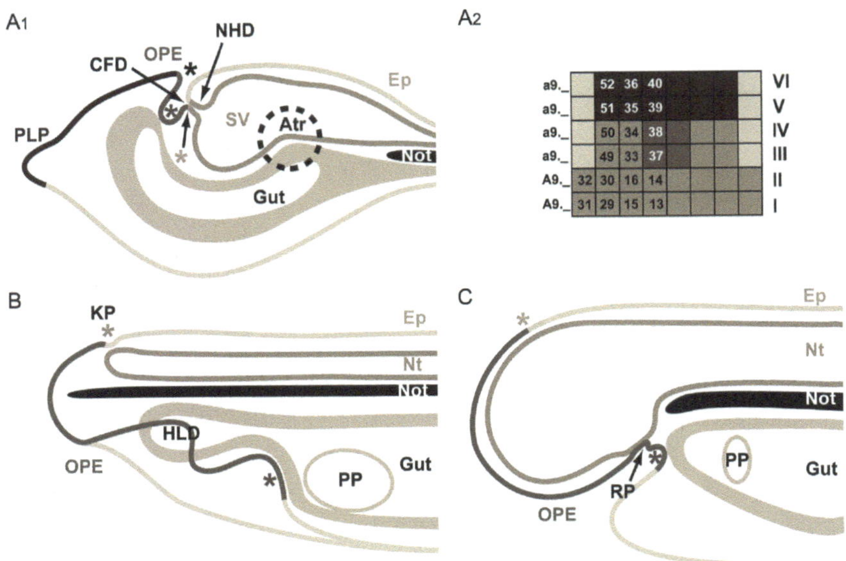

Figure 5 Schematic overview of head region in chordate embryos. Different regions of the ectoderm are distinguished by different colors (green, neural tube; orange, general epidermis; red, oral and preoral part of epidermis; pink, palp-forming region of epidermis in tunicates). The position of anterior neuropore (green asterisk) and mouth is marked by (red asterisk). (A) Tunicates (*Ciona*). Oral and preoral ectoderm (oral siphon primordium) participate in neurulation. As a consequence, the external "neuropore" (purple asterisk) is different from the proper neuropore (green asterisk). The connection between anterior neural tube and the oral siphon primordium may persist giving rise to a neurally derived neurohypophyseal duct (NHD) and a ciliated funnel and duct (CFD) derived from the oral siphon primordium. The atrial siphon primordia (Atr; hatched brown line) invaginate only at late larval stages. The origin of the various ectodermal territories from the so-called neural plate of ascidians at mid-gastrula stage is shown in (A_2) (Nishida, 1987). (B) Amphioxus. Hatschek's left diverticulum (HLD), an endomesodermal pouch will fuse with the preoral ectoderm to give rise to Hatschek's pit. (C) Vertebrates. The preoral ectoderm (including precursors of adenohypophyseal, olfactory, and lens placodes) is expanded due to elaboration of the forebrain. The adenohypophyseal placode buds off the stomodeum as Rathke's pouch (RP) to form the anterior pituitary. Abbreviations: Atr, atrial siphon primordium; CFD, ciliary funnel and duct; Ep, epidermis; HLD, Hatschek's left diverticulum; KP, Kölliker's pit; NHD, neurohypophysial duct: Not, notochord; Nt, neural tube; OPE, oral–preoral ectoderm; PLP, palps; PP, pharyngeal pouches; RP, Rathke's pouch; SV, sensory vesicle. *Panels (A_1), (B), and (C): Modified from Schlosser (2005).* (See the color plate.)

The posterior "neural plate" (derived from the A and b blastomeres at eight-cell stage; Lemaire, Bertrand, & Hudson, 2002; Nishida, 1987) is cell-autonomously specified as neural, while specification of the anterior "neural plate" (derived from the a blastomeres) depends on a sequence of FGF signals from adjacent cells. Initially, an FGF signal from the adjacent

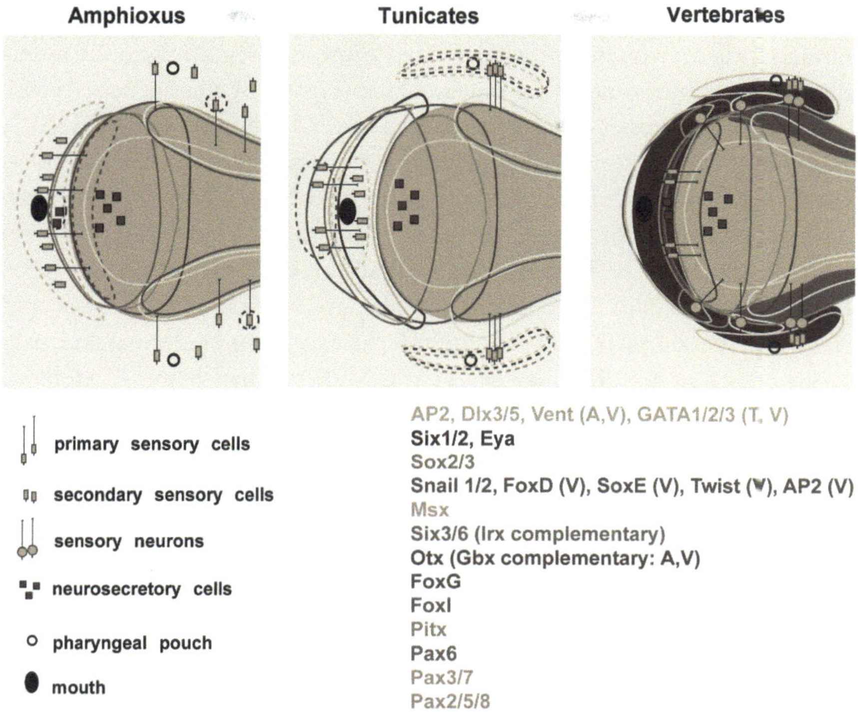

Figure 6 Schematic overview of transcription factor domains and distribution of neurosecretory and sensory cells in ectoderm of chordate embryos. Epidermis is shown in yellow and neural plate in green. Preplacodal ectoderm (expressing Six1/2, Eya) is depicted in red, whereas domains of Six1/2 and Eya expression in tunicates and amphioxus are shown by red outlines. Neural crest (expressing a number of transcription factors) is shown in blue, while domains of Snail1/2 expression in tunicates and amphioxus are shown in blue outlines. Transcription factor expression domains are shown enclosed by colored outlines, with the exception of FoxI in vertebrates and Msx1 in all taxa, which are expressed outside of the colored outlines. Transcription factor domains that are established at later developmental stages are indicated by hatched outlines. Domains of Irx and Gbx expression are not shown but abut the domains of Six3/6 and Otx, respectively. Some expression domains are only present in some taxa (A, amphioxus; T, tunicates; V, vertebrates). *Modified from Schlosser et al. (2014). See text for details.* (See the color plate.)

A-line cells activates expression of Otx in a cell (a6.5) that will form rows III–VI (Bertrand, Hudson, Caillol, Popovici, & Lemaire, 2003; Hudson, Darras, Caillol, Yasuo, & Lemaire, 2003; Hudson & Lemaire, 2001). However, the dedicated neural marker ZicL (the *Ciona* Zic homologue) is actively repressed in these cells until a second FGF signal promotes expression of ZicL exclusively in rows III–IV (Ikeda, Matsuoka, & Satou, 2013; Wagner & Levine, 2012).

Despite the difference in mechanism establishing the neural and nonneural ectoderm, these two territories are defined at the end of gastrulation in ascidians by some of the same TFs as in vertebrates (Fig. 6). While the neural ectoderm in *Ciona* expresses Zic and SoxB1, the Dlx2/3/5 homologue DllB, AP2, and GATA1/2/3 (but not FoxI) are widely expressed in nonneural ectoderm (Christiaen et al., 2002; Imai, Hino, Yagi, Satoh, & Satou, 2004; Imai, Levine, Satoh, & Satou, 2006; Irvine, Cangiano, Millette, & Gutter, 2007; Mazet et al., 2005; Miya & Nishida, 2003; Wada, Katsuyama, & Saiga, 1999; Wada & Saiga, 2002). DllB expression also covers nonneural rows V and VI of the "neural plate" and is required for activating epidermal and palp markers (Imai et al., 2006; Irvine, Vierra, Millette, Blanchette, & Holbert, 2011). It is still unclear how the oral siphon primordium (OSP), which has been fate-mapped to central row III/IV cells in *Halocynthia* (Nishida, 1987), escapes commitment to a neural fate. Upregulation of the Dlx1/4/6 homologue DllA as well as Pitx, and Six1/2 in this domain at early tailbud stages possibly plays a role here (Boorman & Shimeld, 2002; Caracciolo, DiGregorio, Aniello, Dilauro, & Branno, 2000; Christiaen et al., 2002; Irvine et al., 2007; Mazet et al., 2005).

While the gene encoding Vent has been lost from the ascidian genome, Msx in ascidians is also expressed in nonneural ectoderm but extends into the lateral neural plate like in vertebrates, where it overlaps with Pax3/7, Snail, Ets, and elevated Zic expression (Corbo, Erives, DiGregorio, Chang, & Levine, 1997; Gostling & Shimeld, 2003; Imai et al., 2004; Ma et al., 1996; Mazet, Hutt, Millard, & Shimeld, 2003; Mazet, Yu, Liberles, Holland, & Shimeld, 2003; Squarzoni, Parveen, Zanetti, Ristoratore, & Spagnuolo, 2011; Wada, Holland, & Satoh, 1996; Wada & Saiga, 2002; Wagner & Levine, 2012). These TFs may help to define a lateral neural plate identity as in vertebrates. However, apart from Snail, no other TFs acting as neural crest specifiers in vertebrates (SoxE, Twist, FoxD, Myc, and Emc, the *Ciona* Id homologue) are broadly expressed there and no migratory neural crest cells arise from this territory (Imai et al., 2004; Imai, Satoh, & Satou, 2002, 2003; Imai, Stolfi, Levine, & Satou, 2009; Tokuoka, Imai, Satou, & Satoh, 2004; Tokuoka, Satoh, & Satou, 2005). A recent report showed that migratory capacities can be induced by overexpression of Twist (normally expressed in the mesoderm) in the precursor of the ocellus, which develops in the lateral neural plate and expresses FoxD suggesting that the neural crest may have evolved by recruitment of this and other TFs from mesoderm and other tissues into these cells (Abitua, Wagner, Navarrete, & Levine, 2012; Ivashkin & Adameyko, 2013).

Tunicates also show some specializations of the dorsal nonneural ectoderm, which resemble the PPE in vertebrates. Apart from forming various thickenings, this region gives rise to the invaginating primordia of the oral and atrial siphons (Bassham & Postlethwait, 2005; Kourakis, Newman-Smith, & Smith, 2010; Manni, Agnoletto, Zaniolo, & Burighel, 2005; Manni, Lane, et al., 2004; Mazet et al., 2005; Veeman, Newman-Smith, El Nachef, & Smith, 2010). The former develops already in neurula stages from the anterior "neural plate," while the latter only forms at the end of the larval stage. Moreover, Six1/2 and Eya are expressed in the dorsal nonneural ectoderm near the anterior neural plate border and subsequently in and around the oral and atrial siphon primordia (the latter of which also expresses Six4/5) (Bassham & Postlethwait, 2005; Mazet et al., 2005).

Ectodermal patterning along the anteroposterior axis in tunicates has also been greatly modified from an ancestral chordate pattern shared between amphioxus and vertebrates that relies on Wnt and RA gradients (see below). In tunicates, Wnt has largely lost its anteroposterior patterning function and genes encoding various components of the RA signaling pathway are lost from the appendicularian genome (Canestro & Postlethwait, 2007; Nishida, 2005). However, RA still plays some role in regulating *Hox* gene expression along the anteroposterior axis in ascidians and FGF has a posteriorizing function like in vertebrates (Hudson et al., 2003; Hudson, Lotito, & Yasuo, 2007; Kanda, Wada, & Fujiwara, 2009; Nagatomo, Ishibashi, Satou, Satoh, & Fujiwara, 2003; Pasini, Manenti, Rothbächer, & Lemaire, 2012). Despite the degradation of the Wnt patterning system and the loss of the Wnt target *Gbx* from the ascidian genome, many of the TFs which define early anteroposterior domains in the vertebrate ectoderm have similar expression domains in ascidians with Otx, Emx, and Six3/6 expressed anteriorly (Emx in palps; Six3/6 in OSP and anterior sensory vesicle; Otx in each of these domains) and Irx and Hox TFs posteriorly (Ikuta, Yoshida, Satoh, & Saiga, 2004; Imai et al., 2004; Mazet et al., 2005; Moret et al., 2005; Oda & Saiga, 2001; Pasini et al., 2012; Wada, Katsuyama, Sato, Itoh, & Saiga, 1996). Palps (P) and the OSP also express a number of TFs of the extended anterior placodal area in vertebrates such as FoxG (P), Pitx (OSP), and Dmrt, but not Pax6 and FoxE (Boorman & Shimeld, 2002; Christiaen et al., 2002; Glardon, Callaerts, Halder, & Gehring, 1997; Imai et al., 2004; Mazet et al., 2005; Ogasawara & Satou, 2003; Tiozzo et al., 2005). The palps also express neuronal markers such as Atonal, COE, and POU4 TFs (Candiani et al., 2005; Joyce Tang, Chen, & Zeller, 2013; Mazet et al., 2005).

In addition to these similarities in gene expression, the OSP also corresponds in its position to the anteriormost placodes of vertebrates and remains connected to the anterior neural tube of the larva by the so-called neurohypophyseal duct (Fig. 5A). The latter gives rise to the neural gland of the adult, which opens into the oral siphon by a ciliated funnel and duct derived from the posterior wall of the OSP (Manni, Lane, et al., 2004; Veeman et al., 2010). Several types of sensory cells arise from the palps in the larva and from the OSP in the adult (see below).

Further posterior, the atrial siphon primordia in turn express several TFs of the posterior placodal area in vertebrates such as Pax2/5/8 (which is, however, also expressed in the OSP), FoxI, Phox2, and Hox1 (Mazet et al., 2005; Sasakura et al., 2012; Wada, Saiga, Satoh, & Holland, 1998). Like the posterior placodal area, the atrial siphon primordia also are induced by FGF, invaginate and give rise to mechanosensory cells (Bone & Ryan, 1978; Kourakis et al., 2010; Kourakis & Smith, 2007; Mackie & Singla, 2003, 2004) (see below).

Based on these similarities, the oral and atrial siphon primordium have been suggested to be homologous to the adenohypophyseal placode and otic placode of vertebrates, respectively (Boorman & Shimeld, 2002; Christiaen et al., 2002; Graham & Shimeld, 2013; Jefferies, 1986; Manni, Lane, et al., 2004; Mazet et al., 2005; Wada et al., 1998) although they lack key features of placodes and are more likely homologous ectodermal domains from which proper placodes originated in vertebrates.

4.2. Neurosecretory and sensory cell types

Neurosecretory cells in tunicates have been localized in a number of tissues, in particular in the CNS and neural gland of the adult but also the gonads and gut (Schlosser, 2005). However, none of the vertebrate adenohypophyseal hormones (or their specific receptors) were found in the genome of *Ciona*, amphioxus, or any other invertebrate indicating that these hormones are vertebrate innovations (Dehal et al., 2002; Holland et al., 2008; Putnam et al., 2008). Previous reports of cells immunopositive for various adenohypophyseal hormones in tunicates or amphioxus can most likely be attributed to cross-reactivity with other molecules, possibly related hormones, while the isolation of POMC-related peptides from protostomes (Salzet et al., 1997; Stefano, Salzet-Raveillon, & Salzet, 1999) may have been due to contamination with vertebrate tissues.

While the adenohypophyseal hormones of vertebrates are evolutionary innovations, other members of the same hormone families are found throughout bilaterians or even metazoans (Campbell et al., 2004; Jekely, 2013; Mirabeau & Joly, 2013; Roch & Sherwood, 2014). Representatives of the four-helix cytokine-like hormones and their receptors have been identified in protostomes (Huising, Kruiswijk, & Flik, 2006). Genes for the two subunits (GPA2, GPB5) of the heterodimeric glycoprotein thyrostimulin, from which the vertebrate LH/FSH/TSH evolved, and associated GPCR receptors can likewise be traced back at least to the bilaterian ancestor (Dos Santos et al., 2009; Park, Semyonov, Chang, & Hsu, 2005; Sudo, Kuwabara, Park, Hsu, & Hsueh, 2005) as can the neuropeptides of the rhodopsin γ class and their associated GPCRs, which gave rise to the POMC-encoded peptides (opioids, MSH and ACTH) and their receptors in vertebrates (Dores & Baron, 2011; Fredriksson & Schiöth, 2005; Mirabeau & Joly, 2013; Sundström, Dreborg, & Larhammar, 2010). Many other neuropeptides are present in tunicates including six GnRH peptides encoded by two genes (Adams et al., 2003; Roch et al., 2011).

Expression of these various hormones in tunicates is concentrated in the CNS (Hamada et al., 2011), and GPA2 and GPB5, which are not well characterized in tunicates, are expressed in the CNS of both arthropods and amphioxus (Dos Santos, Mazan, Venkatesh, Cohen-Tannoudji, & Querat, 2011; Sellami, Agricola, & Veenstra, 2011; Tando & Kubokawa, 2009). This suggests a possible origin of the adenohypophyseal neurosecretory cells from the CNS. In contrast to vertebrates, there is currently little evidence for a rostral neurosecretory region in ascidians. However, in amphioxus, Hatschek's pit may serve as a rostral neurosecretory organ (see below), suggesting that a corresponding region may have been present in the chordate ancestor but lost in tunicates. Moreover, TFs involved in lineage specification in the vertebrate adenohypophysis are either lost from the *Ciona* genome (Pit1) or are not expressed in OSP or palp regions (Tbx19, GATA2/3, Nr5a1) (Kano, 2010). However, GnRH is expressed in the palps in addition to gonads and CNS where it serves both reproductive and nonreproductive functions (Kavanaugh, Root, & Sower, 2005; Kusakabe et al., 2012; Terakado, 2001). Because no placodal origin of GnRH cells has yet been reported for lampreys, this expression of GnRH in the anterior nonneural ectoderm in tunicates may either have evolved convergently to vertebrates or may reflect a recruitment of GnRH to this domain in the tunicate–vertebrate ancestor with loss in lampreys.

Tunicates also have a diversity of different sensory cells. These are usually ciliated and surrounded by a collar of microvilli, but the number of cilia and microvilli and their arrangement is variable between cells and taxa (Burighel, Caicci, & Manni, 2011; Holland & Holland, 2001). In the absence of direct experimental evidence for the sensory modality mediated by these receptors, it can only be tentatively assigned as chemosensory or mechanosensory based on ultrastructure. The tunicate larva has primary sensory cells associated with the palps or rostral ectoderm and the tail (Bollner, Holmberg, & Olsson, 1986; Caicci et al., 2010; Takamura, 1998; Torrence & Cloney, 1982, 1983). They express Atonal, Ascl, MyT1, POU4f1, COE (palps), and Islet1 (palps) (Candiani et al., 2005; Giuliano, Marino, Pinto, & De Santis, 1998; Joyce Tang et al., 2013; Mazet et al., 2005). Neurogenin and Phox2 are expressed in atrial primordia, but it is not clear whether their expression is restricted to sensory precursors (Mazet et al., 2005). Thus, overall, the differentiation of sensory cells appears to be regulated by the same set of TFs in tunicates as in vertebrates, even though no specific correspondence between particular cell types in tunicates and vertebrates can be established. While the palp neurons have been proposed to be mechanoreceptors, the rostral trunk epidermal neurons in their vicinity are thought to be chemoreceptors which probe the substrate for cues to induce settlement and metamorphosis (Caicci et al., 2010). However, since tunicates have lost all odorant family GPCRs, which mediate olfaction in vertebrates and are also present in amphioxus (Churcher & Taylor, 2009; Niimura, 2009), the underlying chemotransduction mechanism is probably different from olfactory receptor cells.

After metamorphosis, additional sensory cells develop in the primordia of the oral and atrial siphons. The former give rise to the putative photoreceptors of the oral siphon pigment organs, scattered primary sensory cells, and the secondary putatively mechanosensory cells of the coronal (ascidians and thaliaceans) or circumoral organ (appendicularians) which lines the oral tentacles (Auger, Sasakura, Joly, & Jeffery, 2010; Bassham & Postlethwait, 2005; Burighel et al., 2003; Caicci et al., 2013, 2010; Manni, Caicci, Gasparini, Zaniolo, & Burighel, 2004; Manni, Lane, et al., 2004; Manni, Mackie, Caicci, Zaniolo, & Burighel, 2006; Rigon et al., 2013; Takamura, 1998). The latter give rise to the putatively mechanosensory cupular and capsular organs, small sensory organs composed of primary sensory cells (Bone & Ryan, 1978; Mackie & Singla, 2003, 2004). Both the cells of the cupular/capsular organ and the cells of the coronal/circumoral organ have been proposed to be homologues of vertebrate hair cells. However, unlike vertebrate hair cells, the former are primary sensory cells. The latter resemble hair cells in that they are secondary sensory cells with an eccentrically positioned kinocilium, but they develop from a rostral Pitx1 and Six3/6 expressing domain of ectoderm. In vertebrates, this domain gives rise to olfactory and adenohypophyseal placodes that do not form hair cells. While shared lineage and similar gene expression patterns suggest that hair cells and the sensory neurons that innervate them arose as sister cells from a common precursor (Fritzsch, Beisel, & Bermingham, 2000; Fritzsch, Eberl, & Beisel, 2010), it is not clear when this happened. Did hair cells evolve from primary receptor cells in the tunicate–vertebrate ancestor that gave rise to secondary sensory cells and sensory neurons in vertebrates or from secondary receptor cells that were either rostrally confined and later recruited to a more caudal domain in vertebrates or present in rostral and caudal domains with reciprocal loss in tunicates and vertebrates (Patthey et al., 2014)?

Tunicates do not possess lens fiber cells like vertebrates, but the tunicate homologue of vertebrate βγ crystallins is expressed in the palps and the otolith in the neural tube, which is closely associated with the neural tube derived photoreceptor cell (Shimeld et al., 2005). *cis*-regulatory regions of *Ciona* βγ crystallin are able to target gene expression to the vertebrate lens, suggesting that this protein was recruited to the vertebrate lens together with its upstream regulators.

4.3. The last common tunicate–vertebrate ancestor

In spite of the profound divergence and specialization of the tunicate lineage, comparisons between tunicate and vertebrate development allow us to draw some inferences about the tunicate–vertebrate ancestor. Prior to the evolution of lineage-dependent specification of cell fates, different ectodermal cell fates along the dorsoventral and anteroposterior axes were specified by BMP/FGF and Wnt/RA/FGF gradients, respectively. The dorsal non-neural ectoderm was presumably already characterized by the expression of Six1/2 and Eya genes, with anterior and posterior subregions specified by different sets of TFs (Pitx, Six3/6 and Pax2/5/8, FoxI, respectively). These gave rise to special regions of ectoderm around the mouth opening and pharyngeal gill slits, respectively, in which sensory receptor cells were concentrated and which underwent morphogenetic movements. In addition, neurosecretory (neuropeptidergic) cells may have been present in the anterior domain but subsequently disappeared in tunicates. From these, "protoplacodal" ectodermal domains evolved the oral and atrial siphon primordia in tunicates and the extended anterior and the posterior placodal area in vertebrates. However, only in the vertebrate lineage did proper placodes originate from these domains—regions of clustered proliferating progenitors giving rise to complex sense organs with high-density arrays of sensory cells. Rewiring of the gene regulatory network downstream of Six and Eya genes may have played an important role during this process resulting in a shared mechanism for progenitor expansion and neurogenesis in all vertebrate placodes. In the lateral neural plate, the neural crest emerged by similar rewiring leading to recruitment of TFs from mesoderm and other domains to the lateral neural plate.

From the chordate ancestor (see below), the tunicate–vertebrate ancestor inherited a diverse collection of primary and secondary sensory cells. In the vertebrate lineage, some of these existing chemosensory and mechanosensory cell types may have been merely redeployed in new ectodermal domains (with related cell types retained in other parts of the ectoderm or endoderm) and/or developed new specializations (such as the staircase arrangement of microvilli in hair cells), although new cell types such as adenohypophyseal neurosecretory cells and the somato- and viscerosensory neurons have also emerged.

5. CHORDATES

To infer, which of the characters of the tunicate–vertebrate ancestor were primitive chordate traits inherited from the last common chordate ancestor and which one evolved *de novo* in the stem lineage of tunicates and vertebrates, we have to broaden our survey to include cephalochordates (amphioxus) (Figs. 5 and 6). The body plan and development of amphioxus are in many respects more similar to vertebrates than tunicates suggesting that it has deviated much less from the ancestral chordate than tunicates (Holland, 2014).

5.1. Ectodermal patterning

In amphioxus like in vertebrates, different TFs begin to be restricted to dorsal, prospectively neural (SoxB1, Zic), and ventral, prospectively nonneural (AP2, Dlx, Msx, Vent) ectoderm during gastrulation in response to a BMP gradient (Gostling & Shimeld, 2003; Holland, Schubert, Holland, & Neuman, 2000; Holland, Panganiban, Henyey, & Holland, 1996; Kozmik et al., 2007; Kozmikova, Candiani, Fabian, Gurska, & Kozmik, 2013; Kozmikova, Smolikova, Vlcek, & Kozmik, 2011; Meulemans & Bronner-Fraser, 2002, 2007; Sharman, Shimeld, & Holland, 1999; Yu,

Meulemans, McKeown, & Bronner-Fraser, 2008; Yu et al., 2007) (Fig. 6). While AP2 is confined to the nonneural ectoderm, Dlx, Vent, and Msx extend into the lateral or anterior edge of the neural plate together with several other TFs involved in defining lateral neural plate identity in vertebrates such as Pax3/7, Zic, and Snail (Gostling & Shimeld, 2003; Holland, Schubert, Kozmik, & Holland, 1999; Langeland, Tomsa, Jackman, & Kimmel, 1998; Yu et al., 2008). However, none of the other TFs acting as lateral neural plate markers (Myc, Id, Hairy, Irx) or neural crest specifiers (Twist, Ets, FoxD, SoxE) in vertebrates are specifically enriched in this region and no migratory neural crest like cells develop (Kaltenbach, Holland, Holland, & Koop, 2009; Meulemans & Bronner-Fraser, 2004; Meulemans, McCauley, & Bronner-Fraser, 2003; Minguillon, Jimenez-Delgado, Panopoulou, & Garcia-Fernandez, 2003; Van Otterloo et al., 2012; Yasui, Tabata, Ueki, Uemura, & Zhang, 1998; Yu, Holland, & Holland, 2002; Yu et al., 2008). In contrast to vertebrates, amphioxus GATA1/2/3 is not expressed in the early ectoderm (Zhang & Mao, 2009). The nonneural ectoderm gives rise to scattered sensory cells (see below) and shows little evidence of regionalization along the dorsoventral axis, with no concentration of Six1/2 and Eya expression in the dorsal nonneural ectoderm (Kozmik et al., 2007).

Like in vertebrates, Wnt and RA (but not FGF) have been shown to play a posteriorizing role in the amphioxus neural and nonneural ectoderm and several of the Wnt- and RA-dependent TFs of vertebrates have similar distributions along the anteroposterior axis of amphioxus (anterior: Otx, Fezf, and Six3/6; posterior: Gbx, Cdx, Irx, and Hox) (Beaster-Jones et al., 2008; Bertrand et al., 2011; Brooke, Garcia-Fernandez, & Holland, 1998; Castro, Rasmussen, Holland, Holland, & Holland, 2006; Escriva, Holland, Gronemeyer, Laudet, & Holland, 2002; Holland, 2002, 2005; Irimia et al., 2010; Kaltenbach, Holland, et al., 2009; Koop et al., 2010 Kozmik et al., 2007; Onai et al., 2009; Pascual-Anaya et al., 2012; Schubert, Holland, Escriva, Holland, & Laudet, 2004; Schubert, Holland, Laudet, & Holland, 2006; Williams & Holland, 1996; Yu et al., 2007). The TF FoxQ2, which is a Wnt-inhibited anterior marker in many bilaterians but has been lost from some vertebrate genomes, is also anteriorly confined (Yu, Holland, & Holland, 2003).

However, many of the other TFs involved in the subdivision of the PPE along the anteroposterior axis in vertebrates (e.g., Pax3/7, Pax2/5/8, FoxG1, FoxE4) are not expressed in the amphioxus nonneural ectoderm (Holland et al., 1999; Kozmik et al., 1999, 2007; Toresson,

Martinez-Barbera, Bardsley, Caubit, & Krauss, 1998; Yu, Holland, Jamrich, Blitz, & Hollan, 2002) indicating that they either lost a prior role in patterning nonneural ectoderm (e.g., FoxG1) or acquired it later. Several of the TFs defining the posterior placodal area in vertebrates (e.g., Pax2/5/8, Six1/2, Six4/5, Eya, SoxB1, Tbx1/10, Irx) are coexpressed in the pharyngeal pouches of amphioxus and other chordates and, thus, may have been recruited as a network of coregulated genes from the pharyngeal endoderm in the tunicate–vertebrate ancestor (Schlosser, 2005).

Only the preoral ectoderm and Hatschek's pit, which arises from the fusion of an endomesodermal pouch (Hatschek's left diverticulum) with the preoral ectoderm and contacts the brain (Fig. 5B), express TFs of the anterior placodal area (Pax6, Pitx) in addition to Six3/6 (Boorman & Shimeld, 2002; Glardon, Holland, Gehring, & Holland, 1998; Vopalensky et al., 2012; Yasui, Zhang, Uemura, & Saiga, 2000). The rostral ectoderm also harbors primary sensory cells expressing vertebrate type odorant GPCRs (Satoh, 2005), while Hatschek's pit expresses Pit1, Lhx, Islet as well as Six1/2 and Eya and gives rise to neurosecretory cells (Candiani, Holland, Oliveri, Parodi, & Pestarino, 2008; Jackman, Langeland, & Kimmel, 2000; Kozmik et al., 2007; Wang, Zhang, Yasui, & Saiga, 2002). Due to their position and gene expression patterns, the rostral ectoderm has been suggested to be a homologue of the olfactory placode (Glardon et al., 1998; Holland & Holland, 2001) and Hatschek's pit a homologue of the adenohypophyseal placode (Boorman & Shimeld, 2002; Nozaki & Gorbman, 1992; Yasui et al., 2000).

The endomesodermal origin of Hatschek's pit is in conflict with this interpretation since the adenohypophysis in all vertebrates is completely ectodermally derived (Oisi et al., 2013). However, there are several well-documented examples for the evolutionary translocation of developmental programs for cell fate assignment from one germ layer to another in regions of epithelial fusion, e.g., teeth and taste buds, which may form from endo- or ectodermal parts of the mouth cavity (Northcutt, 2004; Soukup, Epperlein, Horacek, & Cerny, 2008; Stone, Finger, Tam, & Tan, 1995). Similarly, expression of TFs specifying cell fates in Hatschek's pit may have shifted from endomesoderm to ectoderm during evolution of the preoral region.

5.2. Neurosecretory and sensory cell types

Many neuropeptides, the glycoprotein hormone thyrostimulin, as well as their associated GPCR receptors were identified in amphioxus

(Mirabeau & Joly, 2013; Roch & Sherwood, 2014). Moreover, neurosecretory cells producing FMRFamide, neuropeptide Y, and many other hormones have been localized to the amphioxus neural tube and Hatschek's pit (see Schlosser, 2005). However, most studies identified such cells based on their immunoreactivity with antibodies against specific vertebrate hormones and, thus, could not rule out the possibility of cross-reactivity with different epitopes. For example, gonadotropins were initially reported to be present in amphioxus based on such immunohistochemical studies (e.g., Nozaki & Gorbman, 1992) but were recently shown to be absent from the amphioxus genome like all other adenohypophyseal hormones (Holland et al., 2008; Putnam et al., 2008). One GnRH peptide previously isolated from amphioxus (Chambery, Parente, Topo, Garcia-Fernandez, & D'Aniello, 2009) could not be confirmed in genomic analyses, but in contrast, another GnRH-like peptide has recently been identified in the amphioxus genome (Roch, Tello, & Sherwood, 2014). Since this peptide activates only one out of four GnRH receptors, additional GnRH peptides are probably present in amphioxus but remain to be identified. GnRH positive cells have been localized to the neural tube but not to Hatschek's pit or preoral ectoderm (Castro, Becerra, Manso, Sherwood, & Anadon, 2006; Roch et al., 2014).

While the nature of the hormones produced in Hatschek's pit is at present still elusive, basally located secretory vesicles identified in ultrastructural studies (Sahlin & Olsson, 1986; Tjoa & Welsch, 1974) provide additional evidence that the latter contains neurosecretory cells (in addition to exocrine cells). Hatschek's pit also expresses the TFs Pit1 and Islet, which play a role in lineage specification of the vertebrate adenohypophysis, but whether they are also involved in specifying neurosecretory cell types in amphioxus is not clear (Candiani et al., 2008; Jackman et al., 2000). Moreover, the cells of Hatschek's pit bear cilia and microvilli resembling chemosensory cells. It has therefore been suggested that they may control endocrine functions such as gonad maturation or gamete release in response to environmental cues (Gorbman, 1995; Nozaki & Gorbman, 1992). However in contrast to chemosensory cells, the cells in Hatschek's pit neither have an axon, nor do they appear to form synaptic contacts with sensory neurons.

In contrast to tunicates, amphioxus contains a large number of different sensory cells throughout the nonneural ectoderm with different types concentrated in different regions. These usually occur as scattered single cells although a few sense organs composed of small clusters of sensory cells (e.g., the rostral corpuscles of de Quatrefages) are also present. Both primary and secondary sensory cells with cilium and microvillar collar have been described and classified as putative chemo- or mechanoreceptors (e.g., Holland & Holland, 2001; Lacalli, 2004; Lacalli & Hou, 1999; Ruppert, 1997).

A rostral population of primary chemosensory cells in amphioxus, which have been discussed as putative homologues of olfactory receptor cells, arises from a region of ectoderm in which the cytodifferentiation TFs Id, Neurogenin, and POU4 are expressed (Candiani, Oliveri, Parodi, Bertini, & Pestarino, 2006; Holland et al., 2000; Meulemans et al., 2003). These sensory neurons also express at least one vertebrate type odorant receptor (Satoh, 2005). Around 50 of these vertebrate type odorant receptors have been identified in the amphioxus genome representing an independent expansion of odorant GPCRs in the amphioxus lineage (Churcher & Taylor, 2009; Niimura, 2009).

Another population of primary mechano- and/or chemosensory cells originates in the ventral nonneural ectoderm, where these cells delaminate, followed by their dorsal migration and reinsertion into the epidermis (Benito-Gutierrez, Nake, Llovera, Comella, & Garcia-Fernandez, 2005; Kaltenbach, Yu, & Holland, 2009). These cells or subsets of them also express the Achaete-scute homologue Ash, Tlx, Hu/Elav, Islet, COE, POU4, SoxB1, Six1/2, Six4/5, Eya, and Delta (Benito-Gutierrez, Illas, Comella, & Garcia-Fernandez, 2005; Candiani et al., 2006; Holland & Holland, 2001; Kaltenbach, Yu, et al., 2009; Kozmik et al., 2007; Lu, Luo, & Yu, 2012; Mazet, Masood, Luke, Holland, & Shimeld, 2004; Meulemans & Bronner-Fraser, 2007; Rasmussen, Holland, Schubert, Beaster-Jones, & Holland, 2007; Satoh, Wang, Zhang, & Satoh, 2001; Schubert et al., 2004), while expression of Atonal has not been described. With this expression profile, these sensory cells resemble various placodally derived sensory cells or sensory neurons, but no one to one correspondence can be established (Patthey et al., 2014). However, unlike vertebrate sensory cells, amphioxus sensory neurons do not originate from placodes—clusters of proliferating progenitors in the dorsal nonneural ectoderm—but from scattered ventral precursors.

5.3. The last common chordate ancestor

Based on the many similarities between amphioxus and vertebrates, the last common ancestor of chordates was probably a quite amphioxus-like filter feeding animal. Dorsoventral and anteroposterior patterning of the ectoderm relied on BMP and Wnt/RA gradients, respectively, while FGF acquired important roles in axial patterning only in the tunicate–vertebrate clade. However, the nonneural ectoderm in the chordate ancestor was not yet regionalized along the dorsoventral axis and Six1/2 and Eya were expressed in pharyngeal pouches and scattered sensory cells. Recruitment of these genes to the dorsalmost part of the nonneural ectoderm thus happened only in tunicate–vertebrate ancestors, when these genes presumably became responsive to inducers from the neural plate or underlying mesoderm. However, some TFs defining the anteriormost (preoral) ectoderm in vertebrates (Six3/6, Pax6, Pitx) were already expressed in the preoral ectoderm, a region which fused with an endomesodermal pouch to give rise to a rostral neurosecretory organ. The expression of some of the TFs expressed in the endomesodermal portion may have later shifted into the adjacent ectoderm forming a rostral protoplacodal ectodermal domain. Further posterior TFs FoxI, Pax2/5/8, Six1/2, Six4/5, and Eya were expressed in pharyngeal pouches and were recruited to the adjacent ectoderm in the tunicate–vertebrate clade. Both primary and secondary chemo- and mechanosensory cells were present, and their cytodifferentiation was regulated by some of the same TFs (e.g., bHLH TFs Ascl, Neurogenin, and probably Atonal; POU4; Six1/2 and Eya) as in vertebrates. However, apart from possibly the chemosensory neurons expressing odorant receptors, there is currently little evidence to suggest that specific subpopulations of these sensory cells evolved into specific placodal cell types. Furthermore, these cells originated throughout the nonneural ectoderm in the chordate ancestor and became concentrated in the rostral and caudal protoplacodal ectodermal domains only in the tunicate–vertebrate ancestor.

6. DEUTEROSTOMES

To determine, which developmental traits of chordates were newly acquired and which ones were inherited from the deuterostome ancestor, we now have to compare development of chordates with their sister taxon, the ambulacrarians (echinoderms and hemichordates). Because the body plan of echinoderms has been drastically reorganized, hemichordates (including enteropneust worms like *Saccoglossus* and *Ptychodera* as well as the pterobranchs) typically provide the more useful model for comparisons although some of their features may also be highly derived and we must occasionally look further afield (i.e., to other bilaterians) to make at least tentative inferences about the deuterostome ancestor.

6.1. Ectodermal patterning

Experimental evidence in hemichordates indicates that BMP plays a conserved role in setting up restricted domains of TF along the dorsoventral axis. However, BMP and TFs promoted by it (e.g., Dlx) are expressed on the dorsal side, i.e., the side opposite to the mouth opening (Lowe et al., 2006), which is different from chordates but similar to protostomes. It has, thus, been suggested that the dorsoventral axis was inverted in the chordate lineage (Arendt & Nübler-Jung, 1994; De Robertis & Sasai, 1996). Different from most protostomes, however, in hemichordates, no prominent centralized nervous system develops from the ventral, BMP-depleted side of the ectoderm. Rather, neurons are found scattered throughout the ectoderm and were reported to form a diffuse nerve net, allowing no obvious distinction between central and peripheral nervous systems (Bullock, 1945; Bullock & Horridge, 1965; Knight-Jones, 1952). Recently, however, several studies have found that both the ventral and dorsal nerve cords of enteropneusts are more complex than previously described and contain different neuron types in addition to axon bundles (Brown, Prendergast, & Swalla, 2008; Kaul & Stach, 2010; Nomaksteinsky et al., 2009). Taken together with findings that similar TFs are involved in the dorsoventral and anteroposterior patterning of the CNS in insects, annelids, and chordates, it has thus been suggested that a centralized neural and a complementary nonneural territory are ancient bilaterian traits, with several independent reductions of neural centralization in some bilaterian phyla (reviewed in Arendt, Denes, Jekely, & Tessmar-Raible, 2008; Holland et al., 2013). In the absence of information about how the enteropneust nerve cords are patterned, it is currently difficult to assess whether any of them is indeed a likely homologue of the chordate neural plate. However, Six1/2 and Eya clearly show no dorsoventrally restricted expression in the nonneural ectoderm in hemichordates or echinoderms but rather are confined to endomesoderm and possibly scattered neuronal cells (Gillis, Fritzenwanker, & Lowe, 2012; Materna, Ransick, Li, & Davidson, 2013; Yankura, Martik, Jennings, & Hinman, 2010). In the pharyngeal pouches, they are coexpressed with other TFs that are also implicated in pharyngeal pouch development in chordates such as Pax1/9 and FoxI (Fritzenwanker, Gerhart, Freeman, & Lowe, 2014; Gillis et al., 2012; Lowe et al., 2003).

In contrast to the established role of a Wnt gradient for anteroposterior ectodermal patterning in both chordates and protostomes (see below), in hemichordates and echinoderms, Wnt is primarily involved in specifying vegetative and endomesodermal fates and (with the exception of the rostralmost ectoderm) only indirectly regulates regionalized TF expression along the anteroposterior axis (Darras, Gerhart, Terasaki, Kirschner, & Lowe, 2011; Logan, Miller, Ferkowicz, & McClay, 1999; Pani et al., 2012; Wikramanayake, Huang, & Klein, 1998). Nevertheless, the anteriorly restricted expression of Six3/6, FoxQ2, Otx, Emx, and FoxG and posteriorly restricted expression of Gbx and Hox are conserved in hemichordates with complementary Otx-Gbx domains and Pax2/5/8 expressed in between (Aronowicz & Lowe, 2006; Fritzenwanker et al., 2014; Lowe et al., 2003; Pani et al., 2012). Pitx is not expressed in the rostral ectoderm but rather in the dorsal proboscis pore, a structure that forms on the left side in some species but on the right side in others and is putative homologue of amphioxus' Hatschek's pit. This suggests that Pitx adopted a role in anterior patterning from its more ancient role in left–right patterning only in chordates (Grande & Patel, 2009; Lowe et al., 2006).

6.2. Neurosecretory and sensory cell types

Not much is known about the neurosecretory and sensory cell types in hemichordates. GnRH was shown to be produced in neurosecretory cells scattered throughout the rostral epidermis, while ciliated chemo- or mechanosensory neurons with a collar of microvilli are found associated with the ciliary bands in larvae and scattered throughout the ectoderm in adults (Cameron, Mackie, Powell, Lescheid, & Sherwood, 1999; Jørgensen, 1989). Vertebrate type odorant receptors have been identified in the echinoderm but not hemichordate genome (Krishnan, Almen, Fredriksson, & Schiöth, 2013; Raible et al., 2006).

6.3. The last common deuterostome ancestor

Due to the uncertainty whether the relatively diffuse nervous system of hemichordate is due to retention of a primitive deuterostome condition or reflects secondary simplification, we cannot currently be sure whether the last common deuterostome ancestor had a ventral CNS or not. However, it probably used BMP to establish different TF domains along the dorsoventral axis with a ventral sink of BMP. Pharyngeal slits developed on the dorsal side in the anterior region, presumably regulated by a network of Six1/2, Eya, FoxI, and Pax1/9. Dorsoventral axis inversion and the formation of a new mouth then occurred in the chordate lineage. A Wnt gradient was used to set up TF domains along the anteroposterior axis with Six3/6 and FoxQ2 confined to the Wnt-depleted anterior.

However, RA-dependent anteroposterior patterning and a role of Pitx in anteroposterior patterning probably evolved later in the ancestors of chordates. Several classes of neurosecretory and chemo- and mechanosensory cells were probably inherited from the bilaterian ancestor, but new types of chemosensory cells using vertebrate type odorant receptors were also present. We do not know how these various cell types were distributed in the deuterostome ancestor but there is no evidence to suggest that neurosecretory cells were clustered in a rostral neurosecretory organ.

7. BILATERIA

The protostomes as sister group of the deuterostomes are a highly diverse group, making the recognition of traits shared with the deuterostomes and the reconstruction of the last common bilaterian ancestor a very challenging task. Moreover, our knowledge of protostomian development is based mostly on model organism in the ecdysozoans (e.g., arthropods such as *Drosophila* and nematodes such as *Caenorhabditis*), with representatives of the lophotrochozoans (e.g., the annelid *Platynereis*) only recently being studied.

7.1. Ectodermal patterning

There is now ample evidence for a BMP gradient mediating dorsoventral patterning in many protostomes (Mizutani & Bier, 2008). Different TFs are activated by low BMP levels on the ventral side and by high BMP levels on the dorsal side, with some of the former (e.g., SoxB1) possibly promoting neural and the latter (e.g., Dlx) nonneural ectodermal competence. Subsequently, a BMP gradient may help to further subdivide the neural or nonneural ectoderm. Some of the TF domains that help regionalize the neural ectoderm along the dorsoventral (medial to lateral: NK2.2, Gsx-NK6-Pax6, Msx-Pax3/7) and anteroposterior (rostral to caudal: Otx, Pax2/5/8, Hox) axes are also conserved between insects, annelids, and chordates. Moreover, similar cell types develop at corresponding positions in this molecular coordinate system (Denes et al., 2007; De Velasco et al., 2007; Marlow et al., 2014; Tessmar-Raible et al., 2007) strengthening the proposal that segregation of the ectoderm into a neural (forming the CNS) and nonneural (forming the epidermis and sensory receptors) domain is an ancient bilaterian trait (reviewed in Arendt et al., 2008; Holland et al., 2013). It has recently been suggested that the mediolateral (dorsoventral) patterning system evolved to specify a part of the nervous system with lateral sensory and medial motorneurons originally associated with the blastopore (Tosches & Arendt, 2013).

However, it still needs to be resolved to what extent the regulatory network establishing distinct neural and nonneural territories is conserved between protostomes and vertebrates, since many TFs involved in this process in vertebrates are either not described for protostomes (FoxI) or not expressed in a corresponding pattern (Zic, GATA1/2/3) (Schlosser et al., 2014). Also, since many bilaterian phyla have nervous systems that are more diffuse and/or have concentrations of neurons in different noncorresponding parts of the body, the similarities between annelids, arthropods, and chordates could also be due to convergently evolved CNSs relying on evolutionarily conserved axial patterning systems (Holland, 2003; Lowe et al., 2006, 2003; Moroz, 2009).

Recently, it has been suggested that some of the TFs defining the dorsalmost nonneural ectoderm in chordates have a corresponding expression domain in protostomes. Expression of Six1/2, Six4/5, and Eya homologues was reported at the anterior border of neural ectoderm in the beetle *Tribolium* and is required for epidermis and sensory bristle formation (Posnien, Koniszewski, & Bucher, 2011). A similar rostral domain of Six1/2 expression also exists in annelids (Arendt, Tessmar, Campos-Baptista, Dorresteijn, & Wittbrodt, 2002). It has been proposed that this "head placode" and vertebrate placodes are derived from a common placode precursor in the bilaterian ancestor (Posnien, Koniszewski, & Bucher, 2011), but because no comparable domain exists in ambulacrarians and amphioxus, it may rather represent a convergent recruitment of these genes to the anterior neural border in some protostomes and the tunicate–vertebrate clade. However, the expression of Six1/2 and Eya in various sensory cells including mechanoreceptors and rhabdomeric photoreceptors together with data from functional studies in protostomes indicate that these proteins rather than acting as ectodermal patterning genes may have an ancient function in regulating sensory differentiation (Arendt et al., 2002; Bonini, Bui, Gray-Board, & Warrick, 1997; Bonini, Leiserson, & Benzer, 1993; Cheyette et al., 1994; Halder et al., 1998; Mannini et al., 2004; Pignoni et al., 1997; Pineda et al., 2000; Serikaku & O'Tousa, 1994; Suzuki & Saigo, 2000) with Atonal being a direct target gene in insects and vertebrates (Ahmed, Wong, et al., 2012; Zhang, Ranade, Cai, Clouser, & Pignoni, 2006).

Similar to dorsoventral patterning, there is also evidence for evolutionarily ancient mechanisms of patterning the anteroposterior axis. Expression data from many protostomes and functional studies in arthropods and planarians support a role for a Wnt morphogen gradient with higher levels in the posterior for anteroposterior patterning throughout bilaterians (reviewed in Holstein, Watanabe, & Ozbek, 2011; Niehrs, 2010; Petersen & Reddien, 2009). Alternate mechanisms, however, have superseded this in some lineages (e.g., the bicoid gradient in *Drosophila*). Some of the TF domains established directly or indirectly in response to this Wnt gradient are also conserved both in the neural and in the nonneural ectoderm with Six3/6, FoxQ2, FoxG1, Fezf, Otx, and Emx expressed anteriorly and Irx, Gbx, and Hox posteriorly (Hirth et al., 2003; Irimia et al., 2010; Marlow et al., 2014; Posnien, Koniszewski, Hein, & Bucher, 2011; Santagata, Resh, Hejnol, Martindale, & Passamaneck, 2012; Sen, Reichert, & VijayRaghavan, 2013; Steinmetz, Kostyuchenko, Fischer, & Arendt, 2011; Steinmetz et al., 2010; Tomer, Denes, Tessmar-Raible, & Arendt, 2010). Complementary domains of Otx–Gbx and Fezf–Irx in insects and some other taxa suggest that cross-repressive interactions between these TFs help to sharpen their expression boundaries like in vertebrates (Hirth et al., 2003; Irimia et al., 2010; Steinmetz et al., 2011). It has recently been shown that some of the anteriorly restricted TFs (Six3/6, FoxQ2, Fezf as well as Rx) are associated with the apical organ in larvae of many bilaterian phyla and suggested that they play a role in patterning the apical nervous system with several types of sensory and neurosecretory cells (Marlow et al., 2014; Tosches & Arendt, 2013).

7.2. Neurosecretory and sensory cell types

As already discussed above, many hormone classes including neuropeptides, glycoprotein hormones, and four-helix cytokine-like hormones are found throughout bilaterians and different types of neurosecretory cells have been identified (Hartenstein, 2006; Tessmar-Raible, 2007). Some neuropeptidergic cells (e.g., producing vasotocin or RFamide) in the anterior CNS of insects and annelids, which have been suggested to originate from the apical organ, have been proposed to be direct homologues of neuroendocrine cells in the vertebrate hypothalamus since they originate from equivalent TF domains along the anteroposterior (Six3/6, Rx) and dorsoventral axis (Nk2.1) and also express Otp, a core regulator of many hypothalamic neuroendocrine cells (De Velasco et al., 2007; Marlow et al., 2014; Tessmar-Raible et al., 2007; Tosches & Arendt, 2013). Based mostly on their innervation from this neuroendocrine center in the CNS, the neurosecretory cells of two peripheral endocrine gland in arthropods, the endomesodemal corpora cardiaca or the ectodermal corpora allata, have in turn been proposed to be homologues of the adenohypophysis (De Velasco, Shen, Go, & Hartenstein, 2004; Wirmer, Bradler, & Heinrich, 2012). However, there is currently no evidence for similarly innervated and positioned neurosecretory cells in other phyla, so these glands may instead have evolved by the convergent recruitment of neurosecretory cells.

Many variants on the common theme of ciliated sensory cells with or without a microvillar collar exist in protostomes. These are mostly primary sensory cells (although some secondary sensory cells have been described) and may function as mechano-, chemo-, or photoreceptors (Arendt, 2008; Budelmann, 1989; Jørgensen, 1989; Laverack, 1988; Schlosser, 2005). The evolutionary relationships between these various receptor cells have recently begun to be better understood, but the emerging relationships are complex and can only be briefly sketched here.

On one hand, similarities in the TFs used for cell specification as well as in the molecular signal transduction pathways have suggested that homologues to many vertebrate sensory cell types are present in protostomes. For example, mechanoreceptors in protostomes as well as deuterostomes are specified by Atonal and POU4 together with Pax2/5/8-related TFs and use myosin VIIa and TRP channels in mechanotransduction (Fritzsch et al., 2007). Moreover, some microRNAs (e.g., miR183) are specifically expressed in sensory cells throughout bilaterians (Christodoulou et al., 2010; Pierce et al., 2008).

On the other hand, it has become increasingly clear that evolutionarily closely related cell types using homologous transduction machinery may mediate different sensory modalities suggesting that these modalities are evolutionarily quite flexible. For example, the bHLH TF Atonal, together with POU4, Pax2/5/8 or Pax6 and Six1/2 TFs as well as TRP channels are also employed in specification of rhabdomeric photoreceptors and some chemoreceptors (e.g., vertebrate vomeronasal receptors), while opsins related to the r-opsins of rhabdomeric photoreceptors have been found in mechanoreceptors (Fritzsch et al., 2005; Plachetzki, Fong, & Oakley, 2010; Senthilan et al., 2012). Similarly, c-opsins of ciliary photoreceptors and CNG channels mediate sensory transduction not only in ciliary photoreceptors but also in some chemoreceptors (e.g., vertebrate olfactory receptors) (Kaupp, 2010; Plachetzki et al., 2010).

In addition, each sensory modality can be mediated by different cell types, which are not closely evolutionarily related and use different TFs and/or transduction machinery. For example, olfactory receptors in insects, nematodes, and vertebrates depend on completely different receptors and signal transduction machinery as do olfactory and vomeronasal receptor cells in vertebrates (Kaupp, 2010). Moreover, other bHLH TFs than Atonal, in particular TFs of the Achaete-scute family, act as sensory determination genes for various types of mechano-, photo-, and chemoreceptor cells (Bertrand et al., 2002). Taken together with the possibility that new cell types may arise not only by divergence from parental cell types but also by co-option of programs from multiple parent cells (Arendt, 2008; Patthey et al., 2014) and that often multiple cell types employing identical or related TFs and sensory transduction pathways develop in multiple locations in the body, this makes it very difficult to trace homologous sensory cell types in different phyla.

7.3. The last common bilaterian ancestor

Given the diversity of bilaterian body plans, it is not surprising that it is still contentious, what the last common bilaterian ancestor (or "urbilaterian") looked like. However, it probably used BMP and Wnt signaling to set up TF domains along the dorsoventral and anteroposterior axis (anterior Six3/6, FoxQ2, FoxG1, Fezf, Otx, Emx; posterior: Irx, Gbx, Hox), respectively. It possibly already formed a centralized nervous system on the ventral side with an anterior center composed of neurosecretory and sensory cells ("apical nervous system") and columns of motorneurons and sensory neurons running along the longitudinal axis ("blastoporal nervous system"). It probably also had several types of neurosecretory and sensory cells in its periphery. The latter, which may have included dedicated photo-, mechano-, and chemoreceptors and/or multimodal cells, were ciliated with a microvillar collar and already existed as distinct cell types specified by different TFs (e.g., bHLH TFs Atonal, Ascl as well as POU4, Pax2/5/8, and Six1/2) and used different sensory transduction mechanisms (e.g., TRP and CNG channels). These cell types later gave rise to different sensory cells in the various bilaterian phyla, but probably with frequent switches in modality and changes in distribution.

8. EUMETAZOA AND METAZOA

Because "urbilaterians" can be inferred to have possessed regulatory networks for patterning the body axes as well as many specialized cell types, these must already have evolved during the early evolution of the metazoans. This period was, thus, one in which many important innovations happened but for space constraints we need to treat the eumetazoan (cnidarians and bilaterians) and metazoan (sponges, ctenophores, and eumetazoans) clades jointly here and can highlight only a few points.

8.1. Ectodermal patterning

Although there is no dorsoventral axis in cnidarians, and neurons are distributed in a diffuse nerve net rather than forming a centralized CNS, BMP gradients seem to play some role in patterning the so-called directive axis (orthogonal to the oral–aboral axis) (Matus, Thomsen, & Martindale, 2006; Rentzsch et al., 2006; Saina, Genikhovich, Renfer, & Technau, 2009). However, several of the TFs subdividing the dorsoventral axis in bilaterians such as Nk2.1 and Dlx are expressed in rings around the blastopore suggesting that many of the TFs involved in dorsoventral patterning in bilaterians may have originally patterned ectoderm centered around the blastopore rather than aligned with the directive axis (Marlow et al., 2014; Tosches & Arendt, 2013).

In cnidarians like in bilaterians, a Wnt gradient has been implicated in anteroposterior (oral–aboral) patterning (Duffy, Plickert, Kuenzel, Tilmann, & Frank, 2010; Marlow, Matus, & Martindale, 2013). It has been disputed whether the oral pole, which forms at high Wnt levels, corresponds to the posterior or anterior pole in bilaterians (Holstein et al., 2011; Martindale & Hejnol, 2009; Niehrs, 2010), but the expression of several anterior TFs such as Six3/6 and FoxQ2 at the aboral pole in association with the apical organ and the expression of Irx in a complementary domain support the former hypothesis (Chevalier, Martin, Leclere, Amiel, & Houliston, 2006; Marlow et al., 2013, 2014; Sinigaglia, Busengdal, Leclere, Technau, & Rentzsch, 2013). However, the expression of other TFs such as Otx and Gbx differs from bilaterians suggesting that they adopted a role in anteroposterior patterning only later (Matus et al., 2006; Mazza, Pang, Martindale, & Finnerty, 2007).

The polarized distribution of Wnt and TGFβ along the anteroposterior axis of sponge larvae suggests that their role in axial patterning may even predate eumetazoans (Adamska et al., 2007; Windsor & Leys, 2010) although ctenophores appear to rely on different patterning mechanisms (Pang, Ryan, Baxevanis, & Martindale, 2011; Pang et al., 2010).

8.2. Neurosecretory and sensory cell types

Many neuropeptide hormones and their receptors are present in eumetazoans but not sponges (even though they possess preneuropeptide processing enzymes), whereas glycoprotein hormones have been identified in all metazoans (Hartenstein, 2006; Jekely, 2013; Roch & Sherwood, 2014). In cnidarians, several types of neurosecretory cells have been described in both endoderm and ectoderm. Thus, the origin of some neurosecretory cells can be dated back to the origin of metazoans with significant diversification in eumetazoans and bilaterians.

However, only cnidarians but not sponges have specialized neurons and sensory cells (Galliot et al., 2009; Jacobs et al., 2007; Watanabe, Fujisawa, & Holstein, 2009). Ctenophores, which have recently been suggested to be the most basal metazoans, have neurons, but many neuron-specific genes found in eumetazoans are lacking (Moroz et al., 2014; Ryan et al., 2013). Assuming that the basal phylogenetic position of ctenophores can be confirmed, this suggests that either sponges have lost a primitive type of neuron or ctenophores evolved neurons convergently to eumetazoans. In either case, many neuron-specific genes only evolved in eumetazoans. Neurons in cnidarians form a diffuse nerve net throughout the body column but can be concentrated in ganglia. Ciliated photo-, mechano-, and chemosensory cells likewise may be scattered throughout the body column or form part of complex sense organs such as the rhopalia of medusae. The nematocytes (stinging cells) of cnidarians are highly specialized mechanosensory cells with a modified cilium, which triggers rapid exocytosis of a venom filled capsule upon stimulation (Holstein, 2012). They are closely associated with other sensory cells, which include mechano- and photoreceptors using TRP- and CNG-channel-dependent sensory transduction, respectively, as well as chemoreceptors (Holstein, 2012; Mahoney, Graugnard, Mire, & Watson, 2011; Plachetzki, Fong, & Oakley, 2012). Achaete-scute like bHLH TFs are expressed in both the nematocyte and associated neurons as well as other neurons or secretory cells, whereas Atonal marks different populations of neurons and sensory cells (Hayakawa, Fujisawa, & Fujisawa, 2004; Seipel, Yanze, & Schmid, 2004). Furthermore, Six1/2 and Eya, PaxB (the precursor of Pax6 and Pax2/5/8), and POU4 are all expressed in putative mechanosensory and photosensory cells (Six1/2, Eya, PaxB) with Six1/2 also expressed in muscle cells (Bebenek, Gates, Morris, Hartenstein, & Jacobs, 2004; Graziussi, Suga, Schmid, & Gehring, 2012; Groger, Callaerts, Gehring, & Schmid, 2000; Hroudova et al., 2012; Kozmik et al., 2003; Matus, Pang, Daly, & Martindale, 2007; Nakanishi, Yuan, Hartenstein, & Jacobs, 2010; Stierwald, Yanze, Bamert, Kammermeier, & Schmid, 2004). Thus, some of the sensory cell types found in bilaterians were probably inherited from their eumetazoan ancestor together with sensory transduction mechanisms and networks of TF involved in cell specification.

Sponges, in contrast, have some components of neurons such as proteins involved in postsynaptic specializations as well as TRP and CNG ion channels but no true neurons or sensory cells (Ludeman, Farrar, Riesgo, Paps, & Leys, 2014; Plachetzki et al., 2012; Sakarya et al., 2007). An Atonal like gene is expressed in globular cells, an unciliated cell with potential sensory functions (Richards et al., 2008), while Six1/2 and PaxB are expressed in contractile pinacocytes (Hill et al., 2010; Rivera et al., 2013). Unless sponges have secondarily lost neurons (see above), this suggests that sensory cells in the eumetazoans may have evolved by co-option and diversification of several molecular components expressed in different sponge cells. The choanocytes of sponges, which bear cilia surrounded by a microvillar collar, resemble both the unicellular choanoflagellates, the closest unicellular relatives of metazoans, and the ciliated sensory cells found in eumetazoans and have been proposed to be potential evolutionary precursor of sensory cells, although there is currently little molecular evidence to support this (Fritzsch et al., 2007; Jacobs et al., 2007; Renard et al., 2009).

8.3. The last common eumetazoan and metazoan ancestors

Taken together, this suggests that while Wnt and TGFβ already played some patterning role in ancestral metazoans, an anteroposterior Wnt-dependent axial patterning system with anterior Six3/6 and FoxQ2 and posterior Irx expression evolved first in the eumetazoan ancestor to pattern the aboral (anterior) pole and was elaborated by recruitment of additional TFs in bilaterians. The eumetazoan ancestor may have used TFs such as NK2.1 and Dlx to regionalize the oral pole and a BMP gradient to pattern the directive axis, and these patterning systems may later have been co-opted to pattern the dorsoventral axis in bilaterians.

While the metazoan ancestor already produced many different hormones and expressed many of the molecular components found in neuronal and sensory cells, specialized neurons and sensory cells evolved (or, alternatively, were significantly elaborated form a primitive ctenophore-like type) only in eumetazoans. The eumetazoan ancestor already had multiple photo-, mechano-, and chemosensory cell types, specified by different bHLH genes (e.g., Atonal, Achaete-scute) and employing different sensory transduction mechanisms. Six1/2, Eya, PaxB, and POU4 TFs probably adopted a role in the specification of some sensory cell types in the eumetazoan ancestor which they maintained in bilaterians, while acquiring additional patterning roles in at least some bilaterian phyla.

9. SUMMARY AND CONCLUSIONS

This review has attempted to gain insights into the evolutionary history of vertebrate placodes and their components by comparing ectodermal patterning mechanisms and cell types in different metazoans in a proper phylogenetic framework. Cranial placodes evolved as novelties in the vertebrate lineage by redeployment and rewiring of regulatory networks involved in ectodermal patterning and cell type specification as summarized in Fig. 7. Many of the molecular building blocks of these networks already evolved in basal metazoans. These were reassembled and diversified to build core networks of axial patterning and sensory and neurosecretory differentiation in eumetazoans. In bilaterians, the former were elaborated into BMP- and Wnt-dependent TF regulatory networks along the dorsoventral and anteroposterior axis, respectively, possibly with a centralized nervous system developing on the BMP-depleted ventral side. The latter diversified into networks driving the differentiation of multiple sensory and neurosecretory cell types.

After dorsoventral inversion in chordates and the recruitment of new TFs for roles in anteroposterior and dorsoventral patterning, a rostral neurosecretory organ developed in the region of fusion between an endomesodermal pouch and anteriormost ectoderm (rostral protoplacodal domain). In the tunicate–vertebrate ancestor, Six1/2 and Eya expression were then recruited to the dorsal nonneural ectoderm and an additional caudal protoplacodal domain was established by redeployment of other TFs to the caudal ectoderm possibly from the adjacent pharyngeal slits. Finally, in vertebrates, a proper neural crest and cranial placodes evolved as novel tissues from the border regions of neural and nonneural ectoderm, respectively. This probably involved rewiring of the regulatory network downstream of neural plate border specifiers (Zic, Pax3, Msx1) and Six1/2 and Eya, respectively, as well as the evolution of new cell types.

This general overview of sensory evolution, however, leaves many questions unanswered. Most importantly, our knowledge of patterning and sensory differentiation in most taxa is based merely on gene expression studies with functional data being available only for a few model species. While this

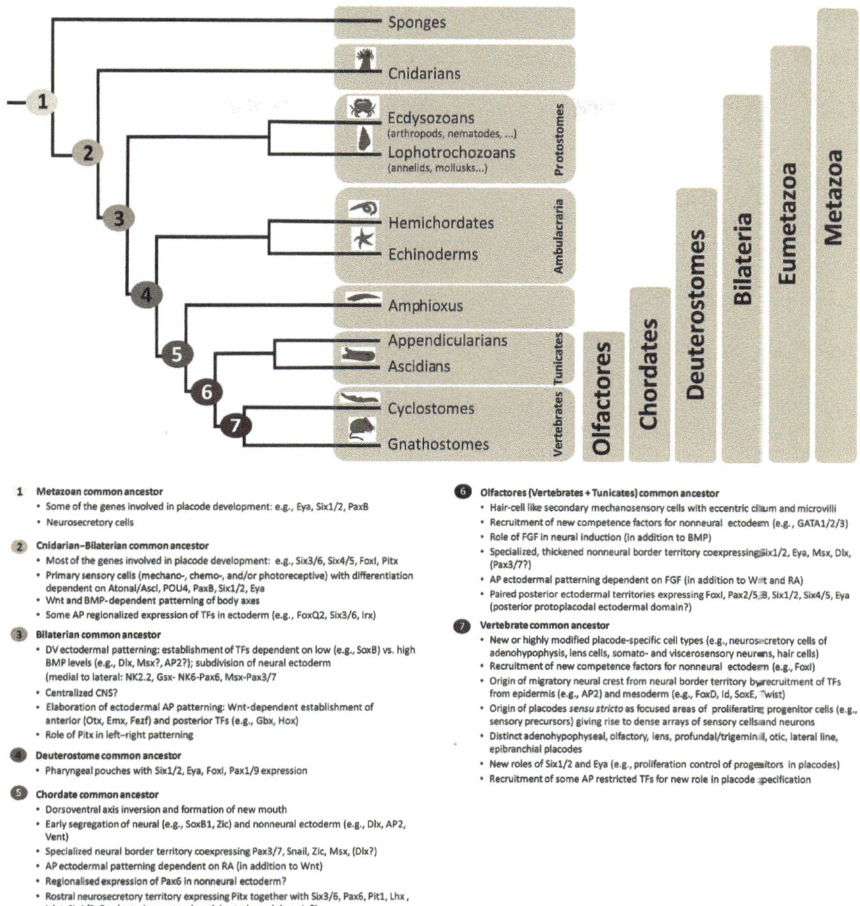

Figure 7 Evolutionary history of cranial placodes and their components. Innovations in ectodermal patterning and neurosecretory and sensory cell types are mapped onto metazoan phylogeny. Some key characters that originated in the ancestors of the various clades (colored ovals) are listed below. Question marks indicate uncertainties about the placement of a character origination event. AP, anteroposterior; DV, dorsoventral; TF, transcription factor. *Modified from Schlosser et al. (2014). See text for details.*

knowledge base allows us to easily recognize conserved patterns of development, it is insufficient to understand how regulatory networks became rewired during the evolution of novelties. Reconstruction of regulatory interactions in experimentally tractable sister taxa (e.g., tunicates and vertebrates) will ultimately be required to overcome this limitation and shed more light on the origin of cranial placodes.

REFERENCES

Abitua, P. B., Wagner, E., Navarrete, I. A., & Levine, M. (2012). Identification of a rudimentary neural crest in a non-vertebrate chordate. *Nature*, *492*, 104–107.

Adams, B. A., Tello, J. A., Erchegyi, J., Warby, C., Hong, D. J., Akinsanya, K. O., et al. (2003). Six novel gonadotropin-releasing hormones are encoded as triplets on each of two genes in the protochordate, Ciona intestinalis. *Endocrinology*, *144*, 1907–1919.

Adamska, M., Degnan, S. M., Green, K. M., Adamski, M., Craigie, A., Larroux, C., et al. (2007). Wnt and TGF-beta expression in the sponge Amphimedon queenslandica and the origin of metazoan embryonic patterning. *PLoS One*, *2*, e1031.

Ahmed, M., Wong, E. Y., Sun, J., Xu, J., Wang, F., & Xu, P. X. (2012). Eya1-six1 interaction is sufficient to induce hair cell fate in the cochlea by activating atoh1 expression in cooperation with sox2. *Developmental Cell*, *22*, 377–390.

Ahmed, M., Xu, J., & Xu, P. X. (2012). EYA1 and SIX1 drive the neuronal developmental program in cooperation with the SWI/SNF chromatin-remodeling complex and SOX2 in the mammalian inner ear. *Development*, *139*, 1965–1977.

Ahrens, K., & Schlosser, G. (2005). Tissues and signals involved in the induction of placodal Six1 expression in Xenopus laevis. *Developmental Biology*, *288*, 40–59.

Alsina, B., Giraldez, F., & Pujades, C. (2009). Patterning and cell fate in ear development. *The International Journal of Developmental Biology*, *53*, 1503–1513.

Andermann, P., Ungos, J., & Raible, D. W. (2002). Neurogenin1 defines zebrafish cranial sensory ganglia precursors. *Developmental Biology*, *251*, 45–58.

Arendt, D. (2008). The evolution of cell types in animals: Emerging principles from molecular studies. *Nature Reviews. Genetics*, *9*, 868–882.

Arendt, D., Denes, A. S., Jekely, G., & Tessmar-Raible, K. (2008). The evolution of nervous system centralization. *Philosophical Transactions of the Royal Society of London. Series B, Biological Sciences*, *363*, 1523–1528.

Arendt, D., & Nübler-Jung, K. (1994). Inversion of dorsoventral axis? *Nature*, *371*, 26.

Arendt, D., Tessmar, K., Campos-Baptista, M. I., Dorresteijn, A., & Wittbrodt, J. (2002). Development of pigment-cup eyes in the polychaete Platynereis dumerilii and evolutionary conservation of larval eyes in Bilateria. *Development*, *129*, 1143–1154.

Aronowicz, J., & Lowe, C. J. (2006). Hox gene expression in the hemichordate Saccoglossus kowalevskii and the evolution of deuterostome nervous systems. *Integrative and Comparative Biology*, *46*, 890–901.

Auger, H., Sasakura, Y., Joly, J. S., & Jeffery, W. R. (2010). Regeneration of oral siphon pigment organs in the ascidian Ciona intestinalis. *Developmental Biology*, *339*, 374–389.

Bailey, A. P., Bhattacharyya, S., Bronner-Fraser, M., & Streit, A. (2006). Lens specification is the ground state of all sensory placodes, from which FGF promotes olfactory identity. *Developmental Cell*, *11*, 505–517.

Baker, C. V. H., & Bronner-Fraser, M. (1997). The origins of the neural crest. Part II: An evolutionary perspective. *Mechanisms of Development*, *69*, 13–29.

Baker, C. V. H., & Bronner-Fraser, M. (2001). Vertebrate cranial placodes. I. Embryonic induction. *Developmental Biology*, *232*, 1–61.

Baker, C. V., O'Neill, P., & McCole, R. B. (2008). Lateral line, otic and epibranchial placodes: Developmental and evolutionary links? *Journal of Experimental Zoology. Part B, Molecular and Developmental Evolution*, *310*, 370–383.

Barraud, P., Seferiadis, A. A., Tyson, L. D., Zwart, M. F., Szabo-Rogers, H. L., Ruhrberg, C., et al. (2010). Neural crest origin of olfactory ensheathing glia. *Proceedings of the National academy of Sciences of the United States of America*, *107*, 21040–21045.

Bassham, S., & Postlethwait, J. H. (2005). The evolutionary history of placodes: A molecular genetic investigation of the larvacean urochordate Oikopleura dioica. *Development*, *132*, 4259–4272.

Beaster-Jones, L., Kaltenbach, S. L., Koop, D., Yuan, S., Chastain, R., & Holland, L. Z. (2008). Expression of somite segmentation genes in amphioxus: A clock without a wavefront? *Development Genes and Evolution*, *218*, 599–611.

Bebenek, I. G., Gates, R. D., Morris, J., Hartenstein, V., & Jacobs, D. K. (2004). sine oculis in basal Metazoa. *Development Genes and Evolution*, *214*, 342–351.

Benito-Gutierrez, E., Illas, M., Comella, J. X., & Garcia-Fernandez, J. (2005). Outlining the nascent nervous system of Branchiostoma floridae (amphioxus) by the pan-neural marker AmphiElav. *Brain Research Bulletin*, *66*, 518–521.

Benito-Gutierrez, E., Nake, C., Llovera, M., Comella, J. X., & Garcia-Fernandez, J. (2005). The single AmphiTrk receptor highlights increased complexity of neurotrophin signalling in vertebrates and suggests an early role in developing sensory neuroepidermal cells. *Development*, *132*, 2191–2202.

Bermingham, N. A., Hassan, B. A., Price, S. D., Vollrath, M. A., Ben Arie, N., Eatock, R. A., et al. (1999). Math1: An essential gene for the generation of inner ear hair cells. *Science*, *284*, 1837–1841.

Bertrand, S., Camasses, A., Somorjai, I., Belgacem, M. R., Chabrol, O., Escande, M. L., et al. (2011). Amphioxus FGF signaling predicts the acquisition of vertebrate morphological traits. *Proceedings of the National academy of Sciences of the United States of America*, *108*, 9160–9165.

Bertrand, N., Castro, D. S., & Guillemot, F. (2002). Proneural genes and the specification of neural cell types. *Nature Reviews. Neuroscience*, *3*, 517–530.

Bertrand, V., Hudson, C., Caillol, D., Popovici, C., & Lemaire, P. (2003). Neural tissue in ascidian embryos is induced by FGF9/16/20, acting via a combination of maternal GATA and Ets transcription factors. *Cell*, *115*, 615–627.

Betancur, P., Bronner-Fraser, M., & Sauka-Spengler, T. (2010). Assembling neural crest regulatory circuits into a gene regulatory network. *Annual Review of Cell and Developmental Biology*, *26*, 581–603.

Bhat, N., Kwon, H. J., & Riley, B. B. (2012). A gene network that coordinates preplacodal competence and neural crest specification in zebrafish. *Developmental Biology*, *373*, 107–117.

Bhattacharyya, S., Bailey, A. P., Bronner-Fraser, M., & Streit, A. (2004). Segregation of lens and olfactory precursors from a common territory: Cell sorting and reciprocity of Dlx5 and Pax6 expression. *Developmental Biology*, *271*, 403–414.

Bicknell, A. B. (2008). The tissue-specific processing of pro-opiomelanocortin. *Journal of Neuroendocrinology*, *20*, 692–699.

Bollner, T., Holmberg, K., & Olsson, R. (1986). A rostral sensory mechanism in Oikopleura dioica (Appendicularia). *Acta Zoologica*, *67*, 235–241.

Bone, Q., & Ryan, K. P. (1978). Cupular sense organs in Ciona (Tunicata: Ascidiacea). *Journal of Zoology (London)*, *186*, 417–429.

Bonini, N. M., Bui, Q. T., Gray-Board, G. L., & Warrick, J. M. (1997). The Drosophila eyes absent gene directs ectopic eye formation in a pathway conserved between flies and vertebrates. *Development*, *124*, 4819–4826.

Bonini, N. M., Leiserson, W. M., & Benzer, S. (1993). The eyes absent gene: Genetic control of cell survival and differentiation in the developing Drosophila eye. *Cell*, *72*, 379–395.

Boorman, C. J., & Shimeld, S. M. (2002). Pitx homeobox genes in Ciona and amphioxus show left-right asymmetry is a conserved chordate character and define the ascidian adenohypophysis. *Evolution & Development*, *4*, 354–365.

Borges, M., Linnoila, R. I., van de Velde, H. J., Chen, H., Nelkin, B. D., Mabry, M., et al. (1997). An achaete-scute homologue essential for neuroendocrine differentiation in the lung. *Nature*, *386*, 852–855.

Bourlat, S. J., Juliusdottir, T., Lowe, C. J., Freeman, R., Aronowicz, J., Kirschner, M., et al. (2006). Deuterostome phylogeny reveals monophyletic chordates and the new phylum Xenoturbellida. *Nature*, *444*, 85–88.

Broccoli, V., Boncinelli, E., & Wurst, W. (1999). The caudal limit of Otx2 expression positions the isthmic organizer. *Nature*, *401*, 164–168.

Brooke, N. M., Garcia-Fernandez, J., & Holland, P. W. H. (1998). The ParaHox cluster is an evolutionary sister of the Hox gene cluster. *Nature*, *392*, 920–922.

Brown, F. D., Prendergast, A., & Swalla, B. J. (2008). Man is but a worm: Chordate origins. *Genesis*, *46*, 605–613.

Brugmann, S. A., Pandur, P. D., Kenyon, K. L., Pignoni, F., & Moody, S. A. (2004). Six1 promotes a placodal fate within the lateral neurogenic ectoderm by functioning as both a transcriptional activator and repressor. *Development*, *131*, 5871–5881.

Budelmann, B. U. (1989). Hydrodynamic receptor systems in invertebrates. In S. Coombs, P. Görner, & H. Münz (Eds.), *The mechanosensory lateral line* (pp. 607–631). New York: Springer.

Bullock, T. H. (1945). The anatomical organization of the nervous system of Enteropneusta. *The Quarterly Journal of Microscopical Science*, *86*, 55–111.

Bullock, T. H., & Horridge, G. A. (1965). *Structure and function in the nervous systems of invertebrates*. San Francisco: Freeman.

Burighel, P., Caicci, F., & Manni, L. (2011). Hair cells in non-vertebrate models: Lower chordates and molluscs. *Hearing Research*, *273*, 14–24.

Burighel, P., Lane, N. J., Fabio, G., Stefano, T., Zaniolo, G., Carnevali, M. D., et al. (2003). Novel, secondary sensory cell organ in ascidians: In search of the ancestor of the vertebrate lateral line. *The Journal of Comparative Neurology*, *461*, 236–249.

Caicci, F., Gasparini, F., Rigon, F., Zaniolo, G., Burighel, P., & Manni, L. (2013). The oral sensory structures of Thaliacea (Tunicata) and consideration of the evolution of hair cells in Chordata. *The Journal of Comparative Neurology*, *521*, 2756–2771.

Caicci, F., Zaniolo, G., Burighel, P., Degasperi, V., Gasparini, F., & Manni, L. (2010). Differentiation of papillae and rostral sensory neurons in the larva of the ascidian Botryllus schlosseri (Tunicata). *The Journal of Comparative Neurology*, *518*, 547–566.

Cameron, C. B., Mackie, G. O., Powell, J. F., Lescheid, D. W., & Sherwood, N. M. (1999). Gonadotropin-releasing hormone in mulberry cells of Saccoglossus and Ptychodera (Hemichordata: Enteropneusta). *General and Comparative Endocrinology*, *114*, 2–10.

Campbell, R. K., Satoh, N., & Degnan, B. M. (2004). Piecing together evolution of the vertebrate endocrine system. *Trends in Genetics*, *20*, 359–366.

Candiani, S., Holland, N. D., Oliveri, D., Parodi, M., & Pestarino, M. (2008). Expression of the amphioxus Pit-1 gene (AmphiPOU1F1/Pit-1) exclusively in the developing preoral organ, a putative homolog of the vertebrate adenohypophysis. *Brain Research Bulletin*, *75*, 324–330.

Candiani, S., Oliveri, D., Parodi, M., Bertini, E., & Pestarino, M. (2006). Expression of AmphiPOU-IV in the developing neural tube and epidermal sensory neural precursors in amphioxus supports a conserved role of class IV POU genes in the sensory cells development. *Development Genes and Evolution*, *216*, 623–633.

Candiani, S., Pennati, R., Oliveri, D., Locascio, A., Branno, M., Castagnola, P., et al. (2005). Ci-POU-IV expression identifies PNS neurons in embryos and larvae of the ascidian Ciona intestinalis. *Development Genes and Evolution*, *215*, 41–45.

Canestro, C., & Postlethwait, J. H. (2007). Development of a chordate anterior-posterior axis without classical retinoic acid signaling. *Developmental Biology*, *305*, 522–538.

Canning, C. A., Lee, L., Luo, S. X., Graham, A., & Jones, C. M. (2008). Neural tube derived Wnt signals cooperate with FGF signaling in the formation and differentiation of the trigeminal placodes. *Neural Development*, *3*, 35.

Caracciolo, A., DiGregorio, A., Aniello, F., Dilauro, R., & Branno, M. (2000). Identification and developmental expression of three Distal-less homeobox containing genes in the ascidian Ciona intestinalis. *Mechanisms of Development*, *99*, 173–176.

Castro, A., Becerra, M., Manso, M. J., Sherwood, N. M., & Anadon, R. (2006). Anatomy of the Hesse photoreceptor cell axonal system in the central nervous system of amphioxus. *The Journal of Comparative Neurology, 494*, 54–62.

Castro, L. F., Rasmussen, S. L., Holland, P. W., Holland, N. D., & Holland, L. Z. (2006). A Gbx homeobox gene in amphioxus: Insights into ancestry of the ANTP class and evolution of the midbrain/hindbrain boundary. *Developmental Biology, 295*, 40–51.

Cau, E., Casarosa, S., & Guillemot, F. (2002). Mash1 and Ngn1 control distinct steps of determination and differentiation in the olfactory sensory neuron lineage. *Development, 129*, 1871–1880.

Cau, E., Gradwohl, G., Fode, C., & Guillemot, F. (1997). Mash1 activates a cascade of bhlh regulators in olfactory neuron progenitors. *Development, 124*, 1611–1621.

Cau, E., & Wilson, S. W. (2003). Ash1a and Neurogenin1 function downstream of Floating head to regulate epiphysial neurogenesis. *Development, 130*, 2455–2466.

Chambery, A., Parente, A., Topo, E., Garcia-Fernandez, J., & D'Aniello, S. (2009). Characterization and putative role of a type I gonadotropin-releasing hormone in the cephalochordate amphioxus. *Endocrinology, 150*, 812–820.

Chen, P., Johnson, J. E., Zoghbi, H. Y., & Segil, N. (2002). The role of Math1 in inner ear development: Uncoupling the establishment of the sensory primordium from hair cell fate determination. *Development, 129*, 2495–2505.

Chen, B., Kim, E. H., & Xu, P. X. (2009). Initiation of olfactory placode development and neurogenesis is blocked in mice lacking both Six1 and Six4. *Developmental Biology, 326*, 75–85.

Chevalier, S., Martin, A., Leclere, L., Amiel, A., & Houliston, E. (2006). Polarised expression of FoxB and FoxQ2 genes during development of the hydrozoan Clytia hemisphaerica. *Development Genes and Evolution, 216*, 709–720.

Cheyette, B. N., Green, P. J., Martin, K., Garren, H., Hartenstein, V., & Zipursky, S. L. (1994). The Drosophila sine oculis locus encodes a homeodomain-containing protein required for the development of the entire visual system. *Neuron, 12*, 977–996.

Christiaen, L., Burighel, P., Smith, W. C., Vernier, P., Bourrat, F., & Joly, J. S. (2002). Pitx genes in Tunicates provide new molecular insight into the evolutionary origin of pituitary. *Gene, 287*, 107–113.

Christodoulou, F., Raible, F., Tomer, R., Simakov, O., Trachana, K., Klaus, S., et al. (2010). Ancient animal microRNAs and the evolution of tissue identity. *Nature, 463*, 1084–1088.

Churcher, A. M., & Taylor, J. S. (2009). Amphioxus (Branchiostoma floridae) has orthologs of vertebrate odorant receptors. *BMC Evolutionary Biology, 9*, 242.

Corbo, J. C., Erives, A., DiGregorio, A., Chang, A., & Levine, M. (1997). Dorsoventral patterning of the vertebrate neural tube is conserved in a protochordate. *Development, 124*, 2335–2344.

Cvekl, A., & Duncan, M. K. (2007). Genetic and epigenetic mechanisms of gene regulation during lens development. *Progress in Retinal and Eye Research, 26*, 555–597.

Darras, S., Gerhart, J., Terasaki, M., Kirschner, M., & Lowe, C. J. (2011). beta-Catenin specifies the endomesoderm and defines the posterior organizer of the hemichordate Saccoglossus kowalevskii. *Development, 138*, 959–970.

Darras, S., & Nishida, H. (2001a). The BMP signaling pathway is required together with the FGF pathway for notochord induction in the ascidian embryo. *Development, 128*, 2629–2638.

Darras, S., & Nishida, H. (2001b). The BMP/CHORDIN antagonism controls sensory pigment cell specification and differentiation in the ascidian embryo. *Developmental Biology, 236*, 271–288.

D'Autreaux, F., Coppola, E., Hirsch, M. R., Birchmeier, C., & Brunet, J. F. (2011). Homeoprotein Phox2b commands a somatic-to-visceral switch in cranial sensory pathways.

Proceedings of the National academy of Sciences of the United States of America, 108, 20018–20023.
Dawkins, R. (2004). *The Ancestor's tale*. London: Weidenfeld and Nicolson.
Decatur, W. A., Hall, J. A., Smith, J. J., Li, W., & Sower, S. A. (2013). Insight from the lamprey genome: Glimpsing early vertebrate development via neuroendocrine-associated genes and shared synteny of gonadotropin-releasing hormone (GnRH). *General and Comparative Endocrinology, 192*, 237–245.
Dehal, P., Satou, Y., Campbell, R. K., Chapman, J., Degnan, B., De Tomaso, A., et al. (2002). The draft genome of Ciona intestinalis: Insights into chordate and vertebrate origins. *Science, 298*, 2157–2167.
Delsuc, F., Brinkmann, H., Chourrout, D., & Philippe, H. (2006). Tunicates and not cephalochordates are the closest living relatives of vertebrates. *Nature, 439*, 965–968.
Denes, A. S., Jekely, G., Steinmetz, P. R., Raible, F., Snyman, H., Prud'homme, B., et al. (2007). Molecular architecture of annelid nerve cord supports common origin of nervous system centralization in bilateria. *Cell, 129*, 277–288.
De Robertis, E. M., & Sasai, Y. (1996). A common plan for dorsoventral patterning in Bilateria. *Nature, 380*, 37–40.
De Velasco, B., Erclik, T., Shy, D., Sclafani, J., Lipshitz, H., McInnes, R., et al. (2007). Specification and development of the pars intercerebralis and pars lateralis, neuroendocrine command centers in the Drosophila brain. *Developmental Biology, 302*, 309–323.
De Velasco, B., Shen, J., Go, S., & Hartenstein, V. (2004). Embryonic development of the Drosophila corpus cardiacum, a neuroendocrine gland with similarity to the vertebrate pituitary, is controlled by sine oculis and glass. *Developmental Biology, 274*, 280–294.
Dores, R. M., & Baron, A. J. (2011). Evolution of POMC: Origin, phylogeny, posttranslational processing, and the melanocortins. *Annals of the New York Academy of Sciences, 1220*, 34–48.
Dos Santos, S., Bardet, C., Bertrand, S., Escriva, H., Habert, D., & Querat, B. (2009). Distinct expression patterns of glycoprotein hormone-alpha2 and -beta5 in a basal chordate suggest independent developmental functions. *Endocrinology, 150*, 3815–3822.
Dos Santos, S., Mazan, S., Venkatesh, B., Cohen-Tannoudji, J., & Querat, B. (2011). Emergence and evolution of the glycoprotein hormone and neurotrophin gene families in vertebrates. *BMC Evolutionary Biology, 11*, 332.
Duffy, D. J., Plickert, G., Kuenzel, T., Tilmann, W., & Frank, U. (2010). Wnt signaling promotes oral but suppresses aboral structures in Hydractinia metamorphosis and regeneration. *Development, 137*, 3057–3066.
Dutta, S., Dietrich, J. E., Aspock, G., Burdine, R. D., Schier, A., Westerfield, M., et al. (2005). pitx3 defines an equivalence domain for lens and anterior pituitary placode. *Development, 132*, 1579–1590.
Dykes, I. M., Tempest, L., Lee, S. I., & Turner, E. E. (2011). Brn3a and Islet1 act epistatically to regulate the gene expression program of sensory differentiation. *The Journal of Neuroscience, 31*, 9789–9799.
Eijkelkamp, N., Quick, K., & Wood, J. N. (2013). Transient receptor potential channels and mechanosensation. *Annual Review of Neuroscience, 36*, 519–546.
Eisthen, H. L. (1997). Evolution of vertebrate olfactory systems. *Brain, Behavior and Evolution, 50*, 222–233.
Elsaesser, R., & Paysan, J. (2007). The sense of smell, its signalling pathways, and the dichotomy of cilia and microvilli in olfactory sensory cells. *BMC Neuroscience, 8*(Suppl. 3), S1.
Eng, S. R., Dykes, I. M., Lanier, J., Fedtsova, N., & Turner, E. E. (2007). POU-domain factor Brn3a regulates both distinct and common programs of gene expression in the spinal and trigeminal sensory ganglia. *Neural Development, 2*, 3.

Escriva, H., Holland, N. D., Gronemeyer, H., Laudet, V., & Holland, L. Z. (2002). The retinoic acid signaling pathway regulates anterior/posterior patterning in the nerve cord and pharynx of amphioxus, a chordate lacking neural crest. *Development, 129*, 2905–2916.

Feledy, J. A., Beanan, M. J., Sandoval, J. J., Goodrich, J. S., Lim, J. H., Matsuo-Takasaki, M., et al. (1999). Inhibitory patterning of the anterior neural plate in Xenopus by homeodomain factors Dlx3 and Msx1. *Developmental Biology, 212*, 455–464

Fode, C., Gradwohl, G., Morin, X., Dierich, A., Lemeur, M., Goridis, C., et al. (1998). The bHLH protein neurogenin 2 is a determination factor for epibranchial placode-derived sensory neurons. *Neuron, 20*, 483–494.

Forni, P. E., Taylor-Burds, C., Melvin, V. S., Williams, T., & Wray, S. (2011). Neural crest and ectodermal cells intermix in the nasal placode to give rise to GnRH-1 neurons, sensory neurons, and olfactory ensheathing cells. *The Journal of Neuroscience, 31*, 6915–6927.

Frederikse, P. H., Kasinathan, C., & Kleiman, N. J. (2012). Parallels between neuron and lens fiber cell structure and molecular regulatory networks. *Developmental Biology, 368*, 255–260.

Fredriksson, R., & Schiöth, H. B. (2005). The repertoire of G-protein-coupled receptors in fully sequenced genomes. *Molecular Pharmacology, 67*, 1414–1425.

Freter, S., Muta, Y., Mak, S. S., Rinkwitz, S., & Ladher, R. K. (2008). Progressive restriction of otic fate: The role of FGF and Wnt in resolving inner ear potential. *Development, 135*, 3415–3424.

Friedle, H., & Knöchel, W. (2002). Cooperative interaction of Xvent-2 and GATA-2 in the activation of the ventral homeobox gene Xvent-1B. *The Journal of Biological Chemistry, 277*, 23872–23881.

Fritzenwanker, J. H., Gerhart, J., Freeman, R. M., Jr., & Lowe, C. J. (2014). The Fox/Forkhead transcription factor family of the hemichordate Saccoglossus kowalevskii. *EvoDevo, 5*, 17.

Fritzsch, B., Beisel, K. W., & Bermingham, N. A. (2000). Developmental evolutionary biology of the vertebrate ear: Conserving mechanoelectric transduction and developmental pathways in diverging morphologies. *Neuroreport, 11*, R35–R44.

Fritzsch, B., Beisel, K. W., Pauley, S., & Soukup, G. (2007). Molecular evolution of the vertebrate mechanosensory cell and ear. *The International Journal of Developmental Biology, 51*, 663–678.

Fritzsch, B., Eberl, D. F., & Beisel, K. W. (2010). The role of bHLH genes in ear development and evolution: Revisiting a 10-year-old hypothesis. *Cellular and Molecular Life Sciences, 67*, 3089–3099.

Fritzsch, B., Piatigorsky, J., Tessmar-Raible, K., Jekely, G., Guy, K., Raible, F., et al. (2005). Ancestry of photic and mechanic sensation? *Science, 308*, 1113–1114.

Galliot, B., Quiquand, M., Ghila, L., de Rosa, R., Miljkovic-Licina, M., & Chera, S. (2009). Origins of neurogenesis, a cnidarian view. *Developmental Biology, 332*, 2–24.

Gillis, J. A., Fritzenwanker, J. H., & Lowe, C. J. (2012). A stem-deuterostome origin of the vertebrate pharyngeal transcriptional network. *Proceedings of the Biological Sciences, 279*, 237–246.

Giuliano, P., Marino, R., Pinto, M. R., & De Santis, R. (1998). Identification and developmental expression of Ci-isl, a homologue of vertebrate islet genes, in the ascidian Ciona intestinalis. *Mechanisms of Development, 78*, 199–202.

Glardon, S., Callaerts, P., Halder, G., & Gehring, W. J. (1997). Conservation of Pax-6 in a lower chordate, the ascidian Phallusia mammillata. *Development, 124*, 817–825.

Glardon, S., Holland, L. Z., Gehring, W. J., & Holland, N. D. (1998). Isolation and developmental expression of the amphioxus pax-6 gene (amphipax-6): Insights into eye and photoreceptor evolution. *Development, 125*, 2701–2710.

Gorbman, A. (1995). Olfactory origins and evolution of the brain-pituitary endocrine system: Facts and speculation. *General and Comparative Endocrinology, 97*, 171–178.

Gostling, N. J., & Shimeld, S. M. (2003). Protochordate Zic genes define primitive somite compartments and highlight molecular changes underlying neural crest evolution. *Evolution & Development, 5*, 136–144.

Graham, A., & Shimeld, S. M. (2013). The origin and evolution of the ectodermal placodes. *Journal of Anatomy, 222*, 32–40.

Grande, C., & Patel, N. H. (2009). Lophotrochozoa get into the game: The nodal pathway and left/right asymmetry in bilateria. *Cold Spring Harbor Symposia on Quantitative Biology, 74*, 281–287.

Graziussi, D. F., Suga, H., Schmid, V., & Gehring, W. J. (2012). The "eyes absent" (eya) gene in the eye-bearing hydrozoan jellyfish Cladonema radiatum: Conservation of the retinal determination network. *Journal of Experimental Zoology. Part B, Molecular and Developmental Evolution, 318*, 257–267.

Grocott, T., Tambalo, M., & Streit, A. (2012). The peripheral sensory nervous system in the vertebrate head: A gene regulatory perspective. *Developmental Biology, 370*, 3–23.

Groger, H., Callaerts, P., Gehring, W. J., & Schmid, V. (2000). Characterization and expression analysis of an ancestor-type Pax gene in the hydrozoan jellyfish Podocoryne carnea. *Mechanisms of Development, 94*, 157–169.

Groves, A. K., & LaBonne, C. (2014). Setting appropriate boundaries: Fate, patterning and competence at the neural plate border. *Developmental Biology, 389*, 2–12.

Halder, G., Callaerts, P., Flister, S., Walldorf, U., Kloter, U., & Gehring, W. J. (1998). Eyeless initiates the expression of both sine oculis and eyes absent during Drosophila compound eye development. *Development, 125*, 2181–2191.

Hallgrimsson, B., Jamniczky, H. A., Young, N. M., Rolian, C., Schmidt-Ott, U., & Marcucio, R. S. (2012). The generation of variation and the developmental basis for evolutionary novelty. *Journal of Experimental Zoology. Part B, Molecular and Developmental Evolution, 318*, 501–517.

Hamada, M., Shimozono, N., Ohta, N., Satou, Y., Horie, T., Kawada, T., et al. (2011). Expression of neuropeptide- and hormone-encoding genes in the Ciona intestinalis larval brain. *Developmental Biology, 352*, 202–214.

Hartenstein, V. (2006). The neuroendocrine system of invertebrates: A developmental and evolutionary perspective. *The Journal of Endocrinology, 190*, 555–570.

Hayakawa, E., Fujisawa, C., & Fujisawa, T. (2004). Involvement of Hydra achaete-scute gene CnASH in the differentiation pathway of sensory neurons in the tentacles. *Development Genes and Evolution, 214*, 486–492.

Hill, A., Boll, W., Ries, C., Warner, L., Osswalt, M., Hill, M., et al. (2010). Origin of Pax and Six gene families in sponges: Single PaxB and Six1/2 orthologs in Chalinula loosanoffi. *Developmental Biology, 343*, 106–123.

Hirth, F., Kammermeier, L., Frei, E., Walldorf, U., Noll, M., & Reichert, H. (2003). An urbilaterian origin of the tripartite brain: Developmental genetic insights from Drosophila. *Development, 130*, 2365–2373.

Holland, L. Z. (2002). Heads or tails? Amphioxus and the evolution of anterior- posterior patterning in deuterostomes. *Developmental Biology, 241*, 209–228.

Holland, N. D. (2003). Early central nervous system evolution: An era of skin brains? *Nature Reviews. Neuroscience, 4*, 617–627.

Holland, L. Z. (2005). Non-neural ectoderm is really neural: Evolution of developmental patterning mechanisms in the non-neural ectoderm of chordates and the problem of sensory cell homologies. *Journal of Experimental Zoology. Part B, Molecular and Developmental Evolution, 304B*, 304–323.

Holland, L. Z. (2014). Genomics, evolution and development of amphioxus and tunicates: The goldilocks principle. *Journal of Experimental Zoology. Part B, Molecular and Developmental Evolution.* http://dx.doi.org/10.1002/jez.b.22569.

Holland, L. Z., Albalat, R., Azumi, K., Benito-Gutierrez, E., Blow, M. J., Bronner-Fraser, M., et al. (2008). The amphioxus genome illuminates vertebrate origins and cephalochordate biology. *Genome Research*, *18*, 1100–1111.

Holland, L. Z., Carvalho, J. E., Escriva, H., Laudet, V., Schubert, M., Shimeld, S. M., et al. (2013). Evolution of bilaterian central nervous systems: A single origin? *EvoDevo*, *4*, 27.

Holland, L. Z., & Holland, N. D. (2001). Evolution of neural crest and placodes: Amphioxus as a model for the ancestral vertebrate? *Journal of Anatomy*, *199*, 85–98.

Holland, N. D., Panganiban, G., Henyey, E. L., & Holland, L. Z. (1996). Sequence and developmental expression of AmphiDll, an amphioxus distalless gene trabscribed in the ectoderm, epidermis and nervous system: Insights into evolution of craniate forebrain and neural crest. *Development*, *122*, 2911–2920.

Holland, L. Z., Schubert, M., Holland, N. D., & Neuman, T. (2000). Evolutionary conservation of the presumptive neural plate markers AmphiSox1/2/3 and AmphiNeurogenin in the invertebrate chordate amphioxus. *Developmental Biology*, *226*, 18–33.

Holland, L. Z., Schubert, M., Kozmik, Z., & Holland, N. D. (1999). AmphiPax3/7, an amphioxus paired box gene: Insights into chordate myogenesis, neurogenesis, and the possible evolutionary precursor of definitive vertebrate neural crest. *Evolution & Development*, *1*, 153–165.

Holstein, T. W. (2012). A view to kill. *BMC Biology*, *10*, 18.

Holstein, T. W., Watanabe, H., & Ozbek, S. (2011). Signaling pathways and axis formation in the lower metazoa. *Current Topics in Developmental Biology*, *97*, 137–177.

Hroudova, M., Vojta, P., Strnad, H., Krejcik, Z., Ridl, J., Paces, J., et al. (2012). Diversity, phylogeny and expression patterns of Pou and Six homeodomain transcription factors in hydrozoan jellyfish Craspedacusta sowerbyi. *PLoS One*, 7, e36420.

Huber, K., Combs, S., Ernsberger, U., Kalcheim, C., & Unsicker, K. (2002). Generation of neuroendocrine chromaffin cells from sympathoadrenal progenitors: Beyond the glucocorticoid hypothesis. *Annals of the New York Academy of Sciences*, *971*, 554–559.

Hudson, C., Darras, S., Caillol, D., Yasuo, H., & Lemaire, P. (2003). A conserved role for the MEK signalling pathway in neural tissue specification and posteriorisation in the invertebrate chordate, the ascidian Ciona intestinalis. *Development*, *130*, 147–159.

Hudson, C., & Lemaire, P. (2001). Induction of anterior neural fates in the ascidian Ciona intestinalis. *Mechanisms of Development*, *100*, 189–203.

Hudson, C., Lotito, S., & Yasuo, H. (2007). Sequential and combinatorial inputs from Nodal, Delta2/Notch and FGF/MEK/ERK signalling pathways establish a grid-like organisation of distinct cell identities in the ascidian neural plate. *Development*, *134*, 3527–3537.

Huising, M. O., Kruiswijk, C. P., & Flik, G. (2006). Phylogeny and evolution of class-I helical cytokines. *The Journal of Endocrinology*, *189*, 1–25.

Ikeda, T., Matsuoka, T., & Satou, Y. (2013). A time delay gene circuit is required for palp formation in the ascidian embryo. *Development*, *140*, 4703–4708.

Ikuta, T., Yoshida, N., Satoh, N., & Saiga, H. (2004). Ciona intestinalis Hox gene cluster: Its dispersed structure and residual colinear expression in development. *Proceedings of the National academy of Sciences of the United States of America*, *101*, 15118–15123.

Imai, K. S., Hino, K., Yagi, K., Satoh, N., & Satou, Y. (2004). Gene expression profiles of transcription factors and signaling molecules in the ascidian embryo: Towards a comprehensive understanding of gene networks. *Development*, *131*, 4047–4058.

Imai, K. S., Levine, M., Satoh, N., & Satou, Y. (2006). Regulatory blueprint for a chordate embryo. *Science*, *312*, 1183–1187.

Imai, K. S., Satoh, N., & Satou, Y. (2002). Region specific gene expressions in the central nervous system of the ascidian embryo. *Mechanisms of Development*, *119*(Suppl. 1), S275–S277.

Imai, K. S., Satoh, N., & Satou, Y. (2003). A Twist-like bHLH gene is a downstream factor of an endogenous FGF and determines mesenchymal fate in the ascidian embryos. *Development, 130*, 4461–4472.

Imai, K. S., Stolfi, A., Levine, M., & Satou, Y. (2009). Gene regulatory networks underlying the compartmentalization of the Ciona central nervous system. *Development, 136*, 285–293.

Irimia, M., Pineiro, C., Maeso, I., Gomez-Skarmeta, J. L., Casares, F., & Garcia-Fernandez, J. (2010). Conserved developmental expression of Fezf in chordates and Drosophila and the origin of the Zona Limitans Intrathalamica (ZLI) brain organizer. *EvoDevo, 1*, 7.

Irvine, S. Q., Cangiano, M. C., Millette, B. J., & Gutter, E. S. (2007). Non-overlapping expression patterns of the clustered Dll-A/B genes in the ascidian Ciona intestinalis. *Journal of Experimental Zoology. Part B, Molecular and Developmental Evolution, 308*, 428–441.

Irvine, S. Q., Vierra, D. A., Millette, B. J., Blanchette, M. D., & Holbert, R. E. (2011). Expression of the Distalless-B gene in Ciona is regulated by a pan-ectodermal enhancer module. *Developmental Biology, 353*, 432–439.

Ivashkin, E., & Adameyko, I. (2013). Progenitors of the protochordate ocellus as an evolutionary origin of the neural crest. *EvoDevo, 4*, 12.

Jackman, W. R., Langeland, J. A., & Kimmel, C. B. (2000). islet reveals segmentation in the amphioxus hindbrain homolog. *Developmental Biology, 220*, 16–26.

Jacob, F. (1977). Evolution and tinkering. *Science, 196*, 1161–1166.

Jacobs, D. K., Nakanishi, N., Yuan, D., Camara, A., Nichols, S. A., & Hartenstein, V. (2007). Evolution of sensory structures in basal metazoa. *Integrative and Comparative Biology, 47*, 712–723.

Jefferies, R. P. S. (1986). *The ancestry of vertebrates*. London: British Museum (Natural History).

Jekely, G. (2013). Global view of the evolution and diversity of metazoan neuropeptide signaling. *Proceedings of the National academy of Sciences of the United States of America, 110*, 8702–8707.

Jemc, J., & Rebay, I. (2007). The eyes absent family of phosphotyrosine phosphatases: Properties and roles in developmental regulation of transcription. *Annual Review of Biochemistry, 76*, 24.21–24.26.

Jørgensen, J. M. (1989). Evolution of octavolateralis cells. In S. Coombs, P. Görner, & H. Münz (Eds.), *The mechanosensory lateral line* (pp. 115–145). New York: Springer.

Joyce Tang, W., Chen, J. S., & Zeller, R. W. (2013). Transcriptional regulation of the peripheral nervous system in Ciona intestinalis. *Developmental Biology, 378*, 183–193.

Kaltenbach, S. L., Holland, L. Z., Holland, N. D., & Koop, D. (2009). Developmental expression of the three iroquois genes of amphioxus (BfIrxA, BfIrxB, and BfIrxC) with special attention to the gastrula organizer and anteroposterior boundaries in the central nervous system. *Gene Expression Patterns, 9*, 329–334.

Kaltenbach, S. L., Yu, J. K., & Holland, N. D. (2009). The origin and migration of the earliest-developing sensory neurons in the peripheral nervous system of amphioxus. *Evolution & Development, 11*, 142–151.

Kanda, M., Wada, H., & Fujiwara, S. (2009). Epidermal expression of Hox1 is directly activated by retinoic acid in the Ciona intestinalis embryo. *Developmental Biology, 335*, 454–463.

Kano, S. (2010). Genomics and developmental approaches to an ascidian adenohypophysis primordium. *Integrative and Comparative Biology, 50*, 35–52.

Karlstrom, R. O., Talbot, W. S., & Schier, A. F. (1999). Comparative synteny cloning of zebrafish you-too: Mutations in the Hedgehog target gli2 affect ventral forebrain patterning. *Genes & Development, 13*, 388–393.

Kaul, S., & Stach, T. (2010). Ontogeny of the collar cord: Neurulation in the hemichordate Saccoglossus kowalevskii. *Journal of Morphology, 271*, 1240–1259.

Kaupp, U. B. (2010). Olfactory signalling in vertebrates and insects: Differences and commonalities. *Nature Reviews. Neuroscience*, *11*, 188–200.

Kavanaugh, S. I., Root, A. R., & Sower, S. A. (2005). Distribution of gonadotropin-releasing hormone (GnRH) by in situ hybridization in the tunicate Ciona intestinalis. *General and Comparative Endocrinology*, *141*, 76–83.

Kawauchi, H., & Sower, S. A. (2006). The dawn and evolution of hormones in the adenohypophysis. *General and Comparative Endocrinology*, *148*, 3–14.

Kelberman, D., Rizzoti, K., Lovell-Badge, R., Robinson, I. C., & Dattani, M. T. (2009). Genetic regulation of pituitary gland development in human and mouse. *Endocrine Reviews*, *30*, 790–829.

Knight-Jones, E. W. (1952). On the nervous system of Saccoglossus cambrensis (Enteropneusta). *Philosophical Transactions of the Royal Society of London. Series B*, *236*, 315–354.

Koop, D., Holland, N. D., Semon, M., Alvarez, S., de Lera, A. R., Laudet, V., et al. (2010). Retinoic acid signaling targets Hox genes during the amphioxus gastrula stage: Insights into early anterior-posterior patterning of the chordate body plan. *Developmental Biology*, *338*, 98–106.

Kourakis, M. J., Newman-Smith, E., & Smith, W. C. (2010). Key steps in the morphogenesis of a cranial placode in an invertebrate chordate, the tunicate Ciona savignyi. *Developmental Biology*, *340*, 134–144.

Kourakis, M. J., & Smith, W. C. (2007). A conserved role for FGF signaling in chordate otic/atrial placode formation. *Developmental Biology*, *312*, 257.

Kozlowski, D. J., Whitfield, T. T., Hukriede, N. A., Lam, W. K., & Weinberg, E. S. (2005). The zebrafish dog-eared mutation disrupts eya1, a gene required for cell survival and differentiation in the inner ear and lateral line. *Developmental Biology*, *277*, 27–41.

Kozmik, Z., Daube, M., Frei, E., Norman, B., Kos, L., Dishaw, L. J., et al. (2003). Role of Pax genes in eye evolution: A cnidarian PaxB gene uniting Pax2 and Pax6 functions. *Developmental Cell*, *5*, 773–785.

Kozmik, Z., Holland, N. D., Kalousova, A., Paces, J., Schubert, M., & Holland, L. Z. (1999). Characterization of an amphioxus paired box gene, AmphiPax2/5/8: Developmental expression patterns in optic support cells, nephridium, thyroid-like structures and pharyngeal gill slits, but not in the midbrain-hindbrain boundary region. *Development*, *126*, 1295–1304.

Kozmik, Z., Holland, N. D., Kreslova, J., Oliveri, D., Schubert, M., Jonasova, K., et al. (2007). Pax-Six-Eya-Dach network during amphioxus development: Conservation in vitro but context-specificity in vivo. *Developmental Biology*, *306*, 143–159.

Kozmikova, I., Candiani, S., Fabian, P., Gurska, D., & Kozmik, Z. (2013). Essential role of Bmp signaling and its positive feedback loop in the early cell fate evolution of chordates. *Developmental Biology*, *382*, 538–554.

Kozmikova, I., Smolikova, J., Vlcek, C., & Kozmik, Z. (2011). Conservation and diversification of an ancestral chordate gene regulatory network for dorsoventral patterning. *PLoS One*, *6*, e14650.

Krishnan, A., Almen, M. S., Fredriksson, R., & Schiöth, H. B. (2013). Remarkable similarities between the hemichordate (Saccoglossus kowalevskii) and vertebrate GPCR repertoire. *Gene*, *526*, 122–133.

Kumar, J. P. (2009). The sine oculis homeobox (SIX) family of transcription factors as regulators of development and disease. *Cellular and Molecular Life Sciences*, *66*, 565–583.

Kusakabe, T. G., Sakai, T., Aoyama, M., Kitajima, Y., Miyamoto, Y., Takigawa, T., et al. (2012). A conserved non-reproductive GnRH system in chordates. *PLoS One*, *7*, e41955.

Kwon, H. J., Bhat, N., Sweet, E. M., Cornell, R. A., & Riley, B. B. (2010). Identification of early requirements for preplacodal ectoderm and sensory organ development. *PLoS Genetics*, *6*, e1001133.

Lacalli, T. C. (2004). Sensory systems in amphioxus: A window on the ancestral chordate condition. *Brain, Behavior and Evolution, 64*, 148–162.

Lacalli, T. C., & Hou, S. (1999). A reexamination of the epithelial sensory cells of amphioxus (Branchiostoma). *Acta Zoologica (Stockholm), 80*, 125–134.

Laclef, C., Souil, E., Demignon, J., & Maire, P. (2003). Thymus, kidney and craniofacial abnormalities in Six1 deficient mice. *Mechanisms of Development, 120*, 669–679.

Ladher, R. K., Anakwe, K. U., Gurney, A. L., Schoenwolf, G. C., & Francis-West, P. H. (2000). Identification of synergistic signals initiating inner ear development. *Science, 290*, 1965–1967.

Ladher, R. K., O'Neill, P., & Begbie, J. (2010). From shared lineage to distinct functions: The development of the inner ear and epibranchial placodes. *Development, 137*, 1777–1785.

Langeland, J. A., Tomsa, J. M., Jackman, W. R., & Kimmel, C. B. (1998). An amphioxus snail gene: Expression in paraxial mesoderm and neural plate suggests a conserved role in patterning the chordate embryo. *Development Genes and Evolution, 208*, 569–577.

Lassiter, R. N., Dude, C. M., Reynolds, S. B., Winters, N. I., Baker, C. V., & Stark, M. R. (2007). Canonical Wnt signaling is required for ophthalmic trigeminal placode cell fate determination and maintenance. *Developmental Biology, 308*, 392–406.

Laverack, M. S. (1988). The diversity of chemoreceptors. In J. Atema, R. R. Fay, A. N. Popper, & W. N. Tavolga (Eds.), *Sensory systems of aquatic animals* (pp. 287–312). New York: Springer.

Lemaire, P. (2009). Unfolding a chordate developmental program, one cell at a time: Invariant cell lineages, short-range inductions and evolutionary plasticity in ascidians. *Developmental Biology, 332*, 48–60.

Lemaire, P., Bertrand, V., & Hudson, C. (2002). Early steps in the formation of neural tissue in ascidian embryos. *Developmental Biology, 252*, 151–169.

Lemaire, P., Smith, W. C., & Nishida, H. (2008). Ascidians and the plasticity of the chordate developmental program. *Current Biology, 18*, R620–R631.

Leonard, J. H., Cook, A. L., Van Gele, M., Boyle, G. M., Inglis, K. J., Speleman, F., et al. (2002). Proneural and proneuroendocrine transcription factor expression in cutaneous mechanoreceptor (Merkel) cells and Merkel cell carcinoma. *International Journal of Cancer, 101*, 103–110.

Li, X., Oghi, K. A., Zhang, J., Krones, A., Bush, K. T., Glass, C. K., et al. (2003). Eya protein phosphatase activity regulates Six1-Dach-Eya transcriptional effects in mammalian organogenesis. *Nature, 426*, 247–254.

Li, H. J., Ray, S. K., Singh, N. K., Johnston, B., & Leiter, A. B. (2011). Basic helix-loop-helix transcription factors and enteroendocrine cell differentiation. *Diabetes, Obesity & Metabolism, 13*(Suppl. 1), 5–12.

Litsiou, A., Hanson, S., & Streit, A. (2005). A balance of FGF, BMP and WNT signalling positions the future placode territory in the head. *Development, 132*, 4051–4062.

Lleras-Forero, L., Tambalo, M., Christophorou, N., Chambers, D., Houart, C., & Streit, A. (2013). Neuropeptides: Developmental signals in placode progenitor formation. *Developmental Cell, 26*, 195–203.

Logan, C. Y., Miller, J. R., Ferkowicz, M. J., & McClay, D. R. (1999). Nuclear beta-catenin is required to specify vegetal cell fates in the sea urchin embryo. *Development, 126*, 345–357.

Lowe, C. J., Terasaki, M., Wu, M., Freeman, R. M., Jr., Runft, L., Kwan, K., et al. (2006). Dorsoventral patterning in hemichordates: Insights into early chordate evolution. *PLoS Biology, 4*, 1602–1619.

Lowe, C. J., Wu, M., Salic, A., Evans, L., Lander, E., Stange-Thomann, N., et al. (2003). Anteroposterior patterning in hemichordates and the origins of the chordate nervous system. *Cell, 113*, 853–865.

Lu, T. M., Luo, Y. J., & Yu, J. K. (2012). BMP and Delta/Notch signaling control the development of amphioxus epidermal sensory neurons: Insights into the evolution of the peripheral sensory system. *Development, 139*, 2020–2030.

Ludeman, D. A., Farrar, N., Riesgo, A., Paps, J., & Leys, S. P. (2014). Evolutionary origins of sensation in metazoans: Functional evidence for a new sensory organ in sponges. *BMC Evolutionary Biology, 14*, 3.

Lumpkin, E. A., & Caterina, M. J. (2007). Mechanisms of sensory transduction in the skin. *Nature, 445*, 858–865.

Ma, Q., Anderson, D. J., & Fritzsch, B. (2000). Neurogenin1 null mutant ears develop fewer, morphologically normal hair cells in smaller sensory epithelia devoid of innervation. *Journal of the Association for Research in Otolaryngology, 1*, 129–143.

Ma, Q. F., Chen, Z. F., Barrantes, I. D., de la Pompa, J. L., & Anderson, D. J. (1998). Neurogenin1 is essential for the determination of neuronal precursors for proximal cranial sensory ganglia. *Neuron, 20*, 469–482.

Ma, L., Swalla, B. J., Zhou, J., Dobias, S. L., Bell, J. R., & Chen, J. (1996). Expression of an msx homeobox gene in ascidians—Insights into the archetypal chordate expression pattern. *Developmental Dynamics, 205*, 308–318.

Mackie, G. O., & Singla, C. L. (2003). The capsular organ of Chelyosoma productum (Ascidiacea: Corellidae): A new tunicate hydrodynamic sense organ. *Brain, Behavior and Evolution, 61*, 45–58.

Mackie, G. O., & Singla, C. L. (2004). Cupular organs in two species of Corella (Tunicata: Ascidiacea). *Invertebrate Biology, 123*, 269–281.

Mahoney, J. L., Graugnard, E. M., Mire, P., & Watson, G. M. (2011). Evidence for involvement of TRPA1 in the detection of vibrations by hair bundle mechanoreceptors in sea anemones. *Journal of Comparative Physiology. A, Neuroethology, Sensory, Neural, and Behavioral Physiology, 197*, 729–742.

Maier, E. C., Saxena, A., Alsina, B., Bronner, M. E., & Whitfield, T. T. (2014). Sensational placodes: Neurogenesis in the otic and olfactory systems. *Developmental Biology, 389*, 50–67.

Manni, L., Agnoletto, A., Zaniolo, G., & Burighel, P. (2005). Stomodeal and neurohypophysial placodes in Ciona intestinalis: Insights into the origin of the pituitary gland. *Journal of Experimental Zoology. Part B, Molecular and Developmental Evolution, 304B*, 324–339.

Manni, L., Caicci, F., Gasparini, F., Zaniolo, G., & Burighel, P. (2004). Hair cells in ascidians and the evolution of lateral line placodes. *Evolution & Development, 6*, 379–381.

Manni, L., Lane, N. J., Joly, J. S., Gasparini, F., Tiozzo, S., Caicci, F., et al. (2004). Neurogenic and non-neurogenic placodes in ascidians. *Journal of Experimental Zoology. Part B, Molecular and Developmental Evolution, 302B*, 483.

Manni, L., Mackie, G. O., Caicci, F., Zaniolo, G., & Burighel, P. (2006). Coronal organ of ascidians and the evolutionary significance of secondary sensory cells in chordates. *The Journal of Comparative Neurology, 495*, 363–373.

Mannini, L., Rossi, L., Deri, P., Gremigni, V., Salvetti, A., Salo, E., et al. (2004). Djeyes absent (Djeya) controls prototypic planarian eye regeneration by cooperating with the transcription factor Djsix-1. *Developmental Biology, 269*, 346–359.

Marchal, L., Luxardi, G., Thome, V., & Kodjabachian, L. (2009). BMP inhibition initiates neural induction via FGF signaling and Zic genes. *Proceedings of the National academy of Sciences of the United States of America, 106*, 17437–17442.

Marlow, H., Matus, D. Q., & Martindale, M. Q. (2013). Ectopic activation of the canonical wnt signaling pathway affects ectodermal patterning along the primary axis during larval development in the anthozoan Nematostella vectensis. *Developmental Biology, 380*, 324–334.

Marlow, H., Tosches, M. A., Tomer, R., Steinmetz, P. R., Lauri, A., Larsson, T., et al. (2014). Larval body patterning and apical organs are conserved in animal evolution. *BMC Biology*, *12*, 7.

Martindale, M. Q., & Hejnol, A. (2009). A developmental perspective: Changes in the position of the blastopore during bilaterian evolution. *Developmental Cell*, *17*, 162–174.

Martinez-Barbera, J. P., Signore, M., Boyl, P. P., Puelles, E., Acampora, D., Gogoi, R., et al. (2001). Regionalisation of anterior neuroectoderm and its competence in responding to forebrain and midbrain inducing activities depend on mutual antagonism between OTX2 and GBX2. *Development*, *128*, 4789–4800.

Materna, S. C., Ransick, A., Li, E., & Davidson, E. H. (2013). Diversification of oral and aboral mesodermal regulatory states in pregastrular sea urchin embryos. *Developmental Biology*, *375*, 92–104.

Matus, D. Q., Pang, K., Daly, M., & Martindale, M. Q. (2007). Expression of Pax gene family members in the anthozoan cnidarian, Nematostella vectensis. *Evolution & Development*, *9*, 25–38.

Matus, D. Q., Thomsen, G. H., & Martindale, M. Q. (2006). Dorso/ventral genes are asymmetrically expressed and involved in germ-layer demarcation during cnidarian gastrulation. *Current Biology*, *16*, 499–505.

Mazet, F., Hutt, J. A., Millard, J., & Shimeld, S. M. (2003). Pax gene expression in the developing central nervous system of Ciona intestinalis. *Gene Expression Patterns*, *3*, 743–745.

Mazet, F., Hutt, J. A., Milloz, J., Millard, J., Graham, A., & Shimeld, S. M. (2005). Molecular evidence from Ciona intestinalis for the evolutionary origin of vertebrate sensory placodes. *Developmental Biology*, *282*, 494–508.

Mazet, F., Masood, S., Luke, G. N., Holland, N. D., & Shimeld, S. M. (2004). Expression of AmphiCoe, an amphioxus COE/EBF gene, in the developing central nervous system and epidermal sensory neurons. *Genesis*, *38*, 58–65.

Mazet, F., Yu, J. K., Liberles, D. A., Holland, L. Z., & Shimeld, S. M. (2003). Phylogenetic relationships of the Fox (Forkhead) gene family in the Bilateria. *Gene*, *316*, 79–89.

Mazza, M. E., Pang, K., Martindale, M. Q., & Finnerty, J. R. (2007). Genomic organization, gene structure, and developmental expression of three clustered otx genes in the sea anemone Nematostella vectensis. *Journal of Experimental Zoology. Part B, Molecular and Developmental Evolution*, *308*, 494–506.

McCabe, K. L., & Bronner-Fraser, M. (2008). Essential role for PDGF signaling in ophthalmic trigeminal placode induction. *Development*, *135*, 1863–1874.

Meulemans, D., & Bronner-Fraser, M. (2002). Amphioxus and lamprey AP-2 genes: Implications for neural crest evolution and migration patterns. *Development*, *129*, 4953–4962.

Meulemans, D., & Bronner-Fraser, M. (2004). Gene-regulatory interactions in neural crest evolution and development. *Developmental Cell*, *7*, 291–299.

Meulemans, D., & Bronner-Fraser, M. (2007). The amphioxus SoxB family: Implications for the evolution of vertebrate placodes. *International Journal of Biological Sciences*, *3*, 356–364.

Meulemans, D., McCauley, D., & Bronner-Fraser, M. (2003). Id expression in amphioxus and lamprey highlights the role of gene cooption during neural crest evolution. *Developmental Biology*, *264*, 430–442.

Milet, C., & Monsoro-Burq, A. H. (2012). Neural crest induction at the neural plate border in vertebrates. *Developmental Biology*, *366*, 22–33.

Millet, S., Campbell, K., Epstein, D. J., Losos, K., Harris, E., & Joyner, A. L. (1999). A role for Gbx2 in repression of Otx-2 and positioning the mid/hindbrain organizer. *Nature*, *401*, 161–164.

Minguillon, C., Jimenez-Delgado, S., Panopoulou, G., & Garcia-Fernandez, J. (2003). The amphioxus hairy family: Differential fate after duplication. *Development*, *130*, 5903–5914.

Mirabeau, O., & Joly, J. S. (2013). Molecular evolution of peptidergic signaling systems in bilaterians. *Proceedings of the National academy of Sciences of the United States of America*, *110*, E2028–E2037.

Miya, T., & Nishida, H. (2003). Expression pattern and transcriptional control of SoxB1 in embryos of the ascidian Halocynthia roretzi. *Zoological Science*, *20*, 59–67.

Mizuseki, K., Kishi, M., Matsui, M., Nakanishi, S., & Sasai, Y. (1998). Xenopus zic-related-1 and sox-2, two factors induced by chordin, have distinct activities in the initiation of neural induction. *Development*, *125*, 579–587.

Mizutani, C. M., & Bier, E. (2008). EvoD/Vo: The origins of BMP signalling in the neuroectoderm. *Nature Reviews. Genetics*, *9*, 663–677.

Moczek, A. P. (2008). On the origins of novelty in development and evolution. *Bioessays*, *30*, 432–447.

Moret, F., Christiaen, L., Deyts, C., Blin, M., Vernier, P., & Joly, J. S. (2005). Regulatory gene expressions in the ascidian ventral sensory vesicle: Evolutionary relationships with the vertebrate hypothalamus. *Developmental Biology*, *277*, 567–579.

Moroz, L. L. (2009). On the independent origins of complex brains and neurons. *Brain, Behavior and Evolution*, *74*, 177–190.

Moroz, L. L., Kocot, K. M., Citarella, M. R., Dosung, S., Norekian, T. P., Povolotskaya, I. S., et al. (2014). The ctenophore genome and the evolutionary origins of neural systems. *Nature*, *510*, 109–114.

Müller, G. B., & Wagner, G. P. (1991). Novelty in evolution: Restructuring the concept. *Annual Review of Ecology and Systematics*, *22*, 229–256.

Murphy, A. E., & Harvey, S. (2001). Extrapituitary beta TSH and GH in early chick embryos. *Molecular and Cellular Endocrinology*, *185*, 161–171.

Nagatomo, K., Ishibashi, T., Satou, Y., Satoh, N., & Fujiwara, S. (2003). Retinoic acid affects gene expression and morphogenesis without upregulating the retinoic acid receptor in the ascidian Ciona intestinalis. *Mechanisms of Development*, *120*, 363–372.

Nakanishi, N., Yuan, D., Hartenstein, V., & Jacobs, D. K. (2010). Evolutionary origin of rhopalia: Insights from cellular-level analyses of Otx and POU expression patterns in the developing rhopalial nervous system. *Evolution & Development*, *12*, 404–415.

Nechiporuk, A., Linbo, T., Poss, K. D., & Raible, D. W. (2007). Specification of epibranchial placodes in zebrafish. *Development*, *134*, 611–623.

Niehrs, C. (2010). On growth and form: A Cartesian coordinate system of Wnt and BMP signaling specifies bilaterian body axes. *Development*, *137*, 845–857.

Niimura, Y. (2009). Evolutionary dynamics of olfactory receptor genes in chordates: Interaction between environments and genomic contents. *Human Genomics*, *4*, 107–118.

Nishida, H. (1987). Cell lineage analysis in ascidian embryos by intracellular injection of a tracer enzyme. III. Up to the tissue restricted stage. *Developmental Biology*, *121*, 526–541.

Nishida, H. (2005). Specification of embryonic axis and mosaic development in ascidians. *Developmental Dynamics*, *233*, 1177–1193.

Nomaksteinsky, M., Röttinger, E., Dufour, H. D., Chettouh, Z., Lowe, C. J., Martindale, M. Q., et al. (2009). Centralization of the deuterostome nervous system predates chordates. *Current Biology*, *19*, 1264–1269.

Northcutt, R. G. (2004). Taste buds: Development and evolution. *Brain, Behavior and Evolution*, *64*, 198–206.

Northcutt, R. G. (2005). The new head hypothesis revisited. *Journal of Experimental Zoology. Part B, Molecular and Developmental Evolution*, *304*, 274–297.

Northcutt, R. G., & Gans, C. (1983). The genesis of neural crest and epidermal placodes: A reinterpretation of vertebrate origins. *The Quarterly Review of Biology*, *58*, 1–28.

Nozaki, M. (2008). The hagfish pituitary gland and its putative adenohypophysial hormones. *Zoological Science*, *25*, 1028–1036.

Nozaki, M., & Gorbman, A. (1992). The question of functional homology of Hatschek's pit of amphioxus (Branchiostoma belcheri) and the vertebrate adenohypophysis. *Zoological Science, 9*, 387–395.
Oda, I., & Saiga, H. (2001). Hremx, the ascidian homologue of ems/emx, is expressed in the anterior and posterior-lateral epidermis but not in the central nervous system during embryogenesis. *Development Genes and Evolution, 211*, 291–298.
Ogasawara, M., & Satou, Y. (2003). Expression of FoxE and FoxQ genes in the endostyle of Ciona intestinalis. *Development Genes and Evolution, 213*, 416–419.
Ohyama, T., Mohamed, O. A., Taketo, M. M., Dufort, D., & Groves, A. K. (2006). Wnt signals mediate a fate decision between otic placode and epidermis. *Development, 133*, 865–875.
Oisi, Y., Ota, K. G., Kuraku, S., Fujimoto, S., & Kuratani, S. (2013). Craniofacial development of hagfishes and the evolution of vertebrates. *Nature, 493*, 175–180.
Onai, T., Lin, H. C., Schubert, M., Koop, D., Osborne, P. W., Alvarez, S., et al. (2009). Retinoic acid and Wnt/beta-catenin have complementary roles in anterior/posterior patterning embryos of the basal chordate amphioxus. *Developmental Biology, 332*, 223–233.
O'Neill, P., Mak, S. S., Fritzsch, B., Ladher, R. K., & Baker, C. V. (2012). The amniote paratympanic organ develops from a previously undiscovered sensory placode. *Nature Communications, 3*, 1041.
Ozair, M. Z., Kintner, C., & Brivanlou, A. H. (2013). Neural induction and early patterning in vertebrates. *Wiley Interdisciplinary Reviews. Developmental Biology, 2*, 479–498.
Pan, B., Geleoc, G. S., Asai, Y., Horwitz, G. C., Kurima, K., Ishikawa, K., et al. (2013). TMC1 and TMC2 are components of the mechanotransduction channel in hair cells of the mammalian inner ear. *Neuron, 79*, 504–515.
Pang, K., Ryan, J. F., Baxevanis, A. D., & Martindale, M. Q. (2011). Evolution of the TGF-beta signaling pathway and its potential role in the ctenophore, Mnemiopsis leidyi. *PLoS One, 6*, e24152.
Pang, K., Ryan, J. F., NISC Comparative Sequencing Program, Mullikin, J. C., Baxevanis, A. D., & Martindale, M. Q. (2010). Genomic insights into Wnt signaling in an early diverging metazoan, the ctenophore Mnemiopsis leidyi. *EvoDevo, 1*, 10.
Pani, A. M., Mullarkey, E. E., Aronowicz, J., Assimacopoulos, S., Grove, E. A., & Lowe, C. J. (2012). Ancient deuterostome origins of vertebrate brain signalling centres. *Nature, 483*, 289–294.
Paps, J., Holland, P. W., & Shimeld, S. M. (2012). A genome-wide view of transcription factor gene diversity in chordate evolution: Less gene loss in amphioxus? *Briefings in Functional Genomics, 11*, 177–186.
Park, J. I., Semyonov, J., Chang, C. L., & Hsu, S. Y. (2005). Conservation of the heterodimeric glycoprotein hormone subunit family proteins and the LGR signaling system from nematodes to humans. *Endocrine, 26*, 267–276.
Pascual-Anaya, J., Adachi, N., Alvarez, S., Kuratani, S., D'Aniello, S., & Garcia-Fernandez, J. (2012). Broken colinearity of the amphioxus Hox cluster. *EvoDevo, 3*, 28.
Pasini, A., Manenti, R., Rothbächer, U., & Lemaire, P. (2012). Antagonizing retinoic acid and FGF/MAPK pathways control posterior body patterning in the invertebrate chordate Ciona intestinalis. *PLoS One, 7*, e46193.
Patthey, C., Schlosser, G., & Shimeld, S. M. (2014). The evolutionary history of vertebrate cranial placodes—I: Cell type evolution. *Developmental Biology, 389*, 82–97.
Pera, E., Stein, S., & Kessel, M. (1999). Ectodermal patterning in the avian embryo: Epidermis versus neural plate. *Development, 126*, 63–73.
Petersen, C. P., & Reddien, P. W. (2009). Wnt signaling and the polarity of the primary body axis. *Cell, 139*, 1056–1068.

Peterson, T., & Müller, G. B. (2013). What is evolutionary novelty? Process versus character based definitions. *Journal of Experimental Zoology. Part B, Molecular and Developmental Evolution, 320*, 345–350.

Philippe, H., Brinkmann, H., Copley, R. R., Moroz, L. L., Nakano, H., Poustka, A. J., et al. (2011). Acoelomorph flatworms are deuterostomes related to Xenoturbella. *Nature, 470*, 255–258.

Philippe, H., Lartillot, N., & Brinkmann, H. (2005). Multigene analyses of bilaterian animals corroborate the monophyly of Ecdysozoa, Lophotrochozoa, and Protostomia. *Molecular Biology and Evolution, 22*, 1246–1253.

Piatigorsky, J. (1998). Multifunctional lens crystallins and corneal enzymes. More than meets the eye. *Annals of the New York Academy of Sciences, 842*, 7–15.

Pieper, M., Ahrens, K., Rink, E., Peter, A., & Schlosser, G. (2012). Differential distribution of competence for panplacodal and neural crest induction to non-neural and neural ectoderm. *Development, 139*, 1175–1187.

Pieper, M., Eagleson, G. W., Wosniok, W., & Schlosser, G. (2011). Origin and segregation of cranial placodes in Xenopus laevis. *Developmental Biology, 360*, 257–275.

Pierce, M. L., Weston, M. D., Fritzsch, B., Gabel, H. W., Ruvkun, G., & Soukup, G. A. (2008). MicroRNA-183 family conservation and ciliated neurosensory organ expression. *Evolution & Development, 10*, 106–113.

Pignoni, F., Hu, B., Zavitz, K. H., Xiao, J., Garrity, P. A., & Zipursky, S. L. (1997). The eye-specification proteins So and Eya form a complex and regulate multiple steps in Drosophila eye development. *Cell, 91*, 881–891.

Pineda, D., Gonzalez, J., Callaerts, P., Ikeo, K., Gehring, W. J., & Salo, E. (2000). Searching for the prototypic eye genetic network: Sine oculis is essential for eye regeneration in planarians. *Proceedings of the National Academy of Sciences of the United States of America, 97*, 4525–4529.

Plachetzki, D. C., Fong, C. R., & Oakley, T. H. (2010). The evolution of phototransduction from an ancestral cyclic nucleotide gated pathway. *Proceedings of the Biological Sciences, 277*, 1963–1969.

Plachetzki, D. C., Fong, C. R., & Oakley, T. H. (2012). Cnidocyte discharge is regulated by light and opsin-mediated phototransduction. *BMC Biology, 10*, 17.

Pogoda, H. M., von der Hardt, S., Herzog, W., Kramer, C., Schwarz, H., & Hammerschmidt, M. (2006). The proneural gene ascl1a is required for endocrine differentiation and cell survival in the zebrafish adenohypophysis. *Development, 133*, 1079–1089.

Posnien, N., Koniszewski, N., & Bucher, G. (2011). Insect Tc-six4 marks a unit with similarity to vertebrate placodes. *Developmental Biology, 350*, 208–216.

Posnien, N., Koniszewski, N. D., Hein, H. J., & Bucher, G. (2011). Candidate gene screen in the red flour beetle Tribolium reveals six3 as ancient regulator of anterior median head and central complex development. *PLoS Genetics*, 7, e1002416.

Putnam, N. H., Butts, T., Ferrier, D. E., Furlong, R. F., Hellsten, U., Kawashima, T., et al. (2008). The amphioxus genome and the evolution of the chordate karyotype. *Nature, 453*, 1064–1071.

Raible, F., Tessmar-Raible, K., Arboleda, E., Kaller, T., Bork, P., Arendt, D., et al. (2006). Opsins and clusters of sensory G-protein-coupled receptors in the sea urchin genome. *Developmental Biology, 300*, 461–475.

Rasmussen, S. L., Holland, L. Z., Schubert, M., Beaster-Jones, L., & Holland, N. D. (2007). Amphioxus AmphiDelta: Evolution of delta protein structure, segmentation, and neurogenesis. *Genesis, 45*, 113–122.

Renard, E., Vacelet, J., Gazave, E., Lapebie, P., Borchiellini, C., & Ereskovsky, A. V. (2009). Origin of the neuro-sensory system: New and expected insights from sponges. *Integrative Zoology, 4*, 294–308.

Rentzsch, F., Anton, R., Saina, M., Hammerschmidt, M., Holstein, T. W., & Technau, U. (2006). Asymmetric expression of the BMP antagonists chordin and gremlin in the sea anemone Nematostella vectensis: Implications for the evolution of axial patterning. *Developmental Biology, 296*, 375–387.

Richards, G. S., Simionato, E., Perron, M., Adamska, M., Vervoort, M., & Degnan, B. M. (2008). Sponge genes provide new insight into the evolutionary origin of the neurogenic circuit. *Current Biology, 18*, 1156–1161.

Rigon, F., Stach, T., Caicci, F., Gasparini, F., Burighel, P., & Manni, L. (2013). Evolutionary diversification of secondary mechanoreceptor cells in tunicata. *BMC Evolutionary Biology, 13*, 112.

Rivera, A., Winters, I., Rued, A., Ding, S., Posfai, D., Cieniewicz, B., et al. (2013). The evolution and function of the Pax/Six regulatory network in sponges. *Evolution & Development, 15*, 186–196.

Roch, G. J., Busby, E. R., & Sherwood, N. M. (2011). Evolution of GnRH: Diving deeper. *General and Comparative Endocrinology, 171*, 1–16.

Roch, G. J., & Sherwood, N. M. (2014). Glycoprotein hormones and their receptors emerged at the origin of metazoans. *Genome Biology and Evolution, 6*, 1466–1479.

Roch, G. J., Tello, J. A., & Sherwood, N. M. (2014). At the transition from invertebrates to vertebrates, a novel GnRH-like peptide emerges in amphioxus. *Molecular Biology and Evolution, 31*, 765–778.

Rodriguez-Seguel, E., Alarcon, P., & Gomez-Skarmeta, J. L. (2009). The Xenopus Irx genes are essential for neural patterning and define the border between prethalamus and thalamus through mutual antagonism with the anterior repressors Fezf and Arx. *Developmental Biology, 329*, 258–268.

Roth, V. L. (1988). The biological basis of homology. In C. J. Humphries (Ed.), *Ontogeny and systematics* (pp. 1–26). New York: Columbia University of Press.

Roth, V. L. (1991). Homology and hierarchies: Problems solved and unresolved. *Journal of Evolutionary Biology, 4*, 167–194.

Ruppert, E. E. (1997). Cephalochordata (Acrania). In F. W. Harrison & E. E. Ruppert (Eds.), *Hemichordata, Chaetognatha, and the invertebrate chordates: Vol. 15. Microscopic anatomy of invertebrates* (pp. 349–504). New York: Wiley-Liss.

Ryan, J. F., Pang, K., Schnitzler, C. E., Nguyen, A. D., Moreland, R. T., Simmons, D. K., et al. (2013). The genome of the ctenophore Mnemiopsis leidyi and its implications for cell type evolution. *Science, 342*, 1242592.

Sabado, V., Barraud, P., Baker, C. V., & Streit, A. (2012). Specification of GnRH-1 neurons by antagonistic FGF and retinoic acid signaling. *Developmental Biology, 362*, 254–262.

Sahlin, K., & Olsson, R. (1986). The wheel organ and Hatschek's groove in the lancelet, Branchiostoma lanceolatum (cephalochordata). *Acta Zoologica, 67*, 201–209.

Saina, M., Genikhovich, G., Renfer, E., & Technau, U. (2009). BMPs and chordin regulate patterning of the directive axis in a sea anemone. *Proceedings of the National academy of Sciences of the United States of America, 106*, 18592–18597.

Saint-Jeannet, J. P., & Moody, S. A. (2014). Establishing the pre-placodal region and breaking it into placodes with distinct identities. *Developmental Biology, 389*, 13–27.

Sakarya, O., Armstrong, K. A., Adamska, M., Adamski, M., Wang, I. F., Tidor, B., et al. (2007). A post-synaptic scaffold at the origin of the animal kingdom. *PLoS One, 2*, e506.

Salzet, M., Salzet-Raveillon, B., Cocquerelle, C., Verger-Bocquet, M., Pryor, S. C., Rialas, C. M., et al. (1997). Leech immunocytes contain proopiomelanocortin: Nitric oxide mediates hemolymph proopiomelanocortin processing. *Journal of Immunology, 159*, 5400–5411.

Santagata, S., Resh, C., Hejnol, A., Martindale, M. Q., & Passamaneck, Y. J. (2012). Development of the larval anterior neurogenic domains of Terebratalia transversa

(Brachiopoda) provides insights into the diversification of larval apical organs and the spiralian nervous system. *EvoDevo, 3*, 3.

Sasakura, Y., Kanda, M., Ikeda, T., Horie, T., Kawai, N., Ogura, Y., et al. (2012). Retinoic acid-driven Hox1 is required in the epidermis for forming the otic/atrial placodes during ascidian metamorphosis. *Development, 139*, 2156–2160.

Sato, T., Sasai, N., & Sasai, Y. (2005). Neural crest determination by co-activation of Pax3 and Zic1 genes in Xenopus ectoderm. *Development, 132*, 2355–2363.

Satoh, G. (2005). Characterization of novel GPCR gene coding locus in amphioxus genome: Gene structure, expression, and phylogenetic analysis with implications for its involvement in chemoreception. *Genesis, 41*, 47–57.

Satoh, T., & Fekete, D. M. (2005). Clonal analysis of the relationships between mechanosensory cells and the neurons that innervate them in the chicken ear. *Development, 132*, 1687–1697.

Satoh, G., Wang, Y., Zhang, P., & Satoh, N. (2001). Early development of amphioxus nervous system with special reference to segmental cell organization and putative sensory cell precursors: A study based on the expression of pan-neuronal marker gene Hu/elav. *The Journal of Experimental Zoology, 291*, 354–364.

Saxena, A., Peng, B. N., & Bronner, M. E. (2013). Sox10-dependent neural crest origin of olfactory microvillous neurons in zebrafish. *eLife, 2*, e00336.

Schlosser, G. (2003). Hypobranchial placodes in Xenopus laevis give rise to hypobranchial ganglia, a novel type of cranial ganglia. *Cell and Tissue Research, 312*, 21–29.

Schlosser, G. (2005). Evolutionary origins of vertebrate placodes: Insights from developmental studies and from comparisons with other deuterostomes. *Journal of Experimental Zoology. Part B, Molecular and Developmental Evolution, 304B*, 347–399.

Schlosser, G. (2006). Induction and specification of cranial placodes. *Developmental Biology, 294*, 303–351.

Schlosser, G. (2010). Making senses: Development of vertebrate cranial placodes. *International Review of Cell and Molecular Biology, 283C*, 129–234.

Schlosser, G. (2014). Early embryonic specification of vertebrate cranial placodes. *WIREs Developmental Biology, 3*, 349–363.

Schlosser, G., Awtry, T., Brugmann, S. A., Jensen, E. D., Neilson, K., Ruan, G., et al. (2008). Eya1 and Six1 promote neurogenesis in the cranial placodes in a SoxB1-dependent fashion. *Developmental Biology, 320*, 199–214.

Schlosser, G., & Northcutt, R. G. (2000). Development of neurogenic placodes in Xenopus laevis. *The Journal of Comparative Neurology, 418*, 121–146.

Schlosser, G., Patthey, C., & Shimeld, S. M. (2014). The evolutionary history of vertebrate cranial placodes. II. Evolution of ectodermal patterning. *Developmental Biology, 389*, 98–119.

Schubert, M., Holland, N. D., Escriva, H., Holland, L. Z., & Laudet, V. (2004). Retinoic acid influences anteroposterior positioning of epidermal sensory neurons and their gene expression in a developing chordate (amphioxus). *Proceedings of the National academy of Sciences of the United States of America, 101*, 10320–10325.

Schubert, M., Holland, N. D., Laudet, V., & Holland, L. Z. (2006). A retinoic acid-Hox hierarchy controls both anterior/posterior patterning and neuronal specification in the developing central nervous system of the cephalochordate amphioxus. *Developmental Biology, 296*, 190–202.

Seipel, K., Yanze, N., & Schmid, V. (2004). Developmental and evolutionary aspects of the basic helix-loop-helix transcription factors Atonal-like 1 and Achaete-scute homolog 2 in the jellyfish. *Developmental Biology, 269*, 331–345.

Sellami, A., Agricola, H. J., & Veenstra, J. A. (2011). Neuroendocrine cells in *Drosophila melanogaster* producing GPA2/GPB5, a hormone with homology to LH, FSH and TSH. *General and Comparative Endocrinology, 170*, 582–588.

Sen, S., Reichert, H., & VijayRaghavan, K. (2013). Conserved roles of ems/Emx and otd/Otx genes in olfactory and visual system development in Drosophila and mouse. *Open Biology*, *3*, 120177.
Senthilan, P. R., Piepenbrock, D., Ovezmyradov, G., Nadrowski, B., Bechstedt, S., Pauls, S., et al. (2012). Drosophila auditory organ genes and genetic hearing defects. *Cell*, *150*, 1042–1054.
Serikaku, M. A., & O'Tousa, J. E. (1994). sine oculis is a homeobox gene required for Drosophila visual system development. *Genetics*, *138*, 1137–1150.
Sharman, A. C., Shimeld, S. M., & Holland, P. W. H. (1999). An amphioxus Msx gene expressed predominantly in the dorsal neural tube. *Development Genes and Evolution*, *209*, 260–263.
Shimeld, S. M., Purkiss, A. G., Dirks, R. P., Bateman, O. A., Slingsby, C., & Lubsen, N. H. (2005). Urochordate betagamma-crystallin and the evolutionary origin of the vertebrate eye lens. *Current Biology*, *15*, 1684–1689.
Shubin, N., Tabin, C., & Carroll, S. (2009). Deep homology and the origins of evolutionary novelty. *Nature*, *457*, 818–823.
Sinigaglia, C., Busengdal, H., Leclere, L., Technau, U., & Rentzsch, F. (2013). The bilaterian head patterning gene six3/6 controls aboral domain development in a cnidarian. *PLoS Biology*, *11*, e1001488.
Sjödal, M., Edlund, T., & Gunhaga, L. (2007). Time of exposure to BMP signals plays a key role in the specification of the olfactory and lens placodes ex vivo. *Developmental Cell*, *13*, 141–149.
So, W. K., Kwok, H. F., & Ge, W. (2005). Zebrafish gonadotropins and their receptors: II. Cloning and characterization of zebrafish follicle-stimulating hormone and luteinizing hormone subunits—Their spatial-temporal expression patterns and receptor specificity. *Biology of Reproduction*, *72*, 1382–1396.
Soukup, V., Epperlein, H. H., Horacek, I., & Cerny, R. (2008). Dual epithelial origin of vertebrate oral teeth. *Nature*, *455*, 795–798.
Sower, S. A., Freamat, M., & Kavanaugh, S. I. (2009). The origins of the vertebrate hypothalamic-pituitary-gonadal (HPG) and hypothalamic-pituitary-thyroid (HPT) endocrine systems: New insights from lampreys. *General and Comparative Endocrinology*, *161*, 20–29.
Sower, S. A., Moriyama, S., Kasahara, M., Takahashi, A., Nozaki, M., Uchida, K., et al. (2006). Identification of sea lamprey GTHbeta-like cDNA and its evolutionary implications. *General and Comparative Endocrinology*, *148*, 22–32.
Squarzoni, P., Parveen, F., Zanetti, L., Ristoratore, F., & Spagnuolo, A. (2011). FGF/MAPK/Ets signaling renders pigment cell precursors competent to respond to Wnt signal by directly controlling Ci-Tcf transcription. *Development*, *138*, 1421–1432.
Stefano, G. B., Salzet-Raveillon, B., & Salzet, M. (1999). Mytilus edulis hemolymph contains pro-opiomelanocortin: LPS and morphine stimulate differential processing. *Brain Research. Molecular Brain Research*, *63*, 340–350.
Steinmetz, P. R., Kostyuchenko, R. P., Fischer, A., & Arendt, D. (2011). The segmental pattern of otx, gbx, and Hox genes in the annelid Platynereis dumerilii. *Evolution & Development*, *13*, 72–79.
Steinmetz, P. R., Urbach, R., Posnien, N., Eriksson, J., Kostyuchenko, R. P., Brena, C., et al. (2010). Six3 demarcates the anterior-most developing brain region in bilaterian animals. *EvoDevo*, *1*, 14.
Steventon, B., Mayor, R., & Streit, A. (2012). Mutual repression between Gbx2 and Otx2 in sensory placodes reveals a general mechanism for ectodermal patterning. *Developmental Biology*, *367*, 55–65.

Stierwald, M., Yanze, N., Bamert, R. P., Kammermeier, L., & Schmid, V. (2004). The Sine oculis/Six class family of homeobox genes in jellyfish with and without eyes: Development and eye regeneration. *Developmental Biology, 274*, 70–81.

Stone, L. M., Finger, T. E., Tam, P. P., & Tan, S. S. (1995). Taste receptor cells arise from local epithelium, not neurogenic ectoderm. *Proceedings of the National Academy of Sciences of the United States of America, 92*, 1916–1920.

Streit, A. (2002). Extensive cell movements accompany formation of the otic placode. *Developmental Biology, 249*, 237–254.

Streit, A. (2004). Early development of the cranial sensory nervous system: From a common field to individual placodes. *Developmental Biology, 276*, 1–15.

Striedter, G. F., & Northcutt, R. G. (1991). Biological hierarchies and the concept of homology. *Brain, Behavior and Evolution, 38*, 177–189.

Sudo, S., Kuwabara, Y., Park, J. I., Hsu, S. Y., & Hsueh, A. J. (2005). Heterodimeric fly glycoprotein hormone-alpha2 (GPA2) and glycoprotein hormone-beta5 (GPB5) activate fly leucine-rich repeat-containing G protein-coupled receptor-1 (DLGR1) and stimulation of human thyrotropin receptors by chimeric fly GPA2 and human GPB5. *Endocrinology, 146*, 3596–3604.

Sundström, G., Dreborg, S., & Larhammar, D. (2010). Concomitant duplications of opioid peptide and receptor genes before the origin of jawed vertebrates. *PLoS One, 5*, e10512.

Suzuki, T., & Saigo, K. (2000). Transcriptional regulation of atonal required for Drosophila larval eye development by concerted action of eyes absent, sine oculis and hedgehog signaling independent of fused kinase and cubitus interruptus. *Development, 127*, 1531–1540.

Suzuki, A., Ueno, N., & Hemmati-Brivanlou, A. (1997). Xenopus msx1 mediates epidermal induction and neural inhibition by BMP4. *Development, 124*, 3037–3044.

Tadjuidje, E., & Hegde, R. S. (2013). The Eyes Absent proteins in development and disease. *Cellular and Molecular Life Sciences, 70*, 1897–1913.

Takamura, K. (1998). Nervous network in larvae of the ascidian Ciona intestinalis. *Development Genes and Evolution, 208*, 1–8.

Tando, Y., & Kubokawa, K. (2009). A homolog of the vertebrate thyrostimulin glycoprotein hormone alpha subunit (GPA2) is expressed in Amphioxus neurons. *Zoological Science, 26*, 409–414.

Terakado, K. (2001). Induction of gamete release by gonadotropin-releasing hormone in a protochordate, Ciona intestinalis. *General and Comparative Endocrinology, 124*, 277–284.

Tessmar-Raible, K. (2007). The evolution of neurosecretory centers in bilaterian forebrains: Insights from protostomes. *Seminars in Cell & Developmental Biology, 18*, 492–501.

Tessmar-Raible, K., Raible, F., Christodoulou, F., Guy, K., Rembold, M., Hausen, H., et al. (2007). Conserved sensory-neurosecretory cell types in annelid and fish forebrain: Insights into hypothalamus evolution. *Cell, 129*, 1389–1400.

Tiozzo, S., Christiaen, L., Deyts, C., Manni, L., Joly, J. S., & Burighel, P. (2005). Embryonic versus blastogenetic development in the compound ascidian Botryllus schlosseri: Insights from Pitx expression patterns. *Developmental Dynamics, 232*, 468–478.

Tjoa, L. T., & Welsch, U. (1974). Electron microscopical observations on Kolliker's and Hatschek's pit and on the wheel organ in the head region of Amphioxus (Branchiostoma lanceolatum). *Cell and Tissue Research, 153*, 175–187.

Tokuoka, M., Imai, K. S., Satou, Y., & Satoh, N. (2004). Three distinct lineages of mesenchymal cells in Ciona intestinalis embryos demonstrated by specific gene expression. *Developmental Biology, 274*, 211–224.

Tokuoka, M., Satoh, N., & Satou, Y. (2005). A bHLH transcription factor gene, Twist-like 1, is essential for the formation of mesodermal tissues of Ciona juveniles. *Developmental Biology, 288*, 387–396.

Tomer, R., Denes, A. S., Tessmar-Raible, K., & Arendt, D. (2010). Profiling by image registration reveals common origin of annelid mushroom bodies and vertebrate pallium. *Cell, 142*, 800–809.

Toresson, H., Martinez-Barbera, J. P., Bardsley, A., Caubit, X., & Krauss, S. (1998). Conservation of BF-1 expression in amphioxus and zebrafish suggests evolutionary ancestry of anterior cell types that contribute to the vertebrate forebrain. *Development Genes and Evolution, 208*, 431–439.

Toro, S., & Varga, Z. M. (2007). Equivalent progenitor cells in the zebrafish anterior preplacodal field give rise to adenohypophysis, lens, and olfactory placodes. *Seminars in Cell & Developmental Biology, 18*, 534–542.

Torrence, S. A., & Cloney, R. A. (1982). Nervous systems of ascidian larvae: Caudal primary sensory neurons. *Zoomorphology, 99*, 103–115.

Torrence, S. A., & Cloney, R. A. (1983). Ascidian larval nervous system: Primary sensory neurons in adhesive papillae. *Zoomorphology, 102*, 111–123.

Tosches, M. A., & Arendt, D. (2013). The bilaterian forebrain: An evolutionary chimaera. *Current Opinion in Neurobiology, 23*, 1080–1089.

Treier, M., Oconnell, S., Gleiberman, A., Price, J., Szeto, D. P., Burgess, R., et al. (2001). Hedgehog signaling is required for pituitary gland development. *Development, 128*(3), 377–386.

True, J. R., & Haag, E. S. (2001). Developmental system drift and flexibility in evolutionary trajectories. *Evolution & Development, 3*, 109–119.

Uchida, K., Moriyama, S., Chiba, H., Shimotani, T., Honda, K., Miki, M., et al. (2010). Evolutionary origin of a functional gonadotropin in the pituitary of the most primitive vertebrate, hagfish. *Proceedings of the National academy of Sciences of the United States of America, 107*, 15832–15837.

Uchida, K., Murakami, Y., Kuraku, S., Hirano, S., & Kuratani, S. (2003). Development of the adenohypophysis in the lamprey: Evolution of epigenetic patterning programs in organogenesis. *Journal of Experimental Zoology. Part B, Molecular and Developmental Evolution, 300*, 32–47.

Van Otterloo, E., Li, W., Garnett, A., Cattell, M., Medeiros, D. M., & Cornell, R. A. (2012). Novel Tfap2-mediated control of soxE expression facilitated the evolutionary emergence of the neural crest. *Development, 139*, 720–730.

Veeman, M. T., Newman-Smith, E., El Nachef, D., & Smith, W. C. (2010). The ascidian mouth opening is derived from the anterior neuropore: Reassessing the mouth/neural tube relationship in chordate evolution. *Developmental Biology, 344*, 138–149.

Vopalensky, P., Pergner, J., Liegertova, M., Benito-Gutierrez, E., Arendt, D., & Kozmik, Z. (2012). Molecular analysis of the amphioxus frontal eye unravels the evolutionary origin of the retina and pigment cells of the vertebrate eye. *Proceedings of the National Academy of Sciences of the United States of America, 109*, 15383–15388.

Wada, H., Holland, P. W. H., & Satoh, N. (1996). Origin of patterning in neural tubes. *Nature, 384*, 123.

Wada, S., Katsuyama, Y., & Saiga, H. (1999). Anteroposterior patterning of the epidermis by inductive influences from the vegetal hemisphere cells in the ascidian embryo. *Development, 126*, 4955–4963.

Wada, S., Katsuyama, Y., Sato, Y., Itoh, C., & Saiga, H. (1996). Hroth, an orthodenticle-related homeobox gene of the ascidian, Halocynthia roretzi: Its expression and putative roles in the axis formation during embryogenesis. *Mechanisms of Development, 60*, 59–71.

Wada, S., & Saiga, H. (2002). HrzicN, a new Zic family gene of ascidians, plays essential roles in the neural tube and notochord development. *Development, 129*, 5597–5608.

Wada, H., Saiga, H., Satoh, N., & Holland, P. W. H. (1998). Tripartite organization of the ancestral chordate brain and the antiquity of placodes: Insights from ascidian Pax-2/5/8, Hox and Otx genes. *Development, 125*, 1113–1122.

Wagner, G. P. (2007). The developmental genetics of homology. *Nature Reviews. Genetics, 8*, 473–479.

Wagner, E., & Levine, M. (2012). FGF signaling establishes the anterior border of the Ciona neural tube. *Development, 139*, 2351–2359.

Wang, Y., Zhang, P. J., Yasui, K., & Saiga, H. (2002). Expression of Bblhx3, a LIM-homeobox gene, in the development of amphioxus Branchiostoma belcheri tsingtauense. *Mechanisms of Development, 117*, 315–319.

Watanabe, H., Fujisawa, T., & Holstein, T. W. (2009). Cnidarians and the evolutionary origin of the nervous system. *Development, Growth & Differentiation, 51*, 167–183.

Whitlock, K. E., Wolf, C. D., & Boyce, M. L. (2003). Gonadotropin-releasing hormone (GnRH) cells arise from cranial neural crest and adenohypophyseal regions of the neural plate in the zebrafish, Danio rerio. *Developmental Biology, 257*, 140–152.

Wikramanayake, A. H., Huang, L., & Klein, W. H. (1998). beta-Catenin is essential for patterning the maternally specified animal-vegetal axis in the sea urchin embryo. *Proceedings of the National academy of Sciences of the United States of America, 95*, 9343–9348.

Williams, N. A., & Holland, P. W. H. (1996). Old head on young shoulders. *Nature, 383*, 490.

Windsor, P. J., & Leys, S. P. (2010). Wnt signaling and induction in the sponge aquiferous system: Evidence for an ancient origin of the organizer. *Evolution & Development, 12*, 484–493.

Wirmer, A., Bradler, S., & Heinrich, R. (2012). Homology of insect corpora allata and vertebrate adenohypophysis? *Arthropod Structure & Development, 41*, 409–417.

Xu, P. X., Adams, J., Peters, H., Brown, M. C., Heaney, S., & Maas, R. (1999). Eya1-deficient mice lack ears and kidneys and show abnormal apoptosis of organ primordia. *Nature Genetics, 23*, 113–117.

Xu, H., Dude, C. M., & Baker, C. V. (2008). Fine-grained fate maps for the ophthalmic and maxillomandibular trigeminal placodes in the chick embryo. *Developmental Biology, 317*, 174–186.

Yankura, K. A., Martik, M. L., Jennings, C. K., & Hinman, V. F. (2010). Uncoupling of complex regulatory patterning during evolution of larval development in echinoderms. *BMC Biology, 8*, 143.

Yasui, K., Tabata, S., Ueki, T., Uemura, M., & Zhang, S. C. (1998). Early development of the peripheral nervous system in a lancelet species. *The Journal of Comparative Neurology, 393*, 415–425.

Yasui, K., Zhang, S., Uemura, M., & Saiga, H. (2000). Left-right asymmetric expression of BbPtx, a Ptx-related gene, in a lancelet species and the developmental left-sidedness in deuterostomes. *Development, 127*, 187–195.

Yu, J. K., Holland, N. D., & Holland, L. Z. (2002). An amphioxus winged helix/forkhead gene, AmphiFoxD: Insights into vertebrate neural crest evolution. *Developmental Dynamics, 225*, 289–297.

Yu, J. K., Holland, N. D., & Holland, L. Z. (2003). AmphiFoxQ2, a novel winged helix/forkhead gene, exclusively marks the anterior end of the amphioxus embryo. *Development Genes and Evolution, 213*, 102–105.

Yu, J. K., Holland, L. Z., Jamrich, M., Blitz, I. L., & Hollan, N. D. (2002). AmphiFoxE4, an amphioxus winged helix/forkhead gene encoding a protein closely related to vertebrate thyroid transcription factor-2: Expression during pharyngeal development. *Evolution & Development, 4*, 9–15.

Yu, J. K., Meulemans, D., McKeown, S. J., & Bronner-Fraser, M. (2008). Insights from the amphioxus genome on the origin of vertebrate neural crest. *Genome Research, 18*, 1127–1132.

Yu, J. K., Satou, Y., Holland, N. D., Shin, I., Kohara, Y., Satoh, N., et al. (2007). Axial patterning in cephalochordates and the evolution of the organizer. *Nature*, *445*, 613–617.

Zhang, Y., & Mao, B. (2009). Developmental expression of an amphioxus (Branchiostoma belcheri) gene encoding a GATA transcription factor. *Zoological Record*, *30*, 137–143.

Zhang, T., Ranade, S., Cai, C. Q., Clouser, C., & Pignoni, F. (2006). Direct control of neurogenesis by selector factors in the fly eye: Regulation of atonal by Ey and So. *Development*, *133*, 4881–4889.

Zheng, W., Huang, L., Wei, Z. B., Silvius, D., Tang, B., & Xu, P. X. (2003). The role of Six1 in mammalian auditory system development. *Development*, *130*, 3989–4000.

Zou, D., Silvius, D., Fritzsch, B., & Xu, P. X. (2004). Eya1 and Six1 are essential for early steps of sensory neurogenesis in mammalian cranial placodes. *Development*, *131*, 5561–5572.

CHAPTER NINE

Transcriptional Regulation of Cranial Sensory Placode Development

Sally A. Moody[*,†,1], **Anthony-Samuel LaMantia**[†,‡]

[*]Department of Anatomy and Regenerative Biology, The George Washington University, School of Medicine and Health Sciences, Washington, DC, USA
[†]George Washington University Institute for Neuroscience, Washington, DC, USA
[‡]Department of Pharmacology and Physiology, The George Washington University, School of Medicine and Health Sciences, Washington, DC, USA
[1]Corresponding author: e-mail address: samoody@gwu.edu

Contents

Abstract

Cranial sensory placodes derive from discrete patches of the head ectoderm and give rise to numerous sensory structures. During gastrulation, a specialized "neural border zone" forms around the neural plate in response to interactions between the neural and nonneural ectoderm and signals from adjacent mesodermal and/or endodermal tissues. This zone subsequently gives rise to two distinct precursor populations of the peripheral nervous system: the neural crest and the preplacodal ectoderm (PPE). The PPE is a common field from which all cranial sensory placodes arise (adenohypophyseal, olfactory, lens, trigeminal, epibranchial, otic). Members of the Six family of transcription factors are major regulators of PPE specification, in partnership with cofactor proteins such as Eya. *Six* gene activity also maintains tissue boundaries between the PPE, neural crest, and epidermis by repressing genes that specify the fates of those

Current Topics in Developmental Biology, Volume 111
ISSN 0070-2153
http://dx.doi.org/10.1016/bs.ctdb.2014.11.009

adjacent ectodermally derived domains. As the embryo acquires anterior–posterior identity, the PPE becomes transcriptionally regionalized, and it subsequently becomes subdivided into specific placodes with distinct developmental fates in response to signaling from adjacent tissues. Each placode is characterized by a unique transcriptional program that leads to the differentiation of highly specialized cells, such as neurosecretory cells, sensory receptor cells, chemosensory neurons, peripheral glia, and supporting cells. In this review, we summarize the transcriptional and signaling factors that regulate key steps of placode development, influence subsequent sensory neuron specification, and discuss what is known about mutations in some of the essential PPE genes that underlie human congenital syndromes.

1. INTRODUCTION

The vertebrate head contains a number of specialized sensory organs that are derived from discrete thickenings in the embryonic ectoderm called cranial sensory placodes (reviewed by Baker & Bronner-Fraser, 2001; Moody & Saint-Jeannet, 2014; Patthey, Schlosser, & Shimeld, 2014; Saint-Jeannet & Moody, 2014; Schlosser, 2006, 2010; Streit, 2004, 2007; Webb & Noden, 1993). Cranial sensory placodes give rise to the anterior pituitary gland, the olfactory epithelium, the lens, the auditory and vestibular organs and their associated sensory ganglia, and in aquatic species the lateral line and electroreceptive organs and their associated sensory ganglia. In addition, placodes give rise to the large neurons in the sensory ganglia of the trigeminal, facial, glossopharyngeal, and vagus cranial nerves. All of these structures are crucial for an animal to successfully interact with other animals and to navigate through its environment. Therefore, it is critically important that these structures develop properly and be replaced, either naturally or by medical intervention, when they are damaged. The focus of this review is to present our current understanding of the transcriptional and signaling pathways that transform a common preplacodal precursor field into these diverse and highly specialized structures.

Experiments performed in a number of animals demonstrate that placode development is highly conserved across vertebrates (reviewed in Patthey et al., 2014; Saint-Jeannet & Moody, 2014). In chick, frog, fish, and mouse, the embryonic ectoderm is divided into neural ectoderm and nonneural ectoderm during gastrulation. As this process is completed, the ectoderm surrounding the nascent neural ectoderm, called the neural border (NB) zone, is specified to give rise to the neural crest and the preplacodal ectoderm (PPE; Fig. 1). During neural tube closure (called neurulation), signals that

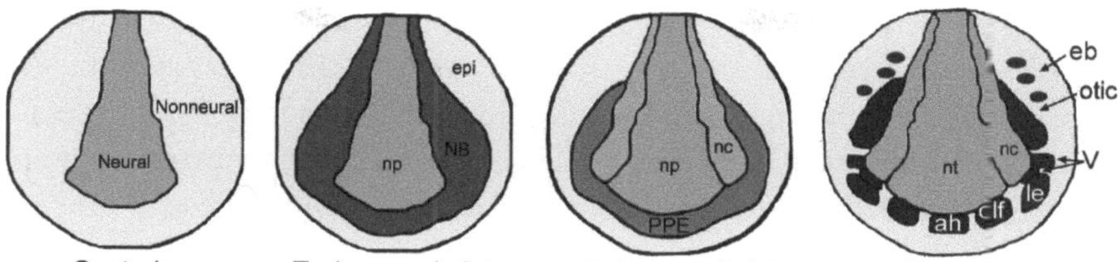

Figure 1 The ectodermal domains depicted at different stages of *Xenopus* embryonic development. At gastrulation, the early embryonic ectoderm is divided into two domains: neural (blue) and nonneural (yellow). Interactions between these two domains and signaling from underlying tissues establish a neural border (NB) zone (green) between the early neural plate (np) and the epidermis (epi). At later neural plate stages, the NB zone divides into the medially located neural crest (nc; light green) and the laterally located preplacodal ectoderm (PPE; dark green). After the PPE acquires initial anterior–posterior patterning and the neural tube (nt) begins to roll up during neurulation, the PPE breaks up into individual placodes in response to local signaling and differential transcription factor expression. These include those derived from the anterior PPE field (the adenohypophyseal (ah), olfactory (olf), and lens (le)); those derived from the intermediate field (V; ophthalmic placode is dorsal to lens placode, and maxillomandibular placode is posterior to it); and those derived from the posterior field (otic and epibranchial (eb)). (See the color plate.)

establish the anterior–posterior (A–P) axis of the embryo also impose regional identity on the PPE, and subsequent signals from adjacent tissues cause the PPE to separate into many discrete placodes that have distinct developmental fates. These placodes will then undergo morphogenetic movements and cellular differentiation processes that result in the numerous specialized sensory structures that characterize the vertebrate head.

One of the key outcomes of placode specification is the genesis of cranial sensory neurons. Chemosensory, auditory, proprioceptive, mechanoreceptive, and nociceptive neurons are generated from placode cells with distinct properties based upon transcriptional specification and subsequent programs of gene expression. These properties include the acquisition of bipolar morphology with a "basal" receptive process (similar to a dendrite) specialized for sensory transduction, and an apical process (an axon) for transmitting information to the central nervous system. In addition, placode-derived sensory neurons are capable of generating action potentials and/or vesicular neurotransmitter release and thus must acquire a full range of molecular regulators of neuronal excitability. We will review data that indicate that these key neuronal properties, and therefore the specific identity of cranial sensory neurons, reflect transcriptional specification of placode precursors, definition of placode-signaling centers, and subsequent interactions with adjacent neural crest-derived mesenchymal cells. We will evaluate this issue by focusing on chemoreceptive olfactory sensory neurons that emerge from the olfactory placode during embryonic development. Olfactory receptor neurons (ORNs) are paradigmatic cranial sensory receptor neurons: they are found in all animals, with remarkably similar cellular and molecular properties (Axel, 2005; Buck, 2000; Hallem & Carlson, 2004).

Over the past two decades, a number of highly conserved transcription factor and signaling genes have been identified that are expressed during these different steps of placode development, allowing us to begin to understand the molecular pathways that induce and specify the fate of these important cells. We will review what is known regarding three steps of placode development: initial induction and specification of the preplacodal field; subdividing the field into region-specific cranial placodes; and differentiation of some of the specialized cells (olfactory receptor cells and cranial ganglion sensory neurons). We will also review the contribution of some of the key genes involved in placode development to human congenital syndromes that include craniofacial and auditory defects.

2. INDUCTION AND SPECIFICATION OF THE PREPLACODAL FIELD

Based on several experimental approaches, it is now widely accepted that the cranial sensory placodes are derived from a common precursor field in the embryonic ectoderm (reviewed in Moody & Saint-Jeannet, 2014; Patthey et al., 2014; Saint-Jeannet & Moody, 2014; Schlosser, 2006, 2010; Streit, 2004, 2007). Classic histological descriptions of cranial sensory placode formation identified the origin of all cranial placodes from a common precursor region called the PPE (also called the "pan-placodal region" or "common placodal field"), which is a U-shaped band of ectoderm that surrounds the anterior margin of the neural plate (Knouff, 1935; Platt, 1896; von Kupffer, 1895; reviewed in Schlosser, 2005; Fig. 1). Fate mapping studies in amphibians and chick also showed that all placodes originate from a common PPE field (Couly & LeDouarin, 1987, 1990; Streit, 2002; reviewed in Pieper, Eagleson, Wosniok, & Schlosser, 2011; Schlosser & Ahrens, 2004; Streit, 2004), and transplantation studies in frog and chick showed that the PPE is initially competent to give rise to all the different types of placodes (Groves & Bronner-Fraser, 2000; Jacobson, 1963). This pan-placodal competence appears to be regulated by a common molecular signature. At early stages of chick, frog, fish, and mouse development, the entire PPE field expresses *Six* and *Eya* genes (Christophorou, Bailey, Hanson, & Streit, 2009; David, Ahrens, Wedlich, & Schlosser, 2001; Esteve & Bovolenta, 1999; Ghanbari, Seo, Fjose, & Brändli, 2001; Kobayashi, Osanai, Kawakami, & Yamamoto, 2000; Litsiou, Hanson, & Streit, 2005; McLarren, Litsiou, & Streit, 2003; Pandur & Moody, 2000; Sato et al., 2010; Schlosser & Ahrens, 2004; Streit, 2004). In both frog and chick, the expression domains of these pan-placodal genes overlap with the fate maps of the placodes (reviewed in Pieper et al., 2011; Schlosser & Ahrens, 2004; Streit, 2004), and these genes are required for the proper development of several placodes and their derivatives (reviewed in Moody & Saint-Jeannet, 2014; Saint-Jeannet & Moody, 2014 and discussed in detail below). Furthermore, studies that explanted the PPE into culture to reveal its developmental potential in the absence of tissue interactions showed that a PPE molecular "ground" state must first be attained before a specific placode identity can be achieved (Bailey, Bhattacharyya, Bronner-Fraser, & Streit, 2006; Martin & Groves, 2005). We will first address how the pan-placodal field arises, and discuss the transcriptional program that both specifies the PPE molecular state and maintains its boundaries.

2.1. Formation of the NB zone

During gastrulation, the embryonic ectoderm is divided into two transcriptionally distinct fields: the neural ectoderm and the nonneural ectoderm (Fig. 1). Interactions between these two fields and signals originating from the underlying mesodermal and endodermal tissues initiate the formation of an intervening zone of ectoderm with the potential to form the neural crest and the PPE, both of which make significant but distinct contributions to the peripheral nervous system. This NB zone was experimentally confirmed by explanting pieces of neural plate into the epidermis. Both neural crest-specific and PPE-specific genes were induced at the border between the neural plate explant and the host epidermis (Ahrens & Schlosser, 2005; Glavic et al., 2004; Litsiou et al., 2005; Mancilla & Mayor, 1996; Selleck & Bronner-Fraser, 1995, 2000; Woda, Pastagia, Mercola, & Artinger, 2003). Several studies have subsequently shown that the NB zone initially expresses a distinct set of genes, called "NB-specifying" genes, that are required for the later expression of neural crest-specific and/or PPE-specific genes (reviewed in Grocott, Tambalo, & Streit, 2012; Meulemans & Bronner-Fraser, 2004; Park & Saint-Jeannet, 2010; Sargent, 2006). These include members of the *Dlx*, *Msx*, *Pax*, and *Zic* families, as well as *TFAP2α*, *GATA*, and *Foxi* genes. In most vertebrates examined, *Dlx*, *Msx1*, *GATA*, *Foxi*, and *TFAP2α* are expressed broadly in the nonneural ectoderm at blastula and gastrula stages, but their expression domains also overlap with the border of the neural ectoderm, which expresses *SoxB1* and *Zic* genes (Fig. 2; reviewed in detail in Grocott et al., 2012; Groves & LaBonne, 2014). As the neural ectoderm thickens into an early neural plate, boundaries form between these expression domains: *SoxB1* genes are confined to the neural plate, *Zic* genes span the lateral neural plate and medial NB zone, *Msx1* and *Pax3* become confined to the NB zone, and *Dlx*, *GATA*, *Foxi*, and *TFAP2α* no longer overlap with neural ectoderm (Fig. 2). At late neural plate/neurula stages, *Msx1* and *Pax3* become confined to the neural crest region, along with neural crest-specifying genes (*Foxd3*, *Sox9/10*, *Snail2*); in addition, some *Dlx* genes and *TFAP2α* are newly expressed in the neural crest. The PPE expresses *Six* and *Eya* genes, and the epidermis expresses *Dlx*, *GATA*, and *Foxi* genes (Fig. 2). These changing expression patterns indicate that NB-specifying genes have dynamic roles in the transcriptional pathway that leads to PPE formation. What is the experimental evidence for their functions?

Distal-less-related homeobox transcription factors, in particular *Dlx3*, *Dlx5*, and *Dlx6*, have two important functions in the NB zone (Fig. 3): they are required for the expression of both neural crest and PPE genes, and they repress neural plate genes (Beanan & Sargent, 2000; Esterberg & Fritz, 2009; Feledy et al., 1999; Kaji & Artinger, 2004; Luo, Matsuo-Takasaki, & Sargent, 2001; McLarren et al., 2003; Solomon & Fritz, 2002; Woda et al., 2003). In the neural plate transplantation experiments described above, both neural crest and PPE markers were induced only when *Dlx* gene activity was intact (Woda et al., 2003). It has been proposed that Dlx factors promote PPE formation by regulating a bone morphogenetic protein (BMP) antagonist in the NB zone (Esterberg & Fritz, 2009) and/or by interacting with GATA factors (McLarren et al., 2003; Phillips et al., 2006; Pieper, Ahrens, Rink, Peter, & Schlosser, 2012; Solomon & Fritz, 2002). However, Dlx and GATA factors are not equivalent in activity; while loss of either Dlx3 or GATA2 significantly reduces PPE gene expression, only Dlx3 can expand PPE gene expression (Pieper et al., 2012). A PPE-specific enhancer in the *Six1* gene contains putative binding sites for both GATA factors and homeodomain (HD)-containing factors such as Dlx (Sato et al., 2010). Mutating the GATA sites reduced expression of a *Six1* enhancer–reporter construct, whereas mutating the HD sites eliminated it.

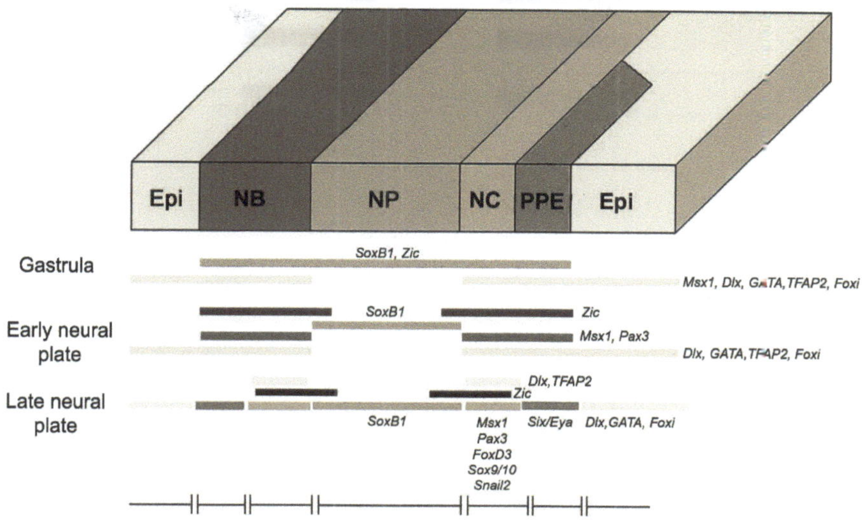

Figure 2 Several steps are involved in forming the preplacodal ectoderm. Top: the embryonic ectoderm of a neurula stage has been flattened into a sheet. On the left is represented the fields apparent at early neural plate stages: epidermis (Epi), neural border (NB) zone, and neural plate (NP). On the right, the NB has divided into its neural crest (NC) and preplacodal ectoderm (PPE) derivatives. Bottom: different sets of transcription factors are differentially expressed in these ectodermal domains over developmental time. At gastrula stages, the expression domains of NP genes (blue (gray in the print version) bar; *SoxB1* and *Zic*) and epidermal genes (yellow (light gray in the print version) bars; *Msx1*, *Dlx*, *GATA*, *TFAP2*, and *Foxi*) overlap in a region that will become the NB zone. At early NP stages, *SoxB1* gene expression recedes from the NB zone, *Zic* genes (purple (light gray in the print version) bars) are no longer expressed in the medial region of the NP, *Msx1* and *Pax3* expression (dark green (dark gray in the print version) bars) is confined to the NB zone, whereas *Dlx*, *GATA*, and *TFAP2* are expressed in both the NB zone and epidermis (yellow (light gray in the print version) bars). At late NP stages, *Six* and *Eya* genes are expressed in the PPE domain (medium green (dark gray in the print version) bars) and "neural crest-specifying" genes (*Msx1*, *Pax3*, *FoxD3*, *Sox9/10*, and *Snail2*) are expressed in the NC domain (light green (gray in the print version) bars). *Dlx3* and *TFAP2* also are expressed in the NC domain, whereas *Dlx5/6*, *GATA*, and *Foxi1* are expressed in the epidermis (yellow (light gray in the print version) bars) and are now excluded from the PPE. Once these four domains are formed, their boundaries appear to be maintained by mutual repression between domain-specific transcription factors (red (dark gray in the print version) bars).

Other experiments indicate that Dlx also functionally interacts with Msx1 (Fig. 3). *Dlx* and *Msx* expression domains partially overlap in the NB zone (Arkell & Beddington, 1997; Feledy et al., 1999; McLarren et al., 2003; Monsoro-Burq, Wang, & Harland, 2005; Schlosser & Ahrens, 2004), and their proteins are known to inhibit each other through

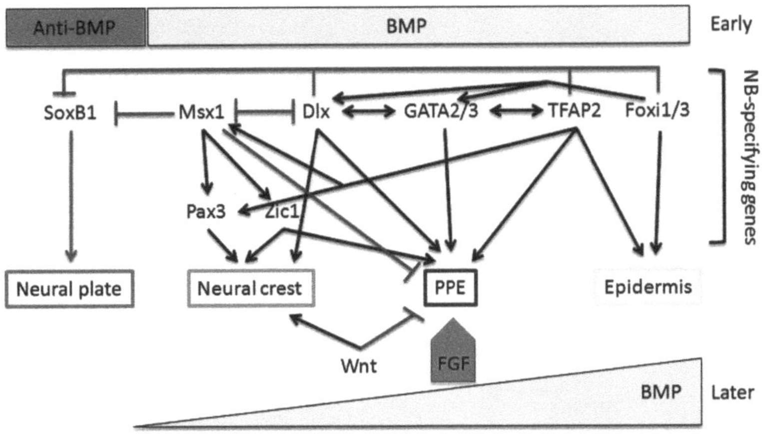

Figure 3 During gastrulation, early, high BMP levels activate several NB-specifying genes and low BMP levels, controlled by the secretion of BMP antagonists, and allow the expression of neural genes (e.g., SoxB1 transcription factors). The NB-specifying genes interact with each other to promote (black arrows) the formation of the NB zone, and they repress (red (dark gray in the print version) bars) the expression of neural genes. Indicated are the various interactions, described in the text, that result in the separation of the NB zone into the neural crest and the preplacodal ectoderm (PPE). Contributing to this separation are several signaling pathways. Later, high levels of BMP promote an epidermal fate, but this pathway must be attenuated to produce neural crest and PPE. An FGF pulse is required for PPE formation. Wnt positively regulates neural crest formation and represses PPE formation.

the formation of heterodimers (Givens et al., 2005; Zhang et al., 1997). Differential knockdown of these genes biases NB zone cells toward either a neural crest fate (*Msx*-high, *Dlx*-low) or a PPE fate (*Msx*-low, *Dlx*-high; Phillips et al., 2006). GST assays with the PPE-specific *Six1* enhancer construct indicates that both Dlx5 and Msx1 can bind to the HD sites, and reporter assays indicate that Dlx5 activates *Six1* whereas Msx1 represses it (Sato et al., 2010). Msx1 activates two other NB-specifying genes: *Pax3* and *Zic1* (Fig. 3). Interactions between Pax3 and Zic1 are well known to initiate neural crest gene expression (Monsoro-Burq et al., 2005; Sato, Sasai, & Sasai, 2005; Tribulo, Ayba, Nguyen, Mullins, & Mayor, 2003). By manipulating the timing of *Pax3* expression with inducible constructs, Hong and Saint-Jeannet (2007) showed that Pax3 has an early role in forming the NB zone and a later role in neural crest specification when coexpressed with Zic1; alternatively, expression of Zic1 in the absence of Pax3 leads to PPE gene expression (Fig. 3).

It needs to be kept in mind, however, that not all murine knockout animals display phenotypes consistent with these results. Although *Msx1*-, *Dlx5*-, and *Pax3*-null mice display neural crest defects that include craniofacial anomalies (www.informatics.jax.org), *Zic1*-null mice do not (Aruga et al., 1998). Instead, they show perinatal growth defects and movement defects consistent with the role of this gene in cerebellar development.

Foxi1 and TFAP2α also are required for both neural crest and PPE formation (Fig. 3). Loss of *Foxi1* in *Xenopus* embryos expands neural plate (*Sox2*) and reduces NB-specifying (*Dlx*), neural crest (*Foxd3*), PPE (*Six1*), and epidermis (*keratin*) genes; in gain-of-function experiments, *Sox2*, *Foxd3*, and *Six1* were reduced whereas *Dlx* and *keratin* were upregulated (Matsuo-Takasaki, Matsumura, & Sasai, 2005). By utilizing a hormone-inducible version of Foxi1, these authors showed that it is the gastrula stage expression of *Foxi1* that regulates NB zone formation. TFAP2α is a key regulator of epidermis genes (Luo, Lee, Saint-Jeannet, & Sargent, 2003; Luo, Matsuo-Takasaki, Thomas, Weeks, & Sargent, 2002; Nguyen et al., 1998; Zhang, Luo, & Sargent, 2006). Loss of TFAP2α downregulates epidermis (*keratin*), NB-specifying (*Msx1*, *Pax3*), neural crest (*Foxd3*, *Slug*, *Sox9*), and PPE (*Six1*, *Eya1*) genes and upregulates neural plate genes; the opposite phenotypes are observed with gain-of-function experiments (de Croze, Maczkowiak, & Monsoro-Burq, 2011; Hoffman, Javier, Campeau, Knight, & Schilling, 2007; Li & Cornell, 2007; Luo et al., 2002). Epistasis analyses show that *TFAP2α* acts upstream of other NB-specifying genes, with *Pax3* being a direct transcriptional target (de Croze et al., 2011; Luo et al., 2002).

These studies indicate that transcriptional interactions between the NB-specifying genes lead to a NB transcriptional state from which either neural crest or PPE cells can arise, depending upon the timing and combinations of genes that are expressed (Figs. 2 and 3). A recent study showed that maintenance of *TFAP2α*, *Foxi1*, and *GATA* expression in the NB zone is required for PPE formation, and this is accomplished by cross-regulation between these factors after local BMP signaling is attenuated (Bhat, Kwon, & Riley, 2013). Likewise, in human embryonic stem cell cultures, the expression of *TFAP2α*, *GATA*, and *DLX* genes must precede the expression of PPE genes for placode cells to differentiate (Leung, Morest, & Li, 2013). Thus, the NB zone transcriptional state appears to set the stage for PPE-specific gene expression.

2.2. Induction of the PPE genes by signaling factors

After the NB zone is established, signals from the local environment induce a distinct set of PPE-specific transcriptional regulators, in particular members of the *Six* and *Eya* gene families (reviewed in Bhattacharyya & Bronner-Fraser, 2004; Grocott et al., 2012; Moody & Saint-Jeannet, 2014; Saint-Jeannet & Moody, 2014; Schlosser & Ahrens, 2004; Streit, 2004). Although the neural plate-grafting experiments discussed earlier indicate that interactions between the neural and nonneural ectoderm are necessary for the induction of PPE genes, ablation and transplantation experiments also demonstrate that they are not sufficient. Experiments in chick and *Xenopus* showed that head mesoderm provides an additional, required, PPE-inducing signal (Ahrens & Schlosser, 2005; Litsiou et al., 2005). Both studies identified fibroblast growth factor (FGF) as the likely additional signal (Fig. 3); FGF also promotes PPE gene expression in fish (Esterberg & Fritz, 2009; Kwon, Bhat, Sweet, Cornell, & Riley, 2010). Consistent with these findings, endogenous FGFs are expressed in the adjacent head mesoderm and anterior neural ridge (Ahrens & Schlosser, 2005; Eagleson & Dempewolf, 2002; Ohuchi, Kimura, Watamoto, & Itoh, 2000; Shamim & Mason, 1999; Shim, Bae, Park, Kim, & Han, 2005). However, several studies also showed that FGF signaling alone is not sufficient to induce PPE genes (Ahrens & Schlosser, 2005; Leger & Brand, 2002; Litsiou et al., 2005; Liu et al., 2003; Maroon et al., 2002; Phillips, Bolding, & Riley, 2001; Solomon, Kwak, & Fritz, 2004). It appears that the level of FGF signaling is important: while low levels of FGF signaling promote PPE fate, high levels inhibit it (Brugmann, Pandur, Kenyon, Pignoni, & Moody, 2004; Hong & Saint-Jeannet, 2007). Grocott et al. (2012) proposed that FGF promotes PPE formation by both inhibiting PPE-repressing factors and upregulating PPE-specific gene expression. However, this transcriptional control may not be direct because two recently characterized PPE-specific enhancers of *Six1*, which are conserved across humans, mouse, chick, and frog, do not contain any obvious FGF pathway binding sites (Sato et al., 2010, 2012). Study of other enhancers is warranted, therefore, to elucidate how the FGF pathway may regulate the expression of the *Six1* gene, as well as other PPE-specific genes.

BMP signaling plays important roles in both NB zone and PPE gene induction (Fig. 3). Most NB-specifying genes are induced by BMP (Bhat et al., 2013; Luo et al., 2003; Luo, Matsuo-Takasaki, Lim, & Sargent, 2001; Luo, Matsuo-Takasaki, & Sargent, 2001; Suzuki,

Ueno, & Hemmati-Brivanlou, 1997; Tribulo et al., 2003). Sustained high levels of BMP promote an epidermal fate, whereas lower levels of BMP signaling, experimentally accomplished by expressing either dominant-negative BMP receptors or BMP antagonists, preferentially expand PPE gene expression domains (Ahrens & Schlosser, 2005; Brugmann et al., 2004; Glavic et al., 2004; Kwon et al., 2010; Litsiou et al., 2005). In fact, in order for FGF8 to induce PPE genes in the nonneural ectoderm, the level of BMP signaling in the epidermis that activates the NB-specifying and epidermal genes must be reduced (Ahrens & Schlosser, 2005). BMP levels also can distinguish between induction of neural crest versus PPE fate in *Xenopus* ectodermal explants (Brugmann et al., 2004). BMP levels that induce a neural crest fate activate *TFAP2α* alone, whereas BMP levels that induce a PPE fate activate *TFAP2α* plus *Foxi1* and *Gata3* (Bhat et al., 2013). The timing of the BMP signaling also is critical (Fig. 3). In fish, high levels of BMP at gastrula stages activate NB-specifying genes, whereas at neural plate stages lower levels of BMP are necessary to activate PPE genes (Bhat et al., 2013; Kwon et al., 2010). Similarly, human embryonic stem cells directed to a placode fate require an early pulse of BMP to activate NB-specifying genes, and a later attenuation of BMP signaling to activate PPE genes (Dincer et al., 2013; Leung et al., 2013).

Wnt signaling appears to antagonize PPE induction (Fig. 3). In both chick and *Xenopus* embryos, elevated Wnt signaling represses PPE genes, whereas reducing it expands the PPE (Brugmann et al., 2004; Hong & Saint-Jeannet, 2007; Litsiou et al., 2005; Matsuo-Takasaki et al., 2005). Secreted anti-Wnt factors are highly expressed in the anterior neural plate and the underlying chordomesoderm (Bradley et al., 2000; Carmona-Fontaine, Acuna, Ellwanger, Niehrs, & Mayor, 2007; Pera & De Robertis, 2000; Takai et al., 2010), which may account for the restriction of PPE gene expression to the ectoderm surrounding the anterior neural plate (Fig. 1). In chick NB zone explants, if Wnt signaling is attenuated in the presence of low BMP, placode markers are expressed, whereas if Wnt signaling persists with low BMP, neural crest markers are expressed (Patthey, Edlund, & Gunhaga, 2009, Patthey, Gunhaga, & Edlund, 2008). In fact, it appears that a pulse of FGF in a BMP low/Wnt low environment is the most effective means of inducing PPE genes (Litsiou et al., 2005). These authors propose that the FGF pulse confers a "neural border" state on the ectoderm, and that those cells within the NB zone that are protected from Wnt signaling become PPE whereas those that are exposed to Wnt signaling become neural crest. It is interesting to note that one of the conserved PPE-specific enhancers of *Six1* contains a Wnt/Lef1 site (Sato et al., 2012); this suggests that Wnt regulation of *Six1* may be direct but this site needs to be functionally tested.

Another important signal in PPE formation is retinoic acid (RA). Raldh2, the RA-synthesizing enzyme, is expressed in a discrete U-shaped ectodermal domain around the anterior neural plate that appears coincident with the PPE (Chen, Pollet, Niehrs, & Pieler, 2001). Decreasing RA signaling during PPE formation expands the posterior limit of a similar U-shaped FGF8 expression domain in the cranial mesoderm (Shiotsugu et al., 2004), suggesting that endogenous RA contributes to limiting the PPE to the head (Figs. 1 and 4). This may be accomplished via the differential expression of two RA-regulated genes: *Tbx1*, a T-box transcription factor, and *Ripply3/*

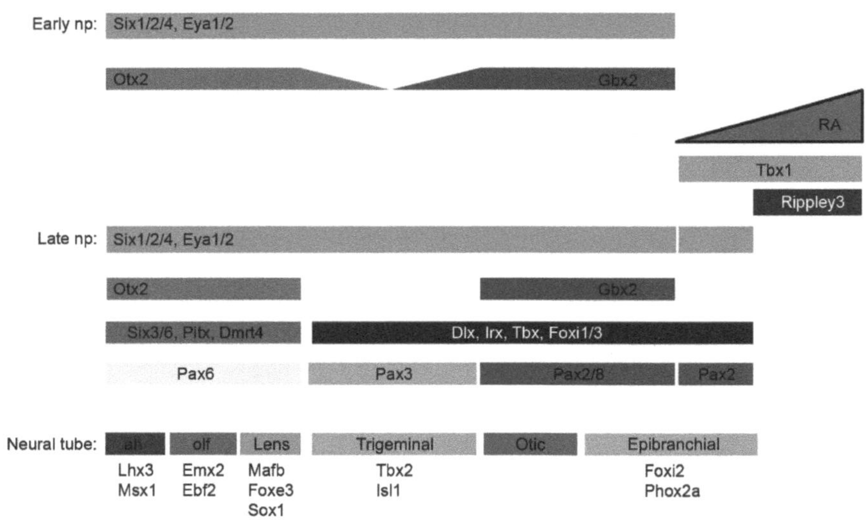

Figure 4 A summary of the anterior–posterior (A–P) patterning of the PPE into anterior, intermediate, and posterior domains. Anterior is to the left and posterior is to the right. At early neural plate (np) stages, the anterior-specific expression of *Otx2* and posterior-specific expression of *Gbx2*, and their mutual repression in the intermediate zone set the stage for A–P patterning of the PPE, which uniformly expresses *Six* and *Eya* genes. Subsequently, RA signaling that locally upregulates *Tbx1* and *Ripply3* likely causes new *Six*/*Eya* expression in a more posterior domain that will give rise to the most posterior epibranchial placodes. At late neural plate stages, other transcription factors become restricted to either the anterior domain or the intermediate/posterior domains. Presumably these transcriptional combinations, together with further local signaling, result in region-specific *Pax* gene expression, leading to the specification of distinct placode fates. Subsequently, each of these placodes (ah, adenohypophyseal; olf, olfactory; Lens; Trigeminal; Otic; Epibranchial) expresses unique combinations of additional genes.

Dscr6, a Groucho-associated corepressor (Arima et al., 2005). *Ripply3* and *Tbx1* expression domains in the PPE partially overlap; in regions where only *Tbx1* is expressed, PPE genes are induced, whereas in regions where *Tbx1* and *Ripply3* overlap, PPE genes are repressed (Janesick, Shiotsugu, Taketani, & Blumberg, 2012). Thus, RA signaling appears to be required for the formation of the posterior, Tbx1-positive part of the PPE and for restricting the posterior boundary of the PPE by coinducing Ripply3 (Fig. 4).

2.3. PPE transcriptional regulators

Six and *Eya* genes were among the first transcriptional regulators to be implicated in placode development because they are expressed in the classically described U-shaped PPE domain that surrounds the anterior neural plate (Fig. 1; Bessarab, Chong, & Korzh, 2004; David et al., 2001; Esteve & Bovolenta, 1999; Ghanbari et al., 2001; Kobayashi et al., 2000; McLarren et al., 2003; Pandur & Moody, 2000; Sato et al., 2010; Schlosser & Ahrens, 2004). Functional studies in several animals have since demonstrated that several members of these two gene families have important roles in the initial specification of the PPE and the formation and differentiation of many of its derivatives.

Drosophila Sine oculis (*SO*), which is essential for fly visual system formation (Cheyette et al., 1994; Serikaku & O'Tousa, 1994), is the founding member of the highly conserved *Six* (*Sine oculis homeobox*) gene family. There are 6 vertebrate *Six* genes that are clustered into two groups (*Six1*/*Six4*/*Six6* and *Six2*/*Six3*/*Six5*) on two separate chromosomes; these are closely syntenic in frog, mouse, and human and slightly rearranged in chick and zebrafish (Moody & Saint-Jeannet, 2014). The proteins are functionally grouped into three subfamilies (Six1/Six2; Six3/Six4; Six5/Six6) based on sequence variations in both the HD, which binds to DNA, and the Six domain (SD), which binds to cofactors that regulate DNA binding specificity and protein activity (Kawakami, Sato, Ozaki, & Ikeda, 2000; Kobayashi, Nishikawa, Suzuki, & Yamamoto, 2001; Pignoni et al., 1997). *Six* genes play major roles in eye, muscle, kidney, genital, limb, and craniofacial development (Brodbeck & Englert, 2004; Kawakami, Ohto, Ikeda, & Roeder, 1996). Although *Six3* plays an important role in lens differentiation (e.g., Oliver, Loosli, Koster, Wittbrodt, & Gruss, 1996), we will not consider it further because it is not expressed in the PPE or in other placodes, and its function in lens development is discussed in detail elsewhere (Ogino, Ochi, Reza, & Yasuda, 2012). Instead, we will focus on the three members of the family (*Six1*, *Six2*, *Six4*) that are strongly expressed in the PPE, placodes, and numerous placode derivatives (reviewed by Brugmann & Moody, 2005).

Several experiments indicate that *Six1* plays a central role in PPE and placode development. *Six1* loss-of-function in *Xenopus* results in reduced expression of other early PPE markers (*Eya1*, *Sox11*, *Sox2/3*; Brugmann et al., 2004; Schlosser et al., 2008). In another study that reduced Six1 activity by repressing Six1 targets, placode-specific gene expression likewise was lost, and the otic placode was malformed (Christophorou et al., 2009); in fish and mouse, *Six1* knockdown also resulted in inner ear defects (Bricaud & Collazo, 2006, 2011; Konishi, Ikeda, Iwakura, & Kawakami, 2006; Laclef, Souil, Demignon, & Maire, 2003; Oliver et al., 1995; Ozaki et al., 2004; Zheng et al., 2003; Zou, Silvius, Fritzsch, & Xu, 2004). Although *Six1* is expressed in the mouse PPE (Sato et al., 2010), its loss only appears to affect later stages of placode development (e.g., Chen, Kim, & Xu, 2009; Ikeda, Kageyama, Suzuki, & Kawakami, 2010; Ikeda et al., 2007). This lack of an early phenotype may be due to gene redundancy because, in contrast, the olfactory placode does not form in *Six1*/*Six4* compound null mice (Chen et al., 2009). Gain-of-function experiments in both chick and frog demonstrate that Six1 upregulates early expressed PPE genes and represses epidermal, neural crest, and neural plate genes (Brugmann et al., 2004; Christophorou et al., 2009). However, ectopic expression of Six1 outside the NB zone is not sufficient to induce other PPE markers (Brugmann et al., 2004; Christophorou et al., 2009; Schlosser et al., 2008), confirming that the appropriate signaling environment of the NB zone (Fig. 3) is necessary to induce PPE gene expression.

Although Six2 and Six4 are frequently used as PPE and placode markers in a number of animals, their functional roles have not yet been described in any detail. Placode deficiencies have not been reported in either *Six2*-null or *Six4*-null mice (Ozaki et al., 2001; Self et al., 2006), but this may be due to redundant functions between the family members. It will be important to elucidate the cooperative roles of Six1, Six2, and Six4 in PPE and placode development, particularly because they are known to have distinct roles in muscle differentiation and kidney development (Brodbeck & Englert, 2004; Fougerousse et al., 2002; Himeda et al., 2004; Ohto et al., 1998; Spitz et al., 1998; Xu et al., 2003).

Six proteins can bind to cofactor proteins that do not bind to DNA on their own, but nonetheless modulate Six transcriptional activity (Bricaud &

Collazo, 2011; Brugmann et al., 2004; Giot et al., 2003; Tessmar, Loosli, & Wittbrodt, 2002; Zhu et al., 2002). The best studied vertebrate Six cofactors are the four members of the Eya family; they are homologous to *Drosophila Eyes absent* (*Eya*), which plays an essential role in fly eye development as a SO cofactor (Bonini, Leiserson, & Benzer, 1993). Like the *Six* genes, vertebrate *Eya* genes are expressed in numerous embryonic tissues, including the eyes, muscles, kidneys, the PPE, and placodes (David et al., 2001; Duncan et al., 1997; Ishihara, Ikeda, Sato, Yajia, & Kawakami, 2008; Modrell & Baker, 2012; Neilson, Pignoni, Yan, & Moody, 2010; Sahly, Andermann, & Petit, 1999; Xu, Cheng, Epstein, & Maas, 1997). In *Xenopus*, chick, and fish, the expression patterns of *Eya1* and *Eya2* are nearly identical to that of *Six1* (Bane, Van Rybroek, Kolker, Weeks, & Manaligod, 2005; David et al., 2001; McLarren et al., 2003; Neilson et al., 2010; Sahly et al., 1999), suggesting that they have important roles in PPE and placode development. Eya proteins contain a highly conserved protein-binding domain called the Eya domain (ED) located at the C-terminus. In *Drosophila*, the ED interacts with the SD of the SO protein (Pignoni et al., 1997). In vertebrates, the interaction between the Six1 SD and the Eya1 ED is essential for Eya1 nuclear translocation and for exerting the transcriptional function of the complex (Ikeda, Watanabe, Ohto, & Kawakami, 2002; Ohto et al., 1999; Patrick et al., 2013). It should be kept in mind, however, that Eya1 can bind to several other proteins, so its function is likely not restricted to the Six1 transcriptional pathway (Giot et al., 2003; Heanue et al., 1999; Kenyon, Li, Clouser, Tran, & Pignoni, 2005; Ohto et al., 1999). Eya also functions as a phosphatase (Rayapureddi et al., 2003; Tootle et al., 2003), which impacts whether the Six1 transcriptional complex activates or represses targets (Li et al., 2003), and as a substrate for mitogen-activated protein kinase (Hsiao, Williams, Davies, & Rebay, 2001). There have been several *Eya1* loss- and gain-of function studies in *Xenopus*, fish, and mouse (Johnson et al., 1999; Li, Manaligod, & Weeks, 2010; Nica et al., 2006; Xu et al., 1999, 2002; Zou et al., 2004; Zou, Silvius, Rodrigo-Blomqvist, Enerback, & Xu, 2006). Although defects are seen in various placode derivatives, to our knowledge no study has addressed the role of this protein in PPE formation. To date, *Eya2*-null mouse phenotypes have not been reported, and while *Eya3*-null mice have behavioral abnormalities consistent with its regulation of neural plate cell survival in *Xenopus* (Kriebel, Muller, & Hollemann, 2007), craniofacial deficits were not noted (Soker et al., 2008). Loss of *Eya4* in fish and mouse results in inner ear defects (Depreux et al., 2008; Schonberger et al., 2005; Wang et al., 2008). Sorting out the distinctive and/or overlapping roles of the different Eya proteins will be important for understanding how these proteins contribute to the transcriptional regulations of placode development.

Another well-described interactor with Six proteins is Groucho. Six1/Six2 can act as both transcriptional activators and repressors, depending on the presence of Eya or Groucho cofactors (Silver, Davies, Doyon, & Rebay, 2003). Using activating (Six1VP16) and repressing (Six1EnR) *Six1* constructs, it was shown that Six1 transcriptionally activates PPE genes in cooperation with Eya1 and transcriptionally represses epidermal and neural crest genes in cooperation with Groucho (Brugmann et al., 2004; Christophorou et al., 2009). In the fish otocyst, Groucho is required (Bajoghli, Aghaallaei, & Czerby, 2005). Further, combined Six1/Eya1 activation of target genes is required for hair cell fate, whereas combined Six1/Groucho repression of target genes is required for neural cell fate (Bricaud & Collazo, 2011). Because both *Eya* and *Groucho* genes are endogenously expressed in the PPE, these data indicate that Six1 functions in PPE development as both a transcriptional activator and a repressor, depending on the cofactor with which it interacts.

Several lines of evidence indicate that there are likely to be additional modifiers of Six transcriptional activity that are developmentally relevant to PPE and placode development. We embarked on a screen of vertebrate homologues of reported novel fly SO interactors (Giot et al., 2003; Kenyon et al., 2005) because the SD of fly SO differs by only four nonconserved amino acids from the SD of *Xenopus* and human Six1/Six2. We identified *Xenopus* genes with amino acid sequences homologous to 20/25 of the fly SO interactors. We analyzed the expression patterns of the homologues of 11 of the fly genes and found that most overlap extensively in the PPE and placodes with *Six1* (Neilson et al., 2010). To date, functional data are only available for one of these: Sine oculis binding protein (Sobp). In fly, this protein is coexpressed with SO in the eye imaginal disc where its misexpression interferes with normal eye neurogenesis (Kenyon et al., 2005). In mouse, a recessive mutation in *Sobp* caused abnormal cochlear development (Chen et al., 2008), and in humans, a truncating mutation of *SOBP* caused craniofacial abnormalities in a family in which one member had hearing loss (Basel-Vanagaite et al., 2007; Birk et al., 2010). There also are likely to be protein regulators that modify Six1 function by binding to or modifying the activity of Eya (or other putative cofactors). For example, Dachshund (Dac), which has an important role in *Drosophila* eye development in cooperation with Eya and SO (Chen, Amoui, Zhang, & Mardon, 1997), binds to both Eya and DNA but does not have a direct interaction with SO. Vertebrate *Dac* is expressed widely in embryonic tissues, including placodes (Grocott et al., 2012; McLarren et al., 2003; Schlosser, 2006), and can regulate the transcriptional effectiveness of Six/Eya complexes (Heanue et al., 1999; Ikeda et al., 2002; Li et al., 2003). However, a specific role for Dac or for other potential members of the Six/Eya transcriptional complex in PPE and placode development remains to be revealed.

Based on studies showing that kidney development is regulated by a transcriptional network that includes *Six*, *Eya*, *Pax*, and *Fox* genes (reviewed in Brodbeck & Englert, 2004), it was proposed that a similar network might regulate sensory placode development (Bhattacharyya & Bronner-Fraser, 2004). At the time, there was only sufficient evidence for the lens placode, so an update is warranted. As discussed above, Pax3 and Foxi1 are important for NB zone formation, and Six and Eya are critical for PPE and placode development. Explant studies suggest that a PPE-wide molecular ground state includes expression of *Pax6* (Bailey et al., 2006), and, as we will discuss below, *Pax* genes are critical in conferring individual placode identity. Foxi1 and Foxi3 are important for otic placode development (Khatri & Groves, 2013; Nissen, Yan, Amsterdam, Hopkins, & Burgess, 2003; Ohyama & Groves, 2004; Solomon, Logsdo, & Fritz, 2003), Foxe3 for lens placode development (Brownell, Dirksen, & Jamrich, 2000; Kenyon, Moody, & Jamrich, 1999; Ogino et al., 2012), and *Foxg1* is expressed in all placodes in the mouse and plays a role in olfactory and auditory development (Duggan, DeMaria, Baudhuin, Stafford, & Ngai, 2008; Hatini, Xin, Balas, & Lai, 1999; Hwang, Simeone, Lai, & Wu, 2009; Kawauchi et al., 2009; Pauley, Lai, & Fritzsch, 2006). Thus, there is a little doubt that *Six/Eya/Pax/Fox* genes are important regulators in PPE and placode development. However, their epistatic and transcriptional relationships have not yet been clearly delineated. Furthermore, there are certain to be numerous additional members of the PPE/placode regulatory network. For example, *Sox*, *Irx*, and *Tbx* genes have been experimentally placed in the pathway (Brugmann et al., 2004; Schlosser & Ahrens, 2004; reviewed in Grocott et al., 2012; Streit, 2004). In *Drosophila*, SO is known to directly regulate several genes required for eye formation (reviewed in Jusiak et al., 2014) that also are involved in placode development. These include *eyeless* (vertebrate *Pax6*; involved in PPE ground state and anterior placodes), *dac* (see above), *atonal* (vertebrate *Atoh1-8*; involved in placode neurogenesis), *prospero* (*Prox1/2*; involved in vertebrate lens development; Mizuno et al., 1999; Wigle, Chowdhury, Gruss, & Oliver, 1999), and *hedgehog* (vertebrate *Shh*, involved in adenohypophyseal placode development; reviewed in Saint-Jeannet & Moody, 2014). A ChIP-Seq analysis of SO binding to DNA isolated from developing fly eye-antennal imaginal discs identified nearly 6000 putative SO target genes, over half of which do not have a described function in eye development (Jusiak et al., 2014). Although flies do not have the equivalent of cranial sensory placodes, we speculate that some of these genes may prove important in the transcriptional regulation of vertebrate placode development because of the significant conservation of many gene regulatory networks between flies and vertebrates. Using a different approach, we recently conducted an expression screen to identify potential new targets of Six1 in *Xenopus* ectodermal explants and identified 72 upregulated and 58 downregulated genes (Yan et al., 2014). Like the fly study, most candidates are of unknown function, but over 30 of these genes are expressed in the PPE and placodes, suggesting they may be part of the placode transcriptional network. Functionally analyzing these new putative SO/Six1 targets in both flies and vertebrates is likely to significantly advance our understanding of the PPE regulatory state.

The PPE transcriptional regulators have two important roles. They set the transcriptional landscape for subsequent development into specific placodes with distinct fates (Fig. 4; to be discussed below). Additionally, they maintain the boundaries between the PPE, neural crest, and epidermis (Fig. 2). Several studies have shown that the expression domains of transcription factors that are initially broad and overlapping become discrete stripes with distinct boundaries (reviewed in Groves & LaBonne, 2014; Moody & Saint-Jeannet, 2014; Schlosser, 2006; Schlosser & Ahrens, 2004; Streit, 2004). The formation of expression domains with sharp boundaries suggests that these factors are mutually repressive. There is a lot of experimental support for this idea. For example, in embryo gain-of-function assays, Dlx5/6 and Zic2 both repress *Six1*, and Six1 represses *Dlx5/6* and *Zic2* (Brugmann et al., 2004; Woda et al., 2003). Epidermal, neural crest, and neural plate genes have mutually repressive interactions with *Six1* (Brugmann et al., 2004; Christophorou et al., 2009; Matsuo-Takasaki et al., 2005). Thus, after the fates of the four major ectodermal subdomains are specified and unique sets of transcription factors are expressed, these same factors appear to continue as maintenance factors to preserve the boundaries between the subdomains (Figs. 2 and 3). However, experiments that control the timing and spatial localization of gene expression are needed to fully understand the molecular interactions that establish and maintain the boundaries between the different ectodermal subdomains.

3. BREAKING THE PPE INTO INDIVIDUAL PLACODES WITH DIFFERENT DEVELOPMENTAL FATES

After the PPE is established as a separate subdomain in the embryonic ectoderm with distinct boundaries from the other subdomains (epidermis, neural crest, neural plate), the PPE undergoes several steps of regionalization (reviewed in Bailey & Streit, 2006; Grocott et al., 2012; Groves & LaBonne, 2014; Saint-Jeannet & Moody, 2014). Classical work in amphibians demonstrated that when pieces of PPE were transplanted to novel sites at early stages, they gave rise to placodes that were appropriate for their *new* A–P position, whereas when transplanted later they gave rise to placodes that were appropriate for their *original* A–P position (Jacobson, 1963). These results indicate that cells within the PPE are initially competent to form any placode, and only after interactions with adjacent tissues at specific A–P addresses do they become specified to particular placode fates (Jacobson, 1966). This idea has since been supported in chick, frog, and fish by similar transplantation studies using molecular markers (Bhattacharyya, Bailey, Bronner-Fraser, & Streit, 2004; Ladher, O'Neill, & Begbie, 2010; Pieper et al., 2011; Streit, 2002; Toro & Varga, 2007; Xu, Dude, & Baker, 2008). Although the early PPE is competent to form all types of placodes and has a unified molecular ground state, a few hours later it begins to acquire separate anterior (adenohypophyseal, olfactory, lens), intermediate (trigeminal), and posterior (otic, lateral line, epibranchial) domains (Fig. 4). Fate mapping studies show that at late PPE stages, precursors of anterior and posterior regions begin to sort out (Bhattacharyya et al., 2004; Dutta et al., 2005; Ladher et al., 2010; Pieper et al., 2011; Streit, 2002; Toro & Varga, 2007; Xu et al., 2008), concomitant with the differential expression of several sets of transcription factors that are placode specific (Fig. 4). What are the transcriptional and signaling factors that regulate this process? Similar to the program that regulates developmental patterning of the neural tube and the neural crest, the PPE appears to first undergo A–P patterning and then subsequent regionalization due to interactions with nearby signaling centers that maintain or repress specific sets of transcription factors.

One potential mechanism that accomplishes A–P regionalization of the PPE is the differential regulation of *Six* gene expression. An in-depth analysis of *Six1* enhancers identified a genomic region that is conserved across tetrapods and activates reporter expression only in the anterior PPE

(Sato et al., 2010). Although a posterior-specific enhancer was not identified, there are other conserved enhancers that are active in combinations of the PPE subdomains: intermediate + posterior; adenohypophyseal + posterior; and anterior + posterior (Sato et al., 2012). It will be very important to identify the factors that bind to these enhancers to determine the direct regulators of region-specific expression of *Six1*. Evidence from *Xenopus* implicates RA signaling for posterior *Six1* expression. First, the most posterior lateral line and epibranchial placodes form posterior to the initial *Six1/Eya1* PPE expression domain, and they only express *Six1/Eya1* during neurulation (Schlosser & Ahrens, 2004). This later, posterior expression is likely to occur in response to RA regulation of *Rippley3/Tbx1* expression described above (Janesick et al., 2012) (Fig. 4). Another potential mechanism for A–P regionalization of the PPE is the differential expression of *Otx2* and *Gbx2* (Steventon, Mayor, & Streit, 2012). *Otx2* is expressed in the anterior PPE domain, *Gbx2* in the posterior PPE domain, and where their expression overlaps in the intermediate PPE domain both genes are downregulated by their mutual inhibition (Fig. 4). How this patterned expression is regulated is not yet understood, but there is evidence that differential responsiveness to BMP signaling may be involved (Sjödal, Edlund, & Gunhaga, 2007). Also, there is evidence for differential responsiveness to Wnt signaling. For example, in *axin1* mutant fish that harbor increased Wnt activity, the most anterior placodes are lost and the posterior placodes are expanded (Heisenberg et al., 2001). In *Xenopus* ectodermal explants, an anterior placode gene (*Dmrt4*) can be induced by coexpression of FGF8 and Noggin (an anti-BMP factor), but adding Wnt to the cocktail only induces a posterior placode gene (*Pax8*; Park & Saint-Jeannet, 2008).

Other genes that are initially expressed broadly in the PPE also gradually become restricted to mostly anterior (e.g., *Six3*, *Six6*, *Pitx3*) versus mostly posterior (e.g., *Dlx*, *Irx*, *Tbx*, *Foxi3*) placodes (Fig. 4; Khatri and Groves, 2013; Schlosser, 2006; Schlosser & Ahrens, 2004; Sjödal & Gunhaga, 2008; reviewed in Grocott et al., 2012). This may be in direct response to differential expression of Otx2 and Gbx2. But it is generally accepted that local signals from underlying mesodermal and endodermal tissues, including specific combinations of BMP, FGF, Wnt, and RA signaling, significantly contribute to placode-specific transcriptional identity (reviewed in detail in Baker & Bronner-Fraser, 2001; Saint-Jeannet & Moody, 2014; Schlosser, 2006; Schlosser & Ahrens, 2004; Streit, 2004). Some signaling pathways appear to be unique for certain placodes, including Sonic hedgehog (adenohypophyseal; Herzog et al., 2003; Lewis et al., 1999; Treier et al.,

2001), Somatostatin/Nociceptin (olfactory, lens; Llleras-Forero et al., 2013), Platelet-derived Growth Factor (trigeminal; McCabe & Bronner-Fraser, 2008), and Notch (otic and epibranchial; Abello, Khatri, Giraldez, & Alsina, 2007; Jayasena, Ohyama, Segli, & Groves, 2008).

One result of local signaling is region-specific induction of different members of the *Pax* gene family (Baker & Bronner-Fraser, 2001; Schlosser & Ahrens, 2004). As shown in Fig. 4, *Pax* genes are differentially expressed: *Pax6* in anterior placodes, *Pax3* in the ophthalmic part of the trigeminal placode, and *Pax2* and *Pax8* in posterior placodes. In fact, the appropriate *Pax* genes are expressed in the PPE domains that fate map to these specific placodes prior to placode separation (Grocott et al., 2012; Pieper et al., 2012), suggesting that they are involved in initiating placode identity. Experimental results support this conclusion: transplantation experiments indicate that the onset of *Pax* expression correlates with the acquisition of placode identity (Baker & Bronner-Fraser, 2000; Baker, Stark, Marcelle, & Bronner-Fraser, 1999); numerous loss-of-function studies show that the *Pax* genes are required for the development of these respective placodes (e.g., Dude et al., 2009; Hans, Liu, & Westerfield, 2004; Mackereth, Kwak, Fritz, & Riley, 2005; Shaham, Menuchin, Farhy, & Ashery-Padan, 2012); and placode domains segregate by mutual repression between the *Pax* genes (Wakamatsu, 2011; reviewed in Grocott et al., 2012).

4. REGULATION OF CELLULAR DIFFERENTIATION

Some genes that are broadly expressed in the PPE (e.g., *Six*, *Eya*, *Dlx*, *Foxi*, *Irx*) or are critical for setting up individual placode identity (*Pax*) have maintained expression in placodes as they differentiate. For example, *Six* and *Eya* genes are expressed in each neurogenic placode but repressed in the nonneural lens placode. This suggests that PPE genes have additional, later roles in regulating the differentiation of placode precursor cells into the distinct cell types found in the mature placode-derived organs. The transcriptional networks for some placodes are dealt with in detail elsewhere (adenohypophysis, Pogoda & Hammerschmidt, 2007; lens, Charlton-Perkins, Brown, & Cook, 2011; Ogino et al., 2012; lateral line, Chitnis & Nogare, 2014). Herein, we focus on two programs of differentiation: the ORNs and the sensory neurons in the ganglia of the trigeminal and epibranchial cranial nerves.

4.1. Specifying ORNs

The transcriptional and signaling mechanisms that establish the PPE, the NB zone, and resulting placode subdomains result in a remarkable dual end point: the definition of organizer regions that will drive subsequent sensory neuron differentiation (Kawauchi et al., 2005; LaMantia, Bhasin, Rhodes, & Heemskerk, 2000; LaMantia, Colbert, & Linney, 1993; Richman & Tickle, 1989, 1992; Shou, Murray, Rim, & Calof, 2000), as well as the precursor populations whose fate will be determined by local signaling from those organizers. This key outcome for placode development is demonstrated dramatically by the olfactory placode, which emerges as a distinct ectodermal domain as initial generation of placodes draws to a close—approximately E9.0 in the mouse (Fig. 5). Once the olfactory placodal ectoderm is established, a series of inductive events follows that drives olfactory epithelial (OE) and ORN differentiation, via signaling within the ectoderm as well as signaling between ectodermal and mesenchymal compartments.

Placode specification alone—establishment of the PPE and subsequent definition of distinct placodes—does not yield a population of neural stem cells capable of generating ORNs and related neuronal classes (including the GnRH neurons that are generated in the olfactory placode and migrate to the hypothalamus during early embryogenesis; Wray, 2010). Separating the olfactory placodal ectoderm from the underlying neural crest-derived mesenchyme at E9.0 in the mouse prior to initial morphogenesis and neurogenesis in the olfactory placode results in a failure of OE and ORN differentiation *in vitro* (Bhasin, Maynard, Gallagher, & LaMantia, 2003; LaMantia et al., 2000; Rawson et al., 2010). Similarly, mutation of *Pax6* disrupts the migration of neural crest-derived mesenchyme as well as the capacity of placodal ectoderm to generate ORNs (Anchan, Drake, Gerwe, Haines, & LaMantia, 1997; Grindley, Davidson, & Hill, 1995; LaMantia et al., 2000). In contrast, when olfactory placodal ectoderm and neural crest-derived frontonasal mesenchyme are apposed to one another, a fully patterned OE with a coherent olfactory nerve emerges (Fig. 5). This patterning includes, critically, the restriction and upregulation of *Sox2* in the placodal ectoderm, as well as the induction of *Meis1*, which identifies a population of slowly dividing precursors in the nascent OE (Fig. 5). Over the next 48 h, the transcriptional profile of precursor cells in the OE diversifies further. Key factors that influence specification of the PPE, including Six1 and Sox2, remain in subsets of OE progenitors. The putative stem cells, found primarily in the lateral portion of the OE, express high levels of Meis1,

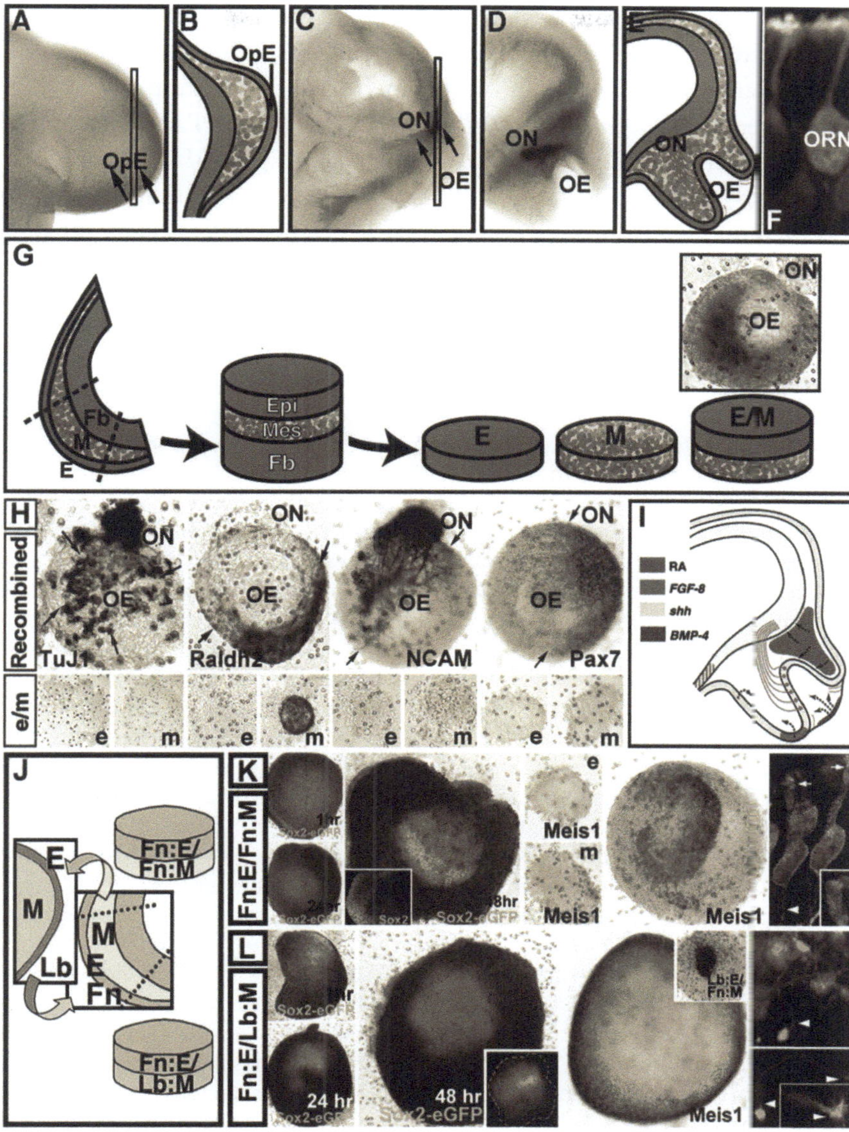

Figure 5 The genesis of olfactory receptor neurons (ORNs) from the olfactory placodal ectoderm (OpE). Top row: whole mouse embryos stained for the neuronal marker NCAM (which labels ORNs and axons in the olfactory nerve—ON), and schematics showing major structures that differentiate from the olfactory placode. (A and B) The undifferentiated OpE at embryonic day (E)9.0 is a slightly thickened epithelial domain on the lateral/ventral surface of the head. (C–E) Within 1.5 days, at E10.5, the OpE invaginates to form the olfactory epithelium (OE). ORNs are generated primarily in the medial OE, and their axons (ON) extend through the frontonasal mesenchyme (gray hatched regions in E) and enter the forebrain. (F) An ORN at E10.5, labeled for the olfactory

(*Continued*)

Pbx1/2/3, and Pax7; low levels of Sox2; and high levels of Six1 (Chen et al., 2009; Ikeda et al., 2007, 2010; Tucker et al., 2010). In contrast, the presumed transit-amplifying cells, found in the medial portion of the OE, express reciprocal levels of Sox2 (high) and Six1 (low). Pax6 is expressed throughout the medial region, and subsets of cells express *bHLH* neurogenic genes (*Ascl1*, *Ngn1*, *NeuroD1*). Differentiating ORNs and GnRH neurons also are found in the medial portion of the OE. Thus, the olfactory placodal

Figure 5—Cont'd marker protein (green, a relatively specific marker), and the ORN-selective adenylcyclase, ACIII (red), which labels the dendritic knob and associated cilia where odor receptor molecules are concentrated, and sensory transduction occurs. (G) A schematic of the approach to isolate the olfactory placode at E9.0; separate the OpE (E), the frontonasal mesenchyme (Mes), and the forebrain neuroepithelium (Fb); and recombine them *in vitro* to assess OpE/ORN differentiation. The inset (right) shows a recombined OpE/frontonasal mesenchyme explant with mesenchyme from a ROSA-26 gene trap mouse that labels all cells with β-galactosidase (blue), and OpE from a WT embryo. There is no movement of cells from the mesenchyme into the OpE over 48 h *in vitro*; however, there is growth of olfactory axons (ON) and cells that migrate with them. (H) The *top row* shows recombined explants (OpE and frontonasal mesenchyme) labeled for molecular markers for differentiating ORNs (TuJ1 and NCAM), a synthetic enzyme for retinoic acid (Raldh2) and Pax7, which is restricted to the lateral mesenchyme. The expression patterns and intensity of each marker are parallel to that seen in the differentiation of the olfactory placode/OE *in vivo* at E11. The *bottom row* shows that the differentiation of the OE, and expression of relevant markers, with the exception of Raldh2, depends upon mesenchymal/epithelial interactions—epithelium (e) nor mesenchyme (m) expresses any of the other markers when isolated. (I) A schematic showing the role of signaling sources of Fgf8, Bmp4, Shh, and RA established via earlier placodal transcriptional mechanisms, in establishing OE neurogenesis and ON trajectory. (J) Schematic of mesenchyme "swapping" experiments *in vitro* to assess specificity of OpE/frontonasal mesenchyme interactions. (K) Isolated E9.0 frontonasal OpE (Fn:E) recombined with frontonasal mesenchyme (Fn:M) leads to the restriction and patterned expression of Sox2 in the OpE/OE over 48 h. The same explant is shown here, imaged at the times indicated. The inset shows the medial (high) to lateral (low) pattern of Sox2 expression facilitated by Fn:E/Fn:M interactions. This pattern is key for specifying ORN stem cells. In addition, Fn:E/Fn:M interactions induce Meis1 in a population of cells that additional evidence (Murdoch, DelConte, & Garcia-Castro., 2010; Tucker et al., 2010) shows are likely OE neural stem cells. At far right, these patterning events yield cells in Fn:E/Fn:M explants that have all of the morphological hallmarks of ORNs. In addition, these cells have physiological properties seen in ORNs (Rawson et al., 2010). (L) When OpE (Fn:E) is recombined with limb bud mesenchyme (Lb:M), Sox2 is not patterned, and *Meis1* expression is seen in the mesenchyme rather than OpE. The OpE still generates neurons (far right) in this heterologous pairing of E and M; however, these neurons lack apical dendrites, their axons cannot enter the limb bud mesenchyme and instead end in elaborate growth cones (arrow, lower right panel), and they lack the physiological hallmarks of ORNs (Rawson et al., 2010). (See the color plate.)

ectoderm established by E9.0 in the mouse as a fairly uniform transcriptional field acquires within 2 days a stunning degree of transcriptional diversity that parallels precursor classes that have the capacity to generate ORNs and other OE neuronal classes.

It seemed likely that the key facilitator of this rapid and dramatic acquisition of cellular diversity and identity were signals exchanged between the placodal ectoderm—especially the "organizer" domains established as the placodal ectoderm is defined—and the adjacent neural crest-derived mesenchyme that accumulates between E8.0 and E9.0 in the mouse. FGF and BMP domains that emerge during early placode formation are potential sources of signals for further cellular differentiation. Their action on the underlying mesenchyme likely drives the capacity of the mesenchyme to support further signaling to the OE. Accordingly, we asked whether the epithelial and mesenchymal tissues that constitute the nascent olfactory primordium (which will form not only the OE and ORNs, but cartilaginous structures of the nose itself) during early gestation in the mouse (E9.0) interact via cardinal morphogenetic signals, particularly FGFs (Balmer & LaMantia, 2005; Kawauchi et al., 2005; LaMantia et al., 2000; Lassiter, Stark, Zhao, & Zhou, 2014; Tucker et al., 2010), BMPs (LaMantia et al., 2000; Shou et al., 2000), and RA (Bhasin et al., 2003; LaMantia et al., 2000, 1993) to facilitate ORN differentiation. Our results are quite clear: these signals, available from distinct zones in the olfactory placodal ectoderm and frontonasal mesenchyme, influence axial patterning and neuronal differentiation in the OE (Fig. 5).

The identity of OE stem cells that generate the initial embryonic OE neuronal lineage remained uncertain until recently. For several decades, the location and cellular characteristics of embryonic OE stem cells—slowly, symmetrically dividing stem cells in a specific location or niche within the OE that have the capacity to generate all differentiated cell types of the mature tissue—were undefined. Initial localization studies indicated that progenitors expressed neuronal bHLH transcription factors, including Ascl1, NeuroG1, and NeuroD1, which were important for OE neurogenesis. From the earliest observations (Guillemot et al., 1993) through several studies of loss-of-function mutants (Cau, Casarosa, & Guillemot, 2002; Cau, Gradwohl, Casarosa, Kageyama, & Guillemot, 2000; Cau, Gradwohl, Fode, & Guillemot, 1997, 2002; Krolewski, Packard, Jang, Wildner, & Schwob, 2012; Murray, Navi, Fesenko, Lander, & Calof, 2003; Tucker et al., 2010), it was clear that *Ascl1* and the other *bHLH* genes were important for the expansion of the numbers of ORNs and other OE cell classes.

Nevertheless, in the absence of Ascl1, ORNs are still produced. Indeed, every neuronal and nonneuronal cell type found in the OE can be identified in these mutants (Guillemot et al., 1993; Tucker et al., 2010), although their numbers appear to be dramatically diminished.

Finally, it was not clear how the specificity of sensory neuron identity was established from distinct placodal domains. For example, trigeminal proprioceptive and mechanoreceptive neurons are also derived from placodal ectodermal precursors that express Six1 and Sox2; however, these cells delaminate from the placode and coalesce with adjacent neural crest cells that become nociceptive sensory neurons, rather than retaining a mesenchymal identity (as is the case for those adjacent to the OE). We found that the A–P identity of the mesenchyme is essential for establishing specific sensory neuron identity. When the limb bud mesenchyme, which contains several of the same molecular constituents found in frontonasal mesenchyme, is apposed to the nascent OE, neurogenesis occurs. These neurons, however, lack the functional properties of ORNs, and their axons cannot penetrate the basal lamina between the anomalously induced OE and the underlying heterologous mesenchyme (Fig. 5; Rawson et al., 2010). Similar frontonasal mesenchymal cues also specify the generation of GnRH neurons (Cariboni et al., 2011; Gamble et al., 2005; Messina et al., 2011; Parkash et al., 2012; Schwarting, Kostek, Bless, Ahmad, & Tobet, 2001); these cells do not differentiate when olfactory placodal ectoderm is induced by limb mesenchyme.

Together, these observations on the initial differentiation of ORNs from the olfactory placodal ectoderm establish two key points. First, through regulated expression of *Six1*, *Sox2*, and other *Six/Sox* genes, the placodal ectoderm acquires the capacity to generate neurons; however, that capacity is only realized, and neuronal identity is specified by interactions with adjacent neural crest-derived mesenchymal cells. Second, the distinct zones of FGF, BMP, and other signaling factors established at placodal ectoderm boundary regions are not only essential for defining placodal limits but also critical for providing signals that drive cranial sensory neuron differentiation, including for special sensory receptor neurons like ORNs. The relationship between the signaling molecules, which are shared by many placodal domains, and the distinctive sensory neuron classes that emerge from each placodal region remains unknown. Nevertheless, it is clear that the transcriptional and signaling mechanisms that define the cranial placodes specify local neural ectodermal progenitors whose subsequent fate depends upon interactions with the underlying neural crest-derived mesenchyme, orchestrated by signaling domains established as placodal identity emerges.

4.2. Cranial ganglion sensory neurons

The trigeminal and epibranchial placodes are specialized niches that produce the large neurons in the sensory ganglia of the trigeminal (V), facial (VII), glossopharyngeal (IX), and vagus (X) cranial nerves. Unlike the neural crest cells that contribute the small neurons and glia to these ganglia, the placode-derived neurons are specified within the placode, delaminate from it, and migrate to the coalescing ganglion as postmitotic neurons (Graham, Blentic, Dusque, & Begbie, 2007). The signaling pathways involved in the neurogenesis in these placodes have recently been reviewed in detail (Lassiter et al., 2014). In general, the Notch pathway plays a critical role in selecting which cells in the neurogenic placode field will become neurons, the FGF pathway regulates delamination, the Wnt pathway contributes to delamination and *Pax* and *bHLH* gene expression, and the BMP pathway regulates *bHLH* and other neural differentiation genes.

Prior to delamination of these neurons, however, PPE factors (e.g., Six1) and SoxB1 neural stem cell factors (e.g., Sox2, Sox3) play important roles in regulating neurogenesis. *Six1* expression is maintained in the outer proliferative layer of the placode and is downregulated as cells delaminate from the epithelium and coalesce into ganglia (Pandur & Moody, 2000; Schlosser et al., 2008). *SoxB1* genes, which are important in maintaining neural stem cells in the central nervous system (reviewed in Bergsland et al., 2011; Moody, Klein, Karpinski, Maynard, & LaMantia, 2013; Thiel, 2013; Wegner, 2013), are expressed in localized regions of the PPE after *Six* and *Eya* genes but before individual placodes segregate (Schlosser & Ahrens, 2004). In the cranial nerve placodes, *SoxB1* genes are expressed at high levels in the inner layer of the placode ectoderm (Schlosser et al., 2008; Fig. 6). bHLH transcription factors (e.g., NeuroG, NeuroD) promote the generation of neural progenitors, cause them to exit the cell cycle, and promote neuronal differentiation in both the central nervous system and in the neurogenic placodes (reviewed in Castro & Guillemot, 2011; Schlosser, 2006). In *Xenopus* cranial nerve placodes, the expression of *NeuroG* and *NeuroD* is complimentary to that of *Six1* (Schlosser & Ahrens, 2004; Schlosser et al., 2008; Schlosser & Northcutt, 2000). *NeuroG* is first expressed in the inner placode layer and later in the delaminating neural progenitor cells; *NeuroG* expression is lost as the coalescing progenitor cells differentiate into neurons. *NeuroD* is expressed later than *NeuroG* in scattered cells within the inner placode layer, and it remains expressed in most of the placode-derived ganglion cells. These expression

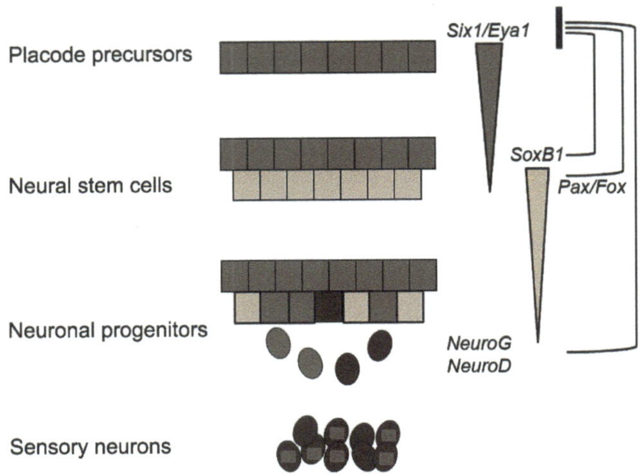

Figure 6 A model of the gene regulatory pathway proposed to control the onset of neurogenesis in neurogenic placodes that form sensory ganglia of cranial nerves V, VII, IX, and X. Six/Eya maintain proliferative, undifferentiated placode precursors (dark green (gray in the print version)). Six/Eya are required for the expression of *SoxB1* genes in neural stem cells (light blue (light gray in the print version)), which are only detected in the deeper positioned cells in the placode once Six/Eya protein levels decrease. *SoxB1* genes are required for the expression of *bHLH* neural progenitor genes (darker blue (dark gray in the print version)), which are only detected once the levels of SoxB1 protein decrease. NeuroG (medium blue (gray in the print version)) is detected in the deeper layer of the placode and in delaminating cells. Later, NeuroD (dark blue (dark gray in the print version)) is detected in these same cells and in the sensory neurons in the coalescing ganglion. These neurons reexpress *Six/Eya* genes (green (gray in the print version) squares), which are required for cell survival. Several studies indicate that the genes expressed at each step of the process, including *Pax* and *Fox* genes that are involved in placode identity, feedback (black bars) to negatively regulate *Six/Eya* genes, thus promoting differentiation.

patterns suggest that Six1 maintains placode cells in a proliferative, undifferentiated "precursor" state, SoxB1 factors transition these cells to a neural stem cell state, and bHLH factors specify neural progenitors (with Notch pathway input; Lassiter et al., 2014) and regulate their differentiation into the sensory neurons of the cranial ganglia (Fig. 6).

What experimental observations place Six1 upstream in this pathway? First, knockdown experiments in *Xenopus* show that Six1 (and Eya1) are required for the expression of *SoxB1* genes, *bHLH* genes, and differentiated neuron markers (Schlosser et al., 2008). In mouse, bHLH factors also appear to be regulated by Six1 and Eya1 in neurogenic placodes. For example, in *Eya1*-null neurogenic placodes, *NeuroG*- and *NeuroD*-positive cells are depleted, delamination is blocked, and sensory neuron differentiation markers (*Phox2a/b*, *SCG10*) are not expressed (Zou et al., 2004). Similar, but weaker phenotypes are seen in *Six1*-null embryos (Zou et al., 2004), perhaps due to the remaining activity of *Six2* and *Six4* (Konishi et al., 2006). Conversely, gain-of-function assays in *Xenopus* indicate that the level of Six1/Eya1 expression is critical for their function in the neural differentiation pathway. Increasing the levels of Six1/Eya1 in the neurogenic placodes increases proliferation, resulting in a larger domain of *SoxB1*-expressing neural stem cells, but fewer *bHLH*-expressing neural progenitors (Schlosser et al., 2008). Lowering the levels of Six1/Eya1 promotes the transition of *SoxB1*-positive cells to *bHLH*-positive cells. These observations indicate that *Six/Eya* genes maintain a proliferative precursor cell, but for neural differentiation to occur they need to be downregulated.

In other systems, *Six* genes also keep precursor cells in a proliferative state prior to cell-type differentiation. The loss of *Six1* in mice appears to decrease proliferation, which results in apoptosis (Li et al., 2003; Ozaki et al., 2004). In humans, SIX1 overexpression was identified in hyperproliferating cell populations (e.g., primary breast cancers and metastatic lesions; Ford, Kabingu, Bump, Mutter, & Pardee, 1998). Six1 overexpression also influences cell proliferation by directly activating the transcription of *CyclinA1* (Coletta et al., 2004), indicating that Six1 may maintain cells in an immature state by influencing cell cycle regulation. In fact, reactivating *Six1* in adult tissues leads to misregulated cell proliferation and a number of human cancers (Li et al., 2013; Patrick et al., 2013; Patrick, Schiemann, Yang, Zhao, & Ford, 2009).

Six genes have an even later role in neurogenesis in the cranial nerve placodes. *Six1* and *Six4* are secondarily expressed in subsets of differentiating sensory neurons in the cranial ganglia (Konishi et al., 2006), and gene knockout studies show that they are required to prevent apoptosis. *Six1*-null, *Eya1*-null, and *Six1/Six4*-null mice have small trigeminal, vestibulocochlear, and epibranchial sensory ganglia (Konishi et al., 2006; Xu et al., 1999; Zheng et al., 2003; Zou et al., 2004). During embryonic stages, these ganglia begin to form, but the loss of *Six1* and *Six4* leads to cell autonomous apoptosis of newly formed cranial ganglion neurons. It is proposed that *Six* genes contribute to sensory neuron survival by regulating the expression of the anti-apoptotic gene *Bcl-x* (Konishi et al., 2006).

The description of the transcriptional path that leads to cranial neuron formation is far from complete. First, there are likely to be other neural stem cell genes involved in this process. For example, neither *Sox2* nor *Sox3* is expressed in *Xenopus* trigeminal placodes (Schlosser & Ahrens, 2004), whereas *Sox11* is expressed in the PPE and is upregulated by Six1 (Brugmann et al., 2004). Second, the critical levels of transcription factors for varying phenotypes have not yet been explored rigorously in ganglion neuron differentiation and the use of null mutants may not be clinically relevant for human syndromes in which only one allele of *Six1* or *Eya1* is affected. Third, studies in mouse indicate that the *Six1*-21 enhancer functions to integrate input from Sox, Pax, Fox, and bHLH factors (Sato et al., 2012), suggesting a transcriptional mechanism by which *Six1* expression can be downregulated to permit differentiation to proceed (Fig. 6). This now needs to be experimentally tested. Finally, not all cranial ganglia are equally affected by the loss of *Six/Eya* genes (Zou et al., 2004), suggesting A–P positional input into the regulatory program. In the future, it will be important to identify all the factors involved in this process and determine precisely how they interact to regulate cranial nerve placode neurogenesis. We expect that the late steps in this process will be very similar to neural plate and neural crest neurogenesis, but the regulatory inputs will obviously be placode specific (e.g., Grocott et al., 2012).

5. NB ZONE, PPE, AND PLACODE GENES INVOLVED IN HUMAN CONGENITAL SYNDROMES

In humans, craniofacial anomalies and congenital hearing loss are among the most common developmental defects. As expected from their roles in placode specification and differentiation, mutations in NB-specifying, PPE, and placode genes often are associated with congenital syndromes characterized by craniofacial and auditory phenotypes (Table 1). The craniofacial defects are likely caused by perturbations in the neural crest derivations of the NB zone, which give rise to the skeleton and connective tissue of the face; the hearing deficits are likely to be caused by perturbations in the otic placode derivative of the NB zone. At first glance, it might be surprising that the phenotypes associated with each syndrome often extend beyond craniofacial and auditory tissues (Table 1), but it is well established that these genes are also expressed in other tissues later during development. Conversely, it is remarkable that the phenotypes are not more severe considering that work from experimental model systems shows that these genes have critical roles in placode development. However, it must be kept in mind that many of the mutations identified in the human patients are heterozygous for the gene involved, whereas experimental gene knockdowns

Table 1 Human congenital syndromes associated with NB-specifying and FPE genes and craniofacial defects

	Name of syndrome	Location	
MSX1	Ectodermal dysplasia	4p16.2	Defects in hair, nails, teeth, and/or sweat glands
	Orofacial cleft		Cleft lip/palate
	Tooth agenesis		Lack of a tooth
PAX3	Cranio-facial-deafness-hand syndrome	2q36.1	Flat facial profile, craniofacial abnormalities, sensorineural hearing loss
	Waardenburg syndrome		Pigment abnormalities, sensorineural hearing loss, craniofacial abnormalities
DLX3	Ameliogenesis imperfect Trichodontoosseous syndrome	17q21.33	Hair, tooth, and bone defects
DLX5	Split-hand/foot malformation 1	7q21.3	Defects in hands and feet; sensorineural hearing loss
GATA2	Emberger syndrome	3q21.3	Congenital deafness, lower limb lymphedema, leukemias
GATA3	Hypoparathyroidism with sensorineural deafness syndrome (HDRS)	10p14	Hypoparathyroidism, sensorineural deafness, renal dysplasia
FOXI1	Enlarged vestibular aqueduct	5q35.1	Sensorineural and mixed hearing loss
TFAP2-α	Branchio-oculo-facial (BOFS)	6p24.3	Branchial clefts, dysmorphic face, external and middle ear anomalies with conductive deafness
SIX1	Branchio-otic syndrome 3 (BOS3)	14q23.1	Branchial arch defects, deafness, lacrimal duct stenosis
	Deafness, autosomal dominant 23 (DFNA23)		Hearing loss
EYA1	Orofacial-cervical syndrome	8q13.3	Facial anomalies, low-set ears, preauricular fistulas, hearing loss, skeletal defects
	Branchio-otic syndrome 1 (BOS1)		Hearing loss, structural defects in outer, middle and inner ear, brachial fistulas
	Branchio-oto-renal syndrome 1 (BOR1)		BOS with renal defects

Continued

Table 1 Human congenital syndromes associated with NB-specifying and PPE genes and craniofacial defects—cont'd

	Name of syndrome	Location	
EYA4	Deafness, autosomal dominant 10	6q23.2	Progressive hearing loss
	Cardiomyopathy, dilated, 1J		Dilated cardiopathy, sensorineural hearing loss
PAX6	Aniridia	11p13	Defects in the iris of the eye
	Various ocular syndromes		Defects in the neural retina
	Peters anomaly		Defects in cornea, lens, and iris
OTX2	Microphthalmia, syndromic 5	14q22.3	Small, dysmorphic retinas
	Pituitary hormone deficiency, combined, 6		Hypoplasia of anterior and/or posterior pituitary
TBX1	Conotruncal anomaly face syndrome	22q11.21	Cardiac anomalies, facial anomalies
	DiGeorge syndrome		Defects in thymus, cardiac outflow tracts, facial anomalies
	Tetralogy of Fallot		Cardiac anomalies
	Velocardiofacial syndrome		Cardiac anomalies, cleft palate, facial anomalies
PAX2	Papillorenal syndrome	10q24.31	Ocular and renal anomalies; high-frequency hearing loss
	Renal hypoplasia		Renal agenesis
PAX8	Congenital hypothyroidism	2q13	Thyroid dysgenesis

Information from the Online Mendelian Inheritance in Man Web site (OMIM.org).

are often complete nulls. It is interesting that hearing loss is the dominant placode deficit identified in humans. This is a phenotype that can easily be clinically tested, whereas the consequences of disruptions in olfactory and cranial sensory neurons might be subtle. Nonetheless, correlating the anatomical phenotypes in humans carrying mutations with those in experimental animals helps confirm the predicted roles of these genes in patients. Furthermore, it is very useful to identify the causative genes for hearing loss, because genetic screening might identify affected infants early enough for intervention. For a full description of the syndromes summarized below, please refer to the Online Mendelian Inheritance in Man web site (OMIM.org).

Mutations in NB-specifying genes often result in defects that likely are caused by perturbations in both neural crest and placode development. Defects in *MSX1* cause ectodermal dysplasia, tooth agenesis, and orofacial clefting, phenotypes that are associated with its dominant role in neural crest development. Mutations in *PAX3* are associated with Waardenburg syndrome and Craniofacial-deafness-hand syndrome. Waardenburg patients have pigment cell defects (neural crest) and both syndromes are characterized by craniofacial defects (neural crest) and hearing loss (placode). Mutations in *DLX3* cause two syndromes that have hair, tooth (neural crest), and bone defects, whereas mutations in *DLX5* result in split-hand/foot malformation 1 syndrome, patients of which have hearing loss (placode). Mutations in *GATA2* are mostly associated with leukemias, as this gene plays a prominent role in hematopoiesis. But one form, called Emberger syndrome, also has hearing loss (placode); mutations in *GATA3* also can cause deafness (placode). Mutations in *FOXI1* cause enlarged vestibular aqueduct syndrome, characterized by hearing loss (placode). Mutations in *TFAP2α* cause Branchio-oculo-facial syndrome; these patients have craniofacial dysmorphologies (neural crest) and hearing deficits (placode).

PPE gene mutations also are causative of a few human congenital syndromes (Table 1). *SIX1* mutations underlie one form of Branchio-otic syndrome (BOS3), whose phenotypes include craniofacial defects and hearing loss (Ruf et al., 2004). Nine mutations in BOS3 patients from 16 unrelated families have been reported to date: 7 are missense mutations in the SD of Six1 and 2 are missense or deletion mutations in the HD of Six1 (Ito, Noguchi, Yashima, & Kitamura, 2006; Kochhar et al., 2008; Noguchi, Ito, Nishio, Honda, & Kitamura, 2011; Ruf et al., 2004; Sanggaard et al., 2007). The mutations either disrupt the Six1–Eya1 interactions or the ability of Six1 to bind to DNA (Patrick et al., 2009). In zebrafish, expression of mutant *Six1* mRNA that carries a BOS patient mutation (R110W) interferes with the Six1/Eya1 interaction that promotes cell proliferation and hair cell formation (Bricaud & Collazo, 2011). The *Catweasel* (*Cwe*) mouse mutant is thought to be a good model for the related Branchio-otic-renal syndrome (BOR), which is similar to BOS but also includes kidney defects, because it harbors a missense mutation in the SD that is similar to at least one BOR family (Bosman, Quint, Fuchs, Hrabe de Angelis, & Steel, 2009; Mosrati et al., 2011). Heterozygous-*Cwe* mice have an ectopic row of hair cells in the cochlea, and homozygous-*Cwe* mice do not have hair cells in the cochlea, semicircular canals, or utricle. To our knowledge, no human syndromes have been assigned to mutations in *SIX2* or *SIX4*, but one affected locus in BOS patients contains the *Six1*, *Six4*, and *Six6* genes (Ruf et al., 2004).

Patients diagnosed with BOS1 and BOR1 harbor *EYA1* mutations (Abdelhak et al., 1997; Kumar, Deffenbacher, Cremers, Van Camp, & Kimberling, 1997; Rodriguez-Soriano, 2003; Spruijt et al., 2006); cataracts of the lens also have been associated with *EYA1* mutations (Azuma, Hirakiyama, Inoue, Asaka, & Yamada, 2000). Often the defects in the protein lie in the ED, where they interfere with the interaction between Eya1 and Six proteins (Buller, Xu, Marquis, Schwanke, & Xu, 2001; Ozaki, Watanabe, Ikeda, & Kawakami, 2002). Partial deletions of *EYA1* that include other genes cause Oto-facio-cervical syndrome, which includes both craniofacial and hearing defects (Estefania et al., 2006; Rickard et al., 2001). To date, no mutations in *EYA2* or *EYA3* have been reported in humans. However, mutations in *EYA4* are involved in autosomal dominant sensorineural deafness 10 (DFNA10; Makishima et al., 2007; Wayne et al., 2001) and dilated cardiomyopathy with sensorineural hearing loss, autosomal dominant (CMD1J; Schonberger et al., 2005), implicating defects in the otic placode. The described mutations, which include small insertions, amino acid substitutions, and large deletions, are all predicted to truncate the EYA4 protein within the ED, which is required for interactions with SIX proteins.

Mutations in genes that lie downstream of SIX/EYA in the placode development pathway have fewer identifiable craniofacial and hearing phenotypes. While *Irx* and *Gbx2* have important roles in placode development (Fig. 4), no human congenital syndromes have been associated with these genes so far. Although Pax6 is an important early marker of the PPR "ground" state (Bailey et al., 2006), mutations in *PAX6* are associated primarily with ocular defects, and a small number of patients have lens defects. Likewise, although patients with *PAX2* mutations have high-frequency hearing deficits, the major abnormalities are found in the kidneys. *PAX8* mutations are associated with thyroid dysgenesis without any reported hearing loss. Mutations of *TBX1* cause a large number of cardiac defects and patients have craniofacial anomalies (neural crest), but no obvious placode-derived defects. *OTX2* mutations cause eye defects as well as defects in both the anterior (placode) and posterior (neural) pituitary. Perhaps because these genes play overlapping roles in placode differentiation, mutation in one is compensated by the activities of the other genes in the regulatory network.

6. CONCLUSIONS

Placodes contribute to important secretory cells and cranial sensory organs that are critical for the normal behavior of an animal. Many of the genes involved in placode development are highly conserved from invertebrates to vertebrates, indicating that the underlying transcriptional pathways are evolutionarily very old (Bassham & Postlthwait, 2005; Gasparini, Degasperi, Shimeld, Burighel, & Manni, 2013; Graham & Shimeld, 2013; Luttrell & Swalla, 2014; Schlosser, 2005, 2007). Mutation of genes involved in placode development leads to a variety of congenital syndromes in humans that share craniofacial dysmorphologies and hearing loss. Future work is needed to identify the specific molecular functions of the *Six* (and other) genes that have key roles in PPE and placode development, and identify all of the potential cofactors; this will reveal further genetic causes of craniofacial syndromes. In addition, identifying and understanding the function of all of the genes involved in PPE specification and placode differentiation pathways will have a major impact on craniofacial tissue repair efforts. Elucidating the basic molecular mechanisms by which PPE cells are induced and transformed from the embryonic ectoderm into numerous differentiated cell types and how the process differs from that described for the closely related neural crest will be critical for designing techniques for sensory organ replacement from various stem and progenitor cell sources. In one case, that of olfactory sensory neurons, the embryonic mechanisms are likely maintained and adapted in the adult to facilitate olfactory sensory neuron replacement throughout life seen in most vertebrates, including mammals (Leung, Coulombe, & Reed, 2007; Schwob, 2002). It is exciting that human embryonic stem cells have been successfully differentiated into placode cells (Chen et al., 2012; Leung et al., 2013), in some instances generating trigeminal sensory neurons capable of *in vivo* engraftment in chick and mouse embryos, mature lens fibers, and anterior pituitary cells capable of producing human growth hormone and adrenocorticotropic hormone *in vivo* (Dincer et al., 2013). Understanding the placode gene regulatory network in the embryo will surely enhance *in vitro* differentiation steps for regenerating placode derivatives, making it possible to repair and regenerate cranial sensory organs.

ACKNOWLEDGMENTS

We apologize to those authors whose work was not cited due to space limitations. Work from the authors' laboratories is supported by NIH R01 Grants DE022065 (S. A. M.), DC011534 (A.-S. L.), and HD29178 (A.-S. L.).

REFERENCES

Abdelhak, S., Kalatzis, V., Heilig, R., Compain, S., Samson, D., Vincent, C., et al. (1997). A human homologue of the *Drosophila eyes absent* gene underlies branchio-oto-renal (BOR) syndrome and identifies a novel gene family. *Nature Genetics*, *15*, 157–164.

Abello, G., Khatri, S., Giraldez, F., & Alsina, B. (2007). Early regionalization of the otic placode and its regulation bu the Notch signaling pathway. *Mechanisms of Development*, *124*, 631–645.

Ahrens, K., & Schlosser, G. (2005). Tissues and signals involved in the induction of placodal Six1 expression in *Xenopus laevis*. *Developmental Biology*, *288*, 40–59.

Anchan, R. M., Drake, D. P., Gerwe, E. A., Haines, C. F., & LaMantia, A.-S. (1997). A failure of retinoid-mediated induction accompanies the loss of the olfactory pathway during mammalian forebrain development. *The Journal of Comparative Neurology*, *379*, 171–184.

Arima, K., Shiotsugu, J., Niu, R., Khandpur, R., Martinez, M., Shin, Y., et al. (2005). Global analysis of RAR-responsive genes in the *Xenopus* neurula using cDNA microarrays. *Developmental Dynamics*, *232*, 414–431.

Arkell, R., & Beddington, R. S. (1997). BMP-7 influences pattern and growth of the developing hindbrain of mouse embryos. *Development*, *124*, 1–12.

Aruga, J., Minowa, O., Yaginuma, H., Kuno, J., Nagai, T., Noda, T., et al. (1998). Mouse Zic1 is involved in cerebellar development. *The Journal of Neuroscience*, *18*, 284–293.

Axel, R. (2005). Scents and sensibility: A molecular logic of olfactory perception (Nobel lecture). *Angewandte Chemie (International Edition in English)*, *44*, 6110–6127.

Azuma, N., Hirakiyama, A., Inoue, T., Asaka, A., & Yamada, M. (2000). Mutations of a human homologue of the *Drosophila eyes absent* gene (EYA1) detected in patients with congenital cataracts and ocular anterior segment anomalies. *Human Molecular Genetics*, *9*, 363–366.

Bailey, A. P., Bhattacharyya, S., Bronner-Fraser, M., & Streit, A. (2006). Lens specification is the ground state of all sensory placodes, from which FGF promotes olfactory identity. *Developmental Cell*, *11*, 505–517.

Bailey, A. P., & Streit, A. (2006). Sensory organs: Making and breaking the pre-placodal region. *Current Topics in Developmental Biology*, *72*, 167–204.

Bajoghli, B., Aghaallaei, N., & Czerby, T. (2005). Groucho corepressor proteins regulate otic vesicle outgrowth. *Developmental Dynamics*, *233*, 760–771.

Baker, C. V. H., & Bronner-Fraser, M. (2000). Establishing neuronal identity in vertebrate neurogenic placodes. *Development*, *127*, 3045–3056.

Baker, C. V., & Bronner-Fraser, M. (2001). Vertebrate cranial placodes I. Embryonic induction. *Developmental Biology*, *232*, 1–61.

Baker, C. V., Stark, M. R., Marcelle, C., & Bronner-Fraser, M. (1999). Competence, specification and induction of Pax-3 in the trigeminal placode. *Development*, *126*, 147–156.

Balmer, C. W., & LaMantia, A.-S. (2005). Noses and neurons: Morphogenesis and neural induction in the olfactory pathway. *Developmental Dynamics*, *234*, 464–481.

Bane, B. C., Van Rybroek, J. M., Kolker, S. J., Weeks, D. L., & Manaligod, J. M. (2005). EYA1 expression in the developing inner ear. *The Annals of Otology, Rhinology, and Laryngology*, *114*, 853–858.

Basel-Vanagaite, L., Rainschtein, L., Inbar, D., Gothelf, D., Hennekam, R., & Straussberg, R. (2007). Autosomal recessive mental retardation syndrome with anterior maxillary protrusion and strabismus: MRAMS syndrome. *American Journal of Medical Genetics, 143A*, 1687–1691.

Bassham, S., & Postlthwait, J. H. (2005). The evolutionary history of placodes: A molecular genetic investigation of the larvacean urochordate *Oikoleura dioica. Development, 132*, 4259–4272.

Beanan, M. J., & Sargent, T. D. (2000). Regulation and function of Dlx3 in vertebrate development. *Developmental Dynamics, 218*, 545–553.

Bergsland, M., Ramskold, D., Zaouter, C., Klum, S., Sandberg, R., & Muhr, J. (2011). Sequentially acting Sox transcription factors in neural lineage development. *Genes & Development, 25*, 2453–2464.

Bessarab, D. A., Chong, S. W., & Korzh, V. (2004). Expression of zebrafish *six1* during sensory organ development and myogenesis. *Developmental Dynamics, 230*, 781–786.

Bhasin, N., Maynard, T. M., Gallagher, P., & LaMantia, A.-S. (2003). Mesenchymal/epithelial interactions regulate retinoid signaling in the olfactory placode. *Developmental Biology, 261*, 82–98.

Bhat, N., Kwon, H. J., & Riley, B. B. (2013). A gene network that coordinates preplacodal competence and neural crest specification in zebrafish. *Developmental Biology, 373*, 107–117.

Bhattacharyya, S., Bailey, A. P., Bronner-Fraser, M., & Streit, A. (2004). Segregation of lens and olfactory precursors from a common territory: Cell sorting and reciprocity of Dlx5 and Pax6 expression. *Developmental Biology, 271*, 403–414.

Bhattacharyya, S., & Bronner-Fraser, M. (2004). Hierarchy of regulatory events in sensory placode development. *Current Opinion in Genetics & Development, 14*, 520–526.

Birk, E., Har-Zahav, A., Manzini, C. M., Pasmanik-Chor, M., Kornreich, L., Walsh, C. A., et al. (2010). SOBP is mutated in syndromic and non-syndromic intellectual disability and is highly expressed in the brain limbic system. *American Journal of Human Genetics, 87*, 694–700.

Bonini, N. M., Leiserson, W. M., & Benzer, S. (1993). The eyes absent gene: Genetic control of cell survival and differentiation in the developing *Drosophila* eye. *Cell, 72*, 379–395.

Bosman, E. A., Quint, E., Fuchs, E., Hrabe de Angelis, M., & Steel, K. P. (2009) Catweasel mice: A novel role for Six1 in sensory patch development and a model for branchio-oticrenal syndrome. *Developmental Biology, 328*, 285–296.

Bradley, L., Sun, B., Collins-Racie, L., LaVallie, E., McCoy, J., & Sive, H. (2000). Different activities of the frizzled-related proteins frzb2 and sizzled2 during *Xenopus* anteroposterior patterning. *Developmental Biology, 227*, 118–132.

Bricaud, O., & Collazo, A. (2006). The transcription factor six1 inhibits neuronal and promotes hair cell fates in the developing zebrafish (*Danio rerio*) inner ear. *The Journal of Neuroscience, 26*, 10438–10451.

Bricaud, O., & Collazo, A. (2011). Balancing cell numbers during organogenesis: Six1a differentially affects neurons and sensory hair cells in the inner ear. *Developmental Biology, 357*, 191–201.

Brodbeck, S., & Englert, C. (2004). Genetic determination of nephrogenesis: The Pax/Eya/Six gene network. *Pediatric Nephrology, 19*, 249–255.

Brownell, I., Dirksen, M., & Jamrich, M. (2000). Forkhead Foxe3 maps to the dysgenetic lens locus and is critical in lens development and differentiation. *Genesis, 27*, 81–93.

Brugmann, S. A., & Moody, S. A. (2005). Induction and specification of the vertebrate ectodermal placodes: Precursors of the cranial sensory organs. *Biology of the Cell, 97*, 303–319.

Brugmann, S. A., Pandur, P. D., Kenyon, K. L., Pignoni, F., & Moody, S. A. (2004). Six1 promotes a placodal fate within the lateral neurogenic ectoderm by functioning as both a transcriptional activator and repressor. *Development, 131*, 5871–5881.

Buck, L. B. (2000). The molecular architecture of odor and pheromone sensing in mammals. *Cell, 100*, 611–618.
Buller, C., Xu, X., Marquis, V., Schwanke, R., & Xu, P. X. (2001). Molecular effects of Eya1 domain mutations causing organ defects in BOR syndrome. *Human Molecular Genetics, 10*, 2775–2781.
Cariboni, A., Davidson, K., Rakic, S., Maggi, R., Parnavelas, J. G., & Ruhrberg, C. (2011). Defective gonadotropin-releasing hormone neuron migration in mice lacking SEMA3A signalling through NRP1 and NRP2: Implications for the aetiology of hypogonadotropic hypogonadism. *Human Molecular Genetics, 20*, 336–344.
Carmona-Fontaine, C., Acuna, G., Ellwanger, K., Niehrs, C., & Mayor, R. (2007). Neural crests are actively precluded from the anterior neural fold by a novel inhibitory mechanism dependent on Dickkopf1 secreted by the prechordal mesoderm. *Developmental Biology, 309*, 208–221.
Castro, D. S., & Guillemot, F. (2011). Old and new functions of proneural factors revealed by the genome-wide characterization of their transcriptional targets. *Cell Cycle, 10*, 4026–4031.
Cau, E., Casarosa, S., & Guillemot, F. (2002). Mash1 and Ngn1 control distinct steps of determination and differentiation in the olfactory sensory neuron lineage. *Development, 129*, 1871–1880.
Cau, E., Gradwohl, G., Casarosa, S., Kageyama, R., & Guillemot, F. (2000). Hes genes regulate sequential stages of neurogenesis in the olfactory epithelium. *Development, 127*, 2323–2332.
Cau, E., Gradwohl, G., Fode, C., & Guillemot, F. (1997). Mash1 activates a cascade of bHLH regulators in olfactory neuron progenitors. *Development, 124*, 1611–1621.
Charlton-Perkins, M., Brown, N. L., & Cook, T. A. (2011). The lens in focus: A comparison of lens development in *Drosophila* and vertebrates. *Molecular Genetics and Genomics, 286*, 798–811.
Chen, R., Amoui, M., Zhang, Z., & Mardon, G. (1997). Dachshund and eyes absent proteins form a complex and function synergistically to induce ectopic eye development in *Drosophila. Cell, 91*, 893–903.
Chen, W., Jongkamonwiwat, N., Abbas, L., Eshtan, S. J., Johnson, S. L., Kuhn, S., et al. (2012). Restoration of auditory evoked response by human ES-derived otic progenitors. *Nature, 490*, 278–282.
Chen, B., Kim, E. H., & Xu, P. X. (2009). Initiation of olfactory placode development and neurogenesis is blocked in mice lacking both Six1 and Six4. *Developmental Biology, 326*, 75–85.
Chen, Z., Montcouquioi, M., Calderon, R., Jenkins, N. A., Copeland, N. G., Kelley, M. W., et al. (2008). Jac1/Sobp, encoding a nuclear zinc finger protein, is critical for cochlear growth, cell fate and patterning of the organ of Corti. *The Journal of Neuroscience, 28*, 6633–6641.
Chen, Y., Pollet, N., Niehrs, C., & Pieler, T. (2001). Increased XRALDH2 activity has a posteriorizing effect on the central nervous system of *Xenopus* embryos. *Mechanisms of Development, 101*, 91–103.
Cheyette, B. N., Green, P. J., Martin, K., Garren, H., Hartenstein, V., & Zipursky, S. L. (1994). The *Drosophila sine oculis* locus encodes a homeodomain-containing protein required for the development of the entire visual system. *Neuron, 12*, 977–996.
Chitnis, A. B., & Nogare, D. D. (2014). Lessons from the zebrafish lateral line system. In S. A. Moody (Ed.), *Principles of developmental genetics* (2nd ed., pp. 266–279). New York, NY: Elsevier.
Christophorou, N. A., Bailey, A. P., Hanson, S., & Streit, A. (2009). Activation of Six1 target genes is required for sensory placode formation. *Developmental Biology, 336*, 327–336.
Coletta, R. D., Christensen, K., Reichenberger, K. J., Lamb, J., Micomonaco, D., Huang, L., et al. (2004). The Six1 homeoprotein stimulates tumorigenesis by

reactivation of cyclin A1. *Proceedings of the National Academy of Sciences of the United States of America, 101*, 6478–6483.

Couly, G. F., & LeDouarin, N. M. (1987). Mapping of the early neural primordium in quail-chick chimeras. II. The prosencephalic neural plate and neural folds: Implications for the genesis of cephalic human congenital abnormalities. *Developmental Biology, 120*, 198–214.

Couly, G. F., & LeDouarin, N. M. (1990). Head morphogenesis in embryonic avian chimeras: Evidence for a segmental pattern in the ectoderm corresponding to the neuromeres. *Development, 108*, 543–558.

David, R., Ahrens, K., Wedlich, D., & Schlosser, G. (2001). *Xenopus* Eya1 demarcates all neurogenic placodes as well as migrating hypaxial muscle precursors. *Mechanisms of Development, 103*, 189–192.

de Croze, N., Maczkowiak, F., & Monsoro-Burq, A. H. (2011). Reiterative AP2a activity controls sequential steps in the neural crest gene regulatory network. *Proceedings of the National Academy of Sciences of the United States of America, 108*, 155–160.

Depreux, F. F., Darrow, K., Conner, D. A., Eavey, R. D., Liberman, M. C., Seidman, C. E., et al. (2008). Eya4-deficient mice are a model for heritable otitis media. *The Journal of Clinical Investigation, 118*, 651–658.

Dincer, Z., Piao, J., Niu, L., Ganat, Y., Kriks, S., Zimmer, B., et al. (2013). Specification of functional cranial placode derivatives from human pluripotent stem cells. *Cell Reports, 5*, 1–16.

Dude, C. M., Kuan, C. Y., Bradshaw, J. R., Greene, N. D., Relaix, F., Stark, M. R., et al. (2009). Activation of Pax3 target genes is necessary but not sufficient for neurogenesis in the ophthalmic trigeminal placode. *Developmental Biology, 326*, 314–326.

Duggan, C. D., DeMaria, S., Baudhuin, A., Stafford, D., & Ngai, J. (2008). Foxg1 is required for development of the vertebrate olfactory system. *The Journal of Neuroscience, 28*, 5229–5239.

Duncan, M. K., Kos, L., Jenkins, N. A., Gilbert, D. J., Copeland, N. G., & Tomarev, S. I. (1997). Eyes absent, a gene family found in several metazoan phyla. *Mammalian Genome, 8*, 479–485.

Dutta, S., Dietrich, J. E., Aspock, G., Burdine, R. D., Schier, A., Westerfield, M., et al. (2005). Pitx3 defines an equivalence domain for lens and anterior pituitary placode. *Development, 132*, 1579–1590.

Eagleson, G. W., & Dempewolf, R. D. (2002). The role of the anterior neural ridge and Fgf-8 in early forebrain patterning and regionalization in *Xenopus laevis*. *Comparative Biochemistry and Physiology Part B: Biochemistry & Molecular Biology, 132*, 179–189.

Estefania, E., Ramirez-Camacho, R., Gomar, M., Trinidad, A., Arellano, B., Garcia-Berrocal, J. R., et al. (2006). Point mutation of an EYA1-gene splice site in a patient with oto-facial-cervical syndrome. *Annals of Human Genetics, 70*, 140–144.

Esterberg, R., & Fritz, A. (2009). Dlx3b/4b are required for the formation of the preplacodal region and otic placode through local modulation of BMP activity. *Developmental Biology, 325*, 189–199.

Esteve, P., & Bovolenta, P. (1999). cSix4, a member of the six gene family of transcription factors, is expressed during placode and somite development. *Mechanisms of Development, 85*, 161–165.

Feledy, J. A., Beanan, M. J., Sandoval, J. J., Goodrich, J. S., Lim, J. H., Matsuo-Takasaki, M., et al. (1999). Inhibitory patterning of the anterior neural plate in *Xenopus* by homeodomain factors Dlx3 and Msx1. *Developmental Biology, 212*, 455–464.

Ford, H. L., Kabingu, E. N., Bump, E. A., Mutter, G. L., & Pardee, A. B. (1998). Abrogation of the G2 cell cycle checkpoint associated with overexpression of HSIX1 A possible mechanism of breast carcinogenesis. *Proceedings of the National Academy of Sciences of the United States of America, 95*, 12608–12613.

Fougerousse, F., Durand, M., Lopez, S., Suel, L., Demignon, J., Thornton, C., et al. (2002). Six and Eya expression during human somitogenesis and MyoD family activation. *Journal of Muscle Research and Cell Motility, 23*, 255–264.

Gamble, J. A., Karunadasa, D. K., Pape, J. R., Skynner, M. J., Todman, M. G., Bicknell, R. J., et al. (2005). Disruption of ephrin signaling associates with disordered axophilic migration of the gonadotropin-releasing hormone neurons. *The Journal of Neuroscience, 25*, 3142–3150.

Gasparini, F., Degasperi, V., Shimeld, S. M., Burighel, P., & Manni, L. (2013). Evolutionary conservation of the placodal transcriptional network during sexual and asexual development of chordates. *Developmental Dynamics, 242*, 752–766.

Ghanbari, H., Seo, H. C., Fjose, A., & Brändli, A. W. (2001). Molecular cloning and embryonic expression of *Xenopus* Six homeobox genes. *Mechanisms of Development, 101*, 271–277.

Giot, L., Bader, J. S., Brouwer, C., Chaudhuri, A., Kuang, B., Li, Y., et al. (2003). A protein interaction map of *Drosophila melanogaster. Science, 302*, 1727–1736.

Givens, M. L., Rave-Harel, N., Goonewardena, V. D., Kurotani, R., Berdy, S. E., Swan, C. H., et al. (2005). Developmental regulation of gonadotropin-releasing hormone gene expression by the MSX and DLX homeodomain protein families. *The Journal of Biological Chemistry, 280*, 19156–19165.

Glavic, A., Maris Honoré, S., Gloria Feijóo, C., Bastidas, F., Allende, M. L., & Mayor, R. (2004). Role of BMP signaling and the homeoprotein Iroquois in the specification of the cranial placodal field. *Developmental Biology, 272*, 89–103.

Graham, A., Blentic, A., Dusque, S., & Begbie, J. (2007). Delamination of cells from neurogenic placodes does not involve an epithelial-to-mesenchymal transition. *Development, 134*, 4141–4145.

Graham, A., & Shimeld, S. M. (2013). The origin and evolution of the ectodermal placodes. *Journal of Anatomy, 222*, 32–40.

Grindley, J. C., Davidson, D. R., & Hill, R. E. (1995). The role of Pax-6 in eye and nasal development. *Development, 121*, 1433–1442.

Grocott, T., Tambalo, M., & Streit, A. (2012). The peripheral sensory nervous system in the vertebrate head: A gene regulatory perspective. *Developmental Biology, 370*, 3–23.

Groves, A. K., & Bronner-Fraser, M. (2000). Competence, specification and commitment in otic placode induction. *Development, 127*, 3489–3499.

Groves, A. K., & LaBonne, C. (2014). Setting appropriate boundaries: Fate, patterning and competence at the neural plate border. *Developmental Biology, 389*, 2–12.

Guillemot, F., Lo, L. C., Johnson, J. E., Auerbach, A., Anderson, D. J., & Joyner, A. L. (1993). Mammalian achaete-scute homolog 1 is required for the early development of olfactory and autonomic neurons. *Cell, 75*, 463–476.

Hallem, E. A., & Carlson, J. R. (2004). The odor coding system of *Drosophila. Trends in Genetics, 20*, 453–459.

Hans, S., Liu, D., & Westerfield, M. (2004). Pax8 and Pax2a function synergistically in otic specification, downstream of the Foxi1 and Dlx3b transcription factors. *Development, 131*, 5091–5102.

Hatini, V., Xin, Y., Balas, G., & Lai, E. (1999). Dynamics of placodal lineage development revealed by targeted transgene expression. *Developmental Dynamics, 215*, 332–343.

Heanue, T. A., Reshef, R., Davis, R. J., Mardon, G., Oliver, G., Tomarev, S., et al. (1999). Synergistic regulation of vertebrate muscle development by Dach2, Eya2, and Six1, homologs of genes required for *Drosophila* eye formation. *Genes & Development, 15*, 3231–3243.

Heisenberg, C. P., Houart, C., Take-Uchi, M., Rauch, G. J., Young, N., Coutinho, P., et al. (2001). A mutation in the Gsk3-binding domain of zebrafish Masterblind/Axin1 leads to a fate transformation of telencephalon and eyes to diencephalon. *Genes & Development, 15*, 1427–1434.

Herzog, W., Zeng, X., Lele, Z., Sonntag, C., Ting, J. W., Chang, C. Y., et al. (2003). Adenohypophysis formation in the zebrafish and its dependence on sonic hedgehog. *Developmental Biology, 245*, 36–49.

Himeda, C. L., Ranish, J. A., Angello, J. C., Maire, P., Aebersold, R., & Hauschka, S. D. (2004). Quantitative proteomic identification of six4 as the trex-binding factor in the muscle creatine kinase enhancer. *Molecular and Cellular Biology, 24*, 2132–2143.

Hoffman, T. L., Javier, A. L., Campeau, S. A., Knight, R. D., & Schilling, T. F. (2007). Tfap2 transcription factors in zebrafish neural crest development and ectodermal evolution. *Journal of Experimental Zoology Part B: Molecular and Developmental Evolution, 308*, 679–691.

Hong, C.-S., & Saint-Jeannet, J.-P. (2007). The activity of Pax3 and Zic1 regulates three distinct cell fates at the neural plate border. *Molecular Biology of the Cell, 18*, 2192–2202.

Hsiao, F. C., Williams, A., Davies, E. L., & Rebay, I. (2001). Eyes absent mediates cross-talk between retinal determination genes and the receptor tyrosine kinase signaling pathway. *Developmental Cell, 1*, 51–61.

Hwang, C. H., Simeone, A., Lai, E., & Wu, D. K. (2009). Foxg1 is required for proper separation and formation of sensory cristae during inner ear development. *Developmental Dynamics, 238*, 2725–2734.

Ikeda, K., Kageyama, R., Suzuki, Y., & Kawakami, K. (2010). Six1 is indispensable for production of functional progenitor cells during olfactory epithelial development. *The International Journal of Developmental Biology, 54*, 1453–1464.

Ikeda, K., Ookawara, S., Sato, S., Ando, Z., Kageyama, R., & Kawakami, K. (2007). Six1 is essential for early neurogenesis in the development of olfactory epithelium. *Developmental Biology, 311*, 53–68.

Ikeda, K., Watanabe, Y., Ohto, H., & Kawakami, K. (2002). Molecular interaction and synergistic activation of a promoter by Six, Eya, and Dach proteins mediated through CREB binding protein. *Molecular and Cellular Biology, 22*, 6759–6766.

Ishihara, T., Ikeda, K., Sato, S., Yajia, H., & Kawakami, K. (2008). Differential expression of Eya1 and Eya2 during chick early embryonic development. *Gene Expression Patterns, 8*, 357–367.

Ito, T., Noguchi, Y., Yashima, T., & Kitamura, K. (2006). SIX1 mutation associated with enlargement of the vestibular aqueduct in a patient with branchio-oto syndrome. *Laryngoscope, 116*, 797–799.

Jacobson, A. G. (1963). The determination and positioning of the nose, lens and ear. III. Effects of reversing the antero-posterior axis of epidermis, neural plate and neural fold. *The Journal of Experimental Zoology, 154*, 293–303.

Jacobson, A. G. (1966). Inductive processes in embryonic development. *Science, 152*, 25–34.

Janesick, A., Shiotsugu, J., Taketani, M., & Blumberg, B. (2012). RIPPLY3 is a retinoic acid-inducible repressor required for setting the borders of the pre-placodal ectoderm. *Development, 139*, 1213–1224.

Jayasena, C. S., Ohyama, T., Segli, N., & Groves, A. K. (2008). Notch signaling augments the canonical Wnt pathway to specify the size of the otic placode. *Development, 135*, 2251–2261.

Johnson, K. R., Cook, S. A., Erway, L. C., Matthews, A. N., Sanford, L. P., Paradies, N. E., et al. (1999). Inner ear and kidney anomalies caused by IAP insertion in an intron of the Eya1 gene in a mouse model of BOR syndrome. *Human Molecular Genetics, 8*, 645–653.

Jusiak, B., Karandikar, U. C., Kwak, S. J., Wang, F., Wang, H., Chen, R., et al. (2014). Regulation of *Drosophila* eye development by the transcription factor Sine oculis. *PLoS One, 9*, e89695.

Kaji, T., & Artinger, K. B. (2004). dlx3b and dlx4b function in the development of Rohon-Beard sensory neurons and trigeminal placode in the zebrafish neurula. *Developmental Biology*, *276*, 523–540.
Kawakami, K., Ohto, H., Ikeda, K., & Roeder, R. G. (1996). Structure, function and expression of a murine homeobox protein AREC3, a homologue of *Drosophila sine oculis* gene product, and implication in development. *Nucleic Acids Research*, *24*, 303–310.
Kawakami, K., Sato, S., Ozaki, H., & Ikeda, K. (2000). Six family genes—Structure and function as transcription factors and their roles in development. *Bioessays*, *22*, 616–626.
Kawauchi, S., Kim, J., Santos, R., Wu, H. H., Lander, A. D., & Calof, A. L. (2009). Foxg1 promotes olfactory neurogenesis by antagonizing Gdf11. *Development*, *136*, 1453–1464.
Kawauchi, S., Shou, J., Santos, R., Hébert, J. M., McConnell, S. K., Mason, I., et al. (2005). Fgf8 expression defines a morphogenetic center required for olfactory neurogenesis and nasal cavity development in the mouse. *Development*, *132*, 5211–5223.
Kenyon, K. L., Li, D. J., Clouser, C., Tran, S., & Pignoni, F. (2005). Fly SIX-type homeodomain proteins Sine oculis and Optix partner with different cofactors during eye development. *Developmental Dynamics*, *234*, 497–504.
Kenyon, K. L., Moody, S. A., & Jamrich, M. (1999). A novel fork head gene mediates early steps during *Xenopus* lens formation. *Development*, *126*, 5107–5116.
Khatri, S. B., & Groves, A. K. (2013). Expression of the Foxi2 and Foxi3 transcription factors during development of chicken sensory placodes and pharyngeal arches. *Gene Expression Patterns*, *13*, 38–42.
Knouff, R. A. (1935). The developmental pattern of ectodermal placodes in *Rana pipiens*. *The Journal of Comparative Neurology*, *62*, 17–71.
Kobayashi, M., Nishikawa, K., Suzuki, T., & Yamamoto, M. (2001). The homeobox protein Six3 interacts with the Groucho corepressor and acts as a transcriptional repressor in eye and forebrain formation. *Developmental Biology*, *232*, 315–326.
Kobayashi, M., Osanai, H., Kawakami, K., & Yamamoto, M. (2000). Expression of three zebrafish Six4 genes in the cranial sensory placodes and the developing somites. *Mechanisms of Development*, *98*, 151–155.
Kochhar, A., Orten, D. J., Sorensen, J. L., Fischer, S. M., Cremers, C. W., Kimberling, W. J., et al. (2008). SIX1 mutation screening in 247 branchio-otic-renal syndrome families: A recurrent missense mutation associated with BOR. *Human Mutation*, *29*, 565.
Konishi, Y., Ikeda, K., Iwakura, Y., & Kawakami, K. (2006). Six1 and Six4 promote survival of sensory neurons during early trigeminal gangliogenesis. *Brain Research*, *1116*, 93–102.
Kriebel, M., Muller, F., & Hollemann, T. (2007). Xeya3 regulates survival and proliferation of neural progenitor cells within the anterior neural plate of *Xenopus* embryos. *Developmental Dynamics*, *236*, 1526–1534.
Krolewski, R. C., Packard, A., Jang, W., Wildner, H., & Schwob, J. E. (2012). Ascl1 (Mash1) knockout perturbs differentiation of nonneuronal cells in olfactory epithelium. *PLoS One*, *7*, e51737.
Kumar, S., Deffenbacher, K., Cremers, C. W., Van Camp, G., & Kimberling, W. J. (1997). Brachio-oto-renal syndrome: Identification of novel mutations, molecular characterization, mutation distribution and prospects for genetic testing. *Genetic Testing*, *1*, 243–251.
Kwon, H. J., Bhat, N., Sweet, E. M., Cornell, R. A., & Riley, B. B. (2010). Identification of early requirements for preplacodal ectoderm and sensory organ development. *PLoS Genetics*, *6*, e1001133.
Laclef, C., Souil, E., Demignon, J., & Maire, P. (2003). Thymus, kidney and craniofacial abnormalities in Six 1 deficient mice. *Mechanisms of Development*, *120*, 669–679.
Ladher, R. K., O'Neill, P., & Begbie, J. (2010). From shared lineage to distinct functions: Development of the inner ear and epibranchial placodes. *Development*, *137*, 1777–1785.
LaMantia, A.-S., Bhasin, N., Rhodes, K., & Heemskerk, J. (2000). Mesenchymal/epithelial induction mediates olfactory pathway formation. *Neuron*, *28*, 411–425.

LaMantia, A.-S., Colbert, M. C., & Linney, E. (1993). Retinoic acid induction and regional differentiation prefigure olfactory pathway formation in the mammalian forebrain. *Neuron, 10*, 1035–1048.

Lassiter, R. N. T., Stark, M. R., Zhao, T., & Zhou, C. J. (2014). Signaling mechanisms controlling cranial placode neurogenesis and delamination. *Developmental Biology, 389*, 39–49.

Leger, S., & Brand, M. (2002). Fgf8 and Fgf3 are required for zebrafish ear placode induction, maintenance and inner ear patterning. *Mechanisms of Development, 119*, 91–108.

Leung, C. T., Coulombe, P. A., & Reed, R. R. (2007). Contribution of olfactory neural stem cells to tissue maintenance and regeneration. *Nature Neuroscience, 10*, 720–726.

Leung, A. W., Morest, K. D., & Li, J. Y. (2013). Differential BMP signaling controls formation and differentiation of multipotent preplacodal progenitors from human embryonic stem cells. *Developmental Biology, 379*, 208–220.

Lewis, K. E., Drossopoulou, G., Paton, I. R., Morrice, D. R., Robertson, K. E., Burt, D. W., et al. (1999). Expression of ptc and gli genes in talpid3 suggests bifurcation in Shh pathway. *Development, 126*, 32397–32407.

Li, W., & Cornell, R. A. (2007). Redundant activities of Tfap2a and Tfap2c are required for neural crest induction and development of other non-neural ectoderm derivatives in zebrafish embryos. *Developmental Biology, 304*, 338–354.

Li, Y., Manaligod, J. M., & Weeks, D. L. (2010). EYA1 mutations associated with the branchio-otic renal syndrome result in defective otic development in *Xenopus laevis*. *Biology of the Cell, 102*, 277–292.

Li, X., Oghi, K. A., Zhang, J., Krones, A., Bush, K. T., Glass, C. K., et al. (2003). Eya protein phosphatase activity regulates Six1-Dach-Eya transcriptional effects in mammalian organogenesis. *Nature, 426*, 247–254.

Li, Z., Tian, T., Lv, F., Chang, Y., Wang, X., Zhang, L., et al. (2013). Six1 promotes proliferation of pancreatic cancer cells via upregulation of Cyclin D1 expression. *PLoS One, 8*, e59203.

Litsiou, A., Hanson, S., & Streit, A. (2005). A balance of FGF, BMP and WNT signalling positions the future placode territory in the head. *Development, 132*, 4051–4062.

Liu, D., Chu, H., Maves, L., Yan, Y. L., Morcos, P. A., Postlethwait, J. H., et al. (2003). Fgf3 and Fgf8 dependent and independent transcription factors are required for otic placode specification. *Development, 130*, 2213–2224.

Lleras-Forero, L., Tambalo, M., Christophorou, N., Chambers, D., Houart, C., & Streit, A. (2013). Neuropeptides: Developmental signals in placode progenitor formation. *Developmental Cell, 26*, 195–203.

Luo, T., Lee, Y.-H., Saint-Jeannet, J.-P., & Sargent, T. D. (2003). Induction of neural crest in *Xenopus* by transcription factor AP2α. *Proceedings of the National Academy of Sciences of the United States of America, 100*, 532–537.

Luo, T., Matsuo-Takasaki, M., Lim, J. H., & Sargent, T. D. (2001). Differential regulation of Dlx gene expression by a BMP morphogenetic gradient. *The International Journal of Developmental Biology, 45*, 681–684.

Luo, T., Matsuo-Takasaki, M., & Sargent, T. D. (2001). Distinct roles for Distal-less genes Dlx3 and Dlx5 in regulating ectodermal development in *Xenopus*. *Molecular Reproduction and Development, 60*, 331–337.

Luo, T., Matsuo-Takasaki, M., Thomas, M. L., Weeks, D. L., & Sargent, T. D. (2002). Transcription factor AP-2 is an essential and direct regulator of epidermal development in *Xenopus*. *Developmental Biology, 245*, 136–144.

Luttrell, S. M., & Swalla, B. J. (2014). Genomic and evolutionary insights into chordate development. In S. A. Moody (Ed.), *Principles of developmental genetics* (2nd ed., pp. 116–128). New York, NY: Elsevier.

Mackereth, M. D., Kwak, S. J., Fritz, A., & Riley, B. B. (2005). Zebrafish pax8 is required for otic placode induction and plays a redundant role with Pax2 genes in the maintenance of the otic placode. *Development, 132*, 371–382.

Makishima, T., Madeo, A. C., Brewer, C. C., Zalewski, C. K., Butman, J. A., Sachdev, V., et al. (2007). Nonsyndromic hearing loss DFNA10 and a novel mutation of EYA4: Evidence for correlation of normal cardiac phenotype with truncating mutations of the Eya domain. *American Journal of Medical Genetics, 143A*, 1592–1598.

Mancilla, A., & Mayor, R. (1996). Neural crest formation in *Xenopus laevis*: Mechanism of Xslug induction. *Developmental Biology, 177*, 580–589.

Maroon, H., Walshe, J., Mahmood, R., Kiefer, P., Dickson, C., & Mason, I. (2002). Fgf3 and Fgf8 are required together for formation of the otic placode and vesicle. *Development, 129*, 2099–2108.

Martin, K., & Groves, A. K. (2005). Competence of cranial ectoderm to respond to Fgf signaling suggests a two-step model of otic placode induction. *Development, 133*, 877–887.

Matsuo-Takasaki, M., Matsumura, M., & Sasai, Y. (2005). As essential role of Foxi1a for ventral specification of the cephalic ectoderm during gastrulation. *Development, 132*, 3885–3894.

McCabe, K. L., & Bronner-Fraser, M. (2008). Essential role for PDGF signaling in ophthalmic trigeminal placode induction. *Development, 135*, 1863–1874.

McLarren, K. W., Litsiou, A., & Streit, A. (2003). DLX5 positions the neural crest and preplacode region at the border of the neural plate. *Developmental Biology, 259*, 34–47.

Messina, A., Ferraris, N., Wray, S., Cagnoni, G., Donohue, D. E., Casoni, F., et al. (2011). Dysregulation of Semaphorin7A/beta1-integrin signaling leads to defective GnRH-1 cell migration, abnormal gonadal development and altered fertility. *Human Molecular Genetics, 20*, 4759–4774.

Meulemans, D., & Bronner-Fraser, M. (2004). Gene-regulatory interactions in neural crest evolution and development. *Developmental Cell, 7*, 291–299.

Mizuno, N., Mochii, M., Yamamoto, T. S., Takahashi, T. C., Eguchi, G., & Okada, T. S. (1999). Pax-6 and Prox-1 expression during lens regeneration from *Cynops* iris and *Xenopus* cornea: Evidence for a genetic program common to embryonic lens development. *Differentiation, 65*, 141–149.

Modrell, M. S., & Baker, C. V. (2012). Evolution of electrosensory ampullary organs: Conservation of Eya4 expression during lateral line development in jawed vertebrates. *Evolution & Development, 14*, 277–285.

Monsoro-Burq, A. H., Wang, E., & Harland, R. (2005). Msx1 and Pax3 cooperate to mediate FGF8 and Wnt signals during *Xenopus* neural crest induction. *Developmental Cell, 8*, 167–178.

Moody, S. A., Klein, S. L., Karpinski, B. A., Maynard, T. M., & LaMantia, A. S. (2013). On becoming neural: What the embryo can tell us about differentiating neural stem cells. *American Journal of Stem Cells, 2*, 74–94.

Moody, S. A., & Saint-Jeannet, J.-P. (2014). Development of the pre-placodal ectoderm and cranial sensory placodes. In S. A. Moody (Ed.), *Principles of developmental genetics* (2nd ed., pp. 331–356). New York, NY: Elsevier.

Mosrati, M. A., Hammami, B., Rebeh, I. B., Ayadi, L., Dhouib, L., Ben Mahfoudh, K., et al. (2011). A novel dominant mutation in SIX1, affecting a highly conserved residue, results in only auditory defects in humans. *European Journal of Medical Genetics, 54*, e484–e488.

Murdoch, B., DelConte, C., & Garcia-Castro, M. I. (2010). Embryonic Pax7-expressing progenitors contribute multiple cell types to the postnatal olfactory epithelium. *Journal of Neuroscience, 30*, 9523–9532.

Murray, R. C., Navi, D., Fesenko, J., Lander, A. D., & Calof, A. L. (2003). Widespread defects in the primary olfactory pathway caused by loss of Mash1 function. *The Journal of Neuroscience*, *23*, 1769–1780.
Neilson, K. M., Pignoni, F., Yan, B., & Moody, S. A. (2010). Developmental expression patterns of candidate co-factors for vertebrate Six family transcription factors. *Developmental Dynamics*, *239*, 3446–3466.
Nguyen, V. H., Schmid, B., Trout, J., Connors, S. A., Ekker, M., & Mullins, M. C. (1998). Ventral and lateral regions of the zebrafish gastrula, including the neural crest progenitors, are established by a bmp2b/swirl pathway of genes. *Developmental Biology*, *199*, 93–110.
Nica, G., Herzog, W., Sonntag, C., Nowak, M., Schwarz, H., Zapata, A. G., et al. (2006). Eya1 is required for lineage-specific differentiation, but not for cell survival in the zebrafish adenohypophysis. *Developmental Biology*, *292*, 189–204.
Nissen, R. M., Yan, J., Amsterdam, A., Hopkins, N., & Burgess, S. M. (2003). Zebrafish foxi modulates cellular responses to Fgf signaling required for the integrity of ear and jaw patterning. *Development*, *130*, 2543–2554.
Noguchi, Y., Ito, T., Nishio, A., Honda, K., & Kitamura, K. (2011). Audiovestibular findings in a branchio-oto syndrome patient with a SIX1 mutation. *Acta Oto-Laryngologica*, *131*, 413–418.
Ogino, H., Ochi, H., Reza, H. M., & Yasuda, K. (2012). Transcription factors involved in lens development from the preplacodal ectoderm. *Developmental Biology*, *363*, 333–347.
Ohto, H., Kamada, S., Tago, K., Tominaga, S., Ozaki, H., Sato, S., et al. (1999). Cooperation of Six and Eya in activation of their target genes through nuclear translocation of Eya. *Molecular and Cellular Biology*, *19*, 6815–6824.
Ohto, H., Takizawa, T., Saito, T., Kobayashi, M., Ikeda, K., & Kawakami, K. (1998). Tissue and developmental distribution of Six family gene products. *The International Journal of Developmental Biology*, *42*, 141–148.
Ohuchi, H., Kimura, S., Watamoto, M., & Itoh, N. (2000). Involvement of fibroblast growth factor (FGF)18-FGF8 signaling in the specification of left-right asymmetry and brain and limb development of the chick embryo. *Mechanisms of Development*, *95*, 55–66.
Ohyama, T., & Groves, A. K. (2004). Expression of mouse Foxi class genes in early craniofacial development. *Developmental Dynamics*, *231*, 640–646.
Oliver, G., Loosli, F., Koster, R., Wittbrodt, J., & Gruss, P. (1996). Ectopic lens induction in fish in response to the murine homeobox gene Six3. *Mechanisms of Development*, *60*, 233–239.
Oliver, G., Mailhos, A., Wehr, R., Copeland, N. G., Jenkins, N. A., & Gruss, P. (1995). *Six3*, a murine homologue of the *sine oculis* gene, demarcates the most anterior border of the developing neural plate and is expressed during eye development. *Development*, *121*, 4045–4055.
Ozaki, H., Nakamura, K., Funahashi, J., Ikeda, K., Yamada, G., Tokano, H., et al. (2004). Six1 controls patterning of the mouse otic vesicle. *Development*, *131*, 551–562.
Ozaki, H., Watanabe, Y., Ikeda, K., & Kawakami, K. (2002). Impaired interactions between mouse Eya1 harboring mutations found in patients with branchio-oto-renal syndrome and Six, Dach, and G proteins. *Journal of Human Genetics*, *47*, 107–116.
Ozaki, H., Watanabe, Y., Takahashi, K., Kitamura, K., Tanaka, A., Urase, K., et al. (2001). Six4, a putative myogenin gene regulator, is not essential for mouse embryonic development. *Molecular and Cellular Biology*, *21*, 3343–3350.
Pandur, P. D., & Moody, S. A. (2000). *Xenopus* Six1 gene is expressed in neurogenic cranial placodes and maintained in the differentiating lateral lines. *Mechanisms of Development*, *96*, 253–257.
Park, B.-Y., & Saint-Jeannet, J.-P. (2008). Hindbrain-derived Wnt and Fgf signals cooperate to specify the otic placode in *Xenopus*. *Developmental Biology*, *324*, 108–121.

Park, B. Y., & Saint-Jeannet, J.-P. (2010). Induction and segregation of the vertebrate cranial placodes. In *Vol. 1. Colloquium series on developmental biology*. San Rafael, CA: Morgan & Claypool Publishers. Available from: *http://www.ncbi.nlm.nih.gov/books/NBK53175/*.

Parkash, J., Cimino, I., Ferraris, N., Casoni, F., Wray, S., Cappy, H., et al. (2012). Suppression of beta1-integrin in gonadotropin-releasing hormone cells disrupts migration and axonal extension resulting in severe reproductive alterations. *The Journal of Neuroscience*, *32*, 16992–17002.

Patrick, A. N., Cabrera, J. H., Smaith, A. L., Chen, X. S., Ford, H. L., & Zhao, R. (2013). Structure function analyses of the human SIX1-EYA2 complex reveal insights into metastasis and BOR syndrome. *Nature Structural & Molecular Biology*, *20*, 447–453.

Patrick, A. N., Schiemann, B. J., Yang, K., Zhao, R., & Ford, H. L. (2009). Biochemical and functional characterization for six SIX1 Branchio-otic-renal syndrome mutations. *The Journal of Biological Chemistry*, *284*, 20781–20790.

Patthey, C., Edlund, T., & Gunhaga, L. (2009). Wnt-regulated temporal control of BMP exposure directs the choice between neural plate border and epidermal fate. *Development*, *136*, 73–83.

Patthey, C., Gunhaga, L., & Edlund, T. (2008). Early development of the central and peripheral nervous systems is coordinated by Wnt and BMP signals. *PLoS One*, *3*, e1625.

Patthey, C., Schlosser, G., & Shimeld, S. M. (2014). The evolutionary history of vertebrate cranial placodes—I: Cell type evolution. *Developmental Biology*, *389*, 82–97.

Pauley, S., Lai, E., & Fritzsch, B. (2006). Foxg1 is required for morphogenesis and histogenesis of the mammalian inner ear. *Developmental Dynamics*, *235*, 2470–2482.

Pera, E. M., & De Robertis, E. M. (2000). A direct screen for secreted proteins in *Xenopus* embryos identifies distinct activities for the Wnt antagonists Crescent and Frzb-1. *Mechanisms of Development*, *96*, 183–195.

Phillips, B. T., Bolding, K., & Riley, B. B. (2001). Zebrafish fgf3 and fgf8 encode redundant functions required for otic placode induction. *Developmental Biology*, *235*, 351–365.

Phillips, B. T., Kwon, H. J., Melton, C., Houghtaling, P., Fritz, A., & Riley, B. B. (2006). Zebrafish msxB, msxC and msxE function together to refine the neural-nonneural border and regulate cranial placodes and neural crest development. *Developmental Biology*, *294*, 376–390.

Pieper, M., Ahrens, K., Rink, E., Peter, A., & Schlosser, G. (2012). Differential distribution of competence for panplacodal and neural crest induction to non-neural and neural ectoderm. *Development*, *139*, 1175–1187.

Pieper, M., Eagleson, G. W., Wosniok, W., & Schlosser, G. (2011). Origin and segregation of cranial placodes in *Xenopus laevis*. *Developmental Biology*, *360*, 257–275.

Pignoni, F., Hu, B., Zavitz, K. H., Xiao, J., Garrity, P. A., & Zipursky, S. L. (1997). The eye-specification proteins So and Eya form a complex and regulate multiple steps in *Drosophila* eye development. *Cell*, *91*, 881–891.

Platt, J. B. (1896). Ontogenetic differentiation of the ectoderm in necturus. II. On the development of the peripheral nervous system. *The Quarterly Journal of Microscopical Science*, *38*, 485–547.

Pogoda, H. M., & Hammerschmidt, M. (2007). Molecular genetics of pituitary development in zebrafish. *Seminars in Cell & Developmental Biology*, *18*, 543–558.

Rawson, N. E., Lischka, F., Yee, K. K., Peters, A. Z., Tucker, E. S., Meechan, D. W., et al. (2010). Specific mesenchymal/epithelial induction of olfactory receptor, vomeronasal, and gonadotropin-releasing hormone (GnRH) neurons. *Developmental Dynamics*, *239*, 1723–1738.

Rayapureddi, J. P., Kattamuri, C., Steinmetz, B. D., Frankfort, B. J., Ostrin, E. J., Mardon, G., et al. (2003). Eyes absent represents a class of protein tyrosine phosphatases. *Nature*, *426*, 295–298.

Richman, J. M., & Tickle, C. (1989). Epithelia are interchangeable between facial primordia of chick embryos and morphogenesis is controlled by the mesenchyme. *Developmental Biology*, *136*, 201–210.
Richman, J. M., & Tickle, C. (1992). Epithelial-mesenchymal interactions in the outgrowth of limb buds and facial primordia in chick embryos. *Developmental Biology*, *154*, 299–308.
Rickard, S., Parker, M., van't Hoff, W., Barnicoat, A., Russell-Eggitt, I., Winter, R. M., et al. (2001). Oto-facial-cervical (OFC) syndrome is a contiguous gene deletion syndrome involving EYA1: Molecular analysis confirms allelism with BOR syndrome and further narrows the Duane syndrome critical regions to 1cM. *Human Genetics*, *108*, 398–403.
Rodriguez-Soriano, J. (2003). Brachio-oto-renal syndrome. *Journal of Nephrology*, *16*, 603–605.
Ruf, R. G., Xu, P. X., Silvius, D., Otto, E. A., Beekmann, F., Muerb, U. T., et al. (2004). SIX1 mutations cause branchio-oto-renal syndrome by disruption of EYA1-SIX1-DNA complexes. *Proceedings of the National Academy of Sciences of the United States of America*, *101*, 8090–8095.
Sahly, I., Andermann, P., & Petit, C. (1999). The zebrafish *eya1* gene and its expression pattern during embryogenesis. *Development Genes and Evolution*, *209*, 399–410.
Saint-Jeannet, J.-P., & Moody, S. A. (2014). Establishing the pre-placodal region and breaking it into placodes with distinct identities. *Developmental Biology*, *389*, 13–27.
Sanggaard, K. M., Rendtorff, N. D., Kjaer, K. W., Eiberg, H., Johnsen, T., Gimsing, S., et al. (2007). Branchio-otic-renal syndrome: Detection of EYA1 and SIX1 mutations in five out of six Danish families by combining linkage, MLPA and sequencing analyses. *European Journal of Human Genetics*, *15*, 1121–1131.
Sargent, T. D. (2006). Transcriptional regulation at the neural plate border. *Advances in Experimental Medicine and Biology*, *589*, 32–44.
Sato, S., Ikeda, K., Ochi, H., Ogino, H., Yajima, H., & Kawakami, K. (2010). Conserved expression of mouse Six1 in the pre-placodal region (PPR) and identification of an enhancer for the rostral PPR. *Developmental Biology*, *344*, 158–171.
Sato, S., Ikeda, K., Shioi, G., Nakao, K., Yajima, H., & Kawakami, K. (2012). Regulation of Six1 expression by evolutionarily conserved enhancers in tetrapods. *Developmental Biology*, *368*, 95–108.
Sato, T., Sasai, N., & Sasai, Y. (2005). Neural crest determination by co-activation of Pax3 and Zic1 genes in *Xenopus* ectoderm. *Development*, *132*, 2355–2363.
Schlosser, G. (2005). Evolutionary origins of vertebrate placodes: Insights from developmental studies and from comparisons with other deuterostomes. *Journal of Experimental Zoology Part B: Molecular and Developmental Evolution*, *304*, 347–399.
Schlosser, G. (2006). Induction and specification of cranial placodes. *Developmental Biology*, *294*, 303–351.
Schlosser, G. (2007). How old genes make a new head: Redeployment of Six and Eya genes during the evolution of vertebrate cranial placodes. *Integrative and Comparative Biology*, *47*, 343–359.
Schlosser, G. (2010). Making senses development of vertebrate cranial placodes. *International Review of Cell and Molecular Biology*, *283*, 129–234.
Schlosser, G., & Ahrens, K. (2004). Molecular anatomy of placode development in *Xenopus laevis*. *Developmental Biology*, *271*, 439–466.
Schlosser, G., Awtry, T., Brugmann, S. A., Jensen, E. D., Neilson, K., Ruan, G., et al. (2008). Eya1 and Six1 promote neurogenesis in the cranial placodes in a SoxB1-dependent fashion. *Developmental Biology*, *320*, 199–214.
Schlosser, G., & Northcutt, R. G. (2000). Development of neurogenic placodes in *Xenopus laevis*. *The Journal of Comparative Neurology*, *418*, 121–146.

Schonberger, J., Wang, L., Shin, J. T., Kim, S. D., Depreux, F. F. S., Zhu, H., et al. (2005). Mutation in the transcriptional coactivator EYA4 causes dilated cardiomyopathy and sensorineural hearing loss. *Nature Genetics, 37*, 418–422.

Schwarting, G. A., Kostek, C., Bless, E. P., Ahmad, N., & Tobet, S. A. (2001). Deleted in colorectal cancer (DCC) regulates the migration of luteinizing hormone-releasing hormone neurons to the basal forebrain. *The Journal of Neuroscience, 21*, 911–919.

Schwob, J. E. (2002). Neural regeneration and the peripheral olfactory system. *The Anatomical Record, 269*, 33–49.

Self, M., Lsagutin, O. V., Bowling, B., Hendrix, J., Cal, Y., Dressler, G. R., et al. (2006). Six2 is required for suppression of nephrogenesis and progenitor renewal in the developing kidney. *The EMBO Journal, 25*, 5214–5228.

Selleck, M. A., & Bronner-Fraser, M. (1995). Origins of the avian neural crest: The role of neural plate-epidermal interactions. *Development, 121*, 525–538.

Selleck, M. A., & Bronner-Fraser, M. (2000). Avian neural crest cell fate decisions: A diffusible signal mediates induction of neural crest by the ectoderm. *International Journal of Developmental Neuroscience, 18*, 621–627.

Serikaku, M. A., & O'Tousa, J. E. (1994). *sine oculis* is a homeobox gene required for *Drosophila* visual system development. *Genetics, 138*, 1137–1150.

Shaham, O., Menuchin, Y., Farhy, C., & Ashery-Padan, R. (2012). Pax6: A multilevel regulator of ocular development. *Progress in Retinal and Eye Research, 31*, 351–376.

Shamim, H., & Mason, I. (1999). Expression of Fgf4 during development of the chick embryo. *Mechanisms of Development, 85*, 189–192.

Shim, S., Bae, N., Park, S. Y., Kim, W.-S., & Han, J.-K. (2005). Isolation of *Xenopus* FGF-8b and comparison with FGF-8a. *Molecules and Cells, 19*, 310–317.

Shiotsugu, J., Katsuyama, Y., Arima, K., Baxter, A., Koide, T., Song, J., et al. (2004). Multiple points of interaction between retinoic acid and FGF signaling during embryonic axis formation. *Development, 131*, 2653–2667.

Shou, J., Murray, R. C., Rim, P. C., & Calof, A. L. (2000). Opposing effects of bone morphogenetic proteins on neuron production and survival in the olfactory receptor neuron lineage. *Development, 127*, 5403–5413.

Silver, S. J., Davies, E. L., Doyon, L., & Rebay, I. (2003). Functional dissection of eyes absent reveals new modes of regulation within the retinal determination gene network. *Molecular and Cellular Biology, 23*, 5989–5999.

Sjödal, M., Edlund, T., & Gunhaga, L. (2007). Time of exposure to BMP signals plays a key role in the specification of the olfactory and lens placodes ex vivo. *Developmental Cell, 13*, 141–149.

Sjödal, M., & Gunhaga, L. (2008). Expression patterns of Shh, Ptc2, Raldh3, Pitx2, Isl1, Lim3 and Pax6 in the developing chick hypophyseal placode and Rathke's pouch. *Gene Expression Patterns, 8*, 481–485.

Soker, T., Dalke, C., Puk, O., Floss, T., Becker, L., Bolle, I., et al. (2008). Pleiotropic effects in Eya3 knockout mice. *BMC Developmental Biology, 8*, 118.

Solomon, K. S., & Fritz, A. (2002). Concerted action of two dlx paralogs in sensory placode formation. *Development, 129*, 3127–3136.

Solomon, K. S., Kwak, S. J., & Fritz, A. (2004). Genetic interactions underlying otic placode induction and formation. *Developmental Dynamics, 230*, 419–433.

Solomon, K. S., Logsdo, J. M., & Fritz, A. (2003). Expression and phylogenetic analyses of three zebrafish Foxi class genes. *Developmental Dynamics, 228*, 301–307.

Spitz, F., Demignon, J., Porteu, A., Kahn, A., Concordet, J. P., Daegelen, D., et al. (1998). Expression of myogenin during embryogenesis is controlled by Six/sine oculis homeoproteins through a conserved MEF3 binding site. *Proceedings of the National Academy of Sciences of the United States of America, 95*, 14220–14225.

Spruijt, L., Hoefsloot, L. H., van Schaijk, G. H., van Waardenburg, D., Kremer, B., Brackel, H. J., et al. (2006). Identification of a novel EYA1 mutation presenting in a newborn with laryngomalacia, glossoptosis, retrognathia, and pectus excavatum. *American Journal of Medical Genetics Part A*, *140*, 1343–1345.

Steventon, B., Mayor, R., & Streit, A. (2012). Mutual repression between Gbx2 and Otx2 in sensory placodes reveals a general mechanism for ectodermal patterning. *Developmental Biology*, *367*, 55–65.

Streit, A. (2002). Extensive cell movements accompany formation of the otic placode. *Developmental Biology*, *249*, 237–254.

Streit, A. (2004). Early development of the cranial sensory nervous system: From a common field to individual placodes. *Developmental Biology*, *276*, 1–15.

Streit, A. (2007). The preplacodal region: An ectodermal domain with multipotential progenitors that contribute to sense organs and cranial sensory ganglia. *The International Journal of Developmental Biology*, *51*, 447–461.

Suzuki, A., Ueno, N., & Hemmati-Brivanlou, A. (1997). *Xenopus* msx1 medicates epidermal induction and neural inhibition by BMP4. *Development*, *124*, 3037–3044.

Takai, A., Inomata, H., Arakawa, A., Yakura, R., Matsuo-Takasaki, M., & Sasai, Y. (2010). Anterior neural development requires Del1, a matrix-associated protein that attenuates canonical Wnt signaling via the Ror2 pathway. *Development*, *137*, 3293–3302.

Tessmar, K., Loosli, F., & Wittbrodt, J. (2002). A screen for co-factors of Six3. *Mechanisms of Development*, *117*, 103–113.

Thiel, G. (2013). How Sox2 maintains neural stem cell identity. *The Biochemical Journal*, *450*, e1–e2.

Tootle, T. L., Silver, S. J., Davies, E. L., Newman, V., Latek, R. R., Mills, I. A., et al. (2003). The transcription factor Eyes absent is a protein tyrosine phosphatase. *Nature*, *426*, 299–302.

Toro, S., & Varga, Z. M. (2007). Equivalent progenitor cells in the zebrafish anterior preplacodal field give rise to adenohypophysis, lens, and olfactory placodes. *Seminars in Cell & Developmental Biology*, *18*, 534–542.

Treier, M., O'Connell, S., Gleiberman, A., Price, J., Szeto, D. P., Burgess, R., et al. (2001). Hedgehog signaling is required for pituitary gland development. *Development*, *128*, 377–386.

Tribulo, C., Ayba, M. J., Nguyen, V. H., Mullins, M. C., & Mayor, R. (2003). Regulation of Msx genes by Bmp gradient is essential for neural crest specification. *Development*, *130*, 6441–6452.

Tucker, E. S., Lehtinen, M. K., Maynard, T. M., Zirlinger, M., Dulac, C., Rawson, N. E., et al. (2010). Proliferative and transcriptional identity of two distinct classes of neural precursors in the mammalian olfactory epithelium. *Development*, *137*, 2471–2481.

von Kupffer, C. (1895). *Studien zur vergleichenden Entwicklungsgeschichte des Kpofes des Cranioten. 3. Heft: Die Entwicklung der Kopfnerven von Ammocoetes planeri*. Munchen: J.H. Lehmann.

Wakamatsu, Y. (2011). Mutual repression between Pax3 and Pax6 is involved in the positioning of ophthalmic trigeminal placode in avian embryo. *Development, Growth & Differentiation*, *53*, 994–1003.

Wang, L., Sewell, W. F., Kim, S. D., Shin, J. T., MacRae, C. A., Zon, L. I., et al. (2008). Eya4 regulation of Na+/K+-ATPase is required for sensory system development in zebrafish. *Development*, *135*, 3425–3434.

Wayne, S., Robertson, N. G., DeClau, F., Chen, N., Verhoeven, K., Prasad, S., et al. (2001). Mutations in the transcriptional activator Eya4 cause late-onset deafness at the DFN10 locus. *Human Molecular Genetics*, *10*, 195–200.

Webb, J. F., & Noden, D. M. (1993). Ectodermal placodes: Contributions to the development of the vertebrate head. *American Zoologist*, *33*, 434–447.

Wegner, M. (2013). SOX after SOX: SOXession regulates neurogenesis. *Genes & Development, 25*, 2423–2428.
Wigle, J. T., Chowdhury, K., Gruss, P., & Oliver, G. (1999). Prox1 function is crucial for mouse lens fibre elongation. *Nature Genetics, 21*, 318–322.
Woda, J. M., Pastagia, J., Mercola, M., & Artinger, K. B. (2003). Dlx proteins position the neural plate border and determine adjacent cell fates. *Development, 130*, 331–342.
Wray, S. (2010). From nose to brain: Development of gonadotrophin-releasing hormone-1 neurons. *Journal of Neuroendocrinology, 22*, 743–753.
Xu, P. X., Adams, J., Peters, H., Brown, M. C., Heaney, S., & Maas, R. (1999). Eya1-deficient mice lack ears and kidneys and show abnormal apoptosis of organ primordia. *Nature Genetics, 23*, 113–117.
Xu, P. X., Cheng, J., Epstein, J. A., & Maas, R. L. (1997). Mouse Eya genes are expressed during limb tendon development and encode a transcriptional activation function. *Proceedings of the National Academy of Sciences of the United States of America, 94*, 11974–11979.
Xu, H., Dude, C. M., & Baker, C. V. (2008). Fine-grained fate maps for the ophthalmic and maxillomandibular trigeminal placodes in the chick embryo. *Developmental Biology, 317*, 174–186.
Xu, P. X., Zheng, W., Huang, L., Maire, P., Laclef, C., & Silvius, D. (2003). Six1 is required for the early organogenesis of mammalian kidney. *Development, 130*, 3085–3094.
Xu, P. X., Zheng, W., Laclef, C., Maire, P., Maas, R. L., Peters, H., et al. (2002). Eya1 is required for the morphogenesis of mammalian thymus, parathyroid and thyroid. *Development, 129*, 3033–3044.
Yan, B., Neilson, K. M., Ranganathan, R., Streit, A., & Moody, S. A. (2014). Microarray identification of novel genes downstream of Six1, a critical factor in cranial placode, somite and kidney development. *Developmental Dynamics*. http://dx.doi.org/10.1002/dvdy.24229.
Zhang, H., Hu, G., Wang, H., Sciavolino, P., Iler, N., Shen, M. M., et al. (1997). Heterodimerization of Msx and Dlx homeoproteins results in functional antagonism. *Molecular and Cellular Biology, 17*, 2920–2932.
Zhang, Y., Luo, T., & Sargent, T. D. (2006). Expression of TFAP2beta and TFAP2gamma genes in *Xenopus laevis*. *Gene Expression Patterns, 6*, 589–595.
Zheng, W., Huang, L., Wei, Z. B., Silvius, D., Tang, B., & Xu, P. X. (2003). The role of Six1 in mammalian auditory system development. *Development, 130*, 3989–4000.
Zhu, C. C., Dyer, M. A., Uchikawa, M., Kondoh, H., Lagutin, O. V., & Oliver, G. (2002). Six3-mediated auto repression and eye development requires its interaction with members of the Groucho-related family of co-repressors. *Development, 129*, 2835–2849.
Zou, D., Silvius, D., Fritzsch, B., & Xu, P. X. (2004). Eya1 and Six1 are essential for early steps of sensory neurogenesis in mammalian cranial placodes. *Development, 131*, 5561–5572.
Zou, D., Silvius, D., Rodrigo-Blomqvist, S., Enerback, S., & Xu, P. X. (2006). Eya1 regulates the growth of otic epithelium and interacts with Pax2 during the development of all sensory areas of the inner ear. *Developmental Biology, 298*, 430–441.

CHAPTER TEN

Neural Crest and Placode Contributions to Olfactory Development

Jun Suzuki*,†, Noriko Osumi*,1

*Department of Developmental Neuroscience, Centers for Neuroscience, Tohoku University Graduate School of Medicine, Sendai, Japan

†Department of Otorhinolaryngology-Head and Neck Surgery, Tohoku University Graduate School of Medicine, Sendai, Japan

[1]Corresponding author: e-mail address: osumi@med.tohoku.ac.jp

Contents

Abstract

Olfaction is the sense of smell that influences many primitive behaviors for survival, e.g., feeding, reproduction, social interaction, and fear response. The olfactory system is an evolutionarily ancient sensory system and composed of the olfactory epithelium (OE), the olfactory bulb (OB), and the olfactory cortex. The OE gives rise to olfactory receptor neurons (ORNs), i.e., primary sensory receptor cells whose axons project directly to the OB. The ORNs are unique in the way that they are continuously replaced during physiological turnover or following injury throughout life. In the OE, horizontal basal cells, i.e., flat and quiescent cells attached to the basal lamina, are now thought to be tissue stem cells. Although OE cells, especially ORNs, were hypothesized to be derived from the olfactory placode (OP), recent genetic fate-mapping studies using Cre reporter mice

Current Topics in Developmental Biology, Volume 111
ISSN 0070-2153
http://dx.doi.org/10.1016/bs.ctdb.2014.11.010

indicate a dual origin, i.e., the OP and neural crest (NC), of the olfactory system. The NC is a transient embryonic tissue that is formed between the dorsal neuroepithelium and epidermis. Neural crest cells (NCCs) are multipotent cells that migrate into various target tissues and differentiate into various cell types, including neurons and glia of the peripheral nervous system, cranial cartilage and bone, and melanocytes. Recent studies have revealed that neural crest-derived cells (NCDCs) are widely distributed in adult tissues, and that a subset of NCDCs still possesses NCC-like multipotency. Here, we review classical and recent studies of the olfactory system, especially focusing on the contribution of the NC and OP to the OE development.

1. INTRODUCTION

The olfactory system is an evolutionarily ancient sensory system used for olfaction, i.e., a sense of smell, and comprises two elements: the primary olfactory pathway, the olfactory epithelium (OE) and the olfactory bulb (OB), and the secondary olfactory pathway, the olfactory cortex (OC) (Treloar, Miller, Ray, & Greer, 2010). Olfaction influences many essential behaviors for survival such as feeding, self-defense, social interaction, and reproduction (Bargmann, 2006; Nomura, Haba, & Osumi, 2007). In vertebrates, peripheral olfactory organs are derived from two specialized embryonic tissue precursors: the neural crest (NC) and the olfactory placode (OP) (Forni & Wray, 2012). The OE is a sensory neuroepithelium and the peripheral component of the olfactory system (Fig. 1). The OE exhibits a unique characteristic in that adult neurogenesis occurs continuously during physiological turnover or following injury throughout life (Graziadei & Graziadei, 1979; Schwob, 2002). In the OE, axons of olfactory receptor neurons (ORNs), or primary sensory receptor cells, project directly to the OB. Typically, individual ORNs express only one odorant gene and characteristically detect a single odorant molecule (Buck & Axel, 1991). Because of these characteristics, the OE is an excellent model for investigating how adult neurogenesis occurs and how neurons acquire diverse fates and organize their connectivity between targets. However, our understanding of the olfactory system, including its origin and development, remains incomplete. Because the olfactory system in vertebrates develops during early embryogenesis, and each component is bilaterally influenced, it has been difficult to decipher the cellular and molecular mechanisms that are required for development of the olfactory system (Nomura et al., 2007).

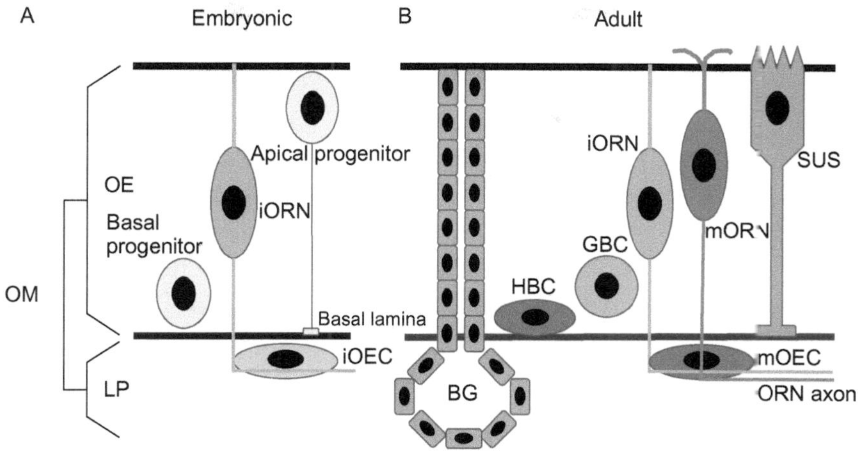

Figure 1 Schematic diagram showing the structure of embryonic and adult olfactory epithelia (OE). The olfactory mucosa (OM) is composed of the OE and the lamina propria (LP) beneath the basal membrane. (A) The embryonic OE contains apical and basal progenitors and immature olfactory receptor neurons (iORNs) provided by apical and basal dividing progenitors. Immature olfactory ensheathing cells (iOECs) that wrap the ORN axons are localized in the developing LP. (B) The adult OE comprises two basal stem/progenitor cells (globose basal cells: GBCs, horizontal basal cells: HBCs), iORNs, mature ORNs (mORNs), sustentacular cells (SUSs), and Bowman's glands (BGs). The LP contains ORN axon bundles and mature OECs (mOECs).

The NC is transiently generated during early vertebrate embryogenesis at the junction between the dorsal neuroepithelium and epidermis (Le Douarin & Kalcheim, 1999). The NC was first described as a "Zwischenstrang" (the intermediate strand) by Wilheim His in 1868 and later renamed NC by Arthur Milnes Marshall (Achilleos & Trainor, 2012; Le Douarin & Kalcheim, 1999). The NC has sometimes been called the fourth germ layer because of its developmental and evolutionary importance (Hall, 2000). Neural crest cells (NCCs) are multipotent cells and leave their original territory through a delamination process that involves epithelial–mesenchymal transition. NCCs migrate to various target tissues throughout the embryo and differentiate into diverse cell types, including neurons and glia of the peripheral nervous system, cranial cartilage and bone, medullary cells of the adrenal gland, and melanocytes (Acloque, Adams, Fishwick, Bronner-Fraser, & Nieto, 2009; Bronner-Fraser, 1995; Le Douarin & Kalcheim, 1999). Recent reports show that neural crest-derived cells (NCDCs) are broadly distributed in adult tissues and organs, and that a subset of NCDCs retain NCC-like multipotency. These multipotent cells are called "neural crest-derived stem cells (NCSCs)" (Stemple & Anderson, 1992) and continue to attract attention as a promising source of cells for regenerative medicine (Achilleos & Trainor, 2012). NCDCs can be identified and isolated from the skin (Fernandes et al., 2004; Sieber-Blum, Grim, Hu, & Szeder, 2004; Toma et al., 2001), heart (Tomita et al., 2005), carotid body (Pardal, Ortega-Saenz, Duran, & Lopez-Barneo, 2007), bone marrow (Morikawa et al., 2009; Nagoshi et al., 2008), dorsal root ganglion (Nagoshi et al., 2008), telencephalon (Yamanishi, Takahashi, Saga, & Osumi, 2012), eye (Kikuchi et al., 2011; Yoshida et al., 2006), gut (Kruger et al., 2002), palatum (Widera et al., 2009), and dental pulp (Janebodin et al., 2011).

The OP is defined as a region of nonneural ectoderm that has undergone focal thickening in response to environmental and intrinsic stimuli (Forni & Wray, 2012). The OP gives rise to both neural and nonneural structures as does the trigeminal, profundal, lateral line, otic, and epibranchial placodes (Maier, Saxena, Alsina, Bronner, & Whitfield, 2014; Schlosser, 2006). Similar to all cranial placodes, the OP is hypothesized to derive exclusively from adjacent preplacodal regions for several decades (Bhattacharyya, Bailey, Bronner-Fraser, & Streit, 2004; Couly & Le Douarin, 1985; Forni & Wray, 2012). However, recent fate-mapping studies using zebrafish and avian models suggest that the OP and otic placode are derived from multiple embryonic cells converging from wide regions, including the NC (Streit, 2002; Whitfield, 2013; Whitlock & Westerfield, 2000). In addition, previous studies using rodents suggested that NCDCs play crucial roles in the development of the front-nasal tissues including the OE (Osumi-Yamashita et al., 1997; Osumi-Yamashita, Ninomiya, Doi, & Eto, 1994). However, the olfactory sensory neurons were thought to be derived exclusively from these placode cells (Streit, 2002; Whitfield, 2013; Whitlock & Westerfield, 2000). In recent years, fate-mapping experiments in chicken embryos have revealed that olfactory ensheathing cells (OECs), i.e., vertebrate glial cells that envelop bundles of ORN axons, are indeed derived from the NC, and not from the OP (Barraud et al., 2010). Moreover, genetic fate-mapping studies using transgenic mice indicate that NC contributes more broadly to the development of olfactory and auditory systems than previously appreciated (Barraud et al., 2010; Forni, Taylor-Burds, Melvin, Williams, & Wray, 2011; Freyer, Aggarwal, & Morrow, 2011; Katoh et al., 2011; Murdoch, DelConte, & Garcia-Castro, 2010; Suzuki, Yoshizaki, Kobayashi, & Osumi, 2013). In particular, NCDCs may potentially serve as a source of cells for neurogenesis and regeneration of the OE (Katoh et al., 2011; Suzuki et al., 2013).

Here, we review our current understanding of the peripheral olfactory organs, particularly with respect to NC and OP contributions to olfactory development.

2. THE OP FORMATION

In mice, the OP arises at the anterior end of the neural plate around 9.5 days of development (E9.5) soon after the closure of the neural tube. The OP invaginates to form the olfactory pit, the beginning of the nasal cavity, at around E10.5 (Chen, Kim, & Xu, 2009; Cuschieri & Bannister, 1975; Forni & Wray, 2012), and some mesenchymal cells delaminate from the placodal epithelium and differentiate into secretory cells or glial cells (Schlosser, 2006; Treloar et al., 2010). By E11, the olfactory pit has deepened and become more complex forming the nostrils and nasal cavity. The vomeronasal organ forms in the medial region of the nasal cavity around the same time, and the main olfactory and accessory olfactory cavities can be distinguished by E11.5 (Treloar et al., 2010).

During OP development, the delineation of neurogenic versus nonneurogenic domains is an initial and critical step that involves many gene networks and signals (Maier et al., 2014). Past studies reveal that the spatiotemporal expressions of several transcription factors, e.g., Pax6, Dlx5, Dlx3 Hes1, Hes5, Sox2, Pou2fl, Msx1/2, and Id3, are very important for OP formation (Bhattacharyya et al., 2004; Bhattacharyya & Bronner-Fraser, 2008; Cau, Gradwohl, Casarosa, Kageyama, & Guillemot, 2000; Donner, Episkopou, & Maas, 2007; Maier & Gunhaga, 2009; Maier et al., 2010). Pax6, which has multiple roles in the developing nervous system (Osumi, Shinohara, Numayama-Tsuruta, & Maekawa, 2008), is expressed in the front-nasal region including the OP (Grindley, Davidson, & Hill, 1995; Matsuo et al., 1993). In spontaneous *Pax6* mutant mice (*Sey/Sey*) and rats (*rSey/rSey*), development of the OP is severely compromised, which results in the absence of ORNs (Grindley et al., 1995; Hill et al., 1991; Matsuo et al., 1993). Although Pax6 and Dlx5 are initially coexpressed by future olfactory cells, Dlx5 expression is upregulated and Pax6 expression is downregulated during acquisition of OP character. In contrast, Dlx3 is initially expressed at low levels, but is upregulated as it commits to an olfactory fate (Bhattacharyya & Bronner-Fraser, 2008). Hes1 is expressed in a broad placodal domain, while expression of Hes5 gene is restricted to clusters of neural progenitor cells. Together, these Hes genes determine the neurogenic and nonneurogenic regions of the OP (Cau et al., 2000; Maier et al., 2014). Sox2 and its cofactor Oct-1 regulate OP induction by maintaining Pax6 expression (Donner et al., 2007). Msx1/2 and Id3 are expressed in future respiratory epithelial cells in chick (Maier et al., 2010). Thus, although some of the signaling pathways are known, the full picture of genetic networks involved in the formation of the OP is incomplete, and further studies are still needed.

Signaling by bone morphogenetic proteins (BMPs), fibroblast growth factors (FGFs), and retinoic acid (RA) modulates OP formation (Kawauchi et al., 2005; LaMantia, Bhasin, Rhodes, & Heemskerk, 2000; Maier et al., 2010, 2014). Cells in the presumptive OP area respond to BMP and FGF signals, respectively, which influence their fate during OP development. BMP signals promote respiratory epithelial character, whereas FGF signals are required for the generation of sensory epithelial cells and the retention of the neurogenic region in chick and mouse (Maier et al., 2010; Sjodal, Edlund, & Gunhaga, 2007). RA signaling is also essential for OE development and olfactory patterning (Bhasin, Maynard, Gallagher, & LaMantia, 2003; LaMantia et al., 2000) and depends on a balance between mesenchymal/epithelial (M/E)-independent and M/E-dependent regulation of RA signaling in the OP. These results suggest that BMP, FGF, and RA signaling coordinately regulate the development and patterning of the OP during early olfactory development (Maier et al., 2014).

3. THE OE DEVELOPMENT

In the OE, neurogenesis occurs throughout life and can be classified into two distinct phases: primary neurogenesis during embryonic development and secondary neurogenesis, i.e., neurogenesis homeostasis in the adult to maintain the mature pseudostratified epithelium (Maier et al., 2014). Primary neurogenesis begins in the invaginating OP at about E10 in mouse and Hamburger–Hamilton stages 14 (HH14) in chick (Kawauchi et al., 2005; Maier & Gunhaga, 2009). Primary neurogenesis forms the fundamental underlying structure of the embryonic OE, and axons of immature ORNs (iORNs) extend to the most rostral part of the telencephalon where the OB will form (Maier et al., 2014; Treloar et al., 2010). During this axon outgrowth process, migratory cells from the OP form the migratory mass together with ORN axons (Miller, Treloar, & Greer, 2010; Valverde, Santacana, & Heredia, 1992). The heterogeneous population of migratory cells include gonadotropin-releasing hormone (GnRH) neurons that migrate to the hypothalamus (Schwanzel-Fukuda & Pfaff, 1989; Whitlock, 2005), OECs (Miller et al., 2010), olfactory marker protein (OMP)-positive cells, and acetylcholine esterase-positive cells (De Carlos, Lopez-Mascaraque, & Valverde, 1995; Miller et al., 2010). Although the migratory mass is supposed to contribute to the axonal projection of ORNs and the induction of the OB, the detailed role of the migratory mass is still to be determined. Most OE cells are proliferative up until E11.5, however by E12.5, the OE begins to organize into a pseudostratified columnar epithelium and commences secondary neurogenesis (Beites, Kawauchi, Crocker, & Calof, 2005; Ikeda et al., 2007).

The olfactory mucosa (OM) is composed of the OE and the lamina propria (LP), the vascularized, NC-derived mesenchymal tissue beneath the OE (Fig. 1). Briefly, the OE proper comprises (1) basal stem/progenitor cells, (2) iORNs, (3) mature ORNs (mORNs), (4) sustentacular cells (SUSs), and (5) secretory ducts of Bowman's glands (BGs). Meanwhile, the LP contains ORN axon bundles, OECs, connective tissues, blood vessels, and acini of BGs (Murdoch & Roskams, 2007). These cells can be distinguished by their morphologic characteristics and gene expression profiles (Murdoch & Roskams, 2007).

The embryonic OE is a neuroepithelium that contains iORNs and two progenitor cell populations: apical progenitor cells with processes that span the height of the OE and basal progenitor cells located closer to the basal lamina (Fig. 1A). These apical and basal progenitor cells express proliferating cell nuclear antigen, Mash1, neurogenin-1 (Ngn1) and nestin, and sandwich the iORNs (Murdoch & Roskams, 2007). At early embryonic stages, most dividing cells are found in the apical part of developing OE. In contrast, at late embryonic stages, cell division is found predominantly near the basal part of the OE (Smart, 1971; Treloar et al., 2010). During early embryogenesis, there are no BGs or SUSs in the OE (Forni & Wray, 2012; Smart, 1971). OEC precursors and immature OECs (iOECs) that ensheathe the ORN axons emerge in the developing LP sometime later. The stem cell/progenitor cell population in the basal layer of the OE continues to produce new ORNs throughout life.

4. THE ADULT OE MAINTENANCE

In the adult OE, its pseudostratified structure is maintained by secondary homeostatic neurogenesis and the replacement of nonneuronal cells based on demand (Fig. 1B). This maintenance is dependent on tissue stem cells, and two subpopulations of basal cells have been proposed as candidates: globose basal cells (GBCs) and horizontal basal cells (HBCs). GBCs are the major proliferating population within the OE. They express Pax6, Sox2, GBC2, and/or Ascl1 (Mash1, an early neuronal marker) depending on their differentiation state and replenish ORNs throughout life (Beites et al., 2005; Caggiano, Kauer, & Hunter, 1994; Chen et al., 2014; Guo et al., 2010; Huard & Schwob, 1995). Although most GBCs are transient amplifying cells under normal physiological conditions, a small population of long-term renewal cells, i.e., stem cell-like cells, can be identified after OE lesioning, and many studies implicate GBCs as multipotent stem/progenitor cells (Chen et al., 2014; Jang, Chen, Flis, Harris, & Schwob, 2014; Packard, Giel-Moloney, Leiter, & Schwob, 2011). In contrast, HBCs are normally quiescent and express the stem cell markers Pax6 and Sox2 and oligodendrocyte progenitor cell and astrocyte markers NG2 and GLAST (Ohnishi et al., 2013), as well as nonneuronal markers Keratin5 (K5) and CD54 (ICAM-1), but are negative for neuronal markers Mash1 and NeuroD (Carter, MacDonald, & Roskams, 2004). Several Cre-loxP lineage-tracing studies have demonstrated that HBCs can regenerate ORNs and nonneuronal cells *in vivo* after extensive OE injury (Duggan & Ngai, 2007; Iwai, Zhou, Roop, & Behringer, 2008; Leung, Coulombe, & Reed, 2007), suggesting the HBCs as tissue stem cells of the OE.

A recent study using *Lgr5-EGFP-IRES-CreERT2* mice revealed that $Lgr5^+$ multipotent GBCs did not give rise to HBCs under normal physiological conditions or following OE injury. This reinforced the idea that GBCs are generated from HBCs after severe epithelial damage (Chen et al., 2014). At the same time, the existence of bromodeoxyuridine or ethynyldeoxyuridine (EdU) label-retaining, quiescent GBCs that are activated in response to epithelial injury and then reestablished after the initial phase of recovery is completed was reported. This indicates that some GBCs satisfy the common criterion for stem cells (Jang et al., 2014). In addition to these two basal cells, SUSs, i.e., "olfactory glia" that function in detoxification, degradation of olfactory stimuli, and regulation of ionic composition, and enhance ORN function, express Pax6 and Sox2 (Guo et al., 2010; Murdoch & Roskams, 2007). However, the role for SUSs as tissue-specific stem cells remains to be determined. As described above, the identification of true tissue stem cells in the OE is not fully resolved and future studies are still needed.

5. THE MULTIPOTENT STEM CELLS IN THE LP

The LP of the OM is a relatively thick layer of connective tissue beneath the OE, separated by the basal lamina. The LP containing connective tissues, cartilages and vesicles, arises from the interaction of cranial NC and the OP, and consists largely of NCDCs as a result. In the LP, multipotent cells with mesenchymal stem cell-like characteristics are derived from embryonic rat OM (Tome, Lindsay, Riddell, & Barnett, 2009). In humans, olfactory stem cells that resemble bone marrow stem cells were observed in the LP and named "olfactory ectomesenchymal stem cells (OE-MSCs)" (Delorme et al., 2010). These stem cells from the LP can give rise to spheres and differentiate into mesenchymal lineages as well as neuronal lineages (Delorme et al., 2010; Tome et al., 2009). Considering the origin of the LP, OE-MSCs are supposed to be of NC origin as NC-derived MSCs are found in the pharyngeal arches and teeth (Delorme et al., 2010; Deng et al., 2004; Ibarretxe et al., 2012; Le Douarin, Calloni, & Dupin, 2008). However, until now, it was unclear whether these stem cells in the LP contribute to the self-renewal of the OE and which cells in the LP they correspond to *in vivo*. Therefore, more sophisticated lineage-tracing studies will be required to identify the OE-MSCs *in vivo* in future.

6. CONTRIBUTION OF NCDCs IN THE OM

6.1. OECs

The OECs are peripheral glial cells that share multiple characteristics with Schwann cells, i.e., NC-derived peripheral glial cells. For example, OEC precursor cells express Sox10 and brain lipid binding protein (BLBP), or fatty acid binding protein 7 (Fabp7). OECs also express glial fibrillary acidic protein (GFAP), low-affinity p75 nerve growth factor receptor ($p75^{NGFR/NTR}$), and S100 similar to Schwann cells (Forni & Wray, 2012). OECs envelope axonal bundles of the olfactory nerve along their course and help axon outgrowth from newly formed ORNs, axonal projection to the OB, and functional synaptic connections throughout life (Ibrahim, Li, Li, Raisman, & El Masry, 2006; Raisman & Li, 2007; Ramon-Cueto & Avila, 1998). Because of this unique property, OECs have attracted attention in regenerative medicine for their possible use in the treatment of spinal cord, brain, and peripheral nerve injury, as well as neurodegenerative diseases like amyotrophic lateral sclerosis (Roet & Verhaagen, 2014).

The debate about the origin of OECs (an ectodermal vs. NC origin) has existed for decades (Couly & Le Douarin, 1985; Wewetzer, Verdu, Angelov, & Navarro, 2002). From traditional morphological studies, OECs were first thought to be olfactory Schwann cells that originate from the NC (Gasser, 1956). Furthermore, because NCCs localize throughout the frontnasal mass (Osumi-Yamashita et al., 1994) and form the LP, it was thought from quail-chick chimeras (quail embryo grafts of the preplacodal anterior neural folds into chick hosts) that the NCCs were the origin of OECs (Couly & Le Douarin, 1985). We have also observed contribution of NCC within the OE during the course of a fate map study of cranial NCC using the lineage tracer DiI (Osumi-Yamashita et al., 1994). However, other studies have proposed a placodal origin of OECs leading to the suggestion that OECs are primarily an OP derivative (Chuah & Au, 1991; Couly & Le Douarin, 1985; Doucette, 1991). Immunohistochemical and ultrastructural observations indicate that OECs are derived from precursor cells in the OE (Chuah & Au, 1991). Moreover, several studies have also reported multipotent progenitor cells from the mouse OE that can produce OEC-like cells as well as ORNs (Carter et al., 2004; Mumm, Shou, & Calof, 1996).

In contrast to these previous studies, recent sophisticated studies suggest an NC origin for OECs (Barraud et al., 2010; Forni et al., 2011; Katoh et al., 2011; Murdoch et al., 2010; Suzuki et al., 2013). Pax7 is an early marker of NCC that is required for NC development (Basch, Bronner-Fraser, & Garcia-Castro, 2006). *Pax7-Cre/Floxed-YFP* mice (Murdoch et al., 2010) indicate that *Pax7*-expressing embryonic progenitors contribute to multiple OE lineages including OECs. Furthermore, fate-mapping techniques in chicken embryos combined with genetic lineage tracing in mice using the *Wnt1-Cre* reporter system show that OECs originate from the NC and hence share a common developmental heritage with Schwann cells (Barraud et al., 2010). They used variety of techniques to reveal OEC origins. First, unilateral isotopic grafting of the anterior neural fold, which generates the OP, at the three to five somites' stage from transgenic ^{GFP} chicken embryos into wild-type host demonstrated that there are no graft-derived GFP-positive OECs in the host LP. Second, isotopic grafting of midbrain-level NCC precursors at the four to seven somites' stage from transgenic ^{GFP} chicken or quail donors into wild-type hosts demonstrated that graft-derived GFP-positive cells expressed OEC markers such as $p75^{NTR}$ and/or myelin protein zero (P0). Finally, genetic NC lineage tracing using *Wnt1-Cre/Floxed-YFP* or *LacZ* mice demonstrated that there were many YFP-positive NCDCs in the front-nasal mass and p75^{NTR}-positive and YFP-positive OECs were found in the LP and the olfactory nerve layer of the OB (Barraud et al., 2010).

Three more reports suggest the NC origin of OECs by using *Wnt1-Cre* and/or *P0-Cre* reporter mice (Forni et al., 2011; Katoh et al., 2011; Suzuki et al., 2013). Forni et al. used *Crect* (an AP-2α ectodermal specific *cis*-regulatory element)-*Cre/Floxed-YFP* or *LacZ* mice that can trace the placodal ectoderm and confirmed the complementary pattern of Crect recombination and Wnt1-Cre recombination in the OM and found no Crect recombination in the OECs (Forni et al., 2011). Although both avian grafting experiments and genetic lineage tracing using Cre reporter mice have some limitations, such as the potential for cell contamination and inappropriate Cre expression, these studies strongly suggest that OECs may have an exclusively NC origin.

6.2. HBCs

The HBCs are relatively quiescent cells that divide every 60 days under normal conditions (Mackay-Sim & Kittel, 1991), and have a capacity to generate neurons, as well as nonneuronal cells *in vitro* (Carter et al., 2004). In addition, K5-Cre lineage-tracing studies demonstrate that HBCs can regenerate all the OE cells, including the GBCs, during normal physiological turn over, as well as during injury-induced regeneration *in vivo* (Iwai et al., 2008; Leung et al., 2007). Recently, two reports reveal the regulation of cellular dynamics of HBCs: (1) ΔNp63 is expressed in HBCs, and a germline mutation in the *p63* gene results in the absence of HBCs in the perinatal OE (Packard, Schnittke, Romano, Sinha, & Schwob, 2011); (2) the transcription factor p63 is a key regulator of postnatal HBCs, in which it functions cell autonomously to maintain olfactory stem cells by promoting self-renewal and inhibiting differentiation into more mature cell types (Fletcher et al., 2011). Interestingly, HBCs do not form in the newborn OE of *p63* knockout mice in spite of the grossly normal appearance of other cellular constituents of the OE (Packard, Schnittke, et al., 2011). This suggests that embryonic HBCs are not necessary for the formation of the OE.

Definitive HBCs, i.e., p63-positive and Keratin 14-positive cells apposed to the basal lamina, appear in the mouse OE at E17.5, and their formation accelerates after birth and becomes confluent by P10 (Holbrook, Szumowski, & Schwob, 1995; Packard, Schnittke, et al., 2011). So what is the origin of the HBCs? Recently, lineage tracing in *P0-Cre* and/or

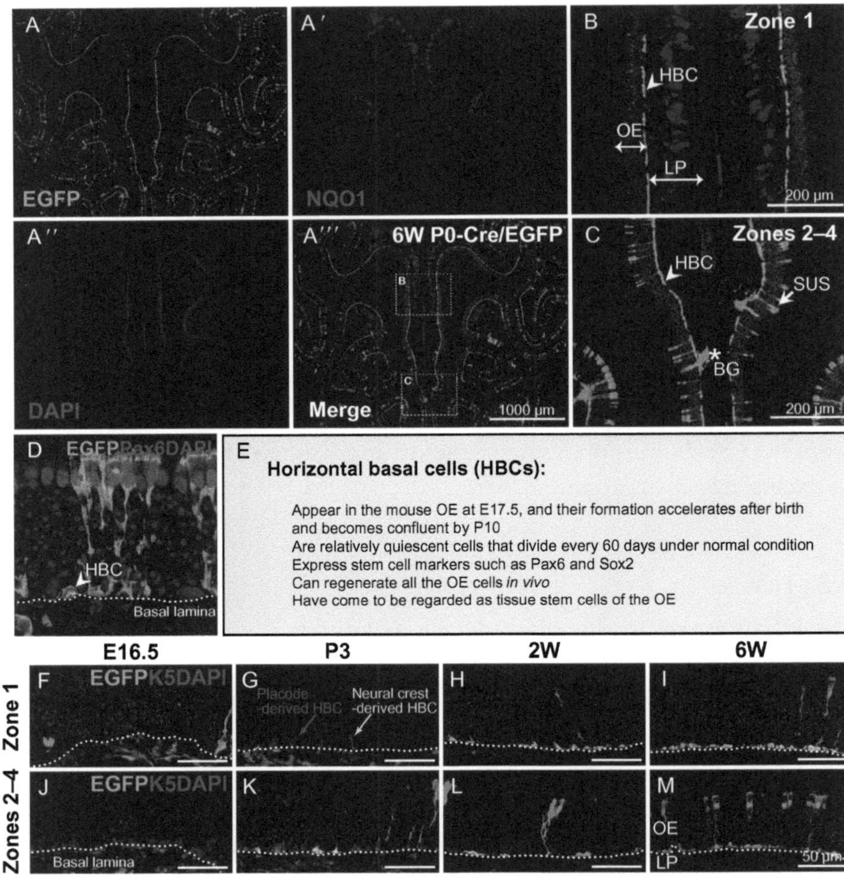

Figure 2 Neural crest-derived cells (NCDCs) in the adult olfactory epithelium (OE). (A–A‴) A coronal section of the adult OE in *P0-Cre/Floxed-EGFP* mice. NQO1 is a marker for zone 1. (B and C) High-magnification images of zone 1 and zones 2–4. Strong EGFP signals are observed in horizontal basal cells (HBCs, arrowhead), sustentacular cells (SUSs, arrow), and Bowman's glands (BGs, *). (D) Confocal image of Pax6-positive HBCs (arrowhead). (E) Summary of HBC characteristics. (F–M) Representative images of the OE at E16.5 (F and J), P3 (G and K), 2W (H and L), and 6W (I and M) in zone 1 (F–I) and zones 2–4 (J–M). Main population of HBCs gradually changes from placode-derived (red arrow) to neural crest-derived (yellow arrow) during postnatal maturation. Abbreviation: LP, lamina propria. *Reproduced from Suzuki et al. (2013) (A–D and F–M).* (See the color plate.)

Wnt1-Cre reporter mice revealed the generation and dynamic changes in two types of HBCs: placode-derived HBCs and NC-derived HBCs (Fig. 2A–M) (Katoh et al., 2011; Suzuki et al., 2013). In zone 1 (a dorsomedial region) of the OE, very few K5-positive HBCs are detected at late embryonic stages (E16.5), and these HBCs increase in number by P3 and continue to do so. In contrast, many K5-positive HBCs were already detected in zones 2–4 (a ventrolateral region) at E16.5. Among these K5-positive HBCs, the ratio of reporter (EGFP)-expressing cells, i.e., NC-derived cells, gradually increased and eventually reached 66.3% in zone 1 and 83.3% in zones 2–4 in 6-week-old mice (Fig. 2F–M) (Suzuki et al., 2013). Similar to previous reports that HBCs can regenerate ORNs and nonneuronal cells *in vivo* after extensive OE injury (Fig. 3A) (Iwai et al., 2008; Leung et al., 2007), these EGFP-expressing HBCs also regenerate

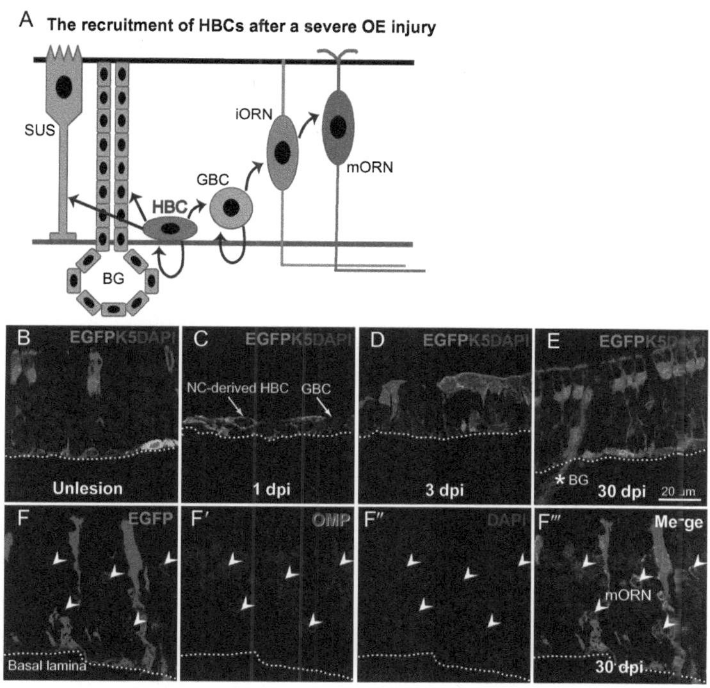

Figure 3 Neural crest-derived cells (NCDCs) contribute to the regeneration of the olfactory epithelium (OE) after methimazole-induced severe OE injury. (A) Schematic representation of the recruitment of horizontal basal cells (HBCs) after a severe OE injury. (B–F‴) An unlesioned OE (B) and injured OE at 1 (C), 3 (D), and 30 days postinjury (dpi) (E–F‴). Following methimazole intraperitoneal injection, widespread cell death is observed within 24 h of drug treatment, but most of the HBCs are not damaged (C). At 30 dpi, the injured OE is nearly completely regenerated, and many EGFP-positive cells are observed in the OE (E). (F–F‴) NCDCs differentiated into OMP-positive mature olfactory receptor neurons (mORNs) (arrowheads). Abbreviations: SUSs, sustentacular cells; BGs, Bowman's glands; GBCs, globose basal cells; iORNs, immature olfactory receptor neurons; mORNs, mature olfactory receptor neurons; NC, neural crest. *Panel (B)–(F‴): Reproduced from Suzuki et al. (2013).* (See the color plate.)

olfactory cells including ORNs *in vivo* after OE injury caused by methimazole injection (Fig. 3B–F‴).

These results demonstrate that (1) HBCs have two different origins and the ratio of NC-derived HBCs to placode-derived HBCs gradually increases during postnatal development; (2) NC-derived HBCs play a backup role to GBCs during the reconstruction of the OE following severe injury in adults; and (3) the different behavior of NC-derived HBCs between zone 1 and zones 2–4 may underpin regional differences in the OE.

HBCs develop from olfactory placode progenitor cells/GBCs that may transiently express Ascl1 in the embryo. Interestingly, *Ascl1* (*Mash1*) knock-out mice exhibit a substantial delay in HBC emergence in the OE. The appearance of HBCs albeit delayed suggests that there is an alternative Ascl1-independent pathway for the development of HBCs (Krolewski, Packard, Jang, Wildner, & Schwob, 2012; Packard, Schnittke, et al., 2011). The presence of NC-derived HBCs could account for this delay in HBC emergence, and NC-derived HBCs may not need Ascl1 for their development and maturation. As to the potential backup function of NC-derived HBCs, there are many reports that reveal the presence of multipotent NCSCs in various tissues (Fernandes et al., 2004; Janebodin et al., 2011; Kikuchi et al., 2011; Kruger et al., 2002; Morikawa et al., 2009; Nagoshi et al., 2008; Pardal et al., 2007; Sieber-Blum et al., 2004; Stemple & Anderson, 1992; Toma et al., 2001; Tomita et al., 2005; Widera et al., 2009; Yoshida et al., 2006). Therefore, perhaps NCDCs remain as quiescent HBCs in the adult OE. An important question that remains is whether the different behavior of NC-derived HBCs between zone 1 and zones 2–4 causes the regional differences in responses to odors (Imai & Sakano, 2007) and/or cellular homeostasis (Vedin, Molander, Bohm, & Berghard, 2009). Olfactory neurons in the dorsomedial OE (zone 1) govern innate responses to odors, whereas the ventrolateral OE (zones 2–4) appears to mediate learned, perceptual functions (Imai & Sakano, 2007). Fewer basal cells exist in S-phase in zone 1 than in the other epithelial parts (Vedin et al., 2009). NCDCs, including HBCs, ORNs, and SUSs, are more frequently observed in zones 2–4 (Fig. 2A–C) (Suzuki et al., 2013), which have a high proliferation rate and a relatively short neuron lifespan. At present, we have no knowledge about the difference between NCDCs and placode-derived cells in the OE, or the effect of ablation of NCDCs in the OE. Similarly, any differences between placode-derived HBCs and NC-derived HBCs remain to be determined and will be the subject of future investigations.

6.3. Globose basal cellss

GBCs are a heterogeneous population that includes immediate neuronal progenitors, transit-amplifying precursors, and multipotent precursors, which are the major proliferating populations in the adult OE (Chen et al., 2014; Jang et al., 2014). NC-derived GBCs are identified as Sox2 and/or Pax6-positive globose cells in the *Wnt1-Cre/Floxed-YFP* mice (Forni et al., 2011), and as Mash1 positive cells in the *P0-Cre/Floxed-LacZ* mice (Katoh et al., 2011). However, we still do not know whether a subtype of GBCs that satisfies the key features of stem cells is derived from the NC or not. Stem cell markers such as Pax6, Sox2, and Mash1 are not specific to GBCs, and GBC2 that is used to label the GBCs in adult rats (Jang, Youngentob, & Schwob, 2003) is not a specific lineage marker for postnatal mouse GBCs (Wang et al., 2011). Future studies using NC reporter mice and thymidine analogue labeling might reveal the presence of mitotic quiescent, label-retaining GBCs.

6.4. Olfactory receptor neurons

ORNs are bipolar transduction cells in the OE that transmit olfactory information to the brain. In the classic view, all the ORNs are derived solely from placode cells (Whitfield, 2013). Recent reports show that the NCCs can give rise to microvillous neurons in the zebrafish olfactory system (Saxena, Peng, & Bronner, 2013). In the zebrafish OE, there are two main types of ORNs: ciliated sensory neurons that detect volatile or easily evaporated substances, and microvillous sensory neurons that detect pheromones and nucleotides (Saxena et al., 2013). Using a Sox10:eGFP line and photoconversion, the authors showed that microvillous neurons are derived from the NC, and that placode cells cannot compensate for the loss of NC-derived neurons. In mice, occasionally labeled ORNs have been identified in *Wnt1-Cre/Floxed-YFP* (Forni et al., 2011) and *P0-Cre/Floxed-LacZ* mice (Katoh et al., 2011) under normal physiological conditions. Furthermore, the percentage of NC-derived mORNs was significantly increased during OE regeneration caused by methimazole administration (Fig. 3) (Suzuki et al., 2013). Methimazole is a drug used in the treatment of hyperthyroidism which functions by inhibiting the enzyme thyroperoxidase that acts in thyroid hormone synthesis. Methimazole also exhibits an olfactory toxicity in rodents and causes OE degeneration and regeneration (Fletcher et al., 2011; Genter, Deamer, Blake, Wesley, & Levi, 1995). Following methimazole administration, most olfactory cells except for the basal cells in the OE die within 24 h. Shortly thereafter, the OE gradually begins to regenerate and the tissue is nearly completely recovered by 30 days post-injury (dpi) (Fig. 3B–E). Interestingly, the incidence of NC-derived ORNs significantly increased in the regenerated OE (Fig. 3F–F‴). This result suggests that NC-derived stem cells have an important role to play in regenerating ORNs. Presently, there are no reports that have investigated the differences between NC-derived ORNs and placode-derived ORNs. Moreover, we still do not know whether regenerated NC-derived ORNs function normally. If NC-derived ORNs function correctly, NC-derived stem cells could be a major target for the treatment of olfaction disorders that currently have no effective treatment.

6.5. GnRH neurons

GnRH neurons in the hypothalamus produce GnRH that stimulates the synthesis and release of gonadotropins, e.g., follicle-stimulating hormone and luteinizing hormone, which are essential for sexual development and reproduction, from the anterior pituitary gland. GnRH neurons are a component of the migratory mass that spans from the nasal area to the forebrain during embryogenesis (Forni et al., 2011; Forni & Wray, 2012). The origin of GnRH neurons has also been controversial for many years, and several regions, e.g., the OP, non-OP, ectoderm, and NC, have been suggested as the origin of GnRH cells (el Amraoui & Dubois, 1993; Schwanzel-Fukuda & Pfaff, 1989; Whitlock, Wolf, & Boyce, 2003; Wray, Grant, & Gainer, 1989). Recently, genetic lineage tracing using *Wnt1-Cre* and *Crect-Cre* reporter mice confirmed that the majority of GnRH-1 neurons are derived from the OP. However, about 30% of the GnRH-1 neurons are derived from the NC. In contrast, avian transplantation experiments suggested that there was no contribution from NC to the formation of GnRH neurons (Sabado, Barraud, Baker, & Streit, 2012). Yet, NC-derived progenitor cells are observed within the OP of the chick as well as of the mouse (Forni et al., 2011; Katoh et al., 2011). Therefore, it is possible that NC-derived progenitor cells are present in OP grafts. In conclusion, GnRH neurons appear to have a dual origin of NC-derived and OP-derived cells.

Kallmann syndrome is a rare genetic disorder in humans that is defined by a delay or absence of puberty onset, along with total or partial anosmia. The pathology of Kallmann syndrome is associated with impaired migration of GnRH-1 neurons (Forni & Wray, 2012; Wray, 2010). SOX10 is a major regulator of NCC development, and recently, *Sox10*-null mutant mice were observed to exhibit and almost complete absence of OECs, together with defasciculation and misrouting of the nerve fibers, and impaired migration of GnRH neurons (Pingault et al., 2013). This further supports the idea that GnRH neuron development is heavily influenced by the NC.

7. THE LIMITATION OF CURRENT TECHNIQUES AND FUTURE PERSPECTIVES

Numerous studies over the past two decades have revealed the contribution of NCC to the tissue development. Classical techniques such as grafting (Couly & Le Douarin, 1985) and dye labeling experiments (Osumi-Yamashita et al., 1994; Serbedzija, Bronner-Fraser, & Fraser, 1989) have limitations, especially with regard to investigations in adult stages. For example, the possibility of cell contamination in grafting experiments cannot be ruled out and neither can the potential for leakage or dilution of the dye could have unintended effects of target cells and tissues. In contrast, genetic lineage tracing using Cre reporter mice has made it possible to study the interaction of NC and OP during morphogenesis of the OE. Although *Wnt1-Cre* and *P0-Cre* are frequently used as NC lineage-tracing mice (Barraud et al., 2010; Forni et al., 2011; Katoh et al., 2011; Suzuki et al., 2013), their promoter activities are not truly "specific" to NCC. Ectopic *Wnt1* expression has been observed in the neural tube and otic placode (Danielian, Muccino, Rowitch, Michael, & McMahon, 1998; Freyer et al., 2011), and several phenotypes affecting midbrain development have been observed in *Wnt1-Cre* transgenic mice because Wnt1 is also expressed in these regions (Lewis, Vasudevan, O'Neill, Soriano, & Bush, 2013). In the same way, *P0* mRNA and protein have been reported to be expressed in the notochord embryonic stages, as well as some endothelial cells in postnatal tissues (Kubota et al., 2011; Lee et al., 2001; Nagoshi et al., 2008). There are also other possibilities such as leaky expression of Cre, lack of reporter signals in all true NC-derived cells, and incomplete penetrance and/or variability of recombination corresponding to the genetic background (Forni & Wray, 2012; Hebert & McConnell, 2000; Suzuki et al., 2013). Although genetic lineage-tracing system using Cre reporter mice has provided many novel insights into the development of the OE, especially with respect to NC contributions, further confirmatory studies will be necessary to evaluate the validity of these new findings.

In addition to the OE, NCDCs contribute to the development of the inner ear (Freyer et al., 2011) and the eye (Kikuchi et al., 2011; Yoshida et al., 2006). These findings indicate that NCDCs may participate in the formation of sensory organs together with placode cells, and furthermore that they may have a backup function as quiescent tissue stem cells in adult sensory organs. NCDCs and NCSCs receive considerable attention as potential sources of cells for regenerative medicine (Achilleos & Trainor, 2012). However, it remains to be fully determined whether there are functional differences between NC-derived olfactory cells and placode-derived olfactory cells. Future investigations will be required to recognize the importance of the dual origin of the OE, not just for our basic understanding of olfactory development but also for any potential benefit in clinical medicine.

CONFLICT OF INTERESTS

The authors have no potential conflict of interests.

ACKNOWLEDGMENTS

The authors thank Dr. Paul Trainor and Dr. Dwayne Simmons for critically reading the chapter.

REFERENCES

Achilleos, A., & Trainor, P. A. (2012). Neural crest stem cells: Discovery, properties and potential for therapy. *Cell Research, 22*, 288–304.

Acloque, H., Adams, M. S., Fishwick, K., Bronner-Fraser, M., & Nieto, M. A. (2009). Epithelial-mesenchymal transitions: The importance of changing cell state in development and disease. *The Journal of Clinical Investigation, 119*, 1438–1449.

Bargmann, C. I. (2006). Comparative chemosensation from receptors to ecology. *Nature, 444*, 295–301.

Barraud, P., Seferiadis, A. A., Tyson, L. D., Zwart, M. F., Szabo-Rogers, H. L., Ruhrberg, C., et al. (2010). Neural crest origin of olfactory ensheathing glia. *Proceedings of the National Academy of Sciences of the United States of America, 107*, 21040–21045.

Basch, M. L., Bronner-Fraser, M., & Garcia-Castro, M. I. (2006). Specification of the neural crest occurs during gastrulation and requires Pax7. *Nature, 441*, 218–222.

Beites, C. L., Kawauchi, S., Crocker, C. E., & Calof, A. L. (2005). Identification and molecular regulation of neural stem cells in the olfactory epithelium. *Experimental Cell Research, 306*, 309–316.

Bhasin, N., Maynard, T. M., Gallagher, P. A., & LaMantia, A. S. (2003). Mesenchymal/epithelial regulation of retinoic acid signaling in the olfactory placode. *Developmental Biology, 261*, 82–98.

Bhattacharyya, S., Bailey, A. P., Bronner-Fraser, M., & Streit, A. (2004). Segregation of lens and olfactory precursors from a common territory: Cell sorting and reciprocity of Dlx5 and Pax6 expression. *Developmental Biology, 271*, 403–414.

Bhattacharyya, S., & Bronner-Fraser, M. (2008). Competence, specification and commitment to an olfactory placode fate. *Development, 135*, 4165–4177.

Bronner-Fraser, M. (1995). Origins and developmental potential of the neural crest. *Experimental Cell Research, 218*, 405–417.

Buck, L., & Axel, R. (1991). A novel multigene family may encode odorant receptors: A molecular basis for odor recognition. *Cell, 65*, 175–187.
Caggiano, M., Kauer, J. S., & Hunter, D. D. (1994). Globose basal cells are neuronal progenitors in the olfactory epithelium: A lineage analysis using a replication-incompetent retrovirus. *Neuron, 13*, 339–352.
Carter, L. A., MacDonald, J. L., & Roskams, A. J. (2004). Olfactory horizontal basal cells demonstrate a conserved multipotent progenitor phenotype. *The Journal of Neuroscience, 24*, 5670–5683.
Cau, E., Gradwohl, G., Casarosa, S., Kageyama, R., & Guillemot, F. (2000). Hes genes regulate sequential stages of neurogenesis in the olfactory epithelium. *Development, 127*, 2323–2332.
Chen, B., Kim, E. H., & Xu, P. X. (2009). Initiation of olfactory placode development and neurogenesis is blocked in mice lacking both Six1 and Six4. *Developmental Biology, 326*, 75–85.
Chen, M., Tian, S., Yang, X., Lane, A. P., Reed, R. R., & Liu, H. (2014). Wnt-responsive lgr5+ globose basal cells function as multipotent olfactory epithelium progenitor cells. *The Journal of Neuroscience, 34*, 8268–8276.
Chuah, M. I., & Au, C. (1991). Olfactory Schwann cells are derived from precursor cells in the olfactory epithelium. *Journal of Neuroscience Research, 29*, 172–180.
Couly, G. F., & Le Douarin, N. M. (1985). Mapping of the early neural primordium in quail-chick chimeras. I. Developmental relationships between placodes, facial ectoderm, and prosencephalon. *Developmental Biology, 110*, 422–439.
Cuschieri, A., & Bannister, L. H. (1975). The development of the olfactory mucosa in the mouse: Light microscopy. *Journal of Anatomy, 119*, 277–286.
Danielian, P. S., Muccino, D., Rowitch, D. H., Michael, S. K., & McMahon, A. P. (1998). Modification of gene activity in mouse embryos in utero by a tamoxifen-inducible form of Cre recombinase. *Current Biology, 8*, 1323–1326.
De Carlos, J. A., Lopez-Mascaraque, L., & Valverde, F. (1995). The telencephalic vesicles are innervated by olfactory placode-derived cells: A possible mechanism to induce neocortical development. *Neuroscience, 68*, 1167–1178.
Delorme, B., Nivet, E., Gaillard, J., Haupl, T., Ringe, J., Deveze, A., et al. (2010). The human nose harbors a niche of olfactory ectomesenchymal stem cells displaying neurogenic and osteogenic properties. *Stem Cells and Development, 19*, 853–866.
Deng, M. J., Jin, Y., Shi, J. N., Lu, H. B., Liu, Y., He, D. W., et al. (2004). Multilineage differentiation of ectomesenchymal cells isolated from the first branchial arch. *Tissue Engineering, 10*, 1597–1606.
Donner, A. L., Episkopou, V., & Maas, R. L. (2007). Sox2 and Pou2f1 interact to control lens and olfactory placode development. *Developmental Biology, 303*, 784–799.
Doucette, R. (1991). PNS-CNS transitional zone of the first cranial nerve. *The Journal of Comparative Neurology, 312*, 451–466.
Duggan, C. D., & Ngai, J. (2007). Scent of a stem cell. *Nature Neuroscience, 10*, 673–674.
el Amraoui, A., & Dubois, P. M. (1993). Experimental evidence for an early commitment of gonadotropin-releasing hormone neurons, with special regard to their origin from the ectoderm of nasal cavity presumptive territory. *Neuroendocrinology, 57*, 991–1002.
Fernandes, K. J., McKenzie, I. A., Mill, P., Smith, K. M., Akhavan, M., Barnabe-Heider, F., et al. (2004). A dermal niche for multipotent adult skin-derived precursor cells. *Nature Cell Biology, 6*, 1082–1093.
Fletcher, R. B., Prasol, M. S., Estrada, J., Baudhuin, A., Vranizan, K., Choi, Y. G., et al. (2011). p63 regulates olfactory stem cell self-renewal and differentiation. *Neuron, 72*, 748–759.
Forni, P. E., Taylor-Burds, C., Melvin, V. S., Williams, T., & Wray, S. (2011). Neural crest and ectodermal cells intermix in the nasal placode to give rise to GnRH-1 neurons, sensory neurons, and olfactory ensheathing cells. *The Journal of Neuroscience, 31* 6915–6927.

Forni, P. E., & Wray, S. (2012). Neural crest and olfactory system: New prospective. *Molecular Neurobiology, 46*, 349–360.

Freyer, L., Aggarwal, V., & Morrow, B. E. (2011). Dual embryonic origin of the mammalian otic vesicle forming the inner ear. *Development, 138*, 5403–5414.

Gasser, H. S. (1956). Olfactory nerve fibers. *The Journal of General Physiology, 39*, 473–496.

Genter, M. B., Deamer, N. J., Blake, B. L., Wesley, D. S., & Levi, P. E. (1995). Olfactory toxicity of methimazole: Dose–response and structure-activity studies and characterization of flavin-containing monooxygenase activity in the Long-Evans rat olfactory mucosa. *Toxicologic Pathology, 23*, 477–486.

Graziadei, P. P., & Graziadei, G. A. (1979). Neurogenesis and neuron regeneration in the olfactory system of mammals. I. Morphological aspects of differentiation and structural organization of the olfactory sensory neurons. *Journal of Neurocytology, 8*, 1–18.

Grindley, J. C., Davidson, D. R., & Hill, R. E. (1995). The role of Pax-6 in eye and nasal development. *Development, 121*, 1433–1442.

Guo, Z., Packard, A., Krolewski, R. C., Harris, M. T., Manglapus, G. L., & Schwob, J. E. (2010). Expression of pax6 and sox2 in adult olfactory epithelium. *The Journal of Comparative Neurology, 518*, 4395–4418.

Hall, B. K. (2000). The neural crest as a fourth germ layer and vertebrates as quadroblastic not triploblastic. *Evolution & Development, 2*, 3–5.

Hebert, J. M., & McConnell, S. K. (2000). Targeting of cre to the Foxg1 (BF-1) locus mediates loxP recombination in the telencephalon and other developing head structures. *Developmental Biology, 222*, 296–306.

Hill, R. E., Favor, J., Hogan, B. L., Ton, C. C., Saunders, G. F., Hanson, I. M., et al. (1991). Mouse small eye results from mutations in a paired-like homeobox-containing gene. *Nature, 354*, 522–525.

Holbrook, E. H., Szumowski, K. E., & Schwob, J. E. (1995). An immunochemical, ultrastructural, and developmental characterization of the horizontal basal cells of rat olfactory epithelium. *The Journal of Comparative Neurology, 363*, 129–146.

Huard, J. M., & Schwob, J. E. (1995). Cell cycle of globose basal cells in rat olfactory epithelium. *Developmental Dynamics, 203*, 17–26.

Ibarretxe, G., Crende, O., Aurrekoetxea, M., Garcia-Murga, V., Etxaniz, J., & Unda, F. (2012). Neural crest stem cells from dental tissues: A new hope for dental and neural regeneration. *Stem Cells International, 2012*, 103503.

Ibrahim, A., Li, Y., Li, D., Raisman, G., & El Masry, W. S. (2006). Olfactory ensheathing cells: Ripples of an incoming tide? *Lancet Neurology, 5*, 453–457.

Ikeda, K., Ookawara, S., Sato, S., Ando, Z., Kageyama, R., & Kawakami, K. (2007). Six1 is essential for early neurogenesis in the development of olfactory epithelium. *Developmental Biology, 311*, 53–68.

Imai, T., & Sakano, H. (2007). Roles of odorant receptors in projecting axons in the mouse olfactory system. *Current Opinion in Neurobiology, 17*, 507–515.

Iwai, N., Zhou, Z., Roop, D. R., & Behringer, R. R. (2008). Horizontal basal cells are multipotent progenitors in normal and injured adult olfactory epithelium. *Stem Cells, 26*, 1298–1306.

Janebodin, K., Horst, O. V., Ieronimakis, N., Balasundaram, G., Reesukumal, K., Pratumvinit, B., et al. (2011). Isolation and characterization of neural crest-derived stem cells from dental pulp of neonatal mice. *PLoS One, 6*, e27526.

Jang, W., Chen, X., Flis, D., Harris, M., & Schwob, J. E. (2014). Label-retaining, quiescent globose basal cells are found in the olfactory epithelium. *The Journal of Comparative Neurology, 522*, 731–749.

Jang, W., Youngentob, S. L., & Schwob, J. E. (2003). Globose basal cells are required for reconstitution of olfactory epithelium after methyl bromide lesion. *The Journal of Comparative Neurology, 460*, 123–140.

Katoh, H., Shibata, S., Fukuda, K., Sato, M., Satoh, E., Nagoshi, N., et al. (2011). The dual origin of the peripheral olfactory system: Placode and neural crest. *Molecular Brain*, *4*, 34.

Kawauchi, S., Shou, J., Santos, R., Hebert, J. M., McConnell, S. K., Mason, I., et al. (2005). Fgf8 expression defines a morphogenetic center required for olfactory neurogenesis and nasal cavity development in the mouse. *Development*, *132*, 5211–5223.

Kikuchi, M., Hayashi, R., Kanakubo, S., Ogasawara, A., Yamato, M., Osumi, N., et al. (2011). Neural crest-derived multipotent cells in the adult mouse iris stroma. *Genes to Cells*, *16*, 273–281.

Krolewski, R. C., Packard, A., Jang, W., Wildner, H., & Schwob, J. E. (2012). *Ascl1 (Mash1)* knockout perturbs differentiation of nonneuronal cells in olfactory epithelium. *PLoS One*, *7*, e51737.

Kruger, G. M., Mosher, J. T., Bixby, S., Joseph, N., Iwashita, T., & Morrison, S. J. (2002). Neural crest stem cells persist in the adult gut but undergo changes in self-renewal, neuronal subtype potential, and factor responsiveness. *Neuron*, *35*, 657–669.

Kubota, Y., Takubo, K., Hirashima, M., Nagoshi, N., Kishi, K., Okuno, Y., et al. (2011). Isolation and function of mouse tissue resident vascular precursors marked by myelin protein zero. *The Journal of Experimental Medicine*, *208*, 949–960.

LaMantia, A. S., Bhasin, N., Rhodes, K., & Heemskerk, J. (2000). Mesenchymal/epithelial induction mediates olfactory pathway formation. *Neuron*, *28*, 411–425.

Le Douarin, N. M., Calloni, G. W., & Dupin, E. (2008). The stem cells of the neural crest. *Cell Cycle*, *7*, 1013–1019.

Le Douarin, N. M., & Kalcheim, C. (1999). *The neural crest*. Cambridge, UK: Cambridge University Press.

Lee, M. J., Calle, E., Brennan, A., Ahmed, S., Sviderskaya, E., Jessen, K. R., et al. (2001). In early development of the rat mRNA for the major myelin protein P(0) is expressed in nonsensory areas of the embryonic inner ear, notochord, enteric nervous system, and olfactory ensheathing cells. *Developmental Dynamics*, *222*, 40–51.

Leung, C. T., Coulombe, P. A., & Reed, R. R. (2007). Contribution of olfactory neural stem cells to tissue maintenance and regeneration. *Nature Neuroscience*, *10*, 720–726.

Lewis, A. E., Vasudevan, H. N., O'Neill, A. K., Soriano, P., & Bush, J. O. (2013). The widely used Wnt1-Cre transgene causes developmental phenotypes by ectopic activation of Wnt signaling. *Developmental Biology*, *379*, 229–234.

Mackay-Sim, A., & Kittel, P. (1991). Cell dynamics in the adult mouse olfactory epithelium: A quantitative autoradiographic study. *The Journal of Neuroscience*, *11*, 979–984.

Maier, E., & Gunhaga, L. (2009). Dynamic expression of neurogenic markers in the developing chick olfactory epithelium. *Developmental Dynamics*, *238*, 1617–1625.

Maier, E. C., Saxena, A., Alsina, B., Bronner, M. E., & Whitfield, T. T. (2014). Sensational placodes: Neurogenesis in the otic and olfactory systems. *Developmental Biology*, *389*, 50–67.

Maier, E., von Hofsten, J., Nord, H., Fernandes, M., Paek, H., Hebert, J. M., et al. (2010). Opposing Fgf and Bmp activities regulate the specification of olfactory sensory and respiratory epithelial cell fates. *Development*, *137*, 1601–1611.

Matsuo, T., Osumi-Yamashita, N., Noji, S., Ohuchi, H., Koyama, E., Myokai, F., et al. (1993). A mutation in the Pax-6 gene in rat small eye is associated with impaired migration of midbrain crest cells. *Nature Genetics*, *3*, 299–304.

Miller, A. M., Treloar, H. B., & Greer, C. A. (2010). Composition of the migratory mass during development of the olfactory nerve. *The Journal of Comparative Neurology*, *518*, 4825–4841.

Morikawa, S., Mabuchi, Y., Niibe, K., Suzuki, S., Nagoshi, N., Sunabori, T., et al. (2009). Development of mesenchymal stem cells partially originate from the neural crest. *Biochemical and Biophysical Research Communications*, *379*, 1114–1119.

Mumm, J. S., Shou, J., & Calof, A. L. (1996). Colony-forming progenitors from mouse olfactory epithelium: Evidence for feedback regulation of neuron production. *Proceedings of the National Academy of Sciences of the United States of America*, *93*, 11167–11172.

Murdoch, B., DelConte, C., & Garcia-Castro, M. I. (2010). Embryonic Pax7-expressing progenitors contribute multiple cell types to the postnatal olfactory epithelium. *The Journal of Neuroscience*, *30*, 9523–9532.

Murdoch, B., & Roskams, A. J. (2007). Olfactory epithelium progenitors: Insights from transgenic mice and in vitro biology. *Journal of Molecular Histology*, *38*, 581–599.

Nagoshi, N., Shibata, S., Kubota, Y., Nakamura, M., Nagai, Y., Satoh, E., et al. (2008). Ontogeny and multipotency of neural crest-derived stem cells in mouse bone marrow, dorsal root ganglia, and whisker pad. *Cell Stem Cell*, *2*, 392–403.

Nomura, T., Haba, H., & Osumi, N. (2007). Role of a transcription factor Pax6 in the developing vertebrate olfactory system. *Development, Growth & Differentiation*, *49*, 683–690.

Ohnishi, Y., Iwatsuki, K., Shinzawa, K., Ishihara, M., Moriwaki, T., Umegaki, M., et al. (2013). Adult olfactory sphere cells are a source of oligodendrocyte and Schwann cell progenitors. *Stem Cell Research*, *11*, 1178–1190.

Osumi, N., Shinohara, H., Numayama-Tsuruta, K., & Maekawa, M. (2008). Concise review: Pax6 transcription factor contributes to both embryonic and adult neurogenesis as a multifunctional regulator. *Stem Cells*, *26*, 1663–1672.

Osumi-Yamashita, N., Kuratani, S., Ninomiya, Y., Aoki, K., Iseki, S., Chareonvit, S., et al. (1997). Cranial anomaly of homozygous rSey rat is associated with a defect in the migration pathway of midbrain crest cells. *Development, Growth & Differentiation*, *39*, 53–67.

Osumi-Yamashita, N., Ninomiya, Y., Doi, H., & Eto, K. (1994). The contribution of both forebrain and midbrain crest cells to the mesenchyme in the frontonasal mass of mouse embryos. *Developmental Biology*, *164*, 409–419.

Packard, A., Giel-Moloney, M., Leiter, A., & Schwob, J. E. (2011). Progenitor cell capacity of NeuroD1-expressing globose basal cells in the mouse olfactory epithelium. *The Journal of Comparative Neurology*, *519*, 3580–3596.

Packard, A., Schnittke, N., Romano, R. A., Sinha, S., & Schwob, J. E. (2011). DeltaNp63 regulates stem cell dynamics in the mammalian olfactory epithelium. *The Journal of Neuroscience*, *31*, 8748–8759.

Pardal, R., Ortega-Saenz, P., Duran, R., & Lopez-Barneo, J. (2007). Glia-like stem cells sustain physiologic neurogenesis in the adult mammalian carotid body. *Cell*, *131*, 364–377.

Pingault, V., Bodereau, V., Baral, V., Marcos, S., Watanabe, Y., Chaoui, A., et al. (2013). Loss-of-function mutations in SOX10 cause Kallmann syndrome with deafness. *American Journal of Human Genetics*, *92*, 707–724.

Raisman, G., & Li, Y. (2007). Repair of neural pathways by olfactory ensheathing cells. *Nature Reviews. Neuroscience*, *8*, 312–319.

Ramon-Cueto, A., & Avila, J. (1998). Olfactory ensheathing glia: Properties and function. *Brain Research Bulletin*, *46*, 175–187.

Roet, K. C., & Verhaagen, J. (2014). Understanding the neural repair-promoting properties of olfactory ensheathing cells. *Experimental Neurology*, *261C*, 594–609.

Sabado, V., Barraud, P., Baker, C. V., & Streit, A. (2012). Specification of GnRH-1 neurons by antagonistic FGF and retinoic acid signaling. *Developmental Biology*, *362*, 254–262.

Saxena, A., Peng, B. N., & Bronner, M. E. (2013). Sox10-dependent neural crest origin of olfactory microvillous neurons in zebrafish. *eLife*, *2*, e00336.

Schlosser, G. (2006). Induction and specification of cranial placodes. *Developmental Biology*, *294*, 303–351.

Schwanzel-Fukuda, M., & Pfaff, D. W. (1989). Origin of luteinizing hormone-releasing hormone neurons. *Nature*, *338*, 161–164.

Schwob, J. E. (2002). Neural regeneration and the peripheral olfactory system. *The Anatomical Record*, *269*, 33–49.

Serbedzija, G. N., Bronner-Fraser, M., & Fraser, S. E. (1989). A vital dye analysis of the timing and pathways of avian trunk neural crest cell migration. *Development, 106*, 809–816.

Sieber-Blum, M., Grim, M., Hu, Y. F., & Szeder, V. (2004). Pluripotent neural crest stem cells in the adult hair follicle. *Developmental Dynamics, 231*, 258–269.

Sjodal, M., Edlund, T., & Gunhaga, L. (2007). Time of exposure to BMP signals plays a key role in the specification of the olfactory and lens placodes ex vivo. *Developmental Cell, 13*, 141–149.

Smart, I. H. (1971). Location and orientation of mitotic figures in the developing mouse olfactory epithelium. *Journal of Anatomy, 109*, 243–251.

Stemple, D. L., & Anderson, D. J. (1992). Isolation of a stem cell for neurons and glia from the mammalian neural crest. *Cell, 71*, 973–985.

Streit, A. (2002). Extensive cell movements accompany formation of the otic placode. *Developmental Biology, 249*, 237–254.

Suzuki, J., Yoshizaki, K., Kobayashi, T., & Osumi, N. (2013). Neural crest-derived horizontal basal cells as tissue stem cells in the adult olfactory epithelium. *Neuroscience Research, 75*, 112–120.

Toma, J. G., Akhavan, M., Fernandes, K. J., Barnabe-Heider, F., Sadikot, A., Kaplan, D. R., et al. (2001). Isolation of multipotent adult stem cells from the dermis of mammalian skin. *Nature Cell Biology, 3*, 778–784.

Tome, M., Lindsay, S. L., Riddell, J. S., & Barnett, S. C. (2009). Identification of non-epithelial multipotent cells in the embryonic olfactory mucosa. *Stem Cells, 27*, 2196–2208.

Tomita, Y., Matsumura, K., Wakamatsu, Y., Matsuzaki, Y., Shibuya, I., Kawaguchi, H., et al. (2005). Cardiac neural crest cells contribute to the dormant multipotent stem cell in the mammalian heart. *The Journal of Cell Biology, 170*, 1135–1146.

Treloar, H. B., Miller, A. M., Ray, A., & Greer, C. A. (2010). Development of the olfactory system. In A. Menini (Ed.), *The neurobiology of olfaction*. Boca Raton, FL: CRC Press.

Valverde, F., Santacana, M., & Heredia, M. (1992). Formation of an olfactory glomerulus: Morphological aspects of development and organization. *Neuroscience, 49*, 255–275.

Vedin, V., Molander, M., Bohm, S., & Berghard, A. (2009). Regional differences in olfactory epithelial homeostasis in the adult mouse. *The Journal of Comparative Neurology, 513*, 375–384.

Wang, Y. Z., Yamagami, T., Gan, Q., Wang, Y., Zhao, T., Hamad, S., et al. (2011). Canonical Wnt signaling promotes the proliferation and neurogenesis of peripheral olfactory stem cells during postnatal development and adult regeneration. *Journal of Cell Science, 124*, 1553–1563.

Wewetzer, K., Verdu, E., Angelov, D. N., & Navarro, X. (2002). Olfactory ensheathing glia and Schwann cells: Two of a kind? *Cell and Tissue Research, 309*, 337–345

Whitfield, T. T. (2013). Shedding new light on the origins of olfactory neurons. *eLife, 2*, e00648.

Whitlock, K. E. (2005). Origin and development of GnRH neurons. *Trends in Endocrinology and Metabolism, 16*, 145–151.

Whitlock, K. E., & Westerfield, M. (2000). The olfactory placodes of the zebrafish form by convergence of cellular fields at the edge of the neural plate. *Development, 127*, 3645–3653.

Whitlock, K. E., Wolf, C. D., & Boyce, M. L. (2003). Gonadotropin-releasing hormone (GnRH) cells arise from cranial neural crest and adenohypophyseal regions of the neural plate in the zebrafish, *Danio rerio*. *Developmental Biology, 257*, 140–152.

Widera, D., Zander, C., Heidbreder, M., Kasperek, Y., Noll, T., Seitz, O., et al. (2009). Adult palatum as a novel source of neural crest-related stem cells. *Stem Cells, 27*, 1899–1910.

Wray, S. (2010). From nose to brain: Development of gonadotrophin-releasing hormone-1 neurones. *Journal of Neuroendocrinology*, *22*, 743–753.
Wray, S., Grant, P., & Gainer, H. (1989). Evidence that cells expressing luteinizing hormone-releasing hormone mRNA in the mouse are derived from progenitor cells in the olfactory placode. *Proceedings of the National Academy of Sciences of the United States of America*, *86*, 8132–8136.
Yamanishi, E., Takahashi, M., Saga, Y., & Osumi, N. (2012). Penetration and differentiation of cephalic neural crest-derived cells in the developing mouse telencephalon. *Development, Growth & Differentiation*, *54*, 785–800.
Yoshida, S., Shimmura, S., Nagoshi, N., Fukuda, K., Matsuzaki, Y., Okano, H., et al. (2006). Isolation of multipotent neural crest-derived stem cells from the adult mouse cornea. *Stem Cells*, *24*, 2714–2722.

CHAPTER ELEVEN

Epithelial Morphogenesis: The Mouse Eye as a Model System[1]

Bharesh Chauhan[*,3], Timothy Plageman[†,3], Ming Lou[‡], Richard Lang[§,2]

*Division of Pediatric Ophthalmology and Strabismus, Department of Ophthalmology, University of Pittsburgh School of Medicine, Pittsburgh, Pennsylvania, USA
†College of Optometry, The Ohio State University, Columbus, Ohio, USA
‡Department of Chemistry and Physics, Lamar University, Beaumont, Texas, USA
§The Visual Systems Group, Abrahamson Pediatric Eye Institute, Division of Pediatric Ophthalmology, Department of Ophthalmology, University of Cincinnati and Children's Hospital Research Foundation, Cincinnati, Ohio, USA
[2]Corresponding author: e-mail address: richard.lang@cchmc.org

Contents

Abstract

Morphogenesis is the developmental process by which tissues and organs acquire the shape that is critical to their function. Here, we review recent advances in our understanding of the mechanisms that drive morphogenesis in the developing eye. These investigations have shown that regulation of the actin cytoskeleton is central to shaping the presumptive lens and retinal epithelia that are the major components of the eye. Regulation of the actin cytoskeleton is mediated by Rho family GTPases, by signaling pathways and indirectly, by transcription factors that govern the expression of critical

[1] This chapter is dedicated to the memory of Yoshiki Sasai, an exceptional scientist whose work contributed critically to many fields of enquiry, including the one reviewed here.
[3] These authors contributed equally to this chapter.

Current Topics in Developmental Biology, Volume 111
ISSN 0070-2153
http://dx.doi.org/10.1016/bs.ctdb.2014.11.011

genes. Changes in the actin cytoskeleton can shape cells through the generation of filopodia (that, in the eye, connect adjacent epithelia) or through apical constriction, a process that produces a wedge-shaped cell. We have also learned that one tissue can influence the shape of an adjacent one, probably by direct force transmission, in a process we term inductive morphogenesis. Though these mechanisms of morphogenesis have been identified using the eye as a model system, they are likely to apply broadly where epithelia influence the shape of organs during development.

1. INTRODUCTION

Though the timescales are rather different, biology and geology both define morphogenesis as the study of mechanisms of shape acquisition. Whether the subject is tissue forms or landforms, the goal is to understand the forces of nature that guide the process. For the biologist, the study of morphogenesis is mostly a study of embryonic development (though regeneration is an option). Biologists interested in morphogenesis frequently combine the tools of developmental biology and cell biology. This gives them the means to establish the relationship between the molecular events at the single cell level and the consequence of those events for organismal shape. Though there will be unifying mechanisms of morphogenesis at the cellular scale, the huge complexity of tissue shapes in different species and different organs teaches us that in morphogenesis, details and subtlety matter (an interesting example is the subtle changes in beak shape that allow "Darwin's finches" to effectively exploit an ecological niche; Mallarino et al., 2011).

Though there are many worthy subjects of study in morphogenesis, in this chapter, we focus on the mouse eye as a model system. Eye morphogenesis in the mouse is very complex and therefore interesting. There are also many tools that can be applied to the study of eye development and so even in these early stages of our understanding, progress has been quite rapid. As a nonessential organ (at least for laboratory mice), the eye also has practical advantages as a subject of study. Finally, perhaps more than any other organ, the shape of the eye is critical for function. When the sensory stimulus is as unforgiving as the photon (just ask anyone over 40 years), the optical characteristics of the tissues and their relative shape and position have to show "inimitable perfection" (Darwin, 1859).

2. EYE MORPHOGENESIS

The basic features of eye morphogenesis that occurs during midgestation in the mouse are shown in Fig. 1. In this chapter, we define

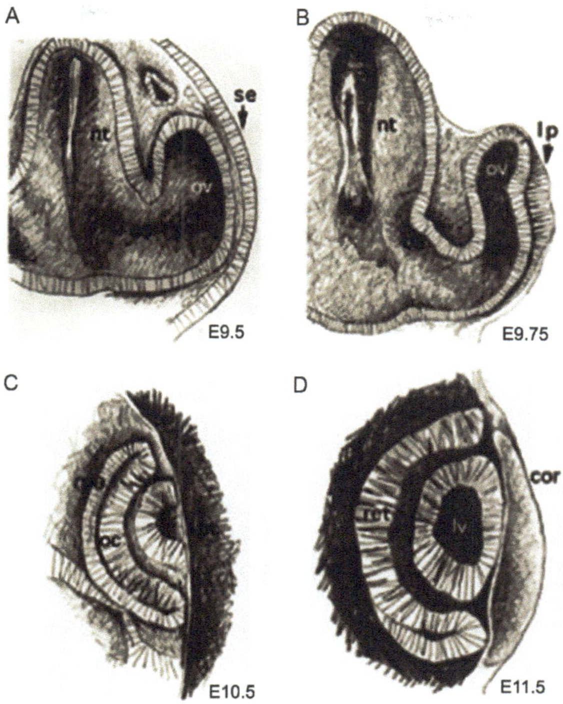

Figure 1 Drawing describing the basic features of eye morphogenesis. (A) At E9.5 in the mouse, the optic stalk (os) and optic vesicle (ov) have evaginated from the diencephalic neural tube (nt) and approached the overlying surface ectoderm (se). (B) By E9.75, the thickened surface ectoderm of the lens placode (lp) has formed. (C) By E1.5, the lens pit (pt) has undergone a coordinated invagination with the bilayered optic cup (oc) that incorporates the forming retinal pigmented epithelium. (D) At E11.5, the lens vesicle (lv) has separated from the surface ectoderm and presumptive corneal epithelium (cor).

two movements of epithelial structures—invagination and evagination—according to the polarity of the epithelium. We define invagination as a coordinated epithelial cell movement in the basal direction and an evagination as an epithelial cell movement in the apical direction. Thus, at E9.5 in the mouse, the optic vesicle (Fig. 1A, ov) is formed through an invagination of the diencephalic neural tube. This invagination brings the distal epithelium of the optic vesicle (the future retina) in close apposition to the embryonic surface ectoderm that will form both the lens and corneal ectoderm. As optic vesicle invagination occurs, the periocular mesenchyme that lies between the optic vesicle and surface ectoderm is excluded. Inductive signaling events (Charlton-Perkins, Brown, & Cook, 2011; Grocott, Tambalo, & Streit, 2012; Gunhaga, 2011; Lang, 2004; Shaham,

Menuchin, Farhy, & Ashery-Padan, 2012) result in the formation of the lens placode (Fig. 1B, lp), a thickened domain of the surface ectoderm that will undergo invagination to form the lens pit (Fig. 1C, lpt). Invagination of the lens pit is coordinated with the formation of the optic cup, a bilayered epithelium that ultimately forms the neural retina and retinal pigmented epithelium (RPE). According to our polarity-based definition of epithelial morphogenesis, the presumptive retina undergoes an evagination and the presumptive RPE undergoes an invagination as the optic cup forms. By E11.5 (Fig. 1D), the lens pit has closed to form the lens vesicle (lv) and the surface ectoderm has separated to form the presumptive corneal ectoderm. At this point in development, the basic layout of the major tissues of the eye is established.

3. DYNAMIC CHANGES IN THE ACTIN CYTOSKELETON DRIVE MORPHOGENESIS

It is now well accepted that regulation of cell migration and cell shape via modulation of the actin cytoskeleton is central to morphogenesis. Central to modulation of the actin cytoskeleton are the Rho family GTPases, molecular switches that transfer information in the cell and gain specificity through a series of effector proteins with a limited set of downstream molecular targets (Bishop & Hall, 2000). The GTPases exist in active (GTP bound) and inactive (GDP bound) states and are regulated in several ways. The Rho GTPase-activating proteins (Tcherkezian & Lamarche-Vane, 2007) and the guanine nucleotide dissociation inhibitors (Dovas & Couchman, 2005) are negative regulators while the Rho guanine-nucleotide exchange factors (RhoGEFs; Rossman, Der, & Sondek, 2005; Schmidt, Diriong, Mery, Fabbrizio, & Debant, 2002) act positively. These are frequently expressed with tissue-specific expression patterns and can themselves be regulated. The Rho family GTPases can regulate morphology, adhesion, differentiation, proliferation, and vesicle trafficking (Heasman & Ridley, 2008; Linseman & Loucks, 2008; Nobes & Hall, 1994, 1999; Villalonga & Ridley, 2006). In some of the first assessments of Rho family GTPase activity, it was noted that cultured cells produced different types of actin and different cell shapes, depending on which active GTPase was expressed. Cdc42 activation resulted in the formation of filopodia, Rac1 in the formation of lamellipodia and RhoA the formation of contractile actin and stress fibers (Jaffe & Hall, 2005). These findings were important for the developmental biologists interested in morphogenesis, because they provided a starting point for the investigation of these pathways *in vivo*.

Based on the experimental analysis that will be summarized below, it is clear that a complex series of physical and signaling interactions between the epithelial and mesenchymal components of the developing eye are required to shape a normal structure. The major theme emerging from this review is that even morphogenetic movements that appear very simple are mechanistically quite complex and require input from many pathways. In addition, it is often the case that each pathway contributes just a little to the overall movement. A further level of complexity emerges from the interaction between adjacent tissues. Though we cannot, as a field, claim to have solved the problem of eye morphogenesis, already many mechanisms have been uncovered. What are these mechanisms? Below, we describe what is known, more-or-less in ontological order. Many of the mechanisms so far identified are directly or indirectly linked to modulation of the actin cytoskeleton.

3.1. What is the function of placode formation in epithelial morphogenesis?

The formation of placodes—regions of thickened epithelium—is associated with the morphogenesis of many structures. Extensive studies on placode specification have been reported (Bailey & Streit, 2006; Streit, 2008), but the mechanisms of epithelial thickening have received less attention. The strong association between placode formation and morphogenesis further suggests that this type of epithelial thickening might be a prerequisite for epithelial bending.

In the case of the lens placode that forms between E9.0 and E9.5 in the mouse, there are several mutant lines in which placodal thickening is compromised. These include the germ line Pax6 mutation present in the *Small eye* mouse (Hill et al., 1991), regulatory mutations that lead to diminished placodal *Pax6* expression (Dimanlig, Faber, Auerbach, Makarenkova, & Lang, 2001), as well as conditional deletion of *Pax6* in the lens placode (Ashery-Padan, Marquardt, Zhou, & Gruss, 2000). Pax6 has a central role in lens development—it is both necessary (Hill et al., 1991) and sufficient (Altmann, Chow, Lang, & Hemmati-Brivanlou, 1997; Chow, Altmann, Lang, & Hemmati-Brivanlou, 1999) for formation of the lens structure from the placode stage—and so it can be difficult to determine how closely this broadly active transcription factor is associated with the morphogenesis machinery. Despite this, it has been shown, with placodal deletion of

Pax6, that there is a loss of expression of Shroom3 (Plageman et al., 2010), an actin scaffold protein with an important role in changing the shape of epithelial cells (Lee, Le, & Wallingford, 2009; Nishimura & Takeichi, 2008; see discussion below). This contrasts with placodal deletion of Sox2, another transcription factor involved in lens development. In the placodal Sox2 deletion, though there are consequences for lens development (Smith, Miller, Radice, Ashery-Padan, & Lang, 2009), Shroom3 expression is retained and placodal invagination initiated (Plageman et al., 2010). Though it is likely that Pax6 regulates many genes after lens induction, these data link the transcription factor to Shroom3 expression and the morphogenesis machinery (Fig. 2).

Some well-known signaling pathways also contribute to formation of the lens placode. When the fibroblast growth factor (Fgf) signaling response is compromised in the preplacodal ectoderm of the mouse, placodal thickening is reduced (Faber et al., 2001), a result confirmed first by mutation of the Fgf receptor adapter protein Frs2α (Gotoh et al., 2004) and later by conditional Fgf receptor deletion (Garcia et al., 2011). Defective Bmp signaling also results in changes to placode formation. Both Bmp4 and Bmp7 germ-line mutant mice exhibit a failure of lens placode formation (Furuta & Hogan, 1998; Wawersik et al., 1999). The defects in lens morphogenesis and changes in epithelial F-actin distribution that arise when the Bmpr1a and Acvr1 receptors are conditionally deleted postinduction (Rajagopal et al., 2009) suggests that this signaling pathway may be quite closely linked to the morphogenesis machinery. Given existing data (Gamell et al., 2008; Konstantinidis, Moustakas, & Stournaras, 2011; Lee-Hoeflich et al., 2004), it is quite likely that there are several levels at which Bmp signaling can regulate cell shape via modulation of the actin cytoskeleton.

The interaction between the presumptive lens and presumptive retina is important for subsequent morphogenesis. In part, this is because of the inductive signaling that takes place, but is also because a physical association is important for coordinated shape formation in the presumptive lens and retina. The importance of the presumptive lens–retina interface is illustrated by experiments that have examined the role of the extracellular matrix (ECM) and more specifically of fibronectin (FN), in lens morphogenesis. In a placode-conditional Pax6 mutant mouse, ECM deposition between the lens and retina is diminished, suggesting that Pax6 directly or indirectly regulates ECM deposition (Huang et al., 2011). The limited ECM at the presumptive lens–retina interface may contribute to the

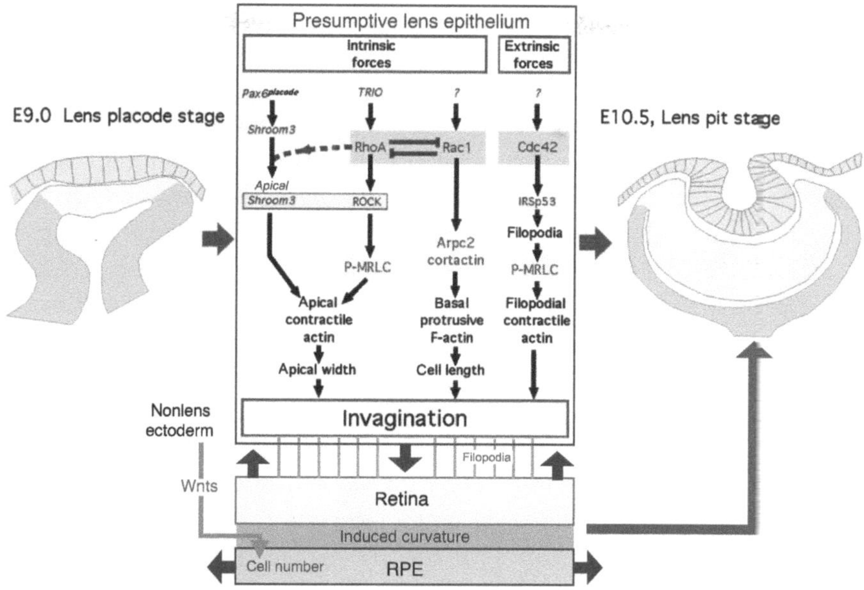

Figure 2 Summary schematic of eye morphogenesis pathways. This schematic describes some of the pathways involved in morphogenesis of the mouse eye from E9.5, when the lens placode overlies the optic vesicle, to E10.5 when the lens pit has invaginated and the optic cup has become bilayered. The surface ectoderm, presumptive retina, and presumptive retina pigmented epithelium are color-coded in red, green, and gray, respectively. Within the red box, we summarize the action of three GTPases that have distinct actions on lens placode invagination. Cdc42 and IRSp53 are required for the generation of filopodia within the presumptive lens that connect to the presumptive retina and permit force transmission between the two epithelia. They are required for a full invagination of the lens pit and for subtle shaping of the optic cup. This is and example where lens pit invagination depends on extrinsic forces. The GTPases RhoA and Rac1 are required for the generation of intrinsic forces for lens pit invagination because they control cell length (Rac1 through an Arpc2/cortactin response) and apical constriction (RhoA through a Rho kinase (ROCK) phospho-myosin pathway). RhoA activity is likely controlled by the TRIO guanine nucleotide exchange factor and in turn, this is required for the apical localization of Shroom3, an actin and ROCK-binding protein required for apical constriction. At the transcriptional level, Shroom3 is positively regulated by Pax6. During lens pit invagination, the coordinated morphogenesis of the optic cup is in part stimulated by Wnt/Catnb pathway ligands that are produced by nonlens surface ectoderm. These ligands stimulate the production of cells within the developing retinal pigmented epithelium (RPE). The RPE thus expands against the relatively static retina and, with the principle of a bimetallic strip, induces curvature in the optic cup. The red arrows indicate the movements of morphogenesis for the each epithelial layer. (See the color plate.)

failure of lens pit invagination in the Pax6 mutant. In addition, when FN1 is conditionally deleted throughout the mouse embryo, there is a failure of both placode thickening and lens pit invagination (Huang et al., 2011). Though this mutant may have many signaling and structural defects, these data suggest that placode thickening may be a prerequisite for invagination. These data are also consistent with descriptive analyses of lens placode thickening, suggesting that an increase in cell density coupled with static cell volumes (Mckeehan, 1951) results in thickening. Additional experimental studies attempting to restrict the outward expansion of presumptive lens placodal cells through physical means resulted in ectopic thickening and/or precocious invagination (Steding, 1967; Wakely, 1984). These data further support the hypothesis that placodal thickening is a prerequisite for invagination. In purely structural terms, this might well make sense: just as a deeper beam can better resist the force of gravity, a thicker epithelium can better generate the forces required for invagination.

3.2. Extrinsic force transmission via filopodia

Events in morphogenesis require precise cell positioning and guidance that can be provided by filopodia. This function can be found in migratory tissue sheets, such as dorsal closure in *Drosophila* (Martin-Blanco, Pastor-Pareja, & Garcia-Bellido, 2000), ventral closure in *Caenorhabditis elegans* (Williams-Masson, Malik, & Hardin, 1997), and wound healing (Wood et al., 2002). Filopodia also feature in developing vasculature (Gallo & Letourneau, 2004), neuronal projections (De Smet, Segura, De Bock, Hohensinner, & Carmeliet, 2009), and the *Drosophila* tracheal system (Wolf, Gerlach, & Schuh, 2002). Studied primarily in fibroblasts and neurite growth cones, filopodia are defined as probing cellular protrusions that aid in migration and extension and that contain bundles of parallel actin filaments (Faix & Rottner, 2006; McClay, 1999; Passey, Pellegrin, & Mellor, 2004; Raich, Agbunag, & Hardin, 1999). Their length differs depending on their tip adhesion; those that are anchored at their tips by cadherins tend to be shorter (Partridge & Marcantonio, 2006; Raich et al., 1999; Wood et al., 2002), whereas the integrin-dependent focal adhesion tips tend to be longer (Partridge & Marcantonio, 2006).

The Rho family GTPase Cdc42 has many functions in the cell, but it is critical for the formation of filopodia (Nobes & Hall, 1995; Ridley, 2006). *In vitro*, it does so through binding in the active state with the ubiquitously expressed effector protein N-WASP (Banzai, Miki, Yamaguchi, & Takenawa, 2000; Egile et al., 1999; Kurisu & Takenawa, 2009; Miki, Sasaki, Takai, & Takenawa, 1998), a Wiskott–Aldrich syndrome gene-family member. N-WASP subsequently activates the actin nucleator Arp2/3, and as a result, fast growing barbed-end actin filaments extend the plasma membrane to form protrusive filopodia. The formation of Cdc42-stimulated filopodia in N-WASP-deficient fibroblasts revealed an alternative pathway for filopodia formation (Snapper et al., 2001). This was later discovered to require the diaphanous-related formins (DRFs; Peng, Wallar, Flanders, Swiatek, & Alberts, 2003), a group of actin nucleators that function downstream of the Rho GTPases. Furthermore, it has been found that Rif forms filopodia through the action of a DRF member, mDia2. A third filopodiogenesis pathway requires the effector protein IRSp53 interacting with either Mena (Krugmann et al., 2001) or Eps8 (Disanza et al., 2006). IRSp53 and Eps8 are also involved in signal transduction, whereas IRSp53 is a substrate for the insulin receptor (Yeh, Ogawa, Danielsen, & Roth, 1996) and Eps8 is a substrate for other tyrosine kinase receptors (Fazioli et al., 1993).

Investigation of filopodia in morphogenesis began with the secondary mesenchymal cells of the sea urchin *Arbacia punctulata* and the sand dollar *Echinarachniu sparma* echinoderm species as early as 1969 (Tilney & Gibbins, 1969). This was later followed by investigation of their contractile function in the tunic cells of the ascidian *Botryllus schlosseri*, where they were found to change cell shape and regulate cell migration (Izzard, 1974). Here, filopodia were observed to be long (reaching lengths of up to 200 μm). Laser ablation studies of the filopodia arising from secondary mesenchymal cells in the sea urchin suggested that they had a contractile function and aided in generating the forces required for invagination of the archenteron during gastrulation (Hardin, 1988). A role for filopodia in connecting epithelium to mesenchyme was found in the examples of the developing avian lung (Gallagher, 1986), and in the dental mesenchyme making contact with the overlying dental epithelium (Fukumoto et al., 2006). Their role in branching morphogenesis was revealed in avian lung development as well as in *Drosophila* tracheal development (Samakovlis et al., 1996). Their role in wound healing was found through the series of studies conducted in the cornea (Crosson, Klyce, & Beuerman, 1986; Pfister, 1975), skin (Pang, Daniels, & Buck, 1978; Vasioukhin & Fuchs, 2001), and then in *Drosophila* dorsal closure (Millard & Martin, 2008; Wood et al., 2002), the latter of which recapitulated a number of embryonic tissue closure mechanisms. The contractile function of filopodia and the specific myosins in the varying filopodia have recently been deciphered (Berg & Cheney, 2002; Chauhan et al., 2009; Jacinto et al., 2002; Moores, Sabry, & Spudich, 1996; Sousa & Cheney, 2005; Tokuo & Ikebe, 2004; Wagner, Barylko, & Albanesi, 1992; Zhang et al., 2004).

Ida Mann's book on human eye development (Mann, 1950) described "protoplasmic processes" that extended between the epithelia of the presumptive lens and retina. Later, the McAvoy group showed, by TEM, that these processes formed a junction-like structure at the tip where they contacted the presumptive retina (McAvoy, 1980). These structures were shown to be actin-rich and devoid of tubulin, suggesting that they were filopodia (Chauhan et al., 2009). A mutant mouse in which the Rho family GTPase Cdc42 was deleted in the lens placode exhibited an absence of interepithelial filopodia further confirming their identity. This mutant also indicated that most of the filopodia originate in the cells of the lens placode and lens pit. A series of mutant mice, including the Cdc42 placode conditional mutant, a germ line mutant for the Cdc42 effector IRSp53, and a conditional mutant for focal adhesion kinase, revealed a spectrum of filopodial loss and an inverse correlation with interepithelial distance. In the Cdc42 conditional mutant where all filopodia were lost, about one-third of the lens pit invagination depth was also lost, suggesting that these structures served to transmit forces between presumptive retina and lens as a means of refining the process of morphogenesis. This suggestion was consistent with the observation that these filopodia contained active phosphomyosin II and were actively contractile. Filopodia-deficient mutants, not surprisingly, show changes in the shape of the lens pit. However, they also show changes in the shape of the adjacent retinal cup (Fig. 3), the structure to which they connect, probably via integrin-mediated adhesion. These nonautonomous epithelial shape changes provide some evidence that one of the functions of these filopodia is to transmit forces from one epithelium to another. Drawing a parallel with the classical concept of inductive fate change, we refer to this phenomenon as inductive morphogenesis. This analysis indicates that the forces required to drive morphogenesis can be extrinsic (Fig. 2).

3.3. Intrinsic force generation through apical constriction and cell elongation

When an epithelium invaginates, the cells within the epithelium must, on average, adopt a wedge-shaped morphology. There is now much evidence

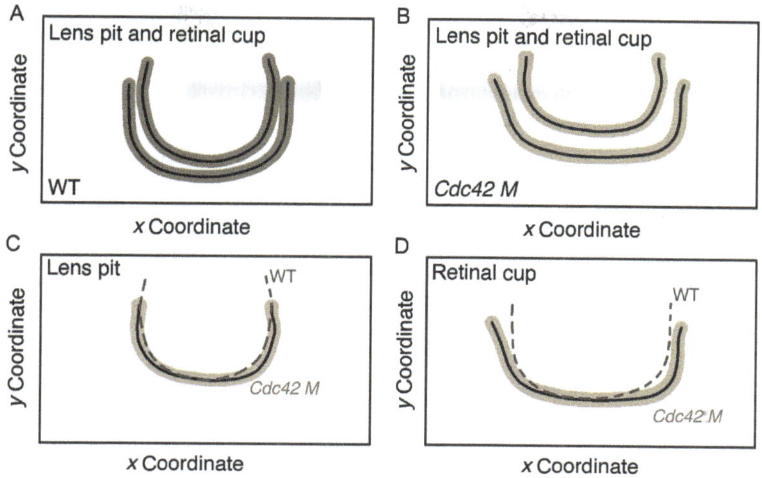

Figure 3 Inductive morphogenesis in the epithelia of the developing eye. (A) The average shape of the basal surfaces of the lens pit and presumptive retina in a wild-type, E10.5 mouse eye. The black line represents that average shape and the green (gray in the print version) shaded area, two standard deviations from the average shape. (B) The average shape of the basal surfaces of the lens pit and presumptive retina in an *Le-cre; Cdc42*$^{fl/fl}$, E10.5 mouse eye. In this mutant (Cdc42 M), there are no filopodia connecting the epithelia. As a consequence, the interepithelial distance has increased. The lens pit (C) has lost about one-third of the invagination distance and the retinal cup (D) is also flatter. Since genetic changes in this mouse line are restricted to the presumptive lens, this provides some evidence for force transmission between the epithelia. These data were generated as described previously (Chauhan et al., 2009).

at both the cellular and molecular levels that this shape change is primarily a result of a reduced apical area that is the result of a process referred to as apical constriction (AC). For the transition between the lens placode and lens pit (Fig. 1) an analysis of cell shape showed that both apical constriction and cell elongation accompanied invagination (Hendrix & Zwaan, 1974). Below, we describe some of the molecular mediators of and epithelial cell elongation and AC and indicate how their activities are integrated.

3.3.1 Apical constriction

Many of the significant advances in our understanding of AC mechanisms have come from analyzing invertebrate morphogenesis including *Drosophila* gastrulation and dorsal closure. During gastrulation, coordinated AC along the ventral embryonic axis drives mesodermal invagination (Leptin, 2005). Mesodermal cells undergo AC through the contraction of an apically positioned network of actomyosin that spans the apical cortex (Dawes-Hoang et al., 2005; Martin, Kaschube, & Wieschaus, 2009). Assembly and contraction of myosin filaments are facilitated through the phosphorylation and activation of myosin regulatory light chain and phosphorylation and deactivation of myosin phosphatases (Bresnick, 1999). The pathway leading to these phosphorylation events in *Drosophila* gastrulation are triggered by a well-understood cascade of events that commences with the mesodermal-specific expression of G protein-coupled receptors that apically recruit a guanine exchange factor (DRhoGEF2). DRhoGEF2 activates the Rho-GTPase, Rho1 and in turn stimulates phosphorylation by activating the serine–threonine kinase Rok (Sawyer et al., 2010). Once active, contraction occurs in a step-wise manner consisting of pulses interspersed by periods of relaxation or apical surface stabilization (Azevedo et al., 2011; Blanchard, Murugesu, Adams, Martinez-Arias, & Gorfinkiel, 2010; David, Tishkina, & Harris, 2010; Martin et al., 2009). A molecular connection bridging actomyosin to junctional complexes are also necessary to draw the apical circumference inward and when disrupted the degree of AC is attenuated (Dawes-Hoang et al., 2005; Roh-Johnson et al., 2012; Sawyer, Harris, Slep, Gaul, & Peifer, 2009).

AC within cells of the lens placode is remarkably similar to that observed in *Drosophila* gastrulation. During lens placode invagination, the vertebrate equivalents of the RhoGEF2-Rho1-Rok pathway of myosin activation are significant contributors. Pharmacological studies inhibiting Rock1/2, the vertebrate homolog of Rok, myosin, or actin dynamics all disrupt lens invagination presumably through the inhibition of AC (Borges, Lamers, Forti, Santos, & Yan, 2011; Plageman et al., 2011). Genetic deletion of RhoA or the RhoA-GEF Trio, functional counterparts of RhoA, and DRhoGEF2, respectively, inhibits AC in the lens and decreased apical levels of activated myosin, indicating a strong molecular conservation of this pathway between species (Chauhan, Lou, Zheng, & Lang, 2011; Plageman et al., 2011). Shroom3, a Rock1/2 and actin-binding cytoskeletal protein, initially identified as a protein required for neural tube AC and morphogenesis, is also necessary for lens placode AC (Haigo, Hildebrand, Harland, & Wallingford, 2003; Hildebrand & Soriano, 1999; Plageman et al., 2010).

Shroom3 appears to play a central role in AC during lens placode invagination. First, its expression domain is regionally restricted to the cells which are undergoing AC. This is likely due in part to lens placode-specific transcription factors, many of which have restricted expression, an idea supported by the observation that Shroom3 expression is dependent upon

Pax6 (Plageman et al., 2010; Fig. 2). Second, Shroom3 is sufficient to induce AC and facilitate the assembly of myosin into sarcomeric-like structures in the apical junctions of cultured epithelial cells (Hildebrand, 2005). Third, the absence of Shroom3 causes a reduction in the apical distribution of both F-actin and myosin in the lens placode (Plageman et al., 2010). Its ability to recruit F-actin is likely dependent on the Mena-family of actin modulators which themselves play a role in AC (Plageman et al., 2010; Roffers-Agarwal, Xanthos, Kragtorp, & Miller, 2008). Given that the function of Shroom3 was also shown to be dependent on RhoA, Rock1, and myosin activity, we posit that Shroom3 acts as a nexus of AC activity in the lens by organizing and/or recruiting most if not all of the necessary components.

Like invertebrate AC, the association of actomyosin with the apical junctions is essential for lens placode AC but unlike *Drosophila* mesodermal cells, much of the actomyosin in the lens placode is found in the circumferential belt positioned near the apical junctional complex (Lang, Herman, Reynolds, Hildebrand, & Plageman, 2014). In a somewhat surprising degree of conservation, actomyosin filaments spanning the apical cortex are indeed present in a subset of lens placodal cells and appear to immediately precede AC. Like invertebrate actomyosin filaments, those in lens placode epithelial cells are connected to deformations in the bicellular junctions (Lang et al., 2014) suggesting that they are contractile. Bicellular deformations and AC are both prevented by the myosin inhibitor blebbistatin (Lang et al., 2014). Further analysis demonstrated that the cadherin-binding modulator of Rho-GTPases, p120-catenin, is required for this junctional displacement and AC in the lens (Lang et al., 2014). This suggests that p120 may be playing a significant role in mediating the connection of actomyosin filaments and/or the circumferential actomyosin belt to the apical junctions. In turn, these mediators are likely critical for the generation of intrinsic forces that can bend an epithelium (Fig. 2).

3.3.2 Cell elongation

In cultured cells, the Rac1 GTPase stimulates the formation of lamellipodia (Jaffe & Hall, 2005). This unique type of cellular process is a consequence of the particular effector molecules that this GTPase activates and in turn, the type of actin complexes that are formed (Jaffe & Hall, 2005). One well-known Rac1 effector is the Wiskott–Aldrich syndrome protein family verprolin homolog WAVE (Kurisu & Takenawa, 2009; Miki, Suetsugu, & Takenawa, 1998) which, through upstream activation of the p21-activated kinase (PAK) effector and binding to the actin nucleator

Arp2/3, encourages polymerization of the Y-branched actin networks that give lamellipodia and membrane ruffles (Burridge & Wennerberg, 2004). It has also been shown that active PAK1 has an additional role in membrane transport of WAVE2 to the leading edge for lamellipodia formation (Takahashi & Suzuki, 2009). Interestingly, it has been shown using fluorescence resonance energy transfer microscopy in cultured cells that RacGTP levels are the highest at the leading edge of a migrating cell (Kraynov et al., 2000), and that integrin–matrix interactions may play a role in localized Rac1 activity.

When *Rac1* is conditionally deleted from the lens placode, the constituent cells exhibit reduced length (Chauhan et al., 2011). This is likely to be the *in vivo* manifestation, in the epithelial cell type, of the activity that forms lamellapodia in migrating cultured cells. From the point of view of tissue structure, it may be that this type of cell elongation is important for epithelial bending in that it extends the length of the levers that can be used to generate force.

3.3.3 Rac1, RhoA pathways integrate to control cell shape in the lens placode

It is clear that in some settings, the Rac1 and RhoA actin modulation pathways are mutually antagonistic (Machacek et al., 2009; Xu et al., 2003). In the invaginating lens placode the basic phenotypic response, where lens pit curvature is reduced in RhoA mutants and increased in Rac1 mutants (Chauhan et al., 2011), also suggested this might be the case *in vivo*. An assessment of cell shapes further showed that, as would be expected, RhoA was required for AC (Chauhan et al., 2011). However, in addition, RhoA mutant lens placode cells also showed increased length. In a reciprocal set of responses, Rac1 mutant placode cells showed reduced length, as would be expected, but also showed increased AC (Chauhan et al., 2011). How can these findings be explained mechanistically?

Using independent markers of the RhoA and Rac1 pathway responses, it was shown that in the *RhoA* mutant lens placode, there were indications of Rac1 pathway gain of function. For example, the Arpc2 actin nucleation complex subunit was significantly over-represented basally in the placodal *RhoA* mutant. This could explain why a *RhoA* mutant epithelial cell had become longer than normal. Similarly, in *Rac1* mutant placode cells, phosphomyosin was over-represented at the cell apex. This suggested that the *Rac1* mutant placodal cell had a RhoA pathway gain of function and was an explanation for the increased AC and the tighter curvature of the lens pit in this mutant. These observations suggest that during lens placode invagination, RhoA and Rac1 are mutually antagonistic. Furthermore, the data suggest their activities are balanced in a way that gives fine control of cell height and apical width (Chauhan et al., 2011). Since, the ratio of these two cellular dimensions controls the angle that an epithelial cell occupies, in aggregate, Rac1 and RhoA mutual antagonism regulates epithelial curvature (Fig. 2).

There are many examples of a dramatic change in epithelial curvature over just a few cell diameters, and so it is also possible that the Rac1 to RhoA activity ratio can also be controlled locally, presumably by the signaling responses known to regulate these GTPases (Raftopoulou & Hall, 2004). These are likely to be general mechanisms for regulating epithelial morphogenesis given the ubiquity of the Rho family GTPases. Studies have shown the mechanism of RhoA–Rac1 mutual antagonism can be mediated through either p190Rho-GAP (Nimnual, Taylor, & Bar-Sagi, 2003), the kinase *sticky* (D'Avino, Savoian, & Glover, 2004), or FilGAP (Ohta, Hartwig, & Stossel, 2006). It remains to be determined whether these mechanisms apply to morphogenesis of the eye.

4. SHAPING OF THE OPTIC CUP

Analysis of the role of GTPases and the AC machinery in the invaginating lens pit has made a strong case that the forces that drive epithelial bending are generated intrinsically through cell shape change and extrinsically, via filopodia that connect the invaginating lens placode to the presumptive retina. This arrangement raises the question of whether there is morphogenetic induction—a change in shape dependent on the shape of the adjacent epithelium—that is polarized or reciprocal.

4.1. Does the optic cup pull on the invaginating lens?

Several lines of evidence indicate that the optic cup influences the shape of the lens pit. In several lines of mutant mice, where interepithelial filpodia are reduced in number, there is a reciprocal relationship between the number of filopodia and interepithelial distance. In other words, as the number of filopodia reduces, so does the depth of lens pit invagination. This suggests quite strongly that the optic cup pulls on the lens pit. There is some subtlety, however, in that even in a mouse in which all filopodia are absent, only about one-third of the lens pit invagination depth is lost (Chauhan et al., 2009). This is consistent with a model in which the lens placode has intrinsic force generation (through cell shape change) and that the extrinsic force generation through connecting filopodia is a fine-tuning mechanism (Fig. 2).

4.2. Does the invaginating lens influence the shape of the optic cup?

Recent experiments in which optic cups are developed in culture from ES cells have suggested that the optic cup develops autonomously and is self-shaping (Eiraku et al., 2011). Though the reality is more complicated (see below), the observation that the optic cups forms in the absence of lens has suggested that the invaginating lens pit does not push on the presumptive retina as the optic cup forms. To extend this line of questioning further, we can ask whether there are any changes in the shape of the optic cup when the invaginating lens placode and presumptive retina are uncoupled. Arguably, an uncoupling occurs in *Cdc42* conditional mutant mice in which the presumptive lens does not generate any connecting filopodia. When we analyze the shape of the optic cup in this mutant, we find that it is changed compared to a control (Fig. 3). Obviously, this suggests that the invaginating lens placode does influence the shape of the optic cup. As in the reciprocal interaction, the changes in the shape are subtle and so again it might be argued that the lens finetunes the shape of the cup. It is also worth noting that when optic cups are generated in culture, very few of them are perfect in shape and a proportion never undergo the evagination of the presumptive retina that generates the bilayered optic cup (Eiraku et al., 2011; Nakano et al., 2012). This observation contrasts with the *in vivo* development of the eye where the structure is perfect nearly all of the time. This contrast suggests that while the optic cup can form its basic shape autonomously some of the time, if it is to do so all the time, the influence of surrounding tissues is required. Below, we review the evidence that the surface ectoderm of the eye produces signals that are critical for morphogenesis of the optic cup.

4.3. The "bimetallic strip" mechanism of optic cup morphogenesis

In the mouse, when the lens placode and adjacent ectoderm are prevented from producing Wnt ligands, there are profound consequences for eye development. One consequence is that the optic cup does not form its normal, near-spherical shape but is flat and saucer shaped (Carpenter et al., 2014). This contrasts with the hypothesis that the optic cup is self-shaping and shows very directly that signaling ligands produced by the ectoderm influence optic cup morphogenesis. With the available data, the most compelling hypothesis to explain this function of Wnt ligands is that they are produced by the surface ectoderm and stimulate a Wnt/β-catenin response in the presumptive RPE adjacent to the retinal rim (Carpenter et al., 2014). It is also understood that the Wnt/β-catenin response upregulates the retinoic acid response pathway, known to be critical for optic cup morphogenesis in the RPE. In turn, the Wnt/β-catenin and retinoic acid responses stimulate the production of cells within the RPE that helps to shape the optic cup.

It has been suggested that the mechanism by which RPE cells shape the optic cup is, in principle, similar to the mechanism by which a bimetallic strip bends when it is heated. A bimetallic strip is a temperature sensing device that consists of two different metals bonded along their length. If these metals have different expansion coefficients, then the bimetallic strip will bend when it is heated (https://www.youtube.com/watch?v=sP5NwEkd3ds). In the same way, as more RPE cells are produced in the bilayered optic cup, they will induce curvature (Fig. 2).

The bimetallic strip model of optic cup morphogenesis is consistent with a number of published observations. First, it has been shown that the presumptive RPE is Wnt/β-catenin responsive according to the expression of Wnt/β-catenin pathway components and reporters (Cho & Cepko, 2006; Fuhrmann, Levine, & Reh, 2000; Liu, Mohamed, Dufort, & Wallace, 2003; Liu, Thurig, Mohamed, Dufort, & Wallace, 2006; Liu et al., 2007; Trimarchi, Cho, & Cepko, 2009) and the expression of Wnt/β-catenin target genes (Eiraku et al., 2011; Fuhrmann, 2008; Westenskow, Piccolo, & Fuhrmann, 2009). Second, assessment of the stiffness of the ocular epithelia using the atomic force microscope has indicated that while the RPE is stiff, the presumptive retina is not (Eiraku et al., 2011). This suggests, as does the bimetallic strip model that the RPE acts as a scaffold and is perhaps the dominant structural determinant of optic cup curvature (Fig. 2).

How do we reconcile the apparently conflicting observations that *in vivo*, optic morphogenesis is dependent on Wnt ligands from the surface ectoderm while in culture, they can form in the absence of the surface ectoderm? The simple reconciliation is that in the ES-derived optic cups, these critical natural signaling ligands are provided as media supplements. Retinoic acid is a standard component of "optic cup medium" and supplementary Wnt ligands have been shown to enhance the developmental process (Eiraku & Sasai, 2012).

5. CONCLUSIONS

A defining message of the current studies in eye morphogenesis is that many small influences can lead to dramatic shape changes when each of those subtle influences are focused on a single structure. The eye has a structure that is more complex than most organs, and so it will be interesting to discover if this principle holds elsewhere.

What does the future hold for studies of morphogenesis, including those focused on the eye? As the field progresses, it will be important to quantify morphogenesis numerically and, perhaps using force sensors (Grashoff et al., 2010), define the forces, in real time, that drive morphogenesis. This, coupled to a more sophisticated understanding of the signaling pathways that drive the morphogenesis machinery, will steadily build a comprehensive model of the process.

ACKNOWLEDGMENTS

We would especially like to thank Dr. Sonya Faber for the drawing of eye morphogenesis shown in Fig. 1. This work was supported by Grant R01 EY016241 from the National Eye Institute of the National Institutes of Health and by funds from the Abrahamson Pediatric Eye Institute of Cincinnati Children's Hospital.

REFERENCES

Altmann, C. R., Chow, R. L., Lang, R. A., & Hemmati-Brivanlou, A. (1997). Lens induction by Pax-6 in Xenopus laevis. *Developmental Biology, 185*(1), 119–123. http://dx.doi.org/10.1006/dbio.1997.8573.

Ashery-Padan, R., Marquardt, T., Zhou, X., & Gruss, P. (2000). Pax6 activity in the lens primordium is required for lens formation and for correct placement of a single retina in the eye. *Genes & Development, 14*(21), 2701–2711.

Azevedo, D., Antunes, M., Prag, S., Ma, X., Hacker, U., Brodland, G. W., et al. (2011). DRhoGEF2 regulates cellular tension and cell pulsations in the Amnioserosa during Drosophila dorsal closure. *PLoS One, 6*(9), e23964. http://dx.doi.org/10.1371/journal.pone.0023964.

Bailey, A. P., & Streit, A. (2006). Sensory organs: Making and breaking the pre-placodal region. *Current Topics in Developmental Biology, 72*, 167–204. http://dx.doi.org/10.1016/S0070-2153(05)72003-2.

Banzai, Y., Miki, H., Yamaguchi, H., & Takenawa, T. (2000). Essential role of neural Wiskott–Aldrich syndrome protein in neurite extension in PC12 cells and rat hippocampal primary culture cells. *Journal of Biological Chemistry, 275*(16), 11987–11992.

Berg, J. S., & Cheney, R. E. (2002). Myosin-X is an unconventional myosin that undergoes intrafilopodial motility. *Nature Cell Biology, 4*(3), 246–250. http://dx.doi.org/10.1038/ncb762.

Bishop, A. L., & Hall, A. (2000). Rho GTPases and their effector proteins. *Biochemical Journal, 348*(Pt. 2), 241–255.

Blanchard, G. B., Murugesu, S., Adams, R. J., Martinez-Arias, A., & Gorfinkiel N. (2010). Cytoskeletal dynamics and supracellular organisation of cell shape fluctuations during dorsal closure. *Development*, *137*(16), 2743–2752. http://dx.doi.org/10.1242/dev.045872.

Borges, R. M., Lamers, M. L., Forti, F. L., Santos, M. F., & Yan, C. Y. (2011). Rho signaling pathway and apical constriction in the early lens placode. *Genesis*, *49*(5), 368–379. http://dx.doi.org/10.1002/dvg.20723.

Bresnick, A. R. (1999). Molecular mechanisms of nonmuscle myosin-II regulation. *Current Opinion in Cell Biology*, *11*(1), 26–33.

Burridge, K., & Wennerberg, K. (2004). Rho and Rac take center stage. *Cell*, *116*(2), 167–179.

Carpenter, A. C., Smith, A. N., Wagner, H., Wallace, V. A., Ashery-Padan, R., & Lang, R. A. (2014). *Wnt ligands from the embryonic surface ectoderm regulate "bimetallic strip" optic cup morphogenesis in the mouse.* Development, in press.

Charlton-Perkins, M., Brown, N. L., & Cook, T. A. (2011). The lens in focus: A comparison of lens development in Drosophila and vertebrates. *Molecular Genetics & Genomics*, *286*(3–4), 189–213. http://dx.doi.org/10.1007/s00438-011-0643-y.

Chauhan, B. K., Disanza, A., Choi, S. Y., Faber, S. C., Lou, M., & Beggs, H. E. (2009). Cdc42- and IRSp53-dependent contractile filopodia tether presumptive lens and retina to coordinate epithelial invagination. *Development*, *136*(21), 3657–3667. http://dx.doi.org/10.1242/dev.042242.

Chauhan, B. K., Lou, M., Zheng, Y., & Lang, R. A. (2011). Balanced Rac1 and RhoA activities regulate cell shape and drive invagination morphogenesis in epithelia. *Proceedings of the National Academy of Sciences of the United States of America*, *108*(45), 18289–18294. http://dx.doi.org/10.1073/pnas.1108993108.

Cho, S. H., & Cepko, C. L. (2006). Wnt2b/beta-catenin-mediated canonical Wnt signaling determines the peripheral fates of the chick eye. *Development*, *133*(16), 3167–3177.

Chow, R. L., Altmann, C. R., Lang, R. A., & Hemmati-Brivanlou, A. (1999). Pax6 induces ectopic eyes in a vertebrate. *Development*, *126*(19), 4213–4222.

Crosson, C. E., Klyce, S. D., & Beuerman, R. W. (1986). Epithelial wound closure in the rabbit cornea. A biphasic process. *Investigative Ophthalmology & Visual Science*, *27*(4), 464–473.

D'Avino, P. P., Savoian, M. S., & Glover, D. M. (2004). Mutations in sticky lead to defective organization of the contractile ring during cytokinesis and are enhanced by Rho and suppressed by Rac. *Journal of Cell Biology*, *166*(1), 61–71. http://dx.doi.org/10.1083/jcb.200402157.

Darwin, C. (1859). *The origin of species*. London: John Murray.

David, D. J., Tishkina, A., & Harris, T. J. (2010). The PAR complex regulates pulsed actomyosin contractions during amnioserosa apical constriction in Drosophila. *Development*, *137*(10), 1645–1655. http://dx.doi.org/10.1242/dev.044107.

Dawes-Hoang, R. E., Parmar, K. M., Christiansen, A. E., Phelps, C. B., Brand, A. H., & Wieschaus, E. F. (2005). *folded gastrulation*, cell shape change and the control of myosin localization. *Development*, *132*(18), 4165–4178.

De Smet, F., Segura, I., De Bock, K., Hohensinner, P. J., & Carmeliet, P. (2009). Mechanisms of vessel branching: Filopodia on endothelial tip cells lead the way. *Arteriosclerosis, Thrombosis, and Vascular Biology*, *29*(5), 639–649. http://dx.doi.org/10.1161/ATVBAHA.109.185165.

Dimanlig, P. V., Faber, S. C., Auerbach, W., Makarenkova, H. P., & Lang, R. A. (2001). The upstream ectoderm enhancer in Pax6 has an important role in lens induction. *Development*, *128*(22), 4415–4424.

Disanza, A., Mantoani, S., Hertzog, M., Gerboth, S., Frittoli, E., Steffen, A., et al. (2006). Regulation of cell shape by Cdc42 is mediated by the synergic actin-bundling activity of the Eps8–IRSp53 complex. *Nature Cell Biology*, *8*(12), 1337–1347.

Dovas, A., & Couchman, J. R. (2005). RhoGDI: Multiple functions in the regulation of Rho family GTPase activities. *Biochemical Journal, 390*(Pt. 1), 1–9. http://dx.doi.org/10.1042/BJ20050104.

Egile, C., Loisel, T. P., Laurent, V., Li, R., Pantaloni, D., Sansonetti, P. J., et al. (1999). Activation of the CDC42 effector N-WASP by the Shigella flexneri IcsA protein promotes actin nucleation by Arp2/3 complex and bacterial actin-based motility. *Journal of Cell Biology, 146*(6), 1319–1332.

Eiraku, M., & Sasai, Y. (2012). Mouse embryonic stem cell culture for generation of three-dimensional retinal and cortical tissues. *Nature Protocols*, 7(1), 69–79. http://dx.doi.org/10.1038/nprot.2011.429.

Eiraku, M., Takata, N., Ishibashi, H., Kawada, M., Sakakura, E., & Okuda, S. (2011). Self-organizing optic-cup morphogenesis in three-dimensional culture. *Nature, 472*(7341), 51–56. http://dx.doi.org/10.1038/nature09941.

Faber, S. C., Dimanlig, P., Makarenkova, H. P., Shirke, S., Ko, K., & Lang, R. A. (2001). Fgf receptor signaling plays a role in lens induction. *Development, 128*(22), 4425–4438.

Faix, J., & Rottner, K. (2006). The making of filopodia. *Current Opinion in Cell Biology, 18*(1), 18–25. http://dx.doi.org/10.1016/j.ceb.2005.11.002.

Fazioli, F., Minichiello, L., Matoska, V., Castagnino, P., Miki, T., Wong, W. T., et al. (1993). Eps8, a substrate for the epidermal growth factor receptor kinase, enhances EGF-dependent mitogenic signals. *The EMBO Journal, 12*(10), 3799–3808.

Fuhrmann, S. (2008). Wnt signaling in eye organogenesis. *Organogenesis, 4*(2), 60–67.

Fuhrmann, S., Levine, E. M., & Reh, T. A. (2000). Extraocular mesenchyme patterns the optic vesicle during early eye development in the embryonic chick. *Development, 127*(21), 4599–4609.

Fukumoto, S., Miner, J. H., Ida, H., Fukumoto, E., Yuasa, K., Miyazaki, H., et al. (2006). Laminin alpha5 is required for dental epithelium growth and polarity and the development of tooth bud and shape. *Journal of Biological Chemistry, 281*(8), 5008–5016. http://dx.doi.org/10.1074/jbc.M509295200.

Furuta, Y., & Hogan, B. L. M. (1998). BMP4 is essential for lens induction in the mouse embryo. *Genes & Development, 12*(23), 3764–3775.

Gallagher, B. C. (1986). Branching morphogenesis in the avian lung: Electron microscopic studies using cationic dyes. *Journal of Embryology and Experimental Morphology, 94*, 189–205.

Gallo, G., & Letourneau, P. C. (2004). Regulation of growth cone actin filaments by guidance cues. *Journal of Neurobiology, 58*(1), 92–102. http://dx.doi.org/10.1002/neu.10282.

Gamell, C., Osses, N., Bartrons, R., Ruckle, T., Camps, M., Rosa, J. L., et al. (2008). BMP2 induction of actin cytoskeleton reorganization and cell migration requires PI3-kinase and Cdc42 activity. *Journal of Cell Science, 121*(Pt. 23), 3960–3970. http://dx.doi.org/10.1242/jcs.031286.

Garcia, C. M., Huang, J., Madakashira, B. P., Liu, Y., Rajagopal, R., Dattilo, L., et al. (2011). The function of FGF signaling in the lens placode. *Developmental Biology, 351*(1), 176–185. http://dx.doi.org/10.1016/j.ydbio.2011.01.001.

Gotoh, N., Ito, M., Yamamoto, S., Yoshino, I., Song, N., & Wang, Y. (2004). Tyrosine phosphorylation sites on FRS2alpha responsible for Shp2 recruitment are critical for induction of lens and retina. *Proceedings of the National Academy of Sciences of the United States of America, 101*(49), 17144–17149. http://dx.doi.org/10.1073/pnas.0407577101.

Grashoff, C., Hoffman, B. D., Brenner, M. D., Zhou, R., Parsons, M., Yang, M. T., et al. (2010). Measuring mechanical tension across vinculin reveals regulation of focal adhesion dynamics. *Nature, 466*(7303), 263–266. http://dx.doi.org/10.1038/nature09198.

Grocott, T., Tambalo, M., & Streit, A. (2012). The peripheral sensory nervous system in the vertebrate head: A gene regulatory perspective. *Developmental Biology, 370*(1), 3–23. http://dx.doi.org/10.1016/j.ydbio.2012.06.028.

Gunhaga, L. (2011). The lens: A classical model of embryonic induction providing new insights into cell determination in early development. *Philosophical Transactions of the Royal Society of London. Series B, Biological Sciences*, *366*(1568), 1193–1203. http://dx.doi.org/10.1098/rstb.2010.0175.

Haigo, S. L., Hildebrand, J. D., Harland, R. M., & Wallingford, J. B. (2003). Shroom induces apical constriction and is required for hingepoint formation during neural tube closure. *Current Biology*, *13*(24), 2125–2137.

Hardin, J. (1988). The role of secondary mesenchyme cells during sea urchin gastrulation studied by laser ablation. *Development*, *103*(2), 317–324.

Heasman, S. J., & Ridley, A. J. (2008). Mammalian Rho GTPases: New insights into their functions from in vivo studies. *Nature Reviews. Molecular Cell Biology*, *9*(9), 690–701. http://dx.doi.org/10.1038/nrm2476.

Hendrix, R. W., & Zwaan, J. (1974). Cell shape regulation and cell cycle in embryonic lens cells. *Nature*, *247*(437), 145–147.

Hildebrand, J. D. (2005). Shroom regulates epithelial cell shape via the apical positioning of an actomyosin network. *Journal of Cell Science*, *118*(Pt. 22), 5191–5203.

Hildebrand, J. D., & Soriano, P. (1999). Shroom, a PDZ domain-containing actin-binding protein, is required for neural tube morphogenesis in mice. *Cell*, *99*(5), 485–497.

Hill, R. E., Favor, J., Hogan, B. L., Ton, C. C., Saunders, G. F., Hanson, I. M., et al. (1991). Mouse small eye results from mutations in a paired-like homeobox-containing gene. *Nature*, *354*(6354), 522–525.

Huang, J., Rajagopal, R., Liu, Y., Dattilo, L. K., Shaham, O., Ashery-Padan, R., et al. (2011). The mechanism of lens placode formation: A case of matrix-mediated morphogenesis. *Developmental Biology*, *355*(1), 32–42. http://dx.doi.org/10.1016/j.ydbio.2011.04.008.

Izzard, C. S. (1974). Contractile filopodia and in vivo cell movement in the tunic of the ascidian, Botryllus schlosseri. *Journal of Cell Science*, *15*(3), 513–535.

Jacinto, A., Wood, W., Woolner, S., Hiley, C., Turner, L., Wilson, C., et al. (2002). Dynamic analysis of actin cable function during Drosophila dorsal closure. *Current Biology*, *12*(14), 1245–1250.

Jaffe, A. B., & Hall, A. (2005). Rho GTPases: Biochemistry and biology. *Annual Review of Cell and Developmental Biology*, *21*, 247–269.

Konstantinidis, G., Moustakas, A., & Stournaras, C. (2011). Regulation of myosin light chain function by BMP signaling controls actin cytoskeleton remodeling. *Cellular Physiology and Biochemistry*, *28*(5), 1031–1044. http://dx.doi.org/10.1159/000335790

Kraynov, V. S., Chamberlain, C., Bokoch, G. M., Schwartz, M. A., Slabaugh, S., & Hahn, K. M. (2000). Localized Rac activation dynamics visualized in living cells. *Science*, *290*(5490), 333–337.

Krugmann, S., Jordens, I., Gevaert, K., Driessens, M., Vandekerckhove, J., & Hall, A. (2001). Cdc42 induces filopodia by promoting the formation of an IRSp53:Mena complex. *Current Biology*, *11*(21), 1645–1655.

Kurisu, S., & Takenawa, T. (2009). The WASP and WAVE family proteins. *Genome Biology*, *10*(6), 226. http://dx.doi.org/10.1186/gb-2009-10-6-226.

Lang, R. A. (2004). Pathways regulating lens induction in the mouse. *International Journal of Developmental Biology*, *48*(8–9), 783–791.

Lang, R. A., Herman, K., Reynolds, A. B., Hildebrand, J. D., & Plageman, T. F., Jr. (2014). p120-Catenin-dependent junctional recruitment of Shroom3 is required for apical constriction during lens pit morphogenesis. *Development*, *141*(16), 3177–3187. http://dx.doi.org/10.1242/dev.107433.

Lee, C., Le, M. P., & Wallingford, J. B. (2009). The shroom family proteins play broad roles in the morphogenesis of thickened epithelial sheets. *Developmental Dynamics*, *238*(6), 1480–1491. http://dx.doi.org/10.1002/dvdy.21942.

Lee-Hoeflich, S. T., Causing, C. G., Podkowa, M., Zhao, X., Wrana, J. L., & Attisano, L. (2004). Activation of LIMK1 by binding to the BMP receptor, BMPRII, regulates BMP-dependent dendritogenesis. *EMBO Journal*, *23*(24), 4792–4801. http://dx.doi.org/10.1038/sj.emboj.7600418.

Leptin, M. (2005). Gastrulation movements: The logic and the nuts and bolts. *Developmental Cell*, *8*(3), 305–320. http://dx.doi.org/10.1016/j.devcel.2005.02.007.

Linseman, D. A., & Loucks, F. A. (2008). Diverse roles of Rho family GTPases in neuronal development, survival, and death. *Frontiers in Bioscience*, *13*, 657–676.

Liu, H., Mohamed, O., Dufort, D., & Wallace, V. A. (2003). Characterization of Wnt signaling components and activation of the Wnt canonical pathway in the murine retina. *Developmental Dynamics*, *227*(3), 323–334. http://dx.doi.org/10.1002/dvdy.10315.

Liu, H., Thurig, S., Mohamed, O., Dufort, D., & Wallace, V. A. (2006). Mapping canonical Wnt signaling in the developing and adult retina. *Investigative Ophthalmology & Visual Science*, *47*(11), 5088–5097. http://dx.doi.org/10.1167/iovs.06-0403.

Liu, H., Xu, S., Wang, Y., Mazerolle, C., Thurig, S., & Coles, B. L. (2007). Ciliary margin transdifferentiation from neural retina is controlled by canonical Wnt signaling. *Developmental Biology*, *308*(1), 54–67. http://dx.doi.org/10.1016/j.ydbio.2007.04.052.

Machacek, M., Hodgson, L., Welch, C., Elliott, H., Pertz, O., Nalbant, P., et al. (2009). Coordination of Rho GTPase activities during cell protrusion. *Nature*, *461*(7260), 99–103. http://dx.doi.org/10.1038/nature08242.

Mallarino, R., Grant, P. R., Grant, B. R., Herrel, A., Kuo, W. P., & Abzhanov, A. (2011). Two developmental modules establish 3D beak-shape variation in Darwin's finches. *Proceedings of the National Academy of Sciences of the United States of America*, *108*(10), 4057–4062. http://dx.doi.org/10.1073/pnas.1011480108.

Mann, I. (1950). *The development of the human eye*. New York: Grune and Stratton.

Martin, A. C., Kaschube, M., & Wieschaus, E. F. (2009). Pulsed contractions of an actin–myosin network drive apical constriction. *Nature*, *457*(7228), 495–499. http://dx.doi.org/10.1038/nature07522.

Martin-Blanco, E., Pastor-Pareja, J. C., & Garcia-Bellido, A. (2000). JNK and decapentaplegic signaling control adhesiveness and cytoskeleton dynamics during thorax closure in Drosophila. *Proceedings of the National Academy of Sciences of the United States of America*, *97*(14), 7888–7893.

McAvoy, J. W. (1980). Cytoplasmic processes interconnect lens placode and optic vesicle during eye morphogenesis. *Experimental Eye Research*, *31*(5), 527–534.

McClay, D. R. (1999). The role of thin filopodia in motility and morphogenesis. *Experimental Cell Research*, *253*(2), 296–301. http://dx.doi.org/10.1006/excr.1999.4723.

Mckeehan, M. S. (1951). Cytological aspects of embryonic lens induction in the chick. *Journal of Experimental Zoology*, *117*, 31–64.

Miki, H., Sasaki, T., Takai, Y., & Takenawa, T. (1998). Induction of filopodium formation by a WASP-related actin-depolymerizing protein N-WASP. *Nature*, *391*(6662), 93–96.

Miki, H., Suetsugu, S., & Takenawa, T. (1998). WAVE, a novel WASP-family protein involved in actin reorganization induced by Rac. *The EMBO Journal*, *17*(23), 6932–6941.

Millard, T. H., & Martin, P. (2008). Dynamic analysis of filopodial interactions during the zippering phase of Drosophila dorsal closure. *Development*, *135*(4), 621–626. http://dx.doi.org/10.1242/dev.014001.

Moores, S. L., Sabry, J. H., & Spudich, J. A. (1996). Myosin dynamics in live Dictyostelium cells. *Proceedings of the National Academy of Sciences of the United States of America*, *93*(1), 443–446.

Nakano, T., Ando, S., Takata, N., Kawada, M., Muguruma, K., Sekiguchi, K., et al. (2012). Self-formation of optic cups and storable stratified neural retina from human ESCs. *Cell Stem Cell*, *10*(6), 771–785. http://dx.doi.org/10.1016/j.stem.2012.05.009.

Nimnual, A. S., Taylor, L. J., & Bar-Sagi, D. (2003). Redox-dependent downregulation of Rho by Rac. *Nature Cell Biology*, *5*(3), 236–241. http://dx.doi.org/10.1038/ncb938.

Nishimura, T., & Takeichi, M. (2008). Shroom3-mediated recruitment of Rho kinases to the apical cell junctions regulates epithelial and neuroepithelial planar remodeling. *Development*, *135*(8), 1493–1502.

Nobes, C., & Hall, A. (1994). Regulation and function of the Rho subfamily of small GTPases. *Current Opinion in Genetics & Development*, *4*(1), 77–81.

Nobes, C. D., & Hall, A. (1995). Rho, rac, and cdc42 GTPases regulate the assembly of multimolecular focal complexes associated with actin stress fibers, lamellipodia, and filopodia. *Cell*, *81*(1), 53–62.

Nobes, C. D., & Hall, A. (1999). Rho GTPases control polarity, protrusion, and adhesion during cell movement. *Journal of Cell Biology*, *144*(6), 1235–1244.

Ohta, Y., Hartwig, J. H., & Stossel, T. P. (2006). FilGAP, a Rho- and ROCK-regulated GAP for Rac binds filamin A to control actin remodelling. *Nature Cell Biology*, *8*(8), 803–814. http://dx.doi.org/10.1038/ncb1437, ncb1437 [pii].

Pang, S. C., Daniels, W. H., & Buck, R. C. (1978). Epidermal migration during the healing of suction blisters in rat skin: A scanning and transmission electron microscopic study. *The American Journal of Anatomy*, *153*(2), 177–191. http://dx.doi.org/10.1002/aja.1001530202.

Partridge, M. A., & Marcantonio, E. E. (2006). Initiation of attachment and generation of mature focal adhesions by integrin-containing filopodia in cell spreading. *Molecular Biology of the Cell*, *17*(10), 4237–4248. http://dx.doi.org/10.1091/mbc.E06-06-0496.

Passey, S., Pellegrin, S., & Mellor, H. (2004). What is in a filopodium? Starfish versus hedgehogs. *Biochemical Society Transactions*, *32*(Pt. 6), 1115–1117. http://dx.doi.org/10.1042/BST0321115.

Peng, J., Wallar, B. J., Flanders, A., Swiatek, P. J., & Alberts, A. S. (2003). Disruption of the diaphanous-related formin Drf1 gene encoding mDia1 reveals a role for Drf3 as an effector for Cdc42. *Current Biology*, *13*(7), 534–545.

Pfister, R. R. (1975). The healing of corneal epithelial abrasions in the rabbit: A scanning electron microscope study. *Investigative Ophthalmology*, *14*(9), 648–661.

Plageman, T. F., Jr., Chauhan, B. K., Yang, C., Jaudon, F., Shang, X., Zheng, Y., et al. (2011). A Trio-RhoA-Shroom3 pathway is required for apical constriction and epithelial invagination. *Development*, *138*(23), 5177–5188. http://dx.doi.org/10.1242/dev.067868.

Plageman, T. F., Jr., Chung, M. I., Lou, M., Smith, A. N., Hildebrand, J. D., Wallingford, J. B., et al. (2010). Pax6-dependent Shroom3 expression regulates apical constriction during lens placode invagination. *Development*, *137*(3), 405–415. http://dx.doi.org/10.1242/dev.045369.

Raftopoulou, M., & Hall, A. (2004). Cell migration: Rho GTPases lead the way. *Developmental Biology*, *265*(1), 23–32, S001216060300544X [pii].

Raich, W. B., Agbunag, C., & Hardin, J. (1999). Rapid epithelial-sheet sealing in the Caenorhabditis elegans embryo requires cadherin-dependent filopodial priming. *Current Biology*, *9*(20), 1139–1146. http://dx.doi.org/10.1016/S0960-9822(00)80015-9.

Rajagopal, R., Huang, J., Dattilo, L. K., Kaartinen, V., Mishina, Y., Deng, C. X., et al. (2009). The type I BMP receptors, Bmpr1a and Acvr1, activate multiple signaling pathways to regulate lens formation. *Developmental Biology*, *335*(2), 305–316. http://dx.doi.org/10.1016/j.ydbio.2009.08.027.

Ridley, A. J. (2006). Rho GTPases and actin dynamics in membrane protrusions and vesicle trafficking. *Trends in Cell Biology*, *16*(10), 522–529.

Roffers-Agarwal, J., Xanthos, J. B., Kragtorp, K. A., & Miller, J. R. (2008). Enabled (Xena) regulates neural plate morphogenesis, apical constriction, and cellular adhesion required for neural tube closure in Xenopus. *Developmental Biology*, *314*(2), 393–403.

Roh-Johnson, M., Shemer, G., Higgins, C. D., McClellan, J. H., Werts, A. D., Tulu, U. S., et al. (2012). Triggering a cell shape change by exploiting preexisting actomyosin contractions. *Science*, *335*(6073), 1232–1235. http://dx.doi.org/10.1126/science.1217869.

Rossman, K. L., Der, C. J., & Sondek, J. (2005). GEF means go: Turning on RHO GTPases with guanine nucleotide-exchange factors. *Nature Reviews. Molecular Cell Biology*, *6*(2), 167–180. http://dx.doi.org/10.1038/nrm1587.

Samakovlis, C., Hacohen, N., Manning, G., Sutherland, D. C., Guillemin, K., & Krasnow, M. A. (1996). Development of the Drosophila tracheal system occurs by a series of morphologically distinct but genetically coupled branching events. *Development*, *122*(5), 1395–1407.

Sawyer, J. K., Harris, N. J., Slep, K. C., Gaul, U., & Peifer, M. (2009). The Drosophila afadin homologue Canoe regulates linkage of the actin cytoskeleton to adherens junctions during apical constriction. *Journal of Cell Biology*, *186*(1), 57–73. http://dx.doi.org/10.1083/jcb.200904001.

Sawyer, J. M., Harrell, J. R., Shemer, G., Sullivan-Brown, J., Roh-Johnson, M., & Goldstein, B. (2010). Apical constriction: A cell shape change that can drive morphogenesis. *Developmental Biology*, *341*(1), 5–19. http://dx.doi.org/10.1016/j.ydbio.2009.09.009.

Schmidt, S., Diriong, S., Mery, J., Fabbrizio, E., & Debant, A. (2002). Identification of the first Rho-GEF inhibitor, TRIPalpha, which targets the RhoA-specific GEF domain of Trio. *FEBS Letters*, *523*(1–3), 35–42.

Shaham, O., Menuchin, Y., Farhy, C., & Ashery-Padan, R. (2012). Pax6: A multi-level regulator of ocular development. *Progress in Retinal and Eye Research*, *31*(5), 351–376. http://dx.doi.org/10.1016/j.preteyeres.2012.04.002.

Smith, A. N., Miller, L. A., Radice, G., Ashery-Padan, R., & Lang, R. A. (2009). Stage-dependent modes of Pax6-Sox2 epistasis regulate lens development and eye morphogenesis. *Development*, *136*(17), 2977–2985. http://dx.doi.org/10.1242/dev.037341.

Snapper, S. B., Takeshima, F., Anton, I., Liu, C. H., Thomas, S. M., Nguyen, D., et al. (2001). N-WASP deficiency reveals distinct pathways for cell surface projections and microbial actin-based motility. *Nature Cell Biology*, *3*(10), 897–904.

Sousa, A. D., & Cheney, R. E. (2005). Myosin-X: A molecular motor at the cell's fingertips. *Trends in Cell Biology*, *15*(10), 533–539. http://dx.doi.org/10.1016/j.tcb.2005.08.006.

Steding, G. (1967). Reasons for embryonic epithelial thickening. *Acta Anatomica (Basel)*, *68*(1), 37–67.

Streit, A. (2008). *The cranial sensory nervous system: Specification of sensory progenitors and placodes*. Cambridge, MA: StemBook.

Takahashi, K., & Suzuki, K. (2009). Membrane transport of WAVE2 and lamellipodia formation require Pak1 that mediates phosphorylation and recruitment of stathmin/Op18 to Pak1-WAVE2-kinesin complex. *Cellular Signalling*, *21*(5), 695–703. http://dx.doi.org/10.1016/j.cellsig.2009.01.007.

Tcherkezian, J., & Lamarche-Vane, N. (2007). Current knowledge of the large RhoGAP family of proteins. *Biology of the Cell*, *99*(2), 67–86. http://dx.doi.org/10.1042/BC20060086.

Tilney, L. G., & Gibbins, J. R. (1969). Microtubules and filaments in the filopodia of the secondary mesenchyme cells of Arbacia punctulata and Echinarachnius parma. *Journal of Cell Science*, *5*(1), 195–210.

Tokuo, H., & Ikebe, M. (2004). Myosin X transports Mena/VASP to the tip of filopodia. *Biochemical and Biophysical Research Communications*, *319*(1), 214–220. http://dx.doi.org/10.1016/j.bbrc.2004.04.167.

Trimarchi, J. M., Cho, S. H., & Cepko, C. L. (2009). Identification of genes expressed preferentially in the developing peripheral margin of the optic cup. *Developmental Dynamics*, *238*(9), 2327–2329. http://dx.doi.org/10.1002/dvdy.21973.

Vasioukhin, V., & Fuchs, E. (2001). Actin dynamics and cell–cell adhesion in epithelia. *Current Opinion in Cell Biology*, *13*(1), 76–84.
Villalonga, P., & Ridley, A. J. (2006). Rho GTPases and cell cycle control. *Growth Factors*, *24*(3), 159–164.
Wagner, M. C., Barylko, B., & Albanesi, J. P. (1992). Tissue distribution and subcellular localization of mammalian myosin I. *Journal of Cell Biology*, *119*(1), 163–170.
Wakely, J. (1984). Observations on the role of ectodermal spreading in the early stages of lens placode invagination in the chick embryo. *Experimental Eye Research*, *38*(6), 627–636.
Wawersik, S., Purcell, P., Rauchman, M., Dudley, A. T., Robertson, E. J., & Maas, R. (1999). BMP7 acts in murine lens placode development. *Developmental Biology*, *207*(1), 176–188.
Westenskow, P., Piccolo, S., & Fuhrmann, S. (2009). Beta-catenin controls differentiation of the retinal pigment epithelium in the mouse optic cup by regulating Mitf and Otx2 expression. *Development*, *136*(15), 2505–2510. http://dx.doi.org/10.1242/dev.032136.
Williams-Masson, E. M., Malik, A. N., & Hardin, J. (1997). An actin-mediated two-step mechanism is required for ventral enclosure of the C. elegans hypodermis. *Development*, *124*(15), 2889–2901.
Wolf, C., Gerlach, N., & Schuh, R. (2002). Drosophila tracheal system formation involves FGF-dependent cell extensions contacting bridge-cells. *EMBO Reports*, *3*(6), 563–568. http://dx.doi.org/10.1093/embo-reports/kvf115.
Wood, W., Jacinto, A., Grose, R., Woolner, S., Gale, J., Wilson, C., et al. (2002). Wound healing recapitulates morphogenesis in Drosophila embryos. *Nature Cell Biology*, *4*(11), 907–912.
Xu, J., Wang, F., Van Keymeulen, A., Herzmark, P., Straight, A., Kelly, K., et al. (2003). Divergent signals and cytoskeletal assemblies regulate self-organizing polarity in neutrophils. *Cell*, *114*(2), 201–214.
Yeh, T. C., Ogawa, W., Danielsen, A. G., & Roth, R. A. (1996). Characterization and cloning of a 58/53-kDa substrate of the insulin receptor tyrosine kinase. *Journal of Biological Chemistry*, *271*(6), 2921–2928.
Zhang, H., Berg, J. S., Li, Z., Wang, Y., Lang, P., Sousa, A. D., et al. (2004). Myosin-X provides a motor-based link between integrins and the cytoskeleton. *Nature Cell Biology*, *6*(6), 523–531. http://dx.doi.org/10.1038/ncb1136.

CHAPTER TWELVE

Developing and Regenerating a Sense of Taste

Linda A. Barlow[*,†,‡,1], **Ophir D. Klein**[§,¶,||]

[*]Department of Cell and Developmental Biology, University of Colorado School of Medicine, Anschutz Medical Campus, Aurora, Colorado, USA
[†]Graduate Program in Cell Biology, Stem Cells and Development, University of Colorado School of Medicine, Anschutz Medical Campus, Aurora, Colorado, USA
[‡]Rocky Mountain Taste and Smell Center, University of Colorado School of Medicine, Anschutz Medical Campus, Aurora, Colorado, USA
[§]Departments of Orofacial Sciences and Pediatrics, University of California San Francisco, San Francisco, California, USA
[¶]Program in Craniofacial and Mesenchymal Biology, University of California San Francisco, San Francisco, California, USA
[||]Institute for Human Genetics, University of California San Francisco, San Francisco, California, USA
[1]Corresponding author: e-mail address: linda.barlow@ucdenver.edu

Contents

Abstract

Taste is one of the fundamental senses, and it is essential for our ability to ingest nutritious substances and to detect and avoid potentially toxic ones. Taste buds, which are clusters of neuroepithelial receptor cells, are housed in highly organized structures called taste papillae in the oral cavity. Whereas the overall structure of the taste periphery is conserved in almost all vertebrates examined to date, the anatomical, histological, and cell biological, as well as potentially the molecular details of taste buds in the oral cavity are diverse across species and even among individuals. In mammals, several types of gustatory papillae reside on the tongue in highly ordered arrangements, and the patterning and distribution of the mature papillae depend on coordinated molecular events in embryogenesis. In this review, we highlight new findings in the field of taste development, including how taste buds are patterned and how taste cell fate is regulated. We discuss whether a specialized taste bud stem cell population exists and how extrinsic signals can define which cell lineages are generated. We also address the

Current Topics in Developmental Biology, Volume 111
ISSN 0070-2153
http://dx.doi.org/10.1016/bs.ctdb.2014.11.012

question of whether molecular regulation of taste cell renewal is analogous to that of taste bud development. Finally, we conclude with suggestions for future directions, including the potential influence of the maternal diet and maternal health on the sense of taste *in utero*.

Taste is important for life. It serves as the gateway to substances that enter the body, allowing us to distinguish nutritious food items from potentially toxic ones. Classically, taste buds in the oral cavity, primarily on the tongue, were shown to detect five basic tastes: sour, salty, bitter, sweet, and umami—savory or "deliciousness" in Japanese. More recently, fatty acids and calcium have emerged as potential tastants that can be sensed by taste bud cells (Iwata, Yoshida, & Ninomiya, 2014; Liman, Zhang, & Montell, 2014; Passilly-Degrace et al., 2014; Tordoff, Reed, & Shao, 2008; Tucker, Mattes, & Running, 2014).

Among humans, taste preferences vary widely, and these preferences in turn influence dietary choices, which impact body weight and therefore health (Mennella, 2014). A key question is what underlies this variability. Not surprisingly, it appears that environmental, genetic, and epigenetic mechanisms are at play. In mammals, including humans, the maternal diet during gestation and postnatal lactation is learned by her offspring. In humans, innervated and differentiated taste buds that are presumably functional are evident by 10–13 weeks of development (Bradley & Stern, 1967; Witt & Reutter, 1996, 1998). Throughout gestation, taste stimuli reach the amniotic fluid, which is continually swallowed by the fetus, and following birth, tastes of the maternal diet are evident in breast milk. This exposure heavily influences the dietary choices of offspring as they discover these new tastes (Beauchamp & Mennella, 2009; Mennella, 2014). However, maternal health also impacts the gestational experience, as it results in fetal metabolic programming via presumed epigenetic mechanisms (Dyer & Rosenfeld, 2011), which, in the case of diabetic or obese mothers, can predispose offspring to diabetes and cardiovascular disease. Although conclusive studies regarding alterations in taste sensitivity in this context have not been performed, it is well known that diabetes and obesity affect taste preferences in adults. For example, in diabetic patients, taste responses, especially to sweet, are blunted (Wasalathanthri, Hettiarachchi, & Prathapan, 2014), and obese individuals also have decreased taste sensitivity (Stewart et al., 2010; Stewart, Feinle-Bisset, & Keast, 2011).

The pattern of taste buds is established during embryogenesis, such that the first functional taste bud cells are specified during gestation. Whereas most sensory epithelia, such as hair cells of the inner ear and photoreceptors of the retina, have limited renewal potential, taste cells are remarkable in their ability to turn over rapidly and continuously throughout adult life (Beidler & Smallman, 1965; Farbman, 1980; Feng, Huang, & Wang, 2014; Hamamichi, Asano-Miyoshi, & Emori, 2006; Perea-Martinez, Nagai, & Chaudhari, 2013). Despite regular sensory cell replacement, the sense of taste is remarkably stable throughout life in healthy individuals. However, taste can be distorted or lost in cancer patients when these individuals are treated with chemotherapeutic drugs, and in head and neck cancer patients following targeted radiotherapy (Berteretche et al., 2004; Hong et al., 2009; Ruo Redda & Allis, 2006; Vissink, Jansma, Spijkervet, Burlage, & Coppes, 2003). These treatments are thought to disrupt taste function by diminishing taste bud cell renewal (Nguyen, Reyland, & Barlow, 2012 and references therein). Thus, we hypothesize that both regulation of taste bud development, including patterning and formation of the proper ratio of taste receptor cell types, and taste bud renewal, i.e., generation of functional taste cell types in the proper ratios with the proper timing, underlie variability in taste function and dysfunction.

In this review, we highlight new data in the context of the important open questions in the field rather than providing an exhaustive survey of the literature; for more comprehensive reviews on taste development, regeneration, and function, see Kapsimali & Barlow (2013), Feng et al. (2014), and Liman et al., (2014), respectively.

1. HOW ARE TASTE BUDS PATTERNED?

Taste bud distribution is highly variable across vertebrate species, including in mammals (Jackowiak, 2006 and references therein), fish, amphibians, and birds (Erdoğan & Iwasaki, 2014; Finger, 1997; Northcutt, 2004). In addition, taste bud pattern varies even within a single species, including in humans (Fischer et al., 2013; Miller & Reedy, 1990) and rodents (mouse inbred lines: Reiner et al., 2008; rat: Miller & Preslar, 1975; Tordoff, Alarcon, & Lawler, 2008).

The initial pattern of taste buds on the tongue is evident at mid-gestation (at embryonic day (E) 12.0 in mice), when bilateral rows of columnar epithelial placodes (taste placodes) form in the otherwise cuboidal epithelium of the developing tongue (Farbman, 1965; Mistretta, 1972). Subsequently, placodes undergo morphogenesis into mushroom-shaped (fungiform) taste papillae, which house taste buds that differentiate at birth. Taste placodes express Sonic hedgehog (*Shh*) from the earliest stages of their development, and expression of *Shh* persists in the apices of papillae through the remainder of embryogenesis (Bitgood & McMahon, 1995; Hall, Hooper, & Finger, 1999; Jung, Oropeza, & Thesleff, 1999). Lineage tracing of the *Shh* + taste placodes, commencing at E12.5 or 13.5, reveals that these cells differentiate directly into the first taste bud cells at birth but do not contribute to the surrounding taste papillae (Thirumangalathu, Harlow, Driskell, Krimm, & Barlow, 2009). Rather, we have proposed that *Shh*-descendent taste bud precursor cells may function as signaling centers to induce adjacent epithelial and mesenchymal cells to form taste papillae (Thirumangalathu et al., 2009).

A number of pathways regulate the initial patterning of taste placodes in rodents, including Wnt/β-catenin (Iwatsuki et al., 2007; Liu et al., 2007), bone morphogenetic proteins (Bmps) (Beites et al., 2009; Zhou, Liu, & Mistretta, 2006), Shh (Hall, Bell, & Finger, 2003; Liu, MacCallum, Edwards, Gaffield, & Mistretta, 2004; Mistretta, Liu, Gaffield, & MacCallum, 2003), epidermal growth factor (Egf) (Liu, Henson, Zhou, D'Silva, & Mistretta, 2008), and fibroblast growth factors (Fgfs) (Kapsimali et al., 2011; Petersen et al., 2011; reviewed in Kapsimali & Barlow, 2013; Table 1). In particular, normal activation of the Wnt/β-catenin pathway

Table 1 Summary of the function of major signaling pathways in embryonic taste bud development

Pathway	Demonstrated functions	References
Wnt/β-catenin	• Promotes taste fate *in vivo* and *in vitro* • Required for taste fate *in vivo*	Iwatsuki et al. (2007) Liu et al. (2007)
BMP	• Prior to taste placode specification, BMPs promote taste fate • Following placode specification, BMPs repress taste fate *in vitro* • Loss of mesenchymal follistatin promotes taste fate *in vivo*	Zhou et al. (2006) Beites et al. (2009)
SHH	• Shh represses taste fate *in vitro* • Inhibition of Shh *in vitro* expands taste fate	Iwatsuki et al. (2007) Hall et al. (2003), Mistretta et al. (2003)
FGF	• Loss of Spry1/2 promotes taste fate • Loss of mesenchymal FGF10 represses taste fate	Petersen et al. (2011)
EGF	• EGF represses taste fate *in vitro* • Inhibition of EGF promotes taste fate *in vitro*	Liu et al. (2008)

within the developing lingual epithelium is required for formation of taste placodes, whereas ectopic activation of the pathway in the entire epithelium drives differentiation into enlarged *Shh*+ taste bud precursors embedded in oversized fungiform papillae (Liu et al., 2007). Interestingly, Wnt ligands are expressed in both epithelial and mesenchymal compartments in the developing tongue (Iwatsuki et al., 2007; Liu et al., 2007, 2012), leaving open the question of which source(s) of Wnt protein are responsible for taste patterning. Recently, conditional deletion of *Wls* (also known as *Gpr177* or *Evi1*) from the early oral endoderm under the $Shh^{GFP\text{-}Cre}$ allele (Harfe et al., 2004) revealed that epithelial Wnt production is required for taste placode initiation (Zhu et al., 2014). The *Wls* gene encodes an intracellular protein that enables Wnt ligand secretion (Banziger et al., 2006), and deletion of epithelial *Wls*, which causes loss of epithelial Wnt ligand secretion, leads to absence of taste placode formation. One caveat to these studies, however, is that *Shh* is expressed within the oral endoderm commencing at E9.5 (Echelard et al., 1993) and thus loss of *Wls* function using the $Shh^{GFP\text{-}Cre}$ is also induced early with respect to tongue and taste placode formation, which begin at E11.5 (Kaufman, 1999) and E12.0 (Hall et al., 2003), respectively. Therefore, it is formally possible that early epithelial WLS function is required for development of epithelial competence to respond to later Wnt signals from the mesenchyme. Nonetheless, these findings are consistent with experimental data from axolotl (salamander) embryos, where early taste bud specification and patterning are governed by mechanisms intrinsic to the epithelium and independent of oral mesenchyme (Barlow, 2001; Barlow & Northcutt, 1997; Parker, Bell, & Barlow, 2004). Thus, numerous pathways can affect taste bud patterning, and subtle differences in timing, competence to receive the signal, and the strength of the signal may be equally influential.

In addition, there are significant distinctions in patterning of different regions of the tongue as well as patterning of taste papillae in different organisms. One area in which this has been explored is the difference between the small anterior fungiform papillae and the large posterior circumvallate papilla (CVP). Most studies of development and patterning have focused on the fungiform papillae; although there is a large literature concerning physiology, anatomy, and cell biology of the adult CVP, only a limited number of developmental studies have been performed on this organ (see Kapsimali & Barlow, 2013 for review). A number of pathways that are known to regulate fungiform papillae pattern have no reported phenotype in the CVP, including the Shh (Mistretta et al., 2003), Bmp (Beites et al., 2009), and Wnt (Iwatsuki et al., 2007) pathways. In contrast, the Fgf pathway has been shown to be a critical regulator of CVP development in mice. A balance between Sprouty (Spry) genes and *Fgf10*, which, respectively, antagonize and activate receptor tyrosine kinase signaling, regulates the number of CVPs (Petersen et al., 2011), such that in wild-type mice, only a single CVP forms. Deletion of a single Sprouty family member, *Spry2*, resulted in duplication of the CVP, as a result of an increase in the size of the CV placode progenitor field. Combined deletion of two Sprouty genes in $Spry1^{-/-};Spry2^{-/-}$ embryos led to the formation of multiple CVPs, demonstrating the redundancy of Sprouty genes in regulating the size of the progenitor field. By contrast, deletion of *Fgf10* led to absence of the CVP, thus identifying FGF10 as an inductive, mesenchyme-derived factor for taste papillae. Recently, the transcription factor Wilms' tumor 1 protein (WT1) was found to have a critical role in CVP development, as deletion of *Wt1* led to failure of CVP development (Gao, Toska, Denmon, Roberts, & Medler, 2014). Several WT1 target genes that are members of canonical signaling pathways were identified, including *Lef1* from the Wnt pathway, *Ptch1* from the Shh pathway, and *Bmp4*. Interestingly, there are also some hints emerging that anterior versus posterior taste papillae may be regulated in opposite ways by specific signaling pathways, such as the Fgf pathway (Petersen et al., 2011). This may be due to origins in different germ layers, as the anterior tongue epithelium is derived from ectoderm, whereas epithelium covering the posterior tongue has an endodermal origin (Adams, 1931; Barlow, 2000; Rothova, Thompson, Lickert, & Tucker, 2012). Of interest, while taste buds are restricted to the oral cavity of most vertebrates, several fishes have evolved external taste buds, including catfishes which have thousands of taste buds distributed in the head and trunk epithelium (Atema, 1971; Landacre, 1907; Northcutt, 2005). Their external location suggests that these taste buds must arise from surface ectoderm (Landacre, 1907), while oral taste buds of fish and amphibians originate primarily from endoderm (Barlow & Northcutt, 1995; Johnston, 1910). Thus, the catfish would be an ideal model to test if and how embryonic origin affects the mode of taste bud development.

2. REGULATION OF TASTE CELL FATE

Taste buds comprise 60 – 100 elongate cells, which are classified into three morphological types and as many as five or more functional categories (Fig. 1) (Feng et al., 2014; Liman et al., 2014). In animal models used in the majority of developmental studies including fish and rodents, taste cells

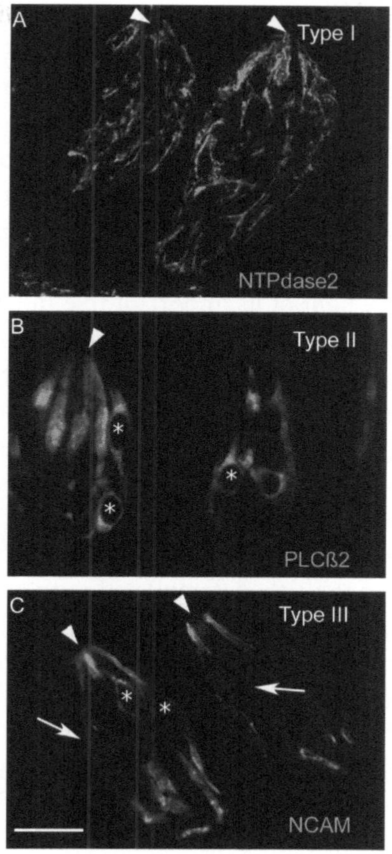

Figure 1 Taste buds comprise three morphological cell types, I, II, and III, which are recognized by their expression of specific marker proteins. (A) Type I cells express NTPdase2, which localizes to cell membranes that tightly wrap other cells within the bud, such that individual NTPDase2+ cells are not discernable (see Miura, Scott, Harada, & Barlow, 2014 for detailed explanation). (B) Type II cells express PLCβ2 in the cytosol, and the protein is excluded from nuclei of PLCβ2+ taste cells (*). (C) NCAM marks the surface of Type III taste cells (*), as well as some nerve fibers extending into buds (arrows). All panels are images of taste buds from the circumvallate papilla of mice. Apical taste pores are indicated with arrowheads. Scale bar is 20 μm. (See the color plate.)

differentiate postnatally (see references in Kapsimali & Barlow (2013)), such that in adults, the different taste cell types occur in proper ratios within each bud (Chaudhari & Roper, 2010; Kim et al., 2003; Ma, Yang, Thomas, & Kinnamon, 2007; Ohtubo & Yoshii, 2010). Generally, the Type I or glial-like cells are most common, followed by Type II sweet/bitter/umami detectors, and least frequent are the Type III sour receptor cells, although these ratios vary with respect to location in the tongue. For example, in rats and mice, Type III cells are more frequent in taste buds in the posterior CVP than in taste buds of the anterior fungiform papillae (Ma et al., 2007; Ohtubo & Yoshii, 2010). In mice, taste buds in the anterior tongue have more sweet cells per bud, whereas in posterior tongue, taste buds have more bitter cells (Kim et al., 2003; Tizzano et al., 2008). How and when these cell fates are assigned in embryos remains an important open question.

Not surprisingly, in light of its important role in fate decisions in many tissues, Notch function has been implicated in taste cell fate decisions during embryonic development (reviewed in Kapsimali & Barlow, 2013). In mice, Notch pathway genes are expressed in and around the developing CV papilla in the posterior tongue, and this expression begins after the papilla is specified (Seta, Seta, & Barlow, 2003). In late gestation embryos, *Mash1/Ascl1*, which is transcriptionally repressed by Notch signaling (Kageyama & Ohtsuka, 1999), is expressed in small numbers of epithelial cells in locations consistent with that of the first differentiated taste buds at early postnatal stages; this expression pattern suggested a role for *Ascl1* in the differentiation of one or more taste cell types. Indeed, genetic deletion of *Ascl1* results in loss of expression of numerous markers of Type III taste cells, supporting the hypothesis that *Ascl1* is required for Type III cell fate (Kito-Shingaki et al., 2014; Seta, Oda, Kataoka, Toyono, & Toyoshima, 2011). Interestingly, Type III cells are the taste cells most similar to neurons, and *Mash1* is a proneural gene that drives expression of Notch ligands cell autonomously to activate Notch signaling in adjacent cells, but *Mash1* also directs fate cell autonomously while keeping neighbors in a stem cell state (Kageyama, Ohtsuka, Hatakeyama, & Ohsawa, 2005). In addition, the transcription factor *Hes1*, considered a primary Notch target gene (Ohtsuka et al., 1999), has been shown to repress differentiation of Type II taste cells, as excess Type II cells differentiate in *Hes1*$^{-/-}$ taste papillae (Ota et al., 2009). Notch function in taste cell fate selection is conserved in zebrafish, where it again plays a role in specifying Type II-like versus Type III-like cell fates (Kapsimali et al., 2011).

In addition to components of the Notch pathway, another transcription factor, *Skn1a/Pou2f3* has been shown to be required for differentiation of Type II sweet/bitter/umami cells in adults, as taste buds in adult *Skn1a*$^{-/-}$ mice lack Type II cells and have excess Type III cells (Matsumoto, Ohmoto, Narukawa, Yoshihara, & Abe, 2011). If and how this transcription factor forms a gene regulatory network with *Ascl1* and/or *Hes1* remains to be explored. Finally, and intriguingly, genetic control of specification of the most common taste cell fate, the Type I glial-like cell, remains a mystery.

3. HOW CAN WE LINK EMBRYONIC DEVELOPMENT AND ADULT TASTE CELL RENEWAL?

As mentioned above, embryonic *Shh*+ placodes are taste bud precursors, which differentiate into the first taste bud cells. As taste bud cells renew, these first taste cells are ultimately replaced over time. However, embryonic *Shh*+ cells do not contribute to the stem cell pool that enables adult taste cell renewal, as all *Shh*-descendent taste bud cells are lost by 4 months postnatally (Thirumangalathu et al., 2009). Rather, in adults, taste receptor cells are renewed from cytokeratin (K) 14+/K5+ basal keratinocytes adjacent to taste buds (Fig. 2) (Okubo, Clark, & Hogan, 2009). K14+ lingual keratinocytes also give rise to the general epithelium of the tongue, which comprises nontaste filiform papillae (Hume & Potten, 1976; Mistretta, 1972). How K14+/K5+ stem cells are regulated to produce both taste buds and general epithelium is poorly understood. However, this population generates only

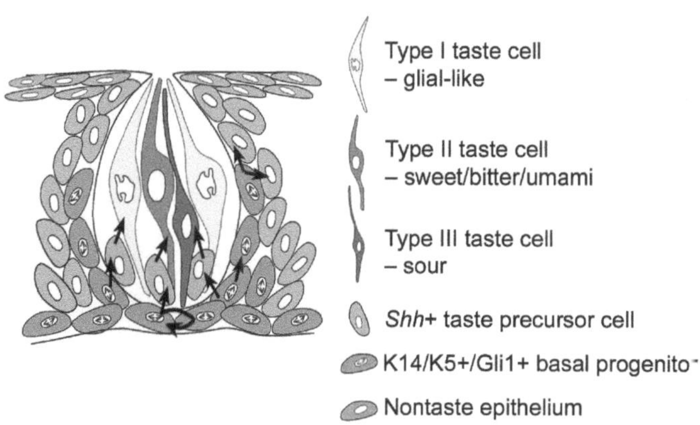

Figure 2 Schematic diagram of taste bud cell types and lineage relationships. Type I (yellow), II (blue), and III (red) taste cells are elongate, postmitotic (open nuclei) cells, which arise from proliferating, K5+/K14+ keratinocytes (green, mitotic nuclei) outside of taste buds. These progenitors also give rise to nontaste epithelial cells (orange). Following their final division, cells destined to become taste cells enter taste buds and become taste bud precursors, which are *Shh*+ ovoid cells (lavender) located in the basal compartment of the taste bud. Shh+ cells are postmitotic, and differentiate into each of the three taste cell types. Black arrows indicate observed lineage relationships among cell types (see text for details), but are not absolute, nor exhaustive. (See the color plate.)

nontaste epithelium *in vitro* (Luo, Okubo, Randell, & Hogan, 2009), suggesting that either taste bud stem cells represent a rare or distinct population, or that key extrinsic signals are required that were absent under the culture conditions employed.

Intriguingly, other lingual epithelial stem populations that contribute to nontaste filiform papillae, but not to taste buds, have been identified. One gene that has been recently studied is *Bmi1*, which together with K14 and K5 labels cells at the base of the interpapillary pit (Tanaka et al., 2013). These cells were reported to be unipotent stem cells for keratinized epithelial cells but not for taste bud cells. *In vitro* organoids could also be generated from single *Bmi1*-positive cells (Hisha et al., 2013). Similarly, lineage tracing using a *Tcf3*^{CreER} knock-in mouse model showed that *Tcf3* marks stem cells as well as transient progenitors and cells undergoing active differentiation in the tongue (Howard, Nuguid, Ngole, & Nguyen, 2014).

4. IS THERE A SPECIALIZED TASTE BUD STEM CELL POPULATION, OR ARE EXTRINSIC SIGNALS RESPONSIBLE FOR DEFINING WHICH CELL LINEAGES ARE GENERATED, AND WHEN?

Recently, a new population of lingual stem cells has been identified that gives rise to both epithelium and taste buds in the large CVP situated at the midline of the posterior tongue. These cells express the leucine-rich repeat-containing G-protein coupled receptor 5 (LGR5), which is also expressed by the stem cells of the intestine and other organs (Barker et al., 2007; Ng et al., 2014; Plaks et al., 2013). In the CVP of mice carrying an *Lgr5* reporter allele (*Lgr5*$^{GFP\text{-}CreER}$; Barker et al., 2007), taste buds themselves are not *Lgr5*^{GFP}+; instead, GFP-bright epithelial cells are situated at the deepest portion of the papilla where taste buds are not present, while low expressing *Lgr5*^{GFP}+ basal epithelial cells are found higher up in the papilla and adjacent to taste buds. Lineage tracing revealed that *Lgr5*+ cells give rise all three taste cell types (Takeda et al., 2013; Yee et al., 2013), and that the deep *Lgr5*^{GFP}+ cells are lineage labeled, and persist for months (Yee et al., 2013). Interestingly, both groups reported that *Lgr5* expression was not detectable in the anterior tongue by GFP expression, or by PCR for *Lgr5* transcripts, indicating that this stem cell population is specific to the posterior CVP. This result highlights again the distinction described above between ectodermally derived anterior taste buds and endodermally derived

CVP taste buds, and it is intriguing in that *Lgr5* also marks a key stem cell population in the endodermally derived gut (see Barker, Tan, & Clevers, 2013 for review).

To date, an analogous stem cell population remains to be identified for fungiform taste buds. However, our recent data point to a different model, where specialized stem cells may not be required in the anterior tongue, but rather extrinsic signals can drive bipotent K14+ cells toward the taste fate (Castillo et al., 2014). Using an inducible gain-of-function approach, we overexpressed SHH (SHHcKI) in K14+ basal keratinocytes and found that ectopic taste buds form throughout the nontaste epithelium and are interspersed among the endogenous fungiform papillae. Thus, SHH signals can induce cell type-replete taste bud differentiation from K14+ progenitors regardless of location. These results indicate that K14+ keratinocytes are broadly competent to generate taste receptor cells in response to SHH, revealing an unappreciated competency of the anterior tongue epithelium to build taste buds outside of fungiform papillae in the adult tongue. Expansion of taste fate in the lingual epithelium is well documented in the embryonic tongue, as all of the following promote the taste fate in lingual regions where taste buds do not form in controls: increased β-catenin *in vitro* or *in vivo* (Iwatsuki et al., 2007; Liu et al., 2007; Okubo, Pevny, & Hogan, 2006), SHH inhibition *in vitro* (Hall et al., 2003; Liu et al., 2004; Mistretta et al., 2003), or genetic deletion of *Follistatin in vivo* (Beites et al., 2009). Interestingly, manipulation of each of these pathways was reported to have no impact on the development of the posterior CVP (Beites et al., 2009; Iwatsuki et al., 2007; Liu et al., 2004; Mistretta et al., 2003). Likewise, in adult mice, despite SHH overexpression, taste buds in the posterior CV papilla were only mildly affected (D. Castillo, O. Klein, & L. Barlow, unpublished), again suggesting that SHH may function differently in ectodermally derived anterior versus endodermally derived posterior lingual epithelium even in adulthood.

5. IS MOLECULAR REGULATION OF TASTE CELL RENEWAL ANALOGOUS TO THAT OF TASTE BUD DEVELOPMENT?

In contrast to development, during which several pathways have been shown to function in taste bud development, only Shh function has been examined thus far in adults. Specifically, ectopic overexpression of SHH promotes differentiation of cell type-replete taste buds (Castillo et al., 2014).

This protaste function contrasts directly with the taste-repressive function of Shh signaling in development. This difference is likely due to differences in cells receiving SHH signals, as *Shh+* cells in adult taste buds and in taste placodes appear quite similar (Miura & Barlow, 2010; Miura, Kusakabe, & Harada, 2006; Nakayama et al., 2008): both are specified, immediate postmitotic precursors of differentiated taste bud cells and both are SHH non-responsive, as they do not express *Ptch1* or *Gli1* (Hall et al., 1999; Liu et al., 2013; Miura et al., 2001; Miura, Scott, Harada, & Barlow, 2014; Thirumangalathu et al., 2009). In embryos, established *Shh+* cells inhibit neighboring *Ptch1* and *Gli1* expressing epithelial cells (Hall et al., 1999) from acquiring a taste fate (Hall et al., 2003; Iwatsuki et al., 2007; Liu et al., 2004; Mistretta et al., 2003). However, in adult epithelium, while postmitotic precursors within taste buds also signal via SHH to surrounding *Gli1*+ and *Ptch1*+ taste papilla epithelial cells (see Fig. 2) (Liu et al., 2013; Miura et al., 2001, 2004), SHH now promotes rather than represses taste fate (Castillo et al., 2014). When this shift in SHH function occurs should shed light on the timing of the transition from initial development to continual taste cell renewal.

In addition to *Ptch1* and *Gli1*, *Gli2* is expressed by basal keratinocytes adjacent to taste buds (a pattern similar to that of *Ptch1* and *Gli1*), as well as more broadly throughout the nontaste epithelium (Liu et al., 2013), suggesting that *Gli2* may play a role in taste cell renewal and underlie in part the ability of SHH to induce ectopic taste buds. Indeed, activation of a hyperactive *Gli2* allele, which is oncogenic in skin (Grachtchouk et al., 2011), abolishes fungiform taste buds in adult mice; however, it is unclear if this is due to oncogenic growth or to a more direct effect on lingual epithelial cell fate (Liu et al., 2013).

Finally, in addition to a role in lingual epithelium, SHH signals are received in the mesenchymal compartment of taste papillae of both embryos and adults (Hall et al., 1999; Liu et al., 2013; Miura et al., 2001). In adults, SHH-receiving cells in the mesenchyme have been proposed to comprise a niche for taste bud cell maintenance (Liu et al., 2013), an idea that was suggested initially based on expression of *Bmp4* in the taste papilla mesenchyme (Nguyen & Barlow, 2010). Likewise, in the embryonic tongue, in addition to signaling to adjacent epithelium, *Shh+* taste placodes signal to the subjacent mesenchymal compartment. This epithelial-to-mesenchymal Shh signaling may function in the extensive morphogenesis of taste papillae, comparable to the role of *Shh* in development of other epithelial appendages, such as teeth, feather, and hair follicles (Chuong, Patel, Lin, Jung, & Widelitz, 2000; Pispa & Thesleff, 2003), although this remains to be tested. As we mentioned above, taste buds in salamander embryos develop independently of oral mesenchyme. Cultured epithelial explants devoid of mesodermal and neural crest-derived cells develop cell type-replete taste buds (Barlow & Northcutt, 1997), but it is important to note that in axolotls taste buds do not reside in papillae, but rather are embedded in the oral epithelium (Fährmann, 1967; Northcutt, Barlow, Braun, & Catania, 2000; Toyoshima, Miyamoto, & Shimamura, 1987). Lingual taste papillae thus appear to be primarily a mammalian innovation, which may have evolved to prevent taste bud desiccation or protect taste buds from abrasive foodstuffs. Thus, we propose that the primary event in taste development is specification of taste bud precursors, and that papillary development is secondary. Specifically, in both amphibians and mammals, we hypothesize that taste bud precursors are specified by epithelium-intrinsic processes, while in mammals, these taste bud precursors in turn organize adjacent epithelium and mesenchyme to build taste papillae around them (Thirumangalathu et al., 2009). It remains to be determined if the canonical Shh signaling pathway, in addition to its role in taste placode patterning, also guides papilla morphogenesis directly and/or indirectly.

In closing, the overall structure of the taste periphery, which is composed of multicellular taste buds that detect primary tastants in the oral cavity, is conserved in all vertebrates examined to date, except the hagfishes (Braun, 1996, 1998; Finger, 1997), whereas the anatomical, histological, cell biological, and molecular details of taste buds in the oral cavity are quite varied (Barlow, 1999; Jiang et al., 2012; Liman, 2012; Northcutt, 2004). These differences in structure and function have evolved as adaptations to the different diets consumed, but also indicate that this sensory system is quite flexible over evolutionary time scales. In addition, the pattern and cellular makeup of taste buds have been shown to vary within mammalian species, such as humans and mice, and in mouse and zebrafish embryos, taste bud pattern and cell complement are easily manipulated experimentally via drug or genetic perturbations. This raises the question of how impressionable the sense of taste is *in utero*, in terms of exposure to maternal diet and to maternal overall health: can the taste bud array be permanently altered anatomically and at the level of taste cell fate decisions by gestational experience? And how might these changes in the taste periphery impact taste function and dietary selections and ultimately the health of offspring?

ACKNOWLEDGMENTS

Thanks to Fernando Giraldez for his good humored comments during the writing of this manuscript, and to members of the Giraldez, Pujades, and Alsina labs at the PRBB for their gracious hosting of LB while on sabbatical at the PRBB, Barcelona, Spain.

This study was supported by DC012675 and DC012383 to L. A. B. and DE021420 to O. D. K.

REFERENCES

Adams, A. E. (1931). Some effects of removal of endoderm from the mouth region of early *Amblystoma punctatum* embryos. *Journal of Experimental Zoology, 58*, 147–163.

Atema, J. (1971). Structures and functions of the sense of taste in the catfish (*Ictalurus natalis*). *Brain, Behavior and Evolution, 4*, 273–294.

Banziger, C., Soldini, D., Schutt, C., Zipperlen, P., Hausmann, G., & Basler, K. (2006). Wntless, a conserved membrane protein dedicated to the secretion of Wnt proteins from signaling cells. *Cell, 125*, 509–522.

Barker, N., Tan, S., & Clevers, H. (2013). Lgr proteins in epithelial stem cell biology. *Development, 140*, 2484–2494.

Barker, N., van Es, J. H., Kuipers, J., Kujala, P., van den Born, M., Cozijnsen, M., et al. (2007). Identification of stem cells in small intestine and colon by marker gene Lgr5. *Nature, 449*, 1003–1007.

Barlow, L. A. (1999). The biology of amphibian taste. In H. Heatwole & E. Dawley (Eds.), *Amphibian biology: Sensory perception* (pp. 743–782).

Barlow, L. A. (2000). Taste buds in ectoderm are induced by endoderm: Implications for mechanisms governing taste bud development. In L. Olsson & C.-O. Jacobson (Eds.), *Regulatory processes in development: The legacy of Sven Hörstadius. Proceedings of the Wenner-Gren international symposium* (pp. 185–190): London, UK: Portland Press.

Barlow, L. A. (2001). Specification of pharyngeal endoderm is dependent on early signals from axial mesoderm. *Development, 128*, 4573–4583.

Barlow, L. A., & Northcutt, R. G. (1995). Embryonic origin of amphibian taste buds. *Developmental Biology, 169*, 273–285.

Barlow, L. A., & Northcutt, R. G. (1997). Taste buds develop autonomously from endoderm without induction by cephalic neural crest or paraxial mesoderm. *Development, 124*, 949–957.

Beauchamp, G. K., & Mennella, J. A. (2009). Early flavor learning and its impact on later feeding behavior. *Journal of Pediatric Gastroenterology and Nutrition, 48*(Suppl. 1), S25–S30.

Beidler, L. M., & Smallman, R. L. (1965). Renewal of cells within taste buds. *The Journal of Cell Biology, 27*, 263–272.

Beites, C. L., Hollenbeck, P. L., Kim, J., Lovell-Badge, R., Lander, A. D., & Calof, A. L. (2009). Follistatin modulates a BMP autoregulatory loop to control the size and patterning of sensory domains in the developing tongue. *Development, 136*, 2187–2197.

Berteretche, M. V., Dalix, A. M., d'Ornano, A. M., Bellisle, F., Khayat, D., & Faurion, A. (2004). Decreased taste sensitivity in cancer patients under chemotherapy. *Support Care Cancer, 12*, 571–576.

Bitgood, M. J., & McMahon, A. P. (1995). Hedgehog and Bmp genes are coexpressed at many diverse sites of cell-cell interaction in the mouse embryo. *Developmental Biology, 172*, 126–138.

Bradley, R. M., & Stern, I. B. (1967). The development of the human taste bud during the foetal period. *Journal of Anatomy, 101*, 743–752.

Braun, C. B. (1996). The sensory biology of the living jawless fishes: A phylogenetic assessment. *Brain, Behavior and Evolution, 48*, 262–276.

Braun, C. B. (1998). Schreiner organs: A new craniate chemosensory modality in hagfishes. *The Journal of Comparative Neurology, 392*, 135–163.

Castillo, D., Seidel, K., Salcedo, E., Ahn, C., de Sauvage, F. J., Klein, O. D., et al. (2014). Induction of ectopic taste buds by SHH reveals the competency and plasticity of adult lingual epithelium. *Development, 141*, 2993–3002.

Chaudhari, N., & Roper, S. D. (2010). The cell biology of taste. *The Journal of Cell Biology, 190*, 285–296.

Chuong, C. M., Patel, N., Lin, J., Jung, H. S., & Widelitz, R. B. (2000). Sonic hedgehog signaling pathway in vertebrate epithelial appendage morphogenesis: Perspectives in development and evolution. *Cellular and Molecular Life Sciences, 57*, 1672–1681.

Dyer, J. S., & Rosenfeld, C. R. (2011). Metabolic imprinting by prenatal, perinatal, and postnatal overnutrition: A review. *Seminars in Reproductive Medicine, 29*, 266–276.

Echelard, Y., Epstein, D. J., St-Jacques, B., Shen, L., Mohler, J., McMahon, J. A., et al. (1993). Sonic hedgehog, a member of a family of putative signaling molecules, is implicated in the regulation of CNS polarity. *Cell, 75*, 1417–1430.

Erdoğan, S., & Iwasaki, S. (2014). Function-related morphological characteristics and specialized structures of the avian tongue. *Annals of Anatomy—Anatomischer Anzeiger: Official Organ of the Anatomische Gesellschaft, 196*, 75–87.

Fährmann, W. (1967). Licht- und electronenmikroskopische Untersuchungen an der Geschmacksknospe des neotenen Axolotls (*Siredon mexicanum* Shaw). *Zeitschrift für Mikroskopisch-Anatomische Forschung, 77*, 117–152.

Farbman, A. I. (1965). Electron microscope study of the developing taste bud in rat fungiform papilla. *Developmental Biology, 11*, 110–135.

Farbman, A. I. (1980). Renewal of taste bud cells in rat circumvallate papillae. *Cell and Tissue Kinetics, 13*, 349–357.

Feng, P., Huang, L., & Wang, H. (2014). Taste bud homeostasis in health, disease, and aging. *Chemical Senses, 39*, 3–16.

Finger, T. E. (1997). Evolution of taste and solitary chemoreceptor cell systems. *Brain, Behavior and Evolution, 50*, 234–243.

Fischer, M. E., Cruickshanks, K. J., Schubert, C. R., Pinto, A., Klein, R., Pankratz, N., et al. (2013). Factors related to fungiform papillae density: The beaver dam offspring study. *Chemical Senses, 38*, 669–677.

Gao, Y., Toska, E., Denmon, D., Roberts, S. G., & Medler, K. F. (2014). WT1 regulates the development of the posterior taste field. *Development, 141*, 2271–2278.

Grachtchouk, M., Pero, J., Yang, S. H., Ermilov, A. N., Michael, L. E., Wang, A., et al. (2011). Basal cell carcinomas in mice arise from hair follicle stem cells and multiple epithelial progenitor populations. *The Journal of Clinical Investigation, 121*, 1768–1781.

Hall, J. M., Bell, M. L., & Finger, T. E. (2003). Disruption of sonic hedgehog signaling alters growth and patterning of lingual taste papillae. *Developmental Biology, 255*, 263–277.

Hall, J. M., Hooper, J. E., & Finger, T. E. (1999). Expression of *Sonic hedgehog*, *Patched* and *Gli1* in developing taste papillae of the mouse. *The Journal of Comparative Neurology, 406*, 143–155.

Hamamichi, R., Asano-Miyoshi, M., & Emori, Y. (2006). Taste bud contains both short-lived and long-lived cell populations. *Neuroscience, 141*, 2129–2138.

Harfe, B. D., Scherz, P. J., Nissim, S., Tian, H., McMahon, A. P., & Tabin, C. J. (2004). Evidence for an expansion-based temporal Shh gradient in specifying vertebrate digit identities. *Cell, 118*, 517–528.

Hisha, H., Tanaka, T., Kanno, S., Tokuyama, Y., Komai, Y., Ohe, S., et al. (2013). Establishment of a novel lingual organoid culture system: Generation of organoids having mature keratinized epithelium from adult epithelial stem cells. *Scientific Reports, 3*, 3224.

Hong, J. H., Omur-Ozbek, P., Stanek, B. T., Dietrich, A. M., Duncan, S. E., Lee, Y. W., et al. (2009). Taste and odor abnormalities in cancer patients. *The Journal of Supportive Oncology*, *7*, 58–65.

Howard, J. M., Nuguid, J. M., Ngole, D., & Nguyen, H. (2014). Tcf3 expression marks both stem and progenitor cells in multiple epithelia. *Development*, *141*, 3143–3152.

Hume, W. J., & Potten, C. S. (1976). The ordered columnar structure of mouse filiform papillae. *Journal of Cell Science*, *22*, 149–160.

Iwata, S., Yoshida, R., & Ninomiya, Y. (2014). Taste transductions in taste receptor cells: Basic tastes and moreover. *Current Pharmaceutical Design*, *20*, 2684–2692.

Iwatsuki, K., Liu, H. X., Gronder, A., Singer, M. A., Lane, T. F., Grosschedl, R., et al. (2007). Wnt signaling interacts with Shh to regulate taste papilla development. *Proceedings of the National Academy of Sciences of the United States of America*, *104*, 2253–2258.

Jackowiak, H. (2006). Scanning electron microscopy study of the lingual papillae in the European mole (Talpa europea, L., Talpidae). *Anatomia, Histologia, Embryologia*, *35*, 190–195.

Jiang, P., Josue, J., Li, X., Glaser, D., Li, W., Brand, J. G., et al. (2012). Major taste loss in carnivorous mammals. *Proceedings of the National Academy of Sciences of the United States of America*, *109*, 4956–4961.

Johnston, J. B. (1910). The limit between the ectoderm and entoderm in the mouth, and the origin of taste buds. I. Amphibians. *American Journal of Anatomy*, *10*, 41–67.

Jung, H. S., Oropeza, V., & Thesleff, I. (1999). Shh, Bmp-2, Bmp-4 and Fgf-8 are associated with initiation and patterning of mouse tongue papillae. *Mechanisms of Development*, *81*, 179–182.

Kageyama, R., & Ohtsuka, T. (1999). The Notch-Hes pathway in mammalian neural development. *Cell Research*, *9*, 179–188.

Kageyama, R., Ohtsuka, T., Hatakeyama, J., & Ohsawa, R. (2005). Roles of bHLH genes in neural stem cell differentiation. *Experimental Cell Research*, *306*, 343–348.

Kapsimali, M., & Barlow, L. A. (2013). Developing a sense of taste. *Seminars in Cell and Developmental Biology*, *24*, 200–209.

Kapsimali, M., Kaushik, A. L., Gibon, G., Dirian, L., Ernest, S., & Rosa, F. M. (2011). Fgf signaling controls pharyngeal taste bud formation through miR-200 and Delta-Notch activity. *Development*, *138*, 3473–3484.

Kaufman, M. H. (1999). *The atlas of mouse development*. London: Academic Press.

Kim, M. R., Kusakabe, Y., Miura, H., Shindo, Y., Ninomiya, Y., & Hino, A. (2003). Regional expression patterns of taste receptors and gustducin in the mouse tongue. *Biochemical and Biophysical Research Communications*, *312*, 500–506.

Kito-Shingaki, A., Seta, Y., Toyono, T., Kataoka, S., Kakinoki, Y., Yanagawa, Y., et al. (2014). Expression of GAD67 and Dlx5 in the taste buds of mice genetically lacking Mash1. *Chemical Senses*, *39*, 403–414.

Landacre, F. L. (1907). On the place of origin and method of distribution of taste buds in Ameirus melas. *The Journal of Comparative Neurology*, *17*, 1–66.

Liman, E. R. (2012). Changing senses: Chemosensory signaling and primate evolution. *Advances in Experimental Medicine and Biology*, *739*, 206–217.

Liman, E. R., Zhang, Y. V., & Montell, C. (2014). Peripheral coding of taste. *Neuron*, *81*, 984–1000.

Liu, H. X., Ermilov, A., Grachtchouk, M., Li, L., Gumucio, D. L., Dlugosz, A. A., et al. (2013). Multiple Shh signaling centers, participate in fungiform papilla and taste bud, formation and maintenance. *Developmental Biology*, *382*, 82–97.

Liu, H. X., Grosse, A. S., Iwatsuki, K., Mishina, Y., Gumucio, D. L., & Mistretta, C. M. (2012). Separate and distinctive roles for Wnt5a in tongue, lingual tissue and taste papilla development. *Developmental Biology*, *361*, 39–56.

Liu, H. X., Henson, B. S., Zhou, Y., D'Silva, N. J., & Mistretta, C. M. (2008). Fungiform papilla pattern: EGF regulates inter-papilla lingual epithelium and decreases papilla number by means of PI3K/Akt, MEK/ERK, and p38 MAPK signaling. *Developmental Dynamics*, *237*, 2378–2393.

Liu, H. X., MacCallum, D. K., Edwards, C., Gaffield, W., & Mistretta, C. M. (2004). Sonic hedgehog exerts distinct, stage-specific effects on tongue and taste papilla development. *Developmental Biology*, *276*, 280–300.

Liu, F., Thirumangalathu, S., Gallant, N. M., Yang, S. H., Stoick-Cooper, C. L., Reddy, S. T., et al. (2007). Wnt-beta-catenin signaling initiates taste papilla development. *Nature Genetics*, *39*, 106–112.

Luo, X., Okubo, T., Randell, S., & Hogan, B. L. (2009). Culture of endodermal stem/progenitor cells of the mouse tongue. In Vitro *Cellular & Developmental Biology. Animal*, *45*, 44–54.

Ma, H., Yang, R., Thomas, S. M., & Kinnamon, J. C. (2007). Qualitative and quantitative differences between taste buds of the rat and mouse. *BMC Neuroscience*, *8*, 5.

Matsumoto, I., Ohmoto, M., Narukawa, M., Yoshihara, Y., & Abe, K. (2011). Skn-1a (Pou2f3) specifies taste receptor cell lineage. *Nature Neuroscience*, *14*, 685–687.

Mennella, J. A. (2014). Ontogeny of taste preferences: Basic biology and implications for health. *The American Journal of Clinical Nutrition*, *99*, 704S–711S.

Miller, I. J., & Preslar, A. J. (1975). Spatial distribution of rat fungiform papillae. *The Anatomical Record*, *181*, 670–684.

Miller, I. J., Jr., & Reedy, F. E., Jr. (1990). Variations in human taste bud density and taste intensity perception. *Physiology and Behavior*, *47*, 1213–1219.

Mistretta, C. M. (1972). Topographical and histological study of the developing rat tongue, palate and taste buds. In J. F. Bosma (Ed.), *Third symposium on oral sensation and perception. The mouth of the infant* (pp. 163–187). Springfield, IL: Charles C. Thomas.

Mistretta, C. M., Liu, H. X., Gaffield, W., & MacCallum, D. K. (2003). Cyclopamine and jervine in embryonic rat tongue cultures demonstrate a role for Shh signaling in taste papilla development and patterning: Fungiform papillae double in number and form in novel locations in dorsal lingual epithelium. *Developmental Biology*, *254*, 1–18.

Miura, H., & Barlow, L. A. (2010). Taste bud regeneration and the search for taste progenitor cells. *Archives Italiennes de Biologie*, *148*, 107–118.

Miura, H., Kato, H., Kusakabe, Y., Tagami, M., Miura-Ohnuma, J., Ninomiya, Y., et al. (2004). A strong nerve dependence of sonic hedgehog expression in basal cells in mouse taste bud and an autonomous transcriptional control of genes in differentiated taste cells. *Chemical Senses*, *29*, 823–831.

Miura, H., Kusakabe, Y., & Harada, S. (2006). Cell lineage and differentiation in taste buds. *Archives of Histology and Cytology*, *69*, 209–225.

Miura, H., Kusakabe, Y., Sugiyama, C., Kawamatsu, M., Ninomiya, Y., Motoyama, J., et al. (2001). Shh and Ptc are associated with taste bud maintenance in the adult mouse. *Mechanisms of Development*, *106*, 143–145.

Miura, H., Scott, J. K., Harada, S., & Barlow, L. A. (2014). Sonic hedgehog-expressing basal cells are general post-mitotic precursors of functional taste receptor cells. *Developmental Dynamics*, *243*, 1286–1297.

Nakayama, A., Miura, H., Shindo, Y., Kusakabe, Y., Tomonari, H., & Harada, S. (2008). Expression of the basal cell markers of taste buds in the anterior tongue and soft palate of the mouse embryo. *The Journal of Comparative Neurology*, *509*, 211–224.

Ng, A., Tan, S., Singh, G., Rizk, P., Swathi, Y., Tan, T. Z., et al. (2014). Lgr5 marks stem/progenitor cells in ovary and tubal epithelia. *Nature Cell Biology*, *16*, 745–757.

Nguyen, H. M., & Barlow, L. A. (2010). Differential expression of a BMP4 reporter allele in anterior fungiform versus posterior circumvallate taste buds of mice. *BMC Neuroscience*, *11*, 129.

Nguyen, H. M., Reyland, M. E., & Barlow, L. A. (2012). Mechanisms of taste bud cell loss after head and neck irradiation. *The Journal of Neuroscience*, *32*, 3474–3484.

Northcutt, R. G. (2004). Taste buds: Development and evolution. *Brain, Behavior and Evolution, 64*, 198–206.
Northcutt, R. G. (2005). Taste bud development in the channel catfish. *The Journal of Comparative Neurology, 482*, 1–16.
Northcutt, R. G., Barlow, L. A., Braun, C. B., & Catania, K. C. (2000). Distribution and innnervation of taste buds in the axolotl. *Brain, Behavior and Evolution, 56*, 123–145.
Ohtsuka, T., Ishibashi, M., Gradwohl, G., Nakanishi, S., Guillemot, F., & Kageyama, R. (1999). Hes1 and Hes5 as notch effectors in mammalian neuronal differentiation. *The EMBO Journal, 18*, 2196–2207.
Ohtubo, Y., & Yoshii, K. (2010). Quantitative analysis of taste bud cell numbers in fungiform and soft palate taste buds of mice. *Brain Research, 1367*, 13–21.
Okubo, T., Clark, C., & Hogan, B. L. (2009). Cell lineage mapping of taste bud cells and keratinocytes in the mouse tongue and soft palate. *Stem Cells, 27*, 442–450.
Okubo, T., Pevny, L. H., & Hogan, B. L. (2006). Sox2 is required for development of taste bud sensory cells. *Genes and Development, 20*, 2654–2659.
Ota, M. S., Kaneko, Y., Kondo, K., Ogishima, S., Tanaka, H., Eto, K., et al. (2009). Combined in silico and *in vivo* analyses reveal role of Hes1 in taste cell differentiation. *PLoS Genetics, 5*, e1000443.
Parker, M. A., Bell, M., & Barlow, L. A. (2004). Cell contact-dependent mechanisms specify taste bud number and size during a critical period early in embryonic development. *Developmental Dynamics, 230*, 630–642.
Passilly-Degrace, P., Chevrot, M., Bernard, A., Ancel, D., Martin, C., & Besnard, P. (2014). Is the taste of fat regulated? *Biochimie, 96*, 3–7.
Perea-Martinez, I., Nagai, T., & Chaudhari, N. (2013). Functional cell types in taste buds have distinct longevities. *PLoS One, 8*, e53399.
Petersen, C. I., Jheon, A. H., Mostowfi, P., Charles, C., Ching, S., Thirumangalathu, S., et al. (2011). FGF signaling regulates the number of posterior taste papillae by controlling progenitor field size. *PLoS Genetics, 7*, e1002098.
Pispa, J., & Thesleff, I. (2003). Mechanisms of ectodermal organogenesis. *Developmental Biology, 262*, 195–362.
Plaks, V., Brenot, A., Lawson, D. A., Linnemann, J. R., Van Kappel, E. C., Wong, K. C., et al. (2013). Lgr5-expressing cells are sufficient and necessary for postnatal mammary gland organogenesis. *Cell Reports, 3*, 70–78.
Reiner, D. J., Jan, T. A., Boughter, J. D., Jr., Li, C. X., Lu, L., Williams, R. W., et al. (2008). Genetic analysis of tongue size and taste papillae number and size in recombinant inbred strains of mice. *Chemical Senses, 33*, 693–707.
Rothova, M., Thompson, H., Lickert, H., & Tucker, A. S. (2012). Lineage tracing of the endoderm during oral development. *Developmental Dynamics, 241*, 1183–1191.
Ruo Redda, M. G., & Allis, S. (2006). Radiotherapy-induced taste impairment. *Cancer Treatment Reviews, 32*, 541–547.
Seta, Y., Oda, M., Kataoka, S., Toyono, T., & Toyoshima, K. (2011). Mash1 is required for the differentiation of AADC-positive type III cells in mouse taste buds. *Developmental Dynamics, 240*, 775–784.
Seta, Y., Seta, C., & Barlow, L. A. (2003). Notch-associated gene expression in embryonic and adult taste papillae and taste buds suggests a role in taste cell lineage decisions. *The Journal of Comparative Neurology, 464*, 49–61.
Stewart, J. E., Feinle-Bisset, C., Golding, M., Delahunty, C., Clifton, P. M., & Keast, R. S. (2010). Oral sensitivity to fatty acids, food consumption and BMI in human subjects. *The British Journal of Nutrition, 104*, 145–152.
Stewart, J. E., Feinle-Bisset, C., & Keast, R. S. (2011). Fatty acid detection during food consumption and digestion: Associations with ingestive behavior and obesity. *Progress in Lipid Research, 50*, 225–233.

Takeda, N., Jain, R., Li, D., Li, L., Lu, M. M., & Epstein, J. A. (2013). Identifies progenitor cells capable of taste bud regeneration after injury. *PLoS One, 8*, e66314.

Tanaka, T., Komai, Y., Tokuyama, Y., Yanai, H., Ohe, S., Okazaki, K., et al. (2013). Identification of stem cells that maintain and regenerate lingual keratinized epithelial cells. *Nature Cell Biology, 15*, 511–518.

Thirumangalathu, S., Harlow, D. E., Driskell, A. L., Krimm, R. F., & Barlow, L. A. (2009). Fate mapping of mammalian embryonic taste bud progenitors. *Development, 136*, 1519–1528.

Tizzano, M., Dvoryanchikov, G., Barrows, J. K., Kim, S., Chaudhari, N., & Finger, T. E. (2008). Expression of Galpha14 in sweet-transducing taste cells of the posterior tongue. *BMC Neuroscience, 9*, 110.

Tordoff, M. G., Alarcon, L. K., & Lawler, M. P. (2008). Preferences of 14 rat strains for 17 taste compounds. *Physiology and Behavior, 95*, 308–332.

Tordoff, M. G., Reed, D. R., & Shao, H. (2008). Calcium taste preferences: Genetic analysis and genome screen of C57BL/6J x PWK/PhJ hybrid mice. *Genes, Brain, and Behavior, 7*, 618–628.

Toyoshima, K., Miyamoto, K., & Shimamura, A. (1987). Fine structure of taste buds in the tongue, palatal mucosa and gill arch of the axolotl, *Ambystoma mexicanum*. *Okajimas Folia Anatomica Japonica, 64*, 99–110.

Tucker, R. M., Mattes, R. D., & Running, C. A. (2014). Mechanisms and effects of "fat taste" in humans. *Biofactors, 40*, 313–326.

Vissink, A., Jansma, J., Spijkervet, F. K., Burlage, F. R., & Coppes, R. P. (2003). Oral sequelae of head and neck radiotherapy. *Critical Reviews in Oral Biology and Medicine, 14*, 199–212.

Wasalathanthri, S., Hettiarachchi, P., & Prathapan, S. (2014). Sweet taste sensitivity in pre-diabetics, diabetics and normoglycemic controls: A comparative cross sectional study. *BMC Endocrine Disorders, 14*, 67.

Witt, M., & Reutter, K. (1996). Embryonic and early fetal development of human taste buds: A transmission electron microscopical study. *The Anatomical Record, 246*, 507–523.

Witt, M., & Reutter, K. (1998). Innervation of developing human taste buds. An immunohistochemical study. *Histochemistry and Cell Biology, 109*, 281–291.

Yee, K. K., Li, Y., Redding, K. M., Iwatsuki, K., Margolskee, R. F., & Jiang, P. (2013). Lgr5-EGFP marks taste bud stem/progenitor cells in posterior tongue. *Stem Cells, 31*, 992–1000.

Zhou, Y., Liu, H. X., & Mistretta, C. M. (2006). Bone morphogenetic proteins and noggin: Inhibiting and inducing fungiform taste papilla development. *Developmental Biology, 297*, 198–213.

Zhu, X., Liu, Y., Zhao, P., Dai, Z., Yang, X., Li, Y., et al. (2014). Gpr177-mediated Wnt signaling is required for fungiform placode initiation. *Journal of Dental Research, 93*, 582–588.

CHAPTER THIRTEEN

Signaling in Tooth, Hair, and Mammary Placodes

Youngwook Ahn[1]
Stowers Institute for Medical Research, Kansas City, Missouri, USA
[1]Corresponding author: e-mail address: yoa@stowers.org

Contents

Abstract

Skin appendages develop from placodes involving reciprocal interactions between the surface ectoderm and the underlying mesenchyme during embryogenesis. Despite their distinct shapes and functions, during early development similar morphological changes are observed among different skin appendages. Previous analyses of genetically modified mice have shown that these skin placodes share many aspects of molecular and cellular programs controlled by a relatively small number of signaling pathways during induction, morphogenesis, and transition to bud stage and beyond This chapter focuses on the major signaling pathways that are reiteratively utilized to control the early developmental processes of placodes for teeth, hair follicles, and mammary glands. I update knowledge on the roles played by individual pathways and cross talk among them in these placodes and discuss similarities as well as differences among the skin appendages.

Current Topics in Developmental Biology, Volume 111
ISSN 0070-2153
http://dx.doi.org/10.1016/bs.ctdb.2014.11.013

1. INTRODUCTION

Teeth and other skin appendages such as hair and mammary glands develop from the embryonic ectoderm and underlying mesenchyme. Despite differences in the final structures and functions, these skin appendages arise through highly similar sets of tissue interactions and morphological processes in the early stages of their development (Biggs & Mikkola, 2014; Mikkola, 2007; Mikkola & Millar, 2006). The initial site of appendage development is characterized by the formation of a placode or thickening of the epithelium, mediated by inductive signals exchanged between the ectoderm and the underlying mesenchyme. Subsequent invagination of the placodal epithelium and condensation of the underlying mesenchymal cells leads to bud formation. The position, number, and size of the skin appendages are primarily determined during the stages of placode and bud formation. Extensive genetic analyses of mouse mutants defective in these structures revealed that signaling pathways (BMP, FGF, Shh, and Wnt) are reiteratively used in multiple stages of development of different skin appendages. Hence, disruption of individual signaling pathways often causes similar developmental defects in multiple skin appendages. This underscores the degree to which many aspects of the early development of skin appendages are controlled by common molecular mechanisms. However, significant differences exist among the skin appendages, particularly in the spatial and temporal dynamics of placode development indicating that there may be unique means for modulation of the signaling pathways in the different contexts. In this chapter, I will summarize current understanding of the molecular and cellular mechanisms controlling early development of teeth, hair, and mammary glands and then discuss similarities and differences among the skin appendages.

2. EARLY PATTERNING AND MORPHOGENESIS OF SKIN APPENDAGES

2.1. Teeth

Teeth are found in all vertebrate groups, and despite the diversity in number, size, and shape, the basic steps of tooth development appear to be similar among species (Jernvall & Thesleff, 2012; Jheon, Seidel, Biehs, & Klein, 2013). In contrast to many other mammals, rodents lack canines and possess either a complete absence or a reduced number of premolars. This results in a large toothless region, called the diastema, between incisors and cheek teeth (premolars and molars) (Peterkova, Lesot, & Peterka, 2006; Fig. 1). In mice, despite lack of any premolars, two vestigial buds develop in each jaw quadrant, then undergo degeneration by the time the first molar arises in the more proximal region of the dental lamina (Peterkova et al., 2006; Peterkova, Lesot, Vonesch, Peterka, & Ruch, 1996; Viriot et al., 2000; Fig. 1). This provides an explanation for the apparent elimination of the premolars in rodents. In a number of mouse mutants with perturbed signaling components, some of these vestigial diastema buds survive and go on to form supernumerary teeth, providing an excellent model system to study the signaling

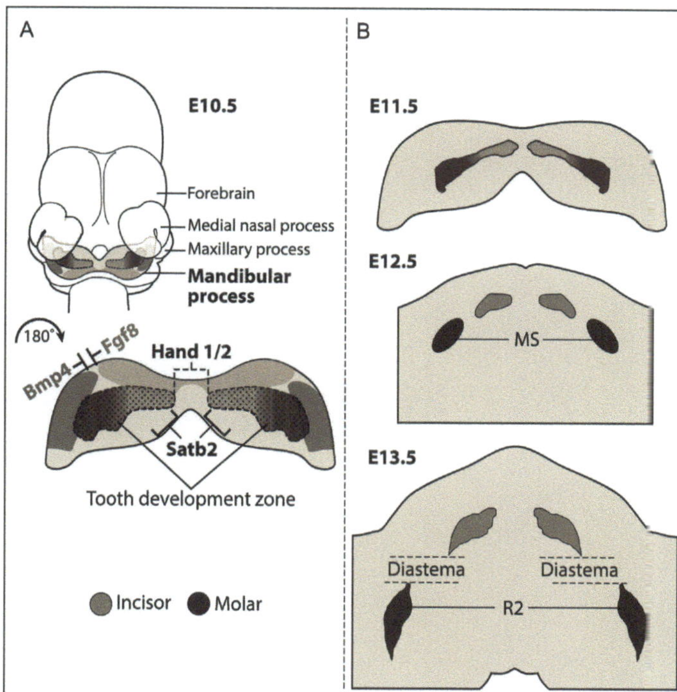

Figure 1 Positioning the sites of tooth formation and their division into the incisor and molar fields. (A) Schematic diagram showing the mandible in the head of an E10.5 embryo. An antagonistic interaction between Bmp4 and Fgf8 is proposed to define the tooth field in the oral ectoderm in each jaw quadrant. Hand1/2 and Satb2 are required for patterning of the distal region. Note that the mandible at the bottom is rotated by 180° for consistency. (B) The tooth field appears to be continuous until E11.5 but subsequently divides into the incisor and molar fields in the distal and proximal regions, respectively. The two fields are separated by a toothless region called diastema. The vestigial tooth buds, MS and R2, form in the molar field but undergo degeneration. (See the color plate.)

mechanisms that control tooth number, size, and growth (Ahn, Sanderson, Klein, & Krumlauf, 2010; Klein et al., 2006; Lan, Jia, & Jiang, 2014; Wang & Fan, 2011; Fig. 2).

To a large degree, much of the current understanding of molecular and cellular mechanisms of tooth development has resulted from genetic analyses in the mouse and tissue recombination experiments (Jussila & Thesleff, 2012; Lan et al., 2014). Teeth develop through sequential and reciprocal interactions between the dental epithelium and the underlying mesenchyme. In mammals, the dental epithelium is derived from oral ectoderm and the mesenchyme from cranial neural crest cells (Chai et al., 2000). During mouse embryogenesis, the first morphological sign of tooth development occurs at embryonic day 11 (E11) when the oral epithelium thickens and becomes the dental lamina. Tooth placodes form within the dental lamina through a further thickening of the epithelium. The placodes

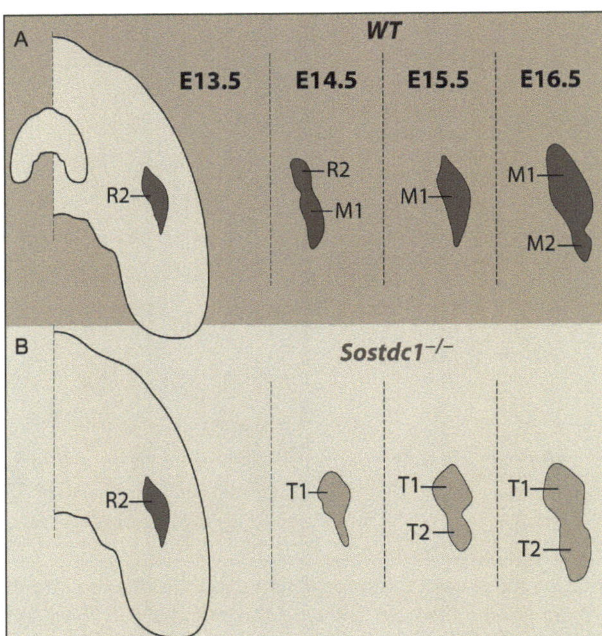

Figure 2 Comparison of the tooth patterns of wild-type and *Sostdc1*-null mice. (A) In the mandible of wild-type (WT) mice, the R2 vestigial degenerates and merges into the first molar (M1), which forms in the more proximal region. The second molar (M2) buds off at the proximal end of M1 around E16.5. Eventually three molars form in each jaw quadrant. (B) In *Sostdc1*-null mice, R2 continues to develop to form the first cheek tooth (T1) leading to formation of four cheek teeth in the mutant. Note that development of the second tooth (T2) is delayed compared to M1.

develop into a bud as the epithelium invaginates, while the dental mesenchymal cells condensate around the invaginating epithelium. During the cap stage (~E14), the dental epithelium invaginates further to surround the condensed mesenchyme, now called the dental papilla. This later gives rise to the dental pulp and dentin-forming odontoblasts. At the surface of the engulfing epithelium, a transient signaling center forms which expresses a number of signaling molecules that control tooth growth and morphogenesis. These signaling centers are called the primary enamel knot at the cap stage, and the secondary enamel knot at the subsequent bell stage. The lack of proliferation in the primary enamel knot and extensive proliferation in the surrounding epithelial cells suggests that the primary enamel knot may have a role in epithelial folding and the bud-to-cap transition (Jernvall & Thesleff, 2000). As tooth development progresses, the epithelial cells adjacent to the dental papilla differentiate into ameloblasts, which deposit enamel alongside dentin.

Bmp4 and Fgf8 are known to be the earliest regulators that specify dental fate in the oral ectoderm. Fgf8 is expressed in the proximal region and required for the development of the proximal mandible and molars (Trumpp, Depew, Rubenstein, Bishop, & Martin, 1999; Tucker, Matthews, & Sharpe, 1998). Before tooth initiation is morphologically apparent, antagonistic interactions between Fgf and Bmp in the oral epithelium play an important role in positioning the sites of tooth formation (Neubuser, Peters, Balling, & Martin, 1997; Vainio, Karavanova, Jowett, & Thesleff, 1993; Fig. 1A). These Fgf–Bmp interactions control expression of *Bmp4*, *Pax9*, *Barx1*, *Msx1*, *Msx2*, *Dlx*, and other genes in the mesenchyme whose combinatorial expression influence tooth type, number, size, and shape (Bei & Maas, 1998; Neubuser et al., 1997; Tucker et al., 1998; Vainio et al., 1993). In addition, *Hand1/2* and *Satb2* are required for patterning of the distal region of the mandible, as loss of these genes results in the fusion or absence of incisors (Barbosa et al., 2007; Britanova et al., 2006; Fig. 1A). During this initiation stage of tooth development, *Shh* and *Wnt* genes are expressed in the oral epithelium and Wnt–Shh interactions are crucial for specification of the dental epithelium (Bitgood & McMahon, 1995; Sarkar et al., 2000; Sarkar & Sharpe, 1999). Once specified, the dental epithelium proliferates to form a horseshoe-like structure called the dental lamina. Classic tissue recombination experiments showed that odontogenic potential initially resides in the dental lamina but shifts to the mesenchyme by E12 (Kollar & Baird, 1970; Lumsden, 1988; Mina & Kollar, 1987). This shift is associated with induction of Bmp4, Fgfs, and many transcription factors in the mesenchyme controlled by the major signaling pathways (Bei, 2009).

Formation of tooth placodes within the dental lamina commences around E11.5 when incisor placodes first appear in the distal region. These placodes contain an epithelial signaling center which is the main source of secreted signaling molecules and their receptors for the major signaling pathways such as Tgf-β/Bmps, Fgfs, Shh, Wnts, and Ectodysplasin-A (Eda). These signaling molecules play a key role in mediating epithelial–mesenchymal interactions and are reiteratively used during multiple steps of tooth development. Early disruption of these signaling pathways results in an arrest at the dental lamina or bud stage (Tummers & Thesleff, 2009). Below, I will summarize the known roles of these signaling pathways in early tooth development.

2.1.1 Tgf-β/Bmp pathway

The *Tgf-β* gene family includes more than 30 members in humans, which are grouped into the *Tgf-β/Activin/Nodal* subfamily and the *Bmp* subfamily. They each initiate activation of signaling by assembling a complex consisting of two type I receptor components and two type II components on the cell membrane (Massague, 2012; Shi & Massague, 2003). During early tooth development, several *Bmp* genes including *Bmp2* and *Bmp4* display dynamic expression patterns in the dental epithelium and mesenchyme (Aberg, Wozney, & Thesleff, 1997; Bitgood & McMahon, 1995). In addition to its role in the determination of tooth-forming sites and tooth types as described above, Bmp signaling is implicated in the odontogenic potential of the mesenchyme and progression from the bud stage to the cap stage via regulation of *Msx1* and *Shh* (Chen, Bei, Woo, Satokata, & Maas, 1996; Zhang et al., 2000). Inactivation of *Bmpr1a* either in the dental epithelium or in the mesenchyme also results in an arrest of tooth development at the bud and cap stages (Andl et al., 2004; Li et al., 2011). The requirement for Bmp signaling in the early tooth development is further supported by supernumerary tooth formation in mice deficient for *Sostdc1* (also known as *Wise*, *ectodin*, and *USAG-1*), which encodes a Wnt/Bmp antagonist, and by an arrest of tooth development at the lamina and early bud stage in mice overexpressing the Bmp antagonist Noggin (Kassai et al., 2005; Murashima-Suginami et al., 2008; Wang et al., 2012).

ActivinβA, another member of the Tgf-β superfamily, is essential for early tooth development. In *ActivinβA* mutant mice incisors and mandibular molars fail to develop beyond the bud stage (Ferguson et al., 1998). In addition, inactivation of *Alk5*, which encodes the Tgf-β type I receptor, in the neural crest cells results in delayed tooth initiation and development (Zhao, Oka, Bringas, Kaartinen, & Chai, 2008).

2.1.2 Fgf pathway

After the initiation stage of tooth development, Fgf signaling plays critical roles in the invagination of the dental epithelium, the bud-to-cap transition, cusp patterning and then differentiation (Li, Prochazka, Goodwin, & Klein, 2014). The function of Fgf signaling at the bud stage was demonstrated by analyses of *Fgfr2* mutant mice. Tooth development failed to progress to the bud stage in concert with reduced epithelial proliferation in *Fgfr2b*-null mice. Furthermore, tooth development was arrested at the early bud stage in mice in which all the *Fgfr2* isoforms were deleted in the dental epithelium (De Moerlooze et al., 2000; Hosokawa et al., 2009; Kettunen et al., 2007). While single mouse mutants for either *Fgf3* or *Fgf10* did not display any early tooth defects, molar development was arrested at the bud stage in *Fgf3;Fgf10* double mutants (Harada et al., 2002; Wang et al., 2007). This indicates a functional overlap between *Fgf3* and *Fgf10* and crucial roles for Fgf signaling in bud morphogenesis and the bud-to-cap transition during tooth development. In line with these roles, inactivation of *Sprouty* genes, which encode negative modulators Fgf signaling, results in formation of supernumerary teeth and ectopic enamel formation (Klein et al., 2008, 2006).

2.1.3 Shh pathway

During the early tooth development, *Shh* expression is restricted to epithelial signaling centers (Bitgood & McMahon, 1995; Dassule & McMahon, 1998). However, expression of downstream target genes including *Ptch1* and *Gli1*, together with tooth defects observed in mice with epithelial deletion of *Smoothened*, which encodes the Shh receptor, collectively suggests that Shh signaling is required both in the dental epithelium and mesenchyme for normal development (Dassule, Lewis, Bei, Maas, & McMahon, 2000; Gritli-Linde et al., 2002; Hardcastle, Mo, Hui, & Sharpe, 1998). Consistent with this idea, conditional inactivation of *Shh* or *Smoothened* in the dental epithelium led to abnormal morphology and asymmetric growth of the tooth bud, suggesting that Shh signaling regulates epithelial invagination, cell proliferation, and buccal–lingual symmetry at the bud and cap stages (Dassule et al., 2000; Gritli-Linde et al., 2002). The lack of tooth development in *Gli2;Gli3* double mutants further supports an early role for Shh signaling in bud morphogenesis (Hardcastle et al., 1998).

2.1.4 *Wnt/β-catenin pathway*

Wnt/β-catenin signaling is dynamically active at all stages of tooth development and plays diverse roles (Liu et al., 2008; Liu & Millar, 2010). Multiple *Wnt* genes are expressed during early tooth development, such as *Wnt4*, *Wnt6*, and *Wnt10b* in the dental epithelium, and these genes are likely to possess overlapping roles in tooth initiation and morphogenesis (Sarkar & Sharpe, 1999). While the source of Wnt ligands and their target cells are still unclear, epithelial inactivation of *Wntless*, which encodes a protein required for sorting and secretion of Wnt ligands, leads to arrest of tooth development at the early cap stage (Zhu et al., 2013). This suggests that Wnt ligands act in an intraepithelial context to regulate the formation of the enamel knot and bud-to-cap transition and that there are additional sources of Wnt ligands.

Transcriptional regulation of target genes of Wnt/β-catenin signaling is mediated by β-catenin and its binding partners, the Tcf/Lef protein family. Loss- and gain-of-function studies of the genes encoding these proteins revealed essential roles for the Wnt signaling pathway during early tooth development. Conditional inactivation of *β-catenin* (formally *Ctnnb1*) in the dental epithelium results in an arrest of tooth development at the bud stage while mesenchymal inactivation of *β-catenin* leads to failure in bud-to-cap transition. This demonstrates roles for Wnt/β-catenin signaling in both the dental epithelium and the mesenchyme at the bud and cap stages of tooth development (Chen, Lan, Baek, Gao, & Jiang, 2009; Liu et al., 2008). Similarly, conditional inactivation of *Lef1* results in arrested tooth development at the bud stage and failure in survival of dental epithelial cells (Kratochwil, Dull, Farinas, Galceran, & Grosschedl, 1996; Sasaki et al., 2005; van Genderen et al., 1994). Conversely, overexpression of *Lef1* in the oral epithelium induces ectopic structures similar to developing teeth (Zhou, Byrne, Jacobs, & Fuchs, 1995).

Dkk1 is a secreted Wnt antagonist and its role in normal tooth development has not been clearly examined due to the severe craniofacial defects and lethality of *Dkk1*-null mice (Mukhopadhyay et al., 2001). Nonetheless, overexpression of *Dkk1* in the dental epithelium results in an arrest of tooth development at the early bud stage, suggesting that Wnt/β-catenin signaling has an earlier role after tooth initiation (Liu et al., 2008).

The strong odontogenic potential of Wnt/β-catenin signaling has been demonstrated by continuous development of ectopic tooth bud-like structures in mice with forced activation of Wnt signaling mediated by expression of a stabilized form of β-catenin in the dental epithelium (Jarvinen et al., 2006; Liu et al., 2008). Ectopic tooth formation was also achieved by inactivation of *Sostdc1*, or of *APC* which encodes a protein required for degradation of β-catenin (Ahn et al., 2010; Kuraguchi et al., 2006; Wang et al., 2009). This provides further support for the crucial role of Wnt/β-catenin signaling in initiation of dental placodes and their progress to the bud and cap stages.

2.1.5 Eda pathway

Ectodysplasin-A (Eda) belongs to the tumor necrosis factor superfamily, and accumulating data suggest that it activates NF-κB signaling through the Eda receptor Edar and the mediator Edardd during skin appendage development (Dickson, Bhakar, & Barker, 2004; Mikkola & Thesleff, 2003; Morlon, Munnich, & Smahi, 2005; Naito et al., 2002; Schmidt-Ullrich et al., 2006). Mice deficient for either *Eda* or *Edar* display no apparent abnormalities in formation of the dental placodes (Kangas, Evans, Thesleff, & Jernvall, 2004; Pispa et al., 1999). However, *Eda* mutant mice exhibit a delay in tooth development with a smaller enamel knot at the cap stage. Furthermore, both *Eda* and *Edar* mutants are characterized by altered number, size, and arrangement of molars (Charles et al., 2009; Kangas et al., 2004; Laurikkala et al., 2001; Peterkova, Lesot, Viriot, & Peterka, 2005). Interestingly, overexpression of *Eda* or *Edar* in the dental epithelium leads to enlargement of tooth placodes and formation of extra tooth buds, suggesting that Eda can promote growth, morphogenesis, and survival of tooth placodes (Kangas et al., 2004; Mustonen et al., 2004, 2003; Tucker, Headon, Courtney, Overbeek, & Sharpe, 2004). It has been proposed that Eda acts downstream of the primary inductive signals for placode initiation and regulates the function of epithelial signaling centers that mediate epithelial–mesenchymal interactions during tooth development (Laurikkala et al., 2001; Mustonen et al., 2004).

2.1.6 Cross talk between the major signaling pathways at the placode and bud stages of tooth development

The dental epithelium is the source of secreted signaling molecules, which control expression of signaling molecules and transcription factors in the underlying mesenchyme (Tummers & Thesleff, 2009). For example, Bmp and Fgf signaling act as epithelial signals that induce expression of *Msx1* and *Pax9* in the mesenchyme (Bei & Maas, 1998; Mandler & Neubuser, 2001; Vainio et al., 1993). Inactivation of either *Msx1* or *Pax9* causes developmental arrest at the bud stage (Peters, Neubuser, Kratochwil, & Balling, 1998; Satokata & Maas, 1994). In return, *Msx1* and *Pax9* are both required for mesenchymal *Bmp4* expression and transgenic *Bmp4* expression driven by an *Msx1* promoter partially rescues tooth development in the *Msx1*-deficient mice (Chen et al., 1996; Peters et al., 1998; Zhao et al., 2000). *Msx1* is also required for expression of *Fgf3* in the dental mesenchyme, and analyses of *Msx1;Msx2* double mutant mice suggest that the two *Msx* genes also have a role in mesenchymal gene expression in the dental lamina (Bei & Maas, 1998). Direct targets of Msx1 include *ink4d*, encoding the cyclin D-dependent kinase inhibitor p19, an important cell cycle regulator (Zhao, Gupta, Raj, Roussel, & Bei, 2013). *Tbx2* is another mesenchymal target of Bmp4 and reduced gene dosage of *Tbx2* partially rescues the bud stage arrest of *Msx1*-deficient mice suggesting a negative feedback mechanism that controls the bud-to-cap transition (Saadi et al., 2013). Together, these data reveal a crucial role for the Bmp4/Msx1 signaling loop in the transition of odontogenic potential from the dental epithelium to the underlying mesenchyme, and this signaling loop was further validated by conditional inactivation of *Bmp4* in the mesenchyme (Jia et al., 2013).

The importance of a Bmp–Wnt feedback loop mediating cross talk between the dental epithelium and mesenchyme was demonstrated by systematic analyses of large-scale gene expression data and mouse genetic experiments during early tooth development (O'Connell et al., 2012). The proposed Bmp–Wnt feedback loop is consistent with previous observations in a number of mouse mutant models. At the early bud stage, Bmp4 and its target Msx1 activate expression of *Lef1*, which is essential for bud formation as tooth development is arrested before the mesenchymal condensation stage in *Lef1*-null mice (Chen et al., 1996; Kratochwil et al., 1996; van Genderen et al., 1994). On the other hand, Bmp4 signaling suppresses expression of *Dkk2* and also *Osr2*, which encodes an inhibitor of tooth development, in dental mesenchyme (Jia et al., 2013). Lef1 controls expression of *Fgf4* in the epithelium signaling center and recombinant FGF proteins can rescue the developmental arrest of tooth development in *Lef1*-null mice. This occurs at least partly through induction of *Fgf3* in the mesenchyme followed by *Shh* in the epithelium, revealing sequential inductive interactions between the epithelium and mesenchyme (Kratochwil, Galceran, Tontsch, Roth, & Grosschedl, 2002). Gain- and loss-of-function studies of *β-catenin* and *Sostdc1* have also identified the components of Fgf and Shh signaling as downstream targets of Wnt/β-catenin signaling in early tooth development (Ahn et al., 2010; Jarvinen et al., 2006; Liu et al., 2008). In addition, Wnt/β-catenin signaling is required for mesenchymal expression of *Bmp4*, which in turn controls *Shh* expression in the dental ectoderm during early incisor development (Fujimori et al., 2010). Further evidence for cross-regulation of *Wnt* genes and *Bmp4* in the developing tooth germs has recently been described (O'Connell et al., 2012).

2.1.7 Control of tooth number, position, and size by integrating signaling activities

Normal patterning and morphogenesis of tooth development requires integration of Bmp, Fgf, Shh, Wnt, and Eda signaling pathways, the inhibition of which leads to an arrest of development at different stages, as described above. Conversely, misregulation, especially ectopic activation, of the same pathways often leads to formation of supernumerary teeth in humans and mice (Cobourne & Sharpe, 2013; Wang & Fan, 2011). For example, forced activation of Wnt/β-catenin signaling leads to continuous development of ectopic teeth as described earlier. In addition, inactivation of *Sp6* (*Epiprofin*), which encodes a transcription factor implicated in modulation of Wnt signaling, results in an excess number of supernumerary teeth in the incisor and molar region (Ibarretxe et al., 2012; Nakamura et al., 2008; Talamillo et al., 2010). *Osr2*-null mice display a rare example for supernumerary teeth developing lingual to the normal position of molars (Zhang, Lan, Chai, & Jiang, 2009). Moreover, it was discovered that Bmp4 negatively regulates expression of *Osr2*, which is required for suppressing expansion of the tooth field mainly by antagonizing the action of Msx1 in the dental mesenchyme (Jia et al., 2013; Zhang, Lan, et al., 2009).

The majority of mouse models for supernumerary teeth represent survival of vestigial tooth buds rather than *de novo* tooth formation Mice have one incisor in each jaw quadrant, but recent gene expression and imaging analyses demonstrated that two successive placodes, marked by *Shh* expression, develop in the prospective incisor region and only the posterior and more recent placode eventually gives rise to a functional incisor (Hovorakova et al., 2011, 2013). The presence of the two successive placodes may be linked to the formation of separate incisors by splitting the large incisor placode as seen with a deficiency of the Sprouty Fgf inhibitor proteins *in vivo*, or inhibition of Bmp signaling in explant culture (Charles et al., 2011; Munne et al., 2010). Rudimentary buds are also observed during later stages of incisor development, and these buds continue to develop to form a supernumerary incisor in mice deficient for the Wnt/Bmp antagonist Sostdc1 or the receptor Lrp4 that interacts with Sostdc1 (Ahn et al., 2010; Murashima-Suginami et al., 2007;

Ohazama et al., 2008). Elevated Wnt and Bmp signaling are causally associated with the incisor phenotype of *Sostdc1*-null mice via *in vivo* genetic rescue and explant culture approaches, respectively (Ahn et al., 2010; Murashima-Suginami et al., 2008). Interestingly, explant culture experiments with wild-type or *Sostdc1*-null incisors suggested that the dental mesenchyme produces additional signaling molecules to limit tooth induction via inhibition of both BMP and Wnt signaling (Munne, Tummers, Jarvinen, Thesleff, & Jernvall, 2009).

Analogous to successive incisor placodes, the placodes for cheek teeth develop sequentially in an anterior to posterior direction in each tooth row, as a posterior extension of the dental epithelium, and degeneration of the two vestigial buds in the diastema leads to the presence of three molars and no premolars (Peterkova et al., 2006; Figs. 1B and 2A). Degeneration of these vestigial buds has been further demonstrated by imaging and lineage tracing experiments (Ahn et al., 2010; Prochazka et al., 2010). Consistent with the proposed role for Wnt/β-catenin, Bmp, Fgf, and Eda signaling in the bud-to-cap transition, elevation of these signaling pathways in the most proximal vestigial bud, namely R2, leads to survival and continuous development to form a supernumerary cheek tooth resulting in four cheek teeth instead of three. For example, mice deficient for *Sostdc1* displayed supernumerary teeth as a result of the survival of the R2 vestigial buds (Ahn et al., 2010; Kassai et al., 2005; Fig. 2B). Genetic interactions of *Sostdc1* with the genes encoding Lrp5/6 Wnt coreceptors indicated that the tooth defects are caused by elevated Wnt/β-catenin signaling (Ahn et al., 2010). Consistent with this observation, mice homozygous for a hypomorphic allele of *Lrp4* displayed similar but somewhat milder defects in molar development compared to *Sostdc1*-null mice (Ohazama et al., 2008). Consistent with the fact that Fgf signaling is the main downstream effector of Wnt signaling during early tooth development, inactivation of *Sprouty* genes, which negatively regulate Fgf signaling during the bud-to-cap transition, also led to survival of the R2 vestigial bud and consequently formation of a supernumerary tooth distal to the first molar (Klein et al., 2006; Peterkova et al., 2009). Lastly, mice overexpressing *Eda* or an activated form of *Edar* also display supernumerary cheek teeth (Mustonen et al., 2003; Tucker et al., 2004).

The exact role for Shh signaling in the survival of the R2 bud and hence in the bud-to-cap transition is confusing. Mice deficient for *growth arrest specific-1* (*Gas1*) displayed supernumerary diastema teeth (Ohazama et al., 2009). This led to a speculation that elevated Shh signaling promotes the bud-to-cap transition and hence survival of the vestigial bud, based on the earlier report that Gas1 acts as an antagonist of Shh signaling during somitogenesis (Lee, Buttitta, & Fan, 2001; Ohazama et al., 2009). However, accumulating evidence supports a role for Gas1 as a facilitator of hedgehog signaling in different tissue contexts (Allen et al., 2011; Allen, Tenzen, & McMahon, 2007; Briscoe & Therond, 2013; Izzi et al., 2011; Martinelli & Fan, 2007). While it is unclear whether elevated Shh signaling in the diastema of the *Gas1* and *Polaris* mutants is causally associated with the tooth phenotype, inhibition of Shh activity *in vivo* using an antibody resulted in similar supernumerary tooth formation supporting a negative role for Shh in the survival of the vestigial tooth bud (Cho et al., 2011). Furthermore, genetic interactions between *Shh* and *Sostdc1*, *Lrp6* and *Ptch1* suggest that Shh normally suppresses survival of the vestigial bud by inhibiting Wnt/β-catenin signaling (Ahn et al., 2010). Supporting this Wnt–Shh negative feedback loop, removal of Kif3a, a motor protein of the primary cilia, from the dental mesenchyme results in an enlarged molar bud and enamel organ associated with loss of Shh signaling and gain of Wnt signaling (Liu, Chen, Cheng, Jing, & Helms, 2014). The formation of supernumerary diastema teeth was also observed in mice lacking the intraflagellar transport protein Polaris, which is a component of primary cilia and believed to be required for Shh signaling (Nozawa, Lin, & Chuang, 2013; Ohazama et al., 2009). While the tooth phenotype of *Polaris* mutants is consistent with the proposed role for Shh in the diastema bud, it is noteworthy to point out that the primary cilium participates in other signaling pathways including Wnt and Fgf signaling (Oh & Katsanis, 2012).

In these mouse models of supernumerary cheek teeth, not only the number but also the relative size and position of teeth were affected. In this regard, a recent study of tooth explant culture provided insights into the mechanisms regulating the relative size and number of molars. The work suggests a model whereby an inhibitory cascade, involving a combination of activators and inhibitors released from an existing tooth germ, determines the timing of initiation and growth of more posterior tooth buds (Kavanagh, Evans, & Jernvall, 2007). This model predicts that continuously developing R2 buds alter the overall pattern of the tooth field. Consistent with this idea, development of the presumptive first molar was delayed leading to repartitioning of the tooth field in the mouse mutants of which R2 buds give rise to supernumerary teeth (Ahn et al., 2010; Kassai et al., 2005; Klein et al., 2006; Ohazama et al., 2008; Fig. 2B). Interestingly, the accelerated growth of the R2 buds in *Sostdc1*-null mice led to an expansion of the entire tooth field into the diastema (Ahn et al., 2010). The relative size of the second, third, and fourth tooth in the mandible appeared to be unchanged in *Sostdc1*-null mice, implying that this Wnt/Bmp antagonist is dispensable for the proposed inhibitory cascade (Ahn et al., 2010; Kavanagh et al., 2007).

2.2. Hair follicles

Similar to teeth, the early stages of hair follicle development proceed via induction of hair fate, placode, bud and peg stage followed by further morphogenesis and differentiation (Schlake, 2007; Schneider, Schmidt-Ullrich, & Paus, 2009; Sennett & Rendl, 2012). During mouse embryogenesis, three consecutive waves of hair follicles develop and each wave gives rise to one or two different hair types (Schlake, 2007). The placodes for the first wave of hair follicles can be visualized at E14 based on the expression of placode markers, such as *β-catenin*, *Dkk4*, *Edar*, and *Lrp4*. These primary hair follicles reach the bud stage by E15.5 and eventually form guard hairs, representing ~1–5% of adult hair coat (Bazzi, Fantauzzo, Richardson, Jahoda, & Christiano, 2007; Fliniaux, Mikkola, Lefebvre, & Thesleff, 2008; Mustonen et al., 2004). The second and third wave of hair follicles are visible around E16 and E18, respectively, giving rise to three different hair types comprising the majority of the hair coat.

Initiation of hair follicle development (~E14) is characterized by reorganization of basal keratinocytes into a placode shortly after the expression of early placode markers and by condensation of underlying mesenchymal cells, the precursor of the dermal papilla (Hardy, 1992; Millar, 2002; Fig. 3). Heterotopic transplant assays, in which dermis and epidermis from different species were recombined at different embryologic stages, suggested that the dermal condensate produces unknown hair-inductive signals required for placode formation (Hardy, 1992). For example, murine E14.5 dermis from back skin induces appendage formation when recombined with nonfeather-forming epidermis from chicken while murine E12.5 dermis lacked the inductive capacity (Dhouailly, 1973). In contrast, epidermis from E14 back skin fails to form hair follicles when recombined with E15 plantar epidermis (Kollar, 1970). The hair-inducing potential of the dermal condensate was further demonstrated by its ability to transform amnion epithelium into skin and hair follicles (Fliniaux, Viallet, Dhouailly, & Jahoda, 2004). Even though the epidermis displayed a limited inductive capacity in these experiments, the placodes produce signals that

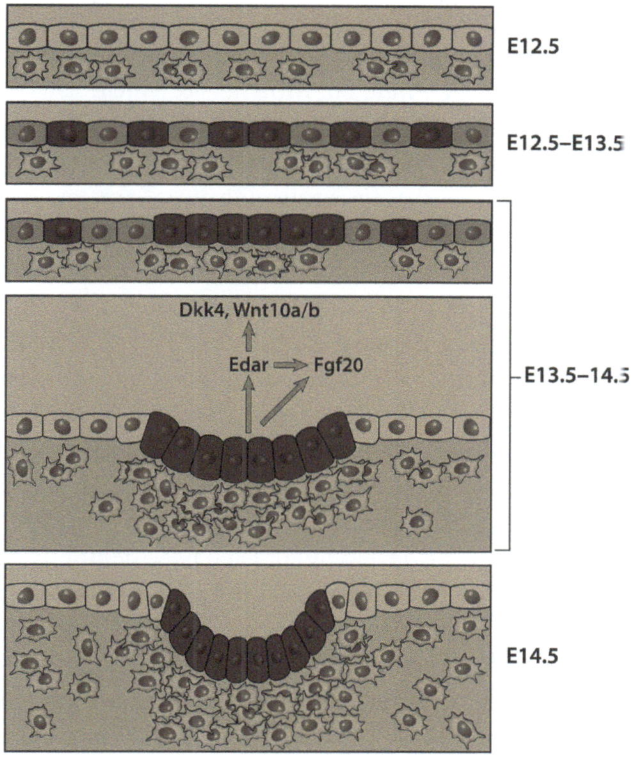

Figure 3 Model for formation of the primary hair follicles. Wnt/β-catenin signaling is irregularly activated in the epithelium between E12.5 and E13.5 and specifies placodal fate. Presumptive placodal cells with high levels of Wnt activity (dark blue; dark gray in the print version) migrate to the site of placode formation while interplacodal cells with low (light blue; light gray in the print version) lose Wnt activity between E13.5 and E14.5. Compaction of the placodal cells contributes to placode morphogenesis, which is accompanied by condensation of dermal fibroblasts underneath the invaginating epithelium.

facilitate dermal condensation and in return signals from the dermal condensate are known to control proliferation and morphogenesis of the placodal epithelium (Hardy, 1992; Millar, 2002; Sennett & Rendl, 2012). Hence, formation of hair follicles requires continuous cross talk between the epidermis and underlying dermal cells and a large part of this cross talk is mediated by the same Bmp, Fgf, Shh, Wnt, and Eda signaling pathways which are essential for tooth development (Biggs & Mikkola, 2014; Sennett & Rendl, 2012).

2.2.1 Wnt/β-catenin pathway

Several *Wnt* and *Frizzled* genes are expressed in developing hair follicles (Reddy et al., 2001; Reddy, Andl, Lu, Morrisey, & Millar, 2004). Reporters for Wnt activity mark the placodal epithelium as well as the underlying mesenchyme as early as E13.5 when the placodes are morphologically indistinguishable from the rest of the epidermis (Zhang, Tomann, et al., 2009). Wnt/β-catenin signaling is first activated throughout the dermis followed by restricted Wnt reporter expression in the emerging placodes (Zhang, Tomann, et al., 2009). Impaired hair follicle induction caused by epidermal inactivation of *Wntless* or *Porcn* suggests that Wnt ligands secreted from the epidermis are important for the initial signaling activity in the dermis (Chen, Jarrell, Guo, Lang, & Atit, 2012; Huang et al., 2012; Liu et al., 2012). A large number of studies have revealed that Wnt/β-catenin signaling is essential for placode induction, dermal condensation, and subsequent morphogenesis and differentiation (Lim & Nusse, 2013). Crucial roles for Wnt/β-catenin signaling in early hair follicle development have been demonstrated by conditional inactivation of *β-catenin* in different cell populations and stages using various *Cre* lines. Mice lacking *β-catenin* in the epidermis from early stages do not form placodes, as stem cells fail to differentiate into follicular keratinocytes (Huelsken, Vogel, Erdmann, Cotsarelis, & Birchmeier, 2001; Zhang, Tomann, et al., 2009). The patterns of Wnt/β-catenin signaling and condensation in the dermis require β-catenin activity in the epithelia (Huelsken et al., 2001; Zhang, Tomann, et al., 2009). Conversely, inactivation of *β-catenin* in the dermis reveals that broad Wnt/β-catenin signaling in dermis is required for fibroblast proliferation and placode induction (Chen et al., 2012). In addition to these early roles, Wnt/β-catenin signaling is required in the dermal condensate for hair follicle formation (Tsai et al., 2014). Impaired development of hair follicles is also observed in mice deficient for *Lef1*, which is required in the dermis for hair growth (Kratochwil et al., 1996; van Genderen et al., 1994).

The inductive role for Wnt/β-catenin signaling in hair follicle development is also revealed by forcing its activation in the epidermis mediated by expression of a stabilized form of β-catenin. These mice display precocious and premature formation of hair follicles and defects in down-growth of follicles and hair shaft production (Gat, DasGupta, Degenstein, & Fuchs, 1998; Narhi et al., 2008; Zhang et al., 2008). This indicates that Wnt/β-catenin signaling promotes the induction of hair follicles at the expense of epidermal fate. Hence, its activity needs to be tightly controlled for proper patterning and morphogenesis of the follicles. Consistent with this role, sustained activation of Wnt/β-catenin signaling in the dermis prior to hair follicle initiation leads to accelerated follicle differentiation and enlargement of both hair placodes and dermal condensates (Chen et al., 2012). This suggests a role for the Wnt signaling in fibroblast proliferation and placode initiation.

2.2.2 Eda pathway

While *Eda* is expressed in the interfollicular epidermis during embryonic hair follicle development, *Edar* is restricted to developing hair placodes. This suggests that Eda released from the interfollicular epidermis regulates the functions of the epithelial signaling centers via its receptor (Headon & Overbeek, 1999; Laurikkala et al., 2002). In mice deficient for *Eda* (*tabby*) and *Edar* (*downless*), placodes for the primary hair follicles initiate but fail to form mature placodes based on the expression patterns of early placode markers (Fliniaux et al., 2008; Laurikkala et al., 2002; Schmidt-Ullrich et al., 2006). This implies that Eda acts through NF-κB signaling downstream of the primary inductive signals to promote further development of hair placodes (Mikkola, 2008; Schmidt-Ullrich et al., 2006). In line with this idea, overexpression of *Eda* leads to increased size and irregular shape of the early hair placodes. However, this does not promote formation of extra hair follicles (Mustonen et al., 2004).

2.2.3 Tgf-β/Bmp pathway

While Wnt/β-catenin and Eda signaling promote induction and morphogenesis of hair follicles, Bmp signaling has an inhibitory role, which needs to be antagonized primarily by Noggin to facilitate placode induction. Expression of *noggin* is detected in dermal condensates together with *Bmp4*, while *Bmpr1a* is expressed in the epidermis (Bitgood & McMahon, 1995; Botchkarev et al., 1999). Mice deficient for *noggin* exhibit a significant delay in hair follicle induction, and embryonic skin organ cultures showed that Noggin neutralizes Bmp4 activity to stimulate hair follicle induction (Botchkarev et al., 1999; Mou, Jackson, Schneider, Overbeek, & Headon, 2006). Analysis of *noggin*-null mice demonstrated that Noggin is necessary for the induction of secondary hair follicles, in addition to its role in advanced stages of development in primary hair follicles (Botchkarev et al., 2002). Conversely, overexpression of *noggin* in the epidermis resulted in thickening of epidermis, increased hair density, and altered hair types (Plikus et al., 2004).

2.2.4 Fgf pathway

Fgf signaling functions downstream of Eda and Wnt/β-catenin signaling to initiate dermal condensation (Huh et al., 2013; Petiot et al., 2003). *Fgfr2b* is initially expressed throughout the epidermis but is downregulated in hair follicles after placode induction (Richardson et al., 2009). Mice deficient for *Fgfr2b* displayed severe epidermal hypoplasia due to a lack of the basal cell proliferation and produced significantly fewer hair follicles in a developmentally delayed manner (Petiot et al., 2003). Individual Fgf ligands showed diverse roles in hair follicle development. *Fgf10*-null mice displayed epidermal hypoplasia and reduced whisker follicles but normal hair follicle development (Ohuchi et al., 2003; Petiot et al., 2003). In contrast, overexpression of *Fgf7* in the epidermis results in *in vivo* suppression of hair follicle induction (Guo, Yu, & Fuchs, 1993). Treatment of embryonic skin organ cultures with FGF7 (KGF) inhibits hair follicle formation and promotes epidermal differentiation (Richardson et al., 2009). *Fgf20* is expressed in hair placodes and is required for formation of primary and secondary dermal condensations but is dispensable for placode formation (Huh et al., 2013). In *Fgf20*-null mice, persistent expression of early placode markers suggests that epithelial patterning is partially independent of underlying mesenchymal patterning during the initial stages of hair follicle development (Huh et al., 2013).

2.2.5 Shh pathway

Shh is expressed in more committed hair placodes, suggesting that it acts after the primary inductive signals (Bitgood & McMahon, 1995; Huh et al., 2013). Analyses of *Shh*- and *Gli2*-null mice demonstrate that Shh signaling is dispensable for placode induction but is required for subsequent stages of hair follicle development, partly by promoting cell proliferation in the epidermis and dermal condensate (Chiang et al., 1999; Mill et al., 2003; St-Jacques et al., 1998). The cell proliferative effects of Shh are mediated by Gli2, which activates expression of cyclin D1 and cyclin D2 in the developing hair follicles (Mill et al., 2003). Experiments involving rescue of the *Gli2*-null skin defects by transgenic *Gli2* expression underscore the importance of Shh signaling within the epidermis (Mill et al., 2003). In contrast, recent analyses of mice with conditional inactivation of *Smoothened* in the dermis show that reception of Shh signaling in the dermal condensate is also essential for dermal papilla formation and hair growth (Woo, Zhen, & Oro, 2012).

2.2.6 *Cross talk between the major signaling pathways during hair placode induction and follicle development*

Loss of hair follicles in mice with epidermal inactivation of *Wntless* suggests that epidermally expressed Wnt ligands have an early role in patterning the underlying dermal fibroblasts, partly by activating expression of *Wnt11*, *Twist2*, *Fgfr1*, *Lef1*, and *Bmp4* (Chen et al., 2012; Huang et al., 2012). In addition, early epidermal Wnt/β-catenin signaling is required for expression of *Dkk4*, *Eda*, *Edar*, and *Wnt10b* in the epidermis (Bazzi et al., 2007; Laurikkala et al., 2002; Zhang, Tomann, et al., 2009). Fgf20 functions downstream of Eda and Wnt/β-catenin signaling as it is required for formation of the dermal condensate induced by forced activation of these two signaling pathways (Huh et al., 2013). Analyses with reporter mice for β-catenin or NF-κB activity demonstrate that Wnt/β-catenin signaling activity precedes Edar activation and that Wnt/β-catenin signaling can be activated in the absence of Eda signaling (Zhang, Tomann, et al., 2009). Furthermore, forced activation of Eda signaling or Noggin treatment fails to rescue the loss of hair follicle formation caused by impaired Wnt/β-catenin signaling in the epidermis (Zhang, Tomann, et al., 2009).

Together with the precocious hair formation mediated by ectopic activation of Wnt/β-catenin signaling, these data place Wnt/β-catenin signaling at the top of the signaling hierarchy that leads to formation of hair follicles (Gat et al., 1998; Narhi et al., 2008; Zhang et al., 2008). After initiation of hair placodes, Eda signaling is required for maintenance of Wnt/β-catenin signaling activity and expression of *Dkk4*, *Lrp4*, *Wnt10a*, and *Wnt10b* (Fliniaux et al., 2008; Zhang, Tomann, et al., 2009). Eda signaling also functions by suppressing BMP activity, possibly through transcriptional activation of Bmp inhibitors (Pummila et al., 2007). Wnt/β-catenin and Eda signaling are required for *Shh* expression in hair placodes while Fgf20 is dispensable (Huelsken et al., 2001; Huh et al., 2013; Pummila et al., 2007). Rescue *Smoothened* mutant mice suggest that Shh promotes dermal papilla formation via regulation of the Bmp inhibitor Noggin (Woo et al., 2012). Bmp signaling suppresses formation of hair placodes in part by downregulating Lef1, the cell-adhesion molecule NCAM and Edar. This suppressive activity is antagonized by Noggin to facilitate hair follicle formation (Botchkarev et al., 1999; Mou et al., 2006). Finally, Wnt/β-catenin signaling and Noggin act in concert to downregulate E-cadherin via Lef1 and promote hair follicle morphogenesis (Jamora, DasGupta, Kocieniewski, & Fuchs, 2003).

2.2.7 Molecular and cellular mechanisms controlling placode initiation and spacing and number of hair follicles

Until recently, the cellular mechanisms that drive the epithelial thickening and invagination of hair placodes have been elusive. The extent of contributions to these processes by cell proliferation, cell shape change, and cell movement has been disputed. However, a recent study utilizing live confocal imaging of embryonic skin whole-mount explants clarified some of the issues (Ahtiainen et al., 2014). Primary hair placodes form through directional cell migration and cell compaction and suppression of cell proliferation is linked to placodal initiation. Interestingly, stimulation of Eda and Wnt/β-catenin pathways, the two major inductive signals for hair follicle formation, leads to reduced proliferation, increased cell movement, and reduced directionality. These findings suggest that cell fate decisions and cell behaviors are coupled during morphogenesis of hair placodes, presumably under the joint control of a common molecular mechanism (Fig. 3).

Since activation of Wnt/β-catenin signaling is tightly linked to induction of hair follicles, it needs to be efficiently blocked in the hairless spaces between developing hair follicles. The Dkk-secreted Wnt antagonists are expressed in hair placodes or dermal condensates and are believed to be involved in patterned activation of Wnt/β-catenin signaling in the epidermis (Monaghan et al., 1999; Sick, Reinker, Timmer, & Schlake, 2006). However, genetic evidence for this role is not yet available. The inhibitory potential of Dkk proteins in hair follicle development can be seen in mice overexpressing *Dkk1* and *Dkk2*, in which placode induction is severely impaired (Andl, Reddy, Gaddapara, & Millar, 2002; Sick et al., 2006). It was hypothesized that Wnt and Dkk proteins control spacing of hair follicles through a reaction–diffusion (Turing) mechanism (Kondo & Miura, 2010; Schlake & Sick, 2007; Sick et al., 2006). In this model, the theoretically higher mobility of smaller Dkk proteins compared to Wnt ligands generates a distinct and stable pattern of activator and inhibitor distribution, leading to patterned morphogenesis of hair placodes (Schlake & Sick, 2007; Sick et al., 2006). In addition to the *Dkks*, *Sostdc1* is highly expressed in the epithelial and mesenchymal cells surrounding hair placodes and its inactivation results in formation of supernumerary vibrissal follicles and enlargement of primary hair placodes (Ahn, Sims, Logue, Weatherbee, & Krumlauf, 2013; Laurikkala, Kassai, Pakkasjarvi, Thesleff, & Itoh, 2003; Narhi et al., 2012). Overexpression of *Sostdc1* leads to severe loss of hair and vibrissal follicles, suggesting that Sostdc1 acts primarily as a Wnt antagonist in this context (Ahn et al., 2013). In addition to the Wnt/Dkk-based model, a recent skin explant culture experiment suggested Edar–Bmp activation–inhibition involvement, in which regulation of Eda signaling by Bmps and Bmp inhibitors determines patterning of primary hair follicles (Mou et al. 2006).

2.3. Mammary glands

In mice, mammary placodes develop along a bilateral structure called a mammary line, which forms in the ventral body between the forelimb and hindlimb buds (Cowin & Wysolmerski, 2010; Propper, Howard, & Veltmaat, 2013; Robinson, 2007; Watson & Khaled, 2008; Fig. 4). While the mammary lines were characterized as stretched elevated ridges on the surface ectoderm in rabbits, murine mammary lines do not appear as a prominent ectodermal ridge but as a thickened epithelium expressing molecular markers such *Wnt10b* and *TopGal* (Chu et al., 2004; Propper, 1978; Veltmaat, Van Veelen, Thiery, & Bellusci, 2004). Analysis of *Wnt10b* expression reveals that three streaks of *Wnt10b* activity initially arise in the axilla, inguen, and the region between forelimb and hindlimb buds. These subsequently become fused into one continuous domain to form a mammary line on each side (Propper et al., 2013; Veltmaat et al., 2004). Starting around E11.5 of mouse embryogenesis, within a day, five mammary placodes sequentially develop along each mammary line in a defined temporal and spatial order while the mammary lines themselves progressively

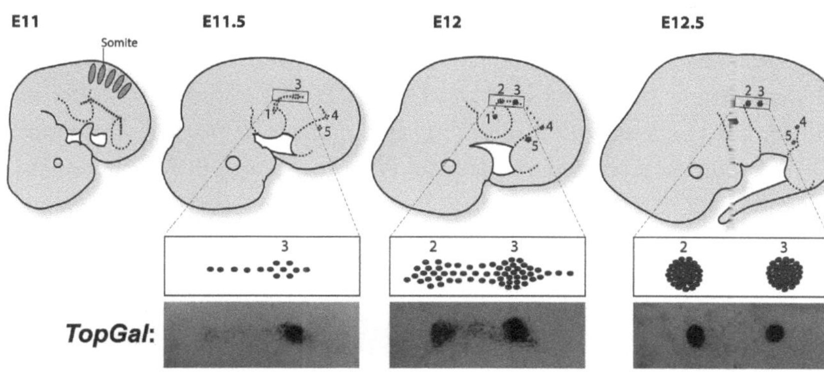

Figure 4 Sequential formation of mammary placodes along the mammary line during mouse embryogenesis. Diagram of an E11 embryo shows the position of the mammary line (dashed blue line) in the axilla, inguen, and the region between forelimb and hindlimb buds. Starting around E11.5, five mammary placodes sequentially develop along the mammary line in a defined temporal and spatial order while the mammary lines themselves progressively disappear. The blowups demonstrate temporal changes in distribution of mammary epithelial cells around placodes #2 and #3 marked by expression of a Wnt reporter, *TopGal* (bottom). (See the color plate.)

disappear (Fig. 4). Mammary placodes grow and invaginate into the underlying mesenchyme, while the mesenchymal cells condense in the region surrounding the epithelium to form mammary buds by E14. Mammary buds remain mostly quiescent until E16 when the buds then begin to sprout and branch into underlying fat pads in female mice.

2.3.1 Signaling pathways important for mammary placode development

Similar to other skin appendages, early mammary gland development requires reciprocal signaling between the mammary epithelium and the underlying mesenchyme (Cowin & Wysolmerski, 2010; Robinson, 2007). Analyses of various mouse mutants indicate that each pair of mammary placodes requires different sets of signals for proper development. This most likely reflects differences in their position along the body axis and timing of initiation.

It has been proposed that Fgf10 generated from somites acts on the Fgfr2b receptor in the surface ectoderm above the somites to determine the dorsoventral position of mammary lines (Mailleux et al., 2002; Veltmaat et al., 2006). It is unclear whether this action by Fgf10 involves a shift in the *Fgf10* expression pattern or delamination/migration of the *Fgf10*-positive mesodermal cells. With the exception of placode #4, both *Fgf10*- and *Fgfr2b*-null mice fail to develop mammary placodes (Mailleux et al., 2002; Veltmaat et al., 2006). A role for Fgf10 produced by somites is supported by early mammary defects in mouse mutants in which the Fgf10 gradient is perturbed (Veltmaat et al., 2006). Although *Gli3* is not expressed in the surface ectoderm until the time of placode formation, *Gli3*-null mice lack mammary placodes #3 and #5 and display hypoplasia of placodes #2 and #4 in association with reduced *Fgf10* expression in somites. This places hedgehog signaling upstream of *Fgf10* in the proposed model for specification of mammary lines (Hatsell & Cowin, 2006; Lee et al., 2011; Veltmaat et al., 2006).

Organ culture experiments suggest that Fgf signaling is required for *Tbx3* expression during mammary placode initiation (Eblaghie et al., 2004), and mice deficient for *Tbx3* display a failure in mammary placode development, mimicking ulnar-mammary syndrome in humans (Bamshad et al., 1997; Davenport, Jerome-Majewska, & Papaioannou, 2003; Jerome-Majewska et al., 2005). The relatively early expression of *Tbx3* and loss of *Wnt10b* and *Lef1* expression in *Tbx3* mutants suggest that *Tbx3* acts upstream of Wnt signaling during early mammary development (Davenport et al.,

2003; Eblaghie et al., 2004). Organ culture experiments involving overexpression of *Tbx3* and *Bmp4* or Noggin treatment led to a hypothesis that mutual interactions between Bmp4 and Tbx3 control expression of *Lef1* and determine the presumptive DV boundary where mammary lines form (Cho et al., 2006). Consistent with the early role for Gli3 as an upstream regulator of *Fgf10* in somites, the expression pattern of *Tbx3* and *Bmp4* is disrupted in *Gli3*-null mice (Chandramouli, Hatsell, Pinderhughes, Koetz, & Cowin, 2013).

As described above, the Wnt/β-catenin pathway appears to be a downstream target of early signaling activities that specify mammary fate and determine the position of mammary lines. Inhibition of initial Wnt/β-catenin signaling by overexpression of *Dkk1* or *Sostdc1* blocks placode formation, while inactivation of *β-catenin* after placode formation leads to hypoplasia of mammary buds (Ahn et al., 2013; Chu et al., 2004). This indicates that Wnt/β-catenin signaling is required for both placode initiation and subsequent growth. Conversely, forced activation of Wnt/β-catenin pathway results in accelerated formation and expansion of mammary placodes (Chu et al., 2004). Furthermore, mice deficient for *Sostdc1* or *Lrp4* display a series of defects in placode patterning, including formation of supernumerary or ectopic placodes (Ahn et al., 2013; Narhi et al., 2012). Rescue of these defects by reducing the gene dosage of *β-catenin* or the Wnt coreceptor genes, *Lrp5* and *Lrp6*, suggests that Sostdc1 and Lrp4 primarily act by antagonizing the Wnt/β-catenin pathway during mammary placode formation (Ahn et al., 2013). Among the Tcf/Lef family of proteins which bind to β-catenin and regulate Wnt target genes, *Lef1* is the member expressed in developing placodes. It is required for placode initiation and subsequent development, as only a subset of placodes form and they eventually degenerate in *Lef1*-null mice (Boras-Granic, Chang, Grosschedl, & Hamel, 2006; van Genderen et al., 1994). This result suggests that other Tcf/Lef family members may play a partly redundant role in mammary specification and placode formation. In addition, mice deficient for *Lrp5* and *Lrp6* display defects in placode development consistent with the critical role of Wnt/β-catenin signaling in early mammary development (Lindvall et al., 2006, 2009).

In contrast to the proposed role in somites, hedgehog signaling was shown to be downregulated in mammary placodes and forced activation leads to a loss of placodes #3 and #5 (Hatsell & Cowin, 2006). This implies that inhibition of hedgehog signaling is necessary for normal placode development. Interestingly, conditional ablation of hedgehog signaling in the epidermis results in ectopic development of mammary gland-like structures in the ventral skin (Gritli-Linde et al., 2007). This further supports the inhibitory role played by hedgehog signaling in specification of mammary fate. Recently, analyses of mice deficient for *Gli3* or possessing a constitutively active Gli1 suggest that continuous repression of hedgehog signaling by Gli3 is necessary not only for early placode patterning but also for later mammary mesenchyme specification and epithelial morphogenesis and proliferation (Chandramouli et al., 2013; Lee et al., 2011).

Edar is expressed during mammary placode development, but neither *Eda*- nor *Edar*-null mice display any defects in the number or position of mammary glands, with the exception that *Eda*-null mice possess nipples with a rudimentary and flat appearance (Mustonen et al., 2003; Pispa, Mikkola, Mustonen, & Thesleff, 2003). This raises the possibility of functional compensation or redundancy. Overexpression of *Eda* in the epidermis results in supernumerary placodes/nipples along the mammary lines, demonstrating its ability to promote placode formation (Mustonen et al., 2004, 2003). Neuregulin 3 (Nrg3), a member of the EGF family, has been proposed to function downstream of Fgf and Tbx3 and upstream of Wnt/β-catenin signaling to promote specification and morphogenesis of mammary placodes (Howard, Panchal, McCarthy, & Ashworth, 2005). Consistent with this idea, mice deficient for *Nrg3* or overexpressing *Nrg3* display altered number and position of mammary glands (Howard et al., 2005; Panchal, Wansbury, Parry, Ashworth, & Howard, 2007).

2.3.2 Molecular and cellular mechanisms controlling mammary placode formation

Development of mammary placodes provides a unique model system to study spatiotemporal patterning of ectodermal organs since five mammary placodes form along the mammary line at relatively fixed positions in a defined order. The molecular and cellular basis for this transition from a stretch of cells expressing the same markers of mammary fate into precisely spaced placodes is still unclear. Morphological studies in rabbit embryos reported spindle-shaped cells on top of the mammary ridge, leading to speculation that these cells represent migrating mammary epithelial cells (Propper, 1978). More recently, BrdU-based cell-tracing experiments in mouse embryos indicate that the formation and growth of mammary placodes involves influx of mammary epithelial cells into growing placodes (Lee et al., 2011). Mammary epithelial cells proliferate at much lower rates compared to neighboring epithelial cells, and very limited cell death occurs along the mammary lines, providing further support for the directional cell migration model (Ahn et al., 2013; Balinsky, 1950; Lee et al., 2011). However, the degree to which cell migration may contribute to placode initiation and/or removal of mammary epithelial cells in the interplacodal space is unclear, as it is unknown how far these cells migrate along the mammary line (Lee et al., 2011). Together with improvements in *in vivo* imaging technology and culture conditions, mouse mutants displaying abnormal patterning of mammary placodes should provide an approach to resolve this issue (Cowin & Wysolmerski, 2010; Veltmaat, Ramsdell, & Sterneck, 2013).

Recent studies of mammary development in *Lrp4* and *Sostdc1* mutant mice provide insights into the mechanisms driving mammary placode initiation (Ahn et al., 2013; Narhi et al., 2012).

First, placode initiation requires assembly or aggregation of epithelial precursor cells. In *Lrp4* mutants, even when a comparable number of these cells are present around the site of placode formation, based on expression of a Wnt reporter, they are loosely assembled with a smaller degree of invagination compared to control mice (Ahn et al., 2013). This delay is rescued by reducing Wnt/β-catenin signaling activity, suggesting that Lrp4 normally facilitates aggregation of the precursor cells by temporarily inhibiting the Wnt/β-catenin pathway (Ahn et al., 2013). Modulation of cell-adhesion molecules has been implicated in morphogenesis of skin appendages, but its role in mammary placode development is currently unknown (Jamora et al., 2003). Second, maintenance of spacing between placodes as well as morphogenesis of individual placodes requires tight control over the number of the precursor cells. The significant increase in the number of Wnt reporter-positive cells along the mammary lines in *Lrp4* and *Sostdc1* mutants is linked to reduced distance and fusion between placodes #2 and #3 (Ahn et al., 2013; Narhi et al., 2012). The presence of Wnt reporter-positive cells in the interplacodal space and subsequent formation of ectopic placodes suggests that suppression of Wnt/β-catenin signaling is necessary to generate placode-free regions along the mammary lines. It is unclear whether this is achieved by preventing the cells in the interplacodal region from maintaining or adopting mammary fate or by facilitating migration of these cells toward placodes.

2.4. Common molecular and cellular mechanisms involved in formation of skin appendage placodes

Genetic manipulations that lead to loss or activation of the Wnt/β-catenin pathway have demonstrated that Wnt/β-catenin signaling activity is not only required but also sufficient for formation of skin appendage placodes (Biggs & Mikkola, 2014; Mikkola, 2007). It is likely that Wnt/β-catenin signaling is also required for specification of the dental lamina and mammary line and induction of hair fate based on observed defects in mice with conditional inactivation of *β-catenin* and overexpression of *Dkk1* (Andl et al., 2002; Chu et al., 2004; Liu et al., 2008; Zhang, Tomann, et al., 2009). Furthermore, the Wnt/β-catenin pathway is critical for growth and morphogenesis of established placodes and their progress to the next stages of development in the skin appendages (Ahn et al., 2013; Liu et al., 2008; van Genderen et al., 1994). Wnt/β-catenin signaling is required in the mesenchyme as well as the epithelium during early hair and tooth development, but it is unknown whether Wnt/β-catenin signaling in the mesenchyme plays a critical role in formation of mammary placodes (Chen et al., 2009, 2012; Tsai et al., 2014).

The Fgf pathway is important for initially specifying the dental lamina via its antagonistic interaction with the Bmp pathway, and later it is required for bud morphogenesis, epithelial proliferation, and the bud-to-cap transition (Li et al., 2014). Fgf signaling is believed to act in the somites to position and specify the mammary lines on the surface ectoderm and furthermore is required for formation of mammary placodes (Cowin & Wysolmerski, 2010). An early role for the Fgf pathway in specification of hair fate remains unclear partly due to differential effects of individual Fgf ligands and their potential redundancy (Sennett & Rendl, 2012). Analyses of mice deficient for *Fgfr2b* and *Fgf20* reveals that Fgf signaling is required for hair placode formation partly by controlling cell proliferation and patterning of dermal condensates (Huh et al., 2013; Petiot et al., 2003).

The Eda pathway appears to act downstream of the primary inductive signals, especially the Wnt/β-catenin pathway, to maintain the established fate of tooth and hair placodes, and promote further morphogenesis. Consequently, loss-of-function defects are generally milder than those of mice deficient for the primary inductive signals (Mikkola, 2008). A role for Eda signaling in mammary development needs further clarification due to the lack of apparent phenotypes in loss-of-function mutants.

While Bmp signaling is required for determining the tooth field and for the bud-to-cap transition in tooth development, inhibition of Bmp signaling activity is necessary for induction of hair placodes (Jussila & Thesleff, 2012; Sennett & Rendl, 2012). During the formation of the mammary lines, Bmp signaling is also implicated in inhibition of mammary fate (Cho et al., 2006).

Shh signaling is essential for epithelial morphogenesis and proliferation and for patterning of underlying mesenchyme during early development of the teeth and hair follicles (Jussila & Thesleff, 2012; Sennett & Rendl, 2012). In contrast, Shh signaling needs to be suppressed for proper mammary placode formation, highlighting differences between mammary glands and other skin appendages (Hatsell & Cowin, 2006). The formation of mammary gland-like structures in mice with epidermal ablation of Shh signaling suggests that one of the roles for Shh is to suppress mammary fate in the surface ectoderm, as mammary placode development precedes the appearance of hair placodes (Gritli-Linde et al., 2007).

Compared to the recent advances in our understanding of signaling pathways controlling early development of the skin appendages, relatively little is known about the cellular mechanisms that drive the complicated processes of placode initiation and subsequent morphogenesis. Changes in cell shape, cell proliferation, cell adhesion, and migration have been implicated in these processes and will be an important focus of future research. In contrast to limited proliferation in mammary epithelial cells, active cell proliferation in the epithelium has been coupled to bud morphogenesis during teeth and hair follicle development (Gritli-Linde et al., 2002; Hosokawa et al., 2009; Mill et al., 2003). While contributions of cell proliferation to placode initiation in the dental epithelium remains to be resolved, cell movement is known to be a major force driving hair placode initiation in a manner analogous to its role in initiation and growth of mammary placodes (Ahn et al., 2013; Ahtiainen et al., 2014; Lee et al., 2011).

3. PERSPECTIVES

The past several decades of classical transplant and mouse genetic experiments have provided significant insight and uncovered a small number of signaling pathways that control development of placodes for skin appendages. These data indicate that skin appendages utilize the same major signaling pathways during placode induction and morphogenesis. Consequently, disruptions of signaling activity often result in similar developmental defects in different appendages. However, we know relatively little about the molecular signatures that drive a developmental program toward tooth, hair, or mammary fate at the expense of epidermal differentiation. Therefore, we need to identify, for example, the mesenchymal signals responsible for odontogenic potential and the inhibitory signals that suppress initiation of a new tooth bud. In hair follicle development, the mesenchymal signals that facilitate formation of placodes are still unclear. Lastly, the mechanisms that specify mammary fate and determine the position of placodes along the mammary line are poorly understood. Uncovering these essential molecular signatures is particularly important in the era of stem cells since reprogramming pluripotent stem cells or somatic cells into a certain cell type requires a thorough understanding of molecular and epigenetic determinants of endogenous cell fate (Hanna, Saha, & Jaenisch, 2010; Hussein & Nagy, 2012). This can be achieved by genomewide chromatin immunoprecipitation analyses looking for binding of transcription factors and epigenetic modifications and expression profiling of epithelial and mesenchymal cells at different stages of placode development. Improvement in tissue dissection and sorting technologies should also be beneficial for this goal. An interesting question is whether the molecular mechanisms governing placode development are similarly used in later embryonic development and postnatal development such as hair cycle and mammary branching morphogenesis or pathological conditions such as metastasis of breast cancer cells. Development of new mouse models for cell type-specific and/or inducible loss- and gain-of-function analyses would facilitate future research on potential later roles for the signaling pathways, which are essential for early development of the skin appendages.

ACKNOWLEDGMENTS

I would like to thank Robb Krumlauf for critical reading of this manuscript and valuable suggestions. I also thank Mark Miller for illustration. This work is supported by funds from the Stowers Institute.

REFERENCES

Aberg, T., Wozney, J., & Thesleff, I. (1997). Expression patterns of bone morphogenetic proteins (Bmps) in the developing mouse tooth suggest roles in morphogenesis and cell differentiation. *Developmental Dynamics*, *210*(4), 383–396.

Ahn, Y., Sanderson, B. W., Klein, O. D., & Krumlauf, R. (2010). Inhibition of Wnt signaling by Wise (Sostdc1) and negative feedback from Shh controls tooth number and patterning. *Development*, *137*(19), 3221–3231.

Ahn, Y., Sims, C., Logue, J. M., Weatherbee, S. D., & Krumlauf, R. (2013). Lrp4 and Wise interplay controls the formation and patterning of mammary and other skin appendage placodes by modulating Wnt signaling. *Development*, *140*(3), 583–593.

Ahtiainen, L., Lefebvre, S., Lindfors, P. H., Renvoise, E., Shirokova, V., Vartiainen, M. K., et al. (2014). Directional cell migration, but not proliferation, drives hair placode morphogenesis. *Developmental Cell*, *28*(5), 588–602.

Allen, B. L., Song, J. Y., Izzi, L., Althaus, I. W., Kang, J. S., Charron, F., et al. (2011). Overlapping roles and collective requirement for the coreceptors GAS1, CDO, and BOC in SHH pathway function. *Developmental Cell, 20*(6), 775–787.

Allen, B. L., Tenzen, T., & McMahon, A. P. (2007). The Hedgehog-binding proteins Gas1 and Cdo cooperate to positively regulate Shh signaling during mouse development. *Genes & Development, 21*(10), 1244–1257.

Andl, T., Ahn, K., Kairo, A., Chu, E. Y., Wine-Lee, L., Reddy, S. T., et al. (2004). Epithelial Bmpr1a regulates differentiation and proliferation in postnatal hair follicles and is essential for tooth development. *Development, 131*(10), 2257–2268.

Andl, T., Reddy, S. T., Gaddapara, T., & Millar, S. E. (2002). WNT signals are required for the initiation of hair follicle development. *Developmental Cell, 2*(5), 643–653.

Balinsky, B. I. (1950). On the prenatal growth of the mammary gland rudiment in the mouse. *Journal of Anatomy, 84*(3), 227–235.

Bamshad, M., Lin, R. C., Law, D. J., Watkins, W. C., Krakowiak, P. A., Moore, M. E., et al. (1997). Mutations in human TBX3 alter limb, apocrine and genital development in ulnar-mammary syndrome. *Nature Genetics, 16*(3), 311–315.

Barbosa, A. C., Funato, N., Chapman, S., McKee, M. D., Richardson, J. A., Olson, E. N., et al. (2007). Hand transcription factors cooperatively regulate development of the distal midline mesenchyme. *Developmental Biology, 310*(1), 154–168.

Bazzi, H., Fantauzzo, K. A., Richardson, G. D., Jahoda, C. A., & Christiano, A. M. (2007). The Wnt inhibitor, Dickkopf 4, is induced by canonical Wnt signaling during ectodermal appendage morphogenesis. *Developmental Biology, 305*(2), 498–507.

Bei, M. (2009). Molecular genetics of tooth development. *Current Opinion in Genetics & Development, 19*(5), 504–510.

Bei, M., & Maas, R. (1998). FGFs and BMP4 induce both Msx1-independent and Msx1-dependent signaling pathways in early tooth development. *Development, 125*(21), 4325–4333.

Biggs, L. C., & Mikkola, M. L. (2014). Early inductive events in ectodermal appendage morphogenesis. *Seminars in Cell & Developmental Biology, 25–26*, 11–21.

Bitgood, M. J., & McMahon, A. P. (1995). Hedgehog and Bmp genes are coexpressed at many diverse sites of cell-cell interaction in the mouse embryo. *Developmental Biology, 172*(1), 126–138.

Boras-Granic, K., Chang, H., Grosschedl, R., & Hamel, P. A. (2006). Lef1 is required for the transition of Wnt signaling from mesenchymal to epithelial cells in the mouse embryonic mammary gland. *Developmental Biology, 295*(1), 219–231.

Botchkarev, V. A., Botchkareva, N. V., Roth, W., Nakamura, M., Chen, L. H., Herzog, W., et al. (1999). Noggin is a mesenchymally derived stimulator of hair-follicle induction. *Nature Cell Biology, 1*(3), 158–164.

Botchkarev, V. A., Botchkareva, N. V., Sharov, A. A., Funa, K., Huber, O., & Gilchrest, B. A. (2002). Modulation of BMP signaling by noggin is required for induction of the secondary (nontylotrich) hair follicles. *Journal of Investigative Dermatology, 118*(1), 3–10.

Briscoe, J., & Therond, P. P. (2013). The mechanisms of Hedgehog signalling and its roles in development and disease. *Nature Reviews. Molecular Cell Biology, 14*(7), 416–429.

Britanova, O., Depew, M. J., Schwark, M., Thomas, B. L., Miletich, I., Sharpe, P., et al. (2006). Satb2 haploinsufficiency phenocopies 2q32-q33 deletions, whereas loss suggests a fundamental role in the coordination of jaw development. *American Journal of Human Genetics, 79*(4), 668–678.

Chai, Y., Jiang, X., Ito, Y., Bringas, P., Jr., Han, J., Rowitch, D. H., et al. (2000). Fate of the mammalian cranial neural crest during tooth and mandibular morphogenesis. *Development, 127*(8), 1671–1679.

Chandramouli, A., Hatsell, S. J., Pinderhughes, A., Koetz, L., & Cowin, P. (2013). Gli activity is critical at multiple stages of embryonic mammary and nipple development. *PLoS One, 8*(11), e79845.

Charles, C., Hovorakova, M., Ahn, Y., Lyons, D. B., Marangoni, P., Churava, S., et al. (2011). Regulation of tooth number by fine-tuning levels of receptor-tyrosine kinase signaling. *Development, 138*(18), 4063–4073.

Charles, C., Pantalacci, S., Tafforeau, P., Headon, D., Laudet, V., & Viriot, L. (2009). Distinct impacts of Eda and Edar loss of function on the mouse dentition. *PLoS One, 4*(4), e4985.

Chen, Y., Bei, M., Woo, I., Satokata, I., & Maas, R. (1996). Msx1 controls inductive signaling in mammalian tooth morphogenesis. *Development, 122*(10), 3035–3044.

Chen, D., Jarrell, A., Guo, C., Lang, R., & Atit, R. (2012). Dermal beta-catenin activity in response to epidermal Wnt ligands is required for fibroblast proliferation and hair follicle initiation. *Development, 139*(8), 1522–1533.

Chen, J., Lan, Y., Baek, J. A., Gao, Y., & Jiang, R. (2009). Wnt/beta-catenin signaling plays an essential role in activation of odontogenic mesenchyme during early tooth development. *Developmental Biology, 334*(1), 174–185.

Chiang, C., Swan, R. Z., Grachtchouk, M., Bolinger, M., Litingtung, Y., Robertson, E. K., et al. (1999). Essential role for Sonic hedgehog during hair follicle morphogenesis. *Developmental Biology, 205*(1), 1–9.

Cho, K. W., Kim, J. Y., Song, S. J., Farrell, E., Eblaghie, M. C., Kim, H. J., et al. (2006). Molecular interactions between Tbx3 and Bmp4 and a model for dorsoventral positioning of mammary gland development. *Proceedings of the National Academy of Sciences of the United States of America, 103*(45), 16788–16793.

Cho, S. W., Kwak, S., Woolley, T. E., Lee, M. J., Kim, E. J., Baker, R. E., et al. (2011). Interactions between Shh, Sostdc1 and Wnt signaling and a new feedback loop for spatial patterning of the teeth. *Development, 138*(9), 1807–1816.

Chu, E. Y., Hens, J., Andl, T., Kairo, A., Yamaguchi, T. P., Brisken, C., et al. (2004). Canonical WNT signaling promotes mammary placode development and is essential for initiation of mammary gland morphogenesis. *Development, 131*(19), 4819–4829.

Cobourne, M. T., & Sharpe, P. T. (2013). Diseases of the tooth: The genetic and molecular basis of inherited anomalies affecting the dentition. *Wiley Interdisciplinary Reviews. Developmental Biology, 2*(2), 183–212.

Cowin, P., & Wysolmerski, J. (2010). Molecular mechanisms guiding embryonic mammary gland development. *Cold Spring Harbor Perspectives in Biology, 2*(6), a003251.

Dassule, H. R., Lewis, P., Bei, M., Maas, R., & McMahon, A. P. (2000). Sonic hedgehog regulates growth and morphogenesis of the tooth. *Development, 127*(22), 4775–4785.

Dassule, H. R., & McMahon, A. P. (1998). Analysis of epithelial-mesenchymal interactions in the initial morphogenesis of the mammalian tooth. *Developmental Biology, 202*(2), 215–227.

Davenport, T. G., Jerome-Majewska, L. A., & Papaioannou, V. E. (2003). Mammary gland, limb and yolk sac defects in mice lacking Tbx3, the gene mutated in human ulnar mammary syndrome. *Development, 130*(10), 2263–2273.

De Moerlooze, L., Spencer-Dene, B., Revest, J. M., Hajihosseini, M., Rosewell, I., & Dickson, C. (2000). An important role for the IIIb isoform of fibroblast growth factor receptor 2 (FGFR2) in mesenchymal-epithelial signalling during mouse organogenesis. *Development, 127*(3), 483–492.

Dhouailly, D. (1973). Dermo-epidermal interactions between birds and mammals: Differentiation of cutaneous appendages. *Journal of Embryology and Experimental Morphology, 30*(3), 587–603.

Dickson, K. M., Bhakar, A. L., & Barker, P. A. (2004). TRAF6-dependent NF-κB transcriptional activity during mouse development. *Developmental Dynamics, 231*(1), 122–127.

Eblaghie, M. C., Song, S. J., Kim, J. Y., Akita, K., Tickle, C., & Jung, H. S. (2004). Interactions between FGF and Wnt signals and Tbx3 gene expression in mammary gland initiation in mouse embryos. *Journal of Anatomy*, *205*(1), 1–13.

Ferguson, C. A., Tucker, A. S., Christensen, L., Lau, A. L., Matzuk, M. M., & Sharpe, P. T. (1998). Activin is an essential early mesenchymal signal in tooth development that is required for patterning of the murine dentition. *Genes & Development*, *12*(16), 2636–2649.

Fliniaux, I., Mikkola, M. L., Lefebvre, S., & Thesleff, I. (2008). Identification of dkk4 as a target of Eda-A1/Edar pathway reveals an unexpected role of ectodysplasin as inhibitor of Wnt signalling in ectodermal placodes. *Developmental Biology*, *320*(1), 60–71.

Fliniaux, I., Viallet, J. P., Dhouailly, D., & Jahoda, C. A. (2004). Transformation of amnion epithelium into skin and hair follicles. *Differentiation*, *72*(9–10), 558–565.

Fujimori, S., Novak, H., Weissenbock, M., Jussila, M., Goncalves, A., Zeller, R., et al. (2010). Wnt/beta-catenin signaling in the dental mesenchyme regulates incisor development by regulating Bmp4. *Developmental Biology*, *348*(1), 97–106.

Gat, U., DasGupta, R., Degenstein, L., & Fuchs, E. (1998). De novo hair follicle morphogenesis and hair tumors in mice expressing a truncated beta-catenin in skin. *Cell*, *95*(5), 605–614.

Gritli-Linde, A., Bei, M., Maas, R., Zhang, X. M., Linde, A., & McMahon, A. P. (2002). Shh signaling within the dental epithelium is necessary for cell proliferation, growth and polarization. *Development*, *129*(23), 5323–5337.

Gritli-Linde, A., Hallberg, K., Harfe, B. D., Reyahi, A., Kannius-Janson, M., Nilsson, J., et al. (2007). Abnormal hair development and apparent follicular transformation to mammary gland in the absence of hedgehog signaling. *Developmental Cell*, *12*(1), 99–112.

Guo, L., Yu, Q. C., & Fuchs, E. (1993). Targeting expression of keratinocyte growth factor to keratinocytes elicits striking changes in epithelial differentiation in transgenic mice. *EMBO Journal*, *12*(3), 973–986.

Hanna, J. H., Saha, K., & Jaenisch, R. (2010). Pluripotency and cellular reprogramming: Facts, hypotheses, unresolved issues. *Cell*, *143*(4), 508–525.

Harada, H., Toyono, T., Toyoshima, K., Yamasaki, M., Itoh, N., Kato, S., et al. (2002). FGF10 maintains stem cell compartment in developing mouse incisors. *Development*, *129*(6), 1533–1541.

Hardcastle, Z., Mo, R., Hui, C. C., & Sharpe, P. T. (1998). The Shh signalling pathway in tooth development: Defects in Gli2 and Gli3 mutants. *Development*, *125*(15), 2803–2811.

Hardy, M. H. (1992). The secret life of the hair follicle. *Trends in Genetics*, *8*(2), 55–61.

Hatsell, S. J., & Cowin, P. (2006). Gli3-mediated repression of Hedgehog targets is required for normal mammary development. *Development*, *133*(18), 3661–3670.

Headon, D. J., & Overbeek, P. A. (1999). Involvement of a novel Tnf receptor homologue in hair follicle induction. *Nature Genetics*, *22*(4), 370–374.

Hosokawa, R., Deng, X., Takamori, K., Xu, X., Urata, M., Bringas, P., Jr., et al. (2009). Epithelial-specific requirement of FGFR2 signaling during tooth and palate development. *Journal of Experimental Zoology. Part B, Molecular and Developmental Evolution*, *312B*(4), 343–350.

Hovorakova, M., Prochazka, J., Lesot, H., Smrckova, L., Churava, S., Boran, T., et al. (2011). Shh expression in a rudimentary tooth offers new insights into development of the mouse incisor. *Journal of Experimental Zoology. Part B, Molecular and Developmental Evolution*, *316*(5), 347–358.

Hovorakova, M., Smrckova, L., Lesot, H., Lochovska, K., Peterka, M., & Peterkova, R. (2013). Sequential Shh expression in the development of the mouse upper functional incisor. *Journal of Experimental Zoology. Part B, Molecular and Developmental Evolution*, *320*(7), 455–464.

Howard, B., Panchal, H., McCarthy, A., & Ashworth, A. (2005). Identification of the scaramanga gene implicates Neuregulin3 in mammary gland specification. *Genes & Development*, *19*(17), 2078–2090.

Huang, S., Zhu, X., Liu, Y., Tao, Y., Feng, G., He, L., et al. (2012). Wls is expressed in the epidermis and regulates embryonic hair follicle induction in mice. *PLoS One*, 7(9), e45904.

Huelsken, J., Vogel, R., Erdmann, B., Cotsarelis, G., & Birchmeier, W. (2001). beta-Catenin controls hair follicle morphogenesis and stem cell differentiation in the skin. *Cell*, *105*(4), 533–545.

Huh, S. H., Narhi, K., Lindfors, P. H., Haara, O., Yang, L., Ornitz, D. M., et al. (2013). Fgf20 governs formation of primary and secondary dermal condensations in developing hair follicles. *Genes & Development*, *27*(4), 450–458.

Hussein, S. M., & Nagy, A. A. (2012). Progress made in the reprogramming field: New factors, new strategies and a new outlook. *Current Opinion in Genetics & Development*, *22*(5), 435–443.

Ibarretxe, G., Aurrekoetxea, M., Crende, O., Badiola, I., Jimenez-Rojo, L., Nakamura, T., et al. (2012). Epiprofin/Sp6 regulates Wnt-BMP signaling and the establishment of cellular junctions during the bell stage of tooth development. *Cell and Tissue Research*, *350*(1), 95–107.

Izzi, L., Levesque, M., Morin, S., Laniel, D., Wilkes, B. C., Mille, F., et al. (2011). Boc and Gas1 each form distinct Shh receptor complexes with Ptch1 and are required for Shh-mediated cell proliferation. *Developmental Cell*, *20*(6), 788–801.

Jamora, C., DasGupta, R., Kocieniewski, P., & Fuchs, E. (2003). Links between signal transduction, transcription and adhesion in epithelial bud development. *Nature*, *422*(6929), 317–322.

Jarvinen, E., Salazar-Ciudad, I., Birchmeier, W., Taketo, M. M., Jernvall, J., & Thesleff, I. (2006). Continuous tooth generation in mouse is induced by activated epithelial Wnt/beta-catenin signaling. *Proceedings of the National Academy of Sciences of the United States of America*, *103*(49), 18627–18632.

Jernvall, J., & Thesleff, I. (2000). Reiterative signaling and patterning during mammalian tooth morphogenesis. *Mechanisms of Development*, *92*(1), 19–29.

Jernvall, J., & Thesleff, I. (2012). Tooth shape formation and tooth renewal: Evolving with the same signals. *Development*, *139*(19), 3487–3497.

Jerome-Majewska, L. A., Jenkins, G. P., Ernstoff, E., Zindy, F., Sherr, C. J., & Papaioannou, V. E. (2005). Tbx3, the ulnar-mammary syndrome gene, and Tbx2 interact in mammary gland development through a p19Arf/p53-independent pathway. *Developmental Dynamics*, *234*(4), 922–933.

Jheon, A. H., Seidel, K., Biehs, B., & Klein, O. D. (2013). From molecules to mastication: The development and evolution of teeth. *Wiley Interdisciplinary Reviews. Developmental Biology*, *2*(2), 165–182.

Jia, S., Zhou, J., Gao, Y., Baek, J. A., Martin, J. F., Lan, Y., et al. (2013). Roles of Bmp4 during tooth morphogenesis and sequential tooth formation. *Development*, *140*(2), 423–432.

Jussila, M., & Thesleff, I. (2012). Signaling networks regulating tooth organogenesis and regeneration, and the specification of dental mesenchymal and epithelial cell lineages. *Cold Spring Harbor Perspectives in Biology*, *4*(4), a008425.

Kangas, A. T., Evans, A. R., Thesleff, I., & Jernvall, J. (2004). Nonindependence of mammalian dental characters. *Nature*, *432*(7014), 211–214.

Kassai, Y., Munne, P., Hotta, Y., Penttila, E., Kavanagh, K., Ohbayashi, N., et al. (2005). Regulation of mammalian tooth cusp patterning by ectodin. *Science*, *309*, 2067–2070.

Kavanagh, K. D., Evans, A. R., & Jernvall, J. (2007). Predicting evolutionary patterns of mammalian teeth from development. *Nature*, *449*(7161), 427–432.

Kettunen, P., Spencer-Dene, B., Furmanek, T., Kvinnsland, I. H., Dickson, C., Thesleff, I., et al. (2007). Fgfr2b mediated epithelial-mesenchymal interactions coordinate tooth morphogenesis and dental trigeminal axon patterning. *Mechanisms of Development*, *124*(11–12), 868–883.

Klein, O. D., Lyons, D. B., Balooch, G., Marshall, G. W., Basson, M. A., Peterka, M., et al. (2008). An FGF signaling loop sustains the generation of differentiated progeny from stem cells in mouse incisors. *Development, 135*(2), 377–385.

Klein, O. D., Minowada, G., Peterkova, R., Kangas, A., Yu, B. D., Lesot, H , et al. (2006). Sprouty genes control diastema tooth development via bidirectional antagonism of epithelial-mesenchymal FGF signaling. *Developmental Cell, 11*(2), 181–190.

Kollar, E. J. (1970). The induction of hair follicles by embryonic dermal papillae. *Journal of Investigative Dermatology, 55*(6), 374–378.

Kollar, E. J., & Baird, G. R. (1970). Tissue interactions in embryonic mouse tooth germs. II. The inductive role of the dental papilla. *Journal of Embryology and Experimental Morphology, 24*(1), 173–186.

Kondo, S., & Miura, T. (2010). Reaction-diffusion model as a framework for understanding biological pattern formation. *Science, 329*(5999), 1616–1620.

Kratochwil, K., Dull, M., Farinas, I., Galceran, J., & Grosschedl, R. (1996). Lef1 expression is activated by BMP-4 and regulates inductive tissue interactions in tooth and hair development. *Genes & Development, 10*(11), 1382–1394.

Kratochwil, K., Galceran, J., Tontsch, S., Roth, W., & Grosschedl, R. (2002). FGF4, a direct target of LEF1 and Wnt signaling, can rescue the arrest of tooth organogenesis in Lef1 (−/−) mice. *Genes & Development, 16*(24), 3173–3185.

Kuraguchi, M., Wang, X. P., Bronson, R. T., Rothenberg, R., Ohene-Baah, N. Y., Lund, J. J., et al. (2006). Adenomatous polyposis coli (APC) is required for normal development of skin and thymus. *PLoS Genetics, 2*(9), e146.

Lan, Y., Jia, S., & Jiang, R. (2014). Molecular patterning of the mammalian dentition. *Seminars in Cell & Developmental Biology, 25–26*, 61–70.

Laurikkala, J., Kassai, Y., Pakkasjarvi, L., Thesleff, I., & Itoh, N. (2003). Identification of a secreted BMP antagonist, ectodin, integrating BMP, FGF, and SHH signals from the tooth enamel knot. *Developmental Biology, 264*(1), 91–105.

Laurikkala, J., Mikkola, M., Mustonen, T., Aberg, T., Koppinen, P., Pispa, J , et al. (2001). TNF signaling via the ligand-receptor pair ectodysplasin and edar controls the function of epithelial signaling centers and is regulated by Wnt and activin during tooth organogenesis. *Developmental Biology, 229*(2), 443–455.

Laurikkala, J., Pispa, J., Jung, H. S., Nieminen, P., Mikkola, M., Wang, X., et al. (2002). Regulation of hair follicle development by the TNF signal ectodysplasin and its receptor Edar. *Development, 129*(10), 2541–2553.

Lee, C. S., Buttitta, L., & Fan, C. M. (2001). Evidence that the WNT-inducible growth arrest-specific gene 1 encodes an antagonist of sonic hedgehog signaling in the somite. *Proceedings of the National Academy of Sciences of the United States of America, 98*(20), 11347–11352.

Lee, M. Y., Racine, V., Jagadpramana, P., Sun, L., Yu, W., Du, T., et al. (2011). Ectodermal influx and cell hypertrophy provide early growth for all murine mammary rudiments, and are differentially regulated among them by Gli3. *PLoS One, 6*(10), e26242.

Li, L., Lin, M., Wang, Y., Cserjesi, P., Chen, Z., & Chen, Y. (2011). BmprIa is required in mesenchymal tissue and has limited redundant function with BmprIb in tooth and palate development. *Developmental Biology, 349*(2), 451–461.

Li, C. Y., Prochazka, J., Goodwin, A. F., & Klein, O. D. (2014). Fibroblast growth factor signaling in mammalian tooth development. *Odontology, 102*(1), 1–13.

Lim, X., & Nusse, R. (2013). Wnt signaling in skin development, homeostasis, and disease. *Cold Spring Harbor Perspectives in Biology, 5*(2), a008029.

Lindvall, C., Evans, N. C., Zylstra, C. R., Li, Y., Alexander, C. M., & Williams, B. O. (2006). The Wnt signaling receptor Lrp5 is required for mammary ductal stem cell activity and Wnt1-induced tumorigenesis. *Journal of Biological Chemistry, 281*(46), 35081–35087.

Lindvall, C., Zylstra, C. R., Evans, N., West, R. A., Dykema, K., Furge, K. A., et al. (2009). The Wnt co-receptor Lrp6 is required for normal mouse mammary gland development. *PLoS One, 4*(6), e5813.

Liu, B., Chen, S., Cheng, D., Jing, W., & Helms, J. A. (2014). Primary cilia integrate hedgehog and Wnt signaling during tooth development. *Journal of Dental Research, 93*(5), 475–482.

Liu, F., Chu, E. Y., Watt, B., Zhang, Y., Gallant, N. M., Andl, T., et al. (2008). Wnt/beta-catenin signaling directs multiple stages of tooth morphogenesis. *Developmental Biology, 313*(1), 210–224.

Liu, F., & Millar, S. E. (2010). Wnt/beta-catenin signaling in oral tissue development and disease. *Journal of Dental Research, 89*(4), 318–330.

Liu, W., Shaver, T. M., Balasa, A., Ljungberg, M. C., Wang, X., Wen, S., et al. (2012). Deletion of Porcn in mice leads to multiple developmental defects and models human focal dermal hypoplasia (Goltz syndrome). *PLoS One,* 7(3), e32331.

Lumsden, A. G. (1988). Spatial organization of the epithelium and the role of neural crest cells in the initiation of the mammalian tooth germ. *Development, 103*(Suppl.), 155–169.

Mailleux, A. A., Spencer-Dene, B., Dillon, C., Ndiaye, D., Savona-Baron, C., Itoh, N., et al. (2002). Role of FGF10/FGFR2b signaling during mammary gland development in the mouse embryo. *Development, 129*(1), 53–60.

Mandler, M., & Neubuser, A. (2001). FGF signaling is necessary for the specification of the odontogenic mesenchyme. *Developmental Biology, 240*(2), 548–559.

Martinelli, D. C., & Fan, C. M. (2007). Gas1 extends the range of Hedgehog action by facilitating its signaling. *Genes & Development, 21*(10), 1231–1243.

Massague, J. (2012). TGFbeta signalling in context. *Nature Reviews. Molecular Cell Biology, 13*(10), 616–630.

Mikkola, M. L. (2007). Genetic basis of skin appendage development. *Seminars in Cell & Developmental Biology, 18*(2), 225–236.

Mikkola, M. L. (2008). TNF superfamily in skin appendage development. *Cytokine & Growth Factor Reviews, 19*(3–4), 219–230.

Mikkola, M. L., & Millar, S. E. (2006). The mammary bud as a skin appendage: Unique and shared aspects of development. *Journal of Mammary Gland Biology and Neoplasia, 11*(3–4), 187–203.

Mikkola, M. L., & Thesleff, I. (2003). Ectodysplasin signaling in development. *Cytokine & Growth Factor Reviews, 14*(3–4), 211–224.

Mill, P., Mo, R., Fu, H., Grachtchouk, M., Kim, P. C., Dlugosz, A. A., et al. (2003). Sonic hedgehog-dependent activation of Gli2 is essential for embryonic hair follicle development. *Genes & Development, 17*(2), 282–294.

Millar, S. E. (2002). Molecular mechanisms regulating hair follicle development. *Journal of Investigative Dermatology, 118*(2), 216–225.

Mina, M., & Kollar, E. J. (1987). The induction of odontogenesis in non-dental mesenchyme combined with early murine mandibular arch epithelium. *Archives of Oral Biology, 32*(2), 123–127.

Monaghan, A. P., Kioschis, P., Wu, W., Zuniga, A., Bock, D., Poustka, A., et al. (1999). Dickkopf genes are co-ordinately expressed in mesodermal lineages. *Mechanisms of Development, 87*(1–2), 45–56.

Morlon, A., Munnich, A., & Smahi, A. (2005). TAB2, TRAF6 and TAK1 are involved in NF-kappaB activation induced by the TNF-receptor, Edar and its adaptator Edaradd. *Human Molecular Genetics, 14*(23), 3751–3757.

Mou, C., Jackson, B., Schneider, P., Overbeek, P. A., & Headon, D. J. (2006). Generation of the primary hair follicle pattern. *Proceedings of the National Academy of Sciences of the United States of America, 103*(24), 9075–9080.

Mukhopadhyay, M., Shtrom, S., Rodriguez-Esteban, C., Chen, L., Tsukui, T., Gomer, L., et al. (2001). Dickkopf1 is required for embryonic head induction and limb morphogenesis in the mouse. *Developmental Cell, 1*(3), 423–434.

Munne, P. M., Felszeghy, S., Jussila, M., Suomalainen, M., Thesleff, I., & Jernvall, J. (2010). Splitting placodes: Effects of bone morphogenetic protein and activin on the patterning and identity of mouse incisors. *Evolution and Development, 12*(4), 383–392.

Munne, P. M., Tummers, M., Jarvinen, E., Thesleff, I., & Jernvall, J. (2009). Tinkering with the inductive mesenchyme: Sostdc1 uncovers the role of dental mesenchyme in limiting tooth induction. *Development, 136*(3), 393–402.

Murashima-Suginami, A., Takahashi, K., Kawabata, T., Sakata, T., Tsukamoto, H., Sugai, M., et al. (2007). Rudiment incisors survive and erupt as supernumerary teeth as a result of USAG-1 abrogation. *Biochemical and Biophysical Research Communications, 359*(3), 549–555.

Murashima-Suginami, A., Takahashi, K., Sakata, T., Tsukamoto, H., Sugai, M., Yanagita, M., et al. (2008). Enhanced BMP signaling results in supernumerary tooth formation in USAG-1 deficient mouse. *Biochemical and Biophysical Research Communications, 369*(4), 1012–1016.

Mustonen, T., Ilmonen, M., Pummila, M., Kangas, A. T., Laurikkala, J., Jaatinen, R., et al. (2004). Ectodysplasin A1 promotes placodal cell fate during early morphogenesis of ectodermal appendages. *Development, 131*(20), 4907–4919.

Mustonen, T., Pispa, J., Mikkola, M. L., Pummila, M., Kangas, A. T., Pakkasjarvi, L., et al. (2003). Stimulation of ectodermal organ development by Ectodysplasin-A1. *Developmental Biology, 259*(1), 123–136.

Naito, A., Yoshida, H., Nishioka, E., Satoh, M., Azuma, S., Yamamoto, T., et al. (2002). TRAF6-deficient mice display hypohidrotic ectodermal dysplasia. *Proceedings of the National Academy of Sciences of the United States of America, 99*(13), 8766–8771.

Nakamura, T., de Vega, S., Fukumoto, S., Jimenez, L., Unda, F., & Yamada, Y. (2008). Transcription factor epiprofin is essential for tooth morphogenesis by regulating epithelial cell fate and tooth number. *Journal of Biological Chemistry, 283*(8), 4825–4833.

Narhi, K., Jarvinen, E., Birchmeier, W., Taketo, M. M., Mikkola, M. L., & Thesleff, I. (2008). Sustained epithelial beta-catenin activity induces precocious hair development but disrupts hair follicle down-growth and hair shaft formation. *Development, 135*(6), 1019–1028.

Narhi, K., Tummers, M., Ahtiainen, L., Itoh, N., Thesleff, I., & Mikkola, M. L. (2012). Sostdc1 defines the size and number of skin appendage placodes. *Developmental Biology, 364*(2), 149–161.

Neubuser, A., Peters, H., Balling, R., & Martin, G. R. (1997). Antagonistic interactions between FGF and BMP signaling pathways: A mechanism for positioning the sites of tooth formation. *Cell, 90*(2), 247–255.

Nozawa, Y. I., Lin, C., & Chuang, P. T. (2013). Hedgehog signaling from the primary cilium to the nucleus: An emerging picture of ciliary localization, trafficking and transduction. *Current Opinion in Genetics & Development, 23*(4), 429–437.

O'Connell, D. J., Ho, J. W., Mammoto, T., Turbe-Doan, A., O'Connell, J. T., Haseley, P. S., et al. (2012). A Wnt-bmp feedback circuit controls intertissue signaling dynamics in tooth organogenesis. *Science Signaling, 5*(206), ra4.

Oh, E. C., & Katsanis, N. (2012). Cilia in vertebrate development and disease. *Development, 139*(3), 443–448.

Ohazama, A., Haycraft, C. J., Seppala, M., Blackburn, J., Ghafoor, S., Cobourne, M., et al. (2009). Primary cilia regulate Shh activity in the control of molar tooth number. *Development, 136*(6), 897–903.

Ohazama, A., Johnson, E. B., Ota, M. S., Choi, H. J., Porntaveetus, T., Oommen, S., et al. (2008). Lrp4 modulates extracellular integration of cell signaling pathways in development. *PLoS One*, *3*(12), e4092.

Ohuchi, H., Tao, H., Ohata, K., Itoh, N., Kato, S., Noji, S., et al. (2003). Fibroblast growth factor 10 is required for proper development of the mouse whiskers. *Biochemical and Biophysical Research Communications*, *302*(3), 562–567.

Panchal, H., Wansbury, O., Parry, S., Ashworth, A., & Howard, B. (2007). Neuregulin3 alters cell fate in the epidermis and mammary gland. *BMC Developmental Biology*, 7, 105.

Peterkova, R., Churava, S., Lesot, H., Rothova, M., Prochazka, J., Peterka, M., et al. (2009). Revitalization of a diastemal tooth primordium in Spry2 null mice results from increased proliferation and decreased apoptosis. *Journal of Experimental Zoology. Part B, Molecular and Developmental Evolution*, *312B*(4), 292–308.

Peterkova, R., Lesot, H., & Peterka, M. (2006). Phylogenetic memory of developing mammalian dentition. *Journal of Experimental Zoology. Part B, Molecular and Developmental Evolution*, *306*(3), 234–250.

Peterkova, R., Lesot, H., Viriot, L., & Peterka, M. (2005). The supernumerary cheek tooth in tabby/EDA mice-a reminiscence of the premolar in mouse ancestors. *Archives of Oral Biology*, *50*(2), 219–225.

Peterkova, R., Lesot, H., Vonesch, J. L., Peterka, M., & Ruch, J. V. (1996). Mouse molar morphogenesis revisited by three dimensional reconstruction. I. Analysis of initial stages of the first upper molar development revealed two transient buds. *International Journal of Developmental Biology*, *40*(5), 1009–1016.

Peters, H., Neubuser, A., Kratochwil, K., & Balling, R. (1998). Pax9-deficient mice lack pharyngeal pouch derivatives and teeth and exhibit craniofacial and limb abnormalities. *Genes & Development*, *12*(17), 2735–2747.

Petiot, A., Conti, F. J., Grose, R., Revest, J. M., Hodivala-Dilke, K. M., & Dickson, C. (2003). A crucial role for Fgfr2-IIIb signalling in epidermal development and hair follicle patterning. *Development*, *130*(22), 5493–5501.

Pispa, J., Jung, H. S., Jernvall, J., Kettunen, P., Mustonen, T., Tabata, M. J., et al. (1999). Cusp patterning defect in Tabby mouse teeth and its partial rescue by FGF. *Developmental Biology*, *216*(2), 521–534.

Pispa, J., Mikkola, M. L., Mustonen, T., & Thesleff, I. (2003). Ectodysplasin, Edar and TNFRSF19 are expressed in complementary and overlapping patterns during mouse embryogenesis. *Gene Expression Patterns*, *3*(5), 675–679.

Plikus, M., Wang, W. P., Liu, J., Wang, X., Jiang, T. X., & Chuong, C. M. (2004). Morpho-regulation of ectodermal organs: Integument pathology and phenotypic variations in K14-noggin engineered mice through modulation of bone morphogenic protein pathway. *American Journal of Pathology*, *164*(3), 1099–1114.

Prochazka, J., Pantalacci, S., Churava, S., Rothova, M., Lambert, A., Lesot, H., et al. (2010). Patterning by heritage in mouse molar row development. *Proceedings of the National Academy of Sciences of the United States of America*, *107*(35), 15497–15502.

Propper, A. Y. (1978). Wandering epithelial cells in the rabbit embryo milk line. A preliminary scanning electron microscope study. *Developmental Biology*, *67*(1), 225–231.

Propper, A. Y., Howard, B. A., & Veltmaat, J. M. (2013). Prenatal morphogenesis of mammary glands in mouse and rabbit. *Journal of Mammary Gland Biology and Neoplasia*, *18*(2), 93–104.

Pummila, M., Fliniaux, I., Jaatinen, R., James, M. J., Laurikkala, J., Schneider, P., et al. (2007). Ectodysplasin has a dual role in ectodermal organogenesis: Inhibition of Bmp activity and induction of Shh expression. *Development*, *134*(1), 117–125.

Reddy, S., Andl, T., Bagasra, A., Lu, M. M., Epstein, D. J., Morrisey, E. E., et al. (2001). Characterization of Wnt gene expression in developing and postnatal hair follicles and identification of Wnt5a as a target of Sonic hedgehog in hair follicle morphogenesis. *Mechanisms of Development, 107*(1–2), 69–82.

Reddy, S. T., Andl, T., Lu, M. M., Morrisey, E. E., & Millar, S. E. (2004). Expression of Frizzled genes in developing and postnatal hair follicles. *Journal of Investigative Dermatology, 123*(2), 275–282.

Richardson, G. D., Bazzi, H., Fantauzzo, K. A., Waters, J. M., Crawford, H., Hynd, P., et al. (2009). KGF and EGF signalling block hair follicle induction and promote interfollicular epidermal fate in developing mouse skin. *Development, 136*(13), 2153–2164.

Robinson, G. W. (2007). Cooperation of signalling pathways in embryonic mammary gland development. *Nature Reviews. Genetics, 8*(12), 963–972.

Saadi, I., Das, P., Zhao, M., Raj, L., Ruspita, I., Xia, Y., et al. (2013). Msx1 and Tbx2 antagonistically regulate Bmp4 expression during the bud-to-cap stage transition in tooth development. *Development, 140*(13), 2697–2702.

Sarkar, L., Cobourne, M., Naylor, S., Smalley, M., Dale, T., & Sharpe, P. T. (2000). Wnt/Shh interactions regulate ectodermal boundary formation during mammalian tooth development. *Proceedings of the National Academy of Sciences of the United States of America, 97*(9), 4520–4524.

Sarkar, L., & Sharpe, P. T. (1999). Expression of Wnt signalling pathway genes during tooth development. *Mechanisms of Development, 85*(1–2), 197–200.

Sasaki, T., Ito, Y., Xu, X., Han, J., Bringas, P., Jr., Maeda, T., et al. (2005). LEF1 is a critical epithelial survival factor during tooth morphogenesis. *Developmental Biology, 278*(1), 130–143.

Satokata, I., & Maas, R. (1994). Msx1 deficient mice exhibit cleft palate and abnormalities of craniofacial and tooth development. *Nature Genetics, 6*(4), 348–356.

Schlake, T. (2007). Determination of hair structure and shape. *Seminars in Cell & Developmental Biology, 18*(2), 267–273.

Schlake, T., & Sick, S. (2007). Canonical WNT signalling controls hair follicle spacing. *Cell Adhesion & Migration, 1*(3), 149–151.

Schmidt-Ullrich, R., Tobin, D. J., Lenhard, D., Schneider, P., Paus, R., & Scheidereit, C. (2006). NF-kappaB transmits Eda A1/EdaR signalling to activate Shh and cyclin D1 expression, and controls post-initiation hair placode down growth. *Development, 133*(6), 1045–1057.

Schneider, M. R., Schmidt-Ullrich, R., & Paus, R. (2009). The hair follicle as a dynamic miniorgan. *Current Biology, 19*(3), R132–R142.

Sennett, R., & Rendl, M. (2012). Mesenchymal-epithelial interactions during hair follicle morphogenesis and cycling. *Seminars in Cell & Developmental Biology, 23*(8), 917–927.

Shi, Y., & Massague, J. (2003). Mechanisms of TGF-beta signaling from cell membrane to the nucleus. *Cell, 113*(6), 685–700.

Sick, S., Reinker, S., Timmer, J., & Schlake, T. (2006). WNT and DKK determine hair follicle spacing through a reaction–diffusion mechanism. *Science, 314*(5804), 1447–1450.

St-Jacques, B., Dassule, H. R., Karavanova, I., Botchkarev, V. A., Li, J., Danielian, P. S., et al. (1998). Sonic hedgehog signaling is essential for hair development. *Current Biology, 8*(19), 1058–1068.

Talamillo, A., Delgado, I., Nakamura, T., de-Vega, S., Yoshitomi, Y., Unda, F., et al. (2010). Role of epiprofin, a zinc-finger transcription factor, in limb development. *Developmental Biology, 337*(2), 363–374.

Trumpp, A., Depew, M. J., Rubenstein, J. L., Bishop, J. M., & Martin, G. R. (1999). Cre-mediated gene inactivation demonstrates that FGF8 is required for cell survival and patterning of the first branchial arch. *Genes & Development, 13*(23), 3136–3148.

Tsai, S. Y., Sennett, R., Rezza, A., Clavel, C., Grisanti, L., Zemla, R., et al. (2014). Wnt/beta-catenin signaling in dermal condensates is required for hair follicle formation. *Developmental Biology, 385*(2), 179–188.

Tucker, A. S., Headon, D. J., Courtney, J. M., Overbeek, P., & Sharpe, P. T. (2004). The activation level of the TNF family receptor, Edar, determines cusp number and tooth number during tooth development. *Developmental Biology, 268*(1), 185–194.

Tucker, A. S., Matthews, K. L., & Sharpe, P. T. (1998). Transformation of tooth type induced by inhibition of BMP signaling. *Science, 282*(5391), 1136–1138.

Tummers, M., & Thesleff, I. (2009). The importance of signal pathway modulation in all aspects of tooth development. *Journal of Experimental Zoology. Part B, Molecular and Developmental Evolution, 312B*(4), 309–319.

Vainio, S., Karavanova, I., Jowett, A., & Thesleff, I. (1993). Identification of BMP-4 as a signal mediating secondary induction between epithelial and mesenchymal tissues during early tooth development. *Cell, 75*(1), 45–58.

van Genderen, C., Okamura, R. M., Farinas, I., Quo, R. G., Parslow, T. G., Bruhn, L., et al. (1994). Development of several organs that require inductive epithelial-mesenchymal interactions is impaired in LEF-1-deficient mice. *Genes & Development, 8*(22), 2691–2703.

Veltmaat, J. M., Ramsdell, A. F., & Sterneck, E. (2013). Positional variations in mammary gland development and cancer. *Journal of Mammary Gland Biology and Neoplasia, 18*(2), 179–188.

Veltmaat, J. M., Relaix, F., Le, L. T., Kratochwil, K., Sala, F. G., van Veelen, W., et al. (2006). Gli3-mediated somitic Fgf10 expression gradients are required for the induction and patterning of mammary epithelium along the embryonic axes. *Development, 133*(12), 2325–2335.

Veltmaat, J. M., Van Veelen, W., Thiery, J. P., & Bellusci, S. (2004). Identification of the mammary line in mouse by Wnt10b expression. *Developmental Dynamics, 229*(2), 349–356.

Viriot, L., Lesot, H., Vonesch, J. L., Ruch, J. V., Peterka, M., & Peterkova, R. (2000). The presence of rudimentary odontogenic structures in the mouse embryonic mandible requires reinterpretation of developmental control of first lower molar histomorphogenesis. *International Journal of Developmental Biology, 44*(2), 233–240.

Wang, X. P., & Fan, J. (2011). Molecular genetics of supernumerary tooth formation. *Genesis, 49*(4), 261–277.

Wang, Y., Li, L., Zheng, Y., Yuan, G., Yang, G., He, F., et al. (2012). BMP activity is required for tooth development from the lamina to bud stage. *Journal of Dental Research, 91*(7), 690–695.

Wang, X. P., O'Connell, D. J., Lund, J. J., Saadi, I., Kuraguchi, M., Turbe-Doan, A., et al. (2009). Apc inhibition of Wnt signaling regulates supernumerary tooth formation during embryogenesis and throughout adulthood. *Development, 136*(11), 1939–1949.

Wang, X. P., Suomalainen, M., Felszeghy, S., Zelarayan, L. C., Alonso, M. T., Plikus, M. V., et al. (2007). An integrated gene regulatory network controls stem cell proliferation in teeth. *PLoS Biology, 5*(6), e159.

Watson, C. J., & Khaled, W. T. (2008). Mammary development in the embryo and adult: A journey of morphogenesis and commitment. *Development, 135*(6), 995–1003.

Woo, W. M., Zhen, H. H., & Oro, A. E. (2012). Shh maintains dermal papilla identity and hair morphogenesis via a Noggin-Shh regulatory loop. *Genes & Development, 26*(11), 1235–1246.

Zhang, Y., Andl, T., Yang, S. H., Teta, M., Liu, F., Seykora, J. T., et al. (2008). Activation of beta-catenin signaling programs embryonic epidermis to hair follicle fate. *Development, 135*(12), 2161–2172.

Zhang, Z., Lan, Y., Chai, Y., & Jiang, R. (2009). Antagonistic actions of Msx1 and Osr2 pattern mammalian teeth into a single row. *Science, 323*(5918), 1232–1234.

Zhang, Y., Tomann, P., Andl, T., Gallant, N. M., Huelsken, J., Jerchow, B , et al. (2009). Reciprocal requirements for EDA/EDAR/NF-kappaB and Wnt/beta-catenin signaling pathways in hair follicle induction. *Developmental Cell, 17*(1), 49–61.

Zhang, Y., Zhang, Z., Zhao, X., Yu, X., Hu, Y., Geronimo, B., et al. (2000). A new function of BMP4: Dual role for BMP4 in regulation of sonic hedgehog expression in the mouse tooth germ. *Development, 127*(7), 1431–1443.

Zhao, M., Gupta, V., Raj, L., Roussel, M., & Bei, M. (2013). A network of transcription factors operates during early tooth morphogenesis. *Molecular and Cellular Biology, 33*(16), 3099–3112.

Zhao, H., Oka, K., Bringas, P., Kaartinen, V., & Chai, Y. (2008). TGF-beta type I receptor Alk5 regulates tooth initiation and mandible patterning in a type II receptor-independent manner. *Developmental Biology, 320*(1), 19–29.

Zhao, X., Zhang, Z., Song, Y., Zhang, X., Zhang, Y., Hu, Y., et al. (2000). Transgenically ectopic expression of Bmp4 to the Msx1 mutant dental mesenchyme restores downstream gene expression but represses Shh and Bmp2 in the enamel knot of wild type tooth germ. *Mechanisms of Development, 99*(1–2), 29–38.

Zhou, P., Byrne, C., Jacobs, J., & Fuchs, E. (1995). Lymphoid enhancer factor 1 directs hair follicle patterning and epithelial cell fate. *Genes & Development, 9*(6), 700–713.

Zhu, X., Zhao, P., Liu, Y., Zhang, X., Fu, J., Ivy Yu, H. M., et al. (2013). Intra-epithelial requirement of canonical Wnt signaling for tooth morphogenesis. *Journal of Biological Chemistry, 288*(17), 12080–12089.

CHAPTER FOURTEEN

The Role of Foxi Family Transcription Factors in the Development of the Ear and Jaw

Renée K. Edlund*, Onur Birol*, Andrew K. Groves*,†,‡,1
*Program in Developmental Biology, Baylor College of Medicine, Houston, Texas, USA
†Department of Molecular and Human Genetics, Baylor College of Medicine, Houston, Texas, USA
‡Department of Neuroscience, Baylor College of Medicine, Houston, Texas, USA
[1]Corresponding author: e-mail address: akgroves@bcm.edu

Contents

Abstract

The mammalian outer, middle, and inner ears have different embryonic origins and evolved at different times in the vertebrate lineage. The outer ear is derived from first and second branchial arch ectoderm and mesoderm, the middle ear ossicles are derived from neural crest mesenchymal cells that invade the first and second branchial arches,

Current Topics in Developmental Biology, Volume 111
ISSN 0070-2153
http://dx.doi.org/10.1016/bs.ctdb.2014.11.014

whereas the inner ear and its associated vestibule-acoustic (VIIIth) ganglion are derived from the otic placode. In this chapter, we discuss recent findings in the development of these structures and describe the contributions of members of a Forkhead transcription factor family, the *Foxi* family to their formation. *Foxi* transcription factors are critical for formation of the otic placode, survival of the branchial arch neural crest, and developmental remodeling of the branchial arch ectoderm.

1. THE ANATOMY AND EMBRYONIC ORIGINS OF THE INNER, MIDDLE, AND OUTER EARS

1.1. Components of the mammalian ear

The mammalian auditory apparatus comprises three compartments, termed the inner, middle, and outer ears. Each compartment arises from a separate embryonic origin, but the final structures work in concert to detect sounds and translate them into electrical signals transmitted by sensory neurons to the brain.

1.1.1 The inner ear

The mature inner ear is found inside a highly calcified bony labyrinth. The soft tissue of the inner ear, termed the epithelial or membranous labyrinth, is composed of three parts containing six sensory patches to detect sound and motion. First, the three semicircular canals, each with a sensory crista housed in an ampulla, serve to detect angular acceleration. Second, the utricle and the saccule, each containing a sensory macula, detect linear acceleration and gravity. Finally, the cochlea serves to detect sound with its sensory patch, the organ of Corti, running the length of the cochlear duct (Groves & Fekete, 2012). The sensory patches consist of highly polarized, mechanically sensitive sensory epithelial cells called hair cells. Hair cells are surrounded by supporting cells that provide physical and trophic support and help to maintain potassium balance in the endolymph bathing each sensory organ.

1.1.2 The middle ear

In aquatic vertebrates, sound vibrations from the surrounding water pass directly into the inner ear where they are translated into neural signals. Land tetrapods, on the other hand, require additional structures to allow impedance matching between air pressure sound waves and fluid vibrations within the epithelial labyrinth. This process is achieved by vibrations of one or more small middle ear bones, or ossicles, against the wall of the cochlea. The ossicles, the air-filled space they occupy, and a vibrating tympanic membrane comprise the vertebrate middle ear. While birds and reptiles possess a single ossicle, the columella, mammals have three, the malleus, the incus, and the stapes. When sound waves strike the tympanic membrane it vibrates, creating pulses that move the malleus. Movements of the malleus cause the incus to pivot, which presses the stapes in and out against the oval window of the inner ear with the same frequency as the original air-borne sound, causing vibrations in the fluid inside the cochlea. The middle ear ossicles occupy an air-filled space that is connected to the external environment by the Eustachian tube, with one end in the middle ear and the other in the throat. The Eustachian tube permits adjustments in air pressure in response to external pressure changes and drainage of any fluid that might accumulate in the middle ear space. The middle ear itself is lined with a ciliated epithelium that covers the entire middle ear cavity in birds, but only the ventral two-thirds in mammals, where the remainder is nonciliated epithelium derived from cranial neural crest cells (Thompson & Tucker, 2013). It is now accepted that the tympanic middle ear evolved independently several times in different vertebrate lineages (Manley, 2010).

1.1.3 The outer ear

The tympanic membrane separates the middle ear from the outer ear, consisting in land mammals of an ear canal and an external pinna. Birds, most reptiles, aquatic mammals, and monotremes also possess ear canals but lack an obvious pinna structure. The pinna assists in selecting directionality of sound input by diffraction of incoming sound waves, and in many mammals, it can be rotated to detect sounds from a specific source. The relationship between the diffractive capacity of an object and sound wavelength suggests that pinnae arose in small mammals with the ability to hear moderately high-frequency sounds (Clack & Allin, 2004).

1.2. Development of the inner ear primordium—from nonneural ectoderm to the otic placode

1.2.1 Complex sequential signaling leads to the formation of the preplacodal region from nonneural ectoderm

The entire inner ear, together with the vestibuloacoustic ganglion that will connect its sensory regions to the hindbrain, develops from a thickened region of embryonic ectoderm next to the hindbrain termed the otic placode (Groves, 2005; Streit, 2001). As discussed in other chapters in this volume, the otic placode is one of several cranial placodes that form the olfactory and lens epithelium, the epibranchial and ophthalmic trigeminal sensory ganglia, the inner ear and in some vertebrate taxa, the lateral line system (Baker & Bronner-Fraser, 2001; Saint-Jeannet & Moody, 2014; Schlosser, 2006, 2010; Streit, 2004, 2007). The transition from early embryonic ectoderm to a definitive placode proceeds through a series of sequential developmental decisions. These can be divided conveniently into three stages: the choice between neural and nonneural ectoderm, the division of nonneural ectoderm into neural crest and early "preplacodal" progenitors in the mediolateral axis, and the singling out of individual placodes from preplacodal progenitors along rostrocaudal axis of the head. These developmental decisions have been well summarized in a number of recent reviews (Grocott, Tambalo, & Streit, 2012; Groves & LaBonne, 2014; Patthey & Gunhaga, 2011, 2013), but we will briefly describe some of these steps below.

An emerging consensus is that early embryonic ectoderm shows evidence of division into presumptive neural and nonneural ectoderm even before the start of gastrulation in vertebrates. The action of FGF signals from tissues underlying early ectoderm (such as hypoblast, future definitive endoderm, or anterior visceral endoderm), together with antagonists of the Wnt and BMP signaling pathways leads to the expression of preneural genes, such as *ERNI*, *Otx2*, *Geminin*, and *Sox3* (Bally-Cuif, Gulisano, Broccoli, & Boncinelli, 1995; Groves & LaBonne, 2014; Kroll, Salic, Evans, & Kirschner, 1998; Papanayotou et al., 2008; Rex et al., 1997; Streit, Berliner, Papanayotou, Sirulnik, & Stern, 2000). Coincident with this early induction, BMP and Wnt signaling in more ventral ectoderm populations specifies nonneural ectoderm, defined by the expression of *Dlx*, *Gata2/3*, and *Foxi* gene families (Brown, Wang, & Groves, 2005; Hans & Westerfield, 2007; Hoffman, Javier, Campeau, Knight, & Schilling, 2007; Knight et al., 2003; Li & Cornell, 2007; Luo, Matsuo-Takasaki, & Sargent, 2001; Matsuo-Takasaki, Matsumura, & Sasai, 2005; McLarren, Litsiou, & Streit, 2003; Ohyama & Groves, 2004a; Papalopulu & Kintner, 1993; Pera & Kessel, 1999; Phillips et al., 2006; Pieper, Ahrens, Rink, Peter, & Schlosser, 2012; Sheng & Stern, 1999; Woda, Pastagia, Mercola, & Artinger, 2003).

The exposure of preneural ectoderm to neural-inducing factors leads to expression of markers of the definitive neural plate, such as *Sox2* (Rex et al., 1997; Streit et al., 1997; Uchikawa, Ishida, Takemoto, Kamachi, & Kondoh, 2003), whereas nonneural ectoderm genes such as *Dlx5/6* and *Gata3* begin to be restricted to the border between neural and nonneural ectoderm (Feledy et al., 1999; Khudyakov & Bronner-Fraser, 2009; Kwon, Bhat, Sweet, Cornell, & Riley, 2010; Pieper et al., 2012; Streit,

2002; Woda et al., 2003). Evidence from a number of vertebrate species suggests that this border is sharpened by cross-repressive interactions between neural and nonneural transcription factors and by positive feedback within each domain (Feledy et al., 1999; Kwon et al., 2010; Linker et al., 2009; Luo et al., 2001; Matsuo-Takasaki et al., 2005; McLarren et al., 2003; Pieper et al., 2012; Tribulo, Aybar, Nguyen, Mullins, & Mayor, 2003; Woda et al., 2003). As ectoderm further differentiates, a thin strip at the neural plate border receives both FGF and Wnt signaling; however, it is also subject to continuing BMP inhibition from secreted antagonists (Ahrens & Schlosser, 2005; Brugmann, Pandur, Kenyon, Pignoni, & Moody, 2004; Litsiou, Hanson, & Streit, 2005). This border region divides into two molecularly distinct populations around the time of head formation: the preplacodal region (PPR) confined to the anterior region of the embryo and premigratory neural crest that is generated from all levels of the embryo other than the forebrain (Betancur, Bronner-Fraser, & Sauka-Spengler, 2010; Grocott et al., 2012; Groves & LaBonne, 2014; McCabe & Bronner-Fraser, 2009; Milet & Monsoro-Burq, 2012; Prasad, Sauka-Spengler, & LaBonne, 2012; Sauka-Spengler & Bronner-Fraser, 2008; Schlosser, 2006, 2010; Steventon, Mayor, & Streit, 2014; Stuhlmiller & Garcia-Castro, 2012). The PPR is defined by expression of members of the *Six* (*sine oculis* homeobox) transcription factors and their *Eya* (*eyes absent*) cofactor partners (Bailey & Streit, 2006; Saint-Jeannet & Moody, 2014; Streit, 2007). This domain lies lateral to the neural crest progenitor domain at the boundary of the neural plate.

To date, no genetic fate mapping studies have been performed to determine the derivatives of *Six*- or *Eya*-expressing progenitors, although dye-labeling studies suggest that this region will contribute to all craniofacial placodes including the otic placode (Bhattacharyya, Bailey, Bronner-Fraser, & Streit, 2004; Kozlowski, Murakami, Ho, & Weinberg, 1997; Streit, 2002; Xu, Dude, & Baker, 2008). In addition to these definitive preplacodal markers, nonneural ectoderm genes such as *foxi1* (in fish), *Foxi3* (in amniotes), and *Dlx5/6* and *Gata3* become restricted to the PPR (Grocott et al., 2012; Groves & LaBonne, 2014). The expression and maintenance of preplacodal markers is dependent on FGF signaling and the inhibition of Wnt and Bmp signaling (Ahrens & Schlosser, 2005; Brugmann et al., 2004; Litsiou et al., 2005), while cells closer to the neural plate respond to FGF, BMP, and Wnt signaling and express early neural crest markers such as *Msx1* and *Pax3/7* (Groves & LaBonne, 2014; Milet & Monsoro-Burq, 2012; Prasad et al., 2012). Figure 1 summarizes the

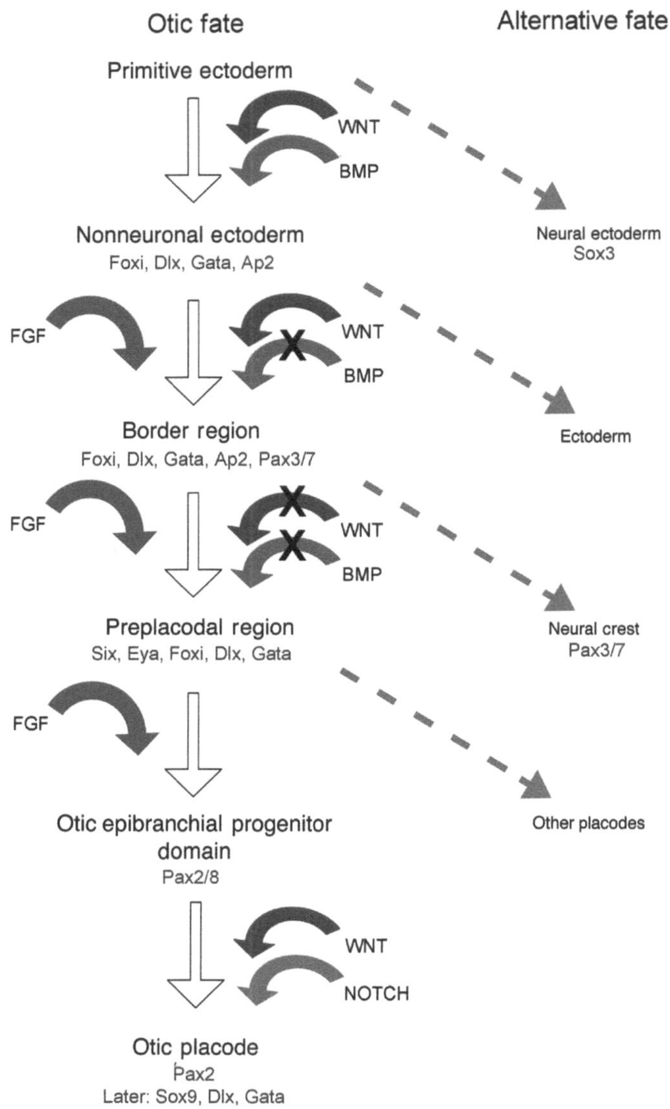

Figure 1 Requirements for otic induction from primitive ectoderm. The figure depicts a consensus scheme of otic induction for amniotes. For each step, the necessary signaling factors for the otic pathway are shown. The alternative pathway is chosen in the absence of the specific combination of signaling and/or with additional signals. Gene expression defining tissue identity for individual steps is indicated. X represents the requirement for an inhibitor of a signaling pathway, such as BMP or Wnt inhibitors. (See the color plate.)

series of signals involved in specification of the PPR from nonneural ectoderm before and after gastrulation.

1.2.2 Induction of the otic placode by FGFs and its refinement by Wnt and Notch signaling

As discussed earlier, the preplacodal ectoderm gives rise to all craniofacial sensory placodes of the vertebrate head (Grocott et al., 2012; Litsiou et al., 2005; Sato et al., 2010; Streit, 2007), and the progenitors for each of the placodes are indistinguishable during early establishment of the PPR at the neural plate boundary. The differentiation of distinct placodal fates requires regional signals, and in the case of the otic placode, data from all major vertebrate groups suggest that FGF signaling plays an early and critical role (Ohyama, Groves, & Martin, 2007; Riley & Phillips, 2003; Solomon, Kwak, & Fritz, 2004). Interestingly, both the location and identity of FGFs involved in otic placode induction vary between species; for example, FGF3 from the hindbrain and FGF10 from the cranial mesoderm are necessary for otic placode induction in mice (Urness, Paxton, Wang, Schoenwolf, & Mansour, 2010; Wright & Mansour, 2003), while in zebrafish, this role is served by Fgf3 and Fgf8 that are both expressed in the hindbrain (Leger & Brand, 2002; Maroon et al., 2002; Phillips, Bolding, & Riley, 2001). The absolute necessity for FGF signaling in otic placode induction has been shown in multiple organisms (Alvarez et al., 2003; Ladher, Wright, Moon, Mansour, & Schoenwolf, 2005; Leger & Brand, 2002; Yang et al., 2013; Zelarayan et al., 2007). The earliest otic markers expressed by preplacodal ectoderm in response to FGF signals are members of the *Pax2/5/8* family (Groves & Bronner-Fraser, 2000; Heller & Brandli, 1999; Ohyama & Groves, 2004b; Pfeffer, Gerster, Lun, Brand, & Busslinger, 1998). Fate mapping studies in mice and chick suggest that the *Pax2*-expressing domain not only contains all the progenitors of the inner ear but also gives rise to the epibranchial placodes and some parts of the epidermis (Ohyama & Groves, 2004b; Ohyama et al., 2007; Streit, 2001). This region of *Pax2*-expressing cells has been referred to as the otic-epibranchial progenitor domain, or OEPD (Freter, Muta, Mak, Rinkwitz, & Ladher, 2008; Ladher, O'Neill, & Begbie, 2010).

Studies in amniotes suggest that one mechanism that refines the broad domain of *Pax2*-expressing cells into the otic placode and epibranchial placodes, and ectoderm is the strength and duration of FGF signaling (Freter et al., 2008; Ladher et al., 2010). Some of the earliest markers of the otic placode are negative regulators of the FGF pathway, such as *Sprouty* genes and dual-specificity phosphatases such as *Dusp6* and *Dusp9* (Chambers & Mason, 2000; Urness, Li, Wang, & Mansour, 2008). The importance of attenuating FGF signaling in otic placode induction was demonstrated by a failure of induction following prolonged ectopic activation of FGF signaling (Freter et al., 2008); extended FGF signaling is instead required to promote epibranchial placode fate (Ladher et al., 2010). Two other mechanisms that demarcate otic from nonotic derivatives within the *Pax2* domain are Wnt and Notch signaling (Groves & Fekete, 2012). A gradient of Wnt signals from the midline and border of the neural plate directs *Pax2*-expressing cells toward an otic fate rather than epidermal derivatives (Ohyama et al., 2007). Disruption of canonical Wnt signaling (for example, by conditional deletion of *β-catenin* in *Pax2*-expressing cells) greatly reduces the size of the otic placode, while activation of Wnt signals by expression of constitutively active forms of *β-catenin* drive the entire *Pax2* domain toward an otic fate (Freter et al., 2008; Ohyama, Mohamed, Taketo, Dufort, & Groves, 2006). The final step in the refinement of otic and nonotic fates is upregulation of ligands of the Notch signaling pathway such as Jagged1 in presumptive otic ectoderm in response to high levels of Wnt signaling (Jayasena, Ohyama, Segil, & Groves, 2008). Jagged1-Notch signaling in this region positively feeds back to further increase the strength of Wnt signaling in regions of *Pax2*-expressing cells close to the neural plate, but not in more lateral regions where Wnt levels are too low to induce *Jag1* expression and activate the Notch pathway (Jayasena et al., 2008). In this way, a continuous gradient of Wnt becomes discontinuous, fixing otic and nonotic fates (Fig. 1).

Although *Pax2/5/8* gene family members are the earliest known markers of the otic placode, these genes do not appear to be necessary for otic placode induction. However, they are necessary for later events in inner ear development (Bouchard, de Caprona, Busslinger, Xu, & Fritzsch, 2010; Burton, Cole, Mulheisen, Chang, & Wu, 2004). To date, no combination of *Six* and *Eya* mouse knockouts, knockdown, or dominant-negative mutants in other vertebrates prevents the induction of the otic placode (Bricaud & Collazo, 2006; Zhang, Knosp, Maconochie, Friedman, & Smith, 2004; Zheng et al., 2003; Zou, Silvius, Rodrigo-Blomqvist, Enerback, & Xu, 2006), although these genes are also individually necessary for later aspects of ear development. Mutation or knockdown of Dlx gene family members can greatly reduce the size of the inner ear (Esterberg & Fritz, 2009; Liu et al., 2003; Solomon & Fritz, 2002; Solomon et al., 2004), but *Dlx* genes again do not appear to be necessary for otic placode induction (Robledo & Lufkin, 2006; Robledo, Rajan, Li, & Lufkin, 2002). To date, the only genes individually necessary for otic placode induction appear to be *foxi1* in fish (Hans, Christison, Liu, & Westerfield, 2007; Hans, Liu, & Westerfield, 2004; Nissen, Yan, Amsterdam, Hopkins, & Burgess, 2003; Solomon, Kudoh, Dawid, & Fritz, 2003) and *Foxi3* in chick and mouse (Edlund, Ohyama, Kantarci, Riley, & Groves, 2014; Khatri, Edlund, & Groves, 2014). We will return to the role of *Foxi* genes in inner ear induction later in the chapter. First, we will discuss the development of the first and second pharyngeal arches that generate much of the middle and external ears, together with the jaw.

1.3. The middle and outer ears develop from the first two branchial arches

During vertebrate embryogenesis, structures in the face and neck develop from transient branchial arches (BAs), also called pharyngeal arches. Following establishment of the embryonic body plan and the development of a basic neural tube, five to six pairs of arches expand from the ventral side of the embryo at the level of the hindbrain and eventually fuse medially (Fig. 2A). The arches are arranged in a simple serial pattern and contain elements of all three embryonic germ layers. The surface of each BA is composed of ectoderm, and each arch is lined with endoderm. Between the epithelial layers lies a mesodermal core surrounded by immigrating neural crest cells from the midbrain and hindbrain. The boundaries between arches are defined by points of direct contact between ectoderm and endoderm, termed pharyngeal pouches (Fig. 2B). The endoderm of the pouches is critical for BA development (Grevellec & Tucker, 2010). The detailed morphology of pharyngeal tissue types has been comprehensively reviewed by Graham, Liu, and colleagues (Graham, Begbie, & McGonnell, 2004; Graham & Richardson, 2012; Graham & Smith, 2001; Szabo-Rogers, Smithers, Yakob, & Liu, 2010).

1.3.1 BA ectoderm, mesoderm, and endoderm contribute to middle and outer ear structures

As BA development progresses, the dorsal territory of the first arch bifurcates into two processes, a posterior mandibular process and an anterior maxillary process. A portion of the outer ear derives from posterior ectoderm and mesoderm of the mandibular process and anterior ectoderm and mesoderm of the second branchial arch (BA2) (Alasti & Van Camp, 2009). Three auricular hillocks arise on the surface of each arch, then expand and remodel,

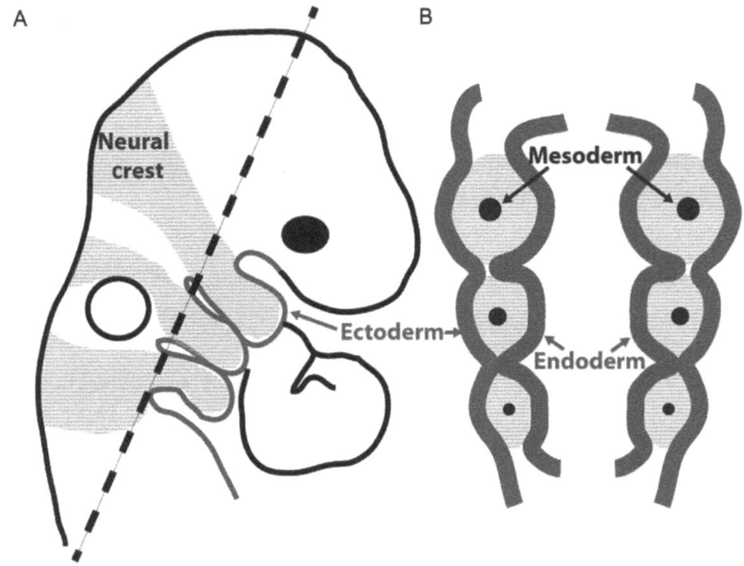

Figure 2 Arrangement of germ layers and neural crest cells in the branchial arches. (A) Sagittal view of an embryonic day 9.5 (E9.5) mouse embryo showing branchial arches (BA) 1–3. The external surface of each arch is ectoderm. Neural crest cells in three distinct streams populate the arches. The most anterior stream contains neural crest cells from the midbrain and rhombomeres 1 and 2 from the hindbrain. Rhombomeres 3 and 5, the white regions between streams of yellow (light gray in the print version) neural crest produce very few neural crest cells. (B) Schematic of a coronal section through the arches of the embryo in (A). Branchial arches consist of external ectoderm, an endodermal lining, a core of mesoderm, and neural crest cells from the midbrain and hindbrain. The boundaries of the arches are defined by points of contact between endoderm and ectoderm: the pharyngeal pouches.

while the ectoderm of the two arches fuses to form the external pinna that is the visible portion of the mammalian ear. Ectoderm from BA1 invaginates and develops into the outer ear canal that leads to the tympanic membrane (Minoux et al., 2013).

Meanwhile, ear structures also develop from pharyngeal pouch endoderm. The Eustachian tube and a portion of the middle ear cavity lining develop from the first pouch. The middle ear consists of an air-filled cavity containing one ossicle in birds and reptiles (the columella) or three ossicles in mammals (the malleus, incus, and stapes) that transmit vibrations from the ear drum to the cochlea. In birds and reptiles, the epithelial lining of this cavity is completely derived from first pouch endoderm. In land tetrapods, however, the three ossicles condense adjacent to the endodermal epithelium of the first pouch (Mallo, 1998, 2001). The endoderm must therefore rupture to line the middle ear cavity, where it comprises only the ventral portion of the cavity epithelium. The dorsal portion is composed of epithelialized neural crest cells, as described below (Thompson & Tucker, 2013). An excellent recent review of middle ear evolution is given by Manley (2010).

1.3.2 Cranial neural crest cells contribute to middle ear structures

One of the developmental innovations unique to vertebrates is the neural crest, a specialized population of cells that exit the neural tube and populate the periphery of the body. Neural crest cells have the potential to differentiate into a variety of cell types and structures. Cranial crest cells derived from the midbrain and hindbrain are particularly plastic, generating bone and cartilage that forms much of the face, jaw, and anterior skull, as well as melanocytes and peripheral neural structures. The process by which cranial neural crest populations adopt their final fate is complicated, involving interactions between all three embryonic primary germ layers and finely tuned signaling, much of which takes place in the BAs.

By the time arch-derived structures begin to take shape, the bulk of mesenchymal cells are cranial neural crest cells that delaminated from the posterior midbrain and the hindbrain. Migration occurs in separate, distinct, and stereotypical streams of cells, with cells delaminating from the posterior midbrain and rhombomeres 1 and 2 of the hindbrain populating BA1, and cells from rhombomere 4 populating BA2. As these cells migrate, they maintain the expression of Hox genes corresponding to their respective rhombomeres of origin, contributing to the establishment of a distinct identity for each BA.

The crest in the mandibular process gives rise to Meckel's cartilage that develops in mammals into two of the three ossicles of the middle ear, the incus, and the malleus, as well as supporting the mandible during ossification (Chai et al., 2000). Neural crest cells in the maxillary process differentiate into the upper jaw, or maxilla, associated connective tissue, and the squamosal bone. BA2 contains neural crest cells that develop into Reichert's cartilage from which the third ear ossicle, the stapes, is derived (Sienknecht, 2013). Additionally, in mammals, the three neural crest-derived middle ear ossicles are surrounded by mesenchymal neural crest cells at the site of the future middle ear cavity. After the incus and malleus separate from Meckel's cartilage and the stapes separates from Reichert's cartilage, the surrounding crest cells retract to create an air-filled space and undergo a mesenchymal-to-epithelial transition to form the dorsal portion of the middle ear cavity lining (Thompson & Tucker, 2013).

1.4. Signals and transcriptional regulators involved in the development of the first and second BAs

The ectoderm and endoderm surrounding the arches produce multiple signaling factors, including FGFs, BMPs, Endothelins (EDNs), and Sonic Hedgehog, that communicate to the underlying mesenchyme—a mixture of mesoderm and neural crest cells (Fig. 3). *Fgfs* are expressed in the epithelial tissues of the BAs and are particularly enriched in the pouch endoderm and overlying ectoderm in the clefts between the arches (Crossley & Martin, 1995; Wall & Hogan, 1995). In the developmental context of the BAs, FGFs function as survival factors for neural crest entering the arches and as signals that induce expression of arch patterning transcription factors in mesenchyme and ectoderm.

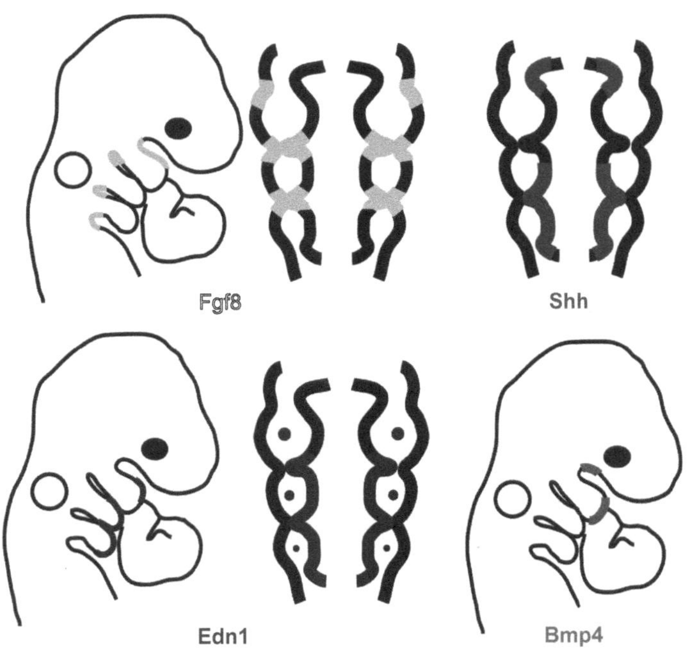

Figure 3 Signaling factors from the pharyngeal ectoderm and endoderm. At embryonic day 9.5 (E9.5), four major signaling factors are secreted from the ectoderm and endoderm. Fgf8 is expressed in the ectoderm between the arches and between the mandibular and maxillary processes of BA1 and in the endoderm at the tips of the pharyngeal pouches. Shh is expressed in endoderm underlying BA2 and BA3. Edn1 is expressed in ventral arch ectoderm, endoderm underlying the mandibular process, BA2, and BA3, and in the mesoderm of each arch. Bmp4 is expressed in the ectoderm overlying the ventral domain and the maxillary process of branchial arch 1 (BA1). (See the color plate.)

One important function of signaling factors from the arch ectoderm is defining the dorsal–ventral signals that pattern postmigratory neural crest cells in the arches (reviewed in Medeiros & Crump, 2012). In order to establish the dorsal–ventral axis in BA1, dorsally expressed *Fgf* in the ectoderm is restricted by bone morphogenetic protein 4 (BMP4), expressed in the ventral portion of the BA1 ectoderm. Ectopic *Bmp4* expression or implantation of a BMP4-soaked bead reduces *Fgf8* expression in the BAs of chick embryos (Shigetani, Nobusada, & Kuratani, 2000). Interestingly, in the arches, BMP4 induces expression of its own antagonists, *Noggin* and *Chordin* (Stottmann, Anderson, & Klingensmith, 2001). In addition to maintaining the boundaries of ectodermal *Fgf8* expression, BMP4 also plays important roles in dorsal–ventral patterning of BA1. A signaling network in which BMP4 and Endothelin 1 (EDN1) overlap to establish the dorsal–ventral axis is conserved from fish to mammals (Alexander et al., 2011; Ozeki, Kurihara, Tonami, Watatani, & Kurihara, 2004; Ruest, Xiang, Lim, Levi, & Clouthier, 2004). BMP4 promotes expression of ventral territory transcription factors in the arch mesenchyme and represses expression of dorsal transcription factors (Liu et al., 2005). Conditional deletion of *Bmp4* from both mouse arch ectoderm and endoderm results in almost complete absence of the mandible and shift toward the midline of the tympanic ring and Meckel's cartilage (Liu et al., 2005).

There is evidence in zebrafish that Bmp4 induces ectodermal expression of *Edn1* (Alexander et al., 2011). As the arches develop, factors induced by Bmp4 come to define the most ventral region of BA1, whereas Edn1 targets occupy a more intermediate domain. One key role of Edn1 signaling is to establish the nested expression pattern of *Dlx* transcription factors that delineate the dorsal–ventral axis of BA1 (Ozeki et al., 2004). Zebrafish and mouse *Edn1*-null mutants transform mandibular arch structures into structures normally derived from the maxillary process, with a mirror duplication observable in skeletal staining (Miller, Schilling, Lee, Parker, & Kimmel, 2000; Ozeki et al., 2004; Ruest et al., 2004).

A fourth signaling pathway, the Hedgehog pathway is also of critical importance for patterning the pharyngeal region. Sonic hedgehog (*Shh*) is expressed in the foregut endoderm prior to arch outgrowth (Wall & Hogan, 1995). Ablation of *Shh*-expressing ectoderm in chick embryos leads to the absence of BA1-derived structures; the defects can be rescued by implantation of Shh-soaked beads in the approximate location of the missing endoderm (Brito, Teillet, & Le Douarin, 2006). These ablation and bead rescue experiments demonstrate a role for Shh in establishing much of the ectodermal patterning required for arch development, including *Fgf8* and *Bmp4* expression (Haworth et al., 2007).

2. FORKHEAD PROTEINS AS TRANSCRIPTION FACTORS AND PIONEER FACTORS

In 1989, a new transcription factor containing a previously uncharacterized protein motif was identified in *Drosophila melanogaster.* Named Forkhead (FKH), the protein was first shown to function in gut development (Weigel, Jurgens, Kuttner, Seifert, & Jackle, 1989). Shortly thereafter, a homolog, HNF-3, was identified in rat, and the motif conserved between the two proteins was determined to be a DNA-binding domain, now called a FKH box (Lai et al., 1990; Weigel & Jackle, 1990). In the ensuing years, an extraordinary number of additional factors containing the FKH box motif have been found across diverse animal species. In an effort to systematize nomenclature of these related genes, a consensus naming strategy was established for chordates in which each gene containing the motif was labeled "Fox" for *f*orkhead b*ox*, sorted into a class lettered A–S based on closest homology, and assigned a unique number within that class (Jackson, Carpenter, Nebert, & Vasiliou, 2010; Kaestner, Knochel, & Martinez, 2000). For example, *Hnf-3* is renamed *Foxa2* to indicate that it is a *Forkhead-box* protein in the *a* class distinct from *Foxa1*.

2.1. FKH proteins are archetypal pioneer factors

While the classical description of transcription factors views these proteins as binding their target sites in DNA and influencing transcription of target genes within a short time after binding, a subset of transcription factors can influence transcriptional activity in a delayed fashion. Often termed *pioneer factors*, such proteins can act through more than one mechanism, both passive and active. Pioneer factors that operate in a passive role are necessary but not sufficient for transcriptional activation (Bossard & Zaret, 1998; Zaret & Carroll, 2011). Rather, they bind target sequences that control gene expression via multiple transcription factors acting in concert, and binding of the pioneer factor merely reduces the number of steps required to initiate gene expression at the appropriate timepoint. On the other hand, many pioneer factors can actively influence gene regulation through manipulation of chromatin structure. They are able to bind the outer face of DNA in regions of tightly compacted chromatin and then open the chromatin to improve accessibility to other transcription factors (Cirillo et al., 2002; Cirillo & Zaret, 1999;

Cuesta, Zaret, & Santisteban, 2007; Perlmann, 1992). Interestingly, some pioneer factors demonstrate the ability to remain bound to chromatin during mitosis (Yan, Xu, Crawford, Wang, & Burgess, 2006; Zaret & Carroll, 2011). Although chromatin remains transcriptionally inactive during this time, the presence of pioneer factors at target sequences upon completion of cell division allows rapid reactivation of target gene transcription (Blobel et al., 2009; Dey, Nishiyama, Karpova, McNally, & Ozato, 2009; Kadauke et al., 2012). This mechanism has been described as "bookmarking" the DNA in a specific cell lineage, as a form of epigenetic mark, to aid in preparing lineage-specific genes for subsequent rapid transcription.

The shared *DNA*-binding domain in FOX proteins is composed of three α helices and two loops resembling insect wings and was thus called a winged-helix domain (Lai, Clark, Burley, & Darnell, 1993). Interestingly, the winged-helix domain bears a distinct resemblance and possibly an evolutionary relationship to the linker histones H5 and H1 (Zaret & Carroll, 2011). The similarity of the FOX winged-helix domain to linker histones suggests that FOX proteins might have access to DNA in regions of compacted chromatin. Indeed, this appears to be the case for more than one member of the FKH family; three FOXA factors, FOXD3, FOXE1, and FOXO, have been shown to be pioneer factors that can actively open chromatin to allow for binding of additional transcription factors (Cuesta et al., 2007; Hatta & Cirillo, 2007; Xu et al., 2009; Zaret & Carroll, 2011). Additionally, these FOX factors and other pioneer factors commonly bind in an unusually tight manner to chromatin (Zaret & Carroll, 2011).

Perhaps the best-characterized FKH pioneer factors are the FOXA proteins, originally identified as early arrivals to mouse albumin regulatory sites during endoderm differentiation (Gualdi et al., 1996). The presence of redundantly expressed FOXA1 or FOXA2 at this albumin regulatory site is required for early liver development (Bossard & Zaret, 1998; Lee, Friedman, Fulmer, & Kaestner, 2005). The FOXA factors, in the cellular context of liver progenitors, bind their target sites in compacted chromatin and open it for further transcription factor binding (Cirillo et al., 2002). FOXA1 protein remains bound to chromatin during mitosis, both at specific target sites and nonspecific sites throughout the compacted chromosomes (Caravaca et al., 2013). Caravaca and colleagues suggest a bookmarking role for FOXA1 at its specific target sites and nonspecific binding as a mechanism for retaining FOXA1 close at hand to jump-start activity at nearby target sites upon the conclusion of telophase. Of note, Yamagishi and colleagues have also suggested a classical transcription factor role for FOXA2 in inducing expression of *Tbx1* in the endoderm (Yamagishi et al., 2003). Overexpression of *Foxa2* alone in HeLa cells was sufficient to activate a luciferase reporter fused to the FOXA-binding site of the *Tbx1* enhancer. While this may be an artifact of the *in vitro* system, it is important not to discount the possibility that a transcription factor may behave as either a pioneer factor or classical transcription factor depending on the target gene and cellular context. Whether acting as classical transcription factors or pioneer factors, with their abundance in chordate genomes, their presence in almost every tissue throughout development, and their ability as pioneer factors to prime genomic regions for rapid activation, FOX transcription factors are critical and universal developmental regulators.

3. THE ROLE OF FOXI FAMILY MEMBERS IN INNER EAR DEVELOPMENT

3.1. The role of *Foxi1* in mammalian ear development

The first Foxi family member to be identified in mammals was *Fkh10*, now renamed *Foxi1* (Hulander, Wurst, Carlsson, & Enerback, 1998). *Foxi1* mutant mice display hyperactivity, circling behavior, and deafness (Hulander et al., 1998), and although the inner ear initially develops normally, much of the epithelial labyrinth becomes swollen and expanded by late embryonic stages, consistent with a failure in endolymphatic fluid regulation in the ear (Hulander et al., 2003). *Foxi1* is first expressed in the primordium of the endolymphatic duct in the developing otocyst (Ohyama & Groves, 2004a) and is later restricted to the epithelium of the endolymphatic duct and sac (Hulander et al., 2003; Raft et al., 2014; Vidarsson et al., 2009). FOXI1 is known to regulate a number of genes involved in ion transport, such as *Slc26A4* which encodes the pendrin anion exchanger implicated in Pendred syndrome (Everett et al., 1997), and human *FOXI1* mutations can also cause Pendred syndrome (Yang et al., 2007). FOXI1 also regulates subunits of other ion transporters such as the vacuolar H^+-ATPases (Vidarsson et al., 2009), and mice with mutations in some of these subunits also have enlarged endolymphatic compartments (Lorente-Canovas et al., 2013). Interestingly, the apparent functional homolog of *Foxi1* in zebrafish is *foxi3*, which also appears to regulate the development and function of transport epithelial cells, the ionocytes of the fish epidermis (Cruz, Chao, & Hwang, 2013; Hsiao et al., 2007; Janicke, Carney, & Hammerschmidt, 2007; Janicke, Renisch, & Hammerschmidt, 2010; Solomon, Logsdon, & Fritz, 2003; Thermes, Lin, & Hwang, 2010).

3.2. Expression, regulation and function of *Foxi1/3* factors in inner ear development—from nonneural ectoderm to otic placode via the preplacodal domain

Both DNA sequence data and phenotypic analysis of mutants suggest that zebrafish *foxi1* and mouse and chicken *Foxi3* should be considered to be homologs (Fig. 4). For convenience, we will refer to the genes as *Foxi1/3* in the following sections to refer to results that have been confirmed in both fish and mouse and/or chicken. *Foxi1/3* transcripts are initially expressed in nonneural ectoderm surrounding the future neural plate (Khatri et al., 2014; Khatri & Groves, 2013; Kwon et al., 2010; Lee, Shen, Fiser, Sali, & Guo, 2003; Nissen et al., 2003; Ohyama & Groves, 2004a; Solomon, Kudoh, Dawid, & Fritz, 2003; Solomon, Logsdon, & Fritz, 2003), overlapping with other transcription factor markers of nonneural ectoderm such as *Gata3*, *Dlx5/6*, and *Ap2a*. In common with these other nonneural ectoderm genes, *Foxi1/3* becomes restricted to the preplacodal domain with definitive PPR markers such as *Six* and *Eya* gene family members(Grocott et al., 2012; Groves & LaBonne, 2014; Kwon et al., 2010; Streit, 2007). After formation of the otic placode, *Foxi1/3* is downregulated in the otic region but continues to be expressed in epibranchial placode ectoderm and ectoderm of the future pharyngeal arches (Edlund et al., 2014; Khatri et al., 2014; Khatri & Groves,

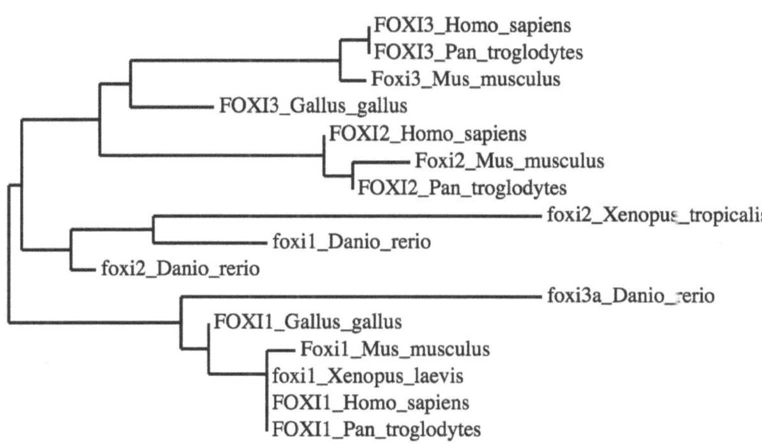

Figure 4 Evolutionary relationship between Foxi genes. The free phylogeny software phylogeny.fr predicts the above relationship between Foxi genes based on protein sequence. Zebrafish foxi1 is more closely related to avian and mammalian Foxi3 genes than to avian and mammalian Foxi1 genes. This relationship reflects naming inconsistencies and supports the hypothesis that zebrafish foxi1 is homologous to avian and mammalian Foxi3 genes in evolutionary history as well as in expression pattern and function.

2013; Ohyama & Groves, 2004a; Solomon, Kudoh, Dawid, & Fritz, 2003; Solomon, Logsdon, & Fritz, 2003). Later in development, *Foxi3* is also expressed in pharyngeal clefts and pouches (Edlund et al., 2014; Ohyama & Groves, 2004a) and in placodes associated with epidermal appendages such as tooth primordia and hair follicles (Drogemuller et al., 2008; Shirokova et al., 2013).

Preplacodal *Foxi1/3* expression is regulated by the same signals that initiate expression of other nonneural ectoderm genes. Zebrafish *foxi1* is dependent on BMP signaling before, but not after, gastrulation (Bhat, Kwon, & Riley, 2013; Hans et al., 2004; Kwon et al., 2010; Solomon, Kudoh, Dawid, & Fritz, 2003). A number of studies have reached differing conclusions on a possible role for Fgf signaling in regulating *foxi1* expression in zebrafish (Hans et al., 2007; Nechiporuk, Linbo, Poss, & Raible, 2007; Padanad, Bhat, Guo, & Riley, 2012; Phillips, Storch, Lekven, & Riley, 2004), although precise time and manner in which Fgf signaling was manipulated may explain some of the differences in these experiments. Another zebrafish study suggests that, in addition to Bmps and Fgfs, *foxi1* is also regulated by retinoic acid (Hans & Westerfield, 2007). Expression of *Pax8*, induced by *foxi1*, as well as expression of *foxi1* itself, expands ectopically in the presence of retinoic acid, whereas *pax8* expression decreases in response to RA inhibitors. This action of retinoic acid may be direct (rather than indirectly affecting anterior–posterior patterning) as *dlx3b* expression in the preplacodal domain is not affected by manipulating retinoic acid levels (Hans & Westerfield, 2007). Finally, in chick, ectodermal *Foxi3* expression in gastrulating embryos is regulated by signals released from the hypoblast, as has also been shown for *Dlx5/6* genes (Khatri et al., 2014; Pera, Stein, & Kessel, 1999). These hypoblast signals appear to be functionally distinct from later signals in cranial mesendoderm, which regulate *Six* and *Eya* family genes in the PPR (Khatri et al., 2014; Litsiou et al., 2005).

In both zebrafish and chick, *Foxi1/3* genes are part of a transcriptional regulatory network that gives the tissue its competence to adopt a preplacodal identity and to respond to FGF signals. In zebrafish, *Foxi1*, *Gata3*, and *Tfap2a* form a complex, cross-regulatory feedback network that is at least partially dependent on Bmp signaling (Bhat et al., 2013; Kwon et al., 2010). Although the same experiments have yet to be performed in chick, recent evidence suggest that amniote *Foxi3*, *Dlx5*, and *Six/Eya* genes can cross-regulate each other. Ectopic expression of *Dlx5* or *Six1* and *Eya2* leads to ectopic *Foxi3* expression in nonneural ectoderm (Khatri et al., 2014). Furthermore, *Foxi3* overexpression at the same developmental stage leads to activation of these three genes in nonneural ectoderm. Together, these fish and chick data support the idea that *Foxi1/3* is part of a transcriptional regulatory network in nonneural ectoderm that is essential for the transition of preplacodal ectoderm into the otic placode.

The only other amniote Foxi homolog, *Foxi2*, does not appear to be expressed in the otic placode, although it is notably expressed in the future epidermis surrounding the otic placode, in progenitors for delaminating epibranchial sensory neurons, and in the BA ectoderm (Freter et al., 2008; Khatri & Groves, 2013; Ohyama & Groves, 2004a). Zebrafish *foxi2* is expressed in chordamesoderm during somitogenesis and is later found in the retina and in the BAs (Solomon, Logsdon, & Fritz, 2003b). Genetic manipulations that reduce the size of the otic placode in amniotes, such as conditional deletion of *β-catenin* in Pax2-expressing progenitors, lead to an expansion of the *Foxi2* ectodermal domain, while manipulations that expand the otic placode lead to a corresponding reduction in *Foxi2* expression (Freter et al., 2008; Jayasena et al., 2008; Ohyama et al., 2006; Urness et al., 2010). Homozygous *Foxi2* mutant mice have been generated in our laboratory but appear to have no obvious defect (T. Ohyama, R. Edlund, O. Birol, & A. Groves, unpublished observations).

3.3. Functional role of *Foxi1/3* in otic placode induction

foxi1 was first identified in two different zebrafish mutagenesis screens as the *hearsay* and *foo* mutants, both of which have significant jaw defects and either very small or completely absent otocysts (Nissen et al., 2003; Solomon, Kudoh, Dawid, & Fritz, 2003). The consensus from a number of studies suggests that *foxi1* acts to provide ectoderm with competence to respond to FGF signaling in otic placode induction (Bhat et al., 2013; Hans et al., 2007; Hans, Irmscher, & Brand, 2013; Hans & Westerfield, 2007; Kwon et al., 2010; Padanad et al., 2012; Padanad & Riley, 2011). Zebrafish *foxi1* is also required for epibranchial placode induction (Lee et al., 2003; Nechiporuk et al., 2007; Nechiporuk, Linbo, & Raible, 2005), consistent with the derivation of both the otic placode and epibranchial placodes from a common *Pax2*-expressing domain. However, an understanding of the precise role of *foxi1* in zebrafish leading from the establishment of the preplacodal domain to the induction of the otic placode in response to FGF signals is complicated by several observations. Analysis of otic placode markers shows that expression of *pax2*, one of the first otic markers, is decreased but still present in *foxi1* mutants (Hans et al., 2007, 2004; Nissen et al., 2003; Solomon, Kudoh,

Dawid, & Fritz, 2003; Solomon et al., 2004). Abolishing otic induction completely in zebrafish requires knocking down *foxi1* together with either three other nonneural ectoderm genes (*tfap2a* and *2c* and *gata3*; Bhat et al., 2013; Kwon et al., 2010) or with *dlx3b* and *dlx4b* (Hans et al., 2007; Solomon et al., 2004). In these embryos, *pax2a* expression is completely abolished and the ear does not form even in the presence of *fgf8* overexpression (Hans et al., 2004, 2007; Solomon et al., 2004). Moreover, in zebrafish unlike mammals, the early otic placode markers *pax2a* and *pax8* appear to be expressed over slightly different time courses (Hans et al., 2004; Mackereth, Kwak, Fritz, & Riley, 2005; Ohyama & Groves, 2004b) and appear to be regulated by different signals. *Pax8* can be regulated independently by either *foxi1* or FGF (Hans et al., 2004; Phillips et al., 2004; Solomon et al., 2004), whereas *pax2a* is regulated by both *dlx3b/4b* and FGF signals (Hans et al., 2007, 2004; Solomon et al., 2004). Further complicating matters, knockdown of *pax2a*, *pax2b*, and *pax8* leads to the disappearance of the inner ear in zebrafish but not in mice (Bouchard et al., 2010; Mackereth et al., 2005). Finally, *dlx3b* and *dlx4b* do not appear to have counterparts in amniotes that share expression pattern and function. *Dlx5* and *Dlx6* are expressed much earlier and more broadly in amniotes than *dlx3b/4b* in fish (McLarren et al., 2003; Pera et al., 1999; Streit, 2002, 2007) and are not necessary for otic placode induction (Robledo & Lufkin, 2006; Robledo et al., 2002). Therefore, the relationship of these genes to *foxi1* is difficult to compare with the interactions between *Foxi3* and *Dlx5/6* in amniotes.

The function of *Foxi3* in otic placode induction in amniotes has received much less attention, and it is not clear at present whether the pathways regulated by *Foxi3* are mechanistically more simple, or simply less well explored. Knockdown of *Foxi3* in chick embryos at preplacodal stages can effectively abolish *Pax2* expression, as well as greatly attenuating the ability of preplacodal ectoderm to express *Pax2* when cultured in the presence of FGFs (Khatri et al., 2014). *Foxi3* mutant mice appear to entirely lack an inner ear (Edlund et al., 2014), although it is not clear whether this represents a complete loss of otic tissue, or whether a small amount of *Pax2*-expressing otic tissue remains as in zebrafish (Hans et al., 2007, 2004; Nissen et al., 2003; Solomon, Kudoh, Dawid, & Fritz, 2003; Solomon et al., 2004). We are currently characterizing the otic placode phenotype of *Foxi3* mutant mice in more detail; our preliminary results suggest that, *like foxi1* mutant zebrafish, *Foxi3* mutant mice also have defects in their epibranchial ganglia in addition to lacking the inner ear (O. Birol, T. Ohyama, R. Edlund, & A. Groves, unpublished observations).

4. THE ROLE OF FOXI FAMILY MEMBERS IN MIDDLE EAR, OUTER EAR, AND JAW DEVELOPMENT

4.1. The role of Foxi1/3 in jaw, middle ear, and outer ear development

In zebrafish, Foxi1 is necessary not only for otic placode induction but also for jaw development (Nissen et al., 2003; Solomon, Kudoh, Dawid, & Fritz, 2003). *foxi1* is expressed in pharyngeal endoderm and ectoderm of zebrafish but not in neural crest cells. Yet, the jaw phenotype in these fish is at least partially attributable to neural crest apoptosis that occurs in the first and second arches, concurrent with decreased *fgf3* expression in the endoderm (Edlund et al., 2014; Nissen et al., 2003). In zebrafish, Foxi1 remains bound to chromatin over multiple cycles of mitosis in cell cultures and may be a chromatin-remodeling pioneer factor that does not itself actively induce expression of many genes, but rather poises genes required for specific developmental processes for rapid expression upon receipt of the appropriate signals (Yan et al., 2006).

Foxi3 is expressed in the pharyngeal region of mouse embryos in a segmented pattern between the BAs (Edlund et al., 2014; Ohyama & Groves, 2004a)). *Foxi3* is expressed in the endoderm of each pharyngeal pouch as well as in the overlying cleft ectoderm, with stronger expression around posterior, later-developing arches and the expression anterior arches gradually becoming fainter. It is important to note that *Foxi3* is expressed neither in BA mesoderm nor in the cranial neural crest cells that populate arch mesenchyme (Edlund et al., 2014; Ohyama & Groves, 2004a). Further supporting homology between the amniote *Foxi3* gene and the zebrafish *foxi1* gene is the catastrophic suite of BA defects present in *Foxi3* mutant mice highly reminiscent of *Hearsay* zebrafish. *Foxi3* mutants are born with completely absent middle and outer ears and severely truncated and deformed jaws; they do not survive after birth (Edlund et al., 2014). The mutant pups are readily identifiable by their lack of mouth, with a continuous ectodermal covering over the lower half of the face, and by the absence of outer ear pinnae. Skeletal preps of the heads of E18.5 *Foxi3* mutant embryos reveal a number of severe defects in bone and cartilage development. In *Foxi3* mutants, only a small anterior portion of the mandible develops and is fused to the distal end of maxilla. The maxilla itself is also malformed (Edlund et al., 2014). In zebrafish *foxi1* mutants, the size of jaw cartilages is significantly reduced and jaw morphology is substantially disrupted (Nissen et al., 2003; Solomon, Kudoh, Dawid, & Fritz, 2003).

The middle ear ossicles are evolutionary derivatives of the two anterior arch cartilages: Meckel's cartilage in BA1 and Reichert's cartilage in BA2 (Sienknecht, 2013). Both cartilages are neural crest derived, thus the severe defects in crest-derived tissue in *Foxi3* mutant mice also affects middle ear structures. There is no sign of the tympanic ring or incus, malleus, or stapes of the middle ear, and the entire temporal bone and inner ear are absent. The absence of the outer ear may be attributable to aberrant ectodermal morphology in *Foxi3* mutant mice, which show a failure to individuate distinct arches (Edlund et al., 2014). In *Foxi3* mutants, cranial neural crest cells undergo apoptosis as they populate the BAs. Since neural crest cells do not express *Foxi3*, this suggests that FOXI3 may regulate the expression of trophic or survival factors in arch ectoderm or endoderm.

Of the four signaling pathways regulating BA development described in Section 1.4, FGF signaling has the clearest role in neural crest cell survival. The presence of strong ectodermal *Fgf* expression promotes postmigratory neural crest survival in the BAs. Increased levels of apoptosis in the arch mesenchyme are observed in *fgf3* knockdown zebrafish, in chicks treated with *Fgf8 RNA*i, and in mouse *Fgf8* hypomorphs and arch ectoderm-conditional *Fgf8* knockouts (Abu-Issa, Smyth, Smoak, Yamamura, & Meyers, 2002; Creuzet, Schuler, Couly, & Le Douarin, 2004; Frank et al., 2002; Nissen et al., 2003; Trumpp, Depew, Rubenstein, Bishop, & Martin, 1999). Experiments with mice carrying hypomorphic alleles of *Fgf8* and with mice in which *Fgf8* is locally knocked out of the arch ectoderm reveal its importance in promoting normal development of the neural crest-derived skeleton. In two studies on compound heterozygote mice, each carrying one null allele and one hypomorphic allele of *Fgf8*, neural crest development is affected in all BAs (Abu-Issa et al., 2002; Frank et al., 2002; Macatee et al., 2003; Trumpp et al., 1999). In BA1, the maxilla, mandible, and middle ear ossicles among other skeletal elements either fail to form or are hypoplastic. This is likely attributable to the substantial increase in cell death in the BA1 mesenchyme.

As apoptosis can be reduced by artificial introduction of FGF when arch morphology is already defective, it is probable that FGF signaling directly promotes survival of migrating neural crest cells (Edlund et al., 2014). FGFs from the ectoderm also play a role in patterning the underlying mesenchyme. Expression of transcription factors, including *Dlx2*, *Gsc*, *Barx1*, *Pax9*, and *Lhx6*, which contribute to defining the dorsal territory of BA1, requires ectodermal signaling (Ferguson, Tucker, & Sharpe, 2000;

Neubuser, Peters, Balling, & Martin, 1997; Tucker, Yamada, Grigoriou, Pachnis, & Sharpe, 1999). *foxi1* and *Foxi3* expression are required for normal pharyngeal pouch morphology in zebrafish and mouse, respectively, FOXI factors establish signaling centers in the developing BAs necessary for crest survival, and the craniofacial phenotype seen in *Foxi3* mouse mutants is due to reduced FGF8 signaling in the pharyngeal region. The activity of FOXI3 in pharyngeal epithelia is required for early expression of *Fgf8* in arch ectoderm; this pathway is conserved in zebrafish where *fgf3* is expressed in BA ectoderm and requires the expression of *foxi1*. In zebrafish *foxi1* morphants, ectopic expression of *fgf3* in pharyngeal ectoderm can reduce neural crest cell death (Edlund et al., 2014).

4.2. Foxi3 has a conserved role in mammalian pharyngeal development

The pharyngeal development function of FOXI3 is conserved in mammals as evidenced by phenotypes seen in three breeds of hairless dogs, all of which possess identical mutations in the *Foxi3* gene (Drogemuller et al., 2008). The mutation, a seven-base pair duplication introduces a premature stop codon prior to the *DNA*-binding domain and is presumed to create a null allele. In addition to a striking lack of hair, possibly related to a role for FOXI3 in regulating the mammalian ectodysplasin pathway (Shirokova et al., 2013), heterozygous dogs have malformed and missing teeth and occasional malformations in the outer ear (Drogemuller et al., 2008; Tassano et al., 2014). A human patient with a deletion in one copy of chromosome 2 encompassing seven genes including *Foxi3* has also been identified. The patient was found to have mandibular asymmetry and unilateral malformation of the outer ear (Tassano et al., 2014).

It is clear from studies in zebrafish, chick, mouse, dog, and from one case study of a human patient that FOXI1/3 homologs are highly conserved developmental regulators that are critical for proper formation of all portions of the ear.

Because of the structural and phenotypic similarities between mammalian FOXI3 and zebrafish Foxi1, it is likely that if Foxi1 is a pioneer factor, so is FOXI3. Their conserved role in mammalian BA development suggests that examination of genes and signaling pathways regulated by FOXI3, perhaps through epigenetic changes and chromatin remodeling, may provide insight into the developmental processes involved with shaping jaws and ears from simple BAs.

5. CONCLUSIONS

5.1. Ear development is a complex process that requires Foxi1/3 in multiple steps

In land vertebrates, the ear consists of three compartments derived from multiple embryonic tissues, while in fish only the inner ear is present. The inner ear arises first during development via sequential refinement of ectodermal identity. Intercellular signaling and subsequent changes in transcriptional profiles of increasingly smaller patches of ectoderm result in conferral of otic identity on a small portion of preplacodal ectoderm. This ectoderm forms the otic placode from which the inner ear develops. Mice- and fish-lacking *Foxi3* or *foxi1*, respectively, fail to form complete otic placodes, a process normally mediated by FGF signaling, and lack all inner ear structures upon completion of embryonic development. In addition, the middle and outer ears are absent in *Foxi3* mutant mice due to failure to form distinct and properly patterned BAs. In the presumptive BAs of *Foxi3* mutant mice, postmigratory neural crest cells undergo apoptosis. This function of *Foxi* is conserved in morphant zebrafish, which also have severe BA-derived skeletal defects, although middle and outer ears are not present in fish. In *Foxi3* mutant mice, *Fgf8* expression in the branchial ectoderm is delayed. Artificial introduction of FGF can rescue neural crest cell apoptosis in zebrafish *Foxi1* morphants, indicating that the proximal cause of cell death in *Foxi* mutants is likely to be reduced FGF signaling.

5.2. Pioneer factors may play a role in morphological diversity

From an evolutionary perspective, it is interesting to speculate about the ways in which minor perturbations of cranial neural crest can lead to the broad spectrum of facial structures present in vertebrates. While apoptosis of neural crest cells in the arches, at least in laboratory genetic experiments, frequently results in severe facial skeletal defects, it is intriguing that induction of apoptosis or even ablation of sizeable populations of precursors to trunk neural crest cells often presents no appreciable obstacle to animal development (Vaglia & Hall, 1999). Trunk neural crest is not usually skeletogenic, with the notable exception of turtle shells. Interestingly, these skeletogenic trunk crest cells migrate in a distinct second wave and express cranial neural crest markers (Cebra-Thomas et al., 2013). Aside from the turtle shell, structures derived from trunk crest, the peripheral nervous system, and melanocytes remain relatively unchanged between classes of vertebrates. Is it possible that the sensitivity of cranial neural crest to changes in total cell number represents a rich substrate for natural selection?

Recent evidence suggests a pivotal role for noncoding DNA in subtle variations of facial shape. While small changes to coding sequences of genes involved with craniofacial development can give rise to disabling phenotypes, sequence changes and even deletions of elements that regulate expression of these same genes can instead generate normal variations in face shape (Attanasio et al., 2013). The authors of the study found that single enhancer deletions resulted in subtle but quantifiable changes to the shape of the mouse skull. It seems then that enhancers may be underappreciated in craniofacial developmental studies. It is possible that some changes in enhancer function are related to chromatin compaction, that is, alterations in the timing or extent of enhancer-containing chromatin in open conformation may permit transcription factor binding in new temporal and spatial patterns. One mechanism potentially at play is changes in pioneer factor activity. Pioneer factors can function to actively open and expose chromatin, priming regions of DNA for transcription factor binding and maintaining cellular lineage throughout multiple rounds of mitosis (Zaret & Carroll, 2011). FOXI1/3 is included among the FKH transcription factors proposed as pioneer factors. Zebrafish Foxi1, the homolog of mammalian FOXI3, has been shown to remain bound to DNA during mitosis in cultured cells (Yan et al., 2006). Yan and colleagues examined chromatin fractions and found Foxi1 both in active chromatin and in chromatin tightly bound to the nuclear matrix, suggesting a dual role for Foxi1 in chromatin organization. Interestingly, the study showed very small changes in gene expression in response to ectopic Foxi1 induction. Although transcription factor overexpression in cultured cells may not produce results representative of *in vivo* activity, the evidence from that study suggests that zebrafish Foxi1 acts not to immediately promote target gene expression, but rather as a pioneer factor that alters the set of genes accessible to other transcription factors necessary for developmental lineage specification. It is tempting to speculate that Foxi1 and FOXI3 could be pioneer factors that influence timing and expression of a suite of pharyngeal patterning factors and might be key in evolution of diverse jaw and ear shapes and sizes. Regardless, when subtle temporal or spatial changes spontaneously occur to BA gene expression, new and potentially adaptive facial structures can arise. Thus, on an individual level the sensitivity of cranial neural crest to genetic perturbations may occasionally result in deformities incompatible with survival, but from a population perspective, accumulation of subtle facial shape changes may advantageous enough that the very same cranial neural crest sensitivity is a powerful force of evolution in vertebrates.

REFERENCES

Abu-Issa, R., Smyth, G., Smoak, I., Yamamura, K., & Meyers, E. N. (2002). Fgf8 is required for pharyngeal arch and cardiovascular development in the mouse. *Development, 129*, 4613–4625.

Ahrens, K., & Schlosser, G. (2005). Tissues and signals involved in the induction of placodal Six1 expression in *Xenopus laevis*. *Developmental Biology, 288*, 40–59.

Alasti, F., & Van Camp, G. (2009). Genetics of microtia and associated syndromes. *Journal of Medical Genetics, 46*, 361–369.

Alexander, C., Zuniga, E., Blitz, I. L., Wada, N., Le Pabic, P., Javidan, Y., et al. (2011). Combinatorial roles for BMPs and Endothelin 1 in patterning the dorsal-ventral axis of the craniofacial skeleton. *Development, 138*, 5135–5146.

Alvarez, Y., Alonso, M. T., Vendrell, V., Zelarayan, L. C., Chamero, P., Theil, T., et al. (2003). Requirements for FGF3 and FGF10 during inner ear formation. *Development, 130*, 6329–6338.

Attanasio, C., Nord, A. S., Zhu, Y., Blow, M. J., Li, Z., Liberton, D. K., et al. (2013). Fine tuning of craniofacial morphology by distant-acting enhancers. *Science, 342*, 1241006.

Bailey, A. P., & Streit, A. (2006). Sensory organs: Making and breaking the pre-placodal region. *Current Topics in Developmental Biology, 72*, 167–204.

Baker, C. V., & Bronner-Fraser, M. (2001). Vertebrate cranial placodes I. Embryonic induction. *Developmental Biology, 232*, 1–61.

Bally-Cuif, L., Gulisano, M., Broccoli, V., & Boncinelli, E. (1995). c-otx2 is expressed in two different phases of gastrulation and is sensitive to retinoic acid treatment in chick embryo. *Mechanisms of Development, 49*, 49–63.

Betancur, P., Bronner-Fraser, M., & Sauka-Spengler, T. (2010). Assembling neural crest regulatory circuits into a gene regulatory network. *Annual Review of Cell and Developmental Biology, 26*, 581–603.

Bhat, N., Kwon, H. J., & Riley, B. B. (2013). A gene network that coordinates preplacodal competence and neural crest specification in zebrafish. *Developmental Biology, 373*, 107–117.

Bhattacharyya, S., Bailey, A. P., Bronner-Fraser, M., & Streit, A. (2004). Segregation of lens and olfactory precursors from a common territory: Cell sorting and reciprocity of Dlx5 and Pax6 expression. *Developmental Biology, 271*, 403–414.

Blobel, G. A., Kadauke, S., Wang, E., Lau, A. W., Zuber, J., Chou, M. M., et al. (2009). A reconfigured pattern of MLL occupancy within mitotic chromatin promotes rapid transcriptional reactivation following mitotic exit. *Molecular Cell, 36*, 970–983.

Bossard, P., & Zaret, K. S. (1998). GATA transcription factors as potentiators of gut endoderm differentiation. *Development, 125*, 4909–4917.

Bouchard, M., de Caprona, D., Busslinger, M., Xu, P., & Fritzsch, B. (2010). Pax2 and Pax8 cooperate in mouse inner ear morphogenesis and innervation. *BMC Developmental Biology, 10*, 89.

Bricaud, O., & Collazo, A. (2006). The transcription factor six1 inhibits neuronal and promotes hair cell fate in the developing zebrafish (Danio rerio) inner ear. *The Journal of Neuroscience, 26*, 10438–10451.

Brito, J. M., Teillet, M. A., & Le Douarin, N. M. (2006). An early role for sonic hedgehog from foregut endoderm in jaw development: Ensuring neural crest cell survival. *Proceedings of the National Academy of Sciences of the United States of America, 103*, 11607–11612.

Brown, S. T., Wang, J., & Groves, A. K. (2005). Dlx gene expression during chick inner ear development. *The Journal of Comparative Neurology, 483*, 48–65.

Brugmann, S. A., Pandur, P. D., Kenyon, K. L., Pignoni, F., & Moody, S. A. (2004). Six1 promotes a placodal fate within the lateral neurogenic ectoderm by functioning as both a transcriptional activator and repressor. *Development, 131*, 5871–5881.

Burton, Q., Cole, L. K., Mulheisen, M., Chang, W., & Wu, D. K. (2004) The role of Pax2 in mouse inner ear development. *Developmental Biology*, *272*, 161–175.

Caravaca, J. M., Donahue, G., Becker, J. S., He, X., Vinson, C., & Zaret, K. S. (2013). Bookmarking by specific and nonspecific binding of FoxA1 pioneer factor to mitotic chromosomes. *Genes & Development*, *27*, 251–260.

Cebra-Thomas, J. A., Terrell, A., Branyan, K., Shah, S., Rice, R., Gyi, L., et al. (2013). Late-emigrating trunk neural crest cells in turtle embryos generate an osteogenic ectomesenchyme in the plastron. *Developmental Dynamics*, *242*, 1223–1235.

Chai, Y., Jiang, X., Ito, Y., Bringas, P., Jr., Han, J., Rowitch, D. H., et al. (2000). Fate of the mammalian cranial neural crest during tooth and mandibular morphogenesis. *Development*, *127*, 1671–1679.

Chambers, D., & Mason, I. (2000). Expression of sprouty2 during early development of the chick embryo is coincident with known sites of FGF signalling. *Mechanisms of Development*, *91*, 361–364.

Cirillo, L. A., Lin, F. R., Cuesta, I., Friedman, D., Jarnik, M., & Zaret, K. S. (2002). Opening of compacted chromatin by early developmental transcription factors HNF3 (FoxA) and GATA-4. *Molecular Cell*, *9*, 279–289.

Cirillo, L. A., & Zaret, K. S. (1999). An early developmental transcription factor complex that is more stable on nucleosome core particles than on free DNA. *Molecular Cell*, *4*, 961–969.

Clack, J. A., & Allin, E. (2004). The evolution of single- and multiple-ossicle ears in fish and tetrapods. In G. A. Manley, A. N. Popper, & R. R. Fay (Eds.), *Evolution of the vertebrate auditory system* (pp. 128–163). New York: Springer Verlag.

Creuzet, S., Schuler, B., Couly, G., & Le Douarin, N. M. (2004). Reciprocal relationships between Fgf8 and neural crest cells in facial and forebrain development. *Proceedings of the National Academy of Sciences of the United States of America*, *101*, 4843–4847.

Crossley, P. H., & Martin, G. R. (1995). The mouse Fgf8 gene encodes a family of polypeptides and is expressed in regions that direct outgrowth and patterning in the developing embryo. *Development*, *121*, 439–451.

Cruz, S. A., Chao, P. L., & Hwang, P. P. (2013). Cortisol promotes differentiation of epidermal ionocytes through Foxi3 transcription factors in zebrafish (Danio rerio). *Comparative Biochemistry and Physiology. Part A, Molecular & Integrative Physiology*, *164*, 249–257.

Cuesta, I., Zaret, K. S., & Santisteban, P. (2007). The forkhead factor FoxE1 binds to the thyroperoxidase promoter during thyroid cell differentiation and modifies compacted chromatin structure. *Molecular and Cellular Biology*, *27*, 7302–7314.

Dey, A., Nishiyama, A., Karpova, T., McNally, J., & Ozato, K. (2009). Brd4 marks select genes on mitotic chromatin and directs postmitotic transcription. *Molecular Biology of the Cell*, *20*, 4899–4909.

Drogemuller, C., Karlsson, E. K., Hytonen, M. K., Perloski, M., Dolf, G., Sainio, K., et al. (2008). A mutation in hairless dogs implicates FOXI3 in ectodermal development. *Science*, *321*, 1462.

Edlund, R. K., Ohyama, T., Kantarci, H., Riley, B. B., & Groves, A. K. (2014). Foxi transcription factors promote pharyngeal arch development by regulating formation of FGF signaling centers. *Developmental Biology*, *390*, 1–13.

Esterberg, R., & Fritz, A. (2009). dlx3b/4b are required for the formation of the preplacodal region and otic placode through local modulation of BMP activity. *Developmental Biology*, *325*, 189–199.

Everett, L. A., Glaser, B., Beck, J. C., Idol, J. R., Buchs, A., Heyman, M., et al. (1997). Pendred syndrome is caused by mutations in a putative sulphate transporter gene (PDS). *Nature Genetics*, *17*, 411–422.

Feledy, J. A., Beanan, M. J., Sandoval, J. J., Goodrich, J. S., Lim, J. H., Matsuo-Takasaki, M., et al. (1999). Inhibitory patterning of the anterior neural plate in Xenopus by homeodomain factors Dlx3 and Msx1. *Developmental Biology*, *212*, 455–464.

Ferguson, C. A., Tucker, A. S., & Sharpe, P. T. (2000). Temporospatial cell interactions regulating mandibular and maxillary arch patterning. *Development, 127*, 403–412.

Frank, D. U., Fotheringham, L. K., Brewer, J. A., Muglia, L. J., Tristani-Firouzi, M., Capecchi, M. R., et al. (2002). An Fgf8 mouse mutant phenocopies human 22q11 deletion syndrome. *Development, 129*, 4591–4603.

Freter, S., Muta, Y., Mak, S. S., Rinkwitz, S., & Ladher, R. K. (2008). Progressive restriction of otic fate: The role of FGF and Wnt in resolving inner ear potential. *Development, 135*, 3415–3424.

Graham, A., Begbie, J., & McGonnell, I. (2004). Significance of the cranial neural crest. *Developmental Dynamics, 229*, 5–13.

Graham, A., & Richardson, J. (2012). Developmental and evolutionary origins of the pharyngeal apparatus. *EvoDevo, 3*, 24.

Graham, A., & Smith, A. (2001). Patterning the pharyngeal arches. *Bioessays, 23*, 54–61.

Grevellec, A., & Tucker, A. S. (2010). The pharyngeal pouches and clefts: Development, evolution, structure and derivatives. *Seminars in Cell & Developmental Biology, 21*, 325–332.

Grocott, T., Tambalo, M., & Streit, A. (2012). The peripheral sensory nervous system in the vertebrate head: A gene regulatory perspective. *Developmental Biology, 370*, 3–23.

Groves, A. K. (2005). The induction of the otic placode. In A. N. Popper, M. W. Kelley, & D. K. Wu (Eds.), *Development of the inner ear* (pp. 10–42). New York: Springer Verlag.

Groves, A. K., & Bronner-Fraser, M. (2000). Competence, specification and commitment in otic placode induction. *Development, 127*, 3489–3499.

Groves, A. K., & Fekete, D. M. (2012). Shaping sound in space: The regulation of inner ear patterning. *Development, 139*, 245–257.

Groves, A. K., & LaBonne, C. (2014). Setting appropriate boundaries: Fate, patterning and competence at the neural plate border. *Developmental Biology, 389*, 2–12.

Gualdi, R., Bossard, P., Zheng, M., Hamada, Y., Coleman, J. R., & Zaret, K. S. (1996). Hepatic specification of the gut endoderm in vitro: Cell signaling and transcriptional control. *Genes & Development, 10*, 1670–1682.

Hans, S., Christison, J., Liu, D., & Westerfield, M. (2007). Fgf-dependent otic induction requires competence provided by Foxi1 and Dlx3b. *BMC Developmental Biology, 7*, 5.

Hans, S., Irmscher, A., & Brand, M. (2013). Zebrafish Foxi1 provides a neuronal ground state during inner ear induction preceding the Dlx3b/4b-regulated sensory lineage. *Development, 140*, 1936–1945.

Hans, S., Liu, D., & Westerfield, M. (2004). Pax8 and Pax2a function synergistically in otic specification, downstream of the Foxi1 and Dlx3b transcription factors. *Development, 131*, 5091–5102.

Hans, S., & Westerfield, M. (2007). Changes in retinoic acid signaling alter otic patterning. *Development, 134*, 2449–2458.

Hatta, M., & Cirillo, L. A. (2007). Chromatin opening and stable perturbation of core histone: DNA contacts by FoxO1. *The Journal of Biological Chemistry, 282*, 35583–35593.

Haworth, K. E., Wilson, J. M., Grevellec, A., Cobourne, M. T., Healy, C., Helms, J. A., et al. (2007). Sonic hedgehog in the pharyngeal endoderm controls arch pattern via regulation of Fgf8 in head ectoderm. *Developmental Biology, 303*, 244–258.

Heller, N., & Brandli, A. W. (1999). Xenopus Pax-2/5/8 orthologues: Novel insights into Pax gene evolution and identification of Pax-8 as the earliest marker for otic and pronephric cell lineages. *Developmental Genetics, 24*, 208–219.

Hoffman, T. L., Javier, A. L., Campeau, S. A., Knight, R. D., & Schilling, T. F. (2007). Tfap2 transcription factors in zebrafish neural crest development and ectodermal evolution. *Journal of Experimental Zoology. Part B, Molecular and Developmental Evolution, 308*, 679–691.

Hsiao, C. D., You, M. S., Guh, Y. J., Ma, M., Jiang, Y. J., & Hwang, P. P. (2007). A positive regulatory loop between foxi3a and foxi3b is essential for specification and differentiation of zebrafish epidermal ionocytes. *PLoS One, 2*, e302.

Hulander, M., Kiernan, A. E., Blomqvist, S. R., Carlsson, P., Samuelsson, E. J., Johansson, B. R., et al. (2003). Lack of pendrin expression leads to deafness and expansion of the endolymphatic compartment in inner ears of Foxi1 null mutant mice. *Development, 130*, 2013–2025.

Hulander, M., Wurst, W., Carlsson, P., & Enerback, S. (1998). The winged helix transcription factor Fkh10 is required for normal development of the inner ear. *Nature Genetics, 20*, 374–376.

Jackson, B. C., Carpenter, C., Nebert, D. W., & Vasiliou, V. (2010). Update of human and mouse forkhead box (FOX) gene families. *Human Genomics, 4*, 345–352.

Janicke, M., Carney, T. J., & Hammerschmidt, M. (2007). Foxi3 transcription factors and Notch signaling control the formation of skin ionocytes from epidermal precursors of the zebrafish embryo. *Developmental Biology, 307*, 258–271.

Janicke, M., Renisch, B., & Hammerschmidt, M. (2010). Zebrafish grainyhead-like1 is a common marker of different non-keratinocyte epidermal cell lineages, which segregate from each other in a Foxi3-dependent manner. *The International Journal of Developmental Biology, 54*, 837–850.

Jayasena, C. S., Ohyama, T., Segil, N., & Groves, A. K. (2008). Notch signaling augments the canonical Wnt pathway to specify the size of the otic placode. *Development, 135*, 2251–2261.

Kadauke, S., Udugama, M. I., Pawlicki, J. M., Achtman, J. C., Jain, D. P., Cheng, Y., et al. (2012). Tissue-specific mitotic bookmarking by hematopoietic transcription factor GATA1. *Cell, 150*, 725–737.

Kaestner, K. H., Knochel, W., & Martinez, D. E. (2000). Unified nomenclature for the winged helix/forkhead transcription factors. *Genes & Development, 14*, 142–146.

Khatri, S. B., Edlund, R. K., & Groves, A. K. (2014). Foxi3 is necessary for the induction of the chick otic placode in response to FGF signaling. *Developmental Biology, 391*, 158–169.

Khatri, S. B., & Groves, A. K. (2013). Expression of the Foxi2 and Foxi3 transcription factors during development of chicken sensory placodes and pharyngeal arches. *Gene Expression Patterns, 13*, 38–42.

Khudyakov, J., & Bronner-Fraser, M. (2009). Comprehensive spatiotemporal analysis of early chick neural crest network genes. *Developmental Dynamics, 238*, 716–723.

Knight, R. D., Nair, S., Nelson, S. S., Afshar, A., Javidan, Y., Geisler, R., et al. (2003). Lockjaw encodes a zebrafish tfap2a required for early neural crest development. *Development, 130*, 5755–5768.

Kozlowski, D. J., Murakami, T., Ho, R. K., & Weinberg, E. S. (1997). Regional cell movement and tissue patterning in the zebrafish embryo revealed by fate mapping with caged fluorescein. *Biochemistry and Cell Biology, 75*, 551–562.

Kroll, K. L., Salic, A. N., Evans, L. M., & Kirschner, M. W. (1998). Geminin, a neuralizing molecule that demarcates the future neural plate at the onset of gastrulation. *Development, 125*, 3247–3258.

Kwon, H. J., Bhat, N., Sweet, E. M., Cornell, R. A., & Riley, B. B. (2010). Identification of early requirements for preplacodal ectoderm and sensory organ development. *PLoS Genetics, 6*(9): e1001133.

Ladher, R. K., O'Neill, P., & Begbie, J. (2010). From shared lineage to distinct functions: The development of the inner ear and epibranchial placodes. *Development, 137*, 1777–1785.

Ladher, R. K., Wright, T. J., Moon, A. M., Mansour, S. L., & Schoenwolf, G. C. (2005). FGF8 initiates inner ear induction in chick and mouse. *Genes & Development, 19*, 603–613.

Lai, E., Clark, K. L., Burley, S. K., & Darnell, J. E., Jr. (1993). Hepatocyte nuclear factor 3/fork head or "winged helix" proteins: A family of transcription factors of diverse biologic function. *Proceedings of the National Academy of Sciences of the United States of America, 90*, 10421–10423.

Lai, E., Prezioso, V. R., Smith, E., Litvin, O., Costa, R. H., & Darnell, J. E., Jr. (1990). HNF-3A, a hepatocyte-enriched transcription factor of novel structure is regulated transcriptionally. *Genes & Development, 4*, 1427–1436.

Lee, C. S., Friedman, J. R., Fulmer, J. T., & Kaestner, K. H. (2005). The initiation of liver development is dependent on Foxa transcription factors. *Nature, 435*, 944–947.

Lee, S. A., Shen, E. L., Fiser, A., Sali, A., & Guo, S. (2003). The zebrafish forkhead transcription factor Foxi1 specifies epibranchial placode-derived sensory neurons. *Development, 130*, 2669–2679.

Leger, S., & Brand, M. (2002). Fgf8 and Fgf3 are required for zebrafish ear placode induction, maintenance and inner ear patterning. *Mechanisms of Development, 119*, 91–108.

Li, W., & Cornell, R. A. (2007). Redundant activities of Tfap2a and Tfap2c are required for neural crest induction and development of other non-neural ectoderm derivatives in zebrafish embryos. *Developmental Biology, 304*, 338–354.

Linker, C., De Almeida, I., Papanayotou, C., Stower, M., Sabado, V., Ghorani, E., et al. (2009). Cell communication with the neural plate is required for induction of neural markers by BMP inhibition: Evidence for homeogenetic induction and implications for xenopus animal cap and chick explant assays. *Developmental Biology, 327*, 478–486.

Litsiou, A., Hanson, S., & Streit, A. (2005). A balance of FGF, BMP and WNT signalling positions the future placode territory in the head. *Development, 132*, 4051–4062.

Liu, D., Chu, H., Maves, L., Yan, Y.-L., Morcos, P. A., Postlethwait, J. H., et al. (2003). Fgf3 and Fgf8 dependent and independent transcription factors are required for otic placode specification. *Development, 130*, 2213–2224.

Liu, W., Selever, J., Murali, D., Sun, X., Brugger, S. M., Ma, L., et al. (2005). Threshold-specific requirements for Bmp4 in mandibular development. *Developmental Biology, 283*, 282–293.

Lorente-Canovas, B., Ingham, N., Norgett, E. E., Golder, Z. J., Karet Frankl, F. E., & Steel, K. P. (2013). Mice deficient in H^+-ATPase a4 subunit have severe hearing impairment associated with enlarged endolymphatic compartments within the inner ear. *Disease Models & Mechanisms, 6*, 434–442.

Luo, T., Matsuo-Takasaki, M., & Sargent, T. D. (2001). Distinct roles for Distal-less genes Dlx3 and Dlx5 in regulating ectodermal development in Xenopus. *Molecular Reproduction and Development, 60*, 331–337.

Macatee, T. L., Hammond, B. P., Arenkiel, B. R., Francis, L., Frank, D. U., & Moon, A. M. (2003). Ablation of specific expression domains reveals discrete functions of ectoderm- and endoderm-derived FGF8 during cardiovascular and pharyngeal development. *Development, 130*, 6361–6374.

Mackereth, M. D., Kwak, S. J., Fritz, A., & Riley, B. B. (2005). Zebrafish pax8 is required for otic placode induction and plays a redundant role with Pax2 genes in the maintenance of the otic placode. *Development, 132*, 371–382.

Mallo, M. (1998). Embryological and genetic aspects of middle ear development. *The International Journal of Developmental Biology, 42*, 11–22.

Mallo, M. (2001). Formation of the middle ear: Recent progress on the developmental and molecular mechanisms. *Developmental Biology, 231*, 410–419.

Manley, G. A. (2010). An evolutionary perspective on middle ears. *Hearing Research, 263*, 3–8.

Maroon, H., Walshe, J., Mahmood, R., Kiefer, P., Dickson, C., & Mason, I. (2002). Fgf3 and Fgf8 are required together for formation of the otic placode and vesicle. *Development, 129*, 2099–2108.

Matsuo-Takasaki, M., Matsumura, M., & Sasai, Y. (2005). An essential role of Xenopus Foxi1a for ventral specification of the cephalic ectoderm during gastrulation. *Development, 132*, 3885–3894.

McCabe, K. L., & Bronner-Fraser, M. (2009). Molecular and tissue interactions governing induction of cranial ectodermal placodes. *Developmental Biology, 332*, 189–195.

McLarren, K. W., Litsiou, A., & Streit, A. (2003). DLX5 positions the neural crest and preplacode region at the border of the neural plate. *Developmental Biology, 259*, 34–47.

Medeiros, D. M., & Crump, J. G. (2012). New perspectives on pharyngeal dorsoventral patterning in development and evolution of the vertebrate jaw. *Developmental Biology, 371*, 121–135.

Milet, C., & Monsoro-Burq, A. H. (2012). Neural crest induction at the neural plate border in vertebrates. *Developmental Biology, 366*, 22–33.

Miller, C. T., Schilling, T. F., Lee, K., Parker, J., & Kimmel, C. B. (2000). Sucker encodes a zebrafish Endothelin-1 required for ventral pharyngeal arch development. *Development, 127*, 3815–3828.

Minoux, M., Kratochwil, C. F., Ducret, S., Amin, S., Kitazawa, T., Kurihara, H., et al. (2013). Mouse Hoxa2 mutations provide a model for microtia and auricle duplication. *Development, 140*, 4386–4397.

Nechiporuk, A., Linbo, T., Poss, K. D., & Raible, D. W. (2007). Specification of epibranchial placodes in zebrafish. *Development, 134*, 611–623.

Nechiporuk, A., Linbo, T., & Raible, D. W. (2005). Endoderm-derived Fgf3 is necessary and sufficient for inducing neurogenesis in the epibranchial placodes in zebrafish. *Development, 132*, 3717–3730.

Neubuser, A., Peters, H., Balling, R., & Martin, G. R. (1997). Antagonistic interactions between FGF and BMP signaling pathways: A mechanism for positioning the sites of tooth formation. *Cell, 90*, 247–255.

Nissen, R. M., Yan, J., Amsterdam, A., Hopkins, N., & Burgess, S. M. (2003). Zebrafish foxi one modulates cellular responses to Fgf signaling required for the integrity of ear and jaw patterning. *Development, 130*, 2543–2554.

Ohyama, T., & Groves, A. K. (2004a). Expression of mouse Foxi class genes in early craniofacial development. *Developmental Dynamics, 231*, 640–646.

Ohyama, T., & Groves, A. K. (2004b). Generation of Pax2-Cre mice by modification of a Pax2 bacterial artificial chromosome. *Genesis, 38*, 195–199.

Ohyama, T., Groves, A. K., & Martin, K. (2007). The first steps towards hearing: Mechanisms of otic placode induction. *The International Journal of Developmental Biology, 51*, 463–472.

Ohyama, T., Mohamed, O. A., Taketo, M. M., Dufort, D., & Groves, A. K. (2006). Wnt signals mediate a fate decision between otic placode and epidermis. *Development, 133*, 865–875.

Ozeki, H., Kurihara, Y., Tonami, K., Watatani, S., & Kurihara, H. (2004). Endothelin-1 regulates the dorsoventral branchial arch patterning in mice. *Mechanisms of Development, 121*, 387–395.

Padanad, M. S., Bhat, N., Guo, B., & Riley, B. B. (2012). Conditions that influence the response to Fgf during otic placode induction. *Developmental Biology, 364*, 1–10.

Padanad, M. S., & Riley, B. B. (2011). Pax2/8 proteins coordinate sequential induction of otic and epibranchial placodes through differential regulation of foxi1, sox3 and fgf24. *Developmental Biology, 351*, 90–98.

Papalopulu, N., & Kintner, C. (1993). Xenopus Distal-less related homeobox genes are expressed in the developing forebrain and are induced by planar signals. *Development*, *117*, 961–975.

Papanayotou, C., Mey, A., Birot, A.-M., Saka, Y., Boast, S., Smith, J. C., et al. (2008). A mechanism regulating the onset of Sox2 expression in the embryonic neural plate. *PLoS Biology*, *6*(1): e2.

Patthey, C., & Gunhaga, L. (2011). Specification and regionalisation of the neural plate border. *The European Journal of Neuroscience*, *34*, 1516–1528.

Patthey, C., & Gunhaga, L. (2013). Signaling pathways regulating ectodermal cell fate choices. *Experimental Cell Research*, *321*, 11–16.

Pera, E., & Kessel, M. (1999). Expression of DLX3 in chick embryos. *Mechanisms of Development*, *89*, 189–193.

Pera, E., Stein, S., & Kessel, M. (1999). Ectodermal patterning in the avian embryo: Epidermis versus neural plate. *Development*, *126*, 63–73.

Perlmann, T. (1992). Glucocorticoid receptor DNA-binding specificity is increased by the organization of DNA in nucleosomes. *Proceedings of the National Academy of Sciences of the United States of America*, *89*, 3884–3888.

Pfeffer, P. L., Gerster, T., Lun, K., Brand, M., & Busslinger, M. (1998). Characterization of three novel members of the zebrafish Pax2/5/8 family: Dependency of Pax5 and Pax8 expression on the Pax2.1 (noi) function. *Development*, *125*, 3063–3074.

Phillips, B. T., Bolding, K., & Riley, B. B. (2001). Zebrafish fgf3 and fgf8 encode redundant functions required for otic placode induction. *Developmental Biology*, *235*, 351–365.

Phillips, B. T., Kwon, H.-J., Melton, C., Houghtaling, P., Fritz, A., & Riley, B. B. (2006). Zebrafish msxB, msxC and msxE function together to refine the neural-nonneural border and regulate cranial placodes and neural crest development. *Developmental Biology*, *294*, 376–390.

Phillips, B. T., Storch, E. M., Lekven, A. C., & Riley, B. B. (2004). A direct role for Fgf but not Wnt in otic placode induction. *Development*, *131*, 923–931.

Pieper, M., Ahrens, K., Rink, E., Peter, A., & Schlosser, G. (2012). Differential distribution of competence for panplacodal and neural crest induction to non-neural and neural ectoderm. *Development*, *139*, 1175–1187.

Prasad, M. S., Sauka-Spengler, T., & LaBonne, C. (2012). Induction of the neural crest state: Control of stem cell attributes by gene regulatory, post-transcriptional and epigenetic interactions. *Developmental Biology*, *366*, 10–21.

Raft, S., Andrade, L. R., Shao, D., Akiyama, H., Henkemeyer, M., & Wu, D. K. (2014). Ephrin-B2 governs morphogenesis of endolymphatic sac and duct epithelia in the mouse inner ear. *Developmental Biology*, *390*, 51–67.

Rex, M., Orme, A., Uwanogho, D., Tointon, K., Wigmore, P. M., Sharpe, P. T., et al. (1997). Dynamic expression of chicken Sox2 and Sox3 genes in ectoderm induced to form neural tissue. *Developmental Dynamics*, *209*, 323–332.

Riley, B. B., & Phillips, B. T. (2003). Ringing in the new ear: Resolution of cell interactions in otic development. *Developmental Biology*, *261*, 289–312.

Robledo, R. F., & Lufkin, T. (2006). Dlx5 and Dlx6 homeobox genes are required for specification of the mammalian vestibular apparatus. *Genesis*, *44*, 425–437.

Robledo, R. F., Rajan, L., Li, X., & Lufkin, T. (2002). The Dlx5 and Dlx6 homeobox genes are essential for craniofacial, axial, and appendicular skeletal development. *Genes & Development*, *16*, 1089–1101.

Ruest, L. B., Xiang, X., Lim, K. C., Levi, G., & Clouthier, D. E. (2004). Endothelin-A receptor-dependent and -independent signaling pathways in establishing mandibular identity. *Development*, *131*, 4413–4423.

Saint-Jeannet, J. P., & Moody, S. A. (2014). Establishing the pre-placodal region and breaking it into placodes with distinct identities. *Developmental Biology*, *389*, 13–27.

Sato, S., Ikeda, K., Shioi, G., Ochi, H., Ogino, H., Yajima, H., et al. (2010). Conserved expression of mouse Six1 in the pre-placodal region (PPR) and identification of an enhancer for the rostral PPR. *Developmental Biology, 344*, 158–171.

Sauka-Spengler, T., & Bronner-Fraser, M. (2008). A gene regulatory network orchestrates neural crest formation. *Nature Reviews. Molecular Cell Biology, 9*, 557–568.

Schlosser, G. (2006). Induction and specification of cranial placodes. *Developmental Biology, 294*, 303–351.

Schlosser, G. (2010). Making senses development of vertebrate cranial placodes. *International Review of Cell and Molecular Biology, 283*, 129–234.

Sheng, G., & Stern, C. D. (1999). Gata2 and Gata3: Novel markers for early embryonic polarity and for non-neural ectoderm in the chick embryo. *Mechanisms of Development, 87*, 213–216.

Shigetani, Y., Nobusada, Y., & Kuratani, S. (2000). Ectodermally derived FGF8 defines the maxillomandibular region in the early chick embryo: Epithelial-mesenchymal interactions in the specification of the craniofacial ectomesenchyme. *Developmental Biology, 228*, 73–85.

Shirokova, V., Jussila, M., Hytonen, M. K., Perala, N., Drogemuller, C., Leeb, T., et al. (2013). Expression of Foxi3 is regulated by ectodysplasin in skin appendage placodes. *Developmental Dynamics, 242*, 593–603.

Sienknecht, U. J. (2013). Developmental origin and fate of middle ear structures. *Hearing Research, 301*, 19–26.

Solomon, K. S., & Fritz, A. (2002). Concerted action of two dlx paralogs in sensory placode formation. *Development, 129*, 3127–3136.

Solomon, K. S., Kudoh, T., Dawid, I. B., & Fritz, A. (2003). Zebrafish foxi1 mediates otic placode formation and jaw development. *Development, 130*, 929–940.

Solomon, K. S., Kwak, S.-J., & Fritz, A. (2004). Genetic interactions underlying otic placode induction and formation. *Developmental Dynamics, 230*, 419–433.

Solomon, K. S., Logsdon, J. M., & Fritz, A. (2003). Expression and phylogenetic analyses of three zebrafish FoxI class genes. *Developmental Dynamics, 228*, 301–307.

Steventon, B., Mayor, R., & Streit, A. (2014). Neural crest and placode interaction during the development of the cranial sensory system. *Developmental Biology, 389*, 28–38.

Stottmann, R. W., Anderson, R. M., & Klingensmith, J. (2001). The BMP antagonists Chordin and Noggin have essential but redundant roles in mouse mandibular outgrowth. *Developmental Biology, 240*, 457–473.

Streit, A. (2001). Origin of the vertebrate inner ear: Evolution and induction of the otic placode. *Journal of Anatomy, 199*, 99–103.

Streit, A. (2002). Extensive cell movements accompany formation of the otic placode. *Developmental Biology, 249*, 237–254.

Streit, A. (2004). Early development of the cranial sensory nervous system: From a common field to individual placodes. *Developmental Biology, 276*, 1–15.

Streit, A. (2007). The preplacodal region: An ectodermal domain with multipotential progenitors that contribute to sense organs and cranial sensory ganglia. *The International Journal of Developmental Biology, 51*, 447–461.

Streit, A., Berliner, A. J., Papanayotou, C., Sirulnik, A., & Stern, C. D. (2000). Initiation of neural induction by FGF signalling before gastrulation. *Nature, 406*, 74–78.

Streit, A., Sockanathan, S., Perez, L., Rex, M., Scotting, P. J., Sharpe, P. T., et al. (1997). Preventing the loss of competence for neural induction: HGF/SF, L5 and Sox-2. *Development, 124*, 1191–1202.

Stuhlmiller, T. J., & Garcia-Castro, M. I. (2012). Current perspectives of the signaling pathways directing neural crest induction. *Cellular and Molecular Life Sciences, 69*, 3715–3737.

Szabo-Rogers, H. L., Smithers, L. E., Yakob, W., & Liu, K. J. (2010). New directions in craniofacial morphogenesis. *Developmental Biology, 341*, 84–94.

Tassano, E., Jagannathan, V., Drögemüller, C., Leoni, M., Hytönen, M. K., Severino, M., et al. (2014). Congenital aural atresia associated with agenesis of internal carotid artery in a girl with FOXI3 deletion. *American Journal of Medical Genetics,* in press.

Thermes, V., Lin, C. C., & Hwang, P. P. (2010). Expression of Ol-foxi3 and Na(+)/K(+)-ATPase in ionocytes during the development of euryhaline medaka (Oryzias latipes) embryos. *Gene Expression Patterns, 10,* 185–192.

Thompson, H., & Tucker, A. S. (2013). Dual origin of the epithelium of the mammalian middle ear. *Science, 339,* 1453–1456.

Tribulo, C., Aybar, M. J., Nguyen, V. H., Mullins, M. C., & Mayor, R. (2003). Regulation of Msx genes by a Bmp gradient is essential for neural crest specification. *Development, 130,* 6441–6452.

Trumpp, A., Depew, M. J., Rubenstein, J. L., Bishop, J. M., & Martin, G. R. (1999). Cre-mediated gene inactivation demonstrates that FGF8 is required for cell survival and patterning of the first branchial arch. *Genes & Development, 13,* 3136–3148.

Tucker, A. S., Yamada, G., Grigoriou, M., Pachnis, V., & Sharpe, P. T. (1999). Fgf-8 determines rostral-caudal polarity in the first branchial arch. *Development, 126,* 51–61.

Uchikawa, M., Ishida, Y., Takemoto, T., Kamachi, Y., & Kondoh, H. (2003). Functional analysis of chicken Sox2 enhancers highlights an array of diverse regulatory elements that are conserved in mammals. *Developmental Cell, 4,* 509–519.

Urness, L. D., Li, C., Wang, X., & Mansour, S. L. (2008). Expression of ERK signaling inhibitors Dusp6, Dusp7, and Dusp9 during mouse ear development. *Developmental Dynamics, 237,* 163–169.

Urness, L. D., Paxton, C. N., Wang, X., Schoenwolf, G. C., & Mansour, S. L. (2010). FGF signaling regulates otic placode induction and refinement by controlling both ectodermal target genes and hindbrain Wnt8a. *Developmental Biology, 340,* 595–604.

Vaglia, J. L., & Hall, B. K. (1999). Regulation of neural crest cell populations: Occurrence, distribution and underlying mechanisms. *The International Journal of Developmental Biology, 43,* 95–110.

Vidarsson, H., Westergren, R., Heglind, M., Blomqvist, S. R., Breton, S., & Enerback, S. (2009). The forkhead transcription factor Foxi1 is a master regulator of vacuolar H-ATPase proton pump subunits in the inner ear, kidney and epididymis. *PLoS One, 4,* e4471.

Wall, N. A., & Hogan, B. L. (1995). Expression of bone morphogenetic protein-4 (BMP-4), bone morphogenetic protein-7 (BMP-7), fibroblast growth factor-8 (FGF-8) and sonic hedgehog (SHH) during branchial arch development in the chick. *Mechanisms of Development, 53,* 383–392.

Weigel, D., & Jackle, H. (1990). The fork head domain: A novel DNA binding motif of eukaryotic transcription factors? *Cell, 63,* 455–456.

Weigel, D., Jurgens, G., Kuttner, F., Seifert, E., & Jackle, H. (1989). The homeotic gene fork head encodes a nuclear protein and is expressed in the terminal regions of the Drosophila embryo. *Cell, 57,* 645–658.

Woda, J. M., Pastagia, J., Mercola, M., & Artinger, K. B. (2003). Dlx proteins position the neural plate border and determine adjacent cell fates. *Development, 130,* 331–342.

Wright, T. J., & Mansour, S. L. (2003). Fgf3 and Fgf10 are required for mouse otic placode induction. *Development, 130,* 3379–3390.

Xu, H., Dude, C. M., & Baker, C. V. H. (2008). Fine-grained fate maps for the ophthalmic and maxillomandibular trigeminal placodes in the chick embryo. *Developmental Biology, 317,* 174–186.

Xu, J., Watts, J. A., Pope, S. D., Gadue, P., Kamps, M., Plath, K., et al. (2009). Transcriptional competence and the active marking of tissue-specific enhancers by defined transcription factors in embryonic and induced pluripotent stem cells. *Genes & Development, 23,* 2824–2838.

Yamagishi, H., Maeda, J., Hu, T., McAnally, J., Conway, S. J., Kume, T., et al. (2003). Tbx1 is regulated by tissue-specific forkhead proteins through a common Sonic hedgehog-responsive enhancer. *Genes & Development, 17*, 269–281.

Yan, J., Xu, L., Crawford, G., Wang, Z., & Burgess, S. M. (2006). The forkhead transcription factor FoxI1 remains bound to condensed mitotic chromosomes and stably remodels chromatin structure. *Molecular and Cellular Biology, 26*, 155–168.

Yang, L., O'Neill, P., Martin, K., Maass, J. C., Vassilev, V., Ladher, R., et al. (2013). Analysis of FGF-dependent and FGF-independent pathways in otic placode induction. *PLoS One, 8*, e55011.

Yang, T., Vidarsson, H., Rodrigo-Blomqvist, S., Rosengren, S. S., Enerback, S., & Smith, R. J. (2007). Transcriptional control of SLC26A4 is involved in Pendred syndrome and nonsyndromic enlargement of vestibular aqueduct (DFNB4). *American Journal of Human Genetics, 80*, 1055–1063.

Zaret, K. S., & Carroll, J. S. (2011). Pioneer transcription factors: Establishing competence for gene expression. *Genes & Development, 25*, 2227–2241.

Zelarayan, L. C., Vendrell, V., Alvarez, Y., Dominguez-Frutos, E., Theil, T., Alonso, M. T., et al. (2007). Differential requirements for FGF3, FGF8 and FGF10 during inner ear development. *Developmental Biology, 308*, 379–391.

Zhang, Y., Knosp, B. M., Maconochie, M., Friedman, R. A., & Smith, R. J. H. (2004). A comparative study of Eya1 and Eya4 protein function and its implication in branchio-oto-renal syndrome and DFNA10. *Journal of the Association for Research in Otolaryngology, 5*, 295–304.

Zheng, W., Huang, L., Wei, Z.-B., Silvius, D., Tang, B., & Xu, P.-X. (2003). The role of Six1 in mammalian auditory system development. *Development, 130*, 3989–4000.

Zou, D., Silvius, D., Rodrigo-Blomqvist, S., Enerback, S., & Xu, P. X. (2006). Eya1 regulates the growth of otic epithelium and interacts with Pax2 during the development of all sensory areas in the inner ear. *Developmental Biology, 298*, 430–441.

CHAPTER FIFTEEN

The Use of Human Pluripotent Stem Cells for the *In Vitro* Derivation of Cranial Placodes and Neural Crest Cells

Bianca E. Borchin*,†, Tiziano Barberi*,‡,1

*Pluripotent Stem Cell Differentiation Laboratory, Southwest National Primate Research Center, Texas Biomedical Research Institute, San Antonio, Texas, USA
†Australian Regenerative Medicine Institute, Monash University, Clayton, Victoria, Australia
‡Department of Anatomy Neuroscience, The University of Melbourne, Parkville, Victoria, Australia
[1]Corresponding author: e-mail address: tbarberi@txbiomed.org

Contents

Abstract

Due to their intrinsic differentiation potential, human pluripotent stem cells (hPSCs) hold remarkable promise for their use in cell-based therapies as well as an *in vitro* model for early human embryogenesis and for modeling disease.

During the development of the human embryo, transient structures such as the neural crest (NC) and the cranial placodes (CPs) are specified in the first 3–4 weeks of gestation. Because of this early occurrence and a scarce availability of embryos for research purposes, these transient structures remain largely unexplored in humans.

Hence, investigators are now exploiting *in vitro* differentiation of hPSC to unveil these early events and to generate NC and CP cells *in vitro*. Derivatives of the NC and CPs will contribute to the formation of very important organs, including most of

Current Topics in Developmental Biology, Volume 111
ISSN 0070-2153
http://dx.doi.org/10.1016/bs.ctdb.2014.11.015

the peripheral nervous system (NC) and the sensory organs of the head (CP). There are many diseases and conditions that affect NC and CP derivatives, thus a better knowledge of how these structures specialize, and the derivation of functional NC and CP cells for therapeutic applications will have an impact on the understanding and treatment of these disorders. Here, we discuss the current state of the art in directing hPSCs into NC or CP cells, which in spite of their importance is still in its infancy.

ABBREVIATIONS

BMP bone morphogenic protein
CNS central nervous system
FACS fluorescence-activated cell sorting
FGF fibroblast growth factor

1. INTRODUCTION

Human pluripotent stem cells (hPSCs) offer an extraordinary tool to investigate the early events taking place during the first 4 weeks of embryonic development. One of these events is the specification of the neural plate border (NPB), an area of cells located at the junction between the neural plate and nonneural ectoderm, which despite existing only temporarily, has a complex diversity in terms of cell fate potential (Patthey & Gunhaga, 2011). The convergence of multiple molecular signals deriving from the neural plate, the nonneural ectoderm, and the underlying mesoderm allow cells located within different areas of the NPB to be specified as neural crest (NC) and, in the most anterior portion, as cranial placode (CP) progenitors (Bhat, Kwon, & Riley, 2013). NC cells are multipotent migratory cells that will differentiate into a variety of specialized cells in different tissues and organs upon reaching their final destinations. NC derivatives include the bone and cartilage of the face, neurons, and glia of the peripheral nervous system (PNS), the pigmented cells of the skin, and some enteric gut cells.

CPs are spatially restricted thickenings of the nonneural ectoderm that form only cranially and lateral to the NPB (Streit, 2007). These localized thickened areas become visible at the end of neurulation and possess specific fate identity (lens, olfactory, hypophyseal, otic, epibranchial, trigeminal, and in aquatic species—lateral line). Overall, they will contribute to the sense organs and sensory ganglia of the head.

In the past two decades, the use of model organisms (zebrafish, frog, chick, and mouse) enabled a comprehensive analysis of the signaling and gene regulatory network underlying NC (Sauka-Spengler & Bronner-Fraser, 2008) and CP induction and specification (Streit, 2007). However, investigating the formation and specialization of NC and CP in humans has been very challenging because of the lack of access to and ethics of working on early embryos. Thus, the recent availability of hPSCs overcomes these limitations, facilitating the modeling of early embryonic development and the etiology and pathogenesis of disease.

2. PLURIPOTENT STEM CELLS

The isolation of human embryonic stem cells (hESCs) by Thomson et al. (1998) was a game changer in the fields of human biology and regenerative medicine. The availability of pluripotent cells that can be directed to become every tissue and cell type of our body transformed our prospective vision about the use of stem cells for therapeutic purposes. In addition, it raised the opportunity to study early human development in a dish, a task that would have been otherwise impossible because of the limitations and scarce availability of human embryos for research purposes. As such, pluripotent stem cell differentiation has been particularly useful to recapitulate the earliest events of human development allowing for the first time to study and model the NPB and its derivatives (NC and CP) in a Petri dish. The discovery of induced pluripotent stem cells (iPSCs) by cellular reprogramming (Takahashi et al., 2007) expanded these possibilities even further. It is now possible not only to derive normal ESC-like cells but also to model human disease *in vitro* by deriving iPSCs from patients. By differentiating iPSCs into the specific cells that are affected by a certain condition, this powerful tool will allow testing for new drugs or compounds and the design of new strategies to correct the molecular lesions causing the disease (Grskovic, Javaherian, Strulovici, & Daley, 2011).

3. DERIVATION OF NC CELLS FROM hESC/hIPSC

It has been shown that neural differentiation in PSCs occurs by default with an anterior and mostly ventral identity (Ozair, Noggle, Warmflash, Krzyspiak, & Brivanlou, 2013; Zeng et al., 2010). Since NC cells are specified at the NPB and migrate out from the dorsal neural tube during neurulation (Wilson, Richards, Ford-Perriss, Panthier, & Murphy, 2004), in order to efficiently obtain bona fide NC cells from PSCs, the system needs to be manipulated. Dorsalization of the emerging neuroectoderm becomes a prerequisite for NC derivation. In some reports, investigators claimed derivation of NC-like cells from PSC-derived neural cultures where Pax6 expression was one of the main neural markers (Cimadamore et al., 2011; Curchoe et al., 2010). Early in development, the transcription factor Pax6 is expressed in the ventral portion of the neural tube (Ericson et al., 1997), in the optic cup/eye field, and in some CP derivatives (Yan et al., 2010). Therefore, its presence together with NC cells is an indication of low efficiency and should encourage alterations to optimize the methods for more efficient neural differentiation.

3.1. Coculture method

A method originally developed to derive neural cells from mouse ESCs, which was based on the coculture of ESCs with bone marrow-derived stromal cell lines [PA6 (Kawasaki et al., 2000) and MS5 (Barberi et al., 2003)], led to the first description of NC-derived peripheral neurons from mouse and nonhuman primate PSCs (Mizuseki et al., 2003). This method centered on what has been called stromal-derived inducing activity (SDIA) (Kawasaki et al., 2000) generated neural cells at a very high efficiency and provided a platform to derive dopamine neurons (Barberi et al., 2003; Kawasaki et al., 2000) and a comprehensive list of central nervous system (CNS) neurons and glia cells (Barberi et al., 2003). To derive NC cells, the SDIA method was used to specify neural cells which were then treated with bone morphogenetic protein (BMP)-4, a morphogen known to play a role in NC specification (Reichert, Randall, & Hill, 2013) and to act as a mild dorsalizing agent (Jessell, 2000). This BMP-4 treatment was sufficient to generate cells expressing markers of sensory and autonomic neurons. Similar results were later obtained by differentiating hESC on PA6 stromal cells but without exposing the cultures to BMP-4 (Pomp, Brokhman, Ben-Dor, Reubinoff, & Goldstein, 2005). Interestingly, Pomp and colleagues did not report any advantage in terms of efficiency when hESCs were treated with BMP-4 versus nontreated cells (SDIA only). However, the overall percentage of NC cells obtained in this study was measured in terms of AP2a-expressing cells, a transcription factor expressed not only by NC cells but also by nonneural ectoderm and CP cells (Li & Cornell, 2007). This raises questions about the real efficiency of this method.

The coculture of hESCs with stromal cells provided the foundation for the first report describing the derivation and FACS purification of putative neural crest stem cells (NCSC) (Lee et al., 2007). Generally, early NC cells are identified by their expression of two surface markers, the low-affinity nerve growth factor receptor p75 (p75) and the carbohydrate moiety CD57/HNK-1 (HNK). Both have been previously shown to be markers of rodent NC cells (p75) (Morrison, White, Zock, & Anderson, 1999; Stemple & Anderson, 1992) and avian NC cells (HNK) (Bronner-Fraser, 1987). In the coculture method, neural rosettes derived from different hESC lines differentiated on MS5 stromal cells were first mechanically isolated and then replated in the absence of stromal cells (Lee et al., 2007). Cells were then treated with (FGF)2 and BMP2 to increase the percentage of cells expressing p75 and HNK. While in a previous report NC cells were isolated using p75 from preimplantation bovine blastocysts induced to acquire a neural fate (Lazzari et al., 2006), this was the first report where NC cells were isolated from hESCs by FACS using also HNK. The hESC-derived NCSC were then differentiated into a variety of different NC fates such as peripheral neurons and glia, smooth muscle, osteocytes, chondrocytes, and adipocytes. The isolation of NCSC from hESCs and their guided differentiation into specific fates marked a substantial advancement in the field. However, the presence of large numbers of Pax6+ neural rosettes in these cultures raised questions about the overall efficiency of this protocol. Subsequently, another study confirmed these findings and relied on p75 for the purification by FACS of NCSC (Jiang et al., 2009).

Two consecutive studies proved that the approach using hPSCs to derive human NC cells is useful for *in vitro* disease modeling (Lee et al., 2009, 2012). First, hiPSC were derived from patients with familiar dysautonomia (FD), a rare but severe peripheral neuropathy, and shown to recapitulate the disease *in vitro* (Lee et al., 2009). Second, a large-scale, small molecule compound screen on NC cells derived from FD hiPSCs uncovered several compounds that positively modulated IKBKAP expression, the gene deficient in FD (Lee et al., 2012). This serves to validate the use of hiPSCs for the identification of candidate drugs for potential clinical applications.

3.2. Embryoid bodies method

A widely used method to differentiate PSCs is based on the formation of embryoid bodies (EBs), spherical aggregates growing in suspension where PSCs spontaneously differentiate into derivatives of all germ layers. Although very popular to differentiate mouse ESCs, this method has not been used extensively with hESCs. Mainly because hESCs do not form EBs very easily unless small aggregates of 50–100 cells are used. Our personal view is also that because of the size and tightness of the aggregates, treatment with growth factors/small molecules to direct differentiation could be ineffective in EBs, as cells at the center would not receive the appropriate dose. Nonetheless, the EB approach has been used to derive NC cells from hPSCs in numerous studies (Bajpai et al., 2010; Cimadamore et al., 2011; Curchoe et al., 2010; Fang et al., 2006; Rada-Iglesias et al., 2012; Zhou & Snead, 2008).

By using the EB method, Fang and colleagues were the first to describe the derivation of melanocytes from hESCs (Fang et al., 2006). Without going through any intermediate state, they demonstrated that exposure of 4-day-old EBs to endothelin-3 (end-3) (Pla & Larue, 2003) was sufficient to generate directly pigmented cells and that the efficiency was enhanced by the addition of Wnt-3a and c-Kit ligand. In an effort to selectively derive cranial NC cells, which comprise the most anterior part of the NC mostly dedicated to provide cells to build the bones and cartilage of the face, Zhou and colleagues attempted to isolate NC cells based on their expression of Frizzled-3 (FZD3) and Cadherin-11 (CAD11) (Zhou & Snead, 2008). FZD3 is a coreceptor for Wnt-ligands (Katsu et al., 2003), and CAD11 is an adhesion molecule that is highly expressed in bone cells (Lecanda et al., 2000). After growing EBs for 10 days, NC cells were purified by FACS based on the expression of these two markers, and although FZD3 and CAD11 are known to be expressed by NC cells, the efficiency in isolating NC cells using this approach was very low (1%). It is not clear whether only cranial NC cells were isolated using this approach since not all cranial NC cell types were obtained using this strategy. However, a more directed differentiation protocol could have probably improved the efficiency of differentiation.

The EB method has subsequently been modified (Bajpai et al., 2010; Cimadamore et al., 2011; Curchoe et al., 2010; Rada-Iglesias et al., 2012), such that forming EBs cultured directly in N_2 medium (Bottenstein & Sato, 1979) differentiate predominantly into neuroepithelial cells. These EBs are then transferred to adherent conditions, and because of the biased neural differentiation, NC cells form and migrate out from the neuroepithelial structures. NC cells can then be harvested and analyzed. Interestingly, using this method, p75 and HNK were not only expressed on the migrating NC-like cells but also in most of the cell types present in the dish, including the neuroepithelial cells (Curchoe et al., 2010). This cautions the use of these two markers for the isolation of NC cells from differentiating PSCs under these particular conditions.

The modified EB method has nonetheless been successfully used for more basic studies of early migrating NC progenitors. Two independent studies support the use of hPSC to study NC function in humans, demonstrating first, that the *CDH7* gene, which encodes for Cadherin-7, plays an active role in the migration ability of human NC. This finding was subsequently confirmed in *CDH7*-mutant zebrafish embryos (Bajpai et al., 2010). Second, hPSC-derived NC cells were used to identify important transcriptional regulators of human NC formation (Rada-Iglesias et al., 2012). Furthermore, analyses of role of *Sox2* in NC-derived sensory ganglia (Cimadamore et al., 2011) revealed that its deficiency affects the number of these cells *in vitro* and *in vivo*.

3.3. Monolayer method

Human PSCs colonies growing on an acellular matrix can be easily induced to differentiate by simply swapping maintenance media with differentiation media. The absence of feeder cells or the need to form complex structures like EBs makes this differentiation method very convenient to work with and yet very efficient. Although neural differentiation can be induced in a monolayer by simply adjusting hPSC colony size and density in the cultures (Mengarelli & Barberi, 2013), a specific treatment based on SMAD inhibition has been described to optimize neural conversion (Chambers et al., 2009). A caveat in all methods used for NC derivation is the tendency of hPSCs to acquire a ventral phenotype during neural induction. This decrease the efficiency in obtaining NC cells, which are a dorsally derived cell type. To overcome this problem, neural conversion can be achieved using two small molecules, an ALK inhibitor (Inman et al., 2002) and a Wnt agonist, the GSK-3 inhibitor (Cohen & Goedert, 2004), to dorsalize the forming neuroepithelium. This provides a more directed approach for efficiently converting hPSCs into NC cells as demonstrated by a marked decrease in Pax6 and Sox2 expression and dramatic increase in the expression of early NC markers (Menendez, Yatskievych, Antin, & Dalton, 2011). NC precursors were then isolated by FACS using p75 and HNK and differentiated into diverse NC lineages via standard protocols. This was the first protocol specifically designed for NC derivation rather than neural differentiation.

The sustained use of a GSK-3 inhibitor for 2 weeks created the conditions for the direct conversion of PSC into dorsal neuroectodermal cells. Subsequent reports confirmed the use of a Wnt agonist for the fast and efficient conversion of hPSC into NC cells (Chambers et al., 2012; Mica, Lee, Chambers, Tomishima, & Studer, 2013). Initially, a combination of small molecules to inhibit several signaling pathways including Notch, FGF, platelet derived growth factor (PDGF), vascular endothelium growth factor (VEGF) in addition to Wnt activation led to a fast and biased differentiation of hPSCs into NC-derived sensory nociceptor neurons (Chambers et al., 2012). Overall, the effect of the small molecule treatment was not just to drive differentiation but also to accelerate the process. Furthermore, the same approach was used together with exogenous BMP-4 and End-3 to selectively bias NC progenitors toward a melanocyte phenotype (Mica et al., 2013). Interestingly, the introduction of a FACS purification strategy based on SOX10 GFP reporter expression helped to improve the identification of bona fide NC cells. To select only for melanoblasts, cells were also labeled with c-Kit antigen, which is highly expressed in pigmented cells (Ito et al., 1999). The high efficiency in derivation of melanocytes enabled the use of this protocol in association with pigmentation disorder patient-specific hiPSCs confirming the disease phenotype in a dish (Mica et al., 2013).

Finally, in a more recent report, NC cells derived from hPSC using a similar small molecule approach centered on a Wnt agonist showed that active Notch signaling is also required for efficient NC derivation (Noisa et al., 2014). It is important to note, however, that this contrasts with studies where Notch inhibition has previously been used to increase the efficiency of NC derivation (Chambers et al., 2012). Thus, further studies will be needed to resolve the role, if any, of the Notch signaling pathway in the *in vitro* generation of NC cells from hPSCs. A summary of the relevant work on NC derivation is described in Table 1, and a schematic representation of the differentiation protocols is illustrated in Fig. 1.

3.4. Comments on NC derivation

The utility of the derivation of NC cells based on coculture with stromal cells, suffered due to widespread variation in the results. The large numbers of cells expressing Pax6 in the cultures even after BMP treatment showed that in contrast with the dorsal identity of NC cells a high proportion of the cells maintained a more ventral CNS phenotype. Furthermore, the use of murine stromal cells for the neural conversion of hPSCs was a limiting factor. As the field became more mechanistic and translational, the use of murine cells influenced the results in an inconsistent manner, which is unacceptable for the generation of cells for preclinical studies.

Table 1 NC derivation

Year	Culture method	FACS	NC derivatives
Stromal (PA6, MS5) coculture			
Pomp et al. (2005)	hESC	No sort	Neurons
Lee et al. (2007)	hESC	p75, HNK-1	NC stem cells, neurons, glia, smooth muscle, bone, cartilage, fat
Lee et al. (2009)	hESC, hiPSC, Disease hiPSC	HNK-1	NC stem cells, neurons
Lee et al. (2012)	hESC, hiPSC, Disease hiPSC	HNK-1	Neurons
Embryoid bodies			
Fang et al. (2006)	hESC	No sort	Melanocytes
Zhou and Snead (2008)	hESC	FZD3, CAD11	NC stem cells, neurons, glia, smooth muscle, bone, cartilage
Bajpai et al. (2010)	hESC	No sort	Migrating NC cells
Rada-Iglesias et al. (2012)	hESC	No sort	Migrating NC cells
Monolayer + small molecules			
Menendez et al. (2011)	hESC, hiPSC	p75, HNK-1	NC stem cells, neurons, smooth muscle, bone, cartilage, fat
Chambers et al. (2012)	hESC, hiPSC	SOX10: GFP	NC stem cells, nociceptors
Mica et al. (2013)	hESC, hiPSC, Disease hiPSC	SOX10: GFP, c-Kit	Melanocytes

An interesting note on the work performed using the EB method is the lack of any growth factor/small molecule treatment required to obtain NC cells. Thus, this method which is heavily used for default neural differentiation may not generate cells of primarily anterior and ventral character as previously reported. However, the lack of a guided stepwise protocol and clear data about the efficiency of this protocol continue to raise questions about its validity and utility.

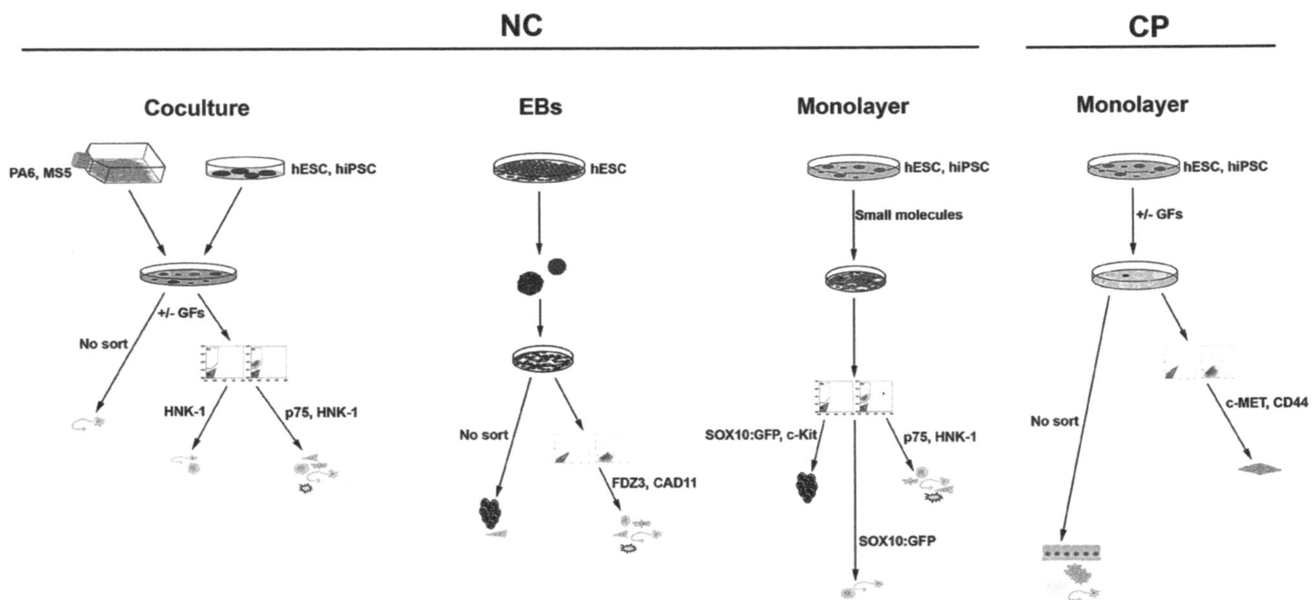

Figure 1 Schematic representation of current methods for NC and CP derivation, illustrating all differentiation steps from undifferentiated hPSCs to specialized NC and CP cells. (See the color Plate.)

In contrast, the monolayer culture method together with small molecule treatment, in particular, Wnt agonists, currently looks the most promising of all. The expression of dorsal markers induced by GSK-3 inhibition creates conditions for the most efficient and controlled derivation of NC cells. Interestingly, the robustness of this strategy has been also confirmed where NC derivation was not the main focus of the study. GSK inhibition appears to be essential to obtain LMX1a expression during early hPSCs differentiation (Borchin, Chen, & Barberi, 2013). *LMX1a* is a dorsal neural tube marker whose expression domain extends to the NPB, encompassing the territory from which NC are derived during early neurulation (Chizhikov & Millen, 2004).

There is also some discordant data related to the sorting strategies used to derive NC cells. The selective markers, p75 and HNK, are widely expressed during early hPSC differentiation. In fact, two p75 + subpopulations, low and high may exist during hPSC differentiation, with the latter enriched in NC cells (Menendez et al., 2011). Furthermore, in GSK-3-inhibitor-treated cells, the majority if not all of the p75 + cells exhibit high levels of p75 and are NC cells. These findings were confirmed in an independent study where only p75 + high cells were identified as NC cells (Mengarelli & Barberi, 2013). Interestingly, in another study where p75 was used as a marker to purify NC cells, the FACS profile revealed the presence of two distinct populations of p75 cells, even though the investigators did not specifically gate these populations (Jiang et al., 2009). Collectively, these results suggest that p75 could be indeed used as a NC marker, but only p75 + high cells are selected for, and above all, only if specific culture conditions for the directed differentiation of NC are met. A histological analysis performed on human embryos confirmed the presence of p75 in pre- and postmigratory NC cells, while HNK expression has been limited to only a few NC cells (Betters, Liu, Kjaeldgaard, Sundstrom, & Garcia-Castro, 2010). This raises questions about the validity and utility of HNK1 as a marker for NC isolation from hPSCs.

In mouse studies, c-Kit has also been used to isolate multipotent NC cells from mESCs (Motohashi, Aoki, Chiba, Yoshimura, & Kunisada, 2007). However, in humans, the use of c-Kit has been restricted to melanoblast isolation (Mica et al., 2013). FDZ3 and CAD11 have been used to isolate NCC from hPSC, but they do so at too low efficiency to be practical (Zhou & Snead, 2008). The lack of other specific markers warrants the investigation of new strategies for the purification of NC cells. Interestingly, while a SOX10 reporter has been utilized to isolate NC cells (Chambers et al.,

2012; Mica et al., 2013), it is unlikely that it would be used for the isolation of NC cells deployed to clinical applications. The green fluorescent protein (GFP), whose expression is also driven in the SOX10 reporter, has been proved to be immunogenic (Stripecke et al., 1999).

4. DERIVATION OF CP CELLS FROM hESC/hIPSC

Human PSCs provide a unique tool to study CPs and their derivatives and for generating placodal cells for therapeutic purposes. Although CPs originate like NC cells from the NPB region and contribute to the formation of essential sensory and endocrine organs, they are less well studies compared to NCC. Investigation of the very early specification and immediate specialization of the placodal territories has proved challenging due to their rapid development. PSCs therefore represent a very promising model to study CP development and differentiation due to their slower rates of development in mammals. Two elegant studies using murine PSCs have proved the potential for pluripotent cells to develop into CP-derived organoids (Koehler, Mikosz, Molosh, Patel, & Hashino, 2013; Suga et al., 2011). In particular, the self-assembly of an anterior adenohypophysis *in vitro* was observed following placodeal ectoderm induction (Suga et al., 2011). In addition, inner ear sensory epithelium was also generated in 3D cultures of mouse PSCs, thus recapitulating otic placode development *in vitro* (Koehler et al., 2013).

4.1. Monolayer method

The first clear evidence describing the directed derivation of human CP cells came from studies growing hPSCs in monolayer culture (Yang et al., 2010). Using sequential growth factor treatment, Noggin was shown to first induce neuroectoderm/ectoderm, which could then be used to derive lens placode and lens epithelium via BMPs. Furthermore, the time-dependent maturation of lens progenitor cells into lentoid bodies and lens cell aggregates secreting specific lens proteins could be observed. These findings were subsequently extended to patient-specific hiPSC derived from age-related cataract lens epithelial cells (LEC) (Qiu et al., 2012). LEC from the capsular bag of patients were first reprogrammed into iPSCs, which were then differentiated into lens cells. The iPSC derived from LEC differentiate into lens cells at a higher efficiency when compared to normal iPSCs.

Subsequently, methods were developed to derive lens epithelium from hPSCs, which could then be isolated by FACS (Mengarelli & Barberi, 2013). Two surface markers, c-MET (Wormstone, Tamiya, Marcantonio, & Reddan, 2000) and CD44 (Saika et al., 1998), known to be expressed in human LEC, facilitated the purification of lens epithelium emerging from differentiating hPSCs and their expansion *in vitro*. LEC were then induced to differentiate and form lentoid bodies (Mengarelli & Barberi, 2013). This was the first report showing isolation by FACS of a CP derivative, which is a fundamental step toward the use of hPSC-derived CP cells in regenerative medicine.

In additional elegant work, the *in vitro* derivation of other anterior CP has also been achieved. Otic progenitors have now been derived from hPSCs, and their functionality proved in a mouse model of auditory neuropathy (Chen et al., 2012). Treating hPSCs with two FGFs, FGF3 and 10, promoted the formation of otic placode cells, which were able to further differentiate into two types of otic progenitors, hair cells, and auditory neurons. The combinatorial transplant of these cells improved the auditory response in the treated animals.

Recently, it was further demonstrated that is possible to generate panplacodal progenitors from hPSCs, thus opening the possibility to obtain not only lens or otic placodal cells but potentially all other CP derivatives (Leung, Kent Morest, & Li, 2013). The study revealed a differential requirement for BMP signaling in the development of CP cells *in vitro*. While the initial panplacodal territory depends on BMP signaling, specialization into anterior CP cells requires BMP inhibition.

In contrast with this work, it has however been reported that during hPSC differentiation, BMP signaling is initially needed for placode induction, and furthermore that there is a requirement for FGFs in the generation of placodal cells. This approach facilitates the differentiation of CP cells into functional trigeminal-type neurons and hormone-releasing pituitary cells (Dincer et al., 2013). Nonetheless, this expands the CP cell types obtained to date from hPSCs. A summary of relevant articles related to CP derivation is presented in Table 2, and a schematic representation of the differentiation protocol is shown in Fig. 1.

4.2. Comments on CP derivation

Even though work in the field on CP cell derivation from hPSCs is still in its infancy, there is a clear indication that hPSCs provide the perfect tool to study the development and specialization of CP cells *in vitro*. The complexity

Table 2 CP derivation

Year	Culture method	FACS	CP derivatives
Monolayer			
Yang et al. (2010)	hESC	No sort	Lens
Qiu et al. (2012)	hESC, hiPSC, Disease hiPSC	No sort	Lens
Chen et al. (2012)	hESC	No sort	Otic progenitors, hair cells, sensory neurons
Mengarelli and Barberi (2013)	hESC	c-MET, CD44	Purified lens epithelium
Leung et al. (2013)	hESC	No sort	Preplacodal ectoderm
Dincer et al. (2013)	hESC, hiPSC	No sort	Adenohypophysis, trygeminal neurons, lens

of CPs and their early commitment into cells with different fates make the use of FACS for cell purification very challenging. On a positive note, the combinatorial use of BMP and FGF treatment during early hPSC differentiation has been sufficient to drive their derivation into placodal ectoderm.

5. CONCLUSIONS

Human PSCs provide the ultimate platform for developmental studies, disease modeling, drug discovery, and regenerative medicine. In the last decade, we have witnessed the start of a revolution in biology and medicine, which encompassed the establishment of hESC lines and the manipulation of cell fate by cellular reprogramming, which may provide the basis for cellular therapies of the future. Although many cell and tissue types have been derived from PSCs *in vitro*, only recently we have begun to see reports of the generation of NC or CP cells. The early formation during embryogenesis of these two specialized areas of the NPB makes PSC an important tool to study NC specification and segregation, and CP specialization in humans. The acquisition of such knowledge will help to address fundamental developmental questions and will provide information about mechanism of actions in normal and diseased cells. As a case in point, NC cells derived from patient-specific hiPSCs have already been successfully used to screen for active molecules that could be used for therapeutic purposes related to familial dysautonomia.

REFERENCES

Bajpai, R., Chen, D. A., Rada-Iglesias, A., Zhang, J., Xiong, Y., Helms, J., et al. (2010). CHD7 cooperates with PBAF to control multipotent neural crest formation. *Nature, 463*(7283), 958–962. http://dx.doi.org/10.1038/nature08733.

Barberi, T., Klivenyi, P., Calingasan, N. Y., Lee, H., Kawamata, H., Loonam, K., et al. (2003). Neural subtype specification of fertilization and nuclear transfer embryonic stem cells and application in Parkinsonian mice. *Nature Biotechnology, 21*(10), 1200–1207. http://dx.doi.org/10.1038/nbt870.

Betters, E., Liu, Y., Kjaeldgaard, A., Sundstrom, E., & Garcia-Castro, M. I. (2010). Analysis of early human neural crest development. *Developmental Biology, 344*(2), 578–592. http://dx.doi.org/10.1016/j.ydbio.2010.05.012.

Bhat, N., Kwon, H. J., & Riley, B. B. (2013). A gene network that coordinates preplacodal competence and neural crest specification in zebrafish. *Developmental Biology, 373*(1), 107–117. http://dx.doi.org/10.1016/j.ydbio.2012.10.012.

Borchin, B., Chen, J., & Barberi, T. (2013). Derivation and FACS-mediated purification of PAX3+/PAX7+ skeletal muscle precursors from human pluripotent stem cells. *Stem Cell Reports, 1*(6), 620–631. http://dx.doi.org/10.1016/j.stemcr.2013.10.007.

Bottenstein, J. E., & Sato, G. H. (1979). Growth of a rat neuroblastoma cell line in serum-free supplemented medium. *Proceedings of the National Academy of Sciences of the United States of America, 76*(1), 514–517.

Bronner-Fraser, M. (1987). Perturbation of cranial neural crest migration by the HNK-1 antibody. *Developmental Biology, 123*(2), 321–331.

Chambers, S. M., Fasano, C. A., Papapetrou, E. P., Tomishima, M., Sadelain, M., & Studer, L. (2009). Highly efficient neural conversion of human ES and iPS cells by dual inhibition of SMAD signaling. *Nature Biotechnology, 27*(3), 275–280. http://dx.doi.org/10.1038/nbt.1529.

Chambers, S. M., Qi, Y., Mica, Y., Lee, G., Zhang, X. J., Niu, L., et al. (2012). Combined small-molecule inhibition accelerates developmental timing and converts human pluripotent stem cells into nociceptors. *Nature Biotechnology, 30*(7), 715–720 http://dx.doi.org/10.1038/nbt.2249.

Chen, W., Jongkamonwiwat, N., Abbas, L., Eshtan, S. J., Johnson, S. L., Kuhn, S., et al. (2012). Restoration of auditory evoked responses by human ES-cell-derived otic progenitors. *Nature, 490*(7419), 278–282. http://dx.doi.org/10.1038/nature11415.

Chizhikov, V. V., & Millen, K. J. (2004). Mechanisms of roof plate formation in the vertebrate CNS. *Nature Reviews Neuroscience, 5*(10), 808–812. http://dx.doi.org/10.1038/nrn1520.

Cimadamore, F., Fishwick, K., Giusto, E., Gnedeva, K., Cattarossi, G., Miller, A., et al. (2011). Human ESC-derived neural crest model reveals a key role for SOX2 in sensory neurogenesis. *Cell Stem Cell, 8*(5), 538–551. http://dx.doi.org/10.1016/j.stem.2011.03.011.

Cohen, P., & Goedert, M. (2004). GSK3 inhibitors: Development and therapeutic potential. *Nature Reviews Drug Discovery, 3*(6), 479–487. http://dx.doi.org/10.1038/nrd1415.

Curchoe, C. L., Maurer, J., McKeown, S. J., Cattarossi, G., Cimadamore, F , Nilbratt, M., et al. (2010). Early acquisition of neural crest competence during hESCs neuralization. *PLoS One, 5*(11), e13890. http://dx.doi.org/10.1371/journal.pone.0013890.

Dincer, Z., Piao, J., Niu, L., Ganat, Y., Kriks, S., Zimmer, B., et al. (2013). Specification of functional cranial placode derivatives from human pluripotent stem cells. *Cell Reports, 5*(5), 1387–1402. http://dx.doi.org/10.1016/j.celrep.2013.10.048.

Ericson, J., Rashbass, P., Schedl, A., Brenner-Morton, S., Kawakami, A., van Heyningen, V., et al. (1997). Pax6 controls progenitor cell identity and neuronal fate in response to graded Shh signaling. *Cell, 90*(1), 169–180.

Fang, D., Leishear, K., Nguyen, T. K., Finko, R., Cai, K., Fukunaga, M., et al. (2006). Defining the conditions for the generation of melanocytes from human embryonic stem cells. *Stem Cells, 24*(7), 1668–1677. http://dx.doi.org/10.1634/stemcells.2005-0414.

Grskovic, M., Javaherian, A., Strulovici, B., & Daley, G. Q. (2011). Induced pluripotent stem cells—Opportunities for disease modelling and drug discovery. *Nature Reviews Drug Discovery*, *10*(12), 915–929. http://dx.doi.org/10.1038/nrd3577.

Inman, G. J., Nicolas, F. J., Callahan, J. F., Harling, J. D., Gaster, L. M., Reith, A. D., et al. (2002). SB-431542 is a potent and specific inhibitor of transforming growth factor-beta superfamily type I activin receptor-like kinase (ALK) receptors ALK4, ALK5, and ALK7. *Molecular Pharmacology*, *62*(1), 65–74.

Ito, M., Kawa, Y., Ono, H., Okura, M., Baba, T., Kubota, Y., et al. (1999). Removal of stem cell factor or addition of monoclonal anti-c-KIT antibody induces apoptosis in murine melanocyte precursors. *The Journal of Investigative Dermatology*, *112*(5), 796–801. http://dx.doi.org/10.1046/j.1523-1747.1999.00552.x.

Jessell, T. M. (2000). Neuronal specification in the spinal cord: Inductive signals and transcriptional codes. *Nature Reviews. Genetics*, *1*(1), 20–29. http://dx.doi.org/10.1038/35049541.

Jiang, X., Gwye, Y., McKeown, S. J., Bronner-Fraser, M., Lutzko, C., & Lawlor, E. R. (2009). Isolation and characterization of neural crest stem cells derived from in vitro-differentiated human embryonic stem cells. *Stem Cells and Development*, *18*(7), 1059–1070. http://dx.doi.org/10.1089/scd.2008.0362.

Katsu, T., Ujike, H., Nakano, T., Tanaka, Y., Nomura, A., Nakata, K., et al. (2003). The human frizzled-3 (FZD3) gene on chromosome 8p21, a receptor gene for Wnt ligands, is associated with the susceptibility to schizophrenia. *Neuroscience Letters*, *353*(1), 53–56.

Kawasaki, H., Mizuseki, K., Nishikawa, S., Kaneko, S., Kuwana, Y., Nakanishi, S., et al. (2000). Induction of midbrain dopaminergic neurons from ES cells by stromal cell-derived inducing activity. *Neuron*, *28*(1), 31–40.

Koehler, K. R., Mikosz, A. M., Molosh, A. I., Patel, D., & Hashino, E. (2013). Generation of inner ear sensory epithelia from pluripotent stem cells in 3D culture. *Nature*, *500*(7461), 217–221. http://dx.doi.org/10.1038/nature12298.

Lazzari, G., Colleoni, S., Giannelli, S. G., Brunetti, D., Colombo, E., Lagutina, I., et al. (2006). Direct derivation of neural rosettes from cloned bovine blastocysts: A model of early neurulation events and neural crest specification in vitro. *Stem Cells*, *24*(11), 2514–2521. http://dx.doi.org/10.1634/stemcells. 2006-0149.

Lecanda, F., Cheng, S. L., Shin, C. S., Davidson, M. K., Warlow, P., Avioli, L. V., et al. (2000). Differential regulation of cadherins by dexamethasone in human osteoblastic cells. *Journal of Cellular Biochemistry*, *77*(3), 499–506.

Lee, G., Kim, H., Elkabetz, Y., Al Shamy, G., Panagiotakos, G., Barberi, T., et al. (2007). Isolation and directed differentiation of neural crest stem cells derived from human embryonic stem cells. *Nature Biotechnology*, *25*(12), 1468–1475. http://dx.doi.org/10.1038/nbt1365.

Lee, G., Papapetrou, E. P., Kim, H., Chambers, S. M., Tomishima, M. J., Fasano, C. A., et al. (2009). Modelling pathogenesis and treatment of familial dysautonomia using patient-specific iPSCs. *Nature*, *461*(7262), 402–406. http://dx.doi.org/10.1038/nature08320.

Lee, G., Ramirez, C. N., Kim, H., Zeltner, N., Liu, B., Radu, C., et al. (2012). Large-scale screening using familial dysautonomia induced pluripotent stem cells identifies compounds that rescue IKBKAP expression. *Nature Biotechnology*, *30*(12), 1244–1248. http://dx.doi.org/10.1038/nbt.2435.

Leung, A. W., Kent Morest, D., & Li, J. Y. (2013). Differential BMP signaling controls formation and differentiation of multipotent preplacodal ectoderm progenitors from human embryonic stem cells. *Developmental Biology*, *379*(2), 208–220. http://dx.doi.org/10.1016/j.ydbio.2013.04.023.

Li, W., & Cornell, R. A. (2007). Redundant activities of Tfap2a and Tfap2c are required for neural crest induction and development of other non-neural ectoderm derivatives in zebrafish embryos. *Developmental Biology, 304*(1), 338–354. http://dx.doi.org/10.1016/j.ydbio.2006.12.042.

Menendez, L., Yatskievych, T. A., Antin, P. B., & Dalton, S. (2011). Wnt signaling and a Smad pathway blockade direct the differentiation of human pluripotent stem cells to multipotent neural crest cells. *Proceedings of the National Academy of Sciences of the United States of America, 108*(48), 19240–19245. http://dx.doi.org/10.1073/pnas.1113746108.

Mengarelli, I., & Barberi, T. (2013). Derivation of multiple cranial tissues and isolation of lens epithelium-like cells from human embryonic stem cells. *Stem Cells Translational Medicine, 2*(2), 94–106. http://dx.doi.org/10.5966/sctm. 2012-0100.

Mica, Y., Lee, G., Chambers, S. M., Tomishima, M. J., & Studer, L. (2013). Modeling neural crest induction, melanocyte specification, and disease-related pigmentation defects in hESCs and patient-specific iPSCs. *Cell Reports, 3*(4), 1140–1152. http://dx.doi.org/10.1016/j.celrep.2013.03.025.

Mizuseki, K., Sakamoto, T., Watanabe, K., Muguruma, K., Ikeya, M., Nishiyama, A., et al. (2003). Generation of neural crest-derived peripheral neurons and floor plate cells from mouse and primate embryonic stem cells. *Proceedings of the National Academy of Sciences of the United States of America, 100*(10), 5828–5833. http://dx.doi.org/10.1073/pnas.1037282100.

Morrison, S. J., White, P. M., Zock, C., & Anderson, D. J. (1999). Prospective identification, isolation by flow cytometry, and in vivo self-renewal of multipotent mammalian neural crest stem cells. *Cell, 96*(5), 737–749.

Motohashi, T., Aoki, H., Chiba, K., Yoshimura, N., & Kunisada, T. (2007). Multipotent cell fate of neural crest-like cells derived from embryonic stem cells. *Stem Cells, 25*(2), 402–410. http://dx.doi.org/10.1634/stemcells. 2006-0323.

Noisa, P., Lund, C., Kanduri, K., Lund, R., Lahdesmaki, H., Lahesmaa, R., et al. (2014). Notch signaling regulates the differentiation of neural crest from human pluripotent stem cells. *Journal of Cell Science, 127*(Pt. 9), 2083–2094. http://dx.doi.org/10.1242/jcs.145755.

Ozair, M. Z., Noggle, S., Warmflash, A., Krzyspiak, J. E., & Brivanlou, A. H. (2013). SMAD7 directly converts human embryonic stem cells to telencephalic fate by a default mechanism. *Stem Cells, 31*(1), 35–47. http://dx.doi.org/10.1002/stem 1246.

Patthey, C., & Gunhaga, L. (2011). Specification and regionalisation of the neural plate border. *The European Journal of Neuroscience, 34*(10), 1516–1528. http://dx.doi.org/10.1111/j.1460-9568.2011.07871.x.

Pla, P., & Larue, L. (2003). Involvement of endothelin receptors in normal and pathological development of neural crest cells. *The International Journal of Developmental Biology, 47*(5), 315–325.

Pomp, O., Brokhman, I., Ben-Dor, I., Reubinoff, B., & Goldstein, R. S. (2005). Generation of peripheral sensory and sympathetic neurons and neural crest cells from human embryonic stem cells. *Stem Cells, 23*(7), 923–930. http://dx.doi.org/10.1634/stemcells.2005-0038.

Qiu, X., Yang, J., Liu, T., Jiang, Y., Le, Q., & Lu, Y. (2012). Efficient generation of lens progenitor cells from cataract patient-specific induced pluripotent stem cells. *PLoS One*, 7(3), e32612. http://dx.doi.org/10.1371/journal.pone.0032612.

Rada-Iglesias, A., Bajpai, R., Prescott, S., Brugmann, S. A., Swigut, T., & Wysocka, J. (2012). Epigenomic annotation of enhancers predicts transcriptional regulators of human neural crest. *Cell Stem Cell, 11*(5), 633–648. http://dx.doi.org/10.1016/j.stem.2012.07.006.

Reichert, S., Randall, R. A., & Hill, C. S. (2013). A BMP regulatory network controls ectodermal cell fate decisions at the neural plate border. *Development, 140*(21), 4435–4444. http://dx.doi.org/10.1242/dev.098707.

Saika, S., Kawashima, Y., Miyamoto, T., Okada, Y., Tanaka, S., Yamanaka, O., et al. (1998). Immunolocalization of hyaluronan and CD44 in quiescent and proliferating human lens epithelial cells. *Journal of Cataract and Refractive Surgery, 24*(9), 1266–1270.

Sauka-Spengler, T., & Bronner-Fraser, M. (2008). A gene regulatory network orchestrates neural crest formation. *Nature Reviews Molecular Cell Biology, 9*(7), 557–568. http://dx.doi.org/10.1038/nrm2428.

Stemple, D. L., & Anderson, D. J. (1992). Isolation of a stem cell for neurons and glia from the mammalian neural crest. *Cell, 71*(6), 973–985.

Streit, A. (2007). The preplacodal region: An ectodermal domain with multipotential progenitors that contribute to sense organs and cranial sensory ganglia. *The International Journal of Developmental Biology, 51*(6–7), 447–461. http://dx.doi.org/10.1387/ijdb.072327as.

Stripecke, R., Carmen Villacres, M., Skelton, D., Satake, N., Halene, S., & Kohn, D. (1999). Immune response to green fluorescent protein: Implications for gene therapy. *Gene Therapy, 6*(7), 1305–1312.

Suga, H., Kadoshima, T., Minaguchi, M., Ohgushi, M., Soen, M., Nakano, T., et al. (2011). Self-formation of functional adenohypophysis in three-dimensional culture. *Nature, 480*(7375), 57–62. http://dx.doi.org/10.1038/nature10637.

Takahashi, K., Tanabe, K., Ohnuki, M., Narita, M., Ichisaka, T., Tomoda, K., et al. (2007). Induction of pluripotent stem cells from adult human fibroblasts by defined factors. *Cell, 131*(5), 861–872. http://dx.doi.org/10.1016/j.cell.2007.11.019.

Thomson, J. A., Itskovitz-Eldor, J., Shapiro, S. S., Waknitz, M. A., Swiergiel, J. J., Marshall, V. S., et al. (1998). Embryonic stem cell lines derived from human blastocysts. *Science, 282*(5391), 1145–1147.

Wilson, Y. M., Richards, K. L., Ford-Perriss, M. L., Panthier, J. J., & Murphy, M. (2004). Neural crest cell lineage segregation in the mouse neural tube. *Development, 131*(24), 6153–6162. http://dx.doi.org/10.1242/dev.01533.

Wormstone, I. M., Tamiya, S., Marcantonio, J. M., & Reddan, J. R. (2000). Hepatocyte growth factor function and c-Met expression in human lens epithelial cells. *Investigative Ophthalmology & Visual Science, 41*(13), 4216–4222.

Yan, Q., Gong, L., Deng, M., Zhang, L., Sun, S., Liu, J., et al. (2010). Sumoylation activates the transcriptional activity of Pax-6, an important transcription factor for eye and brain development. *Proceedings of the National Academy of Sciences of the United States of America, 107*(49), 21034–21039. http://dx.doi.org/10.1073/pnas.1007866107.

Yang, C., Yang, Y., Brennan, L., Bouhassira, E. E., Kantorow, M., & Cvekl, A. (2010). Efficient generation of lens progenitor cells and lentoid bodies from human embryonic stem cells in chemically defined conditions. *The FASEB Journal, 24*(9), 3274–3283. http://dx.doi.org/10.1096/fj.10-157255.

Zeng, H., Guo, M., Martins-Taylor, K., Wang, X., Zhang, Z., Park, J. W., et al. (2010). Specification of region-specific neurons including forebrain glutamatergic neurons from human induced pluripotent stem cells. *PLoS One, 5*(7), e11853. http://dx.doi.org/10.1371/journal.pone.0011853.

Zhou, Y., & Snead, M. L. (2008). Derivation of cranial neural crest-like cells from human embryonic stem cells. *Biochemical and Biophysical Research Communications, 376*(3), 542–547. http://dx.doi.org/10.1016/j.bbrc.2008.09.032.

INDEX

Note: Page numbers followed by "*f*" indicate figures and "*t*" indicate tables.

L

M

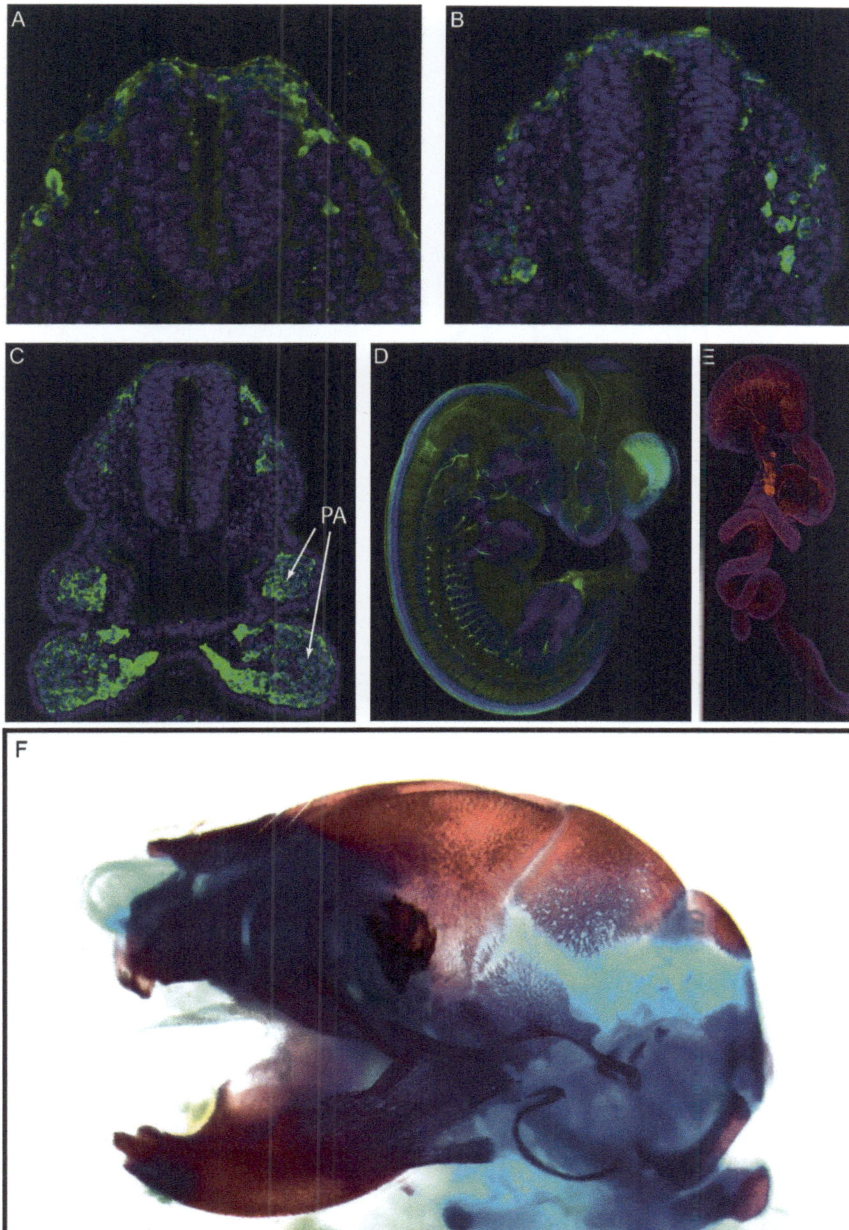

William A. Muñoz and Paul A. Trainor, Figure 1 NCC migration and differentiation in mice. (A–C) Wnt1-Cre YFP mouse section at E10.5. Green labels NCC, blue labels DAPI stained nuclei. (A) NCC in the neural tube and start of emigration from the dorsal neural tube. (B) NCC migrating in streams from the neural tube. (C) NCC populating sites of terminal differentiation including pharyngeal arches (PA). (D) Terminally differentiated NCC-derived neurons in the peripheral nervous system in an E11.5 mouse stained for Tuj1 (green) and DAPI (blue). (E) Terminally differentiated NCC-derived neurons in the enteric nervous system of an E13.5 mouse stained for Tuj1 (red) and DAPI (blue). (F) NCC-derived bone, stained with alizarin red, and cartilage, stained with alcian blue, of the craniofacial skeleton of an E18.5 mouse.

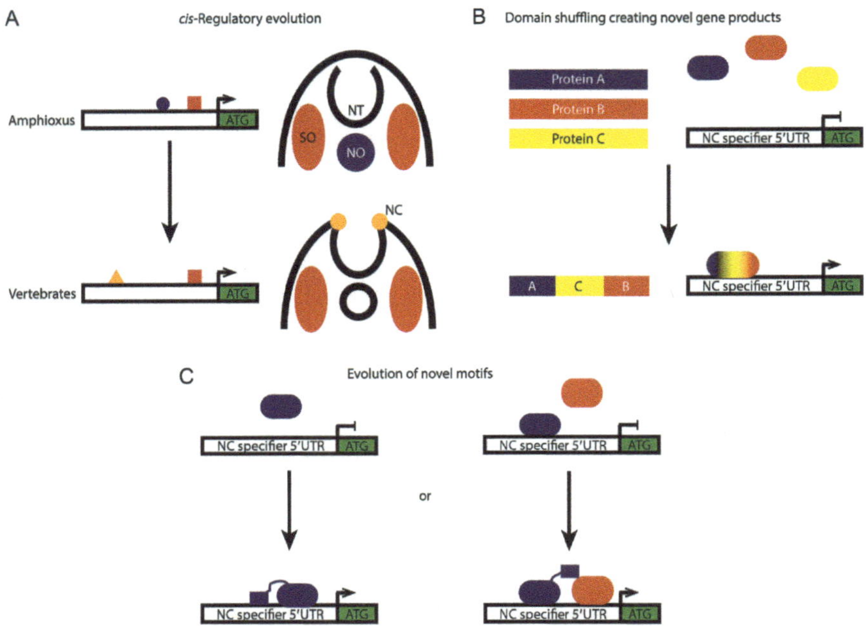

William A. Muñoz and Paul A. Trainor, Figure 3 The potential mechanisms contributing to the molecular evolution of NCC GRN components. (A) Evolution of *cis*-regulatory elements of NCC specifiers. Shapes indicate tissue-specific, *cis*-regulatory elements on the DNA with colors corresponding to expression in the tissues of the embryo section on right. As vertebrates evolved from *Amphioxus*, loss-of-function mutations in regulatory elements (loss of circle) resulted in loss of expression in corresponding tissue. Concurrently, a novel *cis*-regulatory element (triangle) evolved to promote expression of the NCC specifier in the evolving NCC. NT, neural tube; NO, notochord; SO, somites. (B) Evolution of novel gene product via domain shuffling of chromosomes. Several gene products unspliced do not bind to regulatory elements of NCC specifier genes. Following domain shuffling, a novel gene, consisting of regions of each original gene, is created that can bind and activate transcription of the NCC specifier promoting NCC development. (C) Evolution of novel protein motifs in existing molecules. Proteins evolve either a novel motif allowing DNA binding or interactions with additional transcription factors promoting transcription of a NCC specifier.

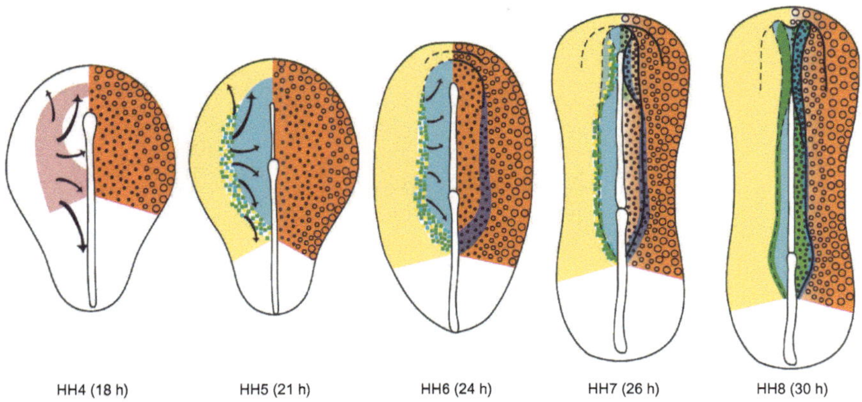

Jean-Loup Duband *et al.*, Figure 1 Integrating space during NCC induction and specification. Hypothetical model depicting how, in the chick embryo, surface movements and cadherin-mediated selective cell adhesion can coordinately contribute to the sorting and partition of the ectodermal, NCC, and neural tube lineages from gastrulation to late neurulation. **Schematic representations of the chick blastoderm viewed dorsally at stages 4–8 of Hamburger and Hamilton (1951) showing: on the left, the expression pattern of the inducing signal BMP-4 at stage 4 and the positions of the different populations of progenitors over time at stages 5–8 and, on the right, the distribution patterns of cadherins as well as the positions of cells according to their sizes. Surface movements combined with the distribution of cells of different sizes in concentric rings result in the formation of folding in the blastoderm, responsible for its transformation into a three-dimensional structure. In a strictly overlapping picture, E- and N-cadherins become distributed in the large/medium cells and in the small cells, respectively, while cadherin-6B appears at the border between medium and small cells where folds develop; this corresponds to the neural plate border where NCC are ultimately generated.

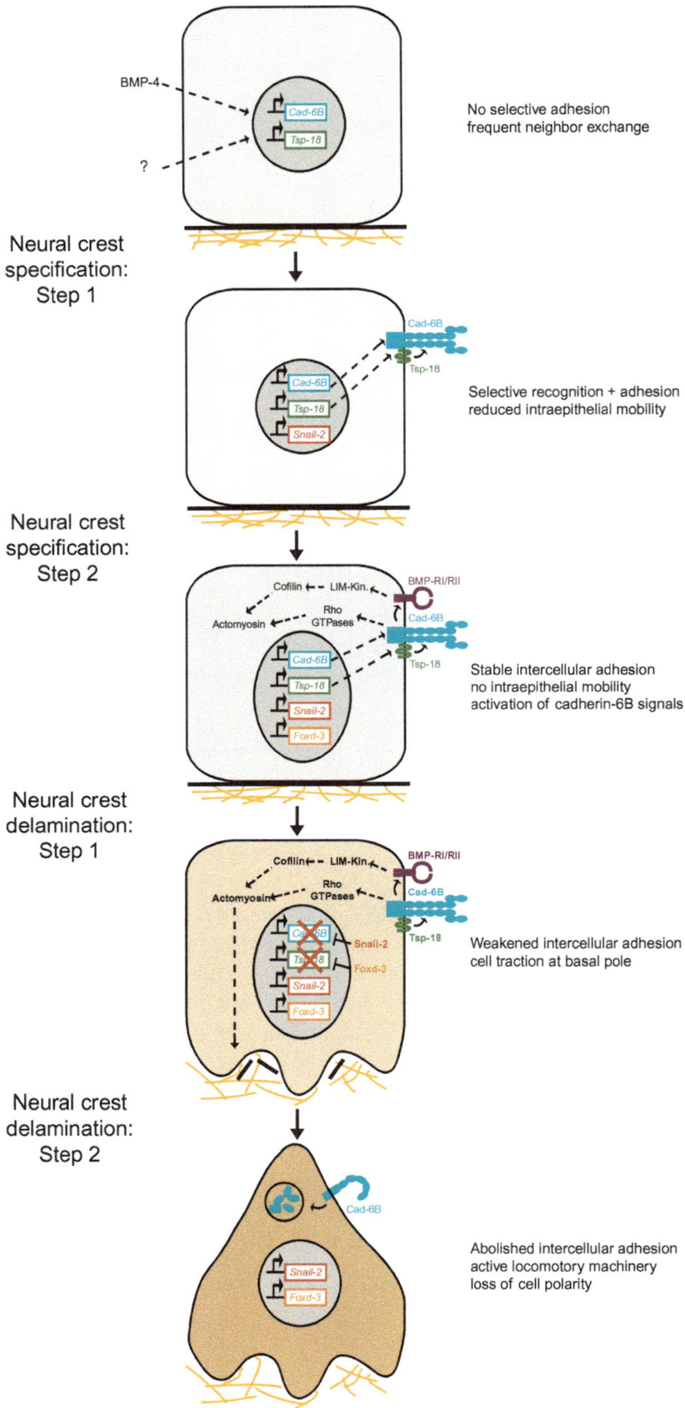

Jean-Loup Duband *et al.*, Figure 2 Dual role of cadherin-6B in NCC specification and delamination. Hypothetical model depicting how cadherin-6B expression is regulated during cranial NCC development in chick and how it may control both cell cohesion and initiation of delamination owing to its dual adhesive and signaling activities. Although this model is illustrated for the cranial region, the same principles apply for trunk NCC except for tetraspanin-18 expression and role.

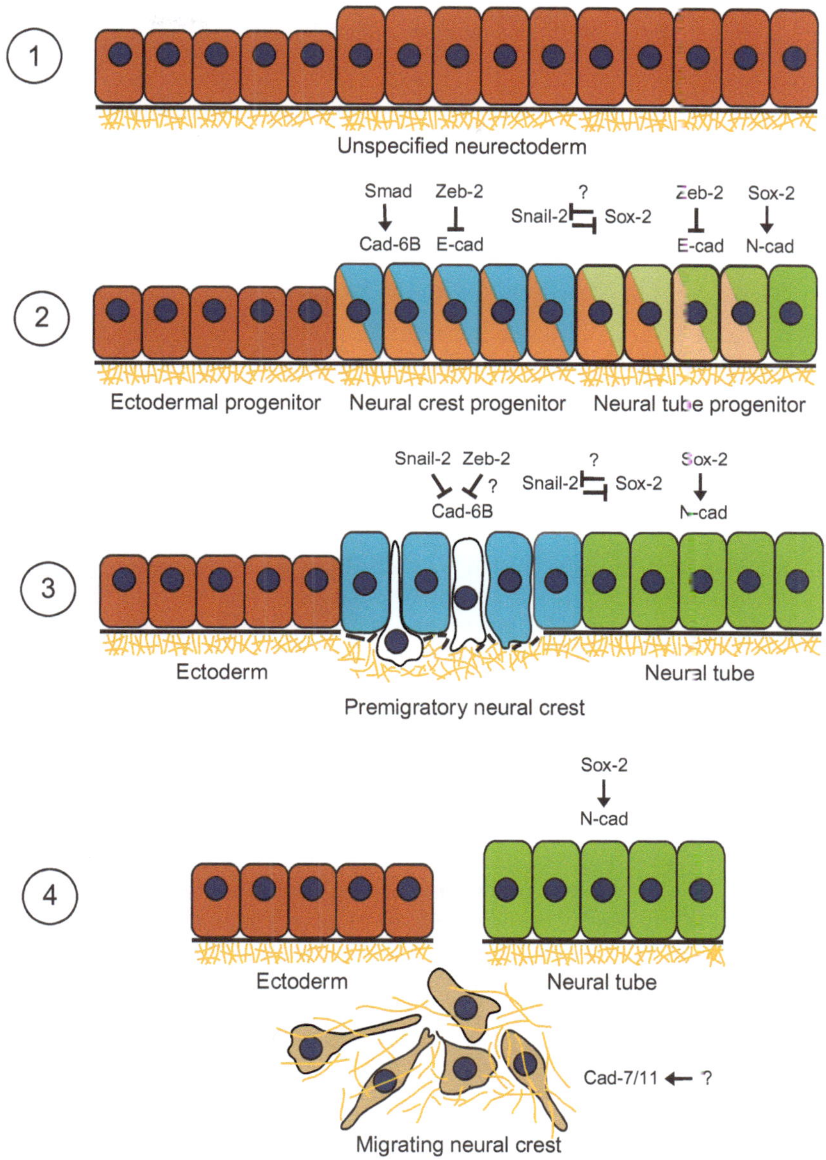

Jean-Loup Duband *et al.*, Figure 3 See legend on next page.

Jean-Loup Duband *et al.*, Figure 3 Coordinated regulation of cadherin expression during NCC specification and delamination. Schematic model depicting the sequence of cadherin expression during NCC development in chick: induction (1), specification (2), delamination (3), and migration (4). The model presented corresponds to the situation found in the hindbrain and anterior trunk. The same scenario applies however to the entire rostral half of the embryo, except that the time at which the E- to N-cadherin switch in the neural tube and in NCC is complete differs according to the level considered, being earlier in the trunk than in the midbrain. In this model, we propose that by the time NCC become induced, the unspecified neurectoderm encompassing all ectodermal, NCC, and neural tube progenitors is an epithelium uniformly expressing E-cadherin (red). Following induction, while E-cadherin expression is maintained at high levels in the ectoderm, it decreases gradually in the rest of the neurectoderm along a mediolateral gradient. Progenitors of the neural tube express progressively higher levels of N-cadherin (green) and lower levels of E-cadherin under the concerted actions of Zeb-2-repressing E-cadherin and Sox-2-activating N-cadherin. NCC also repress *E-cadherin* under the control of Zeb-2 and start expressing *cadherin-6B* (blue) under the control of Smads rather than N-cadherin, due to the putative mutual repression between Snail-2 and Sox-2. This results in N-cadherin confinement to the neural tube. By the time of NCC delamination, cadherin pattern exhibits a clear-cut partition, with E-cadherin only in the ectoderm, cadherin-6B in premigratory NCC, and N-cadherin in the neural tube. NCC undergo EMT after losing cadherin-6B from their surface under the control of Snail-2 and most likely Zeb-2 and transiently express low cadherin levels (white). As NCC initiate migration after delamination, they express a new cadherin content (cadherin-7 or cadherin-11, beige), while the ectoderm and neural tube become entirely segregated.

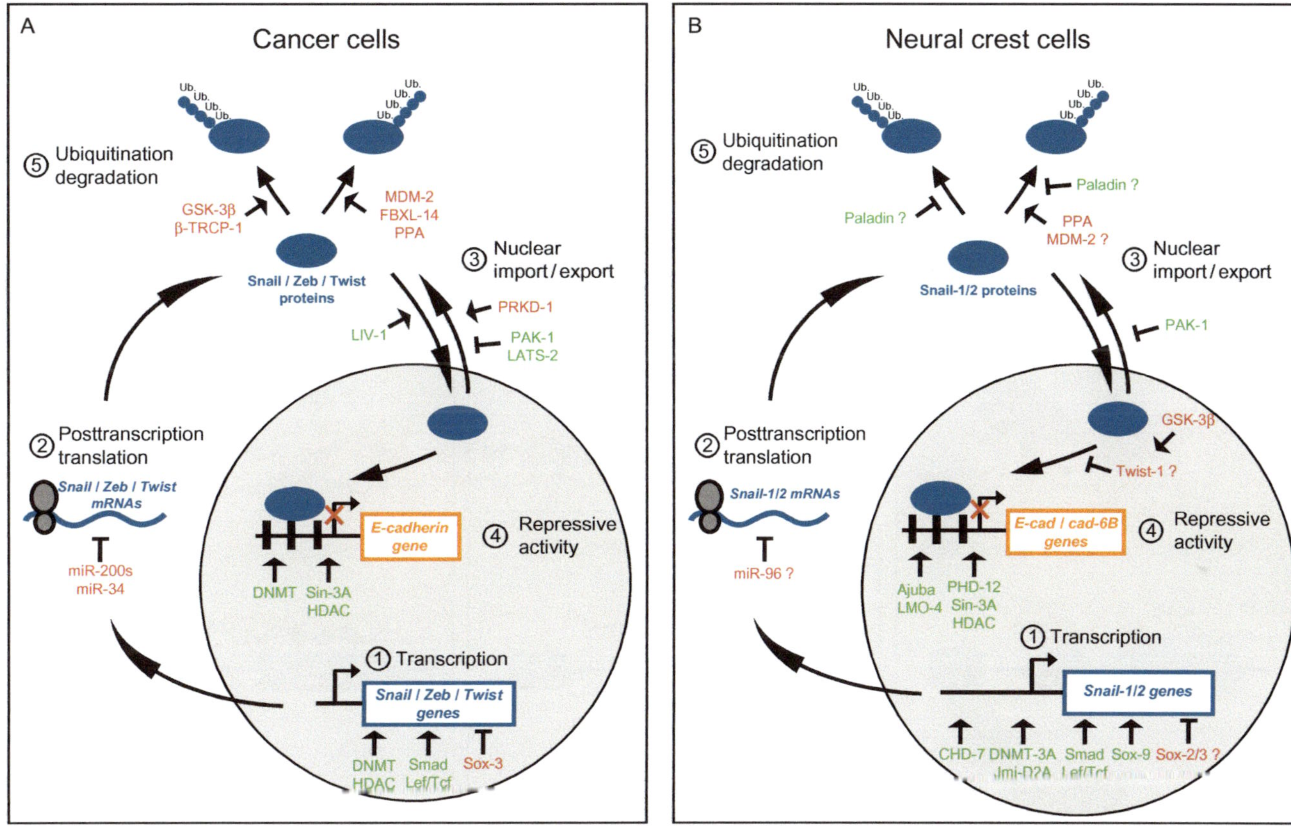

Jean-Loup Duband *et al.*, Figure 4 Control mechanism of the expression and activity of the core EMT regulators. Schematic representation of the molecular cascades controlling (1) transcription and (2) translation of Snail-1/2 messages, (3) the nucleocytoplasmic trafficking of Snail-1/2 proteins, (4) their repressing activity on the E-cadherin gene, and (5) their degradation by the cytoplasmic machinery. The main players identified in cancer cells are depicted in panel (A), while those identified in NCC are shown in panel (B). Players whose action results in increased Snail activity and EMT are indicated in green and those involved in reduced Snail activity are in red. We also propose in this schema that in NCC Twist-1 may reduce Snail-2 activity by titrating Snail-2.

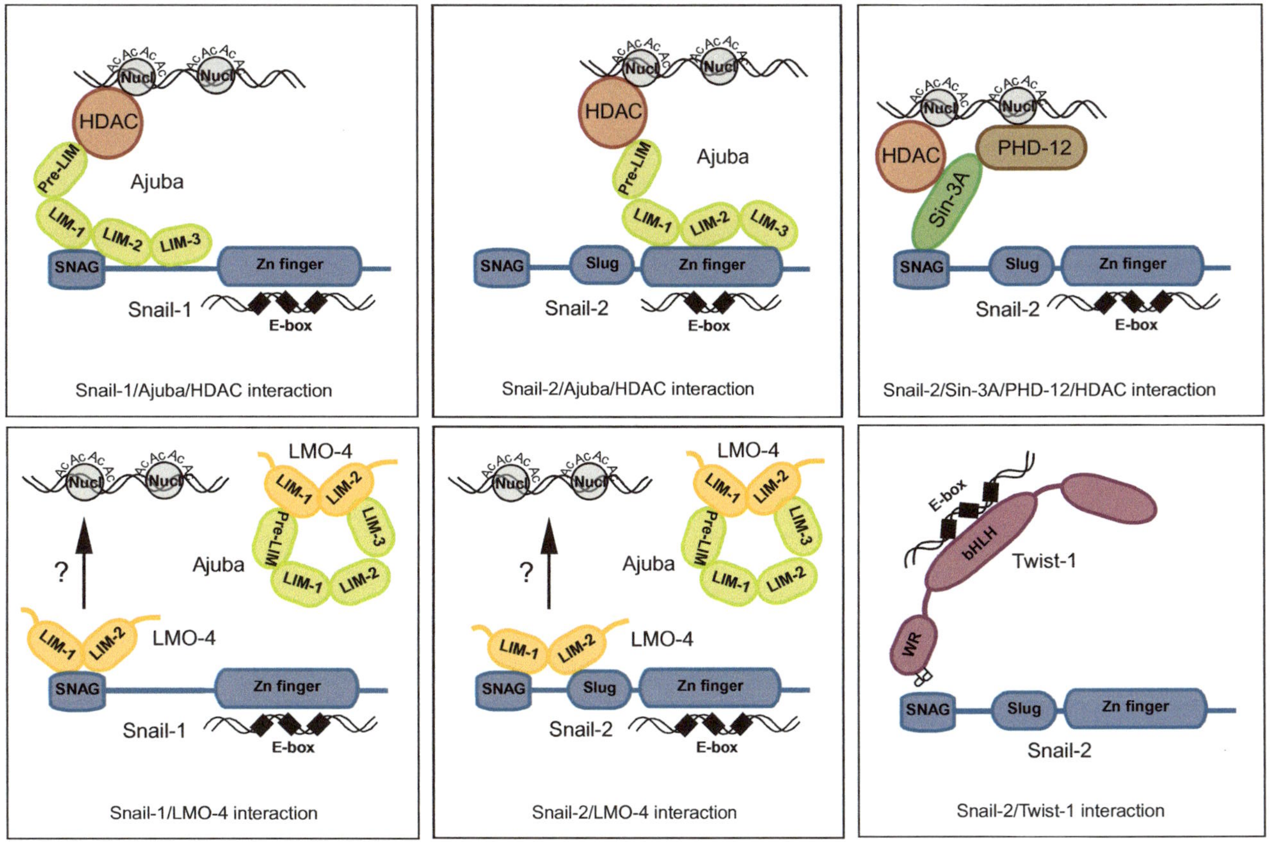

Jean-Loup Duband *et al.*, Figure 5 See legend on opposite page.

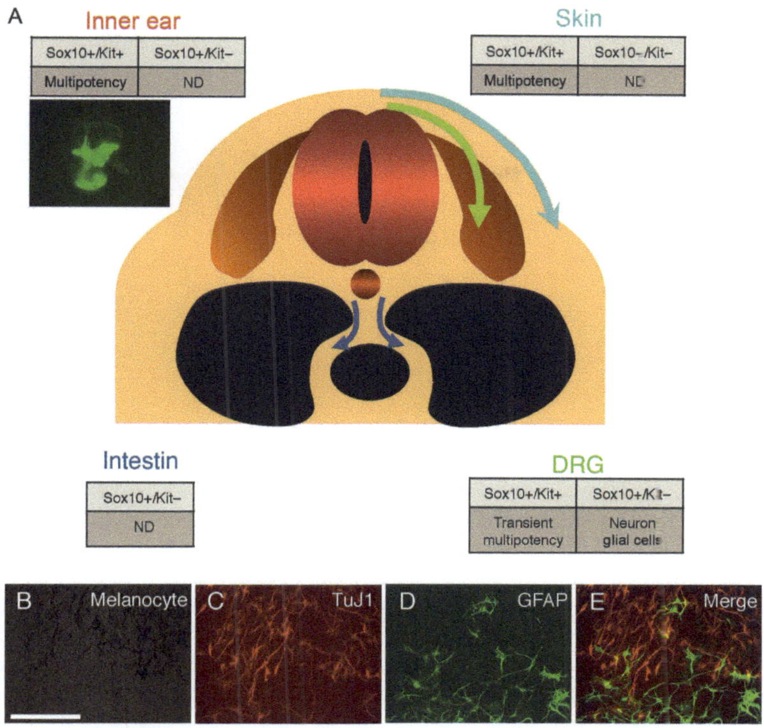

Tsutomu Motohashi and Takahiro Kunisada, Figure 2 NCC-derived cells maintained their multipotency after entry into the target tissues. (A) NCC-derived cells maintain their multipotency after entering into the target tissues. This unexpected differentiation potency of these cells appears to have been restricted by the tissue environment. The scheme shows a transverse section of the embryonic trunk. The arrow shows the NCC migration pathway. The picture shows the inner ear of E12.5 Sox10-IRES-GFP mouse embryo. The table shows the differentiation potencies of Sox10+/Kit+ and Sox10+/Kit- NCC-derived cells in each tissue. ND, not detected. (B–E) Typical multipotent colonies generated from E12.5 skin Sox10+/Kit+ cells. All pictures were taken from a single typical colony. The same visual field is shown in "B–E." Scale bar = 200 μm.

Jean-Loup Duband *et al.*, Figure 5 Snail-1/2 partners in NCC-controlling cadherin gene expression. Snail-1/2 proteins bind to E boxes of the E-cadherin and cadherin-6B promoters via their Zn-finger domain, and their repressing activity involves recruitment of HDAC either through direct binding to the LIM protein Ajuba (as found in *Xenopus*) or via the Sin-3A and PhD-12 adaptors (in chick). Whether Ajuba and PHD-12 exist in both species is not known yet and if so it would be of interest to determine whether they cooperate or compete for Snail activity. Interestingly, Snail-1 and Snail-2 differ from each other by the domain involved in Ajuba binding (the Snail/Gfi (SNAG) domain for Snail-1 and the Zn-finger domain for Snail-2), but whether this may influence the repressing activity of Snail is not known. Snail-1/2 also bind the LIM only protein LMO-4 through their SNAG domain (for Snail-1) or their N-terminal half (for Snail-2). However, as LMO-4 cannot recruit HDAC and that it binds Ajuba, thereby preventing Ajuba–HDAC interaction, it remains to be defined how it functions to mediate Snail-repressing activity. Finally, Snail-2 has been found to bind the GSK-3β-phosphorylated WR domain of Twist-1 and this inhibits its interaction with its target genes. Thus, Twist-1 phosphorylation may function as a negative regulator of Snail-2 activity.

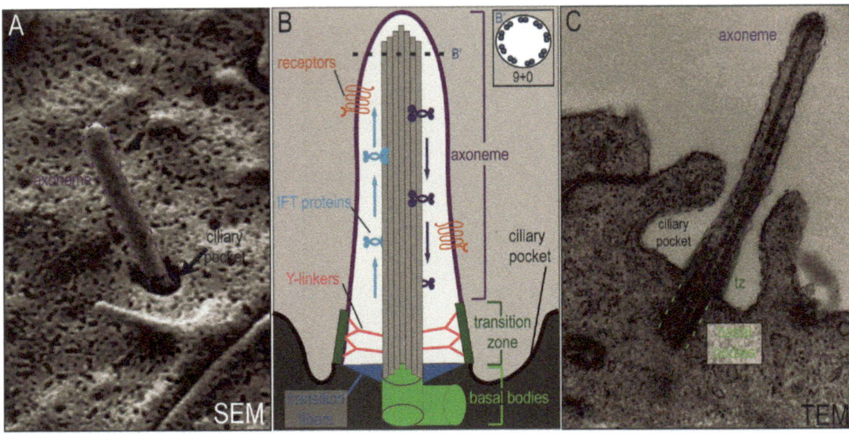

Ching-Fang Chang *et al.*, Figure 1 Structure of the primary cilium. (A) Scanning electron microscopy (SEM) of the primary cilium. The axoneme is outlined in purple and the ciliary pocket is indicated by an arrow. (B) Schematic diagram of the primary cilium. The axoneme (purple) extends out from the basal bodies (light green) into the extracellular space (light gray). Receptors (red) for various signaling pathways localized to the membrane of the axoneme. (B′) Schematic cross-section of the axoneme reveals the microtubules in a 9 + 0 conformation. Intraflagellar transport proteins travel along microtubules, moving cargo up (light blue) and back down (dark blue) the axoneme. The more proximal portion of the primary cilium contains the transition fibers (blue), transition zone (dark green), and Y-linkers (pink). Collectively, these structures are important for ciliogenesis and act as a selective barrier between the intracellular space (dark gray) and the ciliary compartment (white). (C) Transmission electron microscopy (TEM) of the primary cilium. The axoneme (purple dotted lines), transition zone (tz; dark green lines), basal bodies (light green dotted lines), and ciliary pocket (black dotted lines) have all been highlighted in the image.

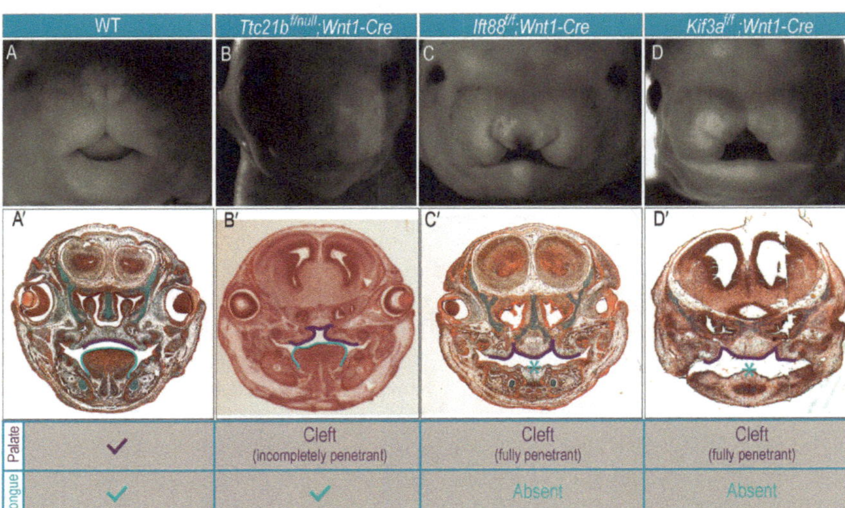

Ching-Fang Chang *et al.*, Figure 3 Craniofacial phenotypes of ciliary mutants. Whole-mount images of e14.5 (A) Wild-type, (B) *Ttc21b*$^{f/null}$;*Wnt1-Cre*, (C) *Ift88*$^{f/f}$;*Wnt1-Cre*, and (D) *Kif3a*$^{f/f}$;*Wnt1-Cre* mice. (A′–D′) Frontal sections of (A–D). (A′) Wild-type mice have a fused palatal shelves (purple outline) and a tongue (blue outline). In contrast, all ciliary mutants have (B′–D′) some degree of cleft palate (purple outline) and *Ift88*$^{f/f}$;*Wnt1-Cre* and *Kif3a*$^{f/f}$;*Wnt1-Cre* mutants have aglossia (C′, D′). Chart indicates craniofacial phenotype for each mouse.

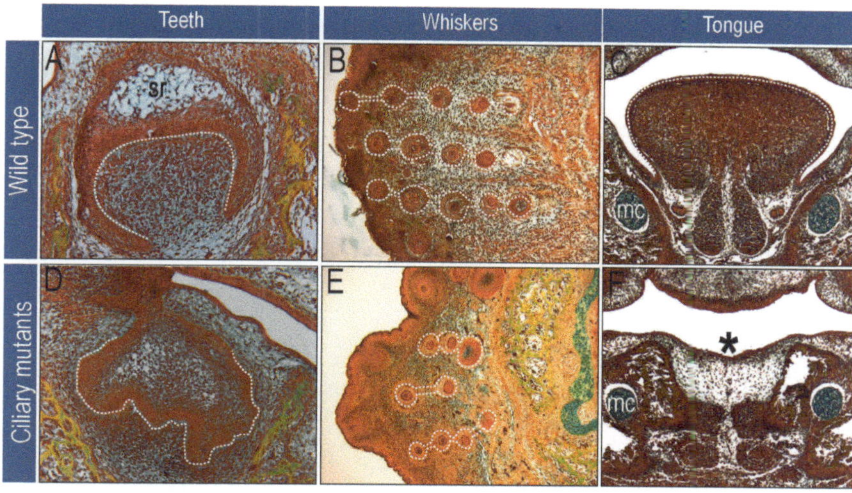

Ching-Fang Chang *et al.*, Figure 5 Defects in craniofacial specialization. Images of wild-type (A–C) and ciliary mutant (D–F) craniofacial structures. Ciliary mutants are unable to properly form teeth (compare dotted white lines in A to D), whiskers are abnormally spaced and have aberrant morphology (compare dotted white lines in B to E) and some ciliary mutants can have aglossia (*) (compare dotted white lines in C to asterisk in F). sr, stellate reticulum; mc, Meckel's cartilage.

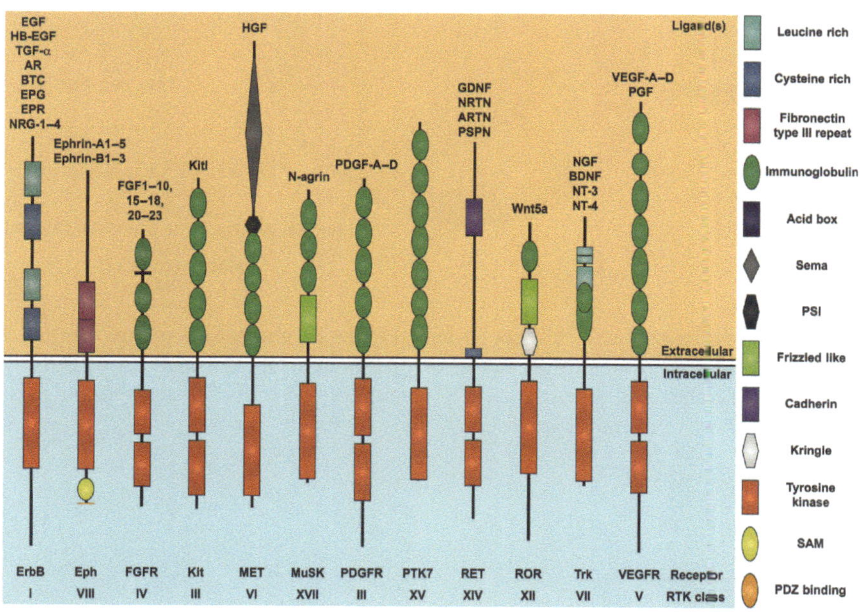

Katherine A. Fantauzzo and Philippe Soriano, Figure 1 Schematic representation of the 12 RTK families involved in mammalian NCC development. The family name and RTK class are listed below each receptor, while the interacting ligands are listed above. Structural domains in the extracellular and intracellular portions of the receptors are depicted according to the key at right.

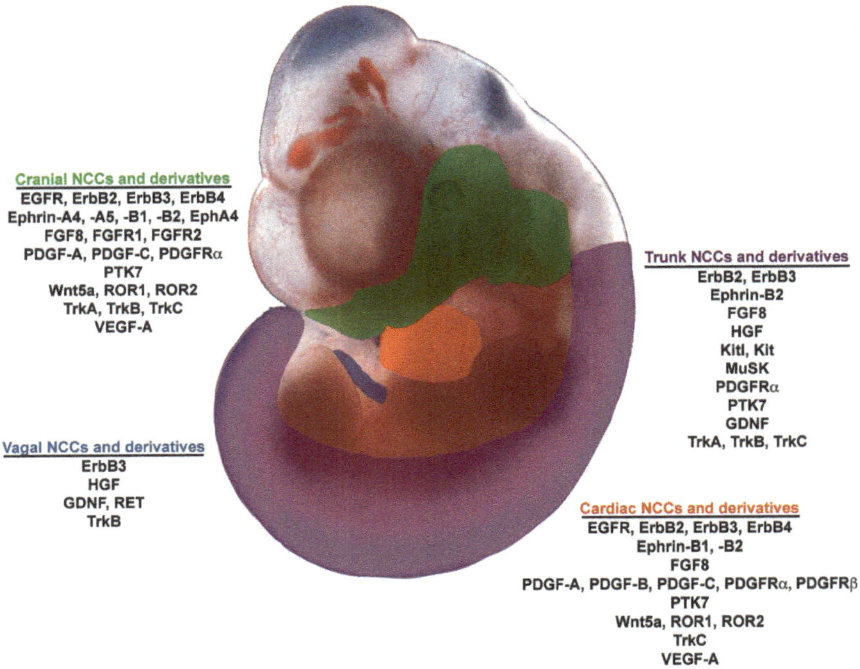

Katherine A. Fantauzzo and Philippe Soriano, Figure 2 Schematic diagram of an E11.5 mouse embryo in which the target tissues of the four axial NCC populations are broadly highlighted. Ligands and corresponding RTKs contributing to the development of each population and its derivatives are listed.

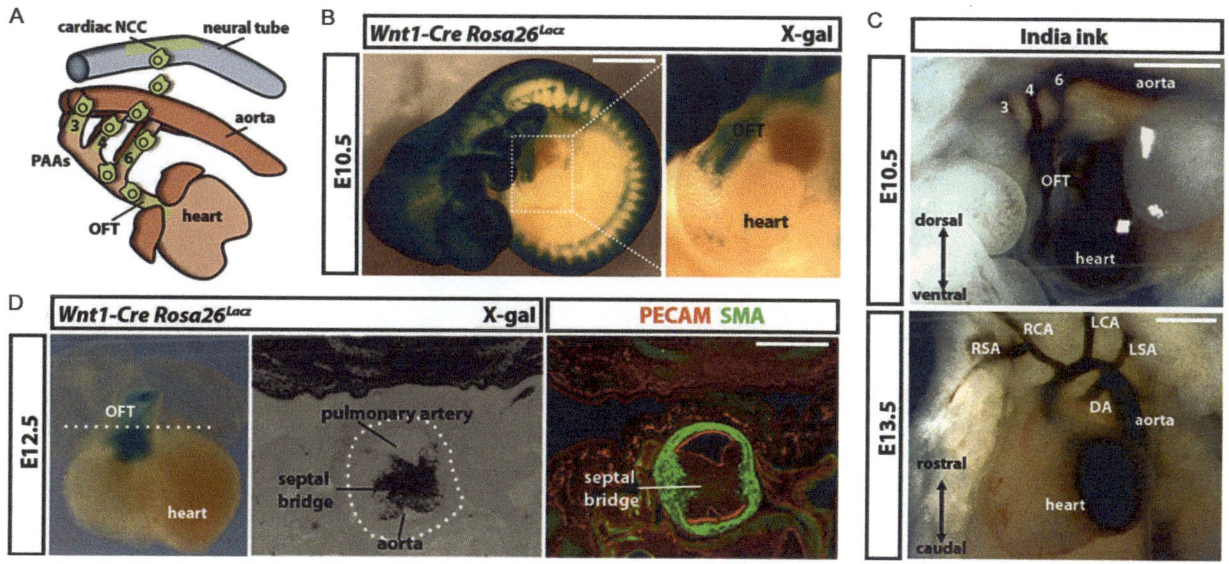

Alice Plein *et al.*, Figure 1 Cardiac NCC contribution to murine OFT and PAA development. (A) Schematic representation of cardiac NCC migration. Cardiac NCCs delaminate from the neural tube at E8.5 and associate with the PAAs to differentiate into the SMCs of the arterial tunica media or continue to the OFT where they contribute to septal bridge formation. (B) Visualization of NCC and their derivatives in an E10.5 *Wnt1-Cre;Rosa26*^{Lacz} mouse embryo. The X-gal stain illustrates NCC contribution to trunk, head, and pharyngeal arch tissues. The higher magnification image shows two prongs of cardiac NCCs in the OFT. (C) PAA remodeling into the aortic arch. Ink injections into embryonic hearts show the third, fourth, and sixth PAAs at E10.5 and their remodeled derivatives at E13.5. (D) Cardiac NCC contribution to the OFT septal bridge. X-gal staining of an E12.5 *Wnt1-Cre;Rosa26*^{Lacz} heart shows accumulation of cells from the cardiac NCC lineage in the OFT; the dotted line indicates the level at which a section was taken to illustrate the position of the cardiac NCCs within the OFT in the adjacent panel. Double staining with the endothelial marker PECAM and the SMC marker SMA illustrates OFT anatomy at E12.5. Scale bars: B—1 mm; C and D—500 μm.

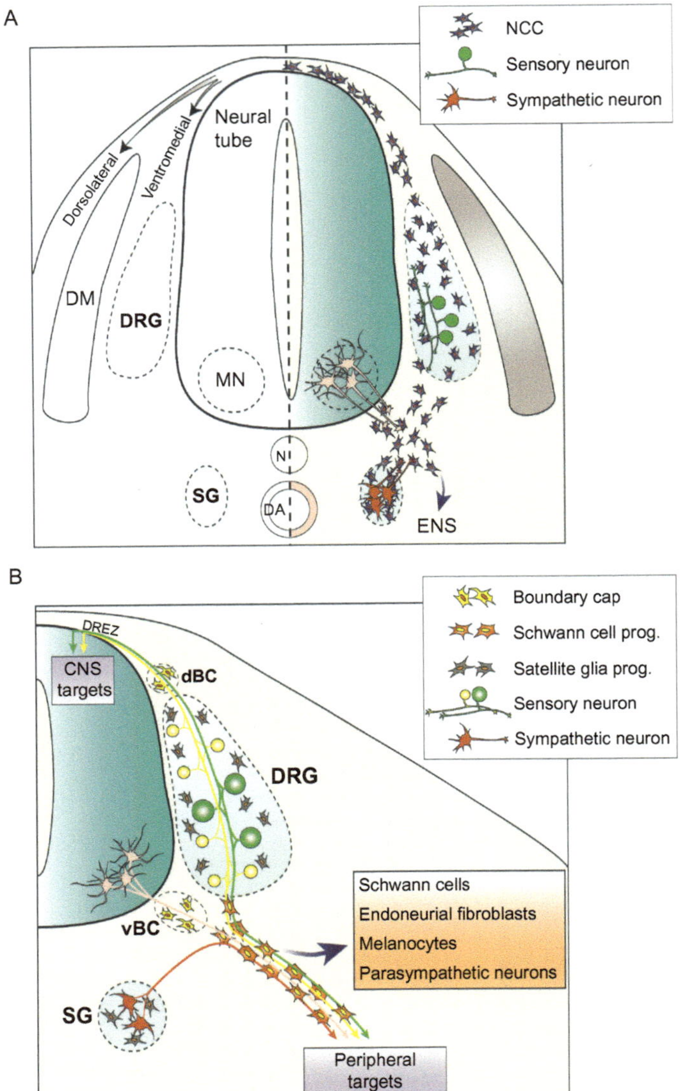

Jason M. Newbern, Figure 1 The initial formation of the PNS from NCCs. (A) After undergoing EMT in the roof plate of the neural tube, migratory trunk NCCs are guided by a combination of attractive, repulsive, and instructive cues derived from the developing dermomyotome (DM), sclerotome, dorsal aorta (DA), and notochord (N). NCCs that generate the PNS migrate ventromedially between the neural tube and developing somite, while the dorsomedial NCCs primarily generate melanocytes. Some NCCs migrate to distant sites in the trunk, such as the enteric nervous system (ENS) and adrenal glands. A population of NCCs cease migration at sites of peripheral ganglia formation and enter a phase of neurogenesis that produces sensory neurons in the dorsal root ganglia (DRG) andsympathetic neurons in the sympathetic ganglia (SG). (B) Axons from NCC-derived SG and DRG neurons and neuroectodermally derived lower motor neurons begin growing into the periphery. The sites where axons enter and exit the spinal cord are populated by the boundary cap, a transient NCC-derived stem-cell niche. The ventral boundary cap (vBC) is localized along outgrowing lower motor neurons axons at CNS/PNS boundary, while the dorsal boundary cap (dBC) is found along

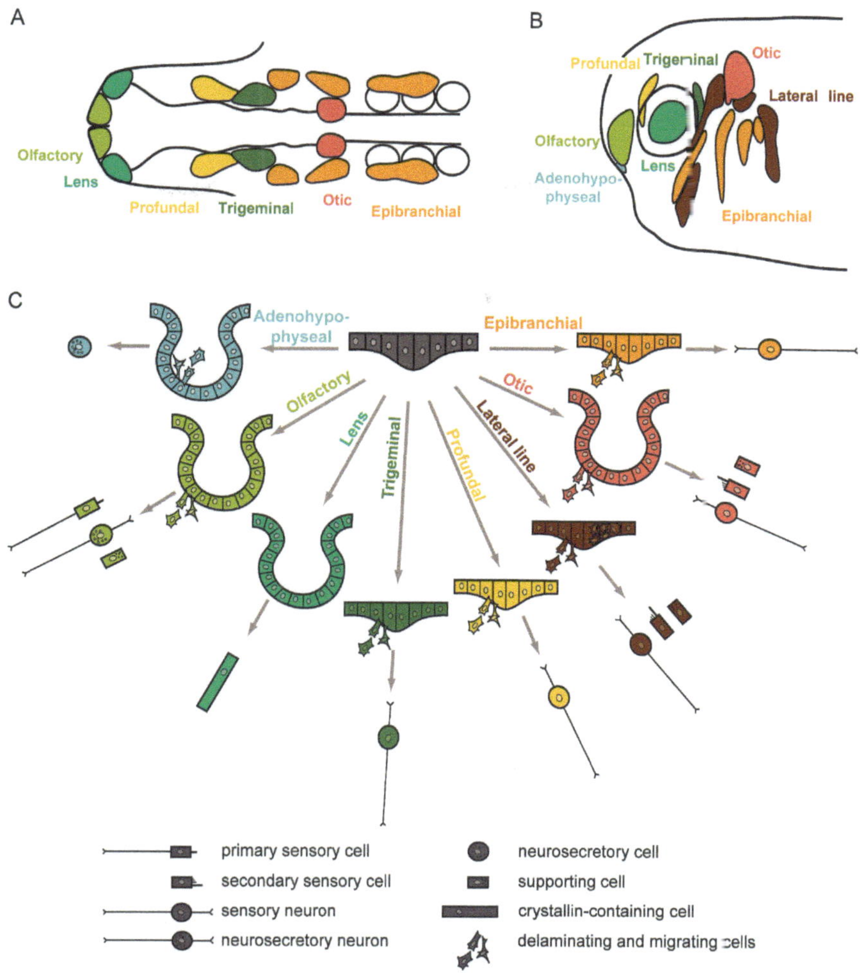

Gerhard Schlosser, Figure 3 The cranial placodes of vertebrates. (A) Chick embryo. (B) *Xenopus* embryo. (C) Developmental fates and derivative cell types of different cranial placodes. *Panel (A): Modified from Streit (2004). Panel (B): Modified from Schlosser and Northcutt (2000). Panel (C): Modified from Schlosser (2005).*

---

fibers near the dorsal root entry zone (DREZ) where sensory afferents enter the spinal cord. TrkC expressing, large-diameter proprioceptive neurons (green) are among the first neurons to be produced in the nascent DRG, followed by small-diameter, TrkA-expressing DRG neurons (red). Subsequent to the onset of neurogenesis, NCCs and boundary cap generate satellite glia progenitors that reside in the ganglia and Schwann cell progenitors (SCPs) that migrate along axons in the developing nerve. SCPs can ultimately differentiate into various cell types that contribute to peripheral nerve function, including myelinating and nonmyelinating Schwann cells and endoneurial fibroblasts. SCPs also generate melanocytes and have even been shown to produce parasympathetic neurons in cranial nerves.

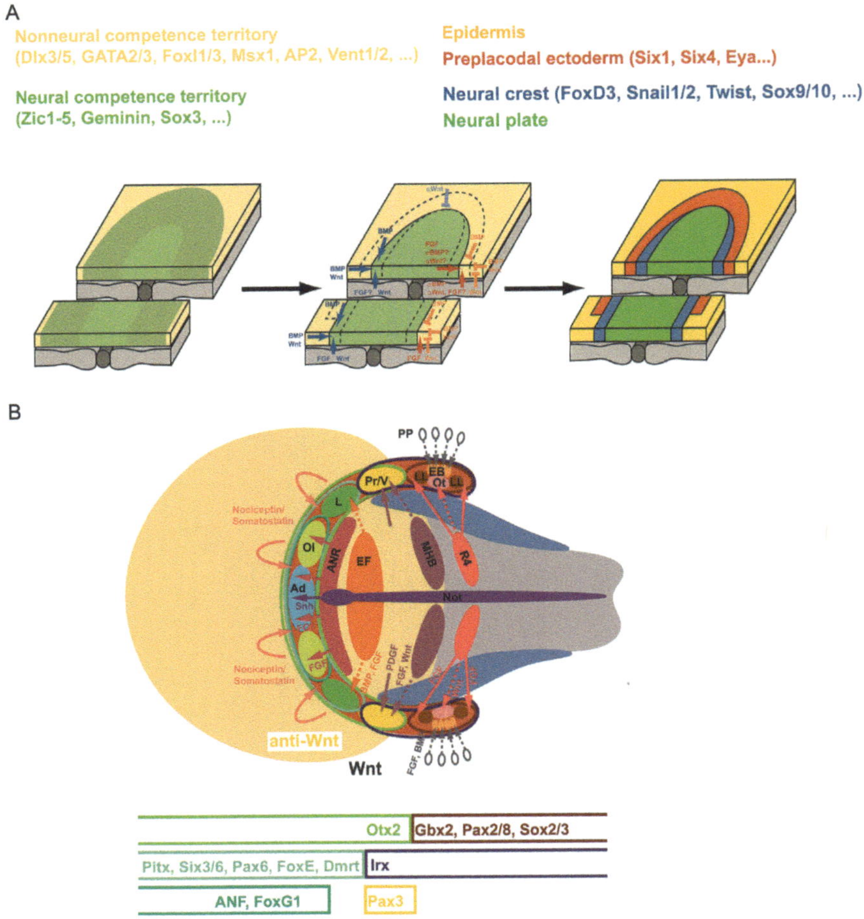

Gerhard Schlosser, Figure 4 Ectodermal patterning and placode induction. (A) Dorsoventral patterning. Dorsally (green) and ventrally (yellow) restricted transcription factors overlap during gastrulation but subsequently resolve into mutually exclusive neural and nonneural competence territories, respectively. The preplacodal ectoderm is then induced at the border of the nonneural territory by FGF, BMP, and Wnt inhibitors (red), while the neural crest is induced at the border of the neural territory by FGF, BMP, and Wnt (blue). (B) Anteroposterior patterning. The preplacodal ectoderm (red) is subdivided into individual placodes by posteriorly restricted Wnt signals and signalling centers in neural plate and mesoderm. These induce transcription factors, which specify multiplacodal areas (colored outlines) and individual placodes (colored ovals). Some examples of transcription factors are listed below. Ad, adenohypophyseal placode; ANR, anterior neural ridge; EB, epibranchial placodes; EF, eye field; L, lens placode; LL, lateral line placodes; MHB, midbrain–hindbrain boundary; Not, notochord; Ol, olfactory placode; Ot, otic placode; PP, pharyngeal pouches; Pr/V, profundal/trigeminal placode; R4, rhombomere 4. *Panel (A): Modified from Schlosser (2006). Panel (B): Modified from Schlosser (2010).*

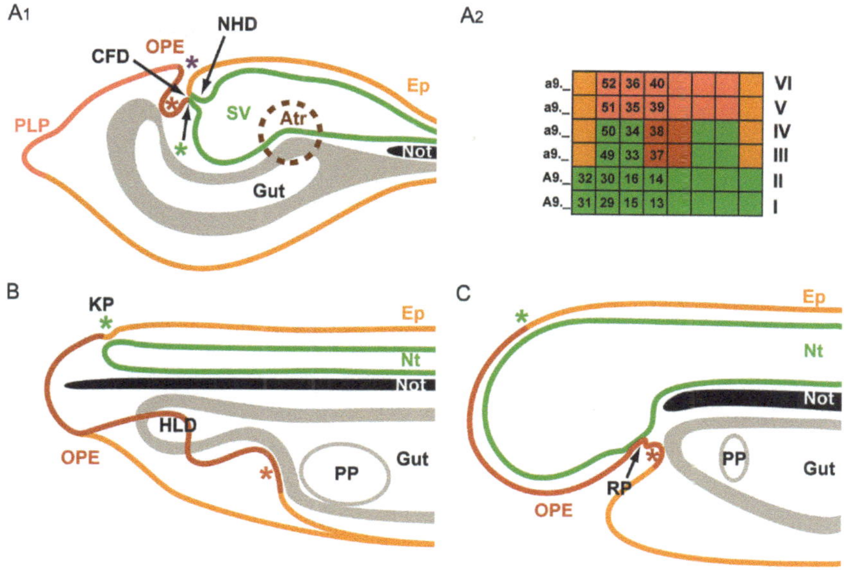

Gerhard Schlosser, Figure 5 Schematic overview of head region in chordate embryos. Different regions of the ectoderm are distinguished by different colors (green, neural tube; orange, general epidermis; red, oral and preoral part of epidermis; pink, palp-forming region of epidermis in tunicates). The position of anterior neuropore (green asterisk) and mouth is marked by (red asterisk). (A) Tunicates (*Ciona*). Oral and preoral ectoderm (oral siphon primordium) participate in neurulation. As a consequence, the external "neuropore" (purple asterisk) is different from the proper neuropore (green asterisk). The connection between anterior neural tube and the oral siphon primordium may persist giving rise to a neurally derived neurohypophyseal duct (NHD) and a ciliated funnel and duct (CFD) derived from the oral siphon primordium. The atrial siphon primordia (Atr; hatched brown line) invaginate only at late larval stages. The origin of the various ectodermal territories from the so-called neural plate of ascidians at mid-gastrula stage is shown in (A_2) (Nishida, 1987). (B) Amphioxus. Hatschek's left diverticulum (HLD), an endomesodermal pouch will fuse with the preoral ectoderm to give rise to Hatschek's pit. (C) Vertebrates. The preoral ectoderm (including precursors of adenohypophyseal, olfactory, and lens placodes) is expanded due to elaboration of the forebrain. The adenohypophyseal placode buds off the stomodeum as Rathke's pouch (RP) to form the anterior pituitary. Abbreviations: Atr, atrial siphon primordium; CFD, ciliary funnel and duct; Ep, epidermis; HLD, Hatschek's left diverticulum; KP, Kölliker's pit; NHD, neurohypophysial duct: Not, notochord; Nt, neural tube; OPE, oral–preoral ectoderm; PLP, palps; PP, pharyngeal pouches; RP, Rathke's pouch; SV, sensory vesicle. *Panels (A_1), (B), and (C): Modified from Schlosser (2005).*

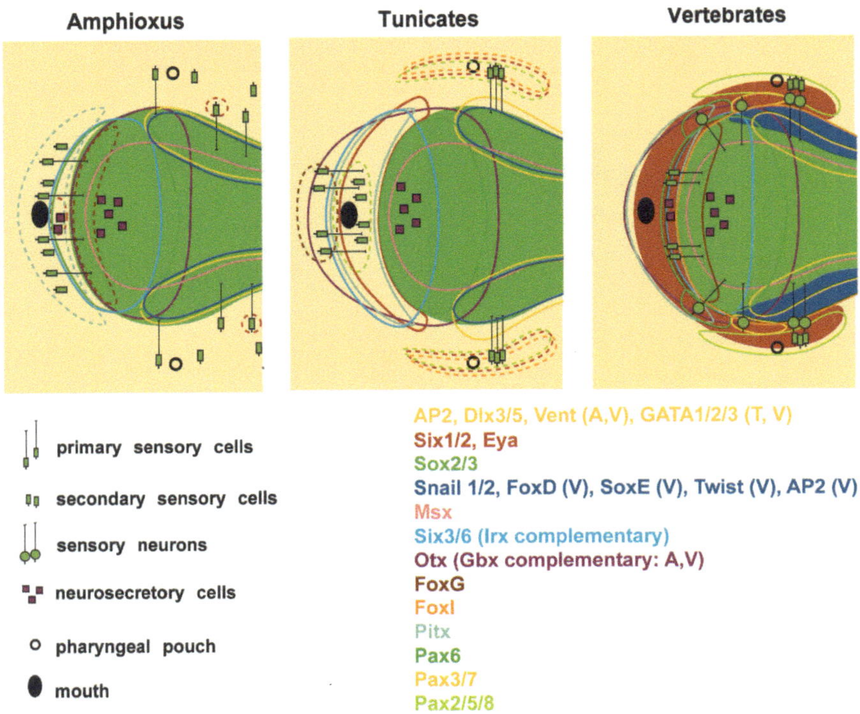

Gerhard Schlosser, Figure 6 Schematic overview of transcription factor domains and distribution of neurosecretory and sensory cells in ectoderm of chordate embryos. Epidermis is shown in yellow and neural plate in green. Preplacodal ectoderm (expressing Six1/2, Eya) is depicted in red, whereas domains of Six1/2 and Eya expression in tunicates and amphioxus are shown by red outlines. Neural crest (expressing a number of transcription factors) is shown in blue, while domains of Snail1/2 expression in tunicates and amphioxus are shown in blue outlines. Transcription factor expression domains are shown enclosed by colored outlines, with the exception of FoxI in vertebrates and Msx1 in all taxa, which are expressed outside of the colored outlines. Transcription factor domains that are established at later developmental stages are indicated by hatched outlines. Domains of Irx and Gbx expression are not shown but abut the domains of Six3/6 and Otx, respectively. Some expression domains are only present in some taxa (A, amphioxus; T, tunicates; V, vertebrates). *Modified from Schlosser et al. (2014). See text for details.*

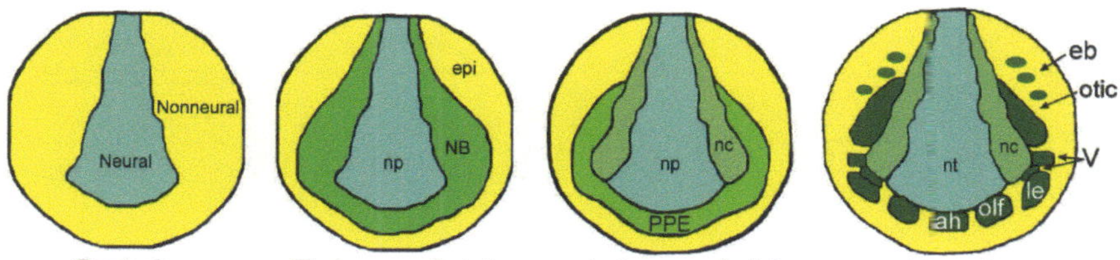

Sally A. Moody and Anthony-Samuel LaMantia, Figure 1 The ectodermal domains depicted at different stages of *Xenopus* embryonic development. At gastrulation, the early embryonic ectoderm is divided into two domains: neural (blue) and nonneural (yellow). Interactions between these two domains and signaling from underlying tissues establish a neural border (NB) zone (green) between the early neural plate (np) and the epidermis (epi). At later neural plate stages, the NB zone divides into the medially located neural crest (nc; light green) and the laterally located preplacodal ectoderm (PPE; dark green). After the PPE acquires initial anterior–posterior patterning and the neural tube (nt) begins to roll up during neurulation, the PPE breaks up into individual placodes in response to local signaling and differential transcription factor expression. These include those derived from the anterior PPE field (the adenohypophyseal (ah), olfactory (olf), and lens (le)); those derived from the intermediate field (V; ophthalmic placode is dorsal to lens placode, and maxillomandibular placode is posterior to it); and those derived from the posterior field (otic and epibranchial (eb)).

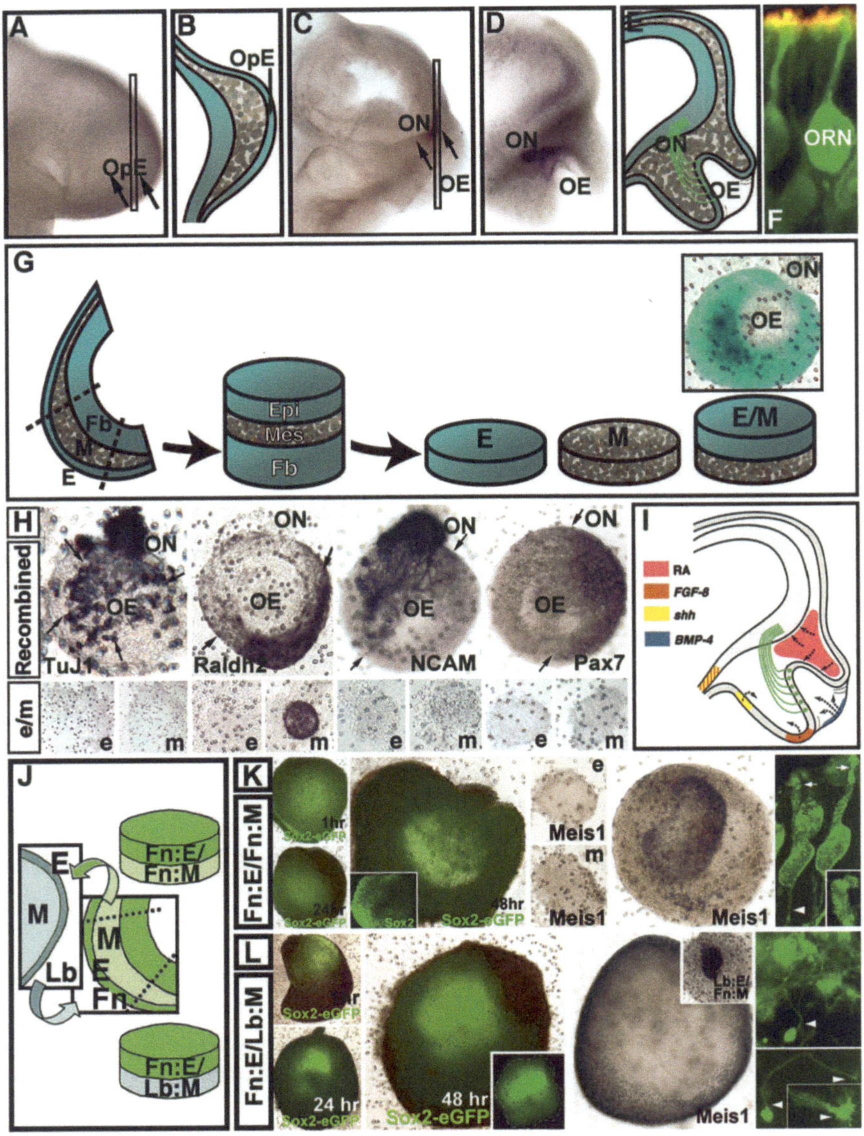

Sally A. Moody and Anthony-Samuel LaMantia, Figure 5 The genesis of olfactory receptor neurons (ORNs) from the olfactory placodal ectoderm (OpE). Top row: whole mouse embryos stained for the neuronal marker NCAM (which labels ORNs and axons in the olfactory nerve—ON), and schematics showing major structures that differentiate from the olfactory placode. (A and B) The undifferentiated OpE at embryonic day (E) 9.0 is a slightly thickened epithelial domain on the lateral/ventral surface of the head. (C–E) Within 1.5 days, at E10.5, the OpE invaginates to form the olfactory epithelium (OE). ORNs are generated primarily in the medial OE, and their axons (ON) extend through the frontonasal mesenchyme (gray hatched regions in E) and enter the forebrain. (F) An ORN at E10.5, labeled for the olfactory marker protein (green, a relatively specific marker), and the ORN-selective adenylcyclase, ACIII (red), which labels the dendritic knob and associated cilia where odor receptor molecules are concentrated, and sensory transduction occurs. (G) A schematic of the approach to isolate the olfactory placode at E9.0; separate the OpE (E), the frontonasal mesenchyme (Mes), and the forebrain neuroepithelium (Fb); and recombine them *in vitro* to assess OpE/ORN differentiation. The inset (right) shows a recombined OpE/frontonasal mesenchyme explant with mesenchyme from a ROSA-26 gene trap mouse that labels all cells with β-galactosidase (blue), and OpE from a WT embryo. There is no movement of cells from the mesenchyme into the OpE over 48 h *in vitro*; however, there is growth of olfactory axons (ON) and cells that migrate with them. (H) The *top row* shows recombined explants (OpE and frontonasal mesenchyme) labeled for molecular markers for differentiating ORNs (TuJ1 and NCAM), a synthetic enzyme for retinoic acid (Raldh2) and Pax7, which is restricted to the lateral mesenchyme. The expression patterns and intensity of each marker are parallel to that seen in the differentiation of the olfactory placode/OE *in vivo* at E11. The *bottom row* shows that the differentiation of the OE, and expression of relevant markers, with the exception of Raldh2, depends upon mesenchymal/epithelial interactions—epithelium (e) nor mesenchyme (m) expresses any of the other markers when isolated. (I) A schematic showing the role of signaling sources of Fgf8, Bmp4, Shh, and RA established via earlier placodal transcriptional mechanisms, in establishing OE neurogenesis and ON trajectory. (J) Schematic of mesenchyme "swapping" experiments *in vitro* to assess specificity of OpE/frontonasal mesenchyme interactions. (K) Isolated E9.0 frontonasal OpE (Fn:E) recombined with frontonasal mesenchyme (Fn:M) leads to the restriction and patterned expression of Sox2 in the OpE/OE over 48 h. The same explant is shown here, imaged at the times indicated. The inset shows the medial (high) to lateral (low) pattern of Sox2 expression facilitated by Fn: E/Fn:M interactions. This pattern is key for specifying ORN stem cells. In addition, Fn: E/Fn:M interactions induce Meis1 in a population of cells that additional evidence (Murdoch, DelConte, & Garcia-Castro., 2010; Tucker et al., 2010) shows are likely OE neural stem cells. At far right, these patterning events yield cells in Fn:E/Fn:M explants that have all of the morphological hallmarks of ORNs. In addition, these cells have physiological properties seen in ORNs (Rawson et al., 2010). (L) When OpE (Fn:E) is recombined with limb bud mesenchyme (Lb:M), Sox2 is not patterned, and *Meis1* expression is seen in the mesenchyme rather than OpE. The OpE still generates neurons (far right) in this heterologous pairing of E and M; however, these neurons lack apical dendrites, their axons cannot enter the limb bud mesenchyme and instead end in elaborate growth cones (arrow, lower right panel), and they lack the physiological hallmarks of ORNs (Rawson et al., 2010).

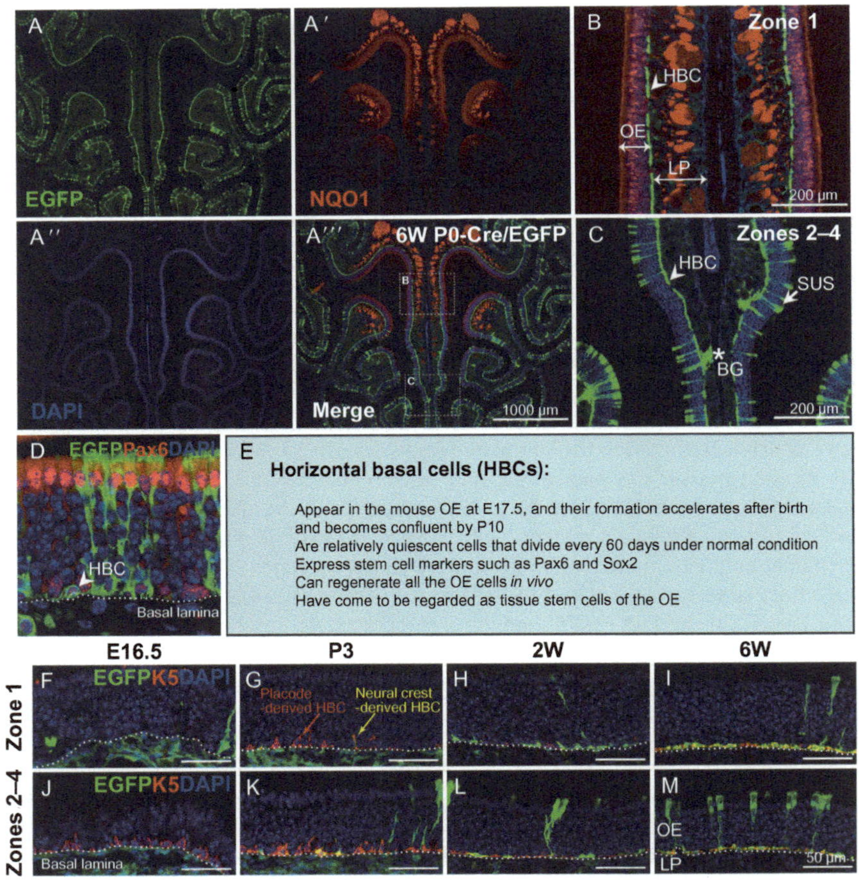

Jun Suzuki and Noriko Osumi, Figure 2 Neural crest-derived cells (NCDCs) in the adult olfactory epithelium (OE). (A–A‴) A coronal section of the adult OE in *P0-Cre/Floxed-EGFP* mice. NQO1 is a marker for zone 1. (B and C) High-magnification images of zone 1 and zones 2–4. Strong EGFP signals are observed in horizontal basal cells (HBCs, arrowhead), sustentacular cells (SUSs, arrow), and Bowman's glands (BGs, *). (D) Confocal image of Pax6-positive HBCs (arrowhead). (E) Summary of HBC characteristics. (F–M) Representative images of the OE at E16.5 (F and J), P3 (G and K), 2W (H and L), and 6W (I and M) in zone 1 (F–I) and zones 2–4 (J–M). Main population of HBCs gradually changes from placode-derived (red arrow) to neural crest-derived (yellow arrow) during postnatal maturation. Abbreviation: LP, lamina propria. *Reproduced from Suzuki et al. (2013) (A–D and F–M).*

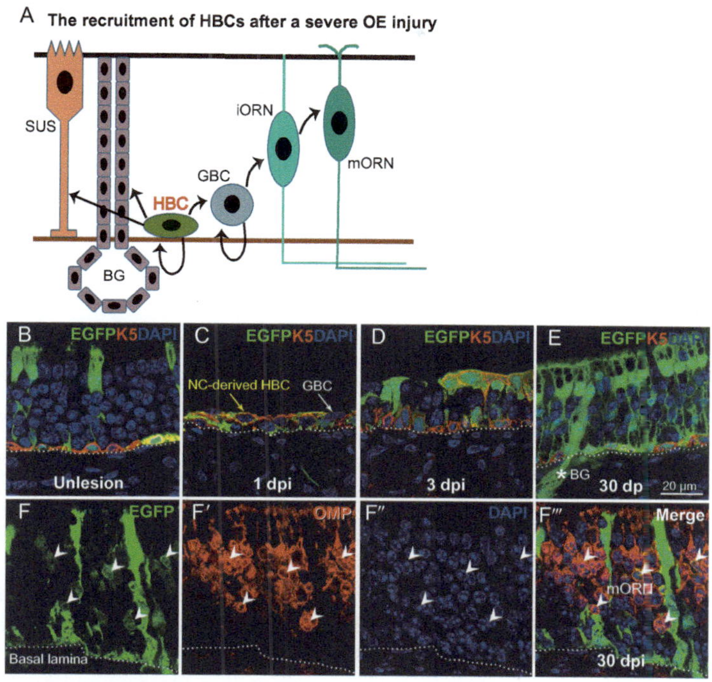

Jun Suzuki and Noriko Osumi, Figure 3 Neural crest-derived cells (NCDCs) contribute to the regeneration of the olfactory epithelium (OE) after methimazole-induced severe OE injury. (A) Schematic representation of the recruitment of horizontal basal cells (HBCs) after a severe OE injury. (B–F‴) An unlesioned OE (B) and injured OE at 1 (C), 3 (D), and 30 days postinjury (dpi) (E–F‴). Following methimazole intraperitoneal injection, widespread cell death is observed within 24 h of drug treatment, but most of the HBCs are not damaged (C). At 30 dpi, the injured OE is nearly completely regenerated, and many EGFP-positive cells are observed in the OE (E). (F–F‴) NCDCs differentiated into OMP-positive mature olfactory receptor neurons (mORNs) (arrowheads). Abbreviations: SUSs, sustentacular cells; BGs, Bowman's glands; GBCs, globose basal cells; iORNs, immature olfactory receptor neurons; mORNs, mature olfactory receptor neurons; NC, neural crest. *Panel (B)–(F‴): Reproduced from Suzuki et al. (2013).*

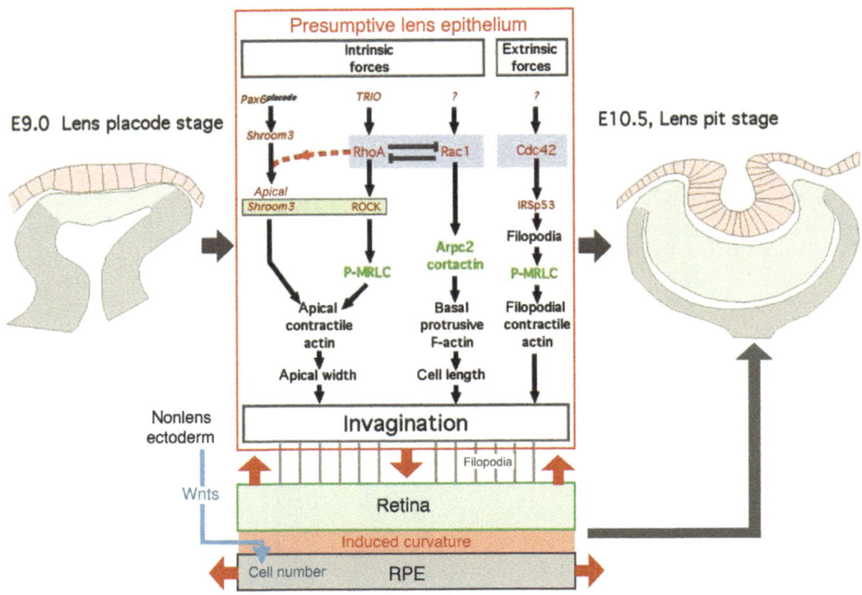

Bharesh Chauhan *et al.*, Figure 2 Summary schematic of eye morphogenesis pathways. This schematic describes some of the pathways involved in morphogenesis of the mouse eye from E9.5, when the lens placode overlies the optic vesicle, to E10.5 when the lens pit has invaginated and the optic cup has become bilayered. The surface ectoderm, presumptive retina, and presumptive retina pigmented epithelium are color-coded in red, green, and gray, respectively. Within the red box, we summarize the action of three GTPases that have distinct actions on lens placode invagination. Cdc42 and IRSp53 are required for the generation of filopodia within the presumptive lens that connect to the presumptive retina and permit force transmission between the two epithelia. They are required for a full invagination of the lens pit and for subtle shaping of the optic cup. This is and example where lens pit invagination depends on extrinsic forces. The GTPases RhoA and Rac1 are required for the generation of intrinsic forces for lens pit invagination because they control cell length (Rac1 through an Arpc2/cortactin response) and apical constriction (RhoA through a Rho kinase (ROCK) phospho-myosin pathway). RhoA activity is likely controlled by the TRIO guanine nucleotide exchange factor and in turn, this is required for the apical localization of Shroom3, an actin and ROCK-binding protein required for apical constriction. At the transcriptional level, Shroom3 is positively regulated by Pax6. During lens pit invagination, the coordinated morphogenesis of the optic cup is in part stimulated by Wnt/Catnb pathway ligands that are produced by nonlens surface ectoderm. These ligands stimulate the production of cells within the developing retinal pigmented epithelium (RPE). The RPE thus expands against the relatively static retina and, with the principle of a bimetallic strip, induces curvature in the optic cup. The red arrows indicate the movements of morphogenesis for the each epithelial layer.

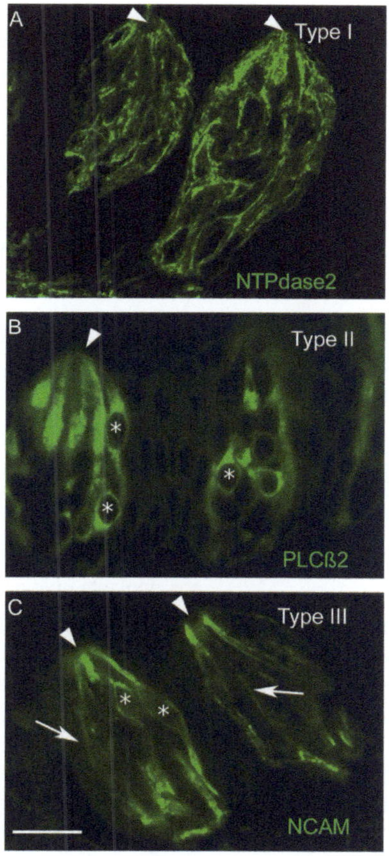

Linda A. Barlow and Ophir D. Klein, Figure 1 Taste buds comprise three morphological cell types, I, II, and III, which are recognized by their expression of specific marker proteins. (A) Type I cells express NTPdase2, which localizes to cell membranes that tightly wrap other cells within the bud, such that individual NTPDase2+ cells are not discernable (see Miura, Scott, Harada, & Barlow, 2014 for detailed explanation). (B) Type II cells express PLCβ2 in the cytosol, and the protein is excluded from nuclei of PLCβ2+ taste cells (*). (C) NCAM marks the surface of Type III taste cells (*), as well as some nerve fibers extending into buds (arrows). All panels are images of taste buds from the circumvallate papilla of mice. Apical taste pores are indicated with arrowheads. Scale bar is 20 μm.

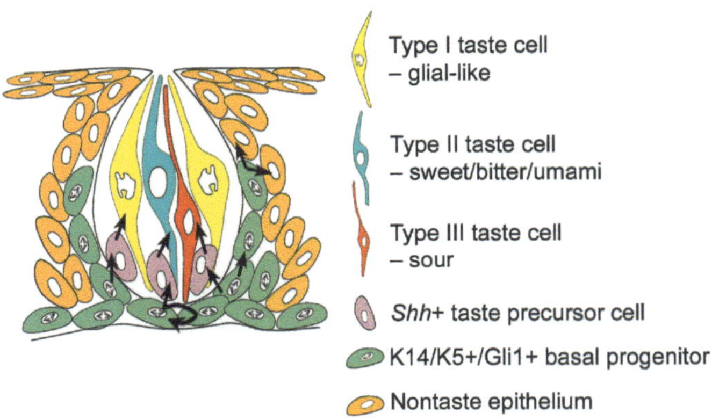

Linda A. Barlow and Ophir D. Klein, Figure 2 Schematic diagram of taste bud cell types and lineage relationships. Type I (yellow), II (blue), and III (red) taste cells are elongate, postmitotic (open nuclei) cells, which arise from proliferating, K5+/K14+ keratinocytes (green, mitotic nuclei) outside of taste buds. These progenitors also give rise to nontaste epithelial cells (orange). Following their final division, cells destined to become taste cells enter taste buds and become taste bud precursors, which are *Shh*+ ovoid cells (lavender) located in the basal compartment of the taste bud. Shh+ cells are postmitotic, and differentiate into each of the three taste cell types. Black arrows indicate observed lineage relationships among cell types (see text for details), but are not absolute, nor exhaustive.

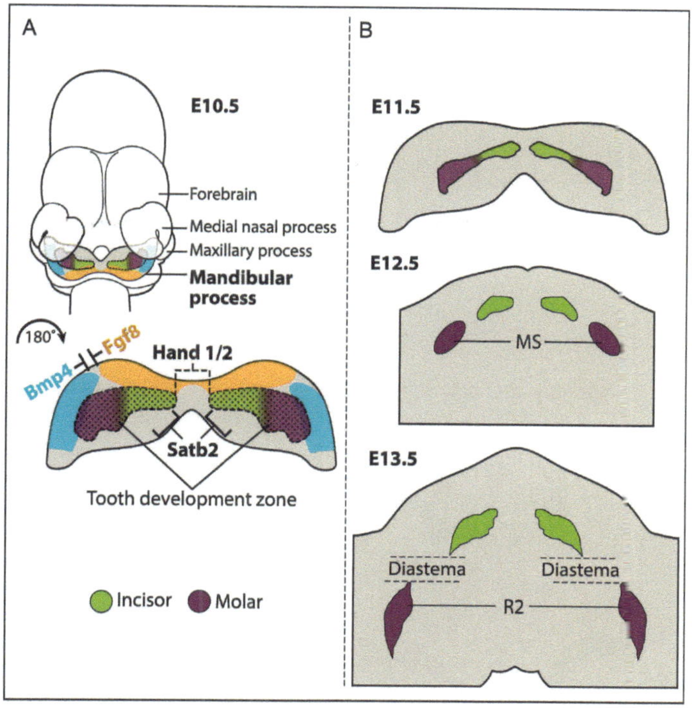

Youngwook Ahn, Figure 1 Positioning the sites of tooth formation and their division into the incisor and molar fields. (A) Schematic diagram showing the mandible in the head of an E10.5 embryo. An antagonistic interaction between Bmp4 and Fgf8 s proposed to define the tooth field in the oral ectoderm in each jaw quadrant. Hand1/2 and Satb2 are required for patterning of the distal region. Note that the mandible at the bottom is rotated by 180° for consistency. (B) The tooth field appears to be continuous until E11.5 but subsequently divides into the incisor and molar fields in the distal and proximal regions, respectively. The two fields are separated by a toothless region called diastema. The vestigial tooth buds, MS and R2, form in the molar field but undergo degeneration.

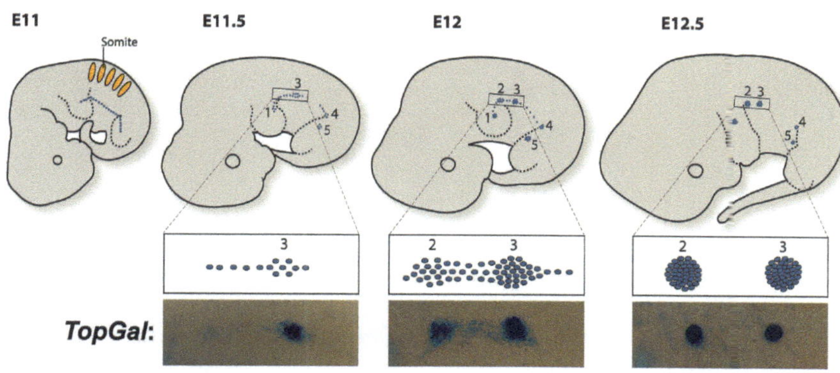

Youngwook Ahn, Figure 4 Sequential formation of mammary placodes along the mammary line during mouse embryogenesis. Diagram of an E11 embryo shows the position of the mammary line (dashed blue line) in the axilla, inguen, and the region between forelimb and hindlimb buds. Starting around E11.5, five mammary placodes sequentially develop along the mammary line in a defined temporal and spatial order while the mammary lines themselves progressively disappear. The blowups demonstrate temporal changes in distribution of mammary epithelial cells around placodes #2 and #3 marked by expression of a Wnt reporter, *TopGal* (bottom).

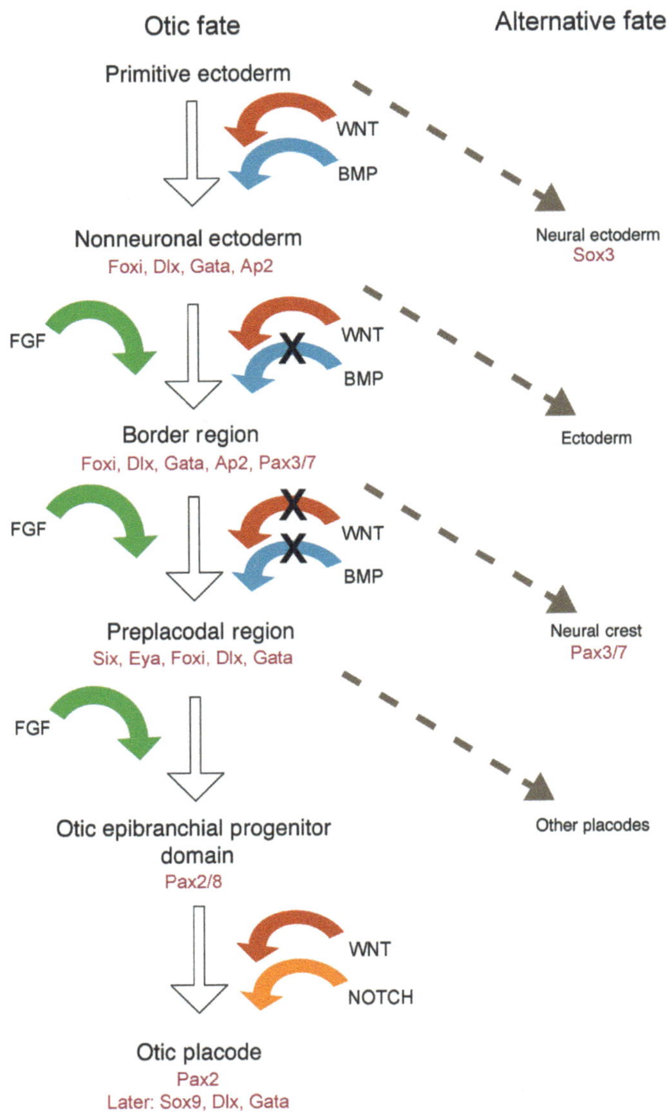

Renée K. Edlund *et al.*, Figure 1 Requirements for otic induction from primitive ectoderm. The figure depicts a consensus scheme of otic induction for amniotes. For each step, the necessary signaling factors for the otic pathway are shown. The alternative pathway is chosen in the absence of the specific combination of signaling and/or with additional signals. Gene expression defining tissue identity for individual steps is indicated. X represents the requirement for an inhibitor of a signaling pathway, such as BMP or Wnt inhibitors.

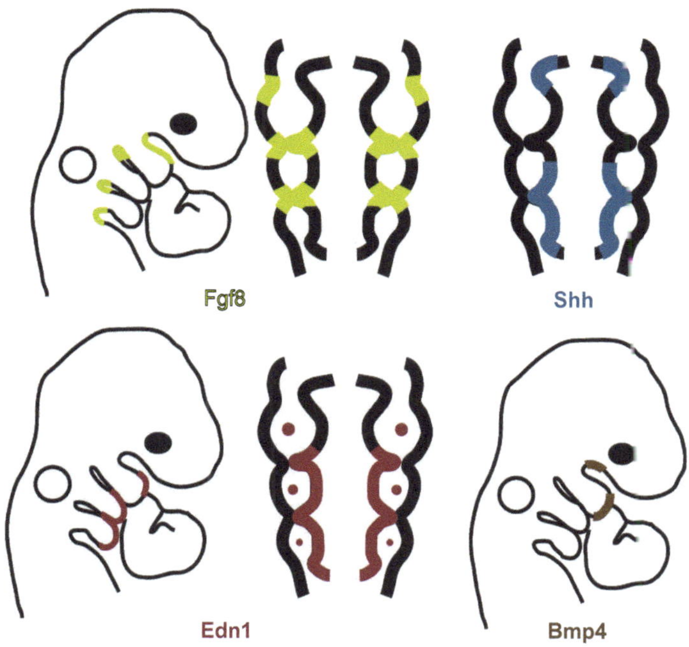

Renée K. Edlund *et al.*, Figure 3 Signaling factors from the pharyngeal ectoderm and endoderm. At embryonic day 9.5 (E9.5), four major signaling factors are secreted from the ectoderm and endoderm. Fgf8 is expressed in the ectoderm between the arches and between the mandibular and maxillary processes of BA1 and in the endoderm at the tips of the pharyngeal pouches. Shh is expressed in endoderm underlying BA2 and BA3. Edn1 is expressed in ventral arch ectoderm, endoderm underlying the mandibular process, BA2, and BA3, and in the mesoderm of each arch. Bmp4 is expressed in the ectoderm overlying the ventral domain and the maxillary process of branchial arch 1 (BA1).

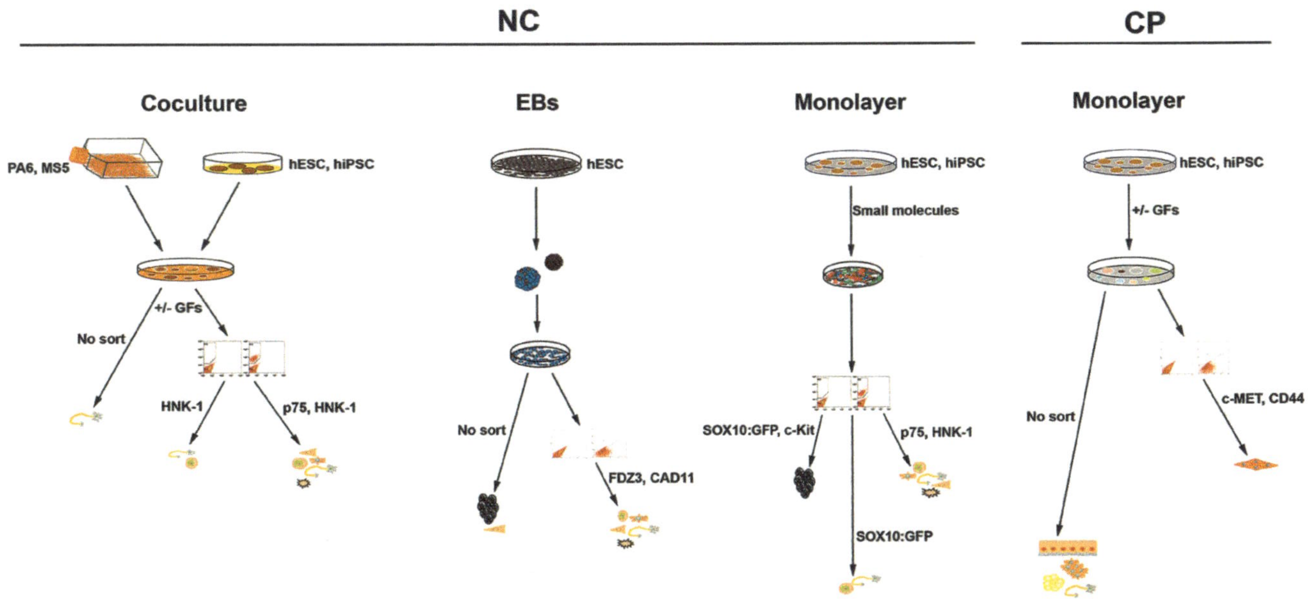

Bianca E. Borchin and Tiziano Barberi, Figure 1 Schematic representation of current methods for NC and CP derivation, illustrating all differentiation steps from undifferentiated hPSCs to specialized NC and CP cells.